CODES AND AUTOMATA

This major revision of Berstel and Perrin's classic *Theory of Codes* has been rewritten with a more modern focus and a much broader coverage of the subject. The concept of unambiguous automata, which is intimately linked with that of codes, now plays a significant role throughout the book, reflecting developments of the last 20 years. This is complemented by a discussion of the connection between codes and transducers, and new material from the field of symbolic dynamics. The authors have also explored links with more practical applications, including data compression and text processing. The treatment remains self-contained: there is background material on discrete mathematics, algebra and theoretical computer science. The wealth of exercises and examples make it ideal for self-study or courses. In sum this is a comprehensive reference on the theory of variable-length codes and their relation to unambiguous automata.

JEAN BERSTEL is Emeritus Professor of Computer Science at the Université Paris-Est.

DOMINIQUE PERRIN is Professor in Computer Science at the Université Paris-Est, and director of ESIEE Paris.

CHRISTOPHE REUTENAUER is Professor of Mathematics in the Combinatorics and Mathematical Computer Science Laboratory (LaCIM) at the University of Québec, Montréal.

ENCYCLOPEDIA OF MATHEMATICS AND ITS APPLICATIONS

All the titles listed below can be obtained from good booksellers or from Cambridge
University Press. For a complete series listing visit

http://www.cambridge.org/uk/series/sSeries.asp?code=EOM

Codes and Automata

JEAN BERSTEL

DOMINIQUE PERRIN
Université Paris-Est

CHRISTOPHE REUTENAUER
Université du Québec à Montréal

CAMBRIDGE
UNIVERSITY PRESS

CAMBRIDGE UNIVERSITY PRESS
Cambridge, New York, Melbourne, Madrid, Cape Town, Singapore, São Paulo, Delhi

Cambridge University Press
The Edinburgh Building, Cambridge CB2 8RU, UK

Published in the United States of America by Cambridge University Press, New York

www.cambridge.org
Information on this title: www.cambridge.org/9780521888318

© Cambridge University Press 2010

First published 2010

Printed in the United Kingdom at the University Press, Cambridge

A catalog record for this publication is available from the British Library

ISBN 978-0-521-88831-8 Hardback

Contents

Contents

Preface

This book presents a comprehensive study of the theory of variable length codes. It is a complete reworking of the book *Theory of Codes* published by the first two authors more than twenty years ago. The present text includes many new results and also contains several additional chapters. Its focus is also broader, in the sense that more emphasis is given to algorithmic questions and to relations with other fields.

The theory of codes takes its origin in the theory of information devised by Shannon in the 1950s. As presented here, it makes use more of combinatorial and algebraic methods than of information theory. Due to the nature of the questions that are raised and solved, this theory has now become clearly a part of theoretical computer science and is strongly related to combinatorics on words, automata theory, formal languages, and the theory of semigroups.

The object of the theory of codes is, from an elementary point of view, the study of the properties concerning factorizations of words into sequences of words taken from a given set. One of the basic techniques used in this book is constructing special automata that perform this kind of parsing. We will show how properties of codes are reflected in combinatorial or algebraic properties of the associated devices.

It is quite remarkable that the problem of encoding as treated here admits a rather simple mathematical formulation: it is the study of embeddings of a free monoid into another. This may be considered to be a basic problem of algebra. There are related problems in other algebraic structures. For instance, if we replace free monoids by free groups, the study of codes reduces to that of subgroups of a free group. However, the situation is quite different at the very beginning since, according to the Nielsen–Schreier theorem, any subgroup of a free group is itself free, whereas the corresponding statement is false for free monoids. Nevertheless the relationship between codes and groups is more than an analogy, and we shall see in this book how the study of a group associated with a code can reveal some of its properties. It was M.-P. Schützenberger's discovery that coding theory is closely related to classical algebra. He has been the main architect of this theory. The main basic results are due to him and most further developments were stimulated by his conjectures.

The aim of the theory of codes is to give a structural description of codes in a way that allows their construction. This is easily accomplished for prefix codes, as shown in Chapter 3. The case of bifix codes is already much more difficult, and the complete structural description given in Chapter 6 is one of the highlights of the theory. However, the structure of general codes (neither prefix nor suffix) still remains unknown to a large extent. For example, no systematic method is known for constructing all finite codes. The result given in Chapter 14 about the factorization of the polynomial of a code must be considered (despite the difficulty of its proof) as an intermediate step toward the understanding of codes.

Many of the results given in this book are concerned with extremal properties, the interest in which comes from the interconnection that appears between different concepts. But it also goes back to the initial investigations on codes considered as communication tools. Indeed, these extremal properties in general reflect some optimization in the encoding process. Thus a maximal code uses, in this sense, the whole capacity of the transmission channel.

Primarily, two types of methods are used in this book: direct methods on words on one hand, and automata and semigroups on the other hand. Direct methods consist of a more or less refined analysis of the sequencing of letters and factors within a word as it occurs in combinatorics on words. Automata and semigroups as used in Chapters 9–14, include the study of special automata associated with codes, called unambiguous automata and of the corresponding monoids of relations (unambiguous monoids of relations).

There are also many connections between the field of codes and automata and the field of symbolic dynamics. This aspect was not covered in *Theory of Codes*, and it is one of the new features of this volume. Symbolic dynamics focuses on the study of symbolic dynamical systems and, in particular of those defined by finite automata. The main point of intersection with codes is the notion of unambiguous automaton which coincides with the notion of *finite-to-one map* between symbolic systems. This relation is spread over several chapters. For example, the solution of the road coloring problem is presented in Chapter 10 and the notion of topological entropy is introduced in Chapter 13. The connections are explained in each chapter in the Notes section.

Codes and automata are related to algorithms on words and graphs. The computational complexity of algorithms related to codes is one of the topics of the book and is considered at various places in the text. We consider in particular algorithms related to tests for codes and to the construction of optimal prefix codes for several criteria.

The degree of generality of the exposition was influenced by the observation that many facts that hold for finite codes remain true for recognizable codes and even for the larger class of thin codes. In general, the transition from finite to recognizable codes does not imply major changes in the proof. However, changing to thin codes may imply some rather delicate computations. This is clearly demonstrated in Chapters 9 and 13, where the summations to be made become infinite when the codes are no longer recognizable. But this approach leads to a greater generality and, as we believe, to a better understanding by focusing attention on the main argument. Moreover, the characterization of the monoids associated with thin codes given in Chapter 9 may be considered to be a justification of our choice.

The organization of the book is as follows: A preliminary chapter (Chapter 1) is intended mainly to fix notation and should be consulted only when necessary. The book is composed of two major parts: part one consisting of Chapters 2–8 and part two formed of Chapters 9–14.

Chapters 2–8 constitute an elementary introduction to the theory of codes in the sense that they primarily make use of direct methods. Chapter 2 contains the definition, the relationship with submonoids, the first results on Bernoulli distributions, and the introduction of the notions of complete, maximal, and thin codes.

Chapter 3 is devoted to a systematic study of prefix codes, developed at an elementary level. Indeed, this is the most intuitive and easy part of the theory of codes and certainly deserves considerable discussion. We believe that its interest largely goes beyond the theory of codes. We consider optimal prefix codes under various constraints. In particular, we give a full proof of the Garsia–Wachs algorithm.

Chapter 4 describes the automata used for representing codes, and for encoding and decoding words. The flower automaton is the basic tool for a syntactic study of codes. It is also helpful in an efficient algorithm for testing whether a rational set of words is a code. Encoders and decoders are transducers. We show how to construct deterministic transducers whenever it is possible.

Chapter 5 introduces the deciphering delay, the family of weakly prefix codes and their relation with weakly deterministic automata. The chapter contains the well-known theorem on maximal codes with finite deciphering delay.

Chapter 6 also is elementary, although it is more dense. Its aims are to describe the structure of maximal bifix codes and to give methods for constructing the finite ones. The use of formal power series is here of great help.

Chapter 7 is combinatorial in nature. It contains a description of length distributions of circular codes which is related to classical enumerative combinatorics. It contains also a systematic theory that leads to the study of the well-known comma-free codes.

Chapter 8 introduces the factorizations of a free monoid and more importantly of the characterization of the codes that may appear as factors. We present complete descriptions of finite factorizations for up to five factors.

The next five chapters contain what is known about codes but can be proved only by syntactic methods.

Chapter 9 is devoted to these techniques, using a more systematic treatment. Instead of the frequently encountered monoids of functions we study unambiguous monoids of relations which do not favor left or right. Chapter 9 contains an important result, already mentioned above: the characterization of thin maximal codes by a finiteness condition on the transition monoid of an unambiguous automaton.

Chapter 10 presents several results linked to the notion of synchronized codes. The notion of locally parsable code is related to that of local automaton. It contains also a proof of the road coloring problem, which has been recently solved. Chapter 11 deals with the groups of codes. It contains in particular the proof of the theorem of synchronization of semaphore codes announced in Chapter 3. Several results on the groups of finite maximal bifix codes are proved.

Chapter 12 presents elements of the theory of factorizations of cyclic groups. Several particular classes of these factorizations are described, such as those due to Hajós and Rédei. The relation with codes is developed.

Chapter 13 starts with a presentation of basics on probability spaces, and contains a proof of Kolmogorov's extension theorem. Next, it shows how to compute the density of the submonoid generated by a code by transferring the computation into the associated unambiguous monoid of relations. The formula of densities, linking together the density of the submonoids, the degree of the code, and the densities of the contexts, is the most striking result.

Chapter 14 contains the proof and discussion of the theorem of the factorization of the polynomial of a finite maximal code. Many of the results of the preceding chapters are used in the proof of this theorem, which contains the most current detailed information about the structure of general codes. The book ends with the connection between maximal bifix codes and semisimple algebras.

In an appendix, we gather, for the convenience of the reader, the conjectures mentioned in the book and present some additional open problems.

The book is written at an elementary level. In particular, the knowledge required is covered by a basic mathematical culture. Complete proofs are given and the necessary results of automata theory or theory of semigroups are presented in Chapter 1. Many examples are given which come from practical applications and illustrate the notions.

Each chapter is followed by a section of exercises. These frequently complement the material covered in the text. Solutions for this set of some 200 exercises are proposed at the end of the book. Each chapter ends with notes containing references, bibliographic discussions, complementary material, and references for the exercises.

It seems impossible to cover the whole text in a one-year course. However, the book contains enough material for several courses, at various levels, in undergraduate or graduate curricula.

A one-semester course at graduate level in discrete mathematics may be composed of Chapter 2, Chapter 3, Chapter 6, and Chapter 4. A one-semester course at undergraduate level may be composed of Chapter 2, Chapter 3 without the last section, and Chapter 4.

Several chapters are largely independent and can be lectured on separately. As an example, a course based solely on Chapter 7 has been taught by one of us. A course based on algorithms may contain the beginning of Chapter 2, the last section of Chapter 3, and Chapter 4.

Because of the extensive use of trees and of the algorithms described there, Chapter 3 by itself might constitute an interesting complement to a programming course.

Chapters 9 and 11, which rely on the structure of unambiguous monoids of relations, are an excellent illustration for a course in algebra. Similarly, Chapter 13 can be used as an adjunct to a course on probability theory.

The present volume is a new version of *Theory of Codes*, for which we have received help and collaboration from many people. It is a pleasure for us to renew our thanks to people who helped us during the preparation of the ancestor book: Aldo De Luca, Georges Hansel, Maurice Nivat, Jean-Eric Pin, Antonio Restivo, Stuart W. Margolis and Paul E. Schupp. The authors are greatly indebted to M.-P. Schützenberger (1920–1996). The project of writing the book stems from him and he has encouraged us constantly in many discussions.

The authors wish to thank, for help and comments on the present text, Marie-Pierre Béal, Jean-Marie Boë, Véronique Bruyère, Arturo Carpi, Christian Choffrut, Clelia De Felice, Sylvain Lavallée, Aaron Lauve, Yun Liu, Roberto Mantaci, Brian H. Marcus, Wojciek Plandowski, Jacques Sakarovitch, Alessandra Savelli, Paul H. Siegel, Sandor Szabó, Stephanie van Willigenburg and Ken Zeger. Special thanks are due to Jean Néraud who has carefully read all exercises and solutions.

1

Preliminaries

In this preliminary chapter, we give an account of some basic notions which will be used throughout the book. This chapter is not designed for a systematic reading but rather as a reference.

The first three sections contain notation and basic vocabulary. Each of the subsequent sections is an introduction to a topic which is not completely treated in this book. These sections are concerned mainly with the theory of automata. Kleene's theorem is given and we show how to construct a minimal automaton from a given automaton. Syntactic monoids are defined. These concepts and results will be discussed in another context in Chapter 9. We introduce formal power series and weighted automata. We give some basic properties and prove parts of Perron–Frobenius theorem.

1.1 Notation

As usual, $\mathbb{N}, \mathbb{Z}, \mathbb{Q}, \mathbb{R}$, and $\mathbb{C}$ denote the sets of nonnegative integers, integers, and rational, real, and complex numbers, respectively. By convention, $0 \in \mathbb{N}$. We set

$$\mathbb{R}_+ = \{x \in \mathbb{R} \mid x \geq 0\}.$$

Next,

$$\binom{n}{p} = \frac{n!}{p!(n-p)!}$$

denotes the binomial coefficient of n and p.

For real numbers $x \leq y$, we denote by $[x, y)$ the set of real numbers z such that $x \leq z$ and $z < y$. In particular, if $x = y$ this set is empty.

Given two subsets X, Y of a set Z, we define

$$X \setminus Y = \{z \in Z \mid z \in X, z \notin Y\}.$$

Frequently, $\overline{X}$ will be used to denote the complement of a subset X of some set Z. An element x and the singleton set $\{x\}$ will usually not be distinguished. The set of all subsets of a set X is denoted by $\mathfrak{P}(X)$.

The function symbols are usually written on the left of their arguments but with some exceptions: When we consider the composition of actions on a set, the action is written on the right. In particular, permutations are written on the right.

A partition of a set X is a family $(X_i)_{i \in I}$ of *nonempty* subsets of X such that

(i) $X = \bigcup_{i \in I} X_i$,
(ii) $X_i \cap X_j = \emptyset, (i \neq j)$.

We usually define a partition as follows: "Let $X = \bigcup_{i \in I} X_i$ be a partition of X". We denote the cardinality of a set X by $\mathrm{Card}(X)$.

1.2 Monoids

A *semigroup* is a set equipped with an associative binary operation. The operation is usually written multiplicatively.

A *monoid* is a semigroup which, in addition, has a neutral element. The neutral element of a monoid M is unique and is denoted by 1_M or simply by 1.

For any monoid M, the set $\mathfrak{P}(M)$ is given a monoid structure by defining, for $X, Y \subset M$,

$$XY = \{xy \mid x \in X, y \in Y\}.$$

The neutral element is $\{1\}$.

A *submonoid* of M is a subset N which is stable under the operation and which contains the neutral element of M, that is $1_M \in N$ and

$$NN \subset N. \tag{1.1}$$

Note that a subset N of M satisfying (1.1) does not always satisfy $1_M = 1_N$ and therefore may be a monoid without being a submonoid of M.

A *morphism* from a monoid M into a monoid N is a function $\varphi : M \to N$ which satisfies, for all $m, m' \in M$,

$$\varphi(mm') = \varphi(m)\varphi(m'),$$

and furthermore

$$\varphi(1_M) = 1_N.$$

The notions of subsemigroup and semigroup morphism are then defined in the same way as the corresponding notions for monoids.

A *congruence* on a monoid M is an equivalence relation θ on M such that, for all $m, m' \in M, u, v \in M$

$$m \equiv m' \bmod \theta \implies umv \equiv um'v \bmod \theta.$$

Let φ be a morphism from M onto N. The equivalence θ defined by $m \equiv m' \bmod \theta$ if and only if $\varphi(m) = \varphi(m')$ is a congruence. It is called the *nuclear congruence* induced by φ. Conversely, if θ is a congruence on the monoid M, the set M/θ of the equivalence classes of θ is equipped with a monoid structure, and the canonical function from M onto M/θ is a monoid morphism.

An *idempotent* of a monoid M is an element e of M such that

$$e = e^2.$$

For each idempotent e of a monoid M, the set eMe is a monoid contained in M. It is easily seen that it is the largest monoid contained in M having e as a neutral element. It is called the *monoid localized* at e.

An element 0 of a monoid M is a *zero* if $0 \neq 1$ and for all $m \in M$

$$0m = m0 = 0.$$

If M contains a zero it is unique.

Let M be a monoid. The set of (left and right) invertible elements of M is a group called the *group of units* of M.

A *cyclic monoid* is a monoid with just one generator, that is,

$$M = \{a^n \mid n \in \mathbb{N}\}$$

with $a^0 = 1$. If M is infinite, it is isomorphic to the additive monoid $\mathbb{N}$ of nonnegative integers. If M is finite, the *index* of M is the smallest integer $i \geq 0$ such that there exists an integer $r \geq 1$ with

$$a^{i+r} = a^i. \tag{1.2}$$

The smallest integer r such that (1.2) holds is called the *period* of M. The pair composed of index i and period p determines a monoid having $i + p$ elements,

$$M_{i,p} = \{1, a, a^2, \ldots, a^{i-1}, a^i, \ldots, a^{i+p-1}\}.$$

Its multiplication is conveniently represented in Figure 1.1.

The monoid $M_{i,p}$ contains two idempotents (provided $i \geq 1$). Indeed, assume that $a^j = a^{2j}$. Then either $j = 0$ or $j \geq i$ and j and $2j$ have the same residue mod p, hence $j \equiv 0 \bmod p$. Conversely, if $j \geq i$ and $j \equiv 0 \bmod p$, then $a^j = a^{2j}$.

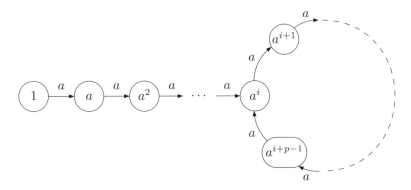

Figure 1.1 The monoid $M_{i,p}$.

Consequently, the unique idempotent $e \neq 1$ in $M_{i,p}$ is $e = a^j$, where j is the unique integer in $\{i, i+1, \ldots, i+p-1\}$ which is a multiple of p.

Let M be a monoid. For $x, y \in M$, we define

$$x^{-1}y = \{z \in M \mid xz = y\} \quad \text{and} \quad xy^{-1} = \{z \in M \mid x = zy\}.$$

For subsets X, Y of M, this notation is extended to

$$X^{-1}Y = \bigcup_{x \in X} \bigcup_{y \in Y} x^{-1}y \quad \text{and} \quad XY^{-1} = \bigcup_{x \in X} \bigcup_{y \in Y} xy^{-1}.$$

The set $X^{-1}Y$ is called a left *residual* of Y. The following identities hold for subsets X, Y, Z of M:

$$(XY)^{-1}Z = Y^{-1}(X^{-1}Z) \quad \text{and} \quad X^{-1}(YZ^{-1}) = (X^{-1}Y)Z^{-1}.$$

The notation $X^{-1}Y$ should not be confused with the product of the inverse of an element with another in some group. There is a case where the confusion could arise, in Chapter 14, where a due "caveat" will be found.

Given a subset X of a monoid M, we define

$$F(X) = M^{-1}XM^{-1}$$

to be the set of *factors* of elements in X. We have

$$F(X) = \{m \in M \mid \exists u, v \in M : umv \in X\}.$$

We sometimes use the notation $\overline{F}(X)$ to denote the complement of $F(X)$ in M,

$$\overline{F}(X) = M \setminus F(X).$$

A *relation m* over a set Q is a subset of $Q \times Q$. The *product* of two relations m and n over Q is the relation mn defined by

$$(p, r) \in mn \iff \exists q \in Q : (p, q) \in m \quad \text{and} \quad (q, r) \in n.$$

The set $\mathfrak{P}(Q \times Q)$ of relations over a set Q is a monoid for this product. Two remarkable relations are the *identity relation* id_Q and the *null relation*, which is the empty subset of $Q \times Q$. The identity relation id_Q is the neutral element of $\mathfrak{P}(Q \times Q)$. The null relation is a zero of this monoid.

A *monoid of relations* over some nonempty set Q is a submonoid of the monoid $\mathfrak{P}(Q \times Q)$. A monoid M of relations over Q is said to be *transitive* if for all $p, q \in Q$, there exists $m \in M$ such that $(p, q) \in m$.

1.3 Words

Let A be a set, which we call an *alphabet*. A *word w* on the alphabet A is a finite sequence of elements of A

$$w = (a_1, a_2, \ldots, a_n), \quad a_i \in A.$$

The set of all words on the alphabet A is denoted by A^* and is equipped with the associative operation defined by the concatenation of two sequences

$$(a_1, a_2, \ldots, a_n)(b_1, b_2, \ldots, b_m) = (a_1, a_2, \ldots, a_n, b_1, b_2, \ldots, b_m).$$

This operation is associative. This allows us to write

$$w = a_1 a_2 \cdots a_n$$

instead of $w = (a_1, a_2, \ldots, a_n)$, by identifying each element $a \in A$ with the sequence (a). An element $a \in A$ is called a *letter*. The empty sequence is called the *empty word* and is denoted by 1 or ε. It is the neutral element for concatenation. Thus the set A^* of words is equipped with the structure of a monoid. The monoid A^* is called the *free monoid* on A. The set of nonempty words on A is denoted by A^+. We therefore have $A^+ = A^* \setminus 1$.

The *length* $|w|$ of the word $w = a_1 a_2 \ldots a_n$ with $a_i \in A$ is the number n of letters in w. Clearly, $|1| = 0$. The function $w \mapsto |w|$ is a morphism from A^* onto the additive monoid $\mathbb{N}$. For $n \geq 0$, we use the notation

$$A^{(n)} = \{w \in A^* \mid |w| \leq n - 1\}$$

and also

$$A^{[n]} = \{w \in A^* \mid |w| \leq n\}.$$

In particular, $A^{(0)} = \emptyset$ and $A^{[0]} = \{1\}$.

For a subset B of A, we denote by $|w|_B$ the number of letters of w which are in B. Thus

$$|w| = \sum_{a \in A} |w|_a.$$

For a word $w \in A^*$, the set

$$\text{alph}(w) = \{a \in A \mid |w|_a > 0\}$$

is the set of all letters occurring at least once in w. For a subset X of A^*, we set

$$\text{alph}(X) = \bigcup_{x \in X} \text{alph}(x).$$

A word $w \in A^*$ is a *factor* of a word $x \in A^*$ if there exist $u, v \in A^*$ such that $x = uwv$. The relation *is a factor of* is a partial order on A^*. A factor w of x is *proper* if $w \neq x$.

A word $w \in A^*$ is a *prefix* of a word $x \in A^*$ if there is a word $u \in A^*$ such that $x = wu$. The factor w is called *proper* if $w \neq x$. The relation *is a prefix of* is again a partial order on A^* called the *prefix order*. We write $w \leq x$ when w is a prefix of x and $w < x$ whenever $w \leq x$ and $w \neq x$. This orderhas the following fundamental property. If, for some x,

$$w \leq x, \quad w' \leq x,$$

then w and w' are comparable, that is, $w \leq w'$ or $w' \leq w$. In other words, if $wu = w'u'$, then either there exists $s \in A^*$ such that $w = w's$ (and also $su = u'$) or there exists $t \in A^*$ such that $w' = wt$ (and then $u = tu'$).

In an entirely symmetric manner, we define a *suffix* w of a word x by $x = vw$ for some $v \in A^*$. A set $P \subset A^*$ is called *prefix-closed* if it contains the prefixes of its elements: $uv \in P \Rightarrow u \in P$. A suffix-closed set is defined symmetrically.

Consider a totally ordered alphabet A. The *lexicographic* or *alphabetic* order on A^* is defined by setting $u \prec v$ if u is a proper prefix of v, or if $u = ras$, $v = rbt$, $a < b$ for $a, b \in A$ and $r, s, t \in A^*$. The lexicographic order has the property

$$u \prec v \Leftrightarrow wu \prec wv$$

for any $u, v, w \in A^*$. Similarly, the *radix order* on A^* is defined by setting $u < v$ if $|u| < |v|$ or if $|u| = |v|$ and $u \prec v$ in the lexicographic order.

The *reversal* w of a word $w = a_1 a_2 \cdots a_n$, with $a_i \in A$, is the word

$$\tilde{w} = a_n \cdots a_2 a_1.$$

The notations $\tilde{w}$ and $w^{\sim}$ are equivalent. Note that for all $u, v \in A^*$,

$$(uv)^{\sim} = \tilde{v}\tilde{u}.$$

The *reversal* $\tilde{X}$ of a set $X \subset A^*$ is the set $\tilde{X} = \{\tilde{x} \mid x \in X\}$.

A *factorization* of a word $w \in A^*$ is a sequence $\{u_1, u_2, \ldots, u_n\}$ of $n \geq 0$ words in A^* such that

$$w = u_1 u_2 \cdots u_n.$$

For a subset X of A^*, we denote by X^* the submonoid generated by X,

$$X^* = \{x_1 x_2 \cdots x_n \mid n \geq 0, x_i \in X\}.$$

Similarly, we denote by X^+ the subsemigroup generated by X,

$$X^+ = \{x_1 x_2 \cdots x_n \mid n \geq 1, x_i \in X\}.$$

We have

$$X^+ = \begin{cases} X^* \setminus 1 & \text{if } 1 \notin X, \\ X^* & \text{otherwise.} \end{cases}$$

By definition, each word w in X^* admits at least one factorization $(x_1, x_2, \ldots, x_n)$ whose elements are all in X. Such a factorization is called an X-*factorization*. We frequently use the pictorial representation of an X-factorization given in Figure 1.2.

Figure 1.2 An X-factorization of w.

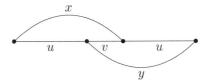

Figure 1.3 Two conjugate words x and y.

A word $x \in A^*$ is called *primitive* if it is not a power of another word. Thus x is primitive if and only if $x = y^n$ with $n \geq 0$ implies $x = y$. Observe that the empty word is not primitive.

Two words x, y are called *conjugate* if there exists words u, v such that $x = uv, y = vu$. (See Figure 1.3.) We frequently say that y is a conjugate of x. Two conjugate words are obtained from each other by a cyclic permutation. More precisely, let γ be the function from A^* into itself defined by

$$\gamma(1) = 1 \quad \text{and} \quad \gamma(av) = va \tag{1.3}$$

for $a \in A$, $v \in A^*$. It is clearly a bijection from A^* onto itself. Two words x and y are conjugate if and only if there exists an integer $n \geq 0$ such that

$$x = \gamma^n(y).$$

This easily implies that the conjugacy relation is an equivalence relation. A *conjugacy class* is a class of this equivalence relation. A conjugacy class is also called a *necklace*. The length of a necklace is the length of the words in the conjugacy class. A necklace is *primitive* if each word in the conjugacy class is primitive.

Proposition 1.3.1 *Each nonempty word is a power of a unique primitive word.*

Proof. Let $x \in A^+$ and δ be the restriction of the function γ defined by (1.3) to the conjugacy class of x. Then $\delta^k = 1$ if and only if x is a power of a word of length dividing k.

Let p be the order of δ, that is, the gcd of the integers k such that $\delta^k = 1$. Since $\delta^p = 1$, there exists a word r of length p such that $x = r^e$ with $e \geq 1$. The word r is primitive, otherwise there would be a word s of length q dividing p such that $r \in s^*$, which in turn implies that $x \in s^*$, contrary to the definition of p. This proves the existence of the primitive word. To show uniqueness, consider a word $t \in A^*$ such that $x \in t^*$ and let $k = |t|$. Since $\delta^k = 1$, the integer k is a multiple of p. Consequently $t \in r^*$. Thus, if t is primitive, we have $t = r$. $\qquad\square$

Table 1.1 *The number $\ell_n(k)$ of primitive conjugacy classes over a k-letter alphabet.*

n	1	2	3	4	5	6	7	8	9	10	11	12
$\ell_n(2)$	2	1	2	3	6	9	18	30	56	99	186	335
$\ell_n(3)$	3	3	8	18	48	116	312	810				
$\ell_n(4)$	4	6	20	60	204	670						
$\ell_n(5)$	5	10	40	150	624							

Let $x \in A^+$. The unique primitive word r such that $x = r^n$ for some integer n is called the *root* of x. The integer n is the *exponent* of x.

Proposition 1.3.2 *Two nonempty conjugate words have the same exponent and their roots are conjugate.*

Proof. Let $x, y \in A^+$ be two conjugate words, and let i be an integer such that $y = \gamma^i(x)$. Set r and s be the roots of x and y respectively and let n be the exponent of x. Then

$$y = \gamma^i(r^n) = (\gamma^i(r))^n.$$

This shows that $\gamma^i(r) \in s^*$. Interchanging the roles of x and y, we have $\gamma^j(s) \in r^*$. It follows that $\gamma^i(r) = s$ and $\gamma^j(s) = r$. Thus r and s are conjugate and consequently x and y have the same exponent. $\qquad\square$

Proposition 1.3.3 *All words in a conjugacy class have the same exponent. If C is a conjugacy class of words of length n with exponent e, then*

$$\mathrm{Card}(C) = n/e.$$

Proof. Let $x \in A^n$ and C be its conjugacy class. Let δ be the restriction of γ to C and p be the order of δ. The root of x is the word r of length p such that $x = r^e$. Thus $n = pe$. Now $C = \{x, \delta(x) \ldots, \delta^{p-1}(x)\}$. These elements are distinct since p is the order of δ. Thus $\mathrm{Card}(C) = p$. $\qquad\square$

We now compute the number of conjugacy classes of words of given length over a finite alphabet. Let A be an alphabet with k letters. For all $n \geq 1$, the number of conjugacy classes of primitive words in A^* of length n is denoted by $\ell_n(k)$. The notation is justified by the fact that this number depends only on k and not on A.

The first values of this function, for $k = 2, 3, 4$, are given in Table 1.1. Clearly $\ell_n(1) = 1$ if $n = 1$, and $\ell_n(1) = 0$ otherwise. Now for $n \geq 1$

$$k^n = \sum_{d|n} d\, \ell_d(k), \tag{1.4}$$

where d runs over the divisors of n. Indeed, every word of length n belongs to exactly one conjugacy class of wordsof length n. Each class has $d = n/e$ elements, where *e* is the exponent of its words. Since there are as many classes whose words have exponent n/e as there are classes of primitive words of length $d = n/e$, the formula follows.

We can obtain an explicit expression for the numbers $\ell_n(k)$ by using the classical technique of Möbius inversion which we now recall.

The *Möbius function* is the function $\mu : \mathbb{N} \setminus 0 \to \mathbb{N}$ defined by $\mu(1) = 1$ and

$$\mu(n) = \begin{cases} (-1)^i & \text{if } n \text{ is the product of } i \text{ distinct prime numbers,} \\ 0 & \text{otherwise.} \end{cases}$$

Proposition 1.3.4 (Möbius inversion formula) *Let α, β be two functions from $\mathbb{N} \setminus 0$ into $\mathbb{N}$. Then*

$$\alpha(n) = \sum_{d|n} \beta(d) \quad (n \geq 1) \tag{1.5}$$

if and only if

$$\beta(n) = \sum_{d|n} \mu(d)\alpha(n/d) \quad (n \geq 1). \tag{1.6}$$

Proof. Let $\mathcal{S}$ be the set of functions from $\mathbb{N} \setminus 0$ into $\mathbb{N}$. Define a product on $\mathcal{S}$ by setting, for $f, g \in \mathcal{S}$

$$f * g(n) = \sum_{n=de} f(d)g(e).$$

It is easily verified that $\mathcal{S}$ is a commutative monoid for this product. Its neutral element is the function I taking the value 1 for $n = 1$ and 0 elsewhere.

Let $\iota \in \mathcal{S}$ be the constant function with value 1. Let us verify that

$$\iota * \mu = I. \tag{1.7}$$

Indeed $\iota * \mu(1) = 1$; for $n \geq 2$, let $n = p_1^{k_1} p_2^{k_2} \cdots p_m^{k_m}$ be the prime decomposition of n. If d divides n, then $\mu(d) \neq 0$ if and only if

$$d = p_1^{\ell_1} p_2^{\ell_2} \cdots p_m^{\ell_m}$$

with all $\ell_i = 0$ or 1. Then $\mu(d) = (-1)^t$ with $t = \sum_{i=1}^m \ell_i$. It follows that

$$\iota * \mu(n) = \sum_{d|n} \mu(d) = \sum_{t=0}^m (-1)^t \binom{m}{t} = 0.$$

Now let α, $\beta \in \mathcal{S}$. Then Formula (1.5) is equivalent to $\alpha = \iota * \beta$ and Formula (1.6) is equivalent to $\beta = \mu * \alpha$. By (1.7) these two formulas are equivalent. $\qquad\square$

Proposition 1.3.5 *The number of conjugacy classes of primitive words of length n over an alphabet with k letters is*

$$\ell_n(k) = \frac{1}{n} \sum_{d \mid n} \mu(n/d) k^d.$$

Proof. This is immediate from Formula (1.4) by Möbius inversion. □

A word $w \in A^+$ is called *unbordered* if no proper nonempty prefix of w is a suffix of w. In other words, w is unbordered if and only if $w \in uA^+ \cap A^+u$ implies $u = 1$. If w is unbordered, then

$$wA^* \cap A^*w = wA^*w \cup w.$$

The following property holds.

Proposition 1.3.6 *Let A be an alphabet with at least two letters. For each word $u \in A^+$, there exists $v \in A^*$ such that uv is unbordered.*

Proof. Let a be the first letter of u, and let $b \in A \setminus a$. Let us verify that the word $w = uab^{|u|}$ is unbordered. A nonempty prefix t of w starts with the letter a. It cannot be a suffix of w unless $|t| > |u|$. But then we have $t = sab^{|u|}$ for some $s \in A^*$, and also $t = uab^{|s|}$. Thus $|s| = |u|$, hence $t = w$. □

Let A be an alphabet. The *free group* $A^{\odot}$ on A is defined as follows: Let $\bar{A}$ be an alphabet in bijection with A and disjoint from A. Denote by $a \mapsto \bar{a}$ the bijection from A onto $\bar{A}$. This notation is extended by setting, for all $a \in A \cup \bar{A}$, $\bar{\bar{a}} = a$. Let δ be the symmetric relation defined for $u, v \in (A \cup \bar{A})^*$ and $a \in A \cup \bar{A}$ by

$$ua\bar{a}v \equiv uv \mod \delta.$$

Let ρ be the reflexive and transitive closure of δ. Then ρ is a congruence. The quotient monoid $A^{\odot} = (A \cup \bar{A})^*/\rho$ is a group. Indeed, for all $a \in A \cup \bar{A}$,

$$a\bar{a} \equiv 1 \mod \rho.$$

Thus the images of the generators are invertible in $A^{\odot}$. This shows that all elements in $A^{\odot}$ are invertible.

Let A be an alphabet. The *free commutative monoid* $A^{\oplus}$ on A is the quotient of A^* by the congruence generated by the pairs (ab, ba) for $a, b \in A$, $a \neq b$. If $A = \{a_1, \ldots, a_k\}$, then the monoid $A^{\oplus}$ can be identified with the additive monoid $\mathbb{N}^k$ through the map $a_1^{n_1} a_2^{n_2} \cdots a_k^{n_k} \mapsto (n_1, n_2, \ldots, n_k)$.

We denote by $\alpha(w)$ the commutative image of a word $w \in A^*$. It is the element of $A^{\oplus}$ defined by

$$\alpha(w) = \prod_{a \in A} a^{|w|_a}.$$

Observe that α is a monoid morphism from A^* onto $A^{\oplus}$.

1.4 Automata

Let A be an alphabet. An *automaton* over A is composed of a set Q (the set of *states*), a subset I of Q (the *initial* states), a subset T of Q (the *terminal* or *final* states), and a set

$$E \subset Q \times A \times Q$$

called the set of *edges*. The automaton is denoted by

$$\mathcal{A} = (Q, I, T).$$

The automaton is *finite* when the set Q is finite.

A *path* in the automaton $\mathcal{A}$ is a sequence $c = (f_1, f_2, \ldots, f_n)$ of consecutive edges

$$f_i = (q_i, a_i, q_{i+1}), \quad 1 \le i \le n.$$

The integer n is called the *length* of the path c. The word $w = a_1 a_2 \cdots a_n$ is the *label* of the path c. The state q_1 is the *origin* of c, and the state q_{n+1} the *end* of c. A useful notation is

$$c : q_1 \xrightarrow{w} q_{n+1}.$$

By convention, there is, for each state $q \in Q$, a path of length 0 from q to q. Its label is the empty word.

A path $c : i \to t$ is *successful* if $i \in I$ and $t \in T$. The set *recognized* by $\mathcal{A}$, denoted by $L(\mathcal{A})$, is defined as the set of labels of successful paths.

A state $q \in Q$ is *accessible* (resp. *coaccessible*) if there exists a path $c : i \to q$ with $i \in I$ (resp. a path $c : q \to t$ with $t \in T$). An automaton is *trim* if each state is both accessible and coaccessible. Let P be the set of accessible and coaccessible states, and let $\mathcal{A}^0 = (P, I \cap P, T \cap P)$. Then it is easy to see that $\mathcal{A}^0$ is trim and $L(\mathcal{A}) = L(\mathcal{A}^0)$. The automaton $\mathcal{A}^0$ is the *trim part* of $\mathcal{A}$.

An automaton can be viewed as a labeled multigraph equipped with two distinguished subset of vertices, the initial and the terminal states. The multigraph having Q as set of vertices, and E as set of edges, is called the *underlying graph* of the automaton. An automaton is called *strongly connected* if its underlying graph is strongly connected, that is if for any pair (p, q) of states (vertices), there is a path from p to q.

Let $\mathcal{A} = (Q, I, T)$ be an automaton over A. For each word w, we denote by $\varphi_{\mathcal{A}}(w)$ the relation over Q defined by

$$(p, q) \in \varphi_{\mathcal{A}}(w) \iff p \xrightarrow{w} q.$$

It follows from the definition that $\varphi_{\mathcal{A}}$ is a morphism from A^* into the monoid of relations over Q. The submonoid $\varphi_{\mathcal{A}}(A^*)$ is called the *transition monoid* of the automaton $\mathcal{A}$.

Clearly, an automaton is strongly connected if and only if its transition monoid is transitive.

An automaton $\mathcal{A} = (Q, I, T)$ is *deterministic* if $\text{Card}(I) = 1$ and if

$$(p, a, q), (p, a, r) \in E \Rightarrow q = r.$$

Thus for each $p \in Q$ and $a \in A$, there is at most one state q in Q such that $p \xrightarrow{a} q$. For $p \in Q$, and $a \in A$, define

$$p \cdot a = \begin{cases} q & \text{if } (p, a, q) \in E, \\ \emptyset & \text{otherwise.} \end{cases}$$

The partial function from $Q \times A$ into Q defined in this way is extended to words by setting $p \cdot 1 = p$ for all $p \in Q$, and, for $w \in A^*$ and $a \in A$,

$$p \cdot wa = (p \cdot w) \cdot a.$$

It follows easily that for words u, v,

$$p \cdot uv = p \cdot u \cdot v. \tag{1.8}$$

This function is called the *transition function* or *next-state* function of $\mathcal{A}$. With this notation, we have with $I = \{i\}$,

$$L(\mathcal{A}) = \{w \in A^* \mid i \cdot w \in T\}.$$

An automaton is *complete* if for all $p \in Q$, $a \in A$, there exists at least one $q \in Q$ such that $p \xrightarrow{a} q$.

Proposition 1.4.1 *For each automaton $\mathcal{A}$, there exists a complete deterministic automaton $\mathcal{B}$ such that*

$$L(\mathcal{A}) = L(\mathcal{B}).$$

If $\mathcal{A}$ is finite, then $\mathcal{B}$ can be chosen to be finite.

Proof. Set $\mathcal{A} = (Q, I, T)$. Define $\mathcal{B} = (R, u, V)$ by setting $R = \mathfrak{P}(Q)$, $u = I$,

$$V = \{S \subset Q \mid S \cap T \neq \emptyset\}.$$

Define the transition function of $\mathcal{B}$, for $S \in R$, $a \in A$ by

$$S \cdot a = \{q \in Q \mid \exists s \in S : s \xrightarrow{a} q\}.$$

The automaton $\mathcal{B}$ is complete and deterministic. It is easily seen that $L(\mathcal{A}) = L(\mathcal{B})$.
□

Example 1.4.2 Figure 1.4 gives, on the left, a nondeterministic automaton recognizing all words over $A = \{a, b\}$ having the suffix aba. The deterministic automaton on the right is obtained by the construction given in the proof of Proposition 1.4.1. It happens that both automata have the same number of states.

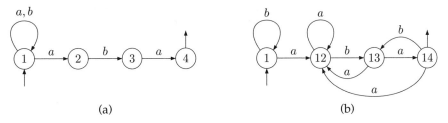

Figure 1.4 (a) A nondeterministic automaton recognizing the set of words $X = \{a, b\}^* aba$, and (b) a deterministic automaton recognizing this set.

Let $\mathcal{A} = (Q, i, T)$ be a deterministic automaton. For each $q \in Q$, let

$$L_q = \{w \in A^* \mid q \cdot w \in T\}.$$

Two states $p, q \in Q$ are called *inseparable* if $L_p = L_q$, and *separable* otherwise. A deterministic automaton is *reduced* if two distinct states are always separable.

Let X be a subset of A^*. We define a special automaton $\mathcal{A}(X)$ in the following way. The states of $\mathcal{A}(X)$ are the nonempty sets $u^{-1}X$ for $u \in A^*$. The initial state is $X = 1^{-1}X$, and the final states are those containing the empty word. The transition function is defined for a state $Y = u^{-1}X$ and a letter $a \in A$ by

$$Y \cdot a = a^{-1}Y.$$

Observe that this defines a partial function. We have

$$L(\mathcal{A}(X)) = X.$$

An easy induction shows that $X \cdot w = w^{-1}X$ for $w \in A^*$. Consequently

$$w \in L(\mathcal{A}(X)) \Leftrightarrow 1 \in X \cdot w \Leftrightarrow 1 \in w^{-1}X \Leftrightarrow w \in X.$$

The automaton $\mathcal{A}(X)$ is reduced. Indeed, for $Y = u^{-1}X$,

$$L_Y = \{v \in A^* \mid Y \cdot v \in T\} = \{v \in A^* \mid uv \in X\}.$$

Thus $L_Y = Y$.

The automaton $\mathcal{A}(X)$ is called the *minimal automaton* of X. This terminology is justified by the following proposition.

Proposition 1.4.3 *Let $\mathcal{A} = (Q, i, T)$ be a trim deterministic automaton and let $X = L(\mathcal{A})$. Let $\mathcal{A}(X) = (P, j, S)$ be the minimal automaton of X. The function φ from Q into P defined by $\varphi(q) = L_q$ is surjective and satisfies $\varphi(i) = j$, $\varphi(T) = S$ and $\varphi(q \cdot a) = \varphi(q) \cdot a$.*

Proof. Let $q \in Q$ and let $u \in A^*$ be such that $i \cdot u = q$. Then

$$L_q = \{w \in A^* \mid q \cdot w \in T\} = u^{-1}X.$$

Since $\mathcal{A}$ is trim, $L_q \neq \emptyset$. This shows that $L_q \in P$. Thus φ is a function from Q into P. Next, let us show that φ is surjective. Let $u^{-1}X \in P$. Then $u^{-1}X \neq \emptyset$. Therefore $i \cdot u \neq \emptyset$ and setting $q = i \cdot u$, we have $L_q = u^{-1}X = \varphi(q)$. Consequently φ is surjective.

Finally, for $q = i \cdot u$, one has $\varphi(q \cdot a) = L_{q \cdot a} = (ua)^{-1}X = (u^{-1}X) \cdot a = L_q \cdot a$.

$\square$

Assume furthermore that the automaton $\mathcal{A}$ in the proposition is reduced. Then the function φ is a bijection, which identifies $\mathcal{A}$ with the minimal automaton. In this sense, there exists just one reduced automaton recognizing a given set.

Let $\mathcal{A} = (Q, i, T)$ be a deterministic automaton. An equivalence relation ρ on the set Q is a *congruence* if for all states p, q and for all letters a, if $p \equiv q \bmod \rho$ and $p \cdot a$ and $q \cdot a$ are defined, then $p \cdot a \equiv q \cdot a \bmod \rho$.

The *quotient automaton* of $\mathcal{A}$ by the congruence ρ, denoted $\mathcal{A}/\rho$, has as states the classes of ρ, its initial state is the class of the initial state of $\mathcal{A}$, its final states are the classes of final states of $\mathcal{A}$. The transition function is defined as follows. If q is a state of $\mathcal{A}/\rho$ and a is a letter, then $q \cdot a$ is defined if there is a state p in the class q such that $p \cdot a$ is defined, and in this case $q \cdot a$ is the class of the state $p \cdot a$. The definition is sound because ρ is a congruence.

For example, the equivalence on the states of a deterministic automaton $\mathcal{A}$ defined by $p \equiv q$ if p and q are inseparable is a congruence. If the automaton is trim, the quotient is the minimal automaton of $L(\mathcal{A})$.

Let $\mathcal{A} = (Q, i, T)$ be a deterministic automaton. Consider the set $\mathcal{F}$ of partial functions from Q into Q. These functions are written on the right: if $q \in Q$ and $m \in \mathcal{F}$, then the image of q by m is denoted by qm. Composition is defined by

$$q(mn) = (qm)n.$$

Thus $\mathcal{F}$ has a monoid structure.

Let φ be the function which to a word $w \in A^*$ associates the partial function from Q into Q defined by

$$q\varphi(w) = q \cdot w.$$

The function φ is a morphism from A^* into the monoid $\mathcal{F}$. The submonoid $\varphi(A^*)$ of $\mathcal{F}$ is called the *transition monoid* of the automaton $\mathcal{A}$. This is consistent with the terminology for general automata since partial functions are a particular case of binary relations.

Observe that, setting $X = L(\mathcal{A})$, we have

$$\varphi^{-1}\varphi(X) = X. \tag{1.9}$$

Indeed $w \in \varphi^{-1}\varphi(X)$ if and only if $\varphi(w) \in \varphi(X)$ which is equivalent to $i\varphi(w) \in T$, that is to $w \in X$.

A morphism φ from a monoid M onto a monoid N is said to *recognize* a subset X of M if

$$\varphi^{-1}\varphi(X) = X.$$

A subset X of M is *recognizable* if it is recognized by a morphism onto a finite monoid.

Let X be a subset of A^*. For $w \in A^*$, a pair (u, v) of words such that $uwv \in X$ is a *context* of w in X. We denote by $\Gamma(w)$ the set of contexts of w, defined by

$$\Gamma(w) = \{(u, v) \in A^* \times A^* \mid uwv \in X\}.$$

The *syntactic congruence* of X is the equivalence relation $\sim_X$ on A^* defined by

$$w \sim_X w' \iff \Gamma(w) = \Gamma(w').$$

It is easily verified that $\sim_X$ is a congruence. The quotient of A^* by $\sim_X$ is, by definition, the *syntactic monoid* of X. We denote it by $\mathcal{M}(X)$, and we denote by φ_X the canonical morphism from A^* onto $\mathcal{M}(X)$. Note that φ_X recognizes X.

Proposition 1.4.4 *Let X be a subset of A^*, and let $\varphi : A^* \to M$ be a surjective morphism. If φ recognizes X, then there exists a morphism ψ from M onto the syntactic monoid $\mathcal{M}(X)$ such that*

$$\varphi_X = \psi \circ \varphi.$$

Proof. It suffices to show that

$$\varphi(w) = \varphi(w') \implies \varphi_X(w) = \varphi_X(w'). \tag{1.10}$$

Indeed, if (1.10) holds, then for an element $m \in M$, $\psi(m)$ is defined as the unique element in $\varphi_X(\varphi^{-1}(m))$. To show (1.10), we consider $(u, v) \in \Gamma(w)$. Then $uwv \in X$. Thus $\varphi(u)\varphi(w)\varphi(v) \in \varphi(X)$. From $\varphi(w) = \varphi(w')$, it follows that $\varphi(u)\varphi(w')\varphi(v) \in \varphi(X)$. Since φ recognizes X, this implies that $uw'v \in X$, showing that $(u, v) \in \Gamma(w')$. $\square$

Proposition 1.4.5 *Let X be a subset of A^*. The syntactic monoid of X is isomorphic to the transition monoid of the minimal automaton $\mathcal{A}(X)$.*

Proof. Let M be the transition monoid of the automaton $\mathcal{A}(X) = (Q, i, T)$ and let $\varphi : A^* \to M$ be the canonical morphism. By (1.9), the morphism φ recognizes X. By Proposition 1.4.4, there exists a morphism ψ from M onto the syntactic monoid $\mathcal{M}(X)$ such that $\varphi_X = \psi \circ \varphi$.

It suffices to show that ψ is injective. For this, consider $m, m' \in M$ such that $\psi(m) = \psi(m')$. Let $w, w' \in A^*$ such that $\varphi(w) = m, \varphi(w') = m'$. Then $\varphi_X(w) = \varphi_X(w')$. To prove that $\varphi(w) = \varphi(w')$, we consider a state $p \in Q$, and let $u \in A^*$ be such that $p = u^{-1}X$. Then

$$p\varphi(w) = p \cdot w = (uw)^{-1}X = \{v \in A^* \mid (u, v) \in \Gamma(w)\}.$$

Since $\Gamma(w) = \Gamma(w')$, we have $p\varphi(w) = p\varphi(w')$. Thus $\varphi(w) = \varphi(w')$, that is $m = m'$. $\square$

We now give a summary of properties which are specific to finite automata.

Theorem 1.4.6 *Let $X \subset A^*$. The following conditions are equivalent.*

(i) *The set X is recognized by a finite automaton.*

(ii) *The minimal automaton $\mathcal{A}(X)$ is finite.*

(iii) *The family of sets $u^{-1}X$, for $u \in A^*$, is finite.*

(iv) *The syntactic monoid $\mathcal{M}(X)$ is finite.*

(v) *The set X is recognizable.*

Proof. (i) $\Rightarrow$ (ii). Let $\mathcal{A}$ be a finite automaton recognizing X. By Proposition 1.4.1, we can assume that $\mathcal{A}$ is deterministic. By Proposition 1.4.3, the minimal automaton $\mathcal{A}(X)$ also is finite.

(ii) $\Leftrightarrow$ (iii) is clear.

(ii) $\Rightarrow$ (iv) holds by Proposition 1.4.5 and by the fact that the transition monoid of a finite automaton is always finite.

(iv) $\Rightarrow$ (v) is clear.

(v) $\Rightarrow$ (i). Let $\varphi : A^* \to M$ be a morphism onto a finite monoid M, and suppose that φ recognizes X. Let $\mathcal{A} = (M, 1, \varphi(X))$ be the deterministic automaton with transition function defined by $m \cdot a = m\varphi(a)$. Then $1 \cdot w \in \varphi(X)$ if and only if $\varphi(w) \in \varphi(X)$, thus if and only if $w \in X$. Consequently $L(\mathcal{A}) = X$. $\qquad\square$

Proposition 1.4.7 *The family of recognizable subsets of A^* is closed under all Boolean operations: union, intersection, complement.*

Proof. Let $X, Y \subset A^*$ be two recognizable subsets of A^*. Let $\mathcal{A} = (P, i, S)$ and $\mathcal{B} = (Q, j, T)$ be complete deterministic automata such that $X = L(\mathcal{A}), Y = L(\mathcal{B})$. Let

$$\mathcal{C} = (P \times Q, (i, j), R)$$

be the complete deterministic automaton defined by

$$(p, q) \cdot a = (p \cdot a, q \cdot a).$$

For $R = (S \times Q) \cup (P \times T)$, we have $L(\mathcal{C}) = X \cup Y$. For $R = S \times T$, we have $L(\mathcal{C}) = X \cap Y$. Finally, for $R = S \times (Q \setminus T)$, we have $L(\mathcal{C}) = X \setminus Y$. $\qquad\square$

Proposition 1.4.8 *Let $\alpha : A^* \to B^*$ be a morphism. If Y is a recognizable subset of B^*, then $X = \alpha^{-1}(Y)$ is a recognizable subset of A^*.*

Proof. Since Y is recognizable, one has $Y = \varphi^{-1}(\varphi(Y))$, where φ is a morphism from B^* onto a finite monoid M. Defining the function ψ from A^* into M by $\psi = \varphi \circ \alpha$, it follows that $X = \psi^{-1}(\psi(X))$. $\qquad\square$

Proposition 1.4.9 *If $X \subset A^*$ is recognizable, then $Y^{-1}X$ is recognizable for any subset Y of A^*.*

Proof. One has $u^{-1}(Y^{-1}X) = \bigcup_{y \in Y}(yu)^{-1}X$. Since X is recognizable, there are finitely many sets of the form $(yu)^{-1}X$, and thus of the form $u^{-1}(Y^{-1}X)$. This shows that $Y^{-1}X$ is recognizable. $\qquad\square$

Consider now a slight generalization of the notion of automaton. An *asynchronous automaton* on A is an automaton $\mathcal{A} = (Q, I, T)$, the edges of which may be labeled by either a letter or the empty word. Therefore the set of its edges satisfies

$$F \subset Q \times (A \cup 1) \times Q.$$

The notions of a path or a successful path extend in a natural way so that the notion of the set recognized by the automaton is clear.

Proposition 1.4.10 *For any finite asynchronous automaton $\mathcal{A}$, there exists a finite automaton $\mathcal{B}$ such that $L(\mathcal{A}) = L(\mathcal{B})$.*

Proof. Let $\mathcal{A} = (Q, I, T)$ be an asynchronous automaton. Let $\mathcal{B}$ be the automaton obtained from $\mathcal{A}$ by replacing its edges by the triples (p, a, q) such that there exists a path $p \xrightarrow{a} q$ in $\mathcal{A}$. We have

$$L(\mathcal{A}) \cap A^+ = L(\mathcal{B}) \cap A^+.$$

If $I \cap T \neq \emptyset$, both sets $L(\mathcal{A})$ and $L(\mathcal{B})$ contain the empty word and are therefore equal. Otherwise, the sets are equal up to the empty word and the result follows from Proposition 1.4.7 since the set $\{1\}$ is recognizable. $\quad\square$

The notion of an asynchronous automaton is useful to prove the following result.

Proposition 1.4.11 *If $X \subset A^*$ is recognizable, then X^* is recognizable. If $X, Y \subset A^*$ are recognizable, then XY is recognizable.*

Proof. Let $\mathcal{A} = (Q, I, T)$ be a finite automaton recognizing X. Let E be the set of its edges. Let $\mathcal{B}$ be the asynchronous automaton obtained from $\mathcal{A}$ by adding to E the triples $(t, 1, i)$, for $t \in T$, $i \in I$. Then $L(\mathcal{B}) = X^+$. In fact, the inclusion $X^+ \subset L(\mathcal{B})$ is clear. Conversely, let $c : i \xrightarrow{w} j$ be a nonempty successful path in $\mathcal{B}$. By the definition of $\mathcal{B}$, this path has the form

$$c : i_1 \xrightarrow{w_1} t_1 \xrightarrow{1} i_2 \xrightarrow{w_2} t_2 \cdots \xrightarrow{1} i_n \xrightarrow{w_n} t_n$$

with $i = i_1$, $j = t_n$ and where no path $c_k : i_k \xrightarrow{w_k} t_k$ contains an edge labeled by the empty word. Then $w_1, w_2, \ldots, w_n \in X$ and therefore $w \in X^+$. This proves that X^+ is recognizable and thus also $X^* = X^+ \cup \{1\}$.

Now let $\mathcal{A} = (P, I, S)$ and $\mathcal{B} = (Q, J, T)$ be two finite automata with sets of edges E and F, respectively. Let $X = L(\mathcal{A})$ and let $Y = L(\mathcal{B})$. One may assume that $P \cap Q = \emptyset$. Let $\mathcal{C} = (P \cup Q, I, T)$ be the asynchronous automaton with edges

$$E \cup F \cup (S \times \{1\} \times J).$$

Then $L(\mathcal{C}) = XY$ as we may easily check. $\quad\square$

We shall now give another characterization of recognizable subsets of A^*. Let M be a monoid. The family of *rational subsets* of M is the smallest family $\mathcal{R}$ of subsets of M such that

(i) any finite subset of M is in $\mathcal{R}$,

(ii) if $X, Y \in \mathcal{R}$, then $X \cup Y \in \mathcal{R}$, and $XY \in \mathcal{R}$,

(iii) if $X \in \mathcal{R}$, then $X^* \in \mathcal{R}$.

The third of these operations, namely $X \mapsto X^*$, is called the *star operation*. Union, product, and star are called the *rational operations*.

Proposition 1.4.12 *Let* $\alpha : A^* \to B^*$ *be a morphism. If* X *is a rational subset of* A^*, *then* $\alpha(X)$ *is a rational subset of* B^*.

Proof. The conclusion clearly holds if X is finite, and if it holds for two subsets X_1 and X_2 of A^*, it holds for their union, their product, and the star. So it holds for every rational subset of A^*.   □

Theorem 1.4.13 (Kleene) *Let* A *be a finite alphabet. A subset of* A^* *is recognizable if and only if it is rational.*

Proof. Denote by $\mathrm{Rec}(A^*)$ the family of recognizable subsets of A^* and by $\mathrm{Rat}(A^*)$ that of rational subsets of A^*. Let us first prove the inclusion $\mathrm{Rat}(A^*) \subset \mathrm{Rec}(A^*)$. In fact, any finite subset X of A^* is clearly recognizable. Moreover, Propositions 1.4.7 and 1.4.11 show that the family $\mathrm{Rec}(A^*)$ satisfies conditions (ii) and (iii) of the definition of $\mathrm{Rat}(A^*)$. This proves the inclusion.

To show that $\mathrm{Rec}(A^*) \subset \mathrm{Rat}(A^*)$, let us consider a recognizable subset X of A^*. Let $\mathcal{A} = (Q, I, T)$ be a finite automaton recognizing X. Set $Q = \{1, 2, \ldots, n\}$ and for $1 \le i, j \le n$,

$$X_{i,j} = \{w \in A^* \mid i \xrightarrow{\ w\ } j\}.$$

We have

$$X = \bigcup_{i \in I} \bigcup_{j \in T} X_{i,j}.$$

It is therefore enough to prove that each $X_{i,j}$ is rational. For $k \in \{0, 1, \ldots, n\}$, denote by $X_{i,j}^{(k)}$ the set of those $w \in A^*$ such that there exists a path $c : i \xrightarrow{\ w\ } j$ passing only through states $\ell \le k$ except perhaps for i, j. In other words we have $w \in X_{i,j}^{(k)}$ if and only if $w = a_1 a_2 \cdots a_m$ with

$$c : i \xrightarrow{\ a_1\ } i_1 \xrightarrow{\ a_2\ } i_2 \to \cdots i_{m-1} \xrightarrow{\ a_m\ } j$$

and $i_1 \le k, \ldots, i_{m-1} \le k$. We have the formulas

$$X_{i,j}^{(0)} \subset A \cup 1, \tag{1.11}$$

$$X_{i,j}^{(n)} = X_{i,j}, \tag{1.12}$$

$$X_{i,j}^{(k+1)} = X_{i,j}^{(k)} \cup X_{i,k+1}^{(k)} (X_{k+1,k+1}^{(k)})^* X_{k+1,j}^{(k)}, \quad (0 \le k < n). \tag{1.13}$$

Since A is finite, $X_{i,j}^{(0)} \in \operatorname{Rat}(A^*)$ by (1.11). Then (1.13) shows by induction on $k \geq 0$ that $X_{i,j}^{(k)} \in \operatorname{Rat}(A^*)$. Therefore $X_{i,j} \in \operatorname{Rat}(A^*)$ by (1.12). □

In the case of an infinite alphabet, recognizable sets need not to be rational: for instance the alphabet itself is recognizable but not rational. However, any recognizable set is the inverse image, by a length preserving morphism, of a recognizable set X over a finite alphabet. Indeed, this morphism identifies letters with the same image in the syntactic monoid of X. The common usage is to call *regular* a recognizable subset of A^*. The previous theorem states that regular sets and rational sets are the same for finite alphabets.

Corollary 1.4.14 *The family of regular sets over finite alphabets is closed under Boolean operations, rational operations, morphisms and inverse morphisms, and left and right quotient by arbitrary sets.* □

A description of a rational set by union, product, and star is called a *rational expression* or a *regular expression*. For instance, the set X of all words over $\{a, b\}$ that contain an even number of occurrences of the letter a has the rational expression $X = (b \cup ab^*a)^*$. Equations (1.11)–(1.13) provide an effective procedure to compute a rational expression for the set recognized by some finite automaton.

Example 1.4.2 (*continued*) The set X of words with suffix aba over the alphabet $A = \{a, b\}$ has the regular expression A^*aba. The equations (1.11)–(1.13), applied to the automaton on the right of Figure 1.4, lead for the same set of words to the regular expression $b^*a(a \cup b(ab)^*a \cup b(ab)^*aa)^*b(ab)^*a$.

1.5 Transducers

A *transducer* $T = (Q, I, T)$ over an input alphabet A and an output alphabet B is composed of a set Q of *states*, together with two distinguished subsets I and T of Q called the sets of *initial* and *terminal* states, and a set E of *edges* which are tuples (p, u, v, q) where p and q are states, u is a word over A and v is a word over B. An edge is also denoted by $p \xrightarrow{u|v} q$. A transducer is *finite* if its set of states is finite.

As in automata, a *path* in a transducer T is a sequence $c = (f_1, f_2, \ldots, f_n)$ of consecutive edges

$$f_i = (q_i, u_i, v_i, q_{i+1}), \quad 1 \leq i \leq n.$$

The integer n is called the *length* of the path c. The word $w = u_1u_2 \cdots u_n$ is the *input label* of the path c and $z = v_1v_2 \cdots v_n$ is its *output label*. The state $p = q_1$ is the *origin* of c, and the state $q = q_{n+1}$ the *end* of c. A useful notation is $c : p \xrightarrow{w|z} q$. A path $i \xrightarrow{x|y} t$ is *successful* if i is an initial state and t is a terminal state.

A transducer T defines a binary relation between words on the two alphabets as follows. A pair (x, y) is in the relation if it is the label of a successful path. This is called the relation *realized* by T. It can be viewed as a multi-valued mapping from

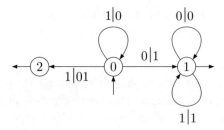

Figure 1.5 A transducer that adds 1 to a number, given by its binary expansion, with bit of highest weight on the right.

the input words into the output words, and also as a multi-valued mapping from the output words into the input words.

A transducer is called *literal* if each input label is a single letter. A transducer is *input-simple* if for any pair of edges (p, u, v, q), (p, u', v', q) with the same origin and the same end, $u = u'$ implies $v = v'$. This guarantees that when the output labels are erased, there are no multiple edges.

A literal transducer which is input-simple defines naturally an automaton over its input alphabet, called its *input automaton*, obtained by forgetting the output labels.

Example 1.5.1 The transducer given in Figure 1.5 has two final states 1 and 2. The only successful paths from 0 to 2 have the labels $(1^n, 0^n 1)$, and the successful paths from 0 to 1 have the labels $(1^n 0w, 0^n 1w)$ for some integer $n \geq 0$ and some word w. Thus the transducer transforms the binary representation of a positive integer N into the binary representation of $N + 1$. This transducer is literal and input-simple.

1.6 Semirings and matrices

A *semiring* K is a set equipped with two operations denoted $+$ and $\cdot$ satisfying the following axioms:

(i) The set K is a commutative monoid for $+$ with a neutral element denoted by 0.
(ii) The set K is a monoid for multiplication with a neutral element denoted by 1.
(iii) Multiplication is distributive on addition.
(iv) For all $x \in K$, $0 \cdot x = x \cdot 0 = 0$.

Clearly, any ring with unit is a semiring. Other examples of semirings are as follows. The set $\mathbb{N}$ of natural integers is a semiring and so is the set $\mathbb{R}_+$ of nonnegative real numbers.

The *Boolean* semiring $\mathcal{B}$ is composed of two elements 0 and 1. The axioms imply

$$0 + 0 = 0, 0 + 1 = 1 + 0 = 1,$$

$$0 \cdot 1 = 1 \cdot 0 = 0 \cdot 0 = 0.$$

The semiring $\mathcal{B}$ is specified by

$$1 + 1 = 1.$$

The other possibility for addition is $1 + 1 = 0$, and it defines the field $\mathbb{Z}/2\mathbb{Z}$.

More generally, for any integer $d \geq 0$, consider the set $\mathcal{B}(d) = \{0, 1, \ldots, d + 1\}$. It becomes a semiring for integer addition and multiplication defined, for $i, j \in \mathcal{B}(d)$, respectively by $\min(i + j, d + 1)$ and $\min(ij, d + 1)$. In particular, $\mathcal{B}(0) = \mathcal{B}$.

For any monoid M, the set $\mathfrak{P}(M)$ is a semiring for the operations of union and set product.

A semiring K is called *ordered* if it is given with a partial order $\leq$ satisfying the following properties:

(i) 0 is the smallest element of K;
(ii) the following implications hold:

$$x \leq y \Rightarrow x + z \leq y + z,$$

$$x \leq y \Rightarrow xz \leq yz, \quad zx \leq zy.$$

The semirings $\mathcal{B}$, $\mathbb{N}$, $\mathbb{R}_+$ are ordered by the usual ordering

$$x \leq y \iff x = y + z.$$

An ordered semiring is said to be *complete* if any subset X of K admits a least upper bound in K. It is the unique element k of K such that

(i) $x \in X \Rightarrow x \leq k$,
(ii) if $x \leq k'$ for all $x \in X$, then $k \leq k'$.

We write $k = \sup(X)$ or $k = \sup\{x \mid x \in X\}$ or $k = \sup_{x \in X}(x)$. The semiring $\mathcal{B}$ is complete. The semirings $\mathbb{N}$, $\mathbb{R}_+$ are not complete, and may be completed as follows. For $K = \mathbb{N}$ or $K = \mathbb{R}_+$, we set

$$\mathcal{K} = K \cup \infty,$$

where $\infty \notin K$. The operations of K are extended to $\mathcal{K}$ by setting for $x \in K$,

(i) $x + \infty = \infty + x = \infty$,
(ii) if $x \neq 0$, then $x \infty = \infty x = \infty$,
(iii) $\infty \infty = \infty$, $\quad 0 \infty = \infty 0 = 0$.

Extending the order of K to $\mathcal{K}$ by $x \leq \infty$ for all $x \in K$, the set $\mathcal{K}$ becomes a totally ordered semiring. It is a complete semiring because any subset has an upper bound and therefore also a least upper bound. We define

$$\mathcal{N} = \mathbb{N} \cup \infty, \quad \mathcal{R}_+ = \mathbb{R}_+ \cup \infty$$

to be the complete semirings obtained by applying this construction to $\mathbb{N}$ and $\mathbb{R}_+$ respectively. If $\mathcal{K}$ is a complete semiring, the sum of an infinite family $(x_i)_{i \in I}$, of elements of $\mathcal{K}$ is defined by

$$\sum_{i \in I} x_i = \sup\left\{ \sum_{j \in J} x_j \mid J \subset I, J \text{ finite} \right\}. \tag{1.14}$$

In the case of the semiring $\mathcal{R}_+$, this gives the usual notion of a summable family: A family $(x_i)_{i \in I}$ of elements in $\mathbb{R}_+$ is summable if the sum (1.14) is finite.

In particular, for a sequence $(x_n)_{n \geq 0}$ of elements of a complete semiring, we have

$$\sum_{n \geq 0} x_n = \sup_{n \geq 0}\left\{ \sum_{i \leq n} x_i \right\}, \tag{1.15}$$

since any finite subset of $\mathbb{N}$ is contained in some interval $\{0, 1, \dots, n\}$. Moreover, if $I = \bigcup_{j \in J} I_j$ is a partition of I, then

$$\sum_{i \in I} x_i = \sum_{j \in J}\left(\sum_{i \in I_j} x_i \right). \tag{1.16}$$

Let P, Q be two sets and let K be a semiring. A $P \times Q$-*matrix* with coefficients in K is a mapping

$$m : P \times Q \to K.$$

We denote indistinctly by

$$(p, m, q) \quad \text{or} \quad m_{p,q}$$

the value of m on $(p, q) \in P \times Q$. We also say that m is a K-*relation* between P and Q. If $P = Q$, we say that it is a K-relation *over* Q. The set of all K-relations between P and Q is denoted by $K^{P \times Q}$.

Let $m \in K^{P \times Q}$ be a K-relation between P and Q. For $p \in P$, the *row* of index p of m is denoted by m_{p*}. It is the element of K^Q defined by

$$(m_{p*})_q = m_{pq}.$$

Similarly, the *column* of index q of m is denoted by m_{*q}. It is an element of K^P. Let P, Q, R be three sets and let K be a complete semiring. For $m \in K^{P \times Q}$ and $n \in K^{Q \times R}$, the product mn is defined as the following element of $K^{P \times R}$. Its value on $(p, r) \in P \times R$ is

$$(mn)_{p,r} = \sum_{q \in Q} m_{p,q} n_{q,r}.$$

When $P = Q = R$, we thus obtain an associative multiplication which turns $K^{Q \times Q}$ into a monoid. Its identity is denoted id_Q or I_Q.

A *monoid* of K-relations over Q is a submonoid of $K^{Q \times Q}$. It contains in particular the identity id_Q.

1.7 Formal series

Let A be an alphabet and let K be a semiring. A *formal series* (or just *series*) over A with coefficients in K is a mapping

$$\sigma : A^* \to K.$$

The value of σ on $w \in A^*$ is denoted (σ, w). We indifferently denote by K^{A^*} or $K \langle\!\langle A \rangle\!\rangle$ the set of formal series over A. We denote by $K \langle A \rangle$ the set of formal series $\sigma \in K \langle\!\langle A \rangle\!\rangle$ such that $(\sigma, w) = 0$ for all but a finite number of $w \in A^*$. An element of $K \langle A \rangle$ is called a *polynomial*. The *degree* of a polynomial $p \neq 0$, denoted $\deg(p)$, is the maximal length of a word w such that $(p, w) \neq 0$. The degree of the null polynomial is $-\infty$.

A series $\sigma \in K \langle\!\langle A \rangle\!\rangle$ can be extended to a linear function from $K \langle A \rangle$ into K by setting, for $p \in K \langle A \rangle$,

$$(\sigma, p) = \sum_{w \in A^*} (\sigma, w)(p, w).$$

This definition makes sense because p is a polynomial. Let $\sigma, \tau \in K \langle\!\langle A \rangle\!\rangle$ and $k \in K$. We define the formal series $\sigma + \tau$, $\sigma\tau$, and $k\sigma$ by

$$(\sigma + \tau, w) = (\sigma, w) + (\tau, w), \tag{1.17}$$

$$(\sigma\tau, w) = \sum_{uv=w} (\sigma, u)(\tau, v), \tag{1.18}$$

$$(k\sigma, w) = k(\sigma, w). \tag{1.19}$$

In (1.18), the sum runs over the $1 + |w|$ pairs (u, v) such that $w = uv$. It is therefore a finite sum. The set $K \langle\!\langle A \rangle\!\rangle$ contains two special elements denoted 0 and 1 defined by

$$(0, w) = 0, \quad (1, w) = \begin{cases} 1 & \text{if } w = 1, \\ 0 & \text{otherwise.} \end{cases}$$

As usual, we denote $\sigma^n = \sigma\sigma \cdots \sigma$ (n times) and $\sigma^0 = 1$. With the operations defined by (1.17) and (1.18) the set $K \langle\!\langle A \rangle\!\rangle$ is a semiring. It may be verified that when K is complete $K \langle\!\langle A \rangle\!\rangle$ is also complete.

The *support* of a series $\sigma \in K \langle\!\langle A \rangle\!\rangle$ is the set

$$\text{supp}(\sigma) = \{w \in A^* \mid (\sigma, w) \neq 0\}.$$

The mapping $\sigma \mapsto \text{supp}(\sigma)$ is an isomorphism from $\mathcal{B} \langle\!\langle A \rangle\!\rangle$ onto $\mathfrak{P}(A^*)$.

A family $(\sigma_i)_{i \in I}$ of series is said to be *locally finite* if for all $w \in A^*$, the set $\{i \in I \mid (\sigma_i, w) \neq 0\}$ is finite. In this case, a series σ denoted

$$\sigma = \sum_{i \in I} \sigma_i$$

can be defined by

$$(\sigma, w) = \sum_{i \in I} (\sigma_i, w). \tag{1.20}$$

This notation makes sense because in the sum (1.20) all but a finite number of terms are different from 0. We easily check that for a locally finite family $(\sigma_i)_{i \in I}$ of elements of $K \langle\!\langle A \rangle\!\rangle$ and any τ in $K \langle\!\langle A \rangle\!\rangle$, we have

$$\tau \left(\sum_{i \in I} \sigma_i \right) = \sum_{i \in I} \tau \sigma_i.$$

Let $\sigma \in K \langle\!\langle A \rangle\!\rangle$ be a series. The *constant term* of σ is the element $(\sigma, 1)$ of K. If σ has zero constant term, then the family $(\sigma^n)_{n \geq 0}$ is locally finite, because the support of σ^n does not contain words of length less than n. We denote by σ^* and by σ^+ the series

$$\sigma^* = \sum_{n \geq 0} \sigma^n, \quad \sigma^+ = \sum_{n \geq 1} \sigma^n.$$

The series σ^* is called *star* of σ. Note that $\sigma^* = 1 + \sigma^+$ and $\sigma^* \sigma = \sigma \sigma^* = \sigma^+$.

Proposition 1.7.1 *Let K be a ring with unit and let $\sigma \in K \langle\!\langle A \rangle\!\rangle$ be a series such that $(\sigma, 1) = 0$. Then $1 - \sigma$ is invertible and*

$$\sigma^* = (1 - \sigma)^{-1}. \tag{1.21}$$

Proof. We have

$$1 = \sigma^* - \sigma^+ = \sigma^* - \sigma^* \sigma = \sigma^* (1 - \sigma).$$

Symmetrically, $1 = (1 - \sigma)\sigma^*$, hence the result. □

For $X \subset A^*$, we denote by $\underline{X}$ the *characteristic series* of X defined by

$$(\underline{X}, x) = \begin{cases} 1 & \text{if } x \in X, \\ 0 & \text{otherwise.} \end{cases}$$

We consider the characteristic series $\underline{X}$ of X as an element of $\mathbb{N}\langle\!\langle A \rangle\!\rangle$. When $X = \{x\}$ we usually write x instead of $\underline{x}$. In particular, since the family $(x)_{x \in X}$ is locally finite, we have $\underline{X} = \sum_{x \in X} x$. More generally, we have for any series $\sigma \in K \langle\!\langle A \rangle\!\rangle$,

$$\sigma = \sum_{w \in A^*} (\sigma, w) w.$$

Proposition 1.7.2 *Let $X, Y \subset A^*$. Then*

$$(\underline{X + Y}, w) = \begin{cases} 0 & \text{if } w \notin X \cup Y, \\ 1 & \text{if } w \in (X \setminus Y) \cup (Y \setminus X), \\ 2 & \text{if } w \in X \cap Y. \end{cases}$$

In particular, with $Z = X \cup Y$,

$$\underline{X} + \underline{Y} = \underline{Z} \quad \textit{if and only if} \quad X \cap Y = \emptyset. \qquad \square$$

Given two sets $X, Y \subset A^*$, the product XY is said to be *unambiguous* if any word $w \in XY$ has only one factorization $w = xy$ with $x \in X$, $y \in Y$.

Proposition 1.7.3 *Let* $X, Y \subset A^*$. *Then*

$$(\underline{X}\,\underline{Y}, w) = \mathrm{Card}\{(x, y) \in X \times Y \mid w = xy\}.$$

In particular, with $Z = XY$,

$$\underline{Z} = \underline{X}\,\underline{Y}$$

if and only if the product XY *is unambiguous.* $\qquad \square$

The following proposition approaches very closely the main subject of this book. It describes the coefficients of the star of a characteristic series.

Proposition 1.7.4 *For* $X \subset A^+$, *we have*

$$((\underline{X})^*, w) = \mathrm{Card}\{(x_1, \ldots, x_n) \mid n \geq 0, x_i \in X, w = x_1 x_2 \cdots x_n\}. \qquad (1.22)$$

Proof. By the definition of $(\underline{X})^*$ we have

$$((\underline{X})^*, w) = \sum_{k \geq 0} ((\underline{X})^k, w).$$

Applying Proposition 1.7.3, we obtain

$$((\underline{X})^k, w) = \mathrm{Card}\{(x_1, x_2, \ldots, x_k) \mid x_i \in X, w = x_l x_2 \ldots x_k\}$$

whence Formula (1.22). $\qquad \square$

Example 1.7.5 The series $\underline{A}^*$ and $\underline{A}^* \underline{A}^*$ satisfy

$$\underline{A}^* = (1 - \underline{A})^{-1} = \sum_{w \in A^*} w, \quad (\underline{A}^* \underline{A}^*, w) = 1 + |w|.$$

We now define the *Hadamard product* of two series $\sigma, \tau \in K \langle\!\langle A \rangle\!\rangle$ as the series $\sigma \odot \tau$ given by

$$(\sigma \odot \tau, w) = (\sigma, w)(\tau, w).$$

This product is distributive over addition, that is $\sigma \odot (\tau + \tau') = \sigma \odot \tau + \sigma \odot \tau'$. If the semiring K satisfies $xy = 0 \Rightarrow x = 0$ or $y = 0$, then

$$\mathrm{supp}(\sigma \odot \tau) = \mathrm{supp}(\sigma) \cap \mathrm{supp}(\tau).$$

In particular, for $X, Y \subset A^*$ and $Z = X \cap Y$,

$$\underline{Z} = \underline{X} \odot \underline{Y}.$$

Given two series $\sigma, \tau \in \mathbb{Z}\langle\langle A \rangle\rangle$ we write $\sigma \leq \tau$ when $(\sigma, w) \leq (\tau, w)$ for all $w \in A^*$.

Let A be an alphabet and let K be a semiring. We denote by $K[[A]]$ the set of formal power series in commutative variables in A with coefficients in K. It is the set of mappings from the free commutative monoid $A^\oplus$ into K.

The canonical morphism α from A^* onto $A^\oplus$ extends by linearity to a morphism from $K\langle\langle A \rangle\rangle$ onto $K[[A]]$. The image by α of a series $\sigma \in K\langle\langle A \rangle\rangle$ is defined, for $w \in A^\oplus$, by

$$(\alpha(\sigma), w) = (\sigma, \alpha^{-1}(w)) = \sum_{\alpha(v)=w} (\sigma, v).$$

The set of commutative polynomials is denoted by $K[A]$.

1.8 Power series

The *power series* in the variable t associated to a sequence a_n of real numbers is the formal sum

$$f(t) = \sum_{n \geq 0} a_n t^n.$$

Given a real number r, the series is said to *converge* for the value r of t if the sum $\sum_{n \geq 0} a_n r^n$ is well defined and finite. Otherwise, $f(t)$ is said to *diverge* for $t = r$. The *radius of convergence* of $f(t)$ is infinite if $f(t)$ converges for all real numbers r. Otherwise, it is the nonnegative real number ρ such that $f(t)$ converges for $0 \leq r < \rho$ and diverges for $r > \rho$. It can be shown that $\rho = \liminf |a_n|^{1/n}$. The series may converge or diverge for $t = \rho$.

For $0 \leq r < \rho$, the series converges. This defines a function from the interval $[0, \rho)$ into the nonnegative reals. For example, $\sum_{n \geq 0} t^n$ defines on the interval $[0, 1)$ the rational function $t \mapsto 1/(1 - t)$.

Example 1.8.1 The series $\sum t^n / n^\alpha$ has radius of convergence 1 for any positive real α. It is known to diverge for $t = 1$ when $\alpha < 2$ and to converge when $\alpha \geq 2$.

Power series, as considered here, are a special case of formal series considered in Section 1.7, when the alphabet is a singleton. In particular, the usual operations of sum, product, and star hold also in this case.

Given a set X of words over an alphabet A, the *generating series* of X is the power series

$$f_X(t) = \sum_{n \geq 0} \text{Card}(X \cap A^n) t^n.$$

Since for all $n \geq 0$, one has $\mathrm{Card}(X \cap A^n) \leq k^n$, with $k = \mathrm{Card}(A)$, it follows that the radius of convergence of f_X is at least $1/k$. The sequence $(u_n)_{n \geq 0}$ where $u_n = \mathrm{Card}(X \cap A^n)$ is called the *length distribution* of the set X.

Proposition 1.8.2 *Let $f(t) = \sum a_n t^n$ be a power series with nonnegative real coefficients, and with finite radius of convergence ρ, and let $g(t) : [0, \rho) \to \mathbb{R}_+$ be the function defined for $r \in [0, \rho)$ by $g(r) = \sum a_n r^n$. Then $f(\rho) = \lim_{r \to \rho, r < \rho} g(r)$. In particular, both quantities are simultaneously finite or infinite.*

Proof. Suppose first that $f(t)$ converges for $t = \rho$, and set $s = f(\rho)$. Given ϵ, there exists an integer N such that $s_N = a_0 + a_1 \rho + \cdots + a_N \rho^N$ satisfies the inequality $s \geq s_N > s - \epsilon/2$. Set $p(t) = a_0 + a_1 t + \cdots + a_n t^N$. There exists a real r with $r < \rho$ such that $s_N \geq p(r) > s_N - \epsilon/2$. For $r \leq x < \rho$, one has $f(\rho) \geq f(x) = g(x) \geq g(r) > p(r) > s_N - \epsilon/2 \geq f(\rho) - \epsilon$. This shows that $g(x)$ tends to $f(\rho)$ when x tends to ρ.

Next, if $f(\rho)$ is infinite, for each $M > 0$ there exists an integer N such that $s_N = a_0 + a_1 \rho + \cdots + a_N \rho^N$ satisfies the inequality $s_N > 2M$. Set again $p(t) = a_0 + a_1 t + \cdots + a_n t^N$. There exists a real r with $r < \rho$ such that $p(r) > s_N/2$. For $r \leq x < \rho$, one has $f(x) = g(x) \geq g(r) > p(r) > s_N/2 \geq M$. This shows that $g(x)$ tends to infinity when x tends to ρ. $\square$

Thus, for a power series $f(t) = \sum_n a_n t^n$ with nonegative coefficients and radius of convergence ρ, we can denote, by the expression $f(r)$, for $0 \leq r \leq \rho$, indifferently the sum $\sum_n a_n r^n$ and the value of the function defined by f for $t = r$, with the property that both values are simultaneously finite or infinite.

Note that this statement only holds because the a_n are nonnegative. Indeed, consider for example $f(t) = \sum (-1)^n t^n$. Here the radius of convergence is 1, and $g(t) = 1/(1 + t)$. We have $g(1) = 1/2$, although $f(t)$ diverges for $t = 1$.

A power series $f(t) = \sum_{n \geq 0} a_n t^n$ with real coefficients can be derivated formally. The result is the series $\sum_{n \geq 0} n a_n t^n$, denoted by $f'(t)$. Let ρ be the radius of convergence of f. For $r < \rho$, $f'(r)$ is equal to the value at r of the derivative of the function defined by f.

Proposition 1.8.3 *Let $f(t)$ be a power series with nonnegative real coefficients. Let ρ be the radius of convergence of f. Then $f'(\rho) = \sum_{n \geq 0} n a_n \rho^n$.*

Proof. This results directly from Proposition 1.8.2. $\square$

The next proposition gives a method for computing the radius of convergence of the star of a power series.

Proposition 1.8.4 *Let $f(t) = \sum_{n \geq 0} a_n t^n$ be a power series with nonnegative real coefficients and with constant term zero. Consider the power series*

$$g(t) = \frac{1}{1 - f(t)} = \sum_{n=0}^{\infty} f(t)^n$$

which is the star of $f(t)$, and denote by ρ_f and ρ_g the radius of convergence of f and g respectively. Then $\rho_g \leq \rho_f$, and if $\rho_g < \rho_f$, then ρ_g is the unique positive real number such that $f(\rho_g) = 1$.

Proof. The coefficients of $g(t)$ are greater than or equal to those of $f(t)$, so $\rho_g \leq \rho_f$. Assume now that $\rho_g < \rho_f$. Then the series $f(t)$ converges for $r = \rho_g$. We use the fact that $f(t)$ defines a continuous function inside its interval of convergence.

Suppose first that $f(r) < 1$. Then there exists a real number s with $r < s < \rho_f$ such that $f(s) < 1$. This implies that $g(s) < \infty$, contradicting the fact that $s > \rho_g$.

Suppose next that $f(r) > 1$. There exists a real number s with $0 < s < r$ such that $f(s) > 1$. This implies that $g(s) = \infty$, contradicting the fact that $s < \rho_g$.

Thus $f(r) = 1$.   □

1.9 Nonnegative matrices

We now consider properties of nonnegative matrices. Let Q be a set of indices. For two Q-vectors v, w with real coordinates, one writes $v \leq w$ if $v_q \leq w_q$ for all $q \in Q$ and $v < w$ if $v_q < w_q$ for all $q \in Q$. A vector v is said to be *nonnegative* (resp. *positive*) if $v \geq 0$ (resp. $v > 0$). Here and below, we denote by 0 the null vector or the null matrix of appropriate size. In the same way, for two $Q \times Q$-matrices M, N with real coefficients, one writes $M \leq N$ when $M_{p,q} \leq N_{p,q}$ for all $p, q \in Q$ and $M < N$ when $M_{p,q} < N_{p,q}$ for all $p, q \in Q$. The $Q \times Q$-matrix M is said to be *nonnegative* (resp. *positive*) if $M \geq 0$ (resp. $M > 0$). We shall use often the elementary fact that if $M > 0$ and $v \geq 0$ with $v \neq 0$, then $Mv > 0$.

A complex number λ is an *eigenvalue* of M if the matrix $\lambda I - M$ is not invertible. In this case there exist vectors $v, w \neq 0$ such that $Mv = \lambda v$ and $wM = \lambda w$. The vectors w, r are left and right *eigenvectors* corresponding to the eigenvalue λ. The *spectral radius* of a matrix is the maximal modulus of its eigenvalues.

A nonnegative matrix M is said to be *stochastic* if the sum of its elements on each row is 1. Equivalently M is stochastic if the vector v with all components equal to 1 is a (right) eigenvector for the eigenvalue 1.

Proposition 1.9.1 *The spectral radius of a stochastic matrix is equal to* 1.

Proof. Let λ be an eigenvalue of the $n \times n$ stochastic matrix M. Let v be a corresponding right eigenvector. Dividing all components of v by the maximum of their modulus, we may assume that $|v_j| \leq 1$ for $1 \leq j \leq n$ and $|v_i| = 1$ for some i. Then $\lambda v_i = \sum_{i=1}^{n} M_{ij} v_j$ implies $|\lambda| \leq \sum_{i=1}^{n} M_{ij}|v_j| \leq \sum_{i=1}^{n} M_{ij} = 1$.   □

The *adjacency matrix* of a finite deterministic automaton $\mathcal{A}$ over the alphabet A with set of states Q is the $Q \times Q$-matrix M with coefficients

$$M_{p,q} = \text{Card}\{a \in A \mid p \cdot a = q\}.$$

Let $k = \text{Card } A$. The matrix M/k is stochastic. A corresponding right eigenvector is the vector with all components equal to 1. It is also an eigenvector of M for the eigenvalue k. By Proposition 1.9.1, the spectral radius of M/k is 1, and therefore the spectral radius of M is k.

If M is the adjacency matrix of a graph G, a useful way to think about an eigenvector v of M is that it assigns a weight v_q to each vertex q. The equality $Mv = \lambda v$ corresponds to the condition that for each vertex p, if we add up the weights of the ends of all edges starting at p, the sum is λ times the weight of p.

A nonnegative matrix M is said to be *irreducible* if for all indices p, q, there is an integer k such that $M_{p,q}^k > 0$, where M^k denotes the k-th power of M. Otherwise, it is called *reducible*. It is easy to verify that M is irreducible if and only if $(I + M)^n > 0$ where n is the dimension of M. It is also easy to prove that M is reducible if there is a reordering of the indices such that M is block triangular, that is of the form

$$M = \begin{bmatrix} U & V \\ 0 & W \end{bmatrix} \tag{1.23}$$

with U, W of dimension > 0.

The following result is part of a theorem known as the Perron–Frobenius theorem. It says in particular that the spectral radius of a nonnegative matrix is an eigenvalue.

Theorem 1.9.2 (Perron–Frobenius) *Any nonnegative matrix M has a real eigenvalue ρ_M such that $|\lambda| \le \rho_M$ for any eigenvalue λ of M, and there corresponds to ρ_M a nonnegative eigenvector v. If M is irreducible, there corresponds to ρ_M a positive eigenvector v.*

Observe that the same result holds both for right and for left eigenvectors.

Before the proof, we state a result of independent interest which will be used in the proof. A sequence $(M_n)_{n \ge 0}$ of real $m \times m$-matrices is said to *converge* if, setting $M_n = (a_{p,q}^{(n)})$, each of the real sequences $(a_{p,q}^{(n)})_{n \ge 0}$ converges. A series $\sum M_n$ of matrices converges if the sequence $(S_m)_{m \ge 0}$ defined by $S_m = \sum_{n \le m} M_n$ converges.

Proposition 1.9.3 *Let M be an $m \times m$-matrix with real coefficients. If the spectral radius ρ of M satisfies $\rho < 1$, then $\sum_n M^n$ converges.*

Proof. Set $N(z) = I - Mz$, where I is the identity matrix and z is a variable. The polynomial $N(z)$ can be considered both as a polynomial with coefficients in the ring of $m \times m$-matrices or as an $m \times m$-matrix with coefficients in the ring of real polynomials in the variable z. The polynomial $N(z)$ is invertible in both structures, and its inverse $N(z)^{-1} = (I - Mz)^{-1}$ can in turn be viewed as a power series with coefficients in the ring of $m \times m$-matrices or as a matrix whose coefficients are rational fractions in the variable z. The radius of convergence of $N(z)^{-1}$, viewed as a power series in z with matrix coefficients, is equal to the minimum of the radius of convergence of the elements of $N(z)^{-1}$, viewed as a matrix of power series expansions of rational fractions. All these rational fractions have denominator $\det(I - Mz)$. Thus the radius of convergence of the expansion of each rational fraction is at least $1/\rho$. Consequently the radius of convergence of $N(z)^{-1}$ is at least $1/\rho$. $\square$

Proof of Theorem 1.9.2. Let us first show that one may reduce to the case where M is irreducible. Indeed, if M is reducible, we may consider a triangular decomposition as in Equation (1.23) above. Applying by induction the theorem to U and W, we obtain nonnegative eigenvectors u and v for the eigenvalues ρ_U and ρ_V of U and V. We prove that $\max(\rho_U, \rho_V)$ is an eigenvalue of M with some nonnegative eigenvector.

If $\rho_U \geq \rho_V$, then ρ_U is an eigenvalue of M with the corresponding eigenvector $\begin{bmatrix} u \\ 0 \end{bmatrix}$. If $\rho_U < \rho_V$, then we show that ρ_V is an eigenvalue of M for the eigenvector $\begin{bmatrix} u' \\ v \end{bmatrix}$, where

$$u' = \left(\sum_{n \geq 0} U^n \rho_V^{-n} \right) v = (I - U/\rho_V)^{-1} v.$$

Since $\rho_U < \rho_V$, the series $\sum_{n \geq 0} U^n \rho_V^{-n}$ converges in view of Proposition 1.9.3, and it converges to a matrix with nonnegative coefficients because each U^n has nonnegative coefficients. If follows that u' has nonnegative coefficients. Moreover

$$Vv = \rho_V v = \rho_V(I - U/\rho_V)u' = \rho_V u' - Uu',$$

showing that $M \begin{bmatrix} u' \\ v \end{bmatrix} = \rho_V \begin{bmatrix} u' \\ v \end{bmatrix}$. This shows that $\rho_M \geq \max(\rho_U, \rho_V)$. Conversely, if λ is an eigenvalue of M with corresponding eigenvector $\begin{bmatrix} u \\ v \end{bmatrix}$, then λ is an eigenvalue of W if $v \neq 0$, and is an eigenvalue of U if $v = 0$. This proves that $\rho_M = \max(\rho_U, \rho_V)$.

We suppose from now on that M is irreducible. For any nonnegative Q-vector $v \neq 0$, let

$$r_M(v) = \min\{(Mv)_i/v_i \mid 1 \leq i \leq n, \ v_i \neq 0\}.$$

Thus $r_M(v)$ is the largest real number r such that $Mv \geq rv$. One has $r_M(\lambda v) = r_M(v)$ for any real number $\lambda \neq 0$. Moreover, the mapping $v \mapsto r_M(v)$ is continuous on the set of nonnegative nonzero vectors.

The set X of nonnegative vectors v such that $\|v\| = 1$ is compact. Define ρ_M by $\rho_M = \max\{r_M(w) \mid w \in X\}$. Since a continuous function on a compact set reaches its maximum on this set, there is an $x \in X$ such that $r_M(x) = \rho_M$. Since $r_M(v) = r_M(\lambda v)$ for $\lambda \neq 0$, we have $\rho_M = \max\{r_M(w) \mid w \geq 0, w \neq 0\}$.

We show that $Mx = \rho_M x$. By the definition of the function r_M, we have $Mx \geq \rho_M x$. Set $y = Mx - \rho_M x$. Then $y \geq 0$. Assume $Mx \neq \rho_M x$. Then $y \neq 0$. Since $(I + M)^n > 0$, this implies that the vector $(I + M)^n y$ is positive. But

$$(I + M)^n y = (I + M)^n (Mx - \rho_M x) = M(I + M)^n x - \rho_M (I + M)^n x = Mz - \rho_M z,$$

with $z = (I + M)^n x$. This shows that $Mz > \rho_M z$, which implies that $r_M(z) > \rho_M$, a contradiction with the definition of ρ_M. This shows that ρ_M is an eigenvalue with a nonnegative eigenvector.

Let us show that $\rho_M \geq |\lambda|$ for each real or complex eigenvalue λ of M. Indeed, let v be an eigenvector corresponding to λ. Then $Mv = \lambda v$. Let $|v|$ be the nonnegative vector with coordinates $|v_i|$. Then $M|v| \geq |\lambda||v|$ by the triangular inequality. By the definition of the function r_M, this implies $r_M(|v|) \geq |\lambda|$ and consequently $\rho_M \geq |\lambda|$.

We have already seen that there corresponds to ρ_M a nonnegative eigenvector x. Let us now verify that $x > 0$. But this is easy since $(I + M)^n x = (1 + \rho_M)^n x$, which implies that $(1 + \rho_M)^n x > 0$ and thus $x > 0$. □

Example 1.9.4 Let $M = \begin{bmatrix} 1 & 1 \\ 1 & 0 \end{bmatrix}$. The eigenvalues of M are $\varphi = \frac{1+\sqrt{5}}{2}$ and $\varphi' = \frac{1-\sqrt{5}}{2}$ which are the root of $z^2 - z - 1 = 0$. There corresponds to φ the nonnegative left eigenvector $[\varphi \ 1]$.

As an example of application of Theorem 1.9.2, we obtain the following result.

Proposition 1.9.5 *Each stochastic matrix has a nonnegative left eigenvector for the eigenvalue 1.*

Proof. Let M be a stochastic matrix. By Proposition 1.9.1, its spectral radius is 1. By Theorem 1.9.2, there exists a corresponding nonnegative left eigenvector. □

Recall that the adjacency matrix of a deterministic automaton over a k-letter alphabet has radius of convergence k and has a corresponding right eigenvector with all components equal to 1. By Theorem 1.9.2, it has also a left eigenvector with nonnegative components corresponding to the eigenvalue k.

Let k be an integer. A k-*approximate eigenvector* of a nonnegative matrix M is, by definition, a vector $v \neq 0$ with integer nonnegative components such that

$$Mv \leq kv.$$

Again, if one assumes that M is the adjacency matrix of a graph G, then an approximate eigenvector of M assigns a nonnegative integer weight v_q to each vertex q and the vector inequality $Mv \leq kv$ corresponds to the condition that for each vertex p, the sum of the weights of the ends of all edges starting at p is at most k times the weight of p. We will use the following result.

Proposition 1.9.6 *An irreducible nonnegative and integral matrix M with spectral radius λ admits a positive k-approximate eigenvector if and only if $k \geq \lambda$.*

Proof. Suppose first that $k > \lambda$. Consider the matrix $N = kI - M$. Since $k > \lambda$, we have $\det(N) > 0$ and therefore N is invertible. Moreover, since $N^{-1} = (I + M/k + M^2/k^2 + \cdots)/k$, and since M is irreducible, the matrix N^{-1} is positive. Let v be a column of N^{-1}. We have $Nv \geq 0$ and thus $Mv \leq kv$. Any column of N^{-1} is then a positive k-approximate eigenvector of M.

If $k = \lambda$, there is by Theorem 1.9.2, a positive vector v such that $Mv = kv$. Since λ is an integer, the coefficients of v can be chosen to be integers.

Let us finally prove that conversely, if M admits a positive k-approximate eigenvector v, then $k \geq \lambda$. Consider the matrix $N = \frac{1}{\lambda} M$. By Theorem 1.9.2, there is a positive vector w such that $Nw = w$. We have $Nv \leq (k/\lambda)v$, implying that $N^n v \leq (k/\lambda)^n v$ for all $n \geq 1$. If $\lambda > k$, the right-hand side tends to 0 as $n \to \infty$, thus N^n tends to the zero matrix, a contradiction with the fact that $N^n w = w$ with $w > 0$. □

Example 1.9.7 Let $M = \begin{bmatrix} 1 & 1 \\ 1 & 0 \end{bmatrix}$. The spectral radius of M is strictly less than 2 and a 2-approximate eigenvector is $\begin{bmatrix} 1 \\ 1 \end{bmatrix}$.

1.10 Weighted automata

Let A be an alphabet. With each automaton $\mathcal{A} = (Q, I, T)$ over A with set of edges E is associated a function denoted by μ_A

$$\mu_A : A \to \mathcal{N}^{Q \times Q}$$

defined by

$$(p, \mu_A(a), q) = \begin{cases} 1 & \text{if } (p, a, q) \in E, \\ 0 & \text{otherwise.} \end{cases}$$

This function extends into a morphism, still denoted μ_A, from A^* into the monoid $\mathcal{N}^{Q \times Q}$ of $\mathcal{N}$-relations over Q (see Section 1.6). In particular, we have

$$\mu_A(1) = I_Q,$$

where I_Q is the identity relation over Q, and for $u, v \in A^*$

$$(p, \mu_A(uv), q) = \sum_{r \in Q} (p, \mu_A(u), r)(r, \mu_A(v), q).$$

The morphism μ_A is called the *representation associated* with $\mathcal{A}$. The correspondence between μ_A and the morphism φ_A defined in Section 1.4 is given by:

$$(p, q) \in \varphi_A(w) \iff (p, \mu_A(w), q) \neq 0.$$

Proposition 1.10.1 *Let $\mathcal{A} = (Q, I, T)$ be an automaton over A. For all $p, q \in Q$ and $w \in A^*$, $(p, \mu_A(w), q)$ is the (possibly infinite) number of paths from p to q with label w.* □

A path $c : i \to t$ is called *successful* if $i \in I$ and $t \in T$. The *behavior* of the automaton $\mathcal{A} = (Q, I, T)$ is the formal power series denoted $|\mathcal{A}|$ and defined by

$$(|\mathcal{A}|, w) = \sum_{i \in I, t \in T} (i, \mu_A(w), t). \tag{1.24}$$

The set *recognized* by $\mathcal{A}$ is the support of $|\mathcal{A}|$. It is just the set of all labels of successful paths. It is denoted by $L(\mathcal{A})$, as in Section 1.4.

Proposition 1.10.2 *Let $\mathcal{A} = (Q, I, T)$ be an automaton over A. For all $w \in A^*$, $(|\mathcal{A}|, w)$ is the (possibly infinite) number of successful paths labeled by w.* □

A more compact writing of Formula (1.24) consists in

$$(|\mathcal{A}|, w) = I \mu_A(w) T. \tag{1.25}$$

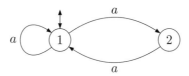

Figure 1.6 The Fibonacci automaton.

Here, the element $I \in \mathcal{N}^Q$ is considered as a row vector and $T \in \mathcal{N}^Q$ as a column vector, both with coefficients 0 and 1.

Example 1.10.3 Let $\mathcal{A}$ be the automaton given by Figure 1.6, with $I = T = \{1\}$. Its behavior is the series

$$|\mathcal{A}| = \sum_{n \geq 0} f_{n+1} a^n,$$

where f_n is the n-th *Fibonacci number*. These numbers are defined by $f_0 = 0$, $f_1 = 1$, and

$$f_{n+1} = f_n + f_{n-1}, \quad (n \geq 1).$$

For $n \geq 1$, we have

$$\mu_\mathcal{A}(a^n) = \begin{bmatrix} f_{n+1} & f_n \\ f_n & f_{n-1} \end{bmatrix}.$$

Proposition 1.10.4 *Let $\mathcal{A} = (Q, I, T)$ be a finite automaton over A. For each integer d, the set $\{w \in A^* \mid (|\mathcal{A}|, w) = d\}$ is regular.*

Proof. Let M be the monoid of $Q \times Q$-matrices over the semiring $\mathcal{B}(d)$. For each word w, let $\alpha(w)$ be the $Q \times Q$-matrix over $\mathcal{B}(d)$ obtained from $\mu_\mathcal{A}(w)$ by replacing each entry $\mu_\mathcal{A}(w)_{p,q}$ by $\min(d + 1, \mu_\mathcal{A}(w)_{p,q})$. Since such a replacement is a morphism from $\mathcal{N}$ onto $\mathcal{B}(d)$, the mapping α is a morphism from A^* into M. The set $\{w \in A^* \mid (|\mathcal{A}|, w) = d\}$ is recognized by α; it is indeed the set of words w such that $I\alpha(w)T$ (computed in $\mathcal{B}(d)$) equals d. □

To each automaton $\mathcal{A} = (Q, I, T)$, we associate an automaton denoted $\mathcal{A}^*$ and called the *star* of the automaton $\mathcal{A}$ by a canonical construction consisting of the two following steps. Let $\omega \notin Q$ be a new state, and let

$$\mathcal{B} = (Q \cup \omega, \omega, \omega) \tag{1.26}$$

be the automaton with edges

$$F = E \cup \widehat{I} \cup \widehat{T} \cup \widehat{O},$$

where E is the set of edges of $\mathcal{A}$, and

$$\widehat{I} = \{(\omega, a, q) \mid \exists i \in I : (i, a, q) \in E\}, \tag{1.27}$$

$$\widehat{T} = \{(q, a, \omega) \mid \exists t \in T : (q, a, t) \in E\}, \tag{1.28}$$

$$\widehat{O} = \{(\omega, a, \omega) \mid \exists i \in I, t \in T : (i, a, t) \in E\}. \tag{1.29}$$

By definition, the automaton $\mathcal{A}^*$ is the trim part of $\mathcal{B}$.

The following terminology is convenient for automata of the form $\mathcal{A} = (Q, 1, 1)$ having just one initial state which is also the unique final state.

A path

$$c : p \xrightarrow{w} q$$

is called *simple* if it is not the null path (that is $w \in A^+$) and if for any factorization

$$c : p \xrightarrow{u} r \xrightarrow{v} q$$

of the path c into two nonnull paths, we have $r \neq 1$.

Any path c from p to q either is the null path or is simple or decomposes in a unique manner as

$$c : p \xrightarrow{u} 1 \xrightarrow{x_1} 1 \xrightarrow{x_2} 1 \cdots 1 \xrightarrow{x_n} 1 \xrightarrow{v} q,$$

where each of these $n + 2$ paths is simple.

Proposition 1.10.5 *Let $X \subset A^+$, and let $\mathcal{A}$ be an automaton such that $|\mathcal{A}| = \underline{X}$. Then*

$$|\mathcal{A}^*| = (\underline{X})^*. \tag{1.30}$$

Proof. Since $\mathcal{A}^*$ is the trim part of the automaton $\mathcal{B}$ defined by Formula (1.26), it suffices to show that $|\mathcal{B}| = |\mathcal{A}|^*$.

Let S be the power series defined as follows: for all $w \in A^*$, (S, w) is the number of simple paths from ω to ω labeled with w. By the preceding remarks, we have

$$|\mathcal{B}| = S^*.$$

Thus it remains to prove that

$$S = \underline{X}.$$

Let $w \in A^*$. If $w = 1$, then

$$(S, 1) = (\underline{X}, 1) = 0,$$

since a simple path is not null. If $w = a \in A$, then $(S, a) = 1$ if and only if $a \in X$, according to Formula (1.29). Assume now $|w| \geq 2$. Set $w = aub$ with $a, b \in A$ and $u \in A^*$. Each simple path $c : \omega \xrightarrow{w} \omega$ factorizes uniquely into

$$c : \omega \xrightarrow{a} p \xrightarrow{u} q \xrightarrow{b} \omega$$

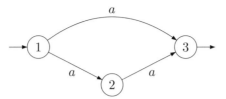

Figure 1.7 An automaton with behavior $\underline{X}$, for $X = \{a, aa\}$.

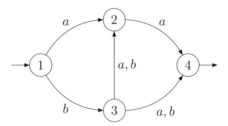

Figure 1.8 An automaton with behavior $\underline{X}$, for $X = \{aa, ba, baa, bb, bba\}$.

for some $p, q \in Q$. There exists at least one successful path

$$i \xrightarrow{a} p \xrightarrow{u} q \xrightarrow{b} t$$

in $\mathcal{A}$. This path is unique because the behavior of $\mathcal{A}$ is a characteristic series. If there is another simple path $c' : \omega \xrightarrow{w} \omega$ in $\mathcal{B}$, then there is also another successful path labeled w in $\mathcal{A}$; this is impossible. Thus there is at most one simple path $c : \omega \xrightarrow{w} \omega$ in $\mathcal{B}$ and such a path exists if and only if $w \in X$. Consequently, $S = \underline{X}$, which was to be proved. □

Example 1.10.6 Let $X = \{a, a^2\}$. Then $\underline{X} = |\mathcal{A}|$ for the automaton given in Figure 1.7, with $I = \{1\}$, $T = \{3\}$. The automaton $\mathcal{A}^*$ is the automaton of Figure 1.6 up to a renaming of ω. Consequently, for $n \geq 0$

$$((\underline{X})^*, a^n) = f_n.$$

Example 1.10.7 Let $X = \{aa, ba, baa, bb, bba\}$. We have $\underline{X} = |\mathcal{A}|$ for the automaton $\mathcal{A}$ of Figure 1.8, with $I = \{1\}$, $T = \{4\}$. The corresponding automaton $\mathcal{A}^*$ is given in Figure 1.9.

We now extend the previous definitions to the more general case where the labels of the edges of an automaton may be weighted. Let A be an alphabet and let K be a semiring. A finite *weighted automaton* $\mathcal{A} = (Q, I, T)$ over the alphabet A and with weights in K is given by a finite set Q with two mappings $I, T : Q \to K$ and by a mapping

$$E : Q \times A \times Q \to K.$$

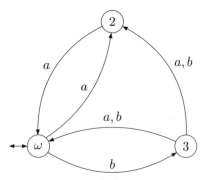

Figure 1.9 An automaton recognizing X^*, for $X = \{aa, ba, baa, bb, bba\}$.

If $E(p, a, q) = k \neq 0$, then we say that (p, a, q) is an edge with label a and weight k and we write $p \xrightarrow{ka} q$. If c is the path

$$p \xrightarrow{k_1 a_1} q_1 \rightarrow \cdots \rightarrow q_{n-1} \xrightarrow{k_n a_n} q$$

then its label is $x = a_1 \cdots a_n$ and its weight is the product $|c| = k_1 \cdots k_n$. We write $c : p \xrightarrow{x} q$ for denoting such a path. The *behavior* of $\mathcal{A}$ is the series denoted $|\mathcal{A}|$ and defined by

$$(|\mathcal{A}|, x) = \sum_{c:p \xrightarrow{x} q} I(p)|c|T(q).$$

Since for each $x \in A^*$, there are only finitely many paths with label x, the sum is well defined. The behavior is also called the series *recognized* by the weighted automaton. A series u is called K-*rational* if it is the behavior of a weighted automaton with weights in the semiring K. We will be particularly interested in $\mathbb{N}$-rational series.

There is an alternative form of the series recognized by a weighted automaton $\mathcal{A} = (Q, I, T)$. Define a morphism μ from A^* into the multiplicative monoid of $Q \times Q$-matrices with coefficients in K by setting, for $a \in A$,

$$\mu(a)_{pq} = E(p, a, q).$$

Then, for any $x \in A^*$, we have

$$(|\mathcal{A}|, x) = I\mu(x)T,$$

with I considered as a row vector and T considered as a column vector. The morphism μ is called the *matrix representation* of $\mathcal{A}$.

Example 1.10.8 Any automaton can be considered as a weighted automaton with weights in the Boolean semiring $\mathcal{B}$, or in the semiring $\mathbb{N}$. In the latter case, the behavior is the number of successful paths.

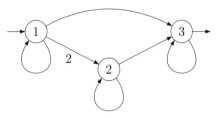

Figure 1.10 A weighted automaton over a single letter alphabet.

Example 1.10.9 The weighted automaton of Figure 1.10 has integer weights and a one letter alphabet. For simplicity, the letter is not specified, and the weight of an edge is not indicated if it is 1. The value of the behavior on the word of length n is n^2.

Indeed, denote by u_n, v_n, w_n the sum of the weights of the paths of length n ending in 3 and starting in 1, 2, 3 respectively. We have $w_n = 1$ for all $n \geq 0$. Next, the form of the automaton shows that $v_{n+1} = v_n + w_n$ for $n \geq 0$, whence $v_n = n$. Finally $u_{n+1} = u_n + 2v_n + w_n$, and thus $u_n = n^2$ for $n \geq 0$.

Let $\mathcal{A} = (Q, I, T)$ be a weighted automaton. When I is a singleton, that is $I(i) = 1$ for some $i \in Q$, and $I(q) = 0$ for $q \neq i$, we write i instead of I. The same convention holds for T.

A weighted automaton $\mathcal{A} = (Q, i, t)$ is said to be *trim* if for each vertex q, there is a path from i to q and a path from q to t. It is said to be *normalized* if no edge enters i, no edge leaves t, and $i \neq t$.

Proposition 1.10.10 *Any $\mathbb{N}$-rational series with zero constant term can be recognized by a normalized weighted automaton.*

Proof. Let $\mathcal{A} = (Q, I, T)$ be a weighted automaton recognizing a series with zero constant term, with edge mapping $E : Q \times A \times \times Q \to K$. Let i and t be two states not in Q, and define a weighted automaton $\mathcal{B} = (Q', i, t)$ with $Q' = Q \cup \{i, t\}$ and edge mapping $F : Q' \times A \times Q' \to K$ by

$$F(p, a, q) = E(p, a, q) \quad \text{for } p, q \in Q,$$

$$F(i, a, q) = \sum_{p \in Q} I(p)E(p, a, q) \quad \text{for } q \in Q,$$

$$F(p, a, t) = \sum_{q \in Q} E(p, a, q)T(q) \quad \text{for } p \in Q,$$

$$F(i, a, t) = \sum_{p, q \in Q} I(p)E(p, a, q)T(q).$$

The matrix representation v of $\mathcal{B}$ is related to the matrix representation μ of $\mathcal{A}$ by

$$v(a) = \begin{bmatrix} 0 & I\mu(a) & I\mu(a)T \\ 0 & \mu(a) & \mu(a)T \\ 0 & 0 & 0 \end{bmatrix}$$

where i and t are reported as the first and the last index respectively. It is easily checked that the same form holds for any word $w \in A^+$, and thus $v(w)_{i,t} = I\mu(w)T$. This holds also for $w = 1$ because $i \neq t$ and $I\mu(w)T = 0$ by assumption. This proves that $\mathcal{A}$ and $\mathcal{B}$ recognize the same series. □

We now consider power series, that is series in one variable.

Proposition 1.10.11 *For any rational subset X of A^*, the generating series $f_X(z)$ is $\mathbb{N}$-rational.*

Proof. Let $\mathcal{A}$ be a deterministic finite automaton recognizing X, and let $\mathcal{B}$ be the weighted automaton obtained by replacing all labels in $\mathcal{A}$ by the symbol z. Clearly $\mathcal{B}$ recognizes the series $\sum_{n \geq 0} \text{Card}(X \cap A^n)z^n$. □

Given a series $u(z) = \sum_{n \geq 0} u_n z^n$ with integer coefficients and with zero constant term $u_0 = 0$, we recall that $u^*(z)$ denotes the series defined by $u^*(z) = 1/(1 - u(z))$.

Proposition 1.10.12 *Let $u(z) = \sum_{n \geq 0} u_n z^n$ be an $\mathbb{N}$-rational series with zero constant term. Let $\mathcal{A} = (Q, i, t)$ be a normalized weighted automaton recognizing $u(z)$. Let $\overline{Q} = Q \setminus t$ and let $\overline{\mathcal{A}} = (\overline{Q}, i, i)$ be the weighted automaton obtained by merging i and t. The behavior of $\overline{\mathcal{A}}$ is the series $u^*(z)$.*

Proof. Recall that a path from i to i is *simple* if it does not go through i inbetween. For each $n > 0$, u_n is the sum of the weights of the simple paths of length n from i to i in $\overline{\mathcal{A}}$. Indeed, since $\mathcal{A}$ is normalized, to each simple path $\bar{c} : i \to i$ in $\overline{\mathcal{A}}$ corresponds a unique path from i to t in $\mathcal{A}$, and conversely.

Next, for $r \geq 1$, let $u_n^{(r)}$ be the sum of the weights of the paths from i to i that go exactly $(r - 1)$ times through i inbetween. Set $u^{(r)}(z) = \sum_{n \geq 0} u_n^{(r)} z^n$ and $u^{(0)}(z) = 1$. The series $u^{(*)}(z) = \sum_{r \geq 0} u^{(r)}(z)$ is the behavior of $\overline{\mathcal{A}}$.

Next, $u^{(r)}(z) = u(z)^r$ for $r \geq 0$. Since $u^*(z) = \sum_{r \geq 0} u(z)^r$, we obtain $u^*(z) = u^{(*)}(z)$. □

Observe that this proposition is related to Proposition 1.10.5 which can be used to give an alternative proof. Indeed, if $\mathcal{A} = (Q, i, t)$ is a normalized automaton, then, in the automaton $\mathcal{A}^*$, state i is no longer accessible and state t is no longer coaccessible. Thus the trimmed automaton is identical with $\overline{\mathcal{A}}$.

Example 1.10.13 Let $u(z) = z + z^2$. The weighted automaton $\mathcal{A}$ with $\mathcal{A}$ given on the left of Figure 1.11 recognizes u with $i = 1$ and $t = 3$. The weighted automaton $\overline{\mathcal{A}}$ is represented on the right.

The following statement relates weighted automata with weights in $\mathbb{N}$ with nonnegative matrices. We extend the definition of *adjacency matrix* to weighted automata. For a weighted automaton $\mathcal{A} = (Q, I, T)$, it is the $Q \times Q$ matrix M defined by

$$M_{p,q} = \sum_{a \in A} E(p, a, q),$$

where $E(p, a, q)$ is the weight of the edge (p, a, q).

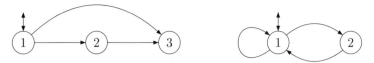

Figure 1.11 Weighted automata recognizing $z + z^2$ and $1/(1 - z - z^2)$.

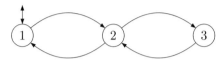

Figure 1.12 A weighted automaton recognizing $(1 - z^2)/(1 - 2z^2)$.

Proposition 1.10.14 *Let $u(z) = \sum_{n \geq 0} u_n z^n$ be an $\mathbb{N}$-rational series recognized by a trim weighted automaton and let M be the adjacency matrix of $\mathcal{A}$. The radius of convergence of the series $u(z)$ is the inverse of the maximal eigenvalue of M.*

Proof. Let λ be the maximal eigenvalue of M, which exists and is positive by the Perron–Frobenius Theorem 1.9.2. Let ρ be the radius of convergence of the series $u(z)$ and, for each $p, q \in Q$, let $\rho_{p,q}$ be the radius of convergence of the series $u_{p,q}(z) = \sum_n M_{p,q}^n z^n$. Then $1/\lambda = \min \rho_{p,q}$ since $\sum_{n \geq 0} M^n z^n$ converges for $|z| < 1/\lambda$. Next, since $\mathcal{A}$ is trim, the series $u_{p,q}(z)$ converges whenever $u(z)$ converges; thus $\rho_{p,q} \geq \rho$ for all $p, q \in Q$. On the other hand $\rho \geq \min \rho_{p,q}$ since u is a nonnegative linear combination of the series $s_{p,q}$. This implies that $\rho = \min \rho_{p,q}$, which concludes the proof. $\square$

Example 1.10.15 The weighted automaton $\mathcal{A}$ of Figure 1.12 recognizes the series

$$u(z) = \frac{1}{1 - \frac{z^2}{1-z^2}} = \frac{1 - z^2}{1 - 2z^2} = 1 + z^2 + 2z^4 + 3z^6 + 4z^8 + \cdots$$

The radius of convergence of $u(z)$ is $\sqrt{2}/2$. The adjacency matrix of $\mathcal{A}$ is

$$\begin{bmatrix} 0 & 1 & 0 \\ 1 & 0 & 1 \\ 0 & 1 & 0 \end{bmatrix}.$$

The eigenvalues are 0 and $\pm\sqrt{2}$.

1.11 Probability distributions

Given an alphabet A, a function $\pi : A^* \to [0, 1]$ such that $\pi(1) = 1$ and

$$\sum_{a \in A} \pi(wa) = \pi(w) \tag{1.31}$$

for all $w \in A^*$ is called a *probability distribution* or *distribution* for short on A^*. Condition (1.31) is called the *coherence condition*. It implies that, for each $n \geq 0$

$$\sum_{x \in A^n} \pi(x) = 1.$$

Indeed, this holds for $n = 0$, and for $n > 0$, one has

$$\sum_{x \in A^n} \pi(x) = \sum_{y \in A^{n-1}} \sum_{a \in A} \pi(ya) = \sum_{y \in A^{n-1}} \pi(y) = 1,$$

where the next-to-last equality holds by the coherence condition and the last equality holds by induction. A distribution is *positive* if $\pi(w) > 0$ for all words w.

These notions are related to usual probability theory. This will be described in Chapter 13. In particular, the coherence condition (1.31) allows to interpret a distribution as a probability corresponding to a sequence of random choices of the letters of a word from left to right.

As a particular case, a *Bernoulli distribution* is a morphism from A^* into $[0, 1]$ such that $\sum_{a \in A} \pi(a) = 1$. Clearly, a Bernoulli distribution is a probability distribution. It is *positive* if and only if $\pi(a) > 0$ for all letters a. A Bernoulli distribution corresponds to a sequence of independent trials all with the same probability. The *uniform Bernoulli distribution* is defined by $\pi(a) = 1/\operatorname{Card}(A)$ for all $a \in A$.

Given a probability distribution π on A^*, we set for any subset X of A^*,

$$\pi(X) = \sum_{x \in X} \pi(x).$$

This may be finite or infinite. The *probability generating series* of a set $X \subset A^*$ is the series

$$F_X(t) = \sum_{n \geq 0} \pi(X \cap A^n) t^n.$$

In particular, $F_X(1) = \pi(X)$. In the case of a uniform Bernoulli distribution, the probability generating series is linked with the (ordinary) generating series by

$$f_X(t) = F_X(kt), \tag{1.32}$$

where $k = \operatorname{Card}(A)$. Indeed, in this case $\operatorname{Card}(X \cap A^n) = k^n \pi(X \cap A^n)$.

A weighted automaton can be used to define a probability distribution on A^*. Recall that the *adjacency matrix* of a weighted automaton $\mathcal{A} = (Q, I, T)$ is the $Q \times Q$-matrix P defined by

$$P_{p,q} = \sum_{a \in A} E(p, a, q).$$

Consider a weighted automaton $\mathcal{A} = (Q, I, T)$ with nonnegative real weights. It is called a *stochastic automaton* if $\sum_{p \in Q} I(p) = 1$ and $T(q) = 1$ for all $q \in Q$ and if its adjacency matrix P is stochastic.

Figure 1.13 A stochastic automaton.

For a stochastic automaton $\mathcal{A}$, the mapping π defined by $\pi(x) = (|\mathcal{A}|, x)$ is a probability distribution, called the probability distribution *defined* by $\mathcal{A}$. Indeed $\pi(1) = \sum_{p \in Q} I(p) = 1$. Next, let μ be the matrix representation of $\mathcal{A}$. The adjacency matrix of $\mathcal{A}$ is $P = \sum_{a \in A} \mu(a)$. Then $PT = T$ and

$$\sum_{a \in A} \pi(xa) = \sum_{a \in A} I\mu(xa)T = I\mu(x)(\sum_{a \in A} \mu(a)T) = I\mu(x)PT = I\mu(x)T = \pi(x),$$

which shows that π satisfies the coherence condition. A probability distribution defined by a stochastic automaton is often called a *hidden Markov chain*.

A particular case of a stochastic automata occurs when the end state of an edge is in bijection with its label. In other terms, this holds if, for edges $E(p, a, q) \neq 0$, $E(p', a', q') \neq 0$

$$a = a' \iff q = q'.$$

In this case, the set of end states of edges can be identified with the alphabet. The probability distribution defined by such a stochastic automaton is called a *Markov chain*.

Example 1.11.1 Let $A = \{a, b\}$. The probability distribution on A^* defined by $\pi(ax) = 2^{-|x|}$, $\pi(bx) = 0$ for all $x \in A^*$ is defined by the stochastic automaton represented in Figure 1.13, with $I = [1\ 0]$. The matrix representation is given by

$$\mu(a) = \begin{bmatrix} 0 & 1 \\ 0 & 1/2 \end{bmatrix}, \quad \mu(b) = \begin{bmatrix} 0 & 0 \\ 0 & 1/2 \end{bmatrix}.$$

It is not a Markov chain because state 2 is the end of edges labeled a and b.

1.12 Ideals in a monoid

Let M be a monoid. A *right ideal* of M is a nonempty subset R of M such that

$$RM \subset R$$

or equivalently such that for all $r \in R$ and all $m \in M$, we have $rm \in R$. Since M is a monoid, we then have $RM = R$ because M contains a neutral element. A *left ideal* of M is a nonempty subset L of M such that $ML \subset L$. A *two-sided ideal* (also called an ideal) is a nonempty subset I of M such that

$$MIM \subset I.$$

A two-sided ideal is therefore both a left and a right ideal. In particular, M itself is an ideal of M.

If M contains a zero, the set $\{0\}$ is a two-sided ideal which is contained in any ideal of M.

An ideal I (resp. a left, right ideal) is called *minimal* if for any ideal J (resp. left, right ideal)

$$J \subset I \Rightarrow J = I.$$

If M contains a minimal two-sided ideal, it is unique because any nonempty intersection of ideals is again an ideal. If M contains a 0, the set $\{0\}$ is the minimal two-sided ideal of M. An ideal $I \neq 0$ (resp. a left, right ideal) is then called 0-*minimal* if for any ideal J (resp. left, right ideal)

$$J \subset I \Rightarrow J = 0 \text{ or } J = I.$$

For any $m \in M$, the set

$$R = mM$$

is a right ideal. It is the smallest right ideal containing m. In the same way, the set $L = Mm$ is the smallest left ideal containing m and the set $I = MmM$ is the smallest two-sided ideal containing m.

We now define in a monoid M four equivalence relations $\mathcal{L}, \mathcal{R}, \mathcal{J}$, and $\mathcal{H}$ as

$$m \mathcal{R} m' \iff mM = m'M,$$
$$m \mathcal{L} m' \iff Mm = Mm',$$
$$m \mathcal{J} m' \iff MmM = Mm'M,$$
$$m \mathcal{H} m' \iff mM = m'M \text{ and } Mm = Mm'.$$

Therefore, we have for instance, $m \mathcal{R} m'$ if and only if there exist $u, u' \in M$ such that

$$m' = mu, \quad m = m'u'.$$

We have $\mathcal{R} \subset \mathcal{J}, \mathcal{L} \subset \mathcal{J}$, and $\mathcal{H} = \mathcal{R} \cap \mathcal{L}$.

Proposition 1.12.1 *The two equivalences $\mathcal{R}$ and $\mathcal{L}$ commute: $\mathcal{R}\mathcal{L} = \mathcal{L}\mathcal{R}$.*

Proof. Let $m, n \in M$ be such that $m \mathcal{R} \mathcal{L} n$. There exists $p \in M$ such that $m \mathcal{R} p, p \mathcal{L} n$ (see Figure 1.14). There exist by the definitions, $u, u', v, v' \in M$ such that $p = mu$, $m = pu', n = vp, p = v'n$. Set $q = vm$. We then have

$$q = vm = v(pu') = (vp)u' = nu', n = vp = v(mu) = (vm)u = qu.$$

This shows that $q \mathcal{R} n$. Furthermore, we have

$$m = pu' = (v'n)u' = v'(nu') = v'q.$$

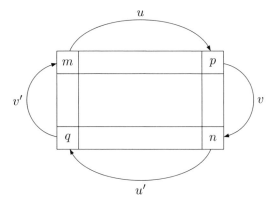

Figure 1.14 The relation $\mathcal{RL} = \mathcal{LR}$.

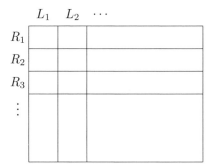

Figure 1.15 A $\mathcal{D}$-class.

Since $q = vm$ by the definition of q, we obtain $m\mathcal{L}q$. Therefore $m\mathcal{L}q\mathcal{R}n$ and consequently $m\mathcal{LR}n$. This proves the inclusion $\mathcal{RL} \subset \mathcal{LR}$. The proof of the converse inclusion is symmetrical. □

Since $\mathcal{R}$ and $\mathcal{L}$ commute, the relation $\mathcal{D}$ defined by

$$\mathcal{D} = \mathcal{RL} = \mathcal{LR}$$

is an equivalence relation. We have the inclusions

$$\mathcal{H} \subset \mathcal{R}, \mathcal{L} \subset \mathcal{D} \subset \mathcal{J}.$$

The classes of the relation $\mathcal{D}$, called $\mathcal{D}$-classes, can be represented by a schema called an "egg-box" as in Figure 1.15.

The $\mathcal{R}$-classes are represented by rows and the $\mathcal{L}$-classes by columns. The squares at the intersection of an $\mathcal{R}$-class and an $\mathcal{L}$-class are the $\mathcal{H}$-classes.

We denote by $L(m), R(m), D(m), H(m)$, respectively, the $\mathcal{L}, \mathcal{R}, \mathcal{D}$, and $\mathcal{H}$-class of an element $m \in M$. We have

$$H(m) = R(m) \cap L(m) \text{ and } R(m), L(m) \subset D(m).$$

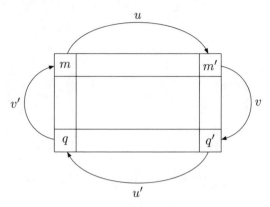

Figure 1.16 The reciprocal bijections.

Proposition 1.12.2 *Let M be a monoid. Let $m, m' \in M$ be $\mathcal{R}$-equivalent. Let $u, u' \in M$ be such that*

$$m = m'u', \quad m' = mu.$$

The mappings

$$\rho_u : q \to qu, \quad \rho_{u'} : q' \to q'u'$$

are bijections from $L(m)$ onto $L(m')$ inverse to each other which map an $\mathcal{R}$-class onto itself.

Proof. We first verify that ρ_u maps $L(m)$ into $L(m')$. If $q \in L(m)$, then $Mq = Mm$ and therefore $Mqu = Mmu = Mm'$. Hence $qu = \rho_u(q)$ is in $L(m')$. Analogously, $\rho_{u'}$ maps $L(m')$ into $L(m)$.

Let $q \in L(m)$ and compute $\rho_{u'}\rho_u(q)$. Since $q \mathcal{L} m$, there exist $v, v' \in M$ such that $q = vm, m = v'q$ (see Figure 1.16). Since $muu' = m'u' = m$, we have

$$\rho_{u'}\rho_u(q) = quu' = vmuu' = vm = q.$$

This proves that $\rho_{u'}\rho_u$ is the identity on $L(m)$. One shows in the same way that $\rho_u\rho_{u'}$ is the identity on $L(m')$.

Finally, since $quu' = q$ for all $q \in L(m)$, the elements q and $\rho_u(q)$ are in the same $\mathcal{R}$-class. $\quad\square$

Proposition 1.12.2 has the following consequence which justifies the regular shape of Figure 1.15.

Proposition 1.12.3 *Any two $\mathcal{H}$-classes contained in the same $\mathcal{D}$-class have the same cardinality.* $\quad\square$

We now address the problem of locating the idempotents in an ideal. The first result describes the $\mathcal{H}$-class of an idempotent.

Proposition 1.12.4 *Let M be a monoid and let $e \in M$ be an idempotent. The $\mathcal{H}$-class of e is the group of units of the monoid eMe.*

Proof. Let $m \in H(e)$. Then, we have for some $u, u', v, v' \in M$

$$e = mu, \quad m = eu', \quad e = vm, \quad m = v'e.$$

Therefore $em = e(eu') = eu' = m$ and in the same way $me = m$. This shows that $m \in eMe$. Since

$$m(eue) = mue = e, \quad (eve)m = evm = e,$$

the element m is both right and left invertible in M. Hence, m belongs to the group of units of eMe. Conversely, if $m \in eMe$ is right and left invertible, we have $mu = vm = e$ for some $u, v \in eMe$. Since $m = em = me$, we obtain $m\mathcal{H}e$. $\square$

Proposition 1.12.5 *An $\mathcal{H}$-class of a monoid M is a group if and only if it contains an idempotent.*

Proof. Let H be an $\mathcal{H}$-class of M. If H contains an idempotent e, then $H = H(e)$ is a group by Proposition 1.12.4. The converse is obvious. $\square$

Proposition 1.12.6 *Let M be a monoid and $m, n \in M$. Then mn is in $R(m) \cap L(n)$ if and only if $R(n) \cap L(m)$ contains an idempotent.*

Proof. If $R(n) \cap L(m)$ contains an idempotent e, then

$$e = nu, \quad n = eu', \quad e = vm, \quad m = v'e$$

for some $u, u', v, v' \in M$. Hence

$$mnu = m(nu) = me = (v'e)e = v'e = m,$$

so that $mn\mathcal{R}m$. We show in the same way that $mn\mathcal{L}n$. Thus $mn \in R(m) \cap L(n)$. Conversely, if $mn \in R(m) \cap L(n)$, then $mn\mathcal{R}m$ and $n\mathcal{L}mn$. By Proposition 1.12.2 the multiplication on the right by n is a bijection from $L(m)$ onto $L(mn)$. Since $n \in L(mn)$, this implies the existence of $e \in L(m)$ such that $en = n$. Since the multiplication by n preserves $\mathcal{R}$-classes, we have additionally $e \in R(n)$. Hence there exists $u \in M$ such that $e = nu$. Consequently

$$nunu = enu = nu$$

and $e = nu$ is an idempotent in $R(n) \cap L(m)$. $\square$

Proposition 1.12.7 *Let M be a monoid and let D be a $\mathcal{D}$-class of M. The following conditions are equivalent.*

(i) *D contains an idempotent.*
(ii) *Each $\mathcal{R}$-class of D contains an idempotent.*
(iii) *Each $\mathcal{L}$-class of D contains an idempotent.*

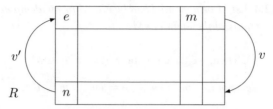

Figure 1.17 Finding an idempotent in R.

Proof. Obviously, only (i) implies (ii) requires a proof. Let $e \in D$ be an idempotent. Let R be an $\mathcal{R}$-class of D. The $\mathcal{H}$-class $H = L(e) \cap R$ is nonempty. Let n be an element of H (See Figure 1.17). Since $n \mathcal{L} e$, there exist $v, v' \in M$ such that

$$n = ve, \quad e = v'n.$$

Let $m = ev'$. Then $mn = e$ because

$$mn = (ev')n = e(v'n) = ee = e.$$

Moreover, we have $m \mathcal{R} e$ since $mn = e$ and $m = ev'$. Therefore, $e = mn$ is in $R(m) \cap L(n)$. This implies, by Proposition 1.12.6, that $R = R(n)$ contains an idempotent. $\square$

A $\mathcal{D}$-class satisfying one of the conditions of Proposition 1.12.7 is called *regular*.

Proposition 1.12.8 *Let M be a monoid and let H be an $\mathcal{H}$-class of M. The two following conditions are equivalent.*

(i) *There exist $h, h' \in H$ such that $hh' \in H$.*
(ii) *H is a group.*

Proof. (i) $\implies$ (ii). If $hh' \in H$, then by Proposition 1.12.6 H contains an idempotent. By Proposition 1.12.5, it is a group. The implication (ii) $\implies$ (i) is obvious. $\square$

We now study the minimal and 0-minimal ideals in a monoid. Recall that if M contains a minimal ideal, it is unique. However, it may contain several 0-minimal ideals.

Let M be a monoid containing a zero. We say that M is *prime* if for any $m, n \in M \setminus 0$, there exists $u \in M$ such that $mun \neq 0$.

Proposition 1.12.9 *Let M be a prime monoid.*

1. *If M contains a 0-minimal ideal, it is unique.*
2. *If M contains a 0-minimal right (resp. left) ideal, then M contains a 0-minimal ideal; this ideal is the union of all 0-minimal right (resp. left) ideals of M.*
3. *If M both contains a 0-minimal right ideal and a 0-minimal left ideal, its 0-minimal ideal is composed of a regular $\mathcal{D}$-class and zero.*

Proof. 1. Let I, J be two 0-minimal ideals of M. Let $m \in I \setminus 0$ and let $n \in J \setminus 0$. Since M is prime, there exist $u \in M$ such that $mun \neq 0$. Then $mun \in J$ implies $I \cap J \neq \{0\}$. Since $I \cap J$ is an ideal, we obtain $I \cap J = I = J$.

2. Let R be a 0-minimal right ideal. We first show that for all $m \in M$, either $mR = \{0\}$ or the set mR is a 0-minimal right ideal. In fact, mR is clearly a right ideal. Suppose $mR \neq \{0\}$ and let $R' \neq \{0\}$ be a right ideal contained in mR. Set $S = \{r \in R \mid mr \in R'\}$. Then $R' = mS$ and $S \neq \{0\}$ since $R' \neq \{0\}$. Moreover, S is a right ideal because R' is a right ideal. Since $S \subset R$, the fact that R is a 0-minimal right ideal implies the equality $S = R$. This shows that $mR = R'$ and consequently that mR is a 0-minimal right ideal.

Let I be the union of all the 0-minimal right ideals. It is a right ideal, and by the preceding discussion, it is also a left ideal. Let $J \neq \{0\}$ be an ideal of M. Then for any 0-minimal right ideal R of M,

$$RJ \subset R \cap J \subset R.$$

We have $RJ \neq \{0\}$ since for any $r \in R \setminus 0$ and $m \in J \setminus 0$, there exists $u \in M$ such that $rum \neq 0$ whence $rum \in RJ \setminus 0$. Since R is a 0-minimal right ideal and $R \cap J$ is a right ideal distinct from $\{0\}$, we have $R \cap J = R$. Thus $R \subset J$. This shows that $I \subset J$. Hence I is contained in any nonzero ideal of M and therefore is the 0-minimal ideal of M.

3. Let I be the 0-minimal ideal of M. Let $m, n \in I \setminus 0$. By 2, the right ideal mM and the left ideal Mn are 0-minimal. Since M is prime, there exists $u \in M$ such that $mun \neq 0$. The right ideal mM being 0-minimal, we have $mM = munM$ and therefore $m\mathcal{R}mun$. In the same way, $mun\mathcal{L}n$. It follows that $m\mathcal{D}n$. This shows that $I \setminus 0$ is contained in a $\mathcal{D}$-class. Conversely, if $m \in I \setminus 0$, $n \in M$ and $m\mathcal{D}n$, there exists a $k \in M$ such that $mM = kM$ and $Mk = Mn$. Consequently $I = MmM = MkM = MnM$ and this implies $n \in I \setminus 0$. This shows that $I \setminus 0$ is a $\mathcal{D}$-class.

Let us show that $I \setminus 0$ is a regular $\mathcal{D}$-class. By Proposition 1.12.7, it is enough to prove that $I \setminus 0$ contains an idempotent. Let $m, n \in I \setminus 0$.

Since M is prime, there exists $u \in M$ such that $mun \neq 0$. Since the right ideal mM is 0-minimal and since $mun \neq 0$, we have $mM = muM = munM$. Thus $mun \in R(m)$. Symmetrically, since Mn is a 0-minimal left ideal, we have $Mn = Mun = Mmun$, whence $mun \in L(n)$. Therefore $mun \in R(m) \cap L(n)$ and by Proposition 1.12.6, this implies that $R(n) \cap L(m)$ contains an idempotent. This idempotent belongs to the $\mathcal{D}$ class of n and therefore to $I \setminus 0$. $\square$

Corollary 1.12.10 *Let M be a prime monoid. If M contains a 0-minimal right ideal and a 0-minimal left ideal, then M contains a unique 0-minimal ideal I which is the union of all the 0-minimal right (resp. left) ideals. This ideal is composed with a regular $\mathcal{D}$ class and 0. Moreover, we have the following computational rules.*

1. *For $m \in I \setminus 0$ and $n \in M$ such that $mn \neq 0$, we have $m\mathcal{R}mn$.*
2. *For $m \in I \setminus 0$ and $n \in M$ such that $nm \neq 0$, we have $m\mathcal{L}nm$.*
3. *For any $\mathcal{H}$ class $H \subset I \setminus 0$ we have $H^2 = H$ or $H^2 = \{0\}$.*

Proof. The first group of statements is an easy consequence of Proposition 1.12.9. Let us prove 1. We have $mnM \subset mM$. Since mM is a 0-minimal right ideal and $mn \neq 0$, this forces the equality $mnM = mM$. The proof of 2 is symmetrical. Finally, to prove

3, let us suppose $H^2 \neq \{0\}$. Let $h, h' \in H$ be such that $hh' \neq 0$. Then, by 1 and 2, $h\mathcal{R}hh'$ and $h'\mathcal{L}hh'$. Since $h\mathcal{L}h'$ and $h'\mathcal{L}hh'$, we have $h\mathcal{L}hh'$. Therefore $hh' \in H$ and H is a group by Proposition 1.12.8.   □

We now give the statements that correspond to Proposition 1.12.9 and Corollary 1.12.10 for minimal ideals instead of 0-minimal ideals. This is of course of interest only in the case where the monoid does not have a zero.

Proposition 1.12.11 *Let M be a monoid.*

1. *If M contains a minimal right (resp. left) ideal, then M contains a minimal ideal which is the union of all the minimal right (resp. left) ideals.*
2. *If M contains a minimal right ideal and a minimal left ideal, its minimal ideal I is a $\mathcal{D}$-class. All the $\mathcal{H}$-classes in I are groups.*

Proof. Let 0 be an element that does not belong to M and let $M_0 = M \cup 0$ be the monoid whose law extends that of M in such a way that 0 is a zero. The monoid M_0 is prime.

An ideal I (resp. a right ideal R, a left ideal L) of M is minimal if and only if $I \cup 0$ (resp. $R \cup 0$, $L \cup 0$) is a 0-minimal ideal (resp. right ideal, left ideal) of M_0. Moreover the restriction to M of the relations $\mathcal{R}, \mathcal{L}, \mathcal{D}, \mathcal{H}$ in M_0 coincide with the corresponding relations in M. Therefore statements 1 and 2 can be deduced from Proposition 1.12.9 and Corollary 1.12.10.   □

Corollary 1.12.12 *Let M be a monoid containing a minimal right ideal and a minimal left ideal. Then M contains a minimal ideal which is the union of all the minimal right (resp. left) ideals. This ideal is a $\mathcal{D}$-class and all its $\mathcal{H}$-classes are groups.*   □

1.13 Permutation groups

In this section we give some elementary results and definitions concerning permutation groups. Let G be a group and let H be a subgroup of G. The *right cosets* of H in G are the sets Hg for $g \in G$. The equality $Hg = Hg'$ holds if and only if $gg'^{-1} \in H$. Hence the right cosets of H in G are a partition of G.

When G is finite, $[G : H]$ denotes the *index* of H in G. This number is both equal to $\mathrm{Card}(G)/\mathrm{Card}(H)$ and to the number of right cosets of H in G.

Let Q be a set. The *symmetric group* over Q composed of all the permutations of Q is denoted by $\mathfrak{S}_Q$. For $Q = \{1, 2, \ldots, n\}$ we write $\mathfrak{S}_n$ instead of $\mathfrak{S}_{\{1,2,\ldots,n\}}$. A permutation is written to the right of its argument. Thus for $g \in \mathfrak{S}_Q$ and $q \in Q$ the image of q by g is denoted by qg.

A *permutation group* over Q is any subgroup of $\mathfrak{S}_Q$. For instance, the *alternating group* over $\{1, 2, \ldots, n\}$, denoted by $\mathfrak{A}_n$ is the permutation group composed of all *even* permutations, that is permutations which are products of an even number of transpositions.

Let G be a permutation group over Q. The *stabilizer* of $q \in Q$ is the subgroup of G composed of all permutations of G fixing q,

$$H = \{h \in G \mid qh = q\}.$$

A *permutation group* over Q is called *transitive* if for all $p, q \in Q$, there exists $g \in G$ such that $pg = q$.

Proposition 1.13.1 *Let G be a group and let H be a subgroup of G. Let Q be the set of right cosets of H in G. Let φ be the mapping from G into $\mathfrak{S}_Q$ defined for $g \in G$ and $Hk \in Q$ by*

$$(Hk)\varphi(g) = H(kg).$$

The mapping φ is a morphism from G into $\mathfrak{S}_Q$ and the permutation group $\varphi(G)$ is transitive. Moreover, the subgroup $\varphi(H)$ is the stabilizer of the point $H \in Q$.

Conversely, let G be a transitive permutation group over Q, let $q \in Q$ and let H be the stabilizer of q. The mapping γ from G into Q defined by

$$\gamma : g \mapsto qg$$

induces a bijection α from the set of right cosets of H onto Q and for all $k \in G$, $g \in G$,

$$\alpha(Hk)g = \alpha(Hkg).$$

Proof. We first prove the direct part. The mapping φ is well defined because $Hk = Hk'$ implies $Hkg = Hk'g$. It is a morphism since $\varphi(1) = 1$ and

$$(Hk)\varphi(g)\varphi(g') = (Hkg)\varphi(g') = Hkgg' = (Hk)\varphi(gg').$$

The permutation group $\varphi(G)$ is transitive since for $k, k' \in G$, we have

$$(Hk)\varphi(k^{-1}k') = Hk'.$$

Finally, for all $h \in H$, $\varphi(h)$ fixes the coset H and conversely, if $\varphi(g)$, with $g \in G$, fixes H, then $Hg = H$, thus $g \in H$.

We now prove the converse. Assume that $Hg = Hg'$. Then $gg'^{-1} \in H$, and therefore $qgg'^{-1} = q$, showing that $qg = qg'$, whence $\gamma(g) = \gamma(g')$. This shows that we can define a function α by setting $\alpha(Hg) = \gamma(g)$. Since G is transitive, γ is surjective and therefore also α is surjective. To show that α is injective, assume that $\alpha(Hg) = \alpha(Hg')$. Then $qg = qg'$, whence $qgg^{-1} = q$. Thus gg^{-1} fixes q. Consequently $gg'^{-1} \in H$, whence $Hg = Hg'$.

The last formula is a direct consequence of the fact that both sides are equal to qkg. ☐

Let G be a transitive permutation group over a finite set Q. By definition, the *degree* of G is the number Card(Q).

Proposition 1.13.2 *Let G be a transitive permutation group over a finite set Q. Let $q \in Q$ and let H be the stabilizer of q. The degree of G is equal to the index of H in G.*

Proof. The function $\alpha : Hg \mapsto qg$ of Proposition 1.13.1(2) is a bijection from the set of right cosets of H onto Q. Consequently $\text{Card}(Q) = [G : H]$.   □

Two permutation groups G over Q and G' over Q' are called *equivalent* if there exists a bijection α from Q onto Q' and an isomorphism φ from G onto G' such that for all $q \in Q$ and $g \in G$,

$$\alpha(qg) = \alpha(q)\varphi(g)$$

or equivalently, for $q' \in Q'$ and $g \in G$,

$$q'\varphi(g) = \alpha((\alpha^{-1}(q'))g).$$

As an example, consider a permutation group G over Q and let H be the stabilizer of some q in Q. According to Proposition 1.13.1(2) this group is equivalent to the permutation group over the set of right cosets of H obtained by the action of G on the cosets of H.

Another example concerns any two stabilizers H and H' of two points q and q' in a transitive permutation group G over Q. Then H and H' are equivalent. Indeed, since G is transitive, there exists $g \in G$ such that $qg = q'$. Then g defines a bijection α from Q onto itself by $\alpha(p) = pg$. The function $\varphi : H \to H'$ given by $\varphi(h) = g^{-1}hg$ is an isomorphism and for all $p \in Q, h \in H$,

$$\alpha(ph) = \alpha(p)\varphi(h).$$

Let G be a transitive permutation group over Q. An *imprimitivity equivalence* of G is an equivalence relation θ over Q that is stable for the action of G. Equivalently, for all $g \in G$,

$$p \equiv q \bmod \theta \Rightarrow pg \equiv qg \bmod \theta.$$

The partition associated with an imprimitivity equivalence is called an *imprimitivity partition*.

Let θ be an imprimitivity equivalence of G. The action of G on the classes of θ defines a transitive permutation group denoted by G_θ called the *imprimitivity quotient* of G for θ.

For any element q in Q, denote by $[q]$ the equivalence class of q mod θ, and let K_q be the transitive permutation group over $[q]$ formed by the restrictions to $[q]$ of the permutations g that globally fix $[q]$, that is verifying $[q]g = [q]$.

The group K_q is the group *induced* by G on the class $[q]$.

We prove that the groups $K_q, q \in Q$ all are equivalent. Indeed let $q, q' \in Q$ and $g \in G$ be such that $qg = q'$. The restriction α of g to $[q]$ is a bijection from $[q]$ onto $[q']$. Clearly, α is injective. It is surjective since if $p \equiv q' \bmod \theta$, then

$pg^{-1} \equiv q \bmod \theta$ and $\alpha(pg^{-1}) = p$. Let φ be the isomorphism from K_q onto $K_{q'}$ defined for $k \in K_q$ by $p'\varphi(k) = \alpha(\alpha^{-1}(p')k)$. This shows that the groups K_q and $K_{q'}$ are equivalent. In particular, all equivalence classes mod θ have the same number of elements.

Any of the equivalent transitive permutation groups K_q is called the *induced group* of G on the classes of θ and is denoted by G^θ.

Let $d = \text{Card}(Q)$ be the degree of G, e the degree of G_θ, and f the degree of G^θ. Then

$$d = ef.$$

Indeed, e is the number of classes of θ and f is the common cardinality of each of the classes mod θ.

Let G be a transitive permutation group over Q. Then G is called *primitive* if the only imprimitivity equivalences of G are the equality relation and the universal relation over Q.

Proposition 1.13.3 *Let G be a transitive permutation group over Q. Let $q \in Q$ and H be the stabilizer of q. Then G is primitive if and only if H is a maximal subgroup of G.*

Proof. Assume first that G is primitive. Let K be a subgroup of G such that $H \subset K \subset G$. Consider the family of subsets of Q having the form qKg for $g \in G$. Any two of these subsets are either disjoint or identical. Suppose indeed that for some $k, k' \in K$ and $g, g' \in G$, we have $qkf = qk'g'$. Then $qkgg'^{-1}k'^{-1} = q$, showing that $kgg'^{-1}k'^{-1} \in H \subset K$. Thus $gg'^{-1} \in K$, whence $Kg = Kg'$ and consequently $qKg = qKg'$. Consequently the sets qKg form a partition of Q which is clearly an imprimitivity partition. Since G is primitive this implies that either $qK = \{q\}$ or $qK = Q$. The first case means that $K = H$. In the second case, $K = G$ since for any $g \in G$ there is some $k \in K$ with $qk = qg$ showing that $gk^{-1} \in H \subset K$ which implies $g \in K$. This proves that H is a maximal subgroup.

Conversely, let H be a maximal subgroup of G and let θ be an imprimitivity equivalence of G. Let K be the subgroup

$$K = \{k \in G \mid qk \equiv q \bmod \theta\}.$$

Then $H \subset K \subset G$, which implies that $K = H$ or $K = G$. If $K = H$, then the class of q is reduced to q and θ is therefore reduced to the equality relation. If $K = G$, then the class of q is equal to Q and θ is the universal equivalence. Thus G is primitive. $\square$

Let G be a transitive permutation group on Q. Then G is said to be *regular* if all elements of $G \setminus 1$ have no fixed point. It is easily verified that in this case $\text{Card}(G) = \text{Card}(Q)$.

Proposition 1.13.4 *Let G be a transitive permutation group over Q and let $q \in Q$. The group G is regular if and only if the stabilizer of q is a singleton.*

Let $k \geq 1$ be an integer. A permutation group G over Q is called *k-transitive* if for all k-tuples $(p_1, p_2, \ldots, p_k) \in Q^k$ and $(q_1, q_2, \ldots, q_k) \in Q^k$ composed of distinct elements, there is a $g \in G$ such that $p_1 g = q_1, p_2 g = q_2, \ldots, p_k g = q_k$.

The 1-transitive groups are just the transitive groups. Any k-transitive group for $k \geq 2$ is clearly also $(k-1)$ transitive. The group $\mathfrak{S}_n$ is n-transitive.

Proposition 1.13.5 *Let $k \geq 2$ be an integer. A permutation group over Q is k-transitive if and only if it is transitive and if the restriction to the set $Q \setminus q$ of the stabilizer of $q \in Q$ is $(k-1)$-transitive.*

Proof. The condition is clearly necessary. Conversely assume that the condition is satisfied by a permutation group G and let $(p_1, p_2, \ldots, p_k) \in Q^k$ and $(q_1, q_2, \ldots, q_k) \in Q^k$ be k-tuples composed of distinct elements. Since G is transitive, there exists a $g \in G$ such that $p_1 g = q_1$. Let H be the stabilizer of q_1. Since the restriction of H to the set $Q \setminus q_1$ is $(k-1)$-transitive, there is an $h \in H$ such that $p_2 gh = q_2, \ldots, p_k gh = q_k$. Since $p_1 gh = q_1$, the permutation $g' = gh$ satisfies $p_1 g' = q_1, p_2 g' = q_2, \ldots, p_k g' = q_k$. This shows that G is k-transitive. $\qquad\square$

A 2-transitive group is also called *doubly transitive*.

Proposition 1.13.6 *A doubly transitive permutation group is primitive.*

Proof. Let G be a doubly transitive permutation group over Q and consider an imprimitivity equivalence θ of G. If θ is not the equality on Q, then there are two distinct elements $q, q' \in Q$ such that $q \equiv q' \bmod \theta$. Let $q'' \in Q$ be distinct from q. Since G is 2-transitive, there exist $g \in G$ such that $qg = q$ and $q'g = q''$. Since θ is an imprimitivity equivalence we have $q \equiv q'' \bmod \theta$. Thus θ is the universal relation on Q. This shows that G is primitive. $\qquad\square$

The converse of Proposition 1.13.6 is false. Indeed, for any prime number p, the cyclic group generated by the permutation $(12 \cdots p)$ is primitive but is not doubly transitive. An interesting case where the converse of Proposition 1.13.6 is true is described in a famous theorem of Schur (Theorem 11.6.7) that will be stated in Chapter 11.

1.14 Notes

Each of the subjects treated in this chapter is part of a theory that we have considered only very superficially. A more complete exposition about words can be found in Lothaire (1997). For automata (Section 1.4) we follow the notation of Eilenberg (1974). Theorem 1.4.13 is due to S. Kleene.

Our definition of a complete semiring is less general than that of Eilenberg (1974) but it will be enough for our purposes. The full statement of the Perron–Frobenius theorem (Theorem 1.9.2) includes additional statements, including the description of the eigenvalues with maximal modulus (see Gantmacher (1959)). The function r_M is sometimes known as the *Wielandt* function.

Our presentation of ideals in monoids (Section 1.12) is developed with more details in Clifford and Preston (1961) or Lallement (1979). The notion of a prime monoid is not classical but it is well fitted to the situation that we shall find in Chapter 9. The 0-minimal ideals of prime monoids are usually called completely 0-simple semigroups. For semirings and formal series see Eilenberg (1974) or Berstel and Reutenauer (1988).

A classical textbook on permutation groups is Wielandt (1964).

Codes

The first two sections contain several equivalent definitions of codes and free submonoids. In Section 2.3 we give a method for verifying that a given set of words is a code.

In Section 2.4 we use Bernoulli distributions to give a necessary condition for a set to be a code (Theorem 2.4.5). The questions about probabilities raised in this and in the following section will be developed in more depth in Chapter 13.

Section 2.5 introduces the concept of a complete set. This is in some sense a notion dual to that of a code. The main result of this chapter (Theorem 2.5.16) describes complete codes by using results on Bernoulli distributions developed previously. In Section 2.6, the operation of composition of codes is introduced and several properties of this operation are established. The last section introduces the prefix graph of a code as a tool for the description of an efficient algorithm testing whether a finite set is a code.

2.1 Definitions

This section contains the definitions of the notions of code, prefix (suffix, bifix) code, maximal code, and coding morphism and gives examples.

Let A be an alphabet. A subset X of the free monoid A^* is a *code* over A if for all $n, m \geq 0$ and $x_1, \ldots, x_n, x_1', \ldots, x_m' \in X$, the condition

$$x_1 x_2 \cdots x_n = x_1' x_2' \cdots x_m' \tag{2.1}$$

implies

$$n = m \quad \text{and} \quad x_i = x_i' \quad \text{for} \quad i = 1, \ldots, n. \tag{2.2}$$

In other words, a set X is a code if any word in X^* can be written *uniquely* as a product of words in X, that is, has a unique factorization in words in X. In particular, a code never contains the empty word 1. It is clear that any subset of a code is a code. In particular, the empty set is a code. An element of a code is sometimes called a *codeword*.

The definition of a code can be rephrased as follows:

Proposition 2.1.1 *If a subset X of A^* is a code, then any bijection from some alphabet B onto X extends to an injective morphism from B^* into A^*. Conversely, if there exists an injective morphism $\beta : B^* \to A^*$ such that $X = \beta(B)$, then X is a code.*

Proof. Let $\beta : B^* \to A^*$ be a morphism such that β is a bijection of B onto X. Let $u, v \in B^*$ be words such that $\beta(u) = \beta(v)$. Set $u = b_1 \cdots b_n, v = b'_1 \cdots b'_m$, with $n, m \geq 0, b_1, \ldots, b_n, b'_1, \ldots, b'_m \in B$. Since β is a morphism, we have

$$\beta(b_1) \cdots \beta(b_n) = \beta(b'_1) \cdots \beta(b'_m).$$

But X is a code and $\beta(b_i), \beta(b'_j) \in X$. Thus $n = m$ and $\beta(b_i) = \beta(b'_i)$ for $i = 1, \ldots, n$. Now β is injective on B. Thus $b_i = b'_i$ for $i = 1, \ldots, n$, and $u = v$. This shows that β is injective.

Conversely, if $\beta : B^* \to A^*$ is an injective morphism, and if

$$x_1 \cdots x_n = x'_1 \cdots x'_m \tag{2.3}$$

for some $n, m \geq 1$ and $x_1, \ldots, x_n, x'_1, \ldots, x'_n \in X = \beta(B)$, then we consider the letters b_i, b'_j in B such that $\beta(b_i) = x_i, \beta(b'_j) = x'_j, i = 1, \ldots, n, j = 1, \ldots, m$. Since β is injective, Equation (2.3) implies that $b_1 \cdots b_n = b_1 \cdots b'_m$. Thus $n = m$ and $b_i = b'_i$, whence $x_i = x'_i$ for $i = 1, \ldots, n$. □

A morphism $\beta : B^* \to A^*$ which is injective and such that $X = \beta(B)$, is called a *coding morphism* for X. For any code $X \subset A^*$, the existence of a coding morphism for X is straightforward: it suffices to take any bijection of a set B onto X and to extend it to a morphism from B^* into A^*. In this context, the alphabet B is called the *source alphabet*, and the alphabet A is the *channel alphabet*.

Proposition 2.1.1 is the origin for the terminology since the words in X encode the letters of the set B. The coding procedure consists of associating to a word $b_1 b_2 \cdots b_n$ ($b_i \in B$) which is the source text an encoded message $\beta(b_1) \cdots \beta(b_n)$ over the channel alphabet by the use of the coding morphism β. The fact that β is injective ensures that the coded text is uniquely decipherable, in order to get the original text back.

Example 2.1.2 For any alphabet A, the set $X = A$ is a code. More generally, if $p \geq 1$ is an integer, then $X = A^p$ is a code called the *uniform code* of words of length p. Indeed, if elements of X satisfy Equation (2.1), then the constant length of words in X implies the conclusion (2.2).

Example 2.1.3 Over an alphabet consisting of a single letter a, a nonempty subset of a^* is a code if and only if it is a singleton distinct from 1.

Example 2.1.4 The set $X = \{aa, baa, ba\}$ over $A = \{a, b\}$ is a code. Indeed, suppose the contrary. Then there exists a word w in X^+, of minimal length, that has two distinct factorizations,

$$w = x_1 x_2 \cdots x_n = x'_1 x'_2 \cdots x'_m$$

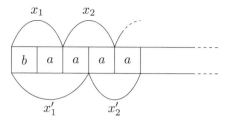

Figure 2.1 A double factorization starting.

$(n, m \geq 1, x_i, x'_j \in X)$. Since w is of minimal length, we have $x_1 \neq x'_1$. Thus x_1 is a proper prefix of x'_1 or vice versa. Assume that x_1 is a proper prefix of x'_1 (see Figure 2.1). By inspection of X, this implies that $x_1 = ba$, $x'_1 = baa$. This in turn implies that $x_2 = aa$, $x'_2 = aa$. Thus $x'_1 = x_1 a$, $x'_1 x'_2 = x_1 x_2 a$, and if we assume that $x'_1 x'_2 \cdot x'_p = x_1 x_2 \cdots x_p a$, it necessarily follows that $x_{p+1} = aa$ and $x'_{p+1} = aa$. Thus $x'_1 x'_2 \cdots x'_{p+1} = x_1 x_2 \cdots x_{p+1} a$. But this contradicts the existence of two factorizations.

Example 2.1.5 The set $X = \{a, ab, ba\}$ is not a code since the word $w = aba$ has two distinct factorizations

$$w = (ab)a = a(ba).$$

The following corollary to Proposition 2.1.1 is useful.

Corollary 2.1.6 *Let $\alpha : A^* \to C^*$ be an injective morphism. If X is a code over A, then $\alpha(X)$ is a code over C. If Y is a code over C, then $\alpha^{-1}(Y)$ is a code over A.*

Proof. Let $\beta : B^* \to A^*$ be a coding morphism for X. Then $\alpha(\beta(B)) = \alpha(X)$ and since $\alpha \circ \beta : B^* \to C^*$ is an injective morphism, Proposition 2.1.1 shows that $\alpha(X)$ is a code.

Conversely, let $X = \alpha^{-1}(Y)$, let $n, m \geq 1, x_1, \ldots, x_n, x'_1, \ldots, x'_m \in X$ be such that

$$x_1 \cdots x_n = x'_1 \cdots x'_m.$$

Then

$$\alpha(x_1) \cdots \alpha(x_n) = \alpha(x'_1) \cdots \alpha(x'_m).$$

Now Y is a code; therefore $n = m$ and $\alpha(x_i) = \alpha(x'_i)$ for $i = 1, \ldots, n$. The injectivity of α implies that $x_i = x'_i$ for $i = 1, \ldots, n$, showing that X is a code. □

Corollary 2.1.7 *If $X \subset A^*$ is a code, then X^n is a code for all integers $n > 0$.*

Proof. Let $\beta : B^* \to A^*$ be a coding morphism for X. Then $X^n = \beta(B^n)$. But B^n is a code. Thus the conclusion follows from Corollary 2.1.6. □

Example 2.1.8 We show that the product of two codes is not a code in general. Consider the sets $X = \{a, ba\}$ and $Y = \{a, ab\}$ which are easily seen to be codes over the alphabet $A = \{a, b\}$. Set $Z = XY$. Then

$$Z = \{aa, aab, baa, baab\}.$$

The word $w = aabaab$ has two distinct factorizations,

$$w = (aa)(baab) = (aab)(aab).$$

Thus Z is not a code.

An important class of codes is the class of prefix codes to be introduced now. A subset X of A^* is *prefix* if no element of X is a proper prefix of another element in X. In an equivalent manner, X is prefix if for all x, x' in X,

$$x \leq x' \Rightarrow x = x'. \tag{2.4}$$

This may be rephrased as: two distinct elements in X are incomparable in the prefix ordering.

It follows immediately from (2.4) that a prefix set X containing the empty word just consists of the empty word. Suffix sets are defined in a symmetric way. A subset X of A^* is *suffix* if no word in X is a proper suffix of another word in X. A set is *bifix* if it is both prefix and suffix. Clearly, a set of words X is suffix if and only if its reversal $\widetilde{X}$ is prefix.

Proposition 2.1.9 *Any prefix (suffix, bifix) set of words $X \neq \{1\}$ is a code.*

Proof. Since $X \neq \{1\}$, it does not contain the empty word. If X is not a code, then there is a word w of minimal length having two factorizations

$$w = x_1 x_2 \cdots x_n = x_1' x_2' \cdots x_m' \quad (x_i, x_j' \in X).$$

Both x_1, x_1' are nonempty, and since w has minimal length, $x_1 \neq x_1'$. But then $x_1 < x_1'$ or $x_1' < x_1$ contradicting the fact that X is prefix. Thus X is a code. The same argument holds for suffix sets. $\qquad\square$

A *prefix code (suffix code, bifix code)* is a prefix set (suffix, bifix set) which is a code, that is distinct from $\{1\}$.

Example 2.1.10 Uniform codes are bifix. The sets X and Y of Example 2.1.8 are a prefix and a suffix code.

Example 2.1.11 The sets $X = a^*b$ and $Y = \{a^n b^n \mid n \geq 1\}$ over $A = \{a, b\}$ are prefix, thus prefix codes. The set Y is suffix, thus bifix, but X is not. This example shows the existence of infinite codes over a finite alphabet.

Example 2.1.12 The *Morse code* associates to each alphanumeric character a sequence of dots and dashes. For instance, A is encoded by " . –" and J is encoded by

". − − −". Provided each codeword is terminated with an additional symbol (usually a space, called a "pause"), the Morse code becomes a prefix code.

A code X is *maximal* over A if X is not properly contained in any other code over A, that is, if

$$X \subset X', \quad X' \text{ code } \Rightarrow X = X'.$$

The maximality of a code depends on the alphabet over which it is given. Indeed, if $X \subset A^*$ and $A \subsetneq B$, then $X \subset B^*$ and X is certainly not maximal over B, even if it is a maximal code over A. The definition of a maximal code gives no algorithm that allows us to verify that it is satisfied. However, maximality is decidable, at least for recognizable codes (see Section 2.5).

Example 2.1.13 Uniform codes A^n are maximal over A. Suppose the contrary. Then there is a word $u \in A^+ \setminus A^n$ such that $Y = A^n \cup \{u\}$ is a code. The word $w = u^n$ belongs to Y^*, and it is also in $(A^n)^*$ because its length is a multiple of n. Thus $w = u^n = x_1 x_2 \cdots x_{|u|}$ for some $x_1, \ldots, x_{|u|} \in A^n$. Now $u \notin A^n$. Thus the two factorizations are distinct, Y is not a code and A^n is maximal.

Proposition 2.1.14 *Any code X over A is contained in some maximal code over A.*

Proof. Let $\mathcal{F}$ be the set of codes over A containing X, ordered by set inclusion. To show that $\mathcal{F}$ contains a maximal element, it suffices to demonstrate, in view of Zorn's lemma, that any chain $\mathcal{C}$ (that is, any totally ordered subset) in $\mathcal{F}$ admits a least upper bound in $\mathcal{F}$.

Consider a chain $\mathcal{C}$ of codes containing X. Then

$$\widehat{Y} = \bigcup_{Y \in \mathcal{C}} Y$$

is the least upper bound of $\mathcal{C}$. It remains to show that $\widehat{Y}$ is a code. For this, let $n, m \geq 1$, and $y_1, \ldots, y_n, y_1', \ldots, y_m' \in \widehat{Y}$ be such that

$$y_1 \cdots y_n = y_1' \cdots y_m'.$$

Each of the y_i, y_j' belongs to a code of the chain $\mathcal{C}$ and this determines $n + m$ elements (not necessarily distinct) of $\mathcal{C}$. One of them, say Z, contains all the others. Thus $y_1, \ldots, y_n, y_1' \ldots, y_m' \in Z$, and since Z is a code, we have $n = m$ and $y_i = y_i'$ for $i = 1, \ldots, n$. This shows that $\widehat{Y}$ is a code. □

Proposition 2.1.14 is no longer true if we restrict ourselves to finite codes. There exist finite codes which are not contained in any finite maximal code. An example of such a code will be given in Section 2.5 (Example 2.5.7).

The fact that a set $X \subset A^*$ is a code admits a very simple expression in the terminology of formal power series.

Proposition 2.1.15 *Let X be a subset of A^+, and let $M = X^*$ be the submonoid generated by X. Then X is a code if and only if $\underline{M} = (\underline{X})^*$ or equivalently $\underline{M} = (1 - \underline{X})^{-1}$.*

Proof. According to Proposition 1.7.4, the coefficient $((\underline{X})^*, w)$ of a word w in $(\underline{X})^*$ is equal to the number of distinct factorizations of w in words in X. By definition, X is a code if and only if this coefficient takes only the values 0 and 1 for any word in A^*. But this is equivalent to saying that $(\underline{X})^*$ is the characteristic series of its support, that is, $(\underline{X})^* = \underline{M}$. □

2.2 Codes and free submonoids

The submonoid X^* generated by a code X is sometimes easier to handle than the code itself. The fact that X is a code (prefix code, bifix code) is equivalent to the property that X^* is a free monoid (a right unitary, biunitary monoid). These properties may be verified directly on the submonoid without any explicit description of its base. Thus we can prove that sets are codes by knowing only the submonoid they generate.

We start with a general property. Let A be an alphabet.

Proposition 2.2.1 *Any submonoid M of A^* has a unique minimal set of generators* $X = (M \setminus 1) \setminus (M \setminus 1)^2$.

Proof. Set $Q = M \setminus 1$. First, we verify that X generates M, that is, that $X^* = M$. Since $X \subset M$, we have $X^* \subset M$. We prove the opposite inclusion by induction on the length of words. Of course, $1 \in X^*$. Let $m \in Q$. If $m \notin Q^2$, then $m \in X$. Otherwise $m = m_1 m_2$ with $m_1, m_2 \in Q$ both strictly shorter than m. Therefore m_1, m_2 belong to X^* by the induction hypothesis and $m \in X^*$.

Now let Y be a set of generators of M. We may suppose that $1 \notin Y$. Then each $x \in X$ is in Y^* and therefore can be written as $x = y_1 y_2 \cdots y_n$ with $y_i \in Y$ and $n \geq 0$. The facts that $x \neq 1$ and $x \notin Q^2$ force $n = 1$ and $x \in Y$. This shows that $X \subset Y$. Thus X is a minimal set of generators and such a set is unique. □

Example 2.2.2 Let $A = \{a, b\}$ and let $M = \{w \in A^* \mid |w|_a \equiv 0 \bmod 2\}$. Then we compute $X = (M \setminus 1) \setminus (M \setminus 1)^2 = b \cup ab^*a$.

We now turn to the study of the submonoid generated by a code. By definition, a submonoid M of A^* is *free* if there exists an isomorphism

$$\alpha : B^* \to M$$

of a free monoid B^* onto M.

Proposition 2.2.3 *If M is a free submonoid of A^*, then its minimal set of generators is a code. Conversely, if $X \subset A^*$ is a code, then the submonoid X^* of A^* is free and X is its minimal set of generators.*

Proof. Let $\alpha : B^* \to M$ be an isomorphism. Then α, considered as morphism from B^* into A^*, is injective. By Proposition 2.1.1, the set $X = \alpha(B)$ is a code. Next $M = \alpha(B^*) = (\alpha(B))^* = X^*$. Thus X generates M. Furthermore $B = B^+ \setminus B^+ B^+$ and $\alpha(B^+) = M \setminus 1$. Consequently $X = (M \setminus 1) \setminus (M \setminus 1)^2$, showing that X is the minimal set of generators of M.

Conversely, assume that $X \subset A^*$ is a code and consider a coding morphism α : $B^* \to A^*$ for X. Then α is injective and α is a bijection from B onto X. Thus α is a bijection from B^* onto $\alpha(B^*) = X^*$. Consequently X^* is free. Now α is a bijection, thus $B = B^+ \setminus B^+ B^+$ implies $X = X^+ \setminus X^+ X^+$, showing by Proposition 2.2.1 that X is the minimal set of generators of M. □

The code X which generates a free submonoid M of A^* is called the *base* of M.

Corollary 2.2.4 *Let X and Y be codes over A. If $X^* = Y^*$, then $X = Y$.*

Example 2.2.2 (*continued*) The set X is a (bifix) code, thus M is a free submonoid of A^*.

According to Proposition 2.2.3, we can distinguish two cases where a set X is not a code. First, when X is not the minimal set of generators of $M = X^*$, that is, there exists an equality

$$x = x_1 x_2 \cdots x_n$$

with $x, x_i \in X$ and $n \geq 2$. Note that despite this fact, M might be free. The other case holds when X is the minimal set of generators, but M is not free (this is the case of Example 2.1.5).

We now give a characterization of free submonoids of A^* which is intrinsic in the sense that it does not rely on the bases. Another slightly different characterization is given in Exercise 2.2.3.

Let M be a monoid. A submonoid N of M is *stable* (in M) if for all $u, v, w \in M$,

$$u, v, uw, wv \in N \Rightarrow w \in N. \tag{2.5}$$

The hypotheses of (2.5) may be written as

$$w \in N^{-1} N \cap N N^{-1},$$

thus the condition for stability becomes

$$N^{-1} N \cap N N^{-1} \subset N$$

or simply

$$N^{-1} N \cap N N^{-1} = N \tag{2.6}$$

since $1 \in N$ and therefore $N \subset N^{-1} N \cap N N^{-1}$.

Figure 2.2 gives a pictorial representation of condition (2.5) when the elements u, v, w are words. The membership in N is represented by an arch.

Stable submonoids appear in almost all of the chapters in this book. A reason for this is Proposition 2.2.5 which gives a remarkable characterization of free submonoids of a free monoid. As a practical application, the proposition is used to prove that some submonoids are free and consequently that their bases are codes.

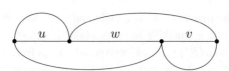

Figure 2.2 Representation of stability.

Proposition 2.2.5 *A submonoid N of A* is stable if and only if it is free.*

Proof. Assume first that N is stable. Set $X = (N \setminus 1) \setminus (N \setminus 1)^2$. To prove that X is a code, suppose the contrary. Then there is a word $z \in N$ of minimal length having two distinct factorizations in words of X,

$$z = x_1 x_2 \cdots x_n = y_1 y_2 \cdots y_m$$

with $x_1, \ldots, x_n, y_1, \ldots, y_m \in X$. We may suppose $|x_1| < |y_1|$. Then $y_1 = x_1 w$ for some nonempty word w. It follows that

$$x_1, \quad y_2 \cdots y_m, \quad x_1 w = y_1, \quad w y_2 \cdots y_m = x_2 \cdots x_n$$

are all in N. Since N is stable, w is in N. Consequently $y_1 = x_1 w \notin X$, which gives the contradiction. Thus X is a code.

Conversely, assume that N is free and let X be its base. Let $u, v, w \in A^*$ and suppose that $u, v, uw, wv \in N$. Set

$$u = x_1 \cdots x_k, \quad wv = x_{k+1} \cdots x_r, \quad uw = y_1 \cdots y_\ell, \quad v = y_{\ell+1} \cdots y_s,$$

with x_i, y_j in X. The equality $u(wv) = (uw)v$ implies

$$x_1 \cdots x_k x_{k+1} \cdots x_r = y_1 \cdots y_\ell y_{\ell+1} \cdots y_s.$$

Thus $r = s$ and $x_i = y_i$ ($i = 1, \ldots, s$) since X is a code. Moreover, $\ell \geq k$ because $|uw| \geq |u|$, showing that

$$uw = x_1 \cdots x_k x_{k+1} \cdots x_\ell = u x_{k+1} \cdots x_\ell,$$

hence $w = x_{k+1} \cdots x_\ell \in N$. Thus N is stable. $\qquad\square$

Submonoids which are generated by prefix codes can also be characterized by a condition which is independent of the base. Let M be a monoid and let N be a submonoid of M. Then N is *right unitary* in M if for all $u, v \in M$,

$$u, uv \in N \Rightarrow v \in N.$$

In a symmetric way, N is *left unitary* if for all $u, v \in M$,

$$u, vu \in N \Rightarrow v \in N.$$

The conditions may be rewritten as follows: N is right unitary if and only if $N^{-1}N = N$, and N is left unitary if and only if $NN^{-1} = N$.

The submonoid N of M is *biunitary* if it is both left and right unitary.

The four properties stable, left unitary, right unitary, and biunitary are of the same nature. Their relationships can be summarized as

$$\text{stable} : N^{-1}N \cap NN^{-1} = N$$

$$\nearrow \qquad\qquad\qquad\qquad \nwarrow$$

$$\text{left unitary} : NN^{-1} = N \qquad\qquad N^{-1}N = N : \text{right unitary}$$

$$\nwarrow \qquad\qquad\qquad\qquad \nearrow$$

$$\text{biunitary} : NN^{-1} = N^{-1}N = N$$

Example 2.2.2 (*continued*) The submonoid M is biunitary. Indeed, if $u, uv \in M$ then $|u|_a$ and $|uv|_a = |u|_a + |v|_a$ are even numbers; consequently $|v|_a$ is even and $v \in M$. Thus M is right unitary.

Example 2.2.6 In group theory, the concepts stable, unitary, and biunitary collapse and coincide with the notion of subgroup. Indeed, let H be a stable submonoid of a group G. For all $h \in H$, both hh^{-1} and $h^{-1}h$ are in H. Stability implies that h^{-1} is in H. Thus H is a subgroup. If H is a subgroup, then conversely $HH^{-1} = H^{-1}H = H$, showing that H is biunitary.

The following proposition shows the relationship between the submonoids we defined and codes.

Proposition 2.2.7 *A submonoid M of A^* is right unitary (resp. left unitary, biunitary) if and only if its minimal set of generators is a prefix code (suffix code, bifix code). In particular, a right unitary (left unitary, biunitary) submonoid of A^* is free.*

Proof. Let $M \subset A^*$ be a submonoid, $Q = M \setminus 1$ and let $X = Q \setminus Q^2$ be its minimal set of generators. Suppose M is right unitary.

To show that X is prefix, let x, xu be in X for some $u \in A^*$. Then $x, xu \in M$ and thus $u \in M$. If $u \neq 1$, then $u \in Q$; but then $xu \in Q^2$ contrary to the assumption. Thus $u = 1$ and X is prefix.

Conversely, suppose that X is prefix. Let $u, v \in A^*$ be such that $u, uv \in M = X^*$. Then

$$u = x_1 \cdots x_n, \qquad uv = y_1 \cdots y_m$$

for some $x_1, \ldots, x_n, y_1, \ldots, y_m \in X$. Consequently

$$x_1 \cdots x_n v = y_1 \cdots y_m.$$

Since X is prefix, neither x_1 nor y_1 is a proper prefix of the other. Thus $x_1 = y_1$, and for the same reason $x_2 = y_2, \ldots, x_n = y_n$. This shows that $m \geq n$ and $v = y_{n+1} \cdots y_m$ belongs to M. Thus M is right unitary. $\qquad\square$

Let M be a free submonoid of A^*. Then M is *maximal* if $M \neq A^*$ and M is not properly contained in any other free submonoid excepted A^*.

Proposition 2.2.8 *If M is a maximal free submonoid of A^*, then its base X is a maximal code.*

Proof. Let Y be a code on A with $X \subsetneq Y$. Then $X^* \subset Y^*$ and $X^* \neq Y^*$ since otherwise $X = Y$ by Corollary 2.2.4. Now X^* is maximal. Thus $Y^* = A^*$ and $Y = A$. Thus $X \subsetneq A$. Let $b \in A \setminus X$. The set $Z = X \cup b^2$ is a code and $M \subsetneq Z^* \subsetneq A^*$. Both inclusions are strict since $b^2 \notin M$ and $b \notin Z^*$. This contradicts the maximality of M. □

Note that the converse of the proposition is false since uniform codes A^n $(n \geq 1)$ are maximal. But if $k, n \geq 2$, we have $(A^{kn})^* \subsetneq (A^n)^* \subsetneq A^*$, showing that $(A^{nk})^*$ is not maximal.

We now introduce a family of bifix codes called group codes which have interesting properties. Before we give the definition, let us consider the following situation.

Let G be a group, H be a subgroup of G, and

$$\varphi : A^* \to G \tag{2.7}$$

be a morphism. The submonoid

$$M = \varphi^{-1}(H) \tag{2.8}$$

is biunitary. Indeed, if, for instance, $p, pq \in M$, then $\varphi(p), \varphi(pq) \in H$, therefore $\varphi(p)^{-1}\varphi(pq) = \varphi(q) \in H$ and $q \in M$. The same proof shows that M is left unitary. Thus the base, say X, of M is a bifix code.

The definition of the submonoid M in (2.8) is equivalent to a description as the intersection of A^* with a subgroup of the free group $A^\odot$ on A. Indeed, the morphism φ in (2.7) factorizes in a unique way in

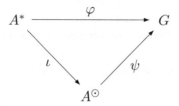

with ι the canonical injection. Setting $Q = \psi^{-1}(H)$, we have

$$M = Q \cap A^*.$$

Conversely if Q is a subgroup of $A^\odot$ and $M = Q \cap A^*$, then

$$M = \iota^{-1}(Q).$$

A *group code* is the base X of a submonoid $M = \varphi^{-1}(H)$, where φ is a morphism given by (2.7) which, moreover, is supposed to be *surjective*. Then X is a bifix code and X is a maximal code. Indeed, if $M = A^*$, then $X = A$ is maximal. Otherwise take $w \in A^* \setminus M$ and setting $Y = X \cup w$, let us verify that Y is not a code. Set $m = \varphi(w)$. Since φ is surjective, there is a word $\bar{w} \in A^*$ such that $\varphi(\bar{w}) = m^{-1}$. The words $u = w\bar{w}, v = \bar{w}w$ both are in M, and $w\bar{w}w = uw = wv \in Y^*$. This word has two distinct factorizations in words in Y, namely, uw formed of words in X followed by a word in Y, and wv which is composed the other way round. Thus Y is not a code and X is maximal.

We give now three examples of group codes.

Example 2.2.9 Let $A = \{a, b\}$ and consider the set

$$M = \{w \in A^* \mid |w|_a \equiv 0 \bmod 2\}$$

of Example 2.2.2. We have $M = \varphi^{-1}(0)$, where

$$\varphi : A^* \to \mathbb{Z}/2\mathbb{Z}$$

is the morphism given by $\varphi(a) = 1, \varphi(b) = 0$. Thus the base of M, namely the code $X = b \cup ab^*a$, is a group code, hence maximal.

Example 2.2.10 The uniform code A^m over A is a group code. The monoid $(A^m)^*$ is indeed the kernel of the morphism of A^* onto $\mathbb{Z}/m\mathbb{Z}$ mapping all letters on the number 1.

Example 2.2.11 Let $A = \{a, b\}$, and consider now the submonoid

$$\{w \in A^* \mid |w|_a = |w|_b\} \tag{2.9}$$

composed of the words on A having as many a's as b's. Let

$$\delta : A^* \to \mathbb{Z}$$

be the morphism defined by $\delta(a) = 1, \delta(b) = -1$. Clearly

$$\delta(w) = |w|_a - |w|_b$$

for all $w \in A^*$. Thus the set (2.9) is equal to $\delta^{-1}(0)$. The base of $\delta^{-1}(0)$ is denoted by D or D_1, the submonoid itself by D^* or D_1^*. Words in D are called *Dyck-primes*, D is the *Dyck code* over A. The set D^* is the *Dyck set* over A.

Example 2.2.12 More generally, let $A = B \cup \bar{B}$ $(B \cap \bar{B} = \emptyset)$ be an alphabet with $2n$ letters, and let $\delta : A^* \to B^\odot$ be the morphism of A^* onto the free group $B^\odot$ defined by $\delta(b) = b, \delta(\bar{b}) = b^{-1}$ for $b \in B, \bar{b} \in \bar{B}$. The base of the submonoid $\delta^{-1}(1)$ is denoted by D_n and is called the *Dyck code* over A or over n letters.

We now turn to a slightly different topic and consider the free submonoids of A^* containing a given submonoid. We start with the following observation which easily follows from Proposition 2.2.5.

Proposition 2.2.13 *The intersection of an arbitrary family of free submonoids of A^* is a free submonoid.*

Proof. Let $(M_i)_{i \in I}$ be a family of free submonoids of A^*, and set $M = \cap_{i \in I} M_i$. Clearly M is a submonoid, and it suffices to show that M is stable. If

$$u, vw, uv, w \in M$$

then these four words belong to each of the M_i. Each M_i being stable, w is in M_i for each $i \in I$. Thus $w \in M$. □

Proposition 2.2.13 leads to the following considerations. Let X be a subset of A^*. As we have just seen, the intersection of all free submonoids of A^* containing X is again a free submonoid. It is the smallest free submonoid of A^* containing X. We call it the *free hull* of X. If X^* is a free submonoid, then it coincides of course with its free hull.

Let X be a subset of A^*, let N be its free hull and let Y be the base of N. If X is not a code, then $X \neq Y$. The following result, known as the *defect theorem* gives an interesting relationship between X and Y.

Theorem 2.2.14 *Let X be a subset of A^*, and let Y be the base of the free hull of X. If X is not a code, then*

$$\mathrm{Card}(Y) \le \mathrm{Card}(X) - 1.$$

The following result is a consequence of the theorem. It can be proved directly as well (Exercise 2.2.1).

Corollary 2.2.15 *Let $X = \{x_1, x_2\}$. Then X is a code if and only if x_1 and x_2 are not powers of the same word.* □

Note that this corollary entirely describes the codes with two elements. The case of sets with three words is already much more complicated. See also Exercises 2.6.2 and 2.6.3.

For the proof of Theorem 2.8, we first show the following result.

Proposition 2.2.16 *Let $X \subset A^*$ and let Y be the base of the free hull of X. Then*

$$Y \subset X(Y^*)^{-1} \cap (Y^*)^{-1} X,$$

that is each word in Y appears as the first (resp. last) factor in the factorization of some word $x \in X$ in words belonging to Y.

Proof. Suppose that a word $y \in Y$ is not in $(Y^*)^{-1} X$. Then $X \subset 1 \cup Y^*(Y \setminus y)$. Setting

$$Z = y^*(Y \setminus y)$$

we have $Z^+ = Y^*(Y \setminus y)$, thus $X \subset Z^*$. Now Z^* is free. Indeed, any word $z \in Z^*$ has a unique factorization

$$z = y_1 y_2 \cdots y_n, \quad y_1, \ldots, y_n \in Y, \quad y_n \neq y$$

and therefore can be written uniquely as

$$z = y^{p_1} z_1 y^{p_2} z_2 \cdots y^{p_r} z_r, \quad z_1, \ldots, z_r \in Y \setminus y, \quad p_i \geq 0.$$

Now $X \subset Z^* \subsetneqq Y^*$, showing that Y^* is not the free hull of X. This gives the contradiction. $\qquad\Box$

Proof of Theorem 2.2.14. If X contains the empty word, then X and $X' = X \setminus 1$ have the same free hull Y^*. If the result holds for X', it also holds for X, since if X' is a code, then $Y = X'$ and $\mathrm{Card}(Y) = \mathrm{Card}(X) - 1$, and otherwise $\mathrm{Card}(Y) \leq \mathrm{Card}(X') - 1 \leq \mathrm{Card}(X) - 2$. Thus we may assume that $1 \notin X$. Let $\alpha : X \to Y$ be the mapping defined by

$$\alpha(x) = y \quad \text{if} \quad x \in yY^*.$$

This mapping is uniquely defined since Y is a code; it is everywhere defined since $X \subset Y^*$. In view of Proposition 2.2.16, the function α is surjective. If X is not a code, then there exists a relation

$$x_1 x_2 \cdots x_n = x_1' x_2' \cdots x_m', \quad x_i, x_j' \in X \tag{2.10}$$

with $x_1 \neq x_1'$. However, Y is a code, and by (2.10) we have

$$\alpha(x_1) = \alpha(x_1').$$

Thus α is not injective. This proves the inequality. $\qquad\Box$

2.3 A test for codes

It is not always easy to verify that a given set of words is a code. The test described in this section is not based on any new property of codes but consists merely in a systematic organization of the computations required to verify that a set of words satisfies the definition of a code.

In the case where X is finite, or more generally if X is recognizable, the amount of computation is finite. In other words, it is effectively decidable whether a finite or recognizable set is a code.

Before starting the description of the algorithm, let us consider an example.

Example 2.3.1 Let $A = \{a, b\}$, and $X = \{b, abb, abbba, bbba, baabb\}$. This set is not a code. For instance $(abb)(baabb) = (abbba)(abb)$. We consider the word

$$w = abbbabbbaabb$$

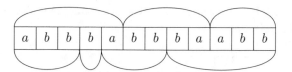

Figure 2.3 Two factorizations of the word *abbbabbbaabb*.

which has the two factorizations (see Figure 2.3)

$$w = (abbba)(bbba)(abb) = (abb)(b)(abb)(baabb).$$

These two factorizations define a sequence of prefixes of w, each one corresponding to an attempt at a double factorization. We give this list, together with the attempt at a double factorization:

$$(abbba) = (abb)\underline{ba}$$

$$(abbba) = (abb)(b)\underline{a}$$

$$(abbba)\underline{bb} = (abb)(b)(abb)$$

$$(abbba)(bbba) = (abb)(b)(abb)\underline{ba}$$

$$(abbba)(bbba)\underline{abb} = (abb)(b)(abb)(baabb)$$

$$(abbba)(bbba)(abb) = (abb)(b)(abb)(baabb).$$

Each but the last one of these attempts fails because of the underlined suffix, which remains after the factorization.

The algorithm presented here computes all the *remainders* in all attempts at a double factorization. It discovers a double factorization by the fact that the empty word is one of the remainders.

Formally, the computations are organized as follows. Let X be a subset of A^+, and let

$$U_1 = X^{-1}X \setminus 1,$$

$$U_{n+1} = X^{-1}U_n \cup U_n^{-1}X \quad (n \geq 1). \tag{2.11}$$

Then we have the following result:

Theorem 2.3.2 *The set $X \subset A^+$ is a code if and only if none of the sets U_n defined above contains the empty word.*

If $X \subset A^+$ is prefix (thus a code), then $U_1 = X^{-1}X \setminus 1 = \emptyset$. Thus the algorithm ends immediately for such codes.

Example 2.3.1 (*continued*) The word ba is in U_1, next $a \in U_2$, then $bb \in U_3$ and $ba \in U_4$, finally $abb \in U_5$ and since $1 \in U_6$, the set X is not a code, according to Theorem 2.3.2

The proof of Theorem 2.3.2 is based on the following lemma.

Lemma 2.3.3 *Let $X \subset A^+$ and let $(U_n)_{n \geq 1}$ be defined as above. For all $n \geq 1$, one has $w \in U_n$ if and only if there exist integers $p, q \geq 1$ with $p + q = n + 1$ and words $x_1, \ldots, x_p, y_1, \ldots, y_q$ in X with $x_1 \neq y_1$ and w suffix of y_q such that*

$$x_1 \cdots x_p w = y_1 \cdots y_q. \tag{2.12}$$

Proof. We show that for $w \in U_n$, words satisfying (2.12) exist, by induction on n. First, if $w \in U_1$, then by definition of U_1, one has $xw = y$ for some $x, y \in X$ with $x \neq y$, and w is a suffix of y, so the assertion holds for $n = 1$.

Let $w \in U_n$, with $n > 1$. Then either $xw = v$ or $vw = x$ for some $x \in X$ and $v \in U_{n-1}$. By induction,

$$x_1 \cdots x_p v = y_1 \cdots y_q,$$

for integers $p, q \geq 1$ with $p + q = n$ and $x_1, \ldots, x_p, y_1, \ldots, y_q$ in X with $x_1 \neq y_1$ and v suffix of y_q. If $xw = v$, then

$$x_1 \cdots x_p x w = y_1 \cdots y_q,$$

showing that the condition is satisfied by $x_1, \ldots, x_p, x_{p+1}, y_1, \ldots, y_q$ with $x_{p+1} = x$, since w is a suffix of y_q. On the other side, if $vw = x$ then

$$x_1 \cdots x_p x = y_1 \cdots y_q w,$$

showing that the condition is satisfied by $y_1, \ldots, y_q, x_1, \ldots, x_p, x_{p+1}$ with $x_{p+1} = x$, since w is a suffix of x.

Conversely, we prove by induction on $n \geq 1$ that if, for $p, q \geq 1$ with $p + q = n + 1$, there are words $x_1, \ldots, x_p, y_1, \ldots, y_q$ in X with $x_1 \neq y_1$ and w suffix of y_q, such that

$$x_1 \cdots x_p w = y_1 \cdots y_q,$$

then $w \in U_n$.

The property is clearly true for $n = 1$. Assume $n > 1$. Since w is a suffix of y_q, we have $y_q = vw$ for some word v, and the equation becomes

$$x_1 \cdots x_p = y_1 \cdots y_{q-1} v.$$

Set $v = v' x_{r+1} \cdots x_p$ with v' suffix of x_r for some r such that $1 \leq r \leq p$. Then $x_1 \cdots x_r = y_1 \cdots y_{q-1} v'$ and thus v' is in U_{r+q-2} by induction hypothesis.

Since $y_q = v' x_{r+1} \cdots x_p w$, one has $x_{r+1} \cdots x_p w \in U_{r+q-2}^{-1} X \subset U_{r+q-1}$. Then we show by induction on i that for $1 \leq i \leq p - r$, we have $x_{r+i} \cdots x_p w \in U_{r+q+i-2}$.

This holds for $i = 1$, and since x_{r+i} is in X, $x_{r+i} \cdots x_p w \in U_{r+q+i-2}$ implies $x_{r+i+1} \cdots x_p w \in U_{r+q+i-1}$. Thus, we obtain $x_p w \in U_{p+q-2}$ and finally $w \in U_{p+q-1}$. This concludes the proof. $\qquad\square$

Proof of Theorem 2.3.2. If X is not a code, then there is a relation

$$x_1 x_2 \cdots x_p = y_1 y_2 \cdots y_q, \quad x_i, y_j \in X, \quad x_1 \neq y_1. \tag{2.13}$$

By the lemma, the empty word is in U_{p+q-1}. Conversely, if $1 \in U_n$, there is a factorization (2.13) with $p + q - 1 = n$, showing that X is not a code. This establishes the theorem.   □

Example 2.3.1 (*continued*) For $X = \{b, abb, abbba, bbba, baabb\}$, we obtain

$$U_1 = \{ba, bba, aabb\}, \qquad X^{-1}U_1 = \{a, ba\}, \qquad U_1^{-1}X = \{abb\},$$
$$U_2 = \{a, ba, abb\}, \qquad X^{-1}U_2 = \{a, 1\}, \qquad U_2^{-1}X = \{bb, bbba, abb, 1, ba\}.$$

Thus $1 \in U_3$ and X is not a code.

Example 2.3.4 Let $X = \{a, ab, ba\}$ and $A = \{a, b\}$. We have

$$U_1 = \{b\}, \quad U_2 = \{a\}, \quad U_3 = \{1, b\}, \quad U_4 = X, \quad U_5 = U_3.$$

The set U_3 contains the empty word. Thus X is not a code.

Example 2.3.5 Let $X = \{aa, ba, bb, baa, bba\}$ and $A = \{a, b\}$. We obtain $U_1 = \{a\}$, $U_2 = U_1$. Thus $U_n = \{a\}$ for all $n \geq 1$ and X is a code.

The next proposition shows that Theorem 2.3.2 provides an algorithm for testing whether a recognizable set is a code.

Proposition 2.3.6 *If $X \subset A^+$ is a recognizable set, then the set of all U_n $(n \geq 1)$ is finite.*

This statement is straightforward when the set X is finite, since each U_n is composed of suffixes of words in X.

Proof. Recall that $\sim_X$ denotes the syntactic congruence of X.

Let μ be the congruence of A^* with the two classes $\{1\}$ and A^+. Let $\iota = \sim_X \cap \mu$. We use the following general fact.

If $L \subset A^*$ is a union of equivalence classes of a congruence θ, then for any subset Y of A^*, $Y^{-1}L$ is a union of congruence classes mod θ. (Indeed, let $z \in Y^{-1}L$ and $z' \equiv z \bmod \theta$. Then $yz \in L$ for some $y \in Y$, whence $yz' \in L$. Thus $z' \in Y^{-1}L$).

We prove that each U_n is a union of equivalence classes of ι by induction on $n \geq 1$. For $n = 1$, X is a union of classes of $\sim_X$, thus $X^{-1}X$ also is a union of classes for $\sim_X$, and finally $X^{-1}X \setminus 1$ is a union of classes of ι. Next, if U_n is a union of classes of ι, then by the previous fact both $U_n^{-1}X$ and $X^{-1}U_n$ are unions of classes of ι. Thus U_{n+1} is a union of classes of ι. The fact that X is recognizable implies that ι has finite index. The result follows.   □

Example 2.3.7 Let $A = \{a, b\}$ and $X = ba^*$. Then X is a recognizable suffix code. Indeed, $U_1 = a^+$ and $U_2 = \emptyset$. Thus the sequence (U_n) has two distinct elements.

2.4 Codes and Bernoulli distributions

In this section, we consider Bernoulli distributions. Recall that for a Bernoulli distribution π on A^* and a set $X \subset A^*$, we set

$$\pi(X) = \sum_{x \in X} \pi(x).$$

The value $\pi(X)$ is a nonnegative number or $+\infty$. For any family $(X_i)_{i \geq 0}$, of subsets of A^*, one has

$$\pi\left(\bigcup_{i \geq 0} X_i\right) \leq \sum_{i \geq 0} \pi(X_i), \tag{2.14}$$

with equality if the sets X_i are pairwise disjoint.

Example 2.4.1 Let $A = \{a, b\}$ and $X = \{a, ba, bb\}$. Let π be a Bernoulli distribution on A^*. Setting $p = \pi(a), q = \pi(b)$, we get $\pi(X) = p + pq + q^2 = p + pq + (1 - p)q = p + q = 1$.

For a Bernoulli distribution π, and a set X, recall that the probability generating series of X is

$$F_X(t) = \sum_{n \geq 0} \pi(X \cap A^n)t^n.$$

Since $\pi(X \cap A^n) \leq 1$, the radius of convergence of $F_X(t)$ is at least 1 and $\pi(X) = F_X(1)$.

Lemma 2.4.2 *Let π be a Bernoulli distribution on A^*. For subsets $X, Y \subset A^+$, one has*

$$F_{X \cup Y}(t) = F_X(t) + F_Y(t) \quad \text{if } X \cap Y = \emptyset,$$

and

$$F_{XY}(t) = F_X(t)F_Y(t) \quad \text{if the product } XY \text{ is unambiguous.}$$

Proof. The first equality is clear. For the second, observe that for all n,

$$XY \cap A^n = \bigcup_{i+j=n} (X \cap A^i)(Y \cap A^j).$$

The above union is disjoint when the product XY is unambiguous. Thus, from the first equality, it follows that

$$\pi(XY \cap A^n) = \sum_{i+j=n} \pi((X \cap A^i)(Y \cap A^j)),$$

and since clearly $\pi((X \cap A^i)(Y \cap A^j)) = \pi(X \cap A^i)\pi(Y \cap A^j)$, the formula follows. $\square$

We observe that

$$F_{X_1 \cdots X_m}(t) = F_{X_1}(t) \cdots F_{X_m}(t)$$

provided every word in $X_1 \cdots X_m$ has a unique factorization as a product of words in $X_1, \ldots, X_m$.

Proposition 2.4.3 *Let $X \subset A^+$ be a code and let π be a Bernoulli distribution on A^*. Then*

$$F_{X^*}(t) = \frac{1}{1 - F_X(t)}.$$

Proof. Since $F_X(0) = 0$, we have $1/(1 - F_X(t)) = \sum_{n \geq 0} F_X(t)^n$. Since X is a code, the products X^n are unambiguous, that is every word in X^n has a unique factorization as a product of n words in X. By Lemma 2.4.2, this implies that $F_{X^n}(t) = F_X(t)^n$. Since moreover the sets X^n are pairwise disjoint, we have $F_{X^*}(t) = F_{\bigcup_{n \geq 0} X^n}(t) = \sum_{n \geq 0} F_{X^n}(t)$. Finally we obtain $1/(1 - F_X(t)) = \sum_{n \geq 0} F_X(t)^n = \sum_{n \geq 0} F_{X^n}(t) = F_{X^*}(t)$. □

In the case of the uniform Bernoulli distribution, we get the following corollary relating the ordinary generating functions $f_X(t)$ and $f_{X^*}(t)$ of X and X^* respectively.

Corollary 2.4.4 *Let X be a code over a finite alphabet A. Then*

$$f_{X^*}(t) = \frac{1}{1 - f_X(t)}.$$

Proof. Indeed, by Equation (1.32) we have, for the uniform Bernoulli distribution, $f_X(t) = F_X(kt)$ and $f_{X^*}(t) = F_{X^*}(kt)$, where $k = \text{Card}(A)$. So the corollary follows from Proposition 2.4.3. □

Theorem 2.4.5 *If X is a code over A, then $\pi(X) \leq 1$ for all Bernoulli distributions π on A^*.*

Proof. Suppose first that X is finite. Then $\pi(X)$ is finite. Assume by contradiction that $\pi(X) > 1$. Then $F_X(1) > 1$, and therefore there is a number $r < 1$ such that $F_X(r) = 1$. Since X is a code, one has $F_{X^*}(t) = 1/(1 - F_X(t))$ by Proposition 2.4.3. Then $F_{X^*}(t)$ diverges for $t = r$ and thus the radius of convergence of $F_{X^*}(t)$ is strictly smaller than 1, a contradiction for probability generating series.

Since $\pi(X)$ is the upper bound of the values for its finite subsets, the result follows. □

In the case where the alphabet A is finite and where the distribution π is uniform, we obtain

Corollary 2.4.6 *Let X be a code over an alphabet with k letters. Then*

$$\sum_{x \in X} k^{-|x|} \leq 1.$$ □

Example 2.4.7 Let $A = \{a, b\}$, and $X = \{b, ab, ba\}$. Define π by $\pi(a) = 1/3$, $\pi(b) = 2/3$. Then

$$\pi(X) = \frac{2}{3} + \frac{2}{9} + \frac{2}{9} = \frac{10}{9}$$

thus X is not a code. Note that for $\pi(a) = \pi(b) = 1/2$, we get $\pi(X) = 1$. Thus it is impossible to conclude that X is not a code from the second distribution.

The following example shows that the converse of Theorem 2.4.5 is false.

Example 2.4.8 Let $A = \{a, b\}$, and $X = \{ab, aba, aab\}$. The set X is not a code since

$$(aba)(ab) = (ab)(aab).$$

However, any Bernoulli distribution π gives $\pi(X) < 1$. Indeed, set $p = \pi(a)$, $q = \pi(b)$. Then

$$\pi(X) = pq + 2p^2q.$$

It is easily seen that we always have $pq \leq \frac{1}{4}$ and also $p^2q \leq \frac{4}{27}$, since $p + q = 1$. Consequently

$$\pi(X) \leq \frac{1}{4} + \frac{8}{27} < 1.$$

This example gives a good illustration of the limits of Theorem 2.4.5 in its use for testing whether a set is a code. Indeed, the set X of Example 2.4.8, where the test fails, is obtained from the set of Example 2.4.7, where the test is successful, simply by replacing b by ab. This shows that the counting argument represented by a Bernoulli distribution takes into account the lengths as well as the number of words. In other terms, Theorem 2.4.5 allows us to conclude that X is not a code only if there are "too many too short words".

Proposition 2.4.9 *Let X be a code over A. If there exists a positive Bernoulli distribution π on A^* such that $\pi(X) = 1$, then the code X is maximal.*

Proof. Suppose that X is not maximal. Then there is some word $y \notin X$ such that $Y = X \cup y$ is a code. By Theorem 2.4.5, we have $\pi(Y) \leq 1$. On the other hand,

$$\pi(Y) = \pi(X) + \pi(y) = 1 + \pi(y).$$

Thus $\pi(y) = 0$, which is impossible since π is positive. $\quad\square$

Proposition 2.4.9 is very useful for proving that a code is maximal. The direct method for proving maximality, based on the definition, indeed is usually much more complicated than the verification of the conditions of the proposition. A more precise statement, holding for a large class of codes, will be given in the next section (Theorem 2.5.16).

Example 2.4.1 (*continued*) Since $\pi(X) = 1$ and X is prefix, X is a maximal code.

Example 2.4.10 We consider again the Dyck code D over $A = \{a, b\}$ described in Example 2.2.11. Let π be a positive Bernoulli distribution on A^*, and set $p = \pi(a)$, $q = \pi(b)$.

Let $D_a = D \cap aA^*$ and $D_b = D \cap bA^*$. Note that D_a is formed of the words x on A such that $|u|_a - |u|_b > 0$ for each nonempty proper prefix u of x or equivalently $|v|_a - |v|_b < 0$ for each nonempty proper suffix v of x. In particular $D_a = \tilde{D}_b$ since the same holds for D_b with b and a interchanged. Let us show that

$$D_a = aD_a^*b, \quad D_b = bD_b^*a. \tag{2.15}$$

Let indeed x be a word of D_a. Clearly $x = ayb$ for some $y \in A^*$. Since $|x|_a = |x|_b$, we have $|y|_a = |y|_b$ and thus $y \in D^*$. Set $y = y_1 y_2 \cdots y_n$ with $y_i \in D$. Then each y_i is in D_a. Indeed, if y_i is in D_b, then $ay_1 \cdots y_{i-1}b$ is a prefix of x which belongs to D_a, a contradiction with the fact that D is a prefix code. Conversely, any word in aD_a^*b is clearly in D_a. This shows that $D_a = aD_a^*b$. The second equality is proved in an analogous way.

Since all products in (2.15) are unambiguous, we obtain $F_{D_a}(t) = F_a(t)F_{D_a^*}(t)F_b(t)$. Since D_a is a code, we have $F_{D_a^*}(t) = 1/(1 - F_{D_a}(t))$. Thus $F_{D_a}(t)$ is one of the two solutions of the quadratic equation

$$Y(t)^2 - Y(t) + pqt^2 = 0.$$

This equation has two solutions $(1 \pm \sqrt{1 - 4pqt^2})/2$. For the series $F_{D_a}(t)$, the correct sign is the minus sign because $F_{D_a}(0) = 0$. Thus

$$F_{D_a}(t) = \frac{1 - \sqrt{1 - 4pqt^2}}{2}.$$

Since $D_a = \tilde{D}_b$, we have $F_{D_a}(t) = F_{D_b}(t)$. Thus $F_D(t) = 2F_{D_a}(t)$ which gives finally

$$F_D(t) = 1 - \sqrt{1 - 4pqt^2}.$$

Thus $\pi(D) = 1 - \sqrt{1 - 4pq}$ or equivalently $\pi(D) = 1 - |p - q|$ since $(p - q)^2 = (p + q)^2 - 4pq = 1 - 4pq$.

For $\pi(a) = \pi(b) = 1/2$, we have $\pi(D) = 1$. This gives another proof that D is a maximal code (Example 2.2.11). Note that $\pi(D) < 1$ for any other Bernoulli distribution.

Example 2.4.11 The set $X = \bigcup_{n \geq 0} a^n b A^n$ is prefix, and therefore is a code over $A = \{a, b\}$. It is a maximal code. Let indeed π be a positive Bernoulli distribution, and set $p = \pi(a)$. Then

$$\pi(a^n b A^n) = p^n(1 - p)$$

hence

$$\pi(X) = \sum_{n \geq 0} p^n(1 - p) = (1/(1 - p))(1 - p) = 1.$$

We now give a statement which proves that the inequality of Corollary 2.4.6 is actually tight.

Theorem 2.4.12 (Kraft–McMillan) *Given a sequence* $(u_n)_{n \geq 1}$ *of integers, there exists a code X over an alphabet A of k symbols such that* $u_n = \mathrm{Card}(X \cap A^n)$ *if and only if*

$$\sum_{n \geq 1} u_n k^{-n} \leq 1. \tag{2.16}$$

Moreover, the code X can be chosen to be prefix.

Inequality (2.16) is called the *Kraft inequality*.

Proof. The necessity of the condition follows from Corollary 2.4.6. Conversely, observe first that by the inequality, one has also $\sum_{1 \leq i \leq n} u_i k^{-i} \leq 1$ or equivalently, multiplying both sides by k^n, $\sum_{1 \leq i \leq n} u_i k^{n-i} \leq k^n$ for all $n \geq 1$. Let us prove by induction on $n \geq 1$ that there exists a prefix code X_n on an alphabet A of k symbols such that $\mathrm{Card}(X_n \cap A^i) = u_i$ for $1 \leq i \leq n$.

This is true for $n = 1$ since $u_1 \leq k$. Next, suppose that the property holds for n. The set of words of length $n + 1$ with a prefix in X_n is $\bigcup_{1 \leq i \leq n} (X_n \cap A^i) A^{n+1-i}$. Consequently, the number of words of length $n + 1$ with a prefix in X_n is

$$s = \sum_{1 \leq i \leq n} u_i k^{n+1-i}.$$

Since $s + u_{n+1} \leq k^{n+1}$, we can choose a set Y of u_{n+1} words of length $n + 1$ without a prefix in X_n. In this way, the set $X_{n+1} = X_n \cup Y$ is a prefix code with length distribution $(u_i)_{1 \leq i \leq n+1}$. □

2.5 Complete sets

Any subset of a code is itself a code. Consequently, it is important to know the structure of maximal codes. Many of the results contained in this book are about maximal codes.

The notion of complete sets introduced in this section is in some sense dual to that of a code. For instance, any set containing a complete set is itself complete. Even if the duality is not perfectly balanced, it allows us to formulate maximality in terms of completeness, thus replacing an extremal property by a combinatorial one.

Let M be a monoid and let P be a subset of M. An element $m \in M$ is *completable* in P if there exist u, v in M such that $umv \in P$. It is equivalent to say that P meets the two-sided ideal MmM,

$$MmM \cap P \neq \emptyset$$

or, in other words, that

$$m \in F(P) = M^{-1} P M^{-1}.$$

A word which is not completable in P is incompletable. The set of words completable in P is of course $F(P)$; the set $\bar{F}(P) = M \setminus F(P)$ of incompletable words is a two-sided ideal of M which is disjoint from P.

A subset P of M is *dense* in M if all elements of M are completable in P, thus if $F(P) = M$ or, in an equivalent way, if P meets all (two-sided) ideals in M. Clearly, each superset of a dense set is dense.

The use of the adjective *dense* is justified by the fact that dense subsets of M are exactly the dense sets relative to some topology on M (see Exercise 2.5.2).

Example 2.5.1 Let $A = \{a\}$. The dense subsets of A^* are the infinite subsets.

Example 2.5.2 In a group G, any nonempty subset is dense, since $GmG = G$ for m in G.

Example 2.5.3 The Dyck code D over $A = \{a, b\}$ is dense in A^*. Indeed, if $w \in A^*$, then $v = a^{2|w|_b} w b^{|w|}$ is easily seen to be in D^*. Furthermore, no proper nonempty prefix of v is in D^*. Thus v is in D, showing that w is completable in D.

It is useful to have a special term for codes X such that the submonoid X^* is dense. A subset P of M is called *complete* in M if the submonoid generated by P is dense. Every dense set is also complete. Next, a subset X of A^* is complete if and only if $F(X^*) = A^*$.

Example 2.5.4 Any nonempty subset of a^+ is complete, since it generates an infinite submonoid.

Theorem 2.5.5 *Any maximal code is complete.*

The theorem is a direct consequence of the following proposition.

Proposition 2.5.6 *Let $X \subset A^+$ be a maximal code. For any word $w \in A^*$, one has*

$$X^* w A^* \cap X^* \neq \emptyset.$$

Proof. The result is clear if $\mathrm{Card}(A) = 1$ or if w is the empty word. Otherwise, by Proposition 1.3.6, there is a word $w' \in A^+$ such that $y = ww'$ is unbordered. Set $Y = X \cup y$. It suffices to prove that $X^* y A^* \cap X^* \neq \emptyset$. Since Y is not a code, we have $x_1 \cdots x_n = y_1 \cdots y_m$ with $n, m \geq 1$, $x_i, y_j \in Y$ and $x_1 \neq y_1$. Since X is a code, at least one of the x_i, y_j is equal to y. Consider the leftmost occurrence of y among the x_i, y_j. We may assume that it occurs among the x_i, say at index k. Thus $x_1, \ldots, x_{k-1} \in X$, $x_k = y$. Let ℓ be the least index such that $x_1 \cdots x_k$ is a prefix of $y_1 \cdots y_\ell$. Set $z = x_1 \cdots x_k u = y_1 \cdots y_\ell$. Clearly $z \in X^* y A^*$ (see Figure 2.4). We prove that $z \in X^*$ by showing that $y_1, \ldots, y_\ell \in X$. Let p be the least index such that $x_1 \cdots x_{k-1}$ is a prefix of $y_1 \cdots y_p$. Set $x_1 \cdots x_{k-1} v = y_1 \cdots y_p$, with v not empty because X is a code. Thus $x_k u = v y_{p+1} \cdots y_\ell$. One has $y_1, \ldots, y_p \in X$ by the minimality of k. Next, $y_{p+1}, \ldots, y_{\ell-1}$ are proper factors of $x_k = y$ and therefore are also in X. Finally, $y_\ell \neq y$ since y is unbordered. So $y_\ell \in X$ and $z \in X^*$. $\qquad\square$

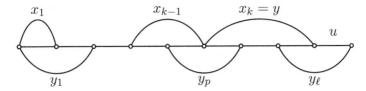

Figure 2.4 Showing that $z \in X^* y A^* \cap X^*$.

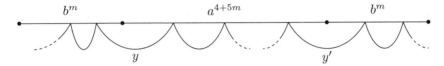

Figure 2.5 The factorization of $b^m a^{4+5m} b^m$ in words in Y.

Example 2.5.7 We are able now to verify one of the claims made in Section 2.1, namely that there do exist finite codes which are not contained in a maximal finite code.

Let $X = \{a^5, ba^2, ab, b\}$. It is a code over $A = \{a, b\}$. Any maximal code containing X is infinite. Indeed, let Y be a maximal code over A containing X, and assume Y finite. Set $m = \max\{|y| \mid y \in Y\}$ and let

$$u = b^m a^{4+5m} b^m.$$

Since Y is maximal, it is complete. Thus u is a factor of a word in Y^*. Neither b^m nor a^{4+5m} can be proper factors of a word in Y. Thus there exist $y, y' \in Y \cup 1$ and integers $p, q, r \geq 0$ such that

$$u = b^p y a^q y' b^r$$

with $a^q \in Y^*$ (see Figure 2.5). The word a^5 is the only word in Y which does not contain b; thus q is a multiple of 5; this implies that $|y|_a + |y'|_a \equiv 4 \bmod 5$. Let $y = b^h a^{5s+i}$ and $y' = a^{j+5t} b^k$ with $0 \leq i, j \leq 4$. We have $i + j \equiv 4 \bmod 5$ whence $i + j = 4$. We will show that any choice of i, j leads to the conclusion that Y is not a code. This yields the contradiction.

If $i = 0$, $j = 4$, then $k \geq 1$ and we have $ba^2 \cdot a^{5t+4} b^k = b \cdot a^{5(t+1)} \cdot ab \cdot b^{k-1}$.
If $i = 1$, $j = 3$, then $b^h a^{5s+1} \cdot b = b^h \cdot a^{5s} \cdot ab$.
If $i = 2$, $j = 2$, then $b \cdot a^{2+5t} b^k = ba^2 \cdot a^{5t} \cdot b^k$.
If $i = 3$, $j = 1$, then $h \geq 1$ and $b^h a^{5s+3} \cdot b = b^{h-1} \cdot ba^2 \cdot a^{5s} \cdot ab$.
Finally, if $i = 4$, $j = 0$, then $b^h a^{5s+4} \cdot ab = b^h \cdot a^{5(s+1)} \cdot b$.

This example is a particular case of a general construction (see Proposition 12.3.3).

The converse of Theorem 2.5.5 is false (see Example 2.5.9). However, it is true under an additional assumption that relies on the following definition.

A subset P of a monoid M which is not dense is called *thin*. If P is thin, there is at least one element m in M which is incompletable in P, that is such that $MmM \cap P = \emptyset$, or equivalently $F(P) \neq M$.

The use of the adjective *thin* is justified by results like Proposition 2.5.8 or 2.5.12.

Proposition 2.5.8 *Let M be a monoid and P, Q, R ⊂ M. Then the set P ∪ Q is thin if and only if P and Q are thin. If R is dense and P is thin, then R \ P is dense.*

Proof. If P and Q are thin, then there exist $m, n \in M$ such that

$$MmM \cap P = \emptyset, \quad MnM \cap Q = \emptyset.$$

Then mn is incompletable in $P \cup Q$ and therefore $P \cup Q$ is thin. Conversely if $P \cup Q$ is thin, there exists $m \in M$ which is incompletable in $P \cup Q$ and therefore incompletable in P and also in Q. Hence P and Q are thin. If R is dense in M and P is thin, then $R \setminus P$ cannot be thin since otherwise $R = (R \setminus P) \cup P$ would also be thin by the above statement.   □

Thin subsets of a free monoid have additional properties. In particular, any finite subset of A^* is clearly thin. Furthermore, if X, Y are thin subsets of A^* then the set XY is thin. In fact, if $u \notin F(X)$, $v \notin F(Y)$, then $uv \notin F(XY)$.

Example 2.5.9 The Dyck code D over $A = \{a, b\}$ is dense (See Example 2.5.3). It is a maximal code since it is a group code (see Example 2.2.11). For each $x \in D$, the code $D \setminus x$ remains dense, in view of Proposition 2.5.8, and thus remains complete. But of course $D \setminus x$ is no more a maximal code. This example shows that the converse of Theorem 2.5.5 does not hold in general.

Theorem 2.5.5 admits a converse in the case of codes which are both thin and complete. Before going on to prove this, we give some useful properties of these sets.

Proposition 2.5.10 *Let $X \subset A^*$ be a thin and complete set. Let w be a word incompletable in X. Then*

$$A^* = \bigcup_{d \in D, g \in G} d^{-1}X^*g^{-1} = D^{-1}X^*G^{-1}, \tag{2.17}$$

where D and G are the sets of suffixes (resp. prefixes) of w.

Proof. Let $z \in A^*$. Since X^* is dense, the word wzw is completable in X^*, thus for some $u, v \in A^*$

$$uwzwv \in X^*.$$

Now w is not a factor of a word in X. Thus there exist two factorizations $w = g_1d = gd_1$ such that

$$ug_1, dzg, d_1v \in X^*.$$

This shows that $z \in d^{-1}X^*g^{-1}$.   □

Proposition 2.5.11 *Let X be a thin and complete subset of A^*. For any positive Bernoulli distribution π on A^*, we have*

$$\pi(X) \geq 1.$$

Proof. We have $\pi(A^*) = \infty$. Since the union in Equation (2.17) is finite, there exists a pair $(d, g) \in D \times G$ such that $\pi(d^{-1}X^*g^{-1}) = \infty$. Now

$$d(d^{-1}X^*g^{-1})g \subset X^*.$$

This implies

$$\pi(d)\pi(d^{-1}X^*g^{-1})\pi(g) \leq \pi(X^*).$$

The positivity of π shows that $\pi(dg) \neq 0$. Thus $\pi(X^*) = \infty$. Now

$$\pi(X^*) \leq \sum_{n\geq 0} \pi(X^n) \leq \sum_{n\geq 0}(\pi(X))^n.$$

Assuming $\pi(X) < 1$, we get $\pi(X^*) < \infty$. Thus $\pi(X) \geq 1$. $\qquad\square$

Note the following property showing, as already claimed before, that a thin set has only *few* words.

Proposition 2.5.12 *Let $X \subset A^*$ be a thin set. For any positive Bernoulli distribution on A^*, we have*

$$\pi(X) < \infty.$$

Proof. Let w be a word which is not a factor of a word in X: $w \notin F(X)$. Set $n = |w|$. We have $n \geq 1$. For $0 \leq i \leq n - 1$, consider

$$X_i = \{x \in X \mid |x| \equiv i \bmod n\}.$$

It suffices to show that $\pi(X_i)$ is finite for $i = 0, \ldots, n - 1$. Now

$$X_i \subset A^i(A^n \setminus w)^*.$$

Since $A^n \setminus w$ is a code, we have

$$\pi[(A^n \setminus w)^*] = \sum_{k\geq 0}(\pi(A^n \setminus w))^k = \sum_{k\geq 0}(1 - \pi(w))^k.$$

The positivity of π implies $\pi(w) > 0$ and consequently

$$\pi[(A^n \setminus w)^*] = \frac{1}{\pi(w)}.$$

Thus $\pi(X_i) \leq 1/\pi(w)$. $\qquad\square$

We are now ready to prove

Theorem 2.5.13 *Any thin and complete code is maximal.*

Proof. Let X be a thin, complete code and let π be a positive Bernoulli distribution. By Proposition 2.5.11, $\pi(X) \geq 1$, and by Theorem 2.4.5, we have $\pi(X) \leq 1$. Thus $\pi(X) = 1$. But then Proposition 2.4.9 shows that X is maximal.   □

Theorems 2.5.5 and 2.5.13 can be grouped together to give

Theorem 2.5.14 *Let X be a code over A. Then X is complete if and only if X is dense or maximal.*

Proof. Assume X is complete. If X is not dense, then it is thin, and consequently X is maximal by the previous theorem. Conversely, a dense set is complete, and a maximal code is complete by Theorem 2.5.5.   □

Before giving other consequences of these statements, let us present a first application of the combinatorial characterization of maximality.

Proposition 2.5.15 *Let $X \subset A^*$ be a finite maximal code. For any nonempty subset B of A, the code $X \cap B^*$ is a maximal code over B. In particular, for each letter $a \in A$, there is an integer n such that $a^n \in X$.*

Proof. The second claim results from the first one by taking $B = \{a\}$. Let $n = \max\{|x| \mid x \in X\}$ be the maximal length of words in X, and let $\emptyset \neq B \subset A$. To show that $Y = X \cap B^*$ is a maximal code over B, it suffices to show, in view of Theorem 2.5.13, that Y is complete (in B^*). Let $w \in B^*$ and $b \in B$. Consider the word

$$w' = b^{n+1} w b^{n+1}.$$

The completeness of X gives words $u, v \in A^*$ such that

$$u w' v = x_1 x_2 \cdots x_k$$

for some $x_1, x_2, \ldots, x_k \in X$. But by the definition of n, there exist two integers i, j $(1 \leq i < j \leq k)$ such that

$$x_i x_{i+1} \cdots x_j = b^r w b^s$$

for some $r, s \in \{1, \ldots, n\}$ (see Fig. 2.6). But then $x_i, x_{i+1}, \ldots, x_j \in X \cap B^* = Y$. This shows that w is completable in Y^*.   □

Let $X \subset A^+$ be a finite maximal code, and let $a \in A$ be a letter. The (unique) integer n such that $a^n \in X$ is called the *order* of a relative to X.

Theorem 2.5.16 *Let X be a thin code. The following conditions are equivalent:*

(i) *X is a maximal code.*
(ii) *There exists a positive Bernoulli distribution π with $\pi(X) = 1$.*
(iii) *For any positive Bernoulli distribution π, we have $\pi(X) = 1$.*
(iv) *X is complete.*

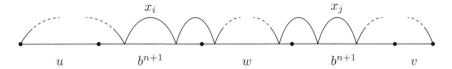

Figure 2.6 The factorization of $ub^{n+1}wb^{n+1}v$.

Proof. (i) $\Rightarrow$ (iv) is Theorem 2.5.5. (iv) $\Rightarrow$ (iii) is a consequence of Theorem 2.4.5 and Proposition 2.5.11. (iii) $\Rightarrow$ (ii) is not very hard, and (ii) $\Rightarrow$ (i) is Proposition 2.4.9. $\square$

Theorem 2.5.16 gives a surprisingly simple method to test whether a thin code X is maximal. It suffices to take any positive Bernoulli distribution π and to check whether $\pi(X) = 1$.

Example 2.5.17 The Dyck code D over $A = \{a, b\}$ is maximal and complete, but satisfies $\pi(D) = 1$ only for one Bernoulli distribution (see Example 2.4.10). Thus the conditions (i) + (ii) + (iv) do not imply (iii) for dense codes.

Example 2.5.18 The prefix code $X = \bigcup_{n \geq 0} a^n b A^n$ over $A = \{a, b\}$ is dense since for all $w \in A^*$, $a^{|w|} b w \in X$. It satisfies (iii), as we have seen in Example 2.4.11. Thus X satisfies the four conditions of the theorem without being thin.

Theorem 2.5.19 *Let X be a thin subset of A^+, and let π be a positive Bernoulli distribution. Any two among the three following conditions imply the third:*

(i) *X is a code.*
(ii) *$\pi(X) = 1$.*
(iii) *X is complete.*

Proof. (i) + (ii) $\Rightarrow$ (iii). The condition $\pi(X) = 1$ implies that X is a maximal code, by Proposition 2.4.9. Thus by Theorem 2.5.5, X is complete.

(i) + (iii) $\Rightarrow$ (ii) Theorem 2.4.5 and condition (i) imply that $\pi(X) \leq 1$. Now X is thin and complete; in view of Proposition 2.5.11, we have $\pi(X) \geq 1$.

(ii) + (iii) $\Rightarrow$ (i) Let $n \geq 1$ be an integer. First, we verify that X^n is thin and complete. To see completeness, let $u \in A^*$, and let $v, w \in A^*$ be such that $vuw \in X^*$. Then $vuw \in X^k$ for some $k \geq 0$. Thus $(vuw)^n \in (X^n)^k \subset (X^n)^*$. This shows that u is completable in $(X^n)^*$. Further, since X is thin and because the product of two thin sets is again thin, the set X^n is thin.

Thus, X^n is thin and complete. Consequently, $\pi(X^n) \geq 1$ by Proposition 2.5.11. On the other hand, we have $\pi(X^n) \leq \pi(X)^n$ and thus $\pi(X^n) \leq 1$. Consequently $\pi(X^n) = 1$. Thus for all $n \geq 1$

$$\pi(X^n) = \pi(X)^n.$$

Proposition 2.4.3 shows that X is a code. $\square$

Thin codes constitute a very important class of codes. They will be characterized by some finiteness condition in Chapter 11. We anticipate these results by proving a particular case which shows that the class of thin codes is quite a large one.

Proposition 2.5.20 *Any recognizable code is thin.*

Proof. Let $X \subset A^*$ be a recognizable code, and let $\mathcal{A} = (Q, i, T)$ be a deterministic complete automaton recognizing X. Associate to a word w, the number

$$\rho(w) = \text{Card}(Q \cdot w) = \text{Card}\{q \cdot w \mid q \in Q\}.$$

We have $\rho(w) \leq \text{Card}(Q)$ and $\rho(uwv) \leq \rho(w)$ for all words u, v.

Let J be the set of words w in A^* with minimal $\rho(w)$. The previous inequality shows that J is a two-sided ideal of A^*.

Let $w \in J$, and let $P = Q \cdot w$. Then $P \cdot w = P$. Indeed $P \cdot w \subset Q \cdot w = P$, and on the other hand, $P \cdot w = Q \cdot w^2$. Thus $\text{Card}(P \cdot w) = \rho(w^2)$. Since $\rho(w)$ is minimal, $\rho(w^2) = \rho(w)$, whence the equality. This shows that the mapping $p \mapsto p \cdot w$ from P onto P is a bijection. It follows that there is some integer n such that the mapping $p \mapsto p \cdot w^n$ is the identity mapping on P.

Since $P = Q \cdot w$, we have $q \cdot w = q \cdot w^{n+1}$ for all $q \in Q$. To show that X is thin, it suffices to show that X does not meet the two-sided ideal J. Assume that $J \cap X \neq \emptyset$ and let $x \in X \cap J$. Then $i \cdot x = t \in T$. Next $x \in J$ and, by the previous discussion, there is some integer $n \geq 1$ such that $i \cdot x^{n+1} = t$. This implies that $x^{n+1} \in X$. But this is impossible, since X is a code.   □

The converse of Proposition 2.5.20 is false, as shown by the following example.

Example 2.5.21 The code $X = \{a^n b^n \mid n \geq 1\}$ is thin (for example, ba is not a factor of X), but X is not recognizable.

Example 2.5.22 In one interesting case, the converse of Proposition 2.5.20 holds: *Any thin group code is recognizable.* Indeed let $X \subset A^*$ be a group code. Let $\varphi : A^* \to G$ be a surjective morphism onto a group G, and let H be a subgroup of G such that $X^* = \varphi^{-1}(H)$. By assumption, X is thin. Let m be a word that is incompletable in X. We show that H has finite index in G, and more precisely that

$$G = \bigcup_{p \leq m} H\varphi(p)^{-1},$$

(where p runs over the prefixes of m). Indeed let $g \in G$ and $w \in \varphi^{-1}(g)$. Let $u \in A^*$ be such that $\varphi(u)$ is the group inverse of $g\varphi(m)$. Then $\varphi(wmu) = g\varphi(m)\varphi(u) = 1$, whence $wmu \in X^*$. Now m is incompletable in X. Thus m is not factor of a word in X and consequently there is a factorization $m = pq$ such that $wp, qu \in X^*$. But then $h = \varphi(wp) \in H$. Since $h = g\varphi(p)$, we have $g \in H\varphi(p)^{-1}$. This proves the formula.

The formula shows that there are finitely many right cosets of H in G. Thus the representation of G by permutations on the right cosets of H is also finite. Denote it by K. Let $\alpha : G \to K$ be the canonical morphism defined by $Hr\alpha(g) = Hrg$

(see Section 1.13). Then, setting $N = \{\sigma \in K \mid H\sigma = H\}$, we have $H = \alpha^{-1}(N) = \alpha^{-1}(\alpha(H))$. Thus $X^* = \psi^{-1}\psi(X^*)$, where $\psi = \alpha \cdot \varphi$. Since K is finite, this shows that X^* is recognizable. Consequently, X is also recognizable (Exercise 2.2.7).

Remark 2.5.23 We have used in the preceding paragraphs arguments which rely basically on two techniques: probabilities on the one hand which allowed us to prove especially Theorem 2.5.13 and direct combinatorial arguments on words on the other (as in the proof of Theorem 2.5.5).

It is interesting to note that some of the proofs can be completed by using just one of the two techniques. A careful analysis shows that all the preceding statements with the exception of those involving maximality can be established by using only arguments on probabilities. As an example, the implication (ii) $\Rightarrow$ (iv) in Theorem 2.5.16 can be proved as follows without using the maximality of X. If X is not complete, then X^* is thin. Thus, by Proposition 2.5.12, $\pi(X^*) < \infty$ which implies $\pi(X) < 1$ by Proposition 2.4.3.

Conversely, there exist, for some of the results given here, combinatorial proofs which do not rely on probabilities. This is the case for Theorem 2.5.13, where the proof given relies heavily on arguments about probabilities. Another proof of this result will be given in Chapter 9 (Corollary 9.4.6). This proof is based on the fact that if $X \subset A^+$ is a thin complete code, then all words $w \in A^*$ satisfy

$$(X^* w X^*)^+ \cap X^* \neq \emptyset.$$

This implies Theorem 2.5.13, because according to this formula, $X \cup w$ is not a code for $w \notin X$ and thus X is a maximal code.

Example 2.5.7 shows that a finite code is not always contained in a finite maximal code. The *inclusion problem*, for a finite code X, is the existence of a finite maximal code containing X. The *inclusion conjecture* claims that the inclusion problem is decidable.

We prove the following remarkable property.

Theorem 2.5.24 (Ehrenfeucht–Rozenberg) *Every rational code is contained in a maximal rational code.*

The proof relies on the following result.

Proposition 2.5.25 *Let $X \subset A^+$ be a code. Let $y \in A^*$ be an unbordered word such that $A^* y A^* \cap X^* = \emptyset$. Let*

$$U = A^* \setminus (X^* \cup A^* y A^*). \tag{2.18}$$

Then the set

$$Y = X \cup y(Uy)^* \tag{2.19}$$

is a complete code.

Proof. Set $V = A^* \setminus A^* y A^*$. Then by assumption $X^* \subset V$ and $U = V \setminus X^*$. Let us first observe that the set $Z = Vy$ is a prefix code.

Assume indeed that $vy < v'y$ for two words v and v' in V. Since y is unbordered, vy must be a prefix of v'. But then v' is in $A^* y A^*$, a contradiction. Thus Z is prefix.

Now we show that Y is a code. Assume the contrary and consider a relation

$$y_1 y_2 \cdots y_n = y'_1 y'_2 \cdots y'_m$$

with $y_1, \ldots, y'_m \in Y$ and $y_1 \neq y'_1$. The set X being a code, one of these words must be in $Y \setminus X$. Assume that one of $y_1, \ldots, y_n$ is in $Y \setminus X$, and let p be the smallest index such that $y_p \in y(Uy)^*$. From $y \notin F(X^*)$ it also follows that $y_p \notin F(X^*)$. Consequently one of $y'_1, \ldots, y'_m$ is in $y(Uy)^*$. Let q be the smallest index that $y'_q \in y(Uy)^*$. Then

$$y_1 \cdots y_{p-1} y, \quad y'_1 y'_2 \cdots y'_{q-1} y \in Z$$

whence $y_1 \cdots y_{p-1} = y'_1 \cdots y'_{q-1}$ since Z is prefix. The set X is a code, thus from $y_1 \neq y'_1$ it follows that $p = q = 1$. Set

$$y_1 = y u_1 y \cdots y u_k y, \quad y'_1 = y u'_1 y \cdots y u'_l y,$$

with $u_1, \ldots, u_k, u'_1, \ldots, u'_l \in U$. Assume $y_1 < y'_1$. Since Z is prefix, the set Z^* is right unitary. From $U \subset V$, it follows that each $u_i y, u'_i y$ is in Z. Consequently

$$u_1 = u'_1, \ldots, u_k = u'_k.$$

Let $t = u'_{k+1} y \cdots y u'_l y$. We have

$$y_2 \cdots y_n = t y'_2 \cdots y'_m.$$

The word y is a factor of t, and thus occurs also in $y_2 \cdots y_n$. This shows that one of $y_2, \ldots, y_n$, say y_r, is in $y(Uy)^*$. Suppose r is chosen minimal. Then $y_2 \cdots y_{r-1} y \in Z$ and $u'_{k+1} y \in Z$ are prefixes of the same word. With the set Z being prefix, we have

$$u'_{k+1} = y_2 \cdots y_{r-1}.$$

Thus $u'_{k+1} \in X^*$, in contradiction with the hypothesis $u'_{k+1} \in U$. This shows that Y is a code.

Finally, let us show that Y is complete. Let $w \in A^*$ and set

$$w = v_1 y v_2 y \cdots y v_{n-1} y v_n$$

with $n \geq 1$ and $v_i \in A^* \setminus A^* y A^*$. Then $ywy \in Y^*$. Indeed let $v_{i_1}, v_{i_2}, \ldots, v_{i_k}$ be those v_i's which are in X^*. Then

$$ywy = (yv_1 y \cdots yv_{i_1-1} y) v_{i_1} (yv_{i_1+1} y \cdots yv_{i_2-1} y) \cdots v_{i_k} (yv_{i_k+1} y \cdots yv_n y).$$

Each of the parenthesized words is in Y. Thus the whole word is in Y^*. □

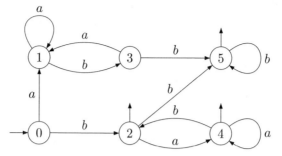

Figure 2.7 An automaton recognizing U.

Proof of Theorem 2.5.24. Since X is rational, the set U defined in Equation (2.18) is also rational. Thus Y is a rational code. By Proposition 2.5.20, the set Y is thin. By Theorem 2.5.13, it follows that Y is a maximal code. □

Example 2.5.26 Let $A = \{a, b\}$ and $X = \{a, ab\}$. The word $y = bba$ is unbordered and is incompletable in X^*. A deterministic automaton recognizing $U = A^* \setminus (X^* \cup A^* y A^*)$ is given in Figure 2.7. Accordingly, we obtain, after some rewriting the expression

$$U = b^+ \cup X^* abb^+ \cup bX^* ab^*.$$

Consider a Bernoulli distribution π on A^* and set $p = \pi(a), q = \pi(b)$. Then an easy computation shows that $\pi(U) = 1/pq$ and thus $\pi(Y) = 1$ for Y defined by (2.18), which implies that Y is maximal.

Example 2.5.27 Let $A = \{a, b\}$ and $X = \{bb, bbab, babb\}$. The word $y = aba$ is incompletable in X^*. However, $X \cup y$ is not a code, since

$$(bb)(aba)(babb) = (bbab)(aba)(bb).$$

This example shows that Proposition 2.5.25 is false without the assumption that y is unbordered.

The following proposition shows how the property of being a complete code is reflected in an automaton.

Proposition 2.5.28 *Let $X \subset A^+$, and let $\mathcal{A} = (Q, 1, 1)$ be a trim automaton recognizing X^*. Then X is complete if and only if the transition monoid of $\mathcal{A}$ does not contain the null relation.*

Proof. If X is complete, then there exist, for each $w \in A^*$, two words $u, v \in A^*$ such that $uwv \in X^*$. Then there exists a path $1 \xrightarrow{u} p \xrightarrow{w} q \xrightarrow{v} 1$. This implies that (p, q) is in $\varphi(w)$ and consequently $\varphi_A(w)$ is not null.

Conversely, if $\varphi_A(A^*)$ does not contain the null relation, then for each $w \in A^*$, there exists at least one path $p \xrightarrow{w} q$. Since $\mathcal{A}$ is trim, there exist two paths $1 \xrightarrow{u} p$ and $q \xrightarrow{v} 1$. Then $uwv \in X^*$. Thus X is complete. $\square$

For a (commutative) polynomial $p \in \mathbb{Q}[A]$, and a Bernoulli distribution π on the alphabet A we denote by $\pi(p)$ the number obtained by substituting $\pi(a)$ to the letter a, for all $a \in A$. More precisely, setting $A = \{a_1, \ldots, a_n\}$ and $p = p(a_1, \ldots, a_n)$, the number $\pi(p)$ is $\pi(p) = p(\pi(a_1), \ldots, \pi(a_n))$.

Proposition 2.5.29 *Let $p \in \mathbb{Q}[A]$ be a polynomial and let $a \in A$ be a letter. The following conditions are equivalent:*

(i) *p is divisible by the polynomial $1 - \sum_{a \in A} a$.*
(ii) *$\pi(p) = 0$ for each positive Bernoulli distribution.*

Proof. The implication (i) $\Rightarrow$ (ii) is clear.

To prove (ii) $\Rightarrow$ (i), fix a letter $a \in A$, and set $B = A \setminus a$. Consider p as a polynomial in the variable a with coefficients in $\mathbb{Q}[B]$. Similarly, consider $\sum_{a \in A} a - 1 = a + u$ as a linear polynomial in a with constant term u where $u = \sum_{b \in B} b - 1$.

The Euclidean division of p by $a + u$ gives $p = q(a + u) + r$ where $q \in \mathbb{Q}[A]$ and $r \in \mathbb{Q}[B]$. Since $\pi(p) = 0$ and $\pi(a + u) = 0$ for each positive Bernoulli distribution π, the polynomial r vanishes at all points $z = (z_1, \ldots, z_{n-1}) \in \mathbb{Q}^{n-1}$ such that $z_i > 0$ and $z_1 + \cdots + z_{n-1} \leq 1$. It follows that r vanishes and consequently $1 - \sum_{a \in A} a$ divides p. $\square$

Recall that α denotes the canonical morphism from $\mathbb{Q}\langle\langle A \rangle\rangle$ onto $\mathbb{Q}[[A]]$.

Theorem 2.5.30 *Let X be a finite maximal code on the alphabet A. Then $\alpha(\underline{X}) - 1$ is divisible by $\alpha(\underline{A}) - 1$.*

Proof. Let π be a positive Bernoulli distribution on A^*. By Theorem 2.5.16, we have $\pi(X) = 1$. By Proposition 2.5.29, this implies the conclusion. $\square$

Example 2.5.31 For the code $X = \{aa, ba, bb, baa, bba\}$ of Example 4.1.7, one has

$$\alpha(\underline{X}) - 1 = (b + 1)(a + b - 1)(a + 1).$$

2.6 Composition

We now introduce a partial binary operation on codes called composition. This operation associates to two codes Y and Z satisfying a certain compatibility condition a third code denoted by $Y \circ Z$.

There is a twofold interest in this operation. First, it gives a useful method for constructing more complicated codes from simple ones. For example, we will see that the composition of a prefix and a suffix code can result in a code that is neither prefix nor suffix.

Second, and this constitutes the main interest for composition, the converse notion of decomposition allows us to study the structure of codes. If a code X decomposes into two codes Y and Z, then these codes are generally simpler.

Let $Z \subset A^*$ and $Y \subset B^*$ be two codes with $B = \text{alph}(Y)$. Then the codes Y and Z are *composable* if there is a bijection from B onto Z. If β is such a bijection, then Y and Z are called composable *through* β. Then β defines a morphism from B^* into A^* which is injective since Z is a code (Proposition 2.1.1). The set

$$X = \beta(Y) \subset Z^* \subset A^* \tag{2.20}$$

is obtained by *composition* of Y and Z (by means of β). We denote it by

$$X = Y \circ_\beta Z,$$

or by $X = Y \circ Z$ when the context permits it. Since β is injective, X and Y are related by bijection, and in particular $\text{Card}(X) = \text{Card}(Y)$. The words in X are obtained just by replacing, in the words of Y, each letter b by the word $\beta(b) \in Z$. The injectivity of β, the Corollary 2.1.6 and (2.20) give the following result.

Proposition 2.6.1 *If Y and Z are two composable codes, then $X = Y \circ Z$ is a code.*

□

Example 2.6.2 Let $A = \{a, b\}$, $B = \{c, d, e\}$ and

$$Z = \{a, ba, bb\} \subset A^*, \quad Y = \{cc, d, dc, e, ec\} \subset B^*.$$

The code Z is prefix, and Y is suffix. Further $\text{Card}(B) = \text{Card}(Z)$. Thus Y and Z are composable, in particular by means of the morphism $\beta : B^* \to A^*$ defined by

$$\beta(c) = a, \quad \beta(d) = ba, \quad \beta(e) = bb.$$

Then $X = Y \circ Z = \{aa, ba, baa, bb, bba\}$. The code X is neither prefix nor suffix. Now define $\beta' : B^* \to A^*$ by

$$\beta'(c) = ba, \quad \beta'(d) = a, \quad \beta'(e) = bb.$$

Then $X' = Y \circ_{\beta'} Z = \{baba, a, aba, bb, bbba\}$. This example shows that the composed code $Y \circ_\beta Z$ depends essentially on the mapping β.

The two expressions $X = X \circ A$ and $X = B \circ X$ are exactly the particular cases obtained by replacing one of the two codes by the alphabet in the expression

$$X = Y \circ Z.$$

Indeed, if $Y = B$, then $Z = \beta(B) = X$; if now $Z = A$, then B can be identified with A, and Y can be identified with X. These examples show that every code is obtained in at least two ways as a composition of codes.

Notice also the formula

$$X = Y \circ_\beta Z \implies X^n = Y^n \circ_\beta Z \quad n \geq 2.$$

Indeed, Y^n is a code (Corollary 2.1.7) and

$$Y^n \circ Z = \beta(Y^n) = X^n.$$

Proposition 2.6.3 *Let $X \subset C^*$, $Y \subset B^*$, and $Z \subset A^*$ be three codes, and assume that X and Y are composable through γ and that Y and Z are composable through β. Then*

$$(X \circ_\gamma Y) \circ_\beta Z = X \circ_{\beta\gamma} (Y \circ_\beta Z).$$

Proof. We may suppose that $C = \mathrm{alph}(X)$, $B = \mathrm{alph}(Y)$. By hypothesis the injective morphisms $\gamma : C^* \to B^*$ and $\beta : B^* \to A^*$ satisfy

$$\gamma(C) = Y, \quad \beta(B) = Z.$$

Let $\delta : D^* \to C^*$ be a coding morphism for X; thus $\delta(D) = X$. Then

$$D^* \xrightarrow{\delta} C^* \xrightarrow{\gamma} B^* \xrightarrow{\beta} A,$$

and $\beta\gamma\delta(D) = \beta\gamma(X) = X \circ_{\beta\gamma} \beta\gamma(C) = X \circ_{\beta\gamma} (Y \circ_\beta Z)$, and also $\beta\gamma\delta(D) = \beta(\gamma\delta(D)) = \gamma\delta(D) \circ_\beta \beta(B) = (X \circ_\gamma Y) \circ Z$. □

Some of the properties of codes are preserved under composition.

Proposition 2.6.4 *Let Y and Z be composable codes, and let $X = Y \circ Z$.*

1. *If Y and Z are prefix (suffix) codes, then X is a prefix (suffix) code.*
2. *If Y and Z are complete, then X is complete.*
3. *If Y and Z are thin, then X is thin.*

The proof of 3 uses Lemma 2.6.5 which cannot be established before Chapter 9 (Lemma 9.4.8), where more powerful tools will be available.

Lemma 2.6.5 *Let Z be a thin complete code over A. For each word $u \in Z^*$ there exists a word $w \in Z^* u Z^*$ having the following property. If $mwn \in Z^*$, then there exists a factorization $w = sut$ with $ms, tn \in Z^*$.*

Proof of Proposition 2.6.4. Let $Y \subset B^*$, $Z \subset A^*$, and let $\beta : B^* \to A^*$ be an injective morphism with $\beta(B) = Z$. Thus $X = \beta(Y) = Y \circ_\beta Z$.

1. Assume Y and Z are prefix codes. Consider $x, xu \in X$ with $u \in A^*$. Since $X \subset Z^*$, we have $x, xu \in Z^*$ and since Z^* is right unitary, this implies $u \in Z^*$. Let $y = \beta^{-1}(x)$, $v = \beta^{-1}(u) \in B^*$. Then $y, yv \in Y$ and Y is prefix; thus $v = 1$ and consequently $u = 1$. This shows that X is prefix. The case of suffix codes is handled in the same way.

2. Let $w \in A^*$. The code Z is complete, thus $uwv \in Z^*$ for some $u, v \in A^*$. Let $h = \beta^{-1}(uwv) \in B^*$. There exist, by the completeness of Y, two words $\bar{u}, \bar{v} \in B^*$ with $\bar{u}h\bar{v} \in Y^*$. But then $\beta(\bar{u})uwv\beta(\bar{v}) \in X^*$. This proves the completeness of X.

3. If Z is not complete, then $F(X) \subset F(Z^*) \neq A^*$ and X is thin. Assume now that Z is complete. The code Y is thin. Consequently $F(Y) \neq B^*$. Let $\bar{u} \in B^* \setminus F(Y)$, and $u = \beta(\bar{u})$. Let w be the word associated to u in Lemma 2.6.5. Then $w \notin F(X)$. Indeed, assuming the contrary, there exist words $m, n \in A^*$ such that

$$x = mwn \in X \subset Z^*.$$

In view of Lemma 2.6.5,

$$x = msutn, \quad \text{with } ms, tn \in Z^* = \beta(B^*).$$

Setting $p = \beta^{-1}(ms), q = \beta^{-1}(tn)$, we have $p\bar{u}q \in Y$. Thus $\bar{u} \in F(Y)$, contrary to the assumption. This shows that w is not in X, and thus X is thin. $\qquad\square$

We now consider the second aspect of the composition operation, namely the decomposition of a code into simpler ones. For this, it is convenient to extend the notation alph in the following way: let $Z \subset A^*$ be a code, and $X \subset A^*$. Then

$$\text{alph}_Z(X) = \{z \in Z \mid \exists u, v \in Z^* : uzv \in X\}.$$

In other words, $\text{alph}_Z(X)$ is the set of words in Z which appear at least once in a factorization of a word in X as a product of words in Z. Of course, $\text{alph}_A = \text{alph}$. The following proposition describes the condition for the existence of a decomposition.

Proposition 2.6.6 *Let $X, Z \subset A^*$ be codes. There exists a code Y such that $X = Y \circ Z$ if and only if*

$$X \subset Z^* \quad \text{and} \quad \text{alph}_Z(X) = Z. \tag{2.21}$$

The second condition in (2.21) means that all words in Z appear in at least one factorization of a word in X as product of words in Z.

Proof. Let $X = Y \circ_\beta Z$, where $\beta : B^* \to A^*$ is an injective morphism, $Y \subset B^*$ and $B = \text{alph}(Y)$. Then $X = \beta(Y) \subset \beta(B^*) = Z^*$ and further $\beta(B) = \text{alph}_{\beta(B)}(\beta(Y))$, that is, $Z = \text{alph}_Z(X)$.

Conversely, let $\beta : B^* \to A^*$ be a coding morphism for Z, and set $Y = \beta^{-1}(X)$. Then $X \subset \beta(B^*) = Z^*$ and $\beta(Y) = X$. By Corollary 2.1.6, Y is a code. Next $\text{alph}(Y) = B$ since $Z = \text{alph}_Z(X)$. Thus Y and Z are composable and $X = Y \circ_\beta Z$. $\qquad\square$

We have already seen that there are two obvious decompositions of a code $X \subset A^*$ as $X = Y \circ Z$, namely $X = B \circ X$ and $X = X \circ A$. They are obtained by taking $Z = X$ and $Z = A$ in Proposition 2.6.6 and assuming $A = \text{alph}(X)$. These decompositions are not interesting. We will call *indecomposable* a code which has no other decompositions. Formally, a code $X \subset A^*$ with $A = \text{alph}(X)$ is called *indecomposable* if $X = Y \circ Z$ and $B = \text{alph}(Y)$ imply $Y = B$ or $Z = A$. If X is decomposable, and if Z is a code such that $X = Y \circ Z$, and $Z \neq X$, $Z \neq A$, then we say that X decomposes *over Z*.

Example 2.6.2 (*continued*) The code X decomposes over Z. On the contrary, the code $Z = \{a, ba, bb\}$ is indecomposable. Indeed, let T be a code such that $Z \subset T^*$, and suppose $T \neq A$. Necessarily, $a \in T$. Thus $b \notin T$. But then $ba, bb \in T$, whence $Z \subset T$. Now Z is a maximal code (Example 2.4.1), thus $Z = T$.

Proposition 2.6.7 *For any finite code X, there exist indecomposable codes $Z_1, \ldots, Z_n$ such that*

$$X = Z_1 \circ \cdots \circ Z_n.$$

To prove this proposition, we introduce a notation. Let X be a finite code, and let

$$\ell(X) = \sum_{x \in X}(|x| - 1) = \sum_{x \in X} |x| - \mathrm{Card}(X).$$

For each $x \in X$, we have $|x| \geq 1$. Thus $\ell(X) \geq 0$, and moreover $\ell(X) = 0$ if and only if X is a subset of the alphabet.

Proposition 2.6.8 *If $X, Z \subset A^*$ and $Y \subset B^*$ are finite codes such that $X = Y \circ Z$, then $\ell(X) \geq \ell(Y) + \ell(Z)$.*

Proof. Let $\beta : B^* \to A^*$ be the injective morphism such that $X = Y \circ_\beta Z$. From $\mathrm{Card}(X) = \mathrm{Card}(Y)$ it follows that

$$\ell(X) - \ell(Y) = \sum_{x \in X} |x| - \sum_{y \in Y} |y| = \sum_{y \in Y}(|\beta(y)| - |y|).$$

Now $|\beta(y)| = \sum_{b \in B} |\beta(b)||y|_b$. Thus

$$\ell(X) - \ell(Y) = \sum_{y \in Y}\left(\sum_{b \in B}(|\beta(b)||y|_b - |y|_b)\right) = \sum_{y \in Y}\left(\sum_{b \in B}(|\beta(b)| - 1)|y|_b\right)$$

$$= \sum_{b \in B}(|\beta(b)| - 1)\left(\sum_{y \in Y} |y|_b\right).$$

By assumption $B = \mathrm{alph}(Y)$, whence $\sum_{y \in Y} |y|_b \geq 1$ for all b in B. Further $|\beta(b)| \geq 1$ for $b \in B$ by the injectivity of β. Thus

$$\ell(X) - \ell(Y) \geq \sum_{b \in B}(|\beta(b)| - 1) = \sum_{z \in Z}(|z| - 1) = \ell(Z). \qquad \square$$

Proof of Proposition 2.6.7. The proof is by induction on $\ell(X)$. If $\ell(X) = 0$, then X is composed of letters, and thus is indecomposable. If $\ell(X) > 0$ and X is decomposable, then $X = Y \circ Z$ for some codes Y, Z. Further Y and Z are not formed of letters only, and thus $\ell(Y) > 0$, $\ell(Z) > 0$. By Proposition 2.6.8, we have $\ell(Y) < \ell(X)$ and $\ell(Z) < \ell(X)$. Thus Y and Z are compositions of indecomposable codes. Thus X also is such a composition. $\qquad \square$

Proposition 2.6.7 shows the existence of a decomposition of codes. This decomposition need not be unique. This is shown in the following example.

Example 2.6.9 Consider the codes

$$X = \{aa, ba, baa, bb, bba\}, \quad Y = \{cc, d, dc, e, ec\}, \quad Z = \{a, ba, bb\}$$

of Example 2.6.2. As we have seen, $X = Y \circ Z$. There is also a decomposition

$$X = Y' \circ_\gamma Z'$$

with

$$Y' = \{cc, d, cd, e, ce\}, \quad Z' = \{aa, b, ba\}$$

and $\gamma : B^* \to A^*$ defined by

$$\gamma(c) = b, \quad \gamma(d) = aa, \quad \gamma(e) = ba.$$

The code Z is indecomposable, the code Z' is obtained from Z by interchanging a and b, and by taking then the reverse. These operations do not change indecomposability.

Example 2.6.10 This example shows that in decompositions of a code in indecomposable codes, even the number of components need not be unique. For $X = \{a^3b\}$, we have

$$X = \{cd\} \circ \{a^2, ab\} = \{cd\} \circ \{u^2, v\} \circ \{a, ab\}$$

and also

$$X = \{cd\} \circ \{a^3, b\}.$$

This gives two decompositions of length 3 and 2, respectively.

The code X in Example 2.6.9 is neither prefix nor suffix, but is composed of such codes. We may ask whether any (finite) code can be obtained by composition of prefix and suffix codes. This is not the case, as shown in the following example, see also Exercise 2.6.3.

Example 2.6.11 The code $X = \{b, ba, a^2b, a^3ba^4\}$ does not decompose over a prefix or a suffix code.

Assume the contrary. Then $X \subset Z^*$ for some prefix (or suffix) code $Z \neq A$. Thus Z^* is right unitary (resp. left unitary). From $b, ba \in Z^*$, it follows that $a \in Z^*$, whence $A = \{a, b\} \subset Z^*$ and $A = Z$. Assuming Z^* left unitary, $b, a^2b \in Z^*$ implies $a^2 \in Z^*$. It follows that $a^3b \in Z^*$, whence $a^3 \in Z^*$ and finally $a \in Z^*$. Thus again $Z = A$.

We now give a list of properties of codes which are inherited by the factors of a decomposition. Proposition 2.6.12 is in some sense dual to Proposition 2.6.4.

Proposition 2.6.12 *Let X, Y, Z be codes with $X = Y \circ Z$*

1. *If X is prefix (suffix), then Y is prefix (suffix).*
2. *If X is maximal, then Y and Z are maximal.*
3. *If X is complete, then Z is complete.*
4. *If X is thin, then Z is thin.*

Proof. We assume that $X, Z \subset A^*$, $Y \subset B^*$, $\beta : B^* \to A^*$ an injective morphism with $\beta(B) = Z$, $\beta(Y) = X$.

1. Let y, $yu \in Y$. Then $\beta(y)$, $\beta(y)\beta(u) \in X$, and since X is prefix, $\beta(u) = 1$. Now β is injective, whence $u = 1$.

2. If Y is not maximal, let $Y' = Y \cup y$ be a code for some $y \notin Y$. Then $\beta(Y') = \beta(Y) \cup \beta(y)$ is a code which is distinct from X by the injectivity of β. Thus X is not maximal.

Assume now that Z is not maximal. Set $Z' = Z \cup z$ for some $z \notin Z$ such that Z' is a code. Extend B to $B' = B \cup b$ ($b \notin B$) and define β over B'^* by $\beta(b) = z$. Then β is injective by Proposition 2.1.1 because Z' is a code. Further $Y' = Y \cup b$ is a code, and consequently $\beta(Y') = X \cup z$ is a code, showing that X is not maximal.

3. is clear from $X^* \subset Z^*$.

4. Any word in Z is a factor of a word in X. Thus $F(Z) \subset F(X)$. By assumption, $F(X) \neq A^*$. Thus $F(Z) \neq A^*$ and Z is thin. $\qquad\square$

Proposition 2.6.13 *Let X, Y, Z be three codes such that $X = Y \circ Z$. Then X is thin and complete if and only if Y and Z are thin and complete.*

Proof. By Proposition 2.6.4, the code X is thin and complete, provided Y and Z are. Assume conversely that X is thin and complete. Proposition 2.6.12 shows that Z is thin and complete. In view of Theorem 2.5.14, X is a maximal code. By Proposition 2.6.12, Y is maximal, and thus Y is complete (Theorem 2.5.5). It remains to show that Y is thin. With the notations of the proof of Proposition 2.6.12, consider a word $u \notin F(X)$. Since Z^* is dense, $sut \in Z^*$ for some words $s, t \in A^*$. Thus $sut = \beta(w)$ for some $w \in B^*$. But now w is not completable in Y, since otherwise $hwk \in Y$ for some $h, k \in B^*$, giving $\beta(h)sut\beta(k) \in X$, whence $u \in F(X)$. Thus Y is thin. $\qquad\square$

By Proposition 2.6.13, for thin codes Y, Z, the code $Y \circ Z$ is maximal if and only if Y and Z are maximal.

Proposition 2.6.14 *Let X be a maximal code over A. For any code $Z \subset A^*$, the code X decomposes over Z if and only if $X^* \subseteq Z^*$. In particular, X is indecomposable if and only if X^* is a maximal free submonoid of A^*.*

Proof. If X decomposes over Z, then $X^* \subset Z^*$. Conversely, if $X^* \subset Z^*$, let $\bar{Z} = \mathrm{alph}_Z(X)$. Then $X \subset \bar{Z}^*$, and of course $\bar{Z} = \mathrm{alph}_{\bar{Z}}(X)$. By Proposition 2.6.6, X decomposes over $\bar{Z}$. In view of Proposition 2.6.12, the code $\bar{Z}$ is maximal. By $\bar{Z} \subset Z$, we have $\bar{Z} = Z$. $\qquad\square$

Example 2.6.15 Let A be an alphabet. We show that the uniform code A^n decomposes over Z if and only if $Z = A^m$ and m divides n. In particular, A^n is indecomposable for n prime and for $n = 1$.

Indeed, let $A^n = X = Y \circ_\beta Z$, where $Y \subset B^*$ and $\beta : B^* \to A^*$. The code X is maximal and bifix, and thus Y also is maximal and bifix and Z is maximal. Let $y \in Y$ be a word of maximal length, and set $y = ub$ with $b \in B$. Then $Y \cup uB$

 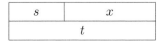

Figure 2.8 The two types of edges in a prefix graph.

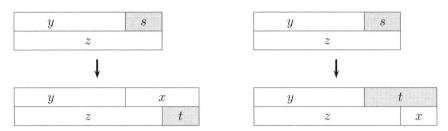

Figure 2.9 The two ways of continuing a double factorization $ys = z$. On the left, it is extended to $yx = zt$, and on the right to $yt = zx$.

is prefix. Let indeed $y' = ub'$, $b' \in B$. Any proper prefix of y' is also a proper prefix of y, and therefore is not in $Y \cup uB$. Next if y' is a prefix of some y'' in $Y \cup uB$, then by the maximality of the length of y, we have $|y'| = |y''|$ and $y' = y''$. Thus $Y \cup uB$ is a code. Hence $Y \cup uB = Y$, because Y is maximal. It follows that $\beta(uB) = \beta(u)Z \subset X$. Now X is a uniform code, thus all words in Z have the same length, say m. Since Z is maximal, $Z = A^m$. It follows that $n = m|y|$.

2.7 Prefix graph of a code

The prefix graph is used to give an efficient test whether a set X is a code. The graph can also answer some other questions on the set X, by applying standard techniques for graph traversal. This will be detailed in later chapters (Exercises 5.1.1 and 5.1.2).

Let X be a finite set of words over some alphabet A. We define a graph G_X for X, called the *prefix graph* of X as follows. The vertices of G_X are the nonempty prefixes of words in X, and there is an edge from s to t if and only if one of the two following situations occurs: either $st \in X$ or $sx = t$ for some $x \in X$, see Figure 2.8.

Edges of the first type are called *crossing*, those of the second type *extending*. A crossing edge (s, t) is labeled with the word t, an extending edge (s, t) with $sx = t$ is labeled with x. As usual, the label of a path is the product of the label of its edges. In the case where $sx = t$ and x, t are in X, then (s, t) is an extending edge labeled with x, and (s, x) is a crossing edge, also labeled with x.

A vertex s is intended to represent a prefix that has been constructed in the process of trying to build a double factorization, say $ys = z$, for $y, z \in X^*$. A crossing edge (s, t), with $st = x \in X$, gives the factorization $yx = zt$, and the prefix t swapped to the other side of the equation, whereas an extending edge (s, t) with $sx = t$ merely replaces the factorization by $yt = zx$, extending the current prefix from s to t. See Figure 2.9.

Example 2.7.1 Let $X = \{a, bb, abbba, babab\}$ over the alphabet $A = \{a, b\}$. The nonempty prefixes, in addition to the words in X, are the words b, ab, ba, abb,

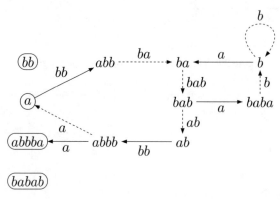

Figure 2.10 The prefix graph G_X for the set $X = \{a, bb, abbba, babab\}$. A crossing edge is drawn dashed, an extending edge is drawn filled. The label of a crossing edge is the name of its endpoint. The label of an extending edge (s, t) is the word x in X for which $sx = t$.

bab, *abbb*, and *baba*, so the graph has 11 vertices. The prefix graph G_X is given in Figure 2.10.

We will prove that the set X is a code if and only if there is no path in the prefix graph G_X from a vertex in X to a vertex in X. In our example, there is a path from a to itself, or to *abbba*, so the set is not a code.

We start with a lemma describing paths in the prefix graph G_X. First, we need a definition. Two factorizations $(x_1, \ldots, x_n)$ and $(y_1, \ldots, y_m)$ of a word are *disjoint* if $x_1 \cdots x_i \neq y_1 \cdots y_j$ for $1 \leq i < n, 1 \leq j < m$. We say simply that

$$x_1 \cdots x_n = y_1 \cdots y_m$$

is a disjoint double factorization when the two factorizations $(x_1, \ldots, x_n)$ and $(y_1, \ldots, y_m)$ of the same word are disjoint.

Lemma 2.7.2 *There is a path of length $n \geq 1$ from s to t in the prefix graph of X if and only if there exist $x_1, \ldots, x_k, y_1, \ldots, y_\ell$ in X such that*

$$sy_1 \cdots y_\ell t = x_1 \cdots x_k \quad or \quad sy_1 \cdots y_\ell = x_1 \cdots x_k t$$

are disjoint factorizations with $k + \ell = n$, and moreover s is a prefix of x_1 (resp. a prefix of t if $k = 0$). The label of the path is $y_1 \cdots y_\ell t$ in the first case and $y_1 \cdots y_\ell$ in the second case. The first (second) case occurs if and only if the path contains an odd (even) number of crossing edges.

Example 2.7.3 Consider as an example the path

$$abb \xrightarrow{ba} ba \xrightarrow{bab} bab \xrightarrow{ab} ab \xrightarrow{bb} abbb$$

in the previous graph. It is represented in the following picture.

b	a	b	a	b

a	b	b	b	a	b	a	b	a	b	b	b

This path has length 4, the first 3 edges are crossing edges, the last one is an extending edge. It corresponds to the disjoint factorizations $abb|babab|abbb = abbba|babab|bb$. Here $\ell = 1$, $k = 3$, and the product of labels is $babababbb$. The path

$$a \xrightarrow{bb} abb \xrightarrow{ba} ba \xrightarrow{bab} bab \xrightarrow{ab} ab \xrightarrow{bb} abbbb \xrightarrow{a} a$$

has two more edges.

a	b	b	b	a	b	a	b	a	b	b	b	a

a	b	b	b	a	b	a	b	a	b	b	b	a

It corresponds to the disjoint factorizations $a|bb|babab|abbba = abbba|babab|bb|a$ which shows that X is not a code.

Proof of Lemma 2.7.2. Assume first that there is a path of length $n \geq 1$ from s to t. If $n = 1$, then either $st = x$, or $sx = t$ with $x \in X$. Thus there is a double factorization of the desired form for $n = 1$.

Assume now $n \geq 1$, and that there is edge from t to u. By induction, $sy_1 \cdots y_\ell t = x_1 \cdots x_k$ or $sy_1 \cdots y_\ell = x_1 \cdots x_k t$, and either $tu = x \in X$ or $tx = u$ for some $x \in X, u \notin X$. So there are four cases to check.

If $sy_1 \cdots y_\ell t = x_1 \cdots x_k$ and $tu = x \in X$, then $sy_1 \cdots y_\ell x = x_1 \cdots x_k u$, and these factorizations are again disjoint because u is a proper suffix of x.

If $sy_1 \cdots y_\ell t = x_1 \cdots x_k$ and $tx = u$ for some $x \in X$, then $sy_1 \cdots y_\ell u = x_1 \cdots x_k x$ and again the factorizations are disjoint because u is a proper suffix of t, so of x_k.

If $sy_1 \cdots y_\ell = x_1 \cdots x_k t$ and $tu = x \in X$, then $sy_1 \cdots y_\ell u = x_1 \cdots x_k x$ and the factorizations are disjoint because u is a proper suffix of x. Moreover, if $k = 0$ then s is a prefix of x because s is a prefix of t and t is a prefix of x.

Finally, if $sy_1 \cdots y_\ell = x_1 \cdots x_k t$ and $tx = u$ for some $x \in X$, then $sy_1 \cdots y_\ell x = x_1 \cdots x_k u$. The factorizations are again disjoint. If $k = 0$, then s is a prefix of t and t is a prefix of u, so the word s is a prefix of u.

Conversely, assume that there is a double factorization $sy_1 \cdots y_\ell t = x_1 \cdots x_k$ or a double factorization $sy_1 \cdots y_\ell = x_1 \cdots x_k t$, with $k + \ell = n$. If $n = 1$, then $k = 1$, $\ell = 0$ in the first case, and $k = 0$, $\ell = 1$ in the second case. Indeed, the value $k = 1$, $\ell = 0$ in the second case is ruled out by the condition that s is a prefix of x_1. Thus, there is a crossing edge (s, t) in the first case, and an extending edge (s, t) in the second case.

Assume $n > 1$ and $sy_1 \cdots y_\ell t = x_1 \cdots x_k$. Since $t \neq x_k$ one of these words is a proper suffix of the other. Suppose first that t is a proper suffix of x_k, and set $x_k = ut$. Then there is an edge from u to t in G_X and moreover $sy_1 \cdots y_\ell = x_1 \cdots x_{k-1} u$. If

$k = 1$, then s is a proper prefix of u, otherwise s remains a proper prefix of x_1. Thus the induction applies and there is a path from s to u of length $n - 1$, whence a path of length n from s to t. Assume next that x_k is a suffix of t and set $t = ux_k$. This defines an extending edge (u, t). Thus $sy_1 \cdots y_\ell u = x_1 \cdots x_{k-1}$. Since the left-hand side is not empty, s is a prefix of x_1. The conclusion again follows by induction.

If the double factorization is $sy_1 \cdots y_\ell = x_1 \cdots x_k t$, then since s is a proper prefix of the right-hand side, one has $\ell > 0$.

If y_ℓ is a proper suffix of t, then $t = uy_\ell$ for some u and there is an extending edge (u, t). Replacing t by uy_ℓ gives $sy_1 \cdots y_{\ell-1} = x_1 \cdots x_k u$. Either s is a prefix of x_1, or $k = 0$, and then s is a proper prefix of u if $\ell > 1$ or $s = u$ if $\ell = 1$. In the first case, there is a path from s to u, in the second case there is just the edge (s, t).

Finally, suppose that t is a proper suffix of y_ℓ. Then $y_\ell = ut$ and thus there is a crossing edge (u, t). Next, $sy_1 \cdots u = x_1 \cdots x_k$, so $k \geq 1$ and s remains a prefix of x_1. There is again a path from s to u of length $n - 1$ by induction. This completes the proof.   □

Theorem 2.7.4 *A set X of nonempty words is a code if and only if there is no path in its prefix graph from a vertex in X to a vertex in X.*

Proof. Assume there is a path from $s \in X$ to $t \in X$ in the prefix graph G_X. Then there exists a disjoint double factorization of one of the forms described in Lemma 2.7.2. In both cases, this gives a double factorization of a word as a product of words in X.

Conversely, assume that X is not a code, and consider a shortest word w in X^+ that has two distinct factorizations

$$w = x_1 \cdots x_n = y_1 \cdots y_m$$

with $x_1, \ldots, x_n, y_1, \ldots, y_m$ in X. We may assume that x_1 is a proper prefix of y_1. Then there exists a path from x_1 to y_m of length $m + n - 2$ in G_X.   □

Given a finite graph G, many properties of G can be checked in linear time with respect to the size of G, where the size is the total number of vertices and edges of G. Among these properties are the existence of cycles, the existence of paths between distinguished sets of nodes, and so on. All properties described in the previous section are of these kind. This requires to estimate the size of the graph G_X of X.

Proposition 2.7.5 *Let X be a finite set of words with n elements, and let $N = \sum_{x \in X} |x|$ be the sum of the lengths of the words in X. The prefix graph G_X has at most N vertices and at most nN edges.*

Proof. The vertices of G_X are the nonempty prefixes of words in X; there are at most $N - 1$ of them. Next, consider a vertex t and an edge (s, t) entering t. If (s, t) is a crossing edge, then $st \in X$ is longer than t, and if $t = sx$ for some $x \in X$, then x is shorter than t. So a word x in X either contributes at most one crossing edge, or it contributes at most one extending edge. So the total number of edges entering t is at most n, and the total number of edges in G_X is at most nN.   □

2.7 Prefix graph of a code 97

Corollary 2.7.6 *Given the prefix graph G_X of a set X of n words of total length N, it can be checked in time $O(nN)$ whether X is a code.*

Proof. This is a direct consequence of the previous discussion. □

It remains to show how to construct the prefix graph G_X of a finite set X in linear time with respect to its size, that is with respect to nN, where n is the number of words in X, and N is the sum of the lengths of the words in X.

The construction is in three steps. First, a simple automaton recognizing X is constructed. This automaton is deterministic but not complete, and has the shape of a tree. Such an automaton is usually called a *trie*. The vertices of G_X are among the states of this automaton. Next, the automaton is converted into what is called a *pattern matching machine*. This is done in equipping the trie with a *failure function*. The role of this function is to provide, in the case a transition does not exist for some letter in some state, another state where one can look for a possible transition. As a result, the pattern matching machine recognizes, with the aid of the failure function, the set A^*X of words ending in a word in X.

These two preliminary steps are used, in the final step, to compute efficiently the edges of the graph G_X.

Given a finite set X of words over the alphabet A, the *trie* of X is the automaton whose set of states is the set P of prefixes of words in X. The initial state is the empty word, the end states are the words in X. The next state function is defined for $p \in P$ and $a \in A$ if and only if pa is in P, and then $p \cdot a = pa$.

The trie of X can be constructed very simply by inserting the words of X into a tree that is initially reduced to the empty word.

Trie(X)
```
 1  T ← New Automaton()
 2  for x ∈ X do
 3        p ← ε
 4        for i ← 1 to |x| do
 5              a ← x[i]
 6              if p · a exists then
 7                    p ← p · a
 8              else  q ← New State()
 9                    p · a ← q
10                    p ← q
11        SetTerminal(p)
12  return T
```

This algorithm clearly computes the trie in time $O(N)$, where N is the sum of the lengths of the words in X.

Example 2.7.7 The trie of $X = \{a, bb, abbbba, babab\}$ is given in Figure 2.11.

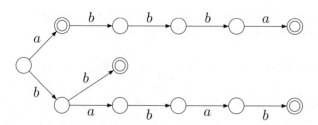

Figure 2.11 The trie of $X = \{a, bb, abbbba, babab\}$. Viewed as an automaton, it accepts words in X.

Given a finite set X of words over the alphabet A, the failure function is intended to be used when the next-state function $p \cdot a$ is undefined in the trie of X. It gives a state q where a new trial for the computation of the next state should be started.

The *failure function* f of X is defined on the set of nonempty prefixes of X. For $p \in P$, $p \neq \varepsilon$, $f(p)$ is the longest proper suffix of p which is in P. For the empty word, $f(\varepsilon) = \varepsilon$.

The *pattern matching machine* of X is the automaton derived from the trie of X by extending the next-state function on P by

$$
p \cdot a = \begin{cases} pa & \text{if } pa \in P, \\ f(p) \cdot a & \text{otherwise.} \end{cases}
$$

Moreover, the state p is terminal if $f(p)$ is terminal. The function COMPUTEFAILURE(T) computes the failure function for the trie T.

COMPUTEFAILURE(T)

```
 1   f(ε) ← ε
 2   F ← NEW QUEUE()
 3   for a ∈ A such that ε · a is defined do
 4         f(ε · a) ← ε
 5         ADD(F, ε · a)
 6   while F ≠ ∅ do
 7         p ← GET(F)
 8         if ISTERMINAL(f(p)) then
 9               SETTERMINAL(p)
10         for a ∈ A such that p · a is defined do
11               q ← f(p)
12               while q · a is undefined do
13                     q ← f(q)
14               f(p · a) ← q · a
15               ADD(F, p · a)
```

The pattern matching machine is obtained by constructing first the trie, and then the failure function.

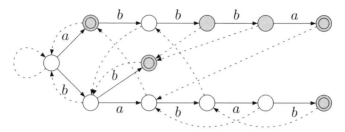

Figure 2.12 The pattern matching machine of $X = \{a, bb, abbbba, babab\}$. Viewed as an automaton, it accepts words in A^*X. Its accepting states are in gray. The failure function is represented by dotted edges.

Example 2.7.8 The pattern matching machine of $X = \{a, bb, abbbba, babab\}$ is given in Figure 2.12.

A state p is terminal for the pattern matching machine if it is a word in A^*X. It appears useful to know the longest suffix of the state p that is in X. Call this $\sigma(p)$. The function σ is undefined on non terminal states, and for terminal states, is is given by

$$\sigma(p) = \begin{cases} f(p) & \text{if } f(p) \text{ is in } X, \\ \sigma(f(p)) & \text{otherwise.} \end{cases}$$

This shows that, provided we remember those states that are in X, is is quite easy, and linear with respect to the number of states, to compute the function σ.

We are now ready to compute the edges of the graph G_X. Each word x in X may produce several crossing edges (s, t). This is a crossing edge provided the suffix t is also a prefix of a word in X. All these suffixes are enumerated by the failure function. Thus one gets the following function for computing the crossing edges:

CROSSINGEDGES(X)

```
1   for x ∈ X do
2       t ← f(x)
3       while t ≠ ε do
4           s ← xt⁻¹
5           ADDCROSSINGEDGE(s, t)
6           t ← f(t)
```

The only tricky line is the computation of the vertex corresponding to the word xt^{-1}. This may be done by maintaining, for each x in X, an array of pointers to the vertices of its prefixes, indexed by their length. So, from the length of x and the length of t one obtains the length of s, thus s in constant time.

The computation of extending edges is quite similar. Given a suffix t, we look for all suffixes x of t. Each of these suffixes gives an extending edge (s, t), with $sx = t$. To loop through the suffixes of t which are in X, one iterates the function σ. Thus the function is

EXTENDINGEDGES(X)

```
1   for t terminal states do
2       x ← σ(t)
3       while x ≠ ε do
4           s ← tx⁻¹
5           ADDEXTENDINGEDGE(s, t)
6           x ← σ(x)
```

Again, the tricky point is the computation of $s = tx^{-1}$. To do this, one maintains for each vertex p a pointer to the longest word in X for which p is a prefix. In the present case, s is a prefix of t, so they share the same longest word in X, and the trick of the array used previously applies again to give the vertex of s in constant time.

Altogether, the following function computes the prefix graph of the set X.

PREFIXGRAPH(X)

```
1   T ← TRIE(X)
2   COMPUTEFAILURE(T)
3   CROSSINGEDGES(X)
4   EXTENDINGEDGES(X)
```

We can finally state the following result as a consequence of the preceding constructions.

Proposition 2.7.9 *Given a set X of n words over some alphabet A, of total length $N = \sum_{x \in X} |x|$, the prefix graph G_X can be constructed in time and space $O(nN)$.*

□

2.8 Exercises

Section 2.1

2.1.1 Let $n \geq 1$ be an integer. Let I, J be two sets of nonnegative integers such that for $i, i' \in I$ and $j, j' \in J$,

$$i + j \equiv i' + j' \bmod n$$

implies $i = i'$, $j = j'$. Let $Y = \{a^i b a^j \mid i \in I, j \in J\}$ and $X = Y \cup a^n$. Show that X is a code.

Section 2.2

2.2.1 Show directly (that is without using Theorem 2.2.14) that a set $X = \{x, y\}$ is a code if and only if x and y are not powers of a single word. (*Hint:* Use induction on $|x| + |y|$.)

2.2.2 Let K be a field and A an alphabet. Let $X \subset A^+$ be a code and let $K\langle X \rangle$ be the subsemiring of $K\langle A \rangle$ generated by the elements of X. Show that $K\langle X \rangle$ is free in the following sense: Let $\beta : B^* \to A^*$ be a coding morphism for X. Extend β by linearity to a morphism from the semiring $K\langle B \rangle$ into $K\langle A \rangle$. Show that β is an isomorphism between $K\langle B \rangle$ and $K\langle X \rangle$.

2.2.3 Show that a submonoid N of a monoid M is stable if and only if for all $m, n \in M$ we have

$$nm, n, mn \in N \Rightarrow m \in N.$$

2.2.4 Let M be a commutative monoid. Show that a submonoid of M is stable if and only if it is biunitary.

2.2.5 For $X \subset A^+$ let Y be the base of the smallest right unitary submonoid containing X.
(a) Show that $Y \subset (Y^*)^{-1} X$.
(b) Deduce that $\mathrm{Card}(Y) \leq \mathrm{Card}(X)$, and give an example showing that equality might hold.

2.2.6 Let X be a subset of A^+. Define a sequence $(S_n)_{n \geq 0}$ of subsets of A^* by setting

$$S_0 = X^*, \quad S_{n+1} = (S_n^{-1} S_n \cap S_n S_n^{-1})^*.$$

Set $S(X) = \bigcup_{n \geq 0} S_n$. Show that $S(X)$ is the free hull of X. Show that when X is recognizable, the free hull of X is recognizable.

2.2.7 Let M be a submonoid of A^* and let $X = (M \setminus 1) \setminus (M \setminus 1)^2$ be its minimal set of generators. Show that X is recognizable if and only if M is recognizable.

2.2.8 Let M be a monoid. Show that M is free if and only if it satisfies the following conditions:
(i) there is a morphism $\lambda : M \to \mathbb{N}$ into the additive monoid $\mathbb{N}$ such that $\lambda^{-1}(0) = 1$,
(ii) for all $x, y, z, t \in M$, the equation $xy = zt$ holds if and only if there exists $u \in M$ such that $xu = z, y = ut$ or $x = zu, uy = t$.

Section 2.3

2.3.1 Let X be a subset of A^+ such that $X \cap XX^+ = \emptyset$. Define a relation $\rho \subset A^* \times A^*$ by $(u, v) \in \rho$ if and only if there exists $x \in X^*$ such that

$$uxv \in X, \quad ux \neq 1, \quad uv \neq 1, \quad xv \neq 1.$$

Show that X is a code if and only if $(1, 1) \notin \rho^+$, where ρ^+ denotes the transitive closure of ρ.

Section 2.4

2.4.1 Let $n \geq 1$ be an integer and I, J be two subsets of $\{0, 1, \ldots, n-1\}$ such that for each integer p in $\{0, 1, \ldots, n-1\}$ there exist a unique pair $(i, j) \in I \times J$ such that

$$p \equiv i + j \bmod n.$$

Let $V = \{i + j - n \mid i \in I, \ j \in J, \ i + j \geq n\}$. For a set K of integers, set $a^K = \{a^k \mid k \in K\}$. Let $X \subset \{a, b\}^*$ be the set defined by

$$X = a^I (ba^V)^* ba^J \cup a^n.$$

Show that X is a maximal code.

a	b	b	a	b	b	b	b	b	b	b	b	a
b	b	b	b	a	b	b	b	b	a	a	b	b

Figure 2.13 This pair of words in U is the product of three words of Y which are $(a, b)(b^2, b^2)(a, b)$, $(b, a)(b^2, b^2)^2(b, a)$ and $(ba)(b, b)(a, b)$.

2.4.2 The *Motzkin code* is the prefix code M on the alphabet $A = \{a, b, c\}$ formed of the words $w \in A^*$ such that $|w|_a - |w|_b = 0$ but $|u|_a - |u|_b > 0$ for any proper nonempty prefix of w. Show that the generating series of M and M^* are

$$f_M(t) = \frac{1 + t - \sqrt{1 - 2t - 3t^2}}{2}, \quad f_{M^*(t)} = \frac{1 - t - \sqrt{1 - 2t - 3t^2}}{2t^2}$$

(*Hint*: Use the fact that $M = c \cup P$ where $P = M \cap aA^*$ and $P = aM^*b$.)

2.4.3 Let $A = \{a_1, \bar{a}_1, \ldots, a_n, \bar{a}_n\}$. Let D be the Dyck code on A. Show that for the uniform Bernoulli distribution on A^*, one has

$$\pi(D) = \frac{1}{2n - 1}.$$

(*Hint*: Set $D_a = D \cap aA^*$ for $a \in A$. Show that $\underline{D}_a = a(\underline{D} - \underline{D}_{\bar{a}})^* \bar{a}$.)

2.4.4 Let $A = \{a, b, c\}$, $B = A \times A$ and $X = \{a, b^2\}$. We identify the set of pairs of words (x, y) of $A^* \times A^*$ of equal length with their representation as words over B, that is we identify $(a_1 a_2 \cdots a_n, b_1 b_2 \cdots b_n)$ with $(a_1, b_1)(a_2, b_2) \cdots (a_n, b_n)$. Here $a_1, \ldots, a_n, b_1, \ldots, b_n \in A$. Show that the set

$$U = \{(x, y) \in X^* \times X^* \mid |x| = |y|\}$$

is a free submonoid of B^* generated by a bifix code Y. See Figure 2.13 for an example. Use this to prove the identity

$$\sum_{n \geq 0} f_{n+1}^2 t^n = \frac{1 - t}{(1 + t)(1 - 3t + t^2)}$$

where f_n is the n-th Fibonacci number defined by $f_0 = 0$, $f_1 = 1$ and $f_{n+1} = f_n + f_{n-1}$ for $n \geq 1$. (*Hint*: Show that U is generated by $Y = (a, a) \cup (b^2, b^2) \cup (a, b)(b^2, b^2)^*(a, b) \cup (a, b)(b^2, b^2)^*(b^2, ba) \cup (b, a)(b^2, b^2)^*(b, a) \cup (b, a)(b^2, b^2)^*(ba, b^2)$.)

Section 2.5

2.5.1 Show that the set $X = \{a^3, b, ab, ba^2, aba^2\}$ is complete and that no proper subset of X is complete. Show that X is not a code.

2.5.2 Let M be a monoid. Let $\mathcal{F}$ be the family of subsets of M which are two-sided ideals of M or empty.

(a) Show that there is a topology on M for which $\mathcal{F}$ is the family of open sets.

(b) Show that a subset P of M is dense in M with respect to this topology if and only if $F(P) = M$, that is if P is dense in the sense of the definition given in Section 2.5.

2.5.3 With the notations of Proposition 2.5.25, and $V = A^* \setminus A^* y A^*$, show successively that

$$\underline{A}^* = (\underline{V}\ y)^* \underline{V} = (\underline{U}\ y)^* (\underline{X}^* y (\underline{U}\ y)^*)^* \underline{V})$$

$$= (\underline{U}\ y)^* \underline{V} + (\underline{U}\ y)^* (\underline{Y})^* y (\underline{U}\ y)^* \underline{V}.$$

(Use the identity $(\sigma + \tau)^* = \tau^* (\sigma \tau^*)^* = (\sigma^* \tau)^* \sigma^*$ for two power series σ, τ having no constant term.) Derive directly from these equations the fact that Y is a code and that Y is complete.

2.5.4 Show that each thin code is contained in a maximal thin code.

Section 2.6

2.6.1 Let $\psi : A^* \to G$ be a morphism from A^* onto a group G. Let H be a subgroup of G and let X the group code defined by $X^* = \psi^{-1}(H)$. Show that X is indecomposable if and only if H is a maximal subgroup of G.

2.6.2 Show that any code $X = \{x, y\}$ with two elements is composed of prefix and suffix codes.

2.6.3 Show that the code $X = \{a, aba, babaab\}$ is not obtained by composition of prefix and suffix codes. Show that it is contained in the finite maximal code Y given by

$$\underline{Y} - 1 = (1 + b + baba(1 + a + b))(a + b - 1)(1 + ba).$$

Show that Y belongs to the family of finite maximal codes defined in Exercise 14.1.7.

2.9 Notes

Codes are frequently called uniquely decipherable codes or UD-codes. The notion of a code originated in the theory of communication initiated by C. Shannon in the late 1940s. The work of Shannon introduced a new scientific domain with many branches and domains of applications. These include data compression, error-correction and cryptography. A comprehensive account of these topics can be found in Pless *et al.* (1998). The development of coding theory lead to a detailed study of constant length codes in connection with problems of error detection and correction. An exposition of this research can be found in MacWilliams and Sloane (1977) or van Lint (1982). The special class of convolution codes, which have close relation with finite automata as presented here, is treated in some detail in McEliece (2004). An early standard book on information and communication theory is Ash (1990).

Variable-length codes were investigated in depth for the first time by Schützenberger (1955) and also by Gilbert and Moore (1959). The direction followed by Schützenberger consists in linking the theory of codes with classical noncommutative algebra. The results presented in this book represent this point of view. An early account of it can be found in Nivat (1966). Since codes are bases of free submonoids of a free monoid, codes are also related with bases of free algebras or of free groups since the free semigroup may be embedded in both structures. For an exposition of free algebras, see Cohn (1985). For an introduction to the theory of free groups, see Magnus *et al.* (2004).

Connections between variable-length codes and automata, and several of the applications mentioned above are presented in Béal (1993) or Béal *et al.* (2009).

The notion of a stable submonoid appears for the first time in Schützenberger (1955) which contains Proposition 2.2.5. The same result is also given in Shevrin (1960), Cohn (1962) and Blum (1965). Proposition 2.2.13 appears in Tilson (1972). The defect theorem (Theorem 2.2.14) has been proved in several formulations in Lentin (1972), Makanin (1976), and Ehrenfeucht and Rozenberg (1978). Some generalizations are discussed in Berstel *et al.* (1979), see also Lothaire (2002). For related questions see also Spehner (1976).

The test for codes given in Section 2.3 goes back to Sardinas and Patterson (1953) and is in fact usually known as the Sardinas and Patterson algorithm. The proof of correctness is surprisingly involved and has motivated a number of papers Bandyopadhyay (1963), Levenshtein (1964), Riley (1967), and de Luca (1976). The problem of testing whether a recognizable set is a code is a special case of a well-known problem in automata theory, namely testing whether a given rational expression is unambiguous. Standard decision procedures exist for this question, see Eilenberg (1974) and Aho *et al.* (1974). These techniques will be used in Chapter 4. The connection between codes and rational expressions has been pointed out in Brzozowski (1967). Further, a characterization of those codes whose coding morphism preserves the star height of rational expressions is given in Hashiguchi and Honda (1976a).

The results of Section 2.4 are well known in information theory. Corollary 2.4.6 with its converse stated in Theorem 2.4.12 are known as the Kraft–McMillan theorem (McMillan (1956)).

The main results of Section 2.5 are from Schützenberger (1955). Our presentation is slightly more general. Proposition 2.5.25 and Theorem 2.5.24 are due to Ehrenfeucht and Rozenberg (1983). They answer a question of Restivo (1977). Theorem 2.5.19 appears in Boë *et al.* (1980). Example 2.5.7 is a special case of a construction due to Restivo (1977), Exercise 2.2.6 is from Berstel *et al.* (1979), Exercise 2.2.8 is known as Levi's lemma (Levi (1944)), Exercise 2.3.1 is from Spehner (1975).

It was shown by Liu (2009) that the composition of maximal codes is not always maximal. Thus the hypothesis that the codes are thin is necessary in Proposition 2.6.13.

We follow Aho and Corasick (1975) for the construction of a trie equipped with a failure function. The resulting structure is called the *pattern matching machine*. The presentation of the algorithm follows closely the description given in Hoffmann (1984), see also Capocelli and Hoffmann (1985). These papers contain the transcription to prefixes of the implementation of Apostolico and Giancarlo (1984). Similar implementation to Hoffmann (1984) are given in Head and Weber (1993, 1995). The implementation proposed in Rodeh (1982) gives the same bounds but is more involved. It is based on the suffix tree, that is a compact tree representing all suffixes of a finite set of words.

The exact complexity of testing unique decipherability is still unknown, see Galil (1985) and Hoffmann (1984) for discussion and partial results.

The basic properties of codes have also been investigated in structures which are more general than free monoids, namely *free partially commutative monoids*. Given a symmetric relation $I \subset A \times A$ over an alphabet A, the free partially commutative monoid $M(A, I)$ is the monoid generated by A subject to the relations $ab = ba$ for all pairs $a, b \in I$. Three problems have been investigated. First, given a homomorphism $f : M \to N$ between free partially commutative monoids M, N, can one decide whether f is injective? This was shown to be undecidable even when M is free (see the survey by Diekert and Muscholl (1996) for details and references). Next, given partially commutative monoids M, N, when does there exist an injective morphism from M into N? This problem, known as the *trace coding problem*, was also shown to be undecidable in general by Kunc (2004). Several particular cases where these problems are decidable were also described by other authors. Finally, the *lifting problem* asking whether any lifting of a trace coding is a word coding was solved positively in Bruyère and De Felice (1996).

The notion of code has also been generalized to codes in symbolic dynamical systems, by Reutenauer (1986) and later by Restivo (1990). Let G be a finite directed multigraph with edges labeled by letters from an alphabet A (the set of labels of bi-infinite paths form what is called the *sofic system* defined by G). Let $S(G)$ be the set of finite words which are the label of a path in G. A subset X of $S(G)$ is a *code over G* if any element of $S(G)$ has at most one factorization in elements of X. Thus a code in the usual sense corresponds to the case where $S(G) = A^*$. A set $X \subset S(G)$ is complete over G if any element of $S(G)$ is a factor of a word of X^*. It is shown in Béal and Perrin (2006) that any maximal code over G is complete over G. This generalizes Theorem 2.5.5. The converse is not true, even for finite codes, and thus there is no generalization of Theorem 2.5.13. Other results for codes over graphs are given in Béal and Perrin (2005) and Reutenauer (1986).

Other structures for which the notion of codes have been considered include trees. Labeled trees are a natural generalization of words, for which the notion of automaton has been introduced a long time ago. Tree codes have been introduced in Nivat (1992) and its study has been further developed, for instance in Mantaci and Restivo (2001).

Dyck codes are named after the German mathematician Walther von Dyck (see also Berstel and Perrin (2007)). Motzkin codes of Exercise 2.4.2 are named after Motzkin paths (see for instance Goulden and Jackson (2004)).

The combinatorial proof for the expression of the generating series of the squares of the Fibonacci numbers given in Exercise 2.4.4 is from Shapiro (1981), see also Stanley (1997), Example 4.7.14, and Foata and Han (1994).

Exercise 2.6.3 is from Derencourt (1996). It is a counterexample to a conjecture in Restivo *et al.* (1989) asserting that every three-word code is composed of prefix and suffix codes. It is not known whether any three-word code is contained in a finite maximal code.

3

Prefix codes

Undoubtedly the prefix codes are the easiest to construct. The verification that a given set of words is a prefix code is straightforward. However, most of the interesting problems on codes can be raised for prefix codes. In this sense, these codes form a family of *models* of codes: frequently, it is easier to gain intuition about prefix codes rather than general codes. However, we can observe that the reasoning behind prefix codes is often valid in the general case.

For this reason we now present a chapter on prefix codes. In the first section, we comment on their definition and give some elementary properties. We also show how to draw the picture of a prefix code as a tree (the literal representation of prefix codes).

In Section 3.2, a construction of the automata associated to prefix codes is given. These automata are deterministic, and we will see in Chapter 9 how to extend their construction to general codes.

The third section deals with maximal prefix codes. Characterizations in terms of completeness are given. Section 3.4 presents the usual operations on prefix codes. Most of them have an easy interpretation as operations on trees.

An important family of prefix codes is introduced in Section 3.5. They have many combinatorial properties which illustrate the notions presented previously. The synchronization of prefix codes is defined in Section 3.6. In fact, this notion will be generalized to arbitrary codes in Chapter 9 where the relationship with groups will be established. The relation between codes and Bernoulli distribution can be extended to probability distributions in the case of prefix codes. This is done in Section 3.7, where the notion of reccurrent event is introduced. The generating series of a rational prefix code is $\mathbb{N}$-rational and satisfies the Kraft inequality. We show in Section 3.8 a converse.

3.1 Prefix codes

This introductory section contains equivalent formulations of the definition of a prefix code together with the description of the tree associated to a prefix code. We then show how any prefix code induces in a natural way a factorization of the free monoid. Of course, all results in this chapter transpose to suffix codes by using the reverse operation.

Recall that for words x, y, we denote by $x \leq y$ (resp. $x < y$) the fact that x is a prefix (resp. a proper prefix) of y. The order defined by $\leq$ is the *prefix order*. We write $x \geq y$ (resp. $x > y$) whenever $y \leq x$ (resp. $y < x$). Two words x, y are *incomparable for the prefix order*, and we write $x \bowtie y$, if neither x is a prefix of y nor y is a prefix of x.

A subset X of A^* is *prefix* if any two distinct words in X are incomparable for the prefix order. If a prefix subset X contains the empty word 1, then $X = \{1\}$. In the other cases, X is a code (Proposition 2.1.9).

Example 3.1.1 The usual binary representation of positive integers is exponentially more succinct than the unary representation, and thus is preferable for efficiency. However, it is not adapted to representation of sequences of integers, since it is not uniquely decipherable: for instance, 11010 may represent the number 26, or the sequence 6, 2, or the sequence 1, 2, 2. The *Elias code* of a positive integer is composed of its binary representation preceded by a number of zeros equal to the length of this representation minus one. For instance, the Elias code of 26 is 000011010. It is easily seen that the set of Elias encodings of positive integers is a prefix code. In fact, it is the same as the code of Example 2.4.11, with a replaced by 0 and b replaced by 1.

It is convenient to have a shorthand for the proper prefixes (resp. proper suffixes) of the words of a set X. For this we use

$$X A^- = X(A^+)^{-1} \text{ and } A^- X = (A^+)^{-1} X.$$

Thus $u \in X A^-$ if and only if $u < x$ for some $x \in X$. Symmetrically, $u \in X A^+$ if and only if $u > x$ for some $x \in X$.

There is a series of equivalent definitions for a set to be prefix, all of which will be useful. The set X is prefix if and only if one of the following properties hold.

(i) $X \cap X A^+ = \emptyset$.
(ii) $X \cap X A^- = \emptyset$.
(iii) $X A^-$, X, $X A^+$ are pairwise disjoint.
(iv) If $x, xu \in X$, then $u = 1$.
(v) If $xu = x'u'$ with $x, x' \in X$, then $x = x'$ and $u = u'$.

The following proposition can be considered as describing a way to construct prefix codes. It also shows a useful relationship between prefix codes and right ideals.

Proposition 3.1.2 *For any subset Y of A^*, the set $X = Y \setminus Y A^+$ is prefix. Moreover $X A^* = Y A^*$, that is X and Y are both empty or generate the same right ideal, and X is the minimal set with this property.*

Proof. Let $X = Y \setminus Y A^+$. From $X \subset Y$, it follows that $X A^+ \subset Y A^+$, whence $X \cap X A^+ \subset X \cap Y A^+ = \emptyset$. This proves that X is a prefix set. Next $X A^* \subset Y A^*$. For the converse, let $u \in Y$ and let v be its shortest prefix in Y. Then $v \in X$, whence $u \in X A^*$. Thus $Y \subset X A^*$ and $Y A^* = X A^*$.

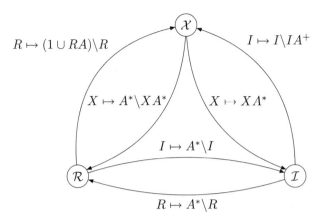

Figure 3.1 The bijections between the three families $\mathcal{X}$, $\mathcal{R}$ and $\mathcal{I}$.

Let Z be a minimal set of generators of YA^*, that is $ZA^* = YA^*$. We show that $X \subset Z$. Let indeed x be a word in X. Then $x = zu$ for some $u \in A^*$ and $z \in Z$. Since X also generates YA^*, $z = x'u'$ for some $x' \in X$, $u' \in A^*$. Thus $x = zu = x'u'u$, and since X is prefix, $uu' = 1$. This shows that $X \subset Z$. Thus $X = Z$. $\qquad\square$

The set $X = Y \setminus YA^+$ is called the *initial part* of Y or also the *base* of the right ideal YA^*.

The following statements describe natural bijections between the following families of subsets of A^*:

1. the family $\mathcal{X}$ of prefix subsets,
2. the family $\mathcal{I}$ composed of the right ideals of A^* together with the empty set,
3. the family $\mathcal{R}$ of prefix-closed subsets.

We describe here these three bijections.

Proposition 3.1.3 *The following bijections hold.*

(i) *The map $X \mapsto XA^*$ is a bijection from $\mathcal{X}$ onto $\mathcal{I}$, and the map $I \mapsto I \setminus IA^+$ is its inverse bijection from $\mathcal{I}$ onto $\mathcal{X}$.*

(ii) *Set complementation maps bijectively $\mathcal{R}$ onto $\mathcal{I}$.*

(iii) *The map $X \mapsto A^* \setminus XA^*$ is a bijection from $\mathcal{X}$ onto $\mathcal{R}$, and the map $R \mapsto (1 \cup RA) \setminus R$ is its inverse bijection from $\mathcal{R}$ onto $\mathcal{X}$.*

Proof. (i) For any nonempty subset X of A^*, the set XA^* is a right ideal. Conversely, for any subset I of A^*, the set $X = I \setminus IA^+$ is prefix. Indeed, a proper prefix of an element of X is not in I and therefore not in X. Thus the two maps are well defined. Let us show that they are inverse to each other.

Let X be a prefix subset of A^* and let $I = XA^*$. Then $X = I \setminus IA^+$. Indeed $I \setminus IA^+ = XA^* \setminus XA^+ = (X \cup XA^+) \setminus XA^+ = X \setminus XA^+ = X$ because $X \cap XA^+ = \emptyset$.

Finally, let I be a right ideal of A^* and let $X = I \setminus IA^+$. By Proposition 3.1.2, $XA^* = IA^* = I$.

(ii) If w is not in the right ideal I, then none of its prefixes is in I. Thus $R = A^* \setminus I$ is prefix-closed. Conversely, the complement of a prefix-closed set is a right ideal or is empty.

(iii) The map sends $\emptyset$ to A^*. For a nonempty prefix code X, the bijection of (i) sends it to the right ideal $I = XA^* \neq A^*$. Taking the complement sends it bijectively to the nonempty prefix-closed set $R = A^* \setminus I = A^* \setminus XA^*$ by (ii). This shows the first assertion.

By (i) and (ii), the inverse maps R to $X = I \setminus IA^+$ with $I = A^* \setminus R = XA^*$. Let $Y = RA \setminus R$. A word x of X is not in R. Set $x = ua$ with $u \in A^*$ and $a \in A$. Since u is not in I, it is in R. Thus x is in Y. Conversely, let y be a word in Y. Then y is not in R and thus y is in I. Since $y \in RA$, any proper prefix of y is in R. Thus y has no proper prefix in I, that is $y \notin IA^+$. This proves that $y \in X$.   □

Note that these bijections, with almost the same proofs, hold in any ordered set.

Example 3.1.4 Let $A = \{a, b\}$ and let $Y = A^*aA^*$ be the set of words containing at least one occurrence of the letter a. Then

$$X = Y \setminus YA^+ = b^*a.$$

Example 3.1.5 Let $A = \{a, b\}$. The set $I = A^*abA^*$ is the set of words containing a factor ab. It is a right ideal. The complement of I is the prefix-closed set $R = b^*a^*$. The prefix code $X = I \setminus IA^+$ is $X = b^*a^*ab$. This code, as the previous one, belongs to the family of semaphore codes studied in Section 3.5.

The preceding bijections have the following counterpart as relations between formal series.

Proposition 3.1.6 *Let X be a prefix code over A and let $R = A^* \setminus XA^*$. Then*

$$\underline{X} - 1 = \underline{R}(\underline{A} - 1), \quad and \quad \underline{A}^* = \underline{X}^*\underline{R}. \tag{3.1}$$

Proof. We show first that the two equations are equivalent. By Proposition 2.6.1, $\underline{X}^* = (1 - \underline{X})^{-1}$. From this and from $(1 - \underline{A})^{-1} = \underline{A}^*$ we get, by multiplying $1 - \underline{X} = \underline{R}(1 - \underline{A})$ on the left by $\underline{X}^*$ and on the right by $\underline{A}^*$ the equation $\underline{A}^* = \underline{X}^*\underline{R}$. The converse operations, that is multiplying on the left by $1 - \underline{X}$ and on the right by $1 - \underline{A}$, give the first equation back.

The product of X and A^* is unambiguous by the property (v) of prefix codes listed above. Thus, $\underline{XA^*} = \underline{X}\,\underline{A}^*$, and

$$\underline{R} = \underline{A^* \setminus XA^*} = \underline{A}^* - \underline{X}\,\underline{A}^* = (1 - \underline{X})\underline{A}^*.$$

Multiplying both sides by $1 - \underline{A}$ on the right, we get $\underline{R}(1 - \underline{A}) = 1 - \underline{X}$. This proves the formula.   □

Note the following combinatorial interpretations of Formulas (3.1). The first can be rewritten as $\underline{R}\,\underline{A} + 1 = \underline{X} + \underline{R}$ and says that a word in R followed by a letter is

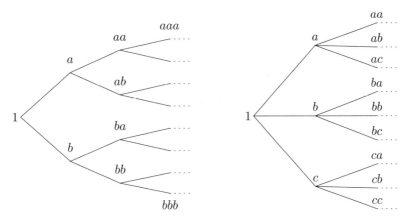

Figure 3.2 The literal representations of $\{a, b\}^*$ and of $\{a, b, c\}^*$.

either in R or in X and that each word in X is composed of a word in R followed by a letter. The second formula says that each word $w \in A^*$ admits a unique factorization

$$w = x_1 x_2 \cdots x_n u, \quad x_1, \ldots, x_n \in X, \quad u \in R.$$

Example 3.1.7 Let $A = \{a, b\}$ and $X = a^*b$ as in Example 3.1.4. Then $R = a^*$. Proposition 3.1.6 gives

$$\underline{X} - 1 = \underline{R}(\underline{A} - 1) = a^*(a + b - 1) = a^*b - 1.$$

We single out the following corollary, which is also contained in Proposition 3.1.3, for ease of reference.

Corollary 3.1.8 *Let X and Y be prefix subsets of A^*. If $XA^* = YA^*$, then $X = Y$.*
□

Observe that there is a straightforward proof by series, since $XA^* = YA^*$ implies $\underline{X}\,\underline{A}^* = \underline{Y}\,\underline{A}^*$, from which the equality follows by simplifying by $\underline{A}^*$.

We now give a useful graphical representation of prefix codes. It consists of associating a tree with each prefix code in such a way that the leaves of the tree represent the words in the code.

First, we associate an infinite tree with the set A^* of words over an alphabet A as follows. The alphabet is totally ordered, and words of equal length are ordered lexicographically. Each node of the tree represents a word in A^*. Words of small length are to the left of words of greater length, and words of equal length are disposed vertically according to lexical ordering. There is an edge from u to v if and only if $v = ua$ for some letter $a \in A$. The tree obtained in this way is the *literal representation* of A^* also called the *Cayley graph* of A^* (see Figure 3.2).

To a given subset X of A^* we associate a subtree of the literal representation of A^* as follows. We keep just the nodes corresponding to the words in X and all the nodes on the paths from the root to these nodes. Nodes corresponding to words in X are marked if necessary. The tree obtained in this way is the *literal representation* of X. Figures 3.3–3.4 give several examples.

Figure 3.3 Literal representations of $X = \{a, ba, baa\}$ with explicit labeling and with implicit labeling.

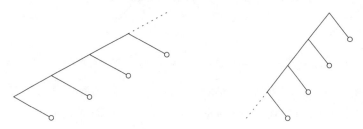

Figure 3.4 Literal representation of $X = a^*b$. On the left, the left-to-right representation, and on the right the top-down drawing.

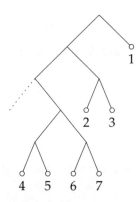

Figure 3.5 The Elias code.

An alternative graphical representation draws the tree from top to bottom instead of from left to right. In this case, words of equal length are disposed horizontally from left to right according to their lexicographic order. See Figure 3.4 for an example.

It is easily seen that a code X is prefix if and only if in the literal representation of X, the nodes corresponding to words in X are all leaves of the tree.

Example 3.1.1 (*continued*) Figure 3.5 is the graphical representation of the Elias code.

The advantage of the literal representation, compared to simple enumeration, lies in the easy readability. Contrary to what might seem to happen, it allows a compact representation of rather big codes (see Figure 3.6).

Example 3.1.9 Let $X = \{a, baa, bab, bb\}$ be the code over $A = \{a, b\}$ represented in Figure 3.7(a). Here $R = \{1, b, ba\} = XA^-$, and $\underline{X} - 1 = (1 + b + ba)(\underline{A} - 1)$. The equality between R and XA^- characterizes maximal prefix codes, as we will see in Section 3.3.

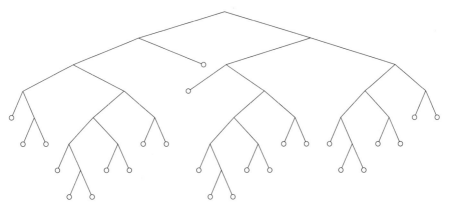

Figure 3.6 A code with 26 elements.

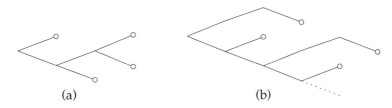

(a)   (b)

Figure 3.7 Two prefix codes: (a) the code $\{a, baa, bab, bb\}$ and (b) the code $(b^2)^*\{a^2b, ba\}$.

Example 3.1.10 Let $X = (b^2)^*\{a^2b, ba\}$, as given in Figure 3.7(b). Here $R = R_1 \cup R_2$, where $R_1 = XA^- = (b^2)^*(1 \cup a \cup b \cup a^2)$ is the set of proper prefixes of X and $R_2 = XA^+ - X - XA^- = (b^2)^*(abA^* \cup a^3A^*)$. Thus Equation (3.1) now gives

$$\underline{X} - 1 = (b^2)^*(1 + a + b + a^2 + ab\underline{A}^* + a^3\underline{A}^*)(\underline{A} - 1).$$

3.2 Automata

The literal representation gives an easy method for verifying whether a word w is in X^* for some fixed prefix code X. It suffices to follow the path starting at the root through the successive letters of w. Whenever a leaf is reached, the corresponding factor of w is split away and the procedure is restarted.

We will consider several automata derived from the literal representation and relate them to the minimal automaton. The particular case of prefix codes is interesting in itself because it is the origin of most of the general results of Chapter 9.

Recall (Chapter 1) that for any subset $X \subset A^*$, we denote by $\mathcal{A}(X)$ the minimal deterministic automaton recognizing X.

Proposition 3.2.1 *Let X be a subset of A^*. The following conditions are equivalent:*

(i) *X is prefix.*
(ii) *The minimal automaton $\mathcal{A}(X)$ is empty or has a single final state t and $t \cdot A = \emptyset$.*
(iii) *There exist a deterministic automaton $\mathcal{A} = (Q, i, T)$ recognizing X with $T \cdot A = \emptyset$.*

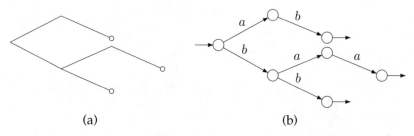

Figure 3.8 (a) Literal representation of X, (b) Literal automaton of X.

Proof. (i) $\implies$ (ii). Suppose that X is nonempty. Set $\mathcal{A}(X) = (Q, i, T)$. First, we claim that for $q \in T$, we have $\{w \in A^* \mid q \cdot w \in T\} = \{1\}$. Indeed let $x \in X$ and $w \in A^*$ be words such that $i \cdot x = q$ (remember that $q \in T$) and $q \cdot w \in T$. Then $xw \in X$, whence $w = 1$. This shows the claim.

Thus, two final states are not separable and from the minimality of $\mathcal{A}(X)$, it follows that $\mathcal{A}(X)$ has just one final state, say t. Assume that $t \cdot A \neq \emptyset$, and that $t \cdot a = p$ for some letter $a \in A$ and some state p. Since p is coaccessible, we have $p \cdot v = t$ for some $v \in A^*$. Thus $t \cdot av = t$, whence $av = 1$, a contradiction.

(ii) $\implies$ (iii) is clear.

(iii) $\implies$ (i). From $T \cdot A = \emptyset$, it follows that $T \cdot A^+ = \emptyset$. Thus, if $x \in X$, and $w \in A^+$ then $i \cdot xw = \emptyset$ and $xw \notin X$. Thus $X \cap XA^+ = \emptyset$. $\square$

It is easy to construct an automaton for a prefix code by starting with the literal representation. This automaton, call it the *literal automaton* of a prefix code X, is the deterministic automaton

$$\mathcal{A} = (XA^- \cup X, 1, X)$$

defined by

$$u \cdot a = \begin{cases} ua & \text{if } ua \in XA^- \cup X, \\ \emptyset & \text{otherwise.} \end{cases}$$

Since $XA^- \cup X$ is prefix-closed, we immediately see that $1 \cdot u \in X$ if and only if $u \in X$, that is $L(\mathcal{A}) = X$. The pictorial representation of a literal automaton corresponds, of course, to the literal representation of the code.

Example 3.2.2 The code $X = \{ab, bab, bb\}$ over $A = \{a, b\}$ has the literal representation given in Figure 3.8(a) and the literal automaton given in Figure 3.8(b).

The literal automaton $\mathcal{A}$ of a prefix code X is trim but is not minimal in general. For infinite codes, it is always infinite. Let us consider two states of $\mathcal{A}$. It is equivalent to consider the two prefixes of words of X, say u and v, leading to these states. These two states are inseparable if and only if

$$u^{-1}X = v^{-1}X.$$

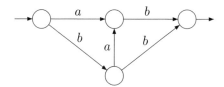

Figure 3.9 The minimal automaton of $X = \{ab, bab, bb\}$.

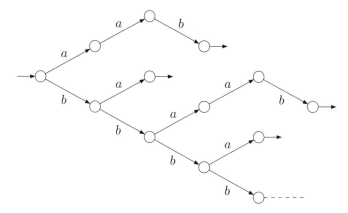

Figure 3.10 The literal automaton of the prefix code $X = (b^2)^*\{a^2b, ba\}$.

Note that this equality means on the literal representation of X that the two subtrees with roots u and v, respectively, are the same. This provides an easy procedure for the computation of the minimal automaton: first, all final states are labeled, say with label 0. If labels up to i are defined we consider subtrees such that all nodes except the roots are labeled. Then roots are labeled identically if the (labeled) subtrees are isomorphic. Taking the labels as states, we obtain the minimal automaton. The procedure is described in Examples 3.2.2–3.2.4.

Example 3.2.2 (*continued*) In view of Proposition 3.2.1, the three terminal states are inseparable. The states a and ba are inseparable because $a^{-1}X = (ba)^{-1}X = b$. No other relation exists. Thus the minimal automaton is given in Figure 3.9.

Example 3.2.3 The literal automaton of $X = (b^2)^*(a^2b \cup ba)$ is given in Figure 3.10. Clearly the final states are equivalent, and also the predecessors of final states and their predecessors. On the main diagonal, however, the states are only equivalent with a step 2. This gives the minimal automaton of Figure 3.11.

Example 3.2.4 In Figure 3.12 the labeling procedure has been carried out for the 26 element code of Figure 3.6. This gives the subsequent minimal automaton of Figure 3.13.

We now consider automata recognizing the submonoid X^* generated by a prefix code X. Recall that X^* is right unitary (Proposition 2.2.7). Proposition 3.2.5 is the analogue of Proposition 3.2.1.

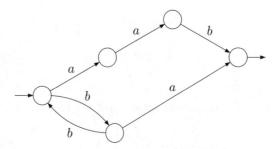

Figure 3.11 Minimal automaton corresponding to Figure 3.10.

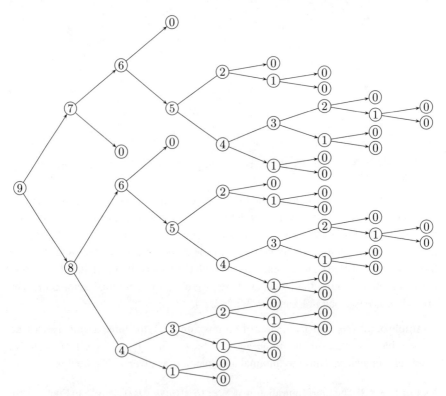

Figure 3.12 The computation of a minimal automaton.

Proposition 3.2.5 *Let P be a subset of A^*. The following conditions are equivalent:*

(i) *P is a right unitary submonoid.*
(ii) *The minimal automaton $\mathcal{A}(P)$ has a unique final state, namely the initial state.*
(iii) *There exists a deterministic automaton recognizing P having the initial state as unique final state.*

Proof. (i) $\implies$ (ii). The states in $\mathcal{A}(P)$ are the nonempty sets $u^{-1}P$, for $u \in A^*$. Now if $u \in P$, then $u^{-1}P = P$ because $uv \in P$ if and only if $v \in P$.

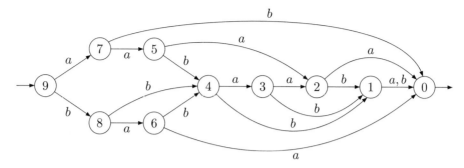

Figure 3.13 A minimal automaton.

Thus, there is only one final state in $\mathcal{A}(P)$, namely P which is also the initial state.

(ii) $\Longrightarrow$ (iii) is clear.

(iii) $\Longrightarrow$ (i). Let $\mathcal{A} = (Q, i, i)$ be the automaton recognizing P. The set P then is a submonoid since the final state and the initial state are the same. Further let $u, uv \in P$. Then $i \cdot u = i$ and $i \cdot uv = i$. This implies that $i \cdot v = i$ because $\mathcal{A}$ is deterministic. Thus, $v \in P$, showing that P is right unitary. □

If $\mathcal{A} = (Q, i, T)$ is any deterministic automaton over A, the *stabilizer* of a state q is the submonoid

$$\text{Stab}(q) = \{w \in A^* \mid q \cdot w = q\}.$$

Proposition 3.2.6 *The stabilizer of a state of a deterministic automaton is a right unitary submonoid. Every right unitary submonoid is the stabilizer of a state of some deterministic automaton.*

Proof. It is an immediate consequence of the proof of Proposition 3.2.5. □

This proposition shows the importance of right unitary submonoids and of prefix codes in automata theory. Proposition 3.2.7 presents a method for deriving the minimal automaton $\mathcal{A}(X^*)$ of X^* from the minimal automata $\mathcal{A}(X)$ of the prefix code X.

Proposition 3.2.7 *Let X be a nonempty prefix code over A, and let $\mathcal{A}(X) = (Q, i, t)$ be the minimal automaton of X. Then the minimal automaton of X^* is*

$$\mathcal{A}(X^*) = \begin{cases} (Q, t, t) & \text{if } \text{Stab}(i) \neq 1, & (3.2) \\ (Q \setminus i, t, t) & \text{if } \text{Stab}(i) = 1. & (3.3) \end{cases}$$

and the action of $\mathcal{A}(X^*)$, denoted by $\circ$, is given by

$$q \circ a = q \cdot a \quad \text{for } q \neq t \tag{3.4}$$

$$t \circ a = i \cdot a \tag{3.5}$$

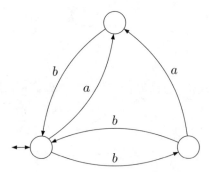

Figure 3.14 The minimal automaton of X^* with $X = \{ab, bab, bb\}$.

Proof. Let $\mathcal{B} = (Q, t, t)$ be the automaton obtained from $\mathcal{A}(X)$, defining the action $\circ$ by (3.4) and (3.5). Then clearly

$$L(\mathcal{B}) = \{w \mid t \circ w = t\} = X^*.$$

Let us verify that the automaton $\mathcal{B}$ is reduced. For this, consider two distinct states p and q. Since $\mathcal{A}(X)$ is reduced, there is a word u in A^* separating p and q, that is such that, say

$$p \cdot u = t, \quad q \cdot u \neq t. \tag{3.6}$$

It follows that $p \circ u = t$, and furthermore $p \circ v \neq t$ for all $v < u$. If $q \circ u \neq t$, then u separates p and q in the automaton $\mathcal{B}$ also. Otherwise, there is a smallest prefix v of u such that $q \circ v = t$. For this v, we have $q \cdot v = t$. In view of (3.6), $v \neq u$. Thus $v < u$. But then $q \circ u = t$ and $p \circ v \neq t$, showing that p and q are separated by v.

Each state in $\mathcal{B}$ is coaccessible because this is the case in $\mathcal{A}(X)$. From $1 \neq X$, we have $i \neq t$. The state i is accessible in $\mathcal{B}$ if and only if the set $\{w \mid t \circ w = i\}$ is nonempty, thus if and only if $\mathrm{Stab}(i) \neq 1$. If this holds, $\mathcal{B}$ is the minimal automaton of X^*. Otherwise, the accessible part of $\mathcal{B}$ is its restriction to $Q \setminus i$. $\qquad\square$

The automaton $\mathcal{A}(X^*)$ always has the form given by (3.3) if X is finite. In this case, it is obtained by identifying the initial and the final state. For a description of the general case, see Exercise 3.2.2.

Example 3.2.2 (*continued*) The minimal automaton of X^* is given in Figure 3.14. The code X is finite and $\mathcal{A}(X^*)$ is given by (3.3).

Example 3.2.3 (*continued*) The automaton $\mathcal{A}(X^*)$ is obtained without removing the initial state of $\mathcal{A}(X)$, and is given by (3.2). See Figure 3.15.

Example 3.2.8 Consider the code $X = ba^*b$ over $A = \{a, b\}$. Its minimal automaton is given in Figure 3.16(a). The stabilizer of the initial state is just the empty word 1. The minimal automaton $A(X^*)$ given in Figure 3.16(b) is derived from Formula (3.3).

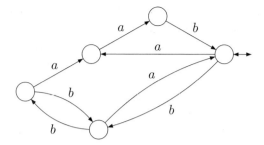

Figure 3.15 The minimal automaton of X^*, with $X = (b^2)^*(a^2b \cup ba)$.

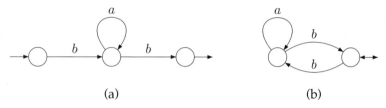

(a)   (b)

Figure 3.16 (a) The minimal automaton of $X = ba^*b$, and (b) the minimal automaton of X^*.

A construction which is analogous to that of Proposition 3.2.7 allows us to define the *literal automaton* of X^* for a prefix code X. It is the automaton

$$\mathcal{A} = (XA^-, 1, 1)$$

whose states are the proper prefixes of words in X, and with the action given by

$$u \cdot a = \begin{cases} ua & \text{if } ua \in XA^-, \\ 1 & \text{if } ua \in X, \\ \emptyset & \text{otherwise.} \end{cases} \tag{3.7}$$

This automaton is obtained from the literal automaton for X by identifying all final states of the latter with the initial state 1. It is immediate that this automaton recognizes X^*.

The following property of rational prefix codes will be useful later (Section 6.6).

Proposition 3.2.9 *For any rational prefix code X over A, there exists an integer N such that the length of any strictly increasing sequence of suffixes of words of X for the prefix order is bounded by N.*

Proof. Let $\mathcal{A} = (Q, i, T)$ be a finite automaton with N states recognizing X, and assume there is a sequence of $N + 1$ suffixes $s_0, \ldots, s_N$ of words of X such that each s_j is a proper prefix of s_{j+1}. Each s_j is the label of a path from some state q_j into a final state t_j in $\mathcal{A}$. Moreover there is, for each j, a word p_j that is the label of a path from i to q_j. Note that $p_j s_j$ is in X for each j. By the definition of N, there exist j, k with $0 \le j < k \le N$ such that $q_j = q_k$. Thus both $p_j s_j$ and $p_j s_k$ are in X, and $p_j s_j$ is a proper prefix of $p_j s_k$, contradicting the fact that X is prefix.   □

Example 3.2.10 Consider the prefix code $X = A^*aba \setminus A^+aba$ over $A = \{a, b\}$. The sequences of maximal length of strictly increasing sequences of suffixes, for the prefix order, are $\varepsilon, a, a^n aba$ with $n \geq 1$. Another sequence is ε, ba.

3.3 Maximal prefix codes

A prefix subset X of A^* is *maximal* if it is not properly contained in any other prefix subset of A^*, that is, if $X \subset Y \subset A^*$ and Y prefix imply $X = Y$.

As for maximal codes, a reference to the underlying alphabet is necessary for the definition to make sense.

The set $\{1\}$ is a maximal prefix set. Every other maximal prefix set is a code. A maximal code which is prefix is always maximal prefix. The converse does not hold: there exist maximal prefix codes which are not maximal as codes. However, under mild assumptions, namely for thin codes, we will show that maximal prefix codes are maximal codes.

The study of maximal prefix codes uses a left-to-right oriented version of dense and complete codes.

Let M be a monoid, and let N be a subset of M. An element $m \in M$ is *right completable* in N if $mw \in N$ for some w in M. It is equivalent to say that N meets the right ideal mM. A subset N is *right dense* if every $m \in M$ is right completable in N, that is if N meets all right ideals. The set N is *right complete* if the submonoid generated by N is right dense. The set N is *right thin* if it is not right dense. Of course, all these definitions make sense if right is replaced by left.

The following implications hold for a subset N of a monoid M:

$$N \text{ right dense} \implies N \text{ dense}$$
$$N \text{ right complete} \implies N \text{ complete}$$
$$N \text{ thin} \implies N \text{ right thin.}$$

In the case of a free monoid A^*, a subset N of A^* is right dense if and only if every word in A^* is a prefix of some word in N. Thus every (nonempty) left ideal is right dense. Similarly, N is right complete if every word w in A^* can be written as

$$w = m_1 m_2 \cdots m_r p$$

for some $r \geq 0, m_1, \ldots, m_r \in N$, and p a prefix of some word in N.

Proposition 3.3.1 *For any subset $X \subset A^*$ the following conditions are equivalent:*

(i) XA^* *is right dense.*

(ii) $A^* = XA^- \cup X \cup XA^+$.

(iii) *For all $w \in A^*$, there exist $u, v \in A^*$, $x \in X$ with $wu = xv$.*

Proof. (i) $\implies$ (iii). Let $w \in A^*$. Since XA^* is right dense, it meets the right ideal wA^*. Thus $wu = xv$ for some $u, v \in A^*$, and $x \in X$.

(iii) $\implies$ (ii). If $wu = xv$, then $w \in XA^-$, $w \in X$ or $w \in XA^+$ according to $w < x$, $w = x$, or $w > x$.

(ii) $\implies$ (i). The set of prefixes of XA^* is $XA^- \cup X \cup XA^+$. $\qquad\square$

Proposition 3.3.2 *Let $X \subset A^+$ be a subset that does not contain the empty word. Then XA^* is right dense if and only if X is right complete.*

Proof. Suppose first that XA^* is right dense and consider a word $w \in A^*$. If $w \in XA^- \cup X$ then $wu \in X$ for some $u \in A^*$. Otherwise $w \in XA^+$ by Proposition 3.3.1. Thus, $w = xw'$ for some $x \in X$, $w' \in A^+$. Since $x \neq 1$, we have $|w'| < |w|$. Arguing by induction, $w'u \in X^*$ for some u in A^*. Thus, w is a prefix of some word in X^*.

Conversely, let $w \in A^*$, and assume that $wu \in X^*$ for some $u \in A^*$. Multiplying if necessary by some word in X, we may assume that $wu \neq 1$. Then $wu \in X^+ \subset XA^*$.

□

Note that Proposition 3.3.2 does not hold for $X = \{1\}$. In this case, $XA^* = A^*$ is right dense, but $X^* = \{1\}$ is, of course, not.

The next statement describes natural bijections between the following families of subsets of A^*:

1. the family $\mathcal{M}$ of maximal prefix sets,
2. the family $\mathcal{D}$ of right ideals which are right dense,
3. the family $\mathcal{P}$ of prefix-closed subsets which do not contain a right ideal.

These bijections are actually restrictions of the bijections of Proposition 3.1.2.

Proposition 3.3.3 *The following bijections hold.*

  (i) *The map $X \mapsto XA^*$ is a bijection from $\mathcal{M}$ onto $\mathcal{D}$, and the map $I \mapsto I \setminus IA^+$ is its inverse.*
 (ii) *Set complementation maps bijectively $\mathcal{P}$ onto $\mathcal{D}$.*
(iii) *The map $X \mapsto XA^-$ is a bijection from $\mathcal{M}$ onto $\mathcal{P}$ and the map $P \mapsto PA \setminus P$ is its inverse.*

Proof. (i) Let X be a maximal prefix set. Any word $u \in A^*$ is comparable with a word of X since otherwise $X \cup u$ would be a prefix, a contradiction with the hypothesis. Thus XA^* is right dense. The converse holds for the same reason.

(ii) is a translation of the fact that a set is right dense if and only if its complement does not contain a right ideal.

(iii) If X is a maximal prefix subset of A^*, then XA^* is right dense. Thus $A^* \setminus XA^* = XA^-$ by Proposition 3.3.1.

□

The following corollary appears to be useful.

Corollary 3.3.4 *Let $L \subset A^+$ and let $X = L \setminus LA^+$. Then L is right complete if and only if X is a maximal prefix code.*

Proof. L is right complete if and only if LA^* is right dense (Proposition 3.3.2). From $XA^* = LA^*$ (Proposition 3.1.2) and from Proposition 3.3.3, the statement follows.

□

A special case of the corollary is the following important statement.

Theorem 3.3.5 *Let $X \subset A^+$ be a prefix code. Then X is right complete if and only if X is a maximal prefix code.*

Proof. This results from the previous corollary by taking for L a prefix code X. $\square$

We now give the statement corresponding to Proposition 3.1.6 for maximal prefix codes.

Theorem 3.3.6 *Let X be a prefix code over A, and let $P = XA^-$ be the set of proper prefixes of words in X. Then X is maximal prefix if and only if one of the following equivalent conditions hold:*

$$\underline{X} - 1 = \underline{P}(\underline{A} - 1), \quad and \quad \underline{A}^* = \underline{X}^*\underline{P}. \tag{3.8}$$

Proof. Set $R = A^* \setminus XA^*$. If X is maximal prefix, then XA^* is right dense and $R = P$ by Proposition 3.3.1. The conclusion then follows directly from Proposition 3.1.6. Conversely, if $\underline{X} - 1 = \underline{P}(\underline{A} - 1)$, then by Equation (3.1)

$$\underline{P}(\underline{A} - 1) = \underline{R}(\underline{A} - 1).$$

Since $\underline{A} - 1$ is invertible we get $P = R$, showing that XA^* is right dense. $\square$

Corollary 3.3.7 *Let X be a finite maximal prefix code with n elements over a k letter alphabet A, let $p = \mathrm{Card}(XA^-)$ be the number of proper prefixes of words in X. Then $n - 1 = p(k - 1)$.* $\square$

In the case of a finite maximal prefix code, the equations of Theorem 3.3.6 give a factorization of $\underline{X} - 1$ into two polynomials. Again, there is a formula derived from Formula (3.8), namely $1 + \underline{P}\,\underline{A} = \underline{P} + \underline{X}$, which has an interpretation on the literal representation of a code X which makes the verification of maximality very easy: if p is a node which is not in X, then for each $a \in A$, there must exist a node pa in the literal representation of X.

We now show that for thin sets, a maximal prefix code is also a maximal code.

Theorem 3.3.8 *Let X be a thin subset of A^+. The following conditions are equivalent.*

(i) *X is a maximal prefix code.*
(ii) *X is prefix and a maximal code.*
(iii) *X is right complete and a code.*

Proof. The implication (ii) $\Longrightarrow$ (i) is clear. (i) $\Longrightarrow$ (iii) follows from Proposition 3.3.3 (i) and Proposition 3.3.2. It remains to prove (iii) $\Longrightarrow$ (ii). Let $Y = X \setminus XA^+$. By Proposition 3.1.2, $YA^* = XA^*$. Thus Y is right complete. Consequently Y is complete. The set Y is also thin, since $Y \subset X$. Thus Y is a maximal code by Theorem 2.5.13. From the inclusion $Y \subset X$, we have $X = Y$. $\square$

The following example shows that Theorem 3.3.8 does not hold without the assumption that the code is thin.

Example 3.3.9 Let $X = \{uba^{|u|} \mid u \in A^*\}$, with $A = \{a, b\}$. This is the reversal of the code given in Example 2.4.11. It is a maximal code which is right dense, whence right complete. However, X is not prefix. From Corollary 3.3.4, it follows that $Y = X \setminus XA^+$ is a maximal prefix code. Of course, $Y \neq X$, and thus, Y is not maximal.

Proposition 3.3.10 *Let X be a thin subset of A^+. The following conditions are equivalent.*

(i) *X is a maximal prefix code.*
(ii) *X is prefix, and there exists a positive Bernoulli distribution π with $\pi(X) = 1$.*
(iii) *X is prefix, and $\pi(X) = 1$ for all positive Bernoulli distributions π.*

Proof. It is an immediate consequence of Theorem 3.3.8 and of Theorem 2.5.16. □

In the previous section, we gave a description of prefix codes by means of the bases of the stabilizers in a deterministic automaton. Now we consider maximal prefix codes. Let us introduce the following definition. A state q of a deterministic automaton $\mathcal{A} = (Q, i, T)$ over A is *recurrent* if for all $u \in A^*$, there is a word $v \in A^*$ such that $q \cdot uv = q$. This implies in particular that $q \cdot u \neq \emptyset$ for all u in A^*.

Proposition 3.3.11 *Let X be a prefix code over A. The following conditions are equivalent.*

(i) *X is maximal prefix.*
(ii) *The minimal automaton of X^* is complete.*
(iii) *All states of the minimal automaton of X^* are recurrent.*
(iv) *The initial state of the minimal automaton of X^* is recurrent.*
(v) *X^* is the stabilizer of a recurrent state in some deterministic automaton.*

Proof. (i) $\implies$ (ii). Let $\mathcal{A}(X^*) = (Q, i, i)$ be the minimal automaton of X^*. Let $q \in Q$, $a \in A$. There is some word $u \in A^*$ such that $i \cdot u = q$. The code X being right complete, $uav \in X^*$ for some word v. Thus $i = i \cdot uav = (q \cdot a) \cdot v$, showing that $q \cdot a \neq \emptyset$. Thus $\mathcal{A}(X^*)$ is complete.

(ii) $\implies$ (iii). Let $q \in Q$, $u \in A^*$; then $q' = q \cdot u \neq \emptyset$ since $\mathcal{A}(X^*)$ is complete. $\mathcal{A}(X^*)$ being minimal, q' is coaccessible, and q is accessible. Thus $q' \cdot v = q$, for some $v \in A^*$, showing that q is recurrent.

The implications (iii) $\implies$ (iv) $\implies$ (v) are clear.

(v) $\implies$ (i). Let $\mathcal{A} = (Q, i, T)$ be a deterministic automaton and $q \in Q$ be a recurrent state such that $X^* = \text{Stab}(q)$. For all $u \in A^*$ there is a word $v \in A^*$ with $q \cdot uv = q$, thus $uv \in X^*$. This shows that X is right complete. The set X being prefix, the result follows from Theorem 3.3.8. □

3.4 Operations on prefix codes

Prefix codes are closed under some simple operations. We start with a general result which will be used several times.

Proposition 3.4.1 *Let X and $(Y_i)_{i \in I}$ be nonempty subsets of A^*, and let $(X_i)_{i \in I}$ be a partition of X. Set*

$$Z = \bigcup_{i \in I} X_i Y_i.$$

1. *If X and the Y_i's are prefix (maximal prefix), then Z is prefix (maximal prefix).*
2. *If Z is prefix, then all Y_i are prefix.*
3. *If X is prefix and Z is maximal prefix, then X and the Y_i's are maximal prefix.*

Proof. 1. Assume that $z, zu \in Z$. Then $z = xy$, $zu = x'y'$ for some $i, j \in I$, $x \in X_i$, $y \in Y_i$, $x' \in X_j$, $y' \in Y_j$. From the relation $xyu = x'y'$ it follows that $x = x'$ because X is prefix, whence $i = j$ and $y = y'$. Thus, $u = 1$ and Z is prefix. Assume now that XA^* and the $Y_i A^*$ are right dense. Let $w \in A^*$. Then $ww' = xv$ for some $w', v \in A^*$, $x \in X$. Let x belong to X_i. Since $Y_i A^*$ is right dense, $vv' \in Y_i A^*$ for some $v' \in A^*$. Thus $ww'v' \in X_i Y_i A^*$, whence $ww'v' \in ZA^*$. Thus Z is maximal prefix.

2. Let $y, yu \in Y_i$ and $x \in X_i$. Then $xy, xyu \in Z$, implying that $u = 1$.

3. From $ZA^* \subset XA^*$ we get that XA^* is right dense. Consequently X is maximal prefix. To show that $Y_i A^*$ is right dense, let $w \in A^*$. For any $x \in X_i$, xw is right-completable in ZA^*. Thus, $xw = zw'$ for some $z \in Z$. Setting $z = x'y'$ with $x' \in X_j$, $y' \in Y_j$ gives $xw = x'y'w'$. The code X being prefix, we get $x = x'$, whence $w = y'w'$, showing that w is in $Y_i A^*$.   □

For $\mathrm{Card}(I) = 1$, we obtain, in particular,

Corollary 3.4.2 *If X and Y are prefix codes (maximal prefix), then XY is a prefix code (maximal prefix).*   □

The converse of Corollary 3.4.2 holds only under rather restrictive conditions and will be given in Proposition 3.4.13.

Example 3.4.3 The *Golomb code* of order $m \geq 1$ over the alphabet $\{0, 1\}$ is the maximal infinite prefix code

$$G_m = 1^* 0 R_m,$$

where $R_1 = \{\epsilon\}$ and, for $m \geq 2$, R_m is the finite maximal prefix code defined below. Thus, each G_m is the product of the maximal prefix codes $1^* 0$ and R_m.

If $m = 2^k$ for some integer k, then R_m is the set of all binary words of length k. Otherwise, the rule is more involved. Set $m = 2^k + \ell$, with $0 < \ell < 2^k$. Setting $n = 2^{k-1}$,

$$R_m = \begin{cases} 0 R_\ell \cup 1 R_{2n} & \text{if } \ell \geq n, \\ 0 R_n \cup 1 R_{n+\ell} & \text{otherwise.} \end{cases}$$

The set R_1 and the codes R_m for $m = 2, \ldots, 7$ are represented in Figure 3.17. Note that, in particular, the lengths of the codewords differ at most by one.

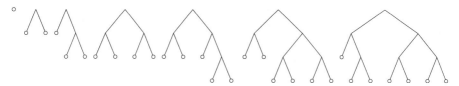

Figure 3.17 The sets R_1 to R_7.

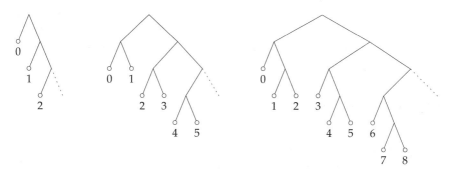

Figure 3.18 The Golomb codes of orders 1, 2, 3.

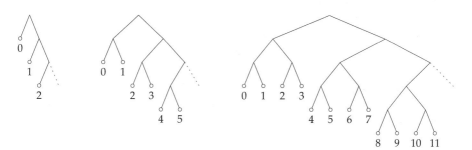

Figure 3.19 The Golomb–Rice codes of orders 0, 1, and 2.

The Golomb codes of order 1, 2, 3 are represented in Figure 3.18. Note that, except possibly for the first level, there are exactly m words of each length. The Golomb codes are used to represent integers as indicated in Figure 3.18. It can be shown that they are optimal for some probability distributions, see Exercise 3.9.1.

Example 3.4.4 The *Golomb–Rice code* of order k is the particular case of the Golomb code for $m = 2^k$. Its structure is especially simple and allows an easy explicit description of the encoding of an integer: The encoding assigns to an integer $n \geq 0$ two binary words, the *base* and the *offset*. The base is the unary expansion of $\lfloor n/2^k \rfloor$ followed by a 0. The offset is the rest of the division written in binary on k bits. Thus, for $k = 2$, the integer $n = 9$ is coded by 110|01. The binary trees representing the Golomb–Rice code of orders 0, 1, 2 are represented in Figure 3.19.

Another expression of the Golomb–Rice code of order k is given by the regular expression

$$GR_k = 1^*0(0 + 1)^k. \tag{3.9}$$

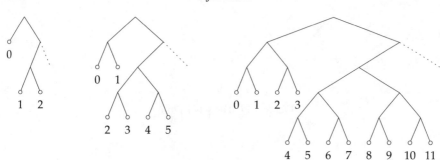

Figure 3.20 The exponential Golomb codes of orders 0, 1, 2.

It expresses the fact that the binary words forming the code are composed of a base of the form 1^i0 for some $i \geq 0$ and an offset which is an arbitrary binary sequence of length k.

Example 3.4.5 The *exponential Golomb codes* form a family depending on an integer k with a length distribution better suited for some probability distributions than the Golomb–Rice codes. The case $k = 0$ is closely related to the *Elias code* already mentioned in Example 3.1.1.

The base of the codeword for an integer n is obtained as follows. Let x be the binary representation of $1 + \lfloor n/2^k \rfloor$ and let i be its length. The base is made of the unary representation of $i - 1$ followed by x with its initial 1 replaced by a 0. The offset is, as before, the binary representation of the rest of the division of n by 2^k, written on k bits. Thus, for $k = 1$, the codeword for 9 is $11001|1$. Figure 3.20 represents the binary trees of the exponential Golomb codes of orders 0, 1, and 2.

An expression describing the exponential Golomb code is

$$EG_k = \bigcup_{i \geq 0} 1^i 0 (0 + 1)^{i+k},$$

and we have the simple relation

$$EG_k = EG_0 (0 + 1)^k.$$

Corollary 3.4.6 *Let $X \subset A^+$, and $n \geq 1$. Then X is (maximal) prefix if and only if X^n is (maximal) prefix.*

Proof. By Corollary 3.4.2, X^n is maximal prefix for a maximal prefix code X. Conversely, setting $Z = X^n = X^{n-1}X$, it follows from Proposition 3.4.1(2) that X is prefix. Writing $Z = XX^{n-1}$, we see by Proposition 3.4.1(3) that X (and X^{n-1}) are maximal prefix if Z is. □

Corollary 3.4.6 is a special case of Proposition 3.4.11, to be proved later.

Corollary 3.4.7 *Let X and Y be prefix codes, and let $X = X_1 \cup X_2$ be a partition. Then $Z = X_1 \cup X_2Y$ is a prefix code and Z is maximal prefix if and only if X and Y are maximal prefix.*

Proof. With $Y' = \{1\}$, we have $Z = X_1Y' \cup X_2Y$. The result follows from Proposition 3.4.1 because Y' is maximal prefix. □

X Y $Z = (X \setminus x) \cup xY$

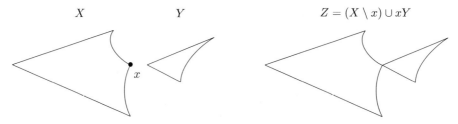

Figure 3.21 Combining codes X and Y.

Z $X = (Z \setminus pY_p) \cup p$

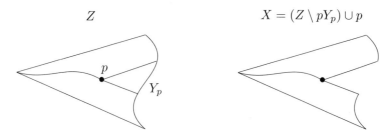

Figure 3.22 Separating Z and Y_p.

There is a special case of this corollary which deserves attention. It constitutes an interesting operation on codes viewed as trees.

Corollary 3.4.8 *Let X and Y be prefix codes, and $x \in X$. Then*

$$Z = (X \setminus x) \cup xY$$

is prefix and Z is maximal prefix if and only if X and Y are.   □

The operation performed on X and Y is sketched in Figure 3.21. We now turn to the converse operation.

Proposition 3.4.9 *Let Z be a prefix code, and let $p \in ZA^-$. Then*

$$Y_p = p^{-1}Z \quad and \quad X = Z \setminus pY_p \cup \{p\} \tag{3.10}$$

are prefix sets. Further if Z is maximal prefix, then Y_p and X are maximal prefix also.

The operation described in (3.10) can be drawn as shown in Figure 3.22. Proposition 3.4.9 is a special case of the following result.

Proposition 3.4.10 *Let Z be a prefix code, and let Q be a prefix subset of ZA^-. For each $p \in ZA^-$, the set $Y_p = p^{-1}Z$ is a prefix code; further*

$$X = Q \cup \left(Z \setminus \bigcup_{p \in Q} pY_p \right)$$

is a prefix set. If Z is maximal prefix, then X and the Y_p ($p \in Q$) are maximal prefix.

Proof. Set $X_0 = Z \setminus \bigcup_{p \in Q} pY_p$, $Y_0 = \{1\}$, $X_p = \{p\}$. Then

$$Z = X_0 Y_0 \cup \bigcup_{p \in Q} X_p Y_p.$$

Thus, to derive the result from Proposition 3.4.1, it suffices to show that X is prefix.

Let x, $xu \in X$ with $u \in A^+$. These words cannot both be in the prefix set Z nor can they both be in the prefix set Q. Since $Q \subset ZA^-$, we have $x \in Q$, $xu \in Z$. Thus $u \in Y_x$ and xu is not in X. $\qquad\square$

Propositions 3.4.1 and 3.4.10 can be used to enumerate maximal prefix sets. Let us illustrate the computation in the case of $A = \{a, b\}$. If Z is maximal prefix and $Z \neq 1$, then both

$$X = a^{-1}Z, \quad Y = b^{-1}Z$$

are maximal prefix and

$$Z = aX \cup bY. \tag{3.11}$$

Conversely, if X and Y are maximal prefix, then so is Z. Thus, Equation (3.11) defines a bijection from maximal prefix codes onto pairs of maximal prefix sets. Further

$$\mathrm{Card}(Z) = \mathrm{Card}(X) + \mathrm{Card}(Y).$$

Let α_n be the number of maximal prefix sets with n elements. Then by Equation (3.11), for $n \geq 2$,

$$\alpha_n = \sum_{k+\ell=n} \alpha_k \alpha_\ell. \tag{3.12}$$

Let $\alpha(t) = \sum_{n \geq 0} \alpha_n t^n$. Then by (3.12)

$$\alpha(t)^2 - \alpha(t) + t = 0.$$

The equation has the solutions $(1 \pm \sqrt{1 - 4t})/2$. Since $\alpha(0) = 0$, one has $\alpha(t) = (1 - \sqrt{1 - 4t})/2$. Using the binomial formula, we get for $n \geq 1$

$$\alpha_n = -\frac{1}{2}(-4)^n \binom{1/2}{n}$$

$$= -\frac{1}{2}(-4)^n \frac{1/2(1/2 - 1) \cdots (1/2 - n + 1)}{n!}$$

$$= -\frac{1}{2}(-4)^n \frac{1}{2^n} \frac{1(1 - 2) \cdots (1 - 2n + 2)}{n!}$$

$$= -\frac{1}{2}(-1)^n 2^n (-1)^{n-1} \frac{1 \cdot 3 \cdots (2n - 3)}{n!}$$

$$= 2^{n-1} \frac{(2n - 2)!}{n!(n - 1)!2^{n-1}} = \frac{1}{n}\binom{2n - 2}{n - 1}.$$

Table 3.1 *The first Catalan numbers.*

n	1	2	3	4	5	6	7	8
α_n	1	1	2	5	14	42	132	429

Thus

$$\alpha_{n+1} = \frac{1}{n+1}\binom{2n}{n}. \tag{3.13}$$

These numbers are called the *Catalan numbers*. See Exercise 3.4.1 for another proof and for the case of more than two letters. No such closed expression is known for the number of finite maximal codes. Table 3.1 gives the first Catalan numbers.

Proposition 3.4.11 *Let Y, Z be composable codes and $X = Y \circ Z$. Then X is a maximal prefix and thin code if and only if Y and Z are maximal prefix and thin codes.*

Proof. Assume first that X is thin and maximal prefix. Then X is right complete by Theorem 3.3.8. Thus X is thin and complete. By Proposition 2.6.13, both Y and Z are thin and complete. Further Y is prefix by Proposition 2.6.12(1). Thus Y, being thin, prefix, and complete, is a maximal prefix code. Next X is right dense and $X \subset Z^*$. Thus Z is right dense. Consequently Z is a right complete, thin code. By Theorem 3.3.8, Z is maximal prefix.

Conversely, Y and Z being prefix, X is prefix by Proposition 2.6.4, and Y, Z being both thin and complete, X is also thin and complete by Proposition 2.6.13. Thus X is a maximal prefix code. □

Proposition 3.4.12 *Let Z be a prefix code over A, and let $Z = X \cup Y$ be a partition. Then $T = X^*Y$ is a prefix code, and further T is maximal prefix if and only if Z is a maximal prefix code.*

Proof. Let B be an alphabet bijectively associated to Z, and let $B = C \cup D$ be the partition of B induced by the partition $Z = X \cup Y$. Then

$$T = C^*D \circ Z.$$

The code C^*D clearly is prefix. Thus, T is prefix by Proposition 2.6.4. Next, $T^* = 1 \cup Z^*Y$ showing that T is right complete if and only if Z is right complete. The second part of the statement thus results from Proposition 3.3.3. □

We conclude this section by the proof of a converse to Corollary 3.4.2.

Proposition 3.4.13 *Let X and Y be finite nonempty subsets of A^* such that the product XY is unambiguous. If XY is a maximal prefix code, then X and Y are maximal prefix codes.*

The following example shows that the conclusion fails for infinite codes.

Example 3.4.14 Consider $X = \{1, a\}$ and $Y = (a^2)^*b$ over $A = \{a, b\}$. Here X is not prefix, and Y is not maximal prefix. However, $XY = a^*b$ is maximal prefix and the product is unambiguous.

Proof of Proposition 3.4.13. Let $Z = XY$ and $n = \max\{|y| \mid y \in Y\}$. The proof is by induction on n. For $n = 0$, we have $Y = \{1\}$ and $Z = X$. Thus, the conclusion clearly holds. Assume $n \geq 1$ and set

$$T = \{y \in Y \mid |y| = n\}, \quad Q = \{q \in YA^- \mid qA \cap T \neq \emptyset\}.$$

By construction, $T \subset QA$. In fact $T = QA$. Indeed, let $q \in Q$, $a \in A$ and let $x \in X$ be a word of maximal length. Then xq is a prefix of a word in Z, and xqa is right-completable in ZA^*. The code Z being prefix, no proper prefix of xqa is in Z. Consequently

$$xqav = x'y'$$

for some $x' \in X$, $y' \in Y$, and $v \in A^*$.

Now $n = |qa| \geq |y'|$, and $|x| \geq |x'|$. Thus $x = x'$, $y' = qa$, $v = 1$. Consequently $qa \in Y$ and $T = QA$. Now let

$$Y' = (Y \setminus T) \cup Q, \quad Z' = XY'.$$

We verify that Z' is prefix. Assume the contrary. Then

$$xy'u = x'y''$$

for some $x, x' \in X$, $y', y'' \in Y'$, $u \neq 1$. Let a be the first letter of u. Then either y' or $y'a$ is in Y. Similarly either y'' or $y''b$ (for any b in A) is in Y. Assume $y' \in Y$. Then $xy' \in Z$ is a proper prefix of $x'y''$ or $x'y''b$, one of them being in Z. This contradicts the fact that Z is prefix. Thus $y'a \in Y$. As before, $xy'a$ is not a proper prefix of $x'y''$ or $x'y''b$. Thus necessarily $u = a$ and $y'' \in Y$, and we have

$$xy'a = x'y''$$

with $y'a, y'' \in Y$. The unambiguity of the product XY shows that $x = x'$, $y'a = y''$. But then $y'' \notin Y'$. This gives the contradiction.

To see that Z' is maximal prefix, observe that $Z \subset Z' \cup Z'A$. Thus $ZA^* \subset Z'A^*$ and the result follows from Proposition 3.3.3. Finally, it is easily seen that the product XY' is unambiguous: if $xy' = x'y''$ with $x, x' \in X$, $y', y'' \in Y'$, then either $y', y'' \in Y \setminus T$ or $y', y'' \in Q$, the third case being ruled out by the prefix character of Z.

Of course, $\max\{|y| \mid y \in Y'\} = n - 1$. By the induction hypothesis, X and Y' are maximal prefix. Since

$$Y = (Y' \setminus Q) \cup QA,$$

the set Y is maximal prefix by Corollary 3.4.7.   □

It is also possible to give a completely different proof of Proposition 3.4.13 using the fact that, under the hypotheses of this proposition, we have $\pi(X)\pi(Y) = 1$ for all Bernoulli distributions π, see Exercise 3.4.2.

3.5 Semaphore codes

This section contains a detailed study of semaphore codes which constitute an interesting subclass of the prefix codes. This investigation also illustrates the techniques introduced in the preceding sections.

Proposition 3.5.1 *For any nonempty subset S of A^+, the set*

$$X = A^*S \setminus A^*SA^+ \tag{3.14}$$

is a maximal prefix code.

Proof. The set $L = A^*S$ is a left ideal, and thus, is right dense. Consequently, L is right complete, and by Corollary 3.3.4, the set $X = L \setminus LA^+$ is maximal prefix. $\quad\square$

A code X of the form given in Equation (3.14) is called a *semaphore code*, the set S being a set of semaphores for X. The terminology stems from the following observation: a word is in X if and only if it ends with a semaphore, but none of its proper prefixes end with a semaphore. Thus, reading a word from left to right, the first appearance of a semaphore gives a "signal" indicating that what has been read up to now is in the code X.

Example 3.5.2 Let $A = \{a, b\}$ and $S = \{a\}$. Then $X = A^*a \setminus A^*aA^+$ whence $X = b^*a$.

Example 3.5.3 For $A = \{a, b\}$ and $S = \{aa, ab\}$, we have $A^*S = A^*aA$. Thus $A^*S \setminus A^*SA^+ = b^*aA$.

The following proposition characterizes semaphore codes among prefix codes.

Proposition 3.5.4 *Let $X \subset A^+$. Then X is a semaphore code if and only if X is prefix and*

$$A^*X \subset XA^*. \tag{3.15}$$

Proof. Let $X = A^*S \setminus A^*SA^+$ be a semaphore code. Then X is prefix and it remains to show (3.15). Let $w \in A^*X$. Since $w \in A^*S$, w has a factor in S. Let w' be the shortest prefix of w which is in A^*S. Then w' is in X. Consequently $w \in XA^*$.

Conversely, assume that a prefix code X satisfies (3.15). Set $M = XA^*$. In view of Proposition 3.1.2 and by the fact that X is prefix, we have $X = M \setminus MA^+$. Equation (3.15) implies that

$$A^*M = A^*XA^* \subset XA^* = M,$$

thus, $M = A^*M$ and $X = A^*M \setminus A^*MA^+$. $\quad\square$

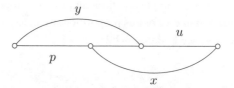

Figure 3.23 Proof of Corollary 3.5.7.

Example 3.5.5 The code $Y = \{a^2, aba, ab^2, b\}$ is a maximal prefix code over A. However, Y is not a semaphore code, since $ab \in A^*Y$ but $ab \notin YA^*$.

A semaphore code is maximal prefix, thus right complete. The following proposition describes those right complete sets which are semaphore codes.

Proposition 3.5.6 *Let $X \subset A^+$. Then X is a semaphore code if and only if X is right complete and*

$$X \cap A^*XA^+ = \emptyset. \tag{3.16}$$

Proof. A semaphore code is maximal prefix, thus also right complete. Further, in view of (3.15),

$$A^*XA^+ \subset XA^+,$$

thus

$$X \cap A^*XA^+ \subset X \cap XA^+ = \emptyset,$$

showing Equation (3.16).

Conversely, if a set X satisfies (3.16), then X is prefix. To show that X is a semaphore code, we verify that (3.15) holds. Let $w = ux \in A^*X$ with $u \in A^*, x \in X$. The code X being right complete, we have $uxv = x'y$ for some $x' \in X$, $y \in X^*$, $v \in A^*$. Now Equation (3.16) shows that ux is not a proper prefix of x'. Thus $ux \in x'A^*$. $\qquad\square$

Corollary 3.5.7 *Let $X \subset A^+$ be a semaphore code and let $P = XA^-$. Then $PX \subset XP \cup X^2$.*

Proof. (See Figure 3.23) Let $p \in P$, $x \in X$. By Equation (3.15), $px = yu$ for some $y \in X$, $u \in A^*$. The code X is prefix, thus $|p| < |y|$. Consequently, u is suffix of x, and by (3.16), $u \notin XA^+$. The code X is maximal prefix, therefore $u \in XA^- \cup X$. $\square$

Formula (3.16) expresses a property of semaphore codes which is stronger than the prefix condition: for a semaphore code X, and two elements $x, x' \in X$, the only possible way for x to occur as a factor in x' is to be a suffix of x'. We now use this fact to characterize semaphore codes among maximal prefix codes.

Proposition 3.5.8 *Let $X \subset A^+$, and let $P = XA^-$ be the set of proper prefixes of words in X. Then X is a semaphore code if and only if X is a maximal prefix code and P is suffix-closed.*

Of course, P is always prefix-closed. Thus P is suffix-closed if and only if it contains the factors of its elements.

Proof. Let X be a semaphore code. Then X is a maximal prefix code (Proposition 3.5.1). Next, let $p = uq \in P$ with $u, q \in A^*$. Let $v \in A^+$ be a word such that $pv \in X$. Then $q \notin XA^*$, since otherwise $pv = uqv \in X \cap A^*XA^+$, violating Proposition 3.5.6. Thus $q \in XA^- = P$.

Conversely assume that X is maximal prefix and that P is suffix-closed. Suppose that $X \cap A^*XA^+ \neq \emptyset$. Let $x \in X \cap A^*XA^+$. Then $x = ux'v$ for some $u \in A^*$, $x' \in X$, $v \in A^+$. It follows that $ux' \in P$, and since P is suffix-closed, also $x' \in P$ which is impossible. Thus X is a semaphore code by Proposition 3.5.6. □

Another consequence of Proposition 3.5.6 is the following result.

Proposition 3.5.9 *Any semaphore code is thin.*

Proof. By Formula (3.16), no word in XA^+ is a factor of a word in X. □

Corollary 3.5.10 *Any semaphore code is a maximal code.*

Proof. A semaphore code is a maximal prefix code and thin by Propositions 3.5.1 and 3.5.9. Thus by Theorem 3.3.8 such a code is a maximal code. □

Now we determine the sets of semaphores giving the same semaphore code.

Proposition 3.5.11 *Two nonempty subsets S and T of A^+ define the same semaphore code if and only if $A^*SA^* = A^*TA^*$. For each semaphore code X, there exists a unique minimal set of semaphores, namely $T = X \setminus A^+X$.*

Proof. Let $X = A^*S \setminus A^*SA^+$, $Y = A^*T \setminus A^*TA^+$. By Proposition 3.1.2, we have $XA^* = A^*SA^*$, $YA^* = A^*TA^*$, and by Corollary 3.1.8, $X = Y$ if and only if $A^*SA^* = A^*TA^*$.

Next, let $X = A^*S \setminus A^*SA^+$ be a semaphore code. By the definition of $T = X \setminus A^+X$, we may apply to T the dual of Proposition 3.1.2. Thus, $A^*T = A^*X$. Since $A^*TA^* = A^*XA^* = A^*SA^*$, the sets S and T define the same semaphore code. Thus $X = A^*T \setminus A^*TA^+$.

Finally, let us verify that $T \subset S$. Let $t \in T$. Since $A^*TA^* = A^*SA^*$, one has $t = usv$ for some $u, v \in A^*$, $s \in S$, and $s = u't'v'$ for some $u', v' \in A^*$, $t' \in T$. Thus, $t = uu't'v'v$. Note that $T \subset X$. Thus, Formula (3.16) applies, showing that $v'v = 1$. Since T is a suffix code, we have $uu' = 1$. Thus, $t = s$ and $t \in S$. □

We now study some operations on semaphore codes.

Proposition 3.5.12 *If X and Y are semaphore codes, then XY is a semaphore code. Conversely, if XY is a semaphore code and if X is a prefix code, then X is a semaphore code.*

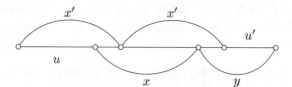

Figure 3.24 Proof of Proposition 3.5.12.

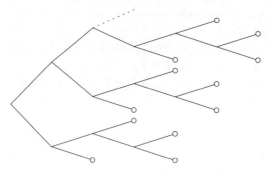

Figure 3.25 The code $a^*b\{a^2, aba, ab^2, b\}$.

Proof. If X, Y are semaphore codes, then by Corollary 3.4.2, XY is a prefix code. Further by Proposition 3.5.4,

$$A^*XY \subset XA^*Y \subset XYA^*,$$

thus XY is a semaphore code.

Assume now that XY is a semaphore code, and that X is a prefix code. We show that $A^*X \subset XA^*$. For this, let $w = ux \in A^*X$, with $u \in A^*$, $x \in X$, and let y be a word in Y of minimal length. Then

$$wy = uxy = x'y'u'$$

for some $x' \in X$, $y' \in Y$, $u' \in A^*$ (see Figure 3.24). By the choice of y, we have $|y| \le |y'| \le |y'u'|$, thus $|ux| \ge |x'|$, showing that $ux \in XA^*$. $\square$

The following example shows that if XY is a semaphore code, then Y need not be semaphore, even if it is maximal prefix.

Example 3.5.13 Over $A = \{a, b\}$, let $X = a^*b$, and $Y = \{a^2, aba, ab^2, b\}$. Then X is a semaphore code, and Y is a maximal prefix code. However, Y is not semaphore (Example 3.5.5). On the other hand the code $Z = XY$ is semaphore. Indeed, Z is maximal prefix, and the set

$$P = ZA^- = a^*\{1, b, ba, bab\}$$

is suffix-closed. The conclusion follows from Proposition 3.5.7 (see Figure 3.25).

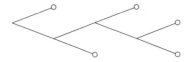

Figure 3.26 The code $X = \{a, baa, baba, bab^2, b^2\}$.

Corollary 3.5.14 *For any $X \subset A^+$ and $n \geq 1$, the set X is a semaphore code if and only if X^n is a semaphore code.*

Proof. If X^n is a semaphore code, then X is a prefix by Corollary 3.4.6 and X is a semaphore code by Proposition 3.5.12. The converse is a direct consequence of Proposition 3.5.12. □

Example 3.5.15 The code $X = \{a, baa, baba, bab^2, b^2\}$ represented in Figure 3.26 is a maximal prefix code but not semaphore. Indeed, the word a has an inner occurrence in bab^2, contradicting Formula (3.16). However, X decomposes into two semaphore codes

$$X = Y \circ Z,$$

with $Y = \{c, dc, d^2, de, e\}$ and $Z = \{a, ba, b^2\}$.

Given a semaphore code

$$X = A^*S \setminus A^*SA^+,$$

it is natural to consider

$$Y = SA^* \setminus A^+SA^*.$$

The code Y is a maximal suffix code. Its reversal $\tilde{Y} = A^*\tilde{S} \setminus A^*\tilde{S}A^+$ is a semaphore code with semaphores $\tilde{S}$. The following result shows a strong relation between X and Y.

Proposition 3.5.16 *Let $S \subset A^+$. There exists a bijection β from $X = A^*S \setminus A^*SA^+$ onto $Y = SA^* \setminus A^+SA^*$ such that, for each $x \in X$, $\beta(x)$ is a conjugate of x.*

Proof. First, consider the two-sided ideal $J = A^*SA^*$. One has

$$X = J \setminus JA^+, \quad Y = J \setminus A^+J.$$

Indeed, $A^*JA^* = A^*SA^*$ and by Proposition 3.5.11, $X = A^*J \setminus JA^+$. The formula for X follows because $A^*J = J$. A symmetric argument holds for Y.

Now we define, for each $x \in X$,

$$D(x) = \{d \in A^+ \mid \text{there is some } g \in A^* \text{ with } x = gd \text{ and } dg \in J\}.$$

Thus, $D(x)$ is composed of nonempty suffixes of x. Further $D(x)$ is nonempty since x is in $D(x)$. Thus, each $D(x)$ contains some shortest element. This will be used to define β as follows. For $x \in X$,

$$\beta(x) = dg, \qquad (3.17)$$

where d is the shortest word in $D(x)$ and g is such that

$$x = gd. \qquad (3.18)$$

Thus, $\beta(x)$ is a conjugate of x, and $\beta(x) \in J$. We show that

$$\beta(x) \in J \setminus A^+ J = Y.$$

Assume the contrary. Then

$$\beta(x) = dg = uj \qquad (3.19)$$

for some $u \in A^+$, $j \in J$.

Next g is a proper prefix of x. Consequently, $g \notin J$. Indeed, if $g \in J$, then g would have a prefix in X, contradicting the fact that X is prefix. This shows that $|g| < |j|$, since otherwise g would belong to the ideal generated by j, thus $g \in J$.

It follows from this and from (3.19) that $|d| > |u|$, thus, $d = ud'$ for some $d' \in A^+$. Moreover $d' \in D(x)$, since $d'(gu) = ju \in J$ and $(gu)d' = gd = x \in X$. This gives a contradiction by the fact that d' is strictly shorter than d. Thus, $\beta(x) \in Y$.

Consider the converse mapping γ from Y into X defined by considering, for y in Y, the set

$$G(y) = \{e \in A^+ \mid y = eh \text{ and } he \in J\},$$

and by setting $\gamma(y) = he$, with $e \in G(y)$ of minimal length.

If $y = \beta(x) = dg$ is given by (3.17) and (3.18) and if $\gamma(y) = he$ with $e \in G(y)$, $eh = y$, then

$$dg = \beta(x) = eh. \qquad (3.20)$$

Note that $gd \in J$. Thus, $d \in G(y)$. Consequently, $|d| \geq |e|$. Now the word e is not a proper prefix of d. Otherwise, setting $d = eu$, $ug = h$ in (3.20) with $u \in A^+$, we get

$$geu = gd = x, \quad uge = he \in J,$$

showing that $u \in D(x)$ and contradicting the minimality of $|d|$. Thus $d = e$, $g = h$, and $\gamma(\beta(x)) = x$. An analogous proof shows that $\beta(\gamma(y)) = y$ for y in Y. Thus, β and γ are reciprocal bijections from X onto Y. $\square$

Example 3.5.17 Let us illustrate the construction of Proposition 3.5.16 by considering, over $A = \{a, b\}$, the set of semaphores $S = \{a^2, ba, b^2\}$. Then

$$X = A^* S \setminus A^* S A^+ = \{a^2, ba, b^2, aba, ab^2\},$$

$$Y = SA^* \setminus A^+ S A^* = \{a^2, a^2 b, ba, bab, b^2\}.$$

Table 3.2 *The correspondence between X and Y.*

X	D	Y
aa	a, aa	aa
aba	a, ba, aba	aab
abb	b, bb, abb	bab
ba	ba	ba
bb	b, bb	bb

Table 3.2 lists on each row an element $x \in X$, the corresponding set $D(x)$ and the element $\beta(x) \in Y$.

Proposition 3.5.16 shows that any semaphore code can be transformed into a suffix code by a bijection which exchanges conjugate words. This property does not hold for arbitrary prefix codes, as shown by the following example.

Example 3.5.18 Let $X = \{ab, ba, c, ac, bca\}$. Assume that there exists a conjugacy preserving bijection β which maps X onto a suffix code Y. Then Y necessarily contains c, and ab, ba. Further Y contains ca (with c and ac, Y would not be suffix!). All the words conjugate to bca now have a suffix equal to one of c, ab, ba, ca. Thus, Y is not suffix.

In fact, X cannot be completed into a semaphore code, since c is a factor of bca.

We end this section with the following result which shows that bifix codes are not usually semaphore codes.

Proposition 3.5.19 *Let X be a bifix semaphore code. Then $X = A^n$ for some $n \geq 1$.*

Proof. It is sufficient to show that $X \subset A^n$ for some n. Let $x, y \in X$. For each suffix q of x, we have $qy \in A^*X \subset XA^*$. Thus there is, in view of Propositions 3.5.4 and 3.5.6, a prefix p of y such that $qp \in X$.

In this way we define a mapping from the set of suffixes of X into the set of prefixes of y. The set X being suffix, the mapping is injective. Indeed, if qp and $q'p$ are in X for two suffixes q, q' of x, then $q = q'$. It follows that $|x| \leq |y|$. Interchanging x and y, we get $|y| \leq |x|$. Thus, all words in X have the same length. □

3.6 Synchronized codes

Let X be a prefix code over A. A word $w \in A^*$ is said to be *synchronizing* for X if for any $u, v \in A^*$, we have

$$uwv \in X^* \implies uw, wv \in X^*.$$

Observe that if this holds, then v also is in X^* since X^* is right unitary. If w is synchronizing, then xwy is synchronizing for any $x, y \in X^*$.

The definition takes a simpler form for a synchronizing word which is in X^*. This is the case we will in general be interested in. A word w of X^* is synchronizing if and only if for any $u, v \in A^*$, we have

$$uwv \in X^* \implies uw \in X^*.$$

A prefix code X is *synchronized* if there exists a word in X^* which is synchronizing for X. We will see later (Chapter 10) a definition of synchronized codes for general codes.

Example 3.6.1 The prefix code $X = \{ab, ba\}$ is synchronized. Indeed, *abba* is a synchronizing word for X, since $uabbav \in X^*$ implies $uab, bav \in X^*$ and thus $uabba \in X^*$.

If X is a maximal prefix code, then w is synchronizing for X if and only if

$$A^*w \subset X^*. \tag{3.21}$$

Indeed, let w be a synchronizing word. For any u in A^*, since X^* is right dense, there exists a word v such that $uwv \in X^*$. Then $uw \in X^*$. This shows that (3.21) holds. Conversely, if (3.21) holds, then $uw \in X^*$ for all $u \in A^*$, and thus w is synchronizing.

Observe that if X is a maximal prefix code, then by (3.21) every synchronizing word is in X^*.

Example 3.6.2 The code $X = b^*a$ is synchronized. Indeed, a is a synchronizing word, since $A^*a \subset X^*$.

Example 3.6.3 A maximal bifix code X over A is never synchronized unless $X = A$. Assume indeed that $w \in A^*$ is synchronizing. For any $u \in A^*$ we have $uw \in X^*$. The monoid X^* being left unitary, it follows that $u \in X^*$. Thus $A^* = X^*$.

The terminology is derived from the following observation: let w be a word which has to be factored into words of some prefix code X. The appearance, in the middle of the word w, of some synchronizing word x in X^*, that is the existence of a factorization

$$w = uxv$$

implies that ux is in X^*. Thus we may start the decoding at the beginning of the word v. Since X^* is right unitary we have indeed $w \in X^*$ if and only if $v \in X^*$. This means that the whole word is in X^* if and only if the final part can be decoded.

Note that any code X over A satisfying (3.21) is maximal prefix. Indeed, let $y, yu \in X$. Then $uw \in X^*$, and $y(uw)$, $(yu)w$ are two X-factorizations which are distinct if $u \neq 1$. Thus $u = 1$. Next, (3.21) shows that X is right complete.

Any synchronized prefix code is thin. Indeed, if x is a nonempty synchronizing word for a prefix code X, then x^2 is not a factor of a word in X, since otherwise $uxxv \in X$ for some $u, v \in A^*$. From $ux \in X^+$, it would follow that X is not prefix.

The fact that a prefix code X is synchronized is well reflected by the automata recognizing X^*. Let us give a definition. Let $\mathcal{A} = (Q, i, T)$ be a deterministic automaton on A. The *rank* of a word $x \in A^*$ in $\mathcal{A}$, denoted by $\text{rank}_\mathcal{A}(x)$, is defined by

$$\text{rank}_\mathcal{A}(x) = \text{Card}(Q \cdot x).$$

It is an integer or $+\infty$. Clearly

$$\text{rank}_\mathcal{A}(uxv) \leq \text{rank}_\mathcal{A}(x).$$

A word $w \in A^*$ is a *synchronizing* in $\mathcal{A}$ if $\text{rank}_\mathcal{A}(w) = 1$. The automaton $\mathcal{A}$ is *synchronized* if there exists a word which is synchronizing in $\mathcal{A}$.

Proposition 3.6.4 *Let X be a prefix code over A. The following conditions are equivalent:*

(i) *X is synchronized.*
(ii) *The literal automaton of X^* is synchronized.*
(iii) *The minimal automaton $\mathcal{A}(X^*)$ is synchronized.*
(iv) *There exists a trim synchronized deterministic automaton recognizing X^*.*

Proof. (i) $\implies$ (ii). Let P be the set of prefixes of X and let $\mathcal{A} = (P, 1, 1)$ be the literal automaton of X^*. Let $x \in X^*$ be a synchronizing word for X. Then 1 is in the set $P \cdot x$, so x has positive rank. Next, let $p \in P$. If $p \cdot x$ exists, there is a word s such $p \cdot xs = 1$. Then $pxs \in X^*$ and $px \in X^*$ since x is synchronizing, showing that $p \cdot x = 1$. This shows that x has rank 1 in $\mathcal{A}$.

(ii) $\implies$ (iii). A synchronizing word in the literal automaton of X^* is also synchronizing in $\mathcal{A}(X^*)$. In fact, any quotient of a synchronized automaton is synchronized.

The implication (iii) $\implies$ (iv) is clear.

(iv) $\implies$ (i). Let $\mathcal{A} = (Q, i, T)$ be trim, let $w \in A^*$ be such that $\text{rank}_\mathcal{A}(w) = 1$. There exists a path $p \xrightarrow{w} q$ in $\mathcal{A}$, and since $\mathcal{A}$ is trim, p is accessible and q is coaccessible. Thus there are words z, y such that $x = zwy \in X^*$. We show that x is a synchronizing word for X.

Let indeed u, v be words such that $uxv \in X^*$. Then $i \cdot ux$ is defined and since x has rank 1, $i \cdot ux = i \cdot x$. Thus $i \cdot ux \in T$ and $ux \in X^*$. □

Two states p, q are said to be *synchronizable* if there exists a word w such that $\text{Card}\{p \cdot w, q \cdot w\} = 1$. The next result is the basis of an algorithm for computing a synchronizing word (see Exercise 3.6.2).

Proposition 3.6.5 *Let $\mathcal{A}$ be a strongly connected deterministic automaton for which there is a word of finite nonnull rank. Then $\mathcal{A}$ is synchronized if and only if any two states of $\mathcal{A}$ are synchronizable.*

Proof. Let Q be the set of states of $\mathcal{A}$. Assume first that $\mathcal{A}$ is synchronized. Let x be a word of rank 1, and let r, s be two states in Q such that $r \cdot x = s$. Let p, q be a pair of states in Q. Since $\mathcal{A}$ is strongly connected, there exists a word y such that

$p \cdot y = r$, whence $p \cdot yx = s$. If $q \cdot yx$ is defined, then it is equal to s, thus p and q are synchronizable.

Conversely, let x be a word of minimal nonzero rank in $\mathcal{A}$. By assumption, this rank is finite. We prove that Card $Q \cdot x = 1$. Assume that there exist $p, q \in Q \cdot x$ with $p \neq q$. Since p and q are synchronizable, there is a word y such that Card$\{p \cdot y, q \cdot y\} = 1$. Then $0 < \text{rank}_A(xy)$ because $p \cdot y$ or $q \cdot y$ is nonempty. Next, $\text{rank}_A(xy) < \text{rank}_A(x)$ because $p \neq q$, a contradiction with the minimality of the rank of the word x. This shows that Card $Q \cdot x = 1$ and thus that $\mathcal{A}$ is synchronized. $\qquad\square$

Proposition 3.6.6 *Let X be a thin maximal prefix code over A, and let $P = XA^-$. Then X is synchronized if and only if for all $p \in P$, there exists $x \in X^*$ such that $px \in X^*$.*

Proof. The condition is necessary. Indeed, let $x \in X^*$ be a synchronizing word for X. Then it follows from Equation (3.21) that $Px \subset X^*$.

The condition is also sufficient. Let $\mathcal{A} = (P, 1, 1)$ be the literal automaton of X^*. The automaton is complete because X is maximal. Since X is thin and maximal, the set $\bar{F}(X) \cap X^*$ is nonempty. Let $w \in \bar{F}(X) \cap X^*$. We show that w has finite positive rank. Clearly, $1 \in P \cdot w$, so this set is nonempty. Next, $P \cdot w$ is composed of suffixes of w. Thus it is finite and w has finite rank.

In view of using Proposition 3.6.5, let p, q be two states in P. There exists a word u such that $pu \in X$. Let $r = q \cdot u$. By hypothesis, there is a word x in X^* such that $rx \in X^*$. Thus $p \cdot ux = 1$ and $q \cdot ux = r \cdot x = 1$, showing that p and q are synchronizable. $\qquad\square$

Proposition 3.6.7 *Let X, Y, Z be maximal prefix codes with $X = Y \circ Z$. Then X is synchronized if and only if Y and Z are synchronized.*

Proof. Let $Y \subset B^*$, $X, Z \subset A^*$, and $\beta : B^* \to A^*$ be such that

$$X = Y \circ_\beta Z.$$

First, assume that Y and Z are synchronized, and let $y \in Y^*$, $z \in Z^*$ be synchronizing words. Then $B^*y \subset Y^*$ and $A^*z \subset Z^*$, whence

$$A^*z\beta(y) \subset Z^*\beta(y) = \beta(B^*y) \subset \beta(Y^*) = X^*,$$

showing that $z\beta(y)$ is a synchronizing word for X. Conversely, assume that $A^*x \subset X^*$ for some $x \in X^*$. Then $x \in Z^*$ and $X^* \subset Z^*$; thus, x is also synchronizing for Z. Next, let $y = \beta^{-1}(x) \in Y^*$. Then

$$\beta(B^*y) = Z^*x \subset A^*x \subset X^* = \beta(Y^*).$$

The mapping β being injective, it follows that $B^*y \subset Y^*$. Consequently Y is synchronized. $\qquad\square$

Example 3.6.8 The code $X = (A^2 \setminus b^2) \cup b^2A^2$ is not synchronized, since it decomposes over the code A^2 which is not synchronized (Example 3.6.3). It is also directly clear that a word $x \in X^*$ can never synchronize words of odd length.

Example 3.6.9 For any maximal prefix code Z and $n \geq 2$, the code $X = Z^n$ is not synchronized. Indeed, such a code has the form $X = B^n \circ Z$ for some alphabet B, and B^n is synchronized only for $n = 1$ (Example 3.6.3).

We now give a result on prefix codes which will be generalized when other techniques will be available (Theorem 9.2.1). The present proof is elementary. Recall from Chapter 2 that for a finite code X, the *order* of a letter a is the integer n such that a^n is in X.

The existence of the order of a results from Proposition 2.5.15. Note that for a finite maximal prefix code, it is an immediate consequence of the inclusion $a^+ \subset X^* P$, with $P = XA^-$.

Theorem 3.6.10 *Let $X \subset A^+$ be a finite maximal prefix code. If the orders of the letters $a \in A$ are relatively prime, then X is synchronized.*

Proof. Let $P = XA^-$ and let $\mathcal{A} = (P, 1, 1)$ be the literal automaton of X^*. This automaton is complete since X is maximal prefix. Recall that its action is given by

$$p \cdot a = \begin{cases} pa & \text{if } pa \in P, \\ 1 & \text{if } pa \in X. \end{cases}$$

For all $w \in A^*$, set $Q(w) = P \cdot w$. Then for $w, w' \in A^*$,

$$Q(w'w) \subset Q(w), \quad \text{Card } Q(w'w) \leq \text{Card } Q(w'). \tag{3.22}$$

Observe that for all $w \in A^*$, $\text{Card}(Q(w)) = \text{rank}_{\mathcal{A}}(w)$.

Let $u \in A^*$ be a word such that $\text{Card}(Q(u))$ is minimal. The code X being right complete, there exists $v \in A^*$ such that $w = uv \in X^+$. By (3.22), $\text{Card}(Q(w))$ is minimal. Further $w \in X^+$ implies

$$1 \in Q(w). \tag{3.23}$$

We will show that $\text{Card}(Q(w)) = 1$. This proves the theorem in view of Proposition 3.6.4.

Let $a \in A$ be a fixed letter, and let n be the positive integer such that $a^n \in X$. We define two sets of integers I and K by

$$I = \{i \in \mathbb{N} \mid Q(w)a^i \cap X \neq \emptyset\},$$

$$K = \{k \in \{0, \ldots, n-1\} \mid a^k w \in X^*\}.$$

First, we show that

$$\text{Card } I = \text{Card } Q(w). \tag{3.24}$$

Indeed, consider a word $p \in Q(w) \subset P$. There is an integer i such that $pa^i \in X$, since X is finite and maximal. This integer is unique since otherwise X would not be prefix. Thus there is a mapping which associates to each p in $Q(w)$ the integer i

such that $pa^i \in X$. This is clearly a surjective mapping onto I. We verify that it is also injective. Assume the contrary. Then $pa^i \in X$ and $p'a^i \in X$ for $p, p' \in Q(w)$, $p \neq p'$. This implies $\text{Card}(Q(wa^i)) < \text{Card}(Q(w))$, contradicting the minimality of $\text{Card}(Q(w))$. Thus the mapping is bijective. This proves (3.24). Next set

$$m = \max\{i + k \mid i \in I, k \in K\}.$$

Clearly $m = \max I + \max K \leq \max I + n - 1$. Let

$$R = \{m, m+1, \ldots, m+n-1\}.$$

We shall find a bijection from $I \times K$ onto R. For this, let $r \in R$ and for each $p \in Q(w)$, let

$$v(p) = p \cdot a^r w.$$

Then

$$v(p) = (p \cdot a^r) \cdot w \in P \cdot w = Q(w).$$

Thus $v(Q(w)) \subset Q(w)$ and $v(Q(w)) = (P \cdot w) \cdot a^r w = P \cdot wa^r w = Q(wa^r w)$, thus $v(Q(w)) = Q(w)$ by the minimality of $Q(w)$. Thus v is a bijection from $Q(w)$ onto itself. It follows by (3.23) that there exists a unique $p_r \in Q(w)$ such that $p_r a^r w \in X^*$. Let i_r be the unique integer such that $p_r a^{i_r} \in X$. Such an integer exists because X is a finite maximal prefix code. Then $i_r \in I$ whence $i_r \leq m \leq r$. Set

$$r = i_r + \lambda n + k_r, \tag{3.25}$$

with $\lambda \in \mathbb{N}$ and $0 \leq k_r < n$. This uniquely defines k_r and we have

$$p_r a^r w = (p_r a^{i_r})(a^n)^\lambda (a^{k_r} w).$$

Since $p_r a^{i_r} \in X$ and X^* is right unitary, we have $(a^n)^\lambda (a^{k_r} w) \in X^*$ and also $a^{k_r} w \in X^*$. Thus, $k_r \in K$. The preceding construction defines a mapping

$$R \to I \times K, \quad r \mapsto (i_r, k_r) \tag{3.26}$$

first by determining i_r, then by computing k_r by means of (3.25). This mapping is injective. Indeed, if $r \neq r'$, then either $i_r \neq i_{r'}$, or it follows from (3.25) and from $r \not\equiv r' \bmod n$ that $k_r \neq k_{r'}$.

  We now show that the mapping (3.26) is surjective. Let $(i, k) \in I \times K$, and let $\lambda \in \mathbb{N}$ be such that

$$r = i + \lambda n + k \in R.$$

By definition of I, there is a unique $q \in Q(w)$ such that $qa^i \in X$, and by the definition of K, we have

$$qa^r w \in X^*.$$

Table 3.3 *The transitions of* $A(X^*)$.

Q	1	2	3	4	5	6	7	8	9
a	2	3	1	1	3	8	9	3	1
b	4	6	7	5	1	4	1	5	1

Thus, $q = p_r$, $i = i_r$, $k = k_r$, showing the surjectivity.

It follows from the bijection that

$$n = \text{Card}(R) = \text{Card}(I)\,\text{Card}(K).$$

This in turn implies, by (3.24), that Card $Q(w)$ divides the integer n. Thus Card $Q(w)$ divides the order of each letter in the alphabet. Since these orders are relatively prime, necessarily $\text{Card}(Q(w)) = 1$. The proof is complete. □

Example 3.6.11 Let $A = \{a, b\}$ and let $X = (A^2 \setminus b^2) \cup b^2 A$. The order of A is 2 and the order of b is 3. Thus X is synchronized by Theorem 3.6.10 and indeed the word *abba* is synchronizing.

We will prove later (Section 11.2) the following important theorem.

Theorem 3.6.12 (Schützenberger) *Let X be a semaphore code. Then there exists a synchronized semaphore code Z and an integer d such that*

$$X = Z^d.$$

This result admits Proposition 3.5.19 as a special case. Consider indeed a bifix semaphore code $X \subset A^+$. Then according to Theorem 3.6.12, we have $X = Z^d$ with Z synchronized. The code X being bifix, Z is also bifix (Proposition 3.4.12); but a bifix synchronized code is trivial by Example 3.6.3. Thus, $Z = A$ and $X = A^d$.

Theorem 3.6.12 describes in a simple manner the structure of semaphore codes which are not synchronized.

We may ask whether such a description exists for general maximal prefix codes: is it true that an indecomposable maximal prefix code X is either bifix or synchronized? Unfortunately, it is not the case, even when X is finite, as shown by the following example.

Example 3.6.13 Let $A = \{a, b\}$, and let X be the prefix code with automaton $A(X) = (Q, 1, 1)$ whose transitions are given in Table 3.3. The automaton $A(X^*)$ is complete, thus X is maximal prefix. In fact, X is finite and it is given in Figure 3.27.

To show that X is not synchronized, observe that the action of the letters a and b preserves globally the sets of states

$$\{1, 2, 3\}, \quad \{1, 4, 5\}, \quad \{4, 6, 7\}, \quad \{1, 8, 9\}$$

as shown in Figure 3.28. This implies that X is not synchronized. Assume indeed that $x \in X^*$ is a synchronizing word. Then by definition $A^*x \subset X^*$, whence $q \cdot x = 1$ for all states $q \in Q$. Thus for each three element subset I, we would have $I \cdot x = \{1\}$.

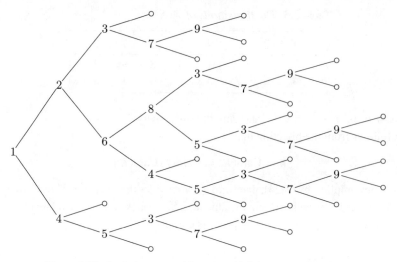

Figure 3.27 An indecomposable code which is not synchronized.

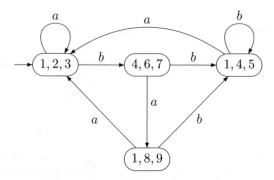

Figure 3.28 The action of the letters a and b.

Further X is not bifix since $b^3, ab^4 \in X$. Finally, the inspection of Figure 3.27 shows that X is indecomposable.

We define a canonical decomposition of a prefix code called its *maximal decomposition*. This is used to show in Chapter 11 that only maximal prefix codes may produce nontrivial groups by composition.

Proposition 3.6.14 *Let $X \subset A^+$ be a prefix code. Let $D = X^*(A^*)^{-1}$ be the set of prefixes of X^*. The set*

$$U = \{u \in A^* \mid u^{-1}D = D\}$$

is a right unitary submonoid of A^*. Let Z be the prefix code generating U. The code X decomposes as

$$X = Y \circ Z \qquad (3.27)$$

where Y is a maximal prefix code.

Proof. Note first that $U \subset D$: Let $u \in U$. Since $1 \in D$, we have $1 \in u^{-1}D$, whence $u \in D$.

The set U is a submonoid. Let indeed $u, v \in U$. Then $(uv)^{-1}D = v^{-1}u^{-1}D = v^{-1}D = D$ showing that $uv \in U$. Assume next that $u, uv \in U$. Then $u^{-1}D = D$, and $v^{-1}D = v^{-1}u^{-1}D = (uv)^{-1}D = D$. Thus U is right unitary.

We have $X^* \subset Z^* = U$. Indeed, X^* is right unitary. Thus for all $x \in X^*, x^{-1}X^* = X^*$. It follows that

$$x^{-1}D = x^{-1}(X^*(A^*)^{-1}) = (x^{-1}X^*)(A^*)^{-1}$$
$$= X^*(A^*)^{-1} = D.$$

We verify that for $u \in U$, there exists $v \in U$ such that $uv \in X^*$. Indeed, let $u \in U$. Then $u \in D$, and therefore $uv \in X^*$ for some $v \in A^*$. Since $X^* \subset U$, we have $u, uv \in U$, and consequently $v \in U$ (U is right unitary). The claim shows that X decomposes over Z. Let Y be such that $X = Y \circ Z$. Then Y is prefix by Proposition 2.6.12. The claim also shows that Y is right complete, hence Y is prefix maximal. $\square$

It can be shown (Exercise 3.6.5) that for any other decomposition $X = Y' \circ Z'$ with Z' prefix and Y' maximal prefix, we have $Z'^* \subset Z^*$. This justifies the name of *maximal decomposition* of the prefix code X given to the decomposition (3.27).

In the case where X is a maximal prefix code, the set D defined above is A^*. Thus $U = A^*$ and $Z = A$ in (3.27). Thus the maximal decomposition, in this case, is trivial.

Example 3.6.15 Let $A = \{a, b\}$ and $X = \{aa, aba, ba\}$. The maximal decomposition of X is $X = Y \circ Z$, with $Y = \{uu, uv, v\} \subset B^+$, $B = \{u, v\}$ and $Z = \{a, ba\}$.

3.7 Recurrent events

The results of Chapter 2 concerning Bernoulli distributions apply of course to prefix codes. However, for these codes, considerable extensions exist in two directions. First, the properties proved in Chapter 2 hold for probability distributions which are much more general than Bernoulli distributions. Second, there exists a remarkable combinatorial interpretation of the average length of a prefix code by means of the sum of the probabilities of its proper prefixes (Proposition 3.7.11).

The following result shows that for prefix codes, Theorem 2.4.5 holds for arbitrary probability distributions.

Proposition 3.7.1 *Let π be a probability distribution on A^*. For any prefix code X, we have $\pi(X) \le 1$.*

Proof. Recall that $A^{[n]}$ denotes the set of words of length at most n. For $x \in X \cap A^{[n]}$, one has $\pi(x) = \pi(xA^{n-|x|})$ by the coherence condition. Next, the sets $xA^{n-|x|}$ for $x \in X \cap A^{[n]}$ are pairwise disjoint because X is prefix. Consequently

$$\sum_{x \in X \cap A^{[n]}} \pi(xA^{n-|x|}) = \pi\left(\bigcup_{x \in X \cap A^{[n]}} xA^{n-|x|} \right) \le \pi(A^n) = 1.$$

It follows that for $n \geq 0$, we have

$$\pi(X \cap A^{[n]}) = \sum_{x \in X \cap A^{[n]}} \pi(x) = \sum_{x \in X \cap A^{[n]}} \pi(x A^{n-|x|}) \leq \pi(A^n) = 1.$$

Thus $\pi(X \cap A^{[n]}) \leq 1$ for all $n \geq 0$. Taking the limit for $n \to \infty$, we obtain $\pi(X) \leq 1$. $\qquad\square$

Proposition 3.7.2 *Let π be a probability distribution on A^*. For any finite maximal prefix code X, we have $\pi(X) = 1$.*

Proof. Let n be greater than the maximal length of the words in X. Since X is maximal, it is right complete, and thus any word of length n has a unique prefix in X. It follows that

$$\pi(X) = \sum_{x \in X} \pi(x) = \sum_{x \in X} \pi(x A^{n-|x|}) = \pi(A^n) = 1. \qquad\square$$

The following computation rule appears to be useful.

Lemma 3.7.3 *Let $X \subset A^+$ be a prefix code. For any probability distribution π on A^* such that $\sum_{x \in X} \pi(x) = 1$, and for any prefix p of a word of X, one has $\pi(p) = \pi(p A^* \cap X)$.*

Proof. Suppose first that $\pi(p) = 0$. Then, using the coherence condition, we obtain that $\pi(x) = 0$ for each $x \in p A^* \cap X$. Thus the conclusion holds. Otherwise, set $Y = p^{-1} X$ and $Z = X \setminus pY$. It is easy to verify that the function ρ defined on A^* by $\rho(u) = \pi(pu)/\pi(p)$ is a probability distribution. Since Y and $Z \cup p$ are prefix codes, we have $\rho(Y) \leq 1$ and $\pi(p) + \pi(Z) \leq 1$, by Proposition 3.7.1. Since $X = pY \cup Z$, we have $1 = \pi(pY) + \pi(Z) \leq \pi(p) + \pi(Z) \leq 1$. Thus $\pi(pY) = \pi(p)$. $\qquad\square$

A *recurrent event* on the alphabet A is a pair composed of a prefix code X on the alphabet A and a probability distribution π on A^* which is multiplicative on X^*, that is such that $\pi(xy) = \pi(x)\pi(y)$ for all $x, y \in X^*$. For example, the pair of a prefix code and a Bernoulli distribution is a recurrent event.

The terminology comes from probability theory. The event considered is the membership in X^* of the prefixes of a word obtained by a succession of trials defining its letters from left to right according to the probability π. A more precise formulation will be given in Chapter 13.

A recurrent event (X, π) is called *persistent* if $\pi(X) = 1$ and *transient* otherwise. In terms of probability, the event is persistent if it occurs at least once with probability 1.

Proposition 3.7.2 shows that (X, π) is persistent whenever X is a finite maximal prefix code.

Example 3.7.4 Let π be a positive Bernoulli distribution on A^* and let X be a thin maximal prefix code. Then (X, π) is persistent by Theorem 2.5.16.

Example 3.7.5 Let D be the Dyck code of Example 2.4.10 and let π be a Bernoulli distribution on $\{a, b\}^*$. Set $p = \pi(a)$ and $q = \pi(b)$. Then $\pi(X) = 1 - |p - q|$. Thus (D, π) is transient when $p \neq q$ and is persistent for $p = q$.

Let $\beta : B \to X$ be a coding morphism for a prefix code X, that is a bijection between a source alphabet B and the code X extended to a injective morphism from B^* into A^*. A persistent recurrent event (X, π) defines a Bernoulli distribution μ on B^* by setting $\mu(b) = \pi(\beta(b))$ for any $b \in B$. Since π is multiplicative on X^*, we then have $\mu(w) = \pi(\beta(w))$ for any $w \in B^*$. The following result shows that conversely, a Bernoulli distribution on the source alphabet defines in a unique way a recurrent event.

Proposition 3.7.6 *Let X be a prefix code and let $\sigma : X \to [0, 1]$ be a mapping such that $\sum_{x \in X} \sigma(x) = 1$. Then there exists a unique probability distribution π on A^* which coincides with σ on X and such that the pair (X, π) is a recurrent event. Moreover, we have $\pi(xw) = \pi(x)\pi(w)$ for any $x \in X^*$ and $w \in A^*$.*

Proof. Let $P = A^* \setminus XA^*$. We first prove the existence of π. For $x_1, \ldots, x_n$ in X and $p \in P$, we set $\pi(x_1 \cdots x_n p) = \sigma(x_1) \cdots \sigma(x_n)\sigma(pA^* \cap X)$. Since $A^* = X^*P$ and the factorization is unambiguous, this defines a function π on A^*. The last two formulas are a direct consequence of the definition, since for $w = yp$ with $y \in X^*$ and $p \in P$, one has $\pi(xw) = \pi(xyp) = \pi(x)\pi(y)\pi(p) = \pi(x)\pi(w)$.

Then π is by definition multiplicative on X^* and coincides with σ on X. We prove now that π satisfies the coherence condition. For any p in P, we have $pA^* \cap X = pAA^* \cap X = \bigcup_{a \in A} paA^* \cap X$ because p is not in X, and thus $\pi(p) = \sigma(pA^* \cap X) = \sum_{a \in A} \sigma(paA^* \cap X) = \sum_{a \in A} \pi(pa)$. This shows that $\pi(w) = \sum_{a \in A} \pi(wa)$ for any $w \in A^*$. This proves that π is a probability distribution.

To prove uniqueness, let π' be a probability distribution such that $\pi'(x) = \sigma(x)$ for all $x \in X$ and which is multiplicative on X^*. Observe first that π and π' coincide on X^* since both are multiplicative on X^* and coincide on X.

Consider a word $w \in A^*$ and let $w = xp$ with $x \in X^*$ and $p \in P$. Let $n \geq 0$ be such that $x \in X^n$. Then, applying Lemma 3.7.3 to the prefix code X^{n+1} and the probability distribution π', we obtain $\pi'(wA^* \cap X^{n+1}) = \pi'(w)$. Since $\pi'(wA^* \cap X^{n+1}) = \pi(wA^* \cap X^{n+1}) = \pi(w)$, we conclude that $\pi(w) = \pi'(w)$. □

Example 3.7.7 Let $A = \{a, b\}$ and $X = \{a, ba\}$. Let $p, q \geq 0$ be such that $p + q = 1$ and let σ be defined by $\sigma(a) = p$ and $\sigma(ba) = q$. The unique probability distribution which is multiplicative on X^* and coincides with σ on X satisfies $\pi(aw) = p\pi(w)$, $\pi(baw) = q\pi(w)$ and $\pi(b^2w) = 0$ for all $w \in A^*$. Note that $\pi(b) = q$ since $\pi(bA^* \cap X) = \pi(ba)$.

Proposition 3.7.8 *For any persistent recurrent event (X, π) over A such that $\pi(x) > 0$ for $x \in X$, there exists a stochastic automaton whose set of states is the set of prefixes of X which defines π.*

Proof. Let Q be the set of proper prefixes of X, and let $\mathcal{A} = (Q, 1, 1)$ be the literal automaton of X^*. We convert it into a weighted automaton (Q, I, T) by setting

$I(1) = 1$ and $I(q) = 0$ for $q \neq 1$ and $T(q) = 1$ for all $q \in Q$. The associated matrix representation is defined by

$$\mu(a)_{p,q} = \begin{cases} \pi(pa)/\pi(p) & \text{if } p \cdot a = q \\ 0 & \text{otherwise.} \end{cases}$$

One has $\sum_{a \in A} \mu(a)_{p,q} = \frac{1}{\pi(p)} \sum_{a \in A} \pi(pa) = 1$ by the coherence condition. Thus the automaton is stochastic. We prove that

$$\mu(w)_{p,q} = \begin{cases} \pi(pw)/\pi(p) & \text{if } p \cdot w = q, \\ 0 & \text{otherwise,} \end{cases}$$

by induction on the length of w. The case of $|w| = 0$ is clear. Next, let $a \in A$ and $w \in A^*$. For $p \in Q$ such that $p \cdot aw$ is defined, set $r = p \cdot a$ and $q = r \cdot w$. Then $\mu(aw)_{p,q} = \mu(a)_{p,r}\mu(w)_{r,q}$. Consequently

$$\mu(aw)_{p,q} = \frac{\pi(pa)}{\pi(p)} \frac{\pi(rw)}{\pi(r)}.$$

If $r \neq 1$, one has $r = pa$ and $\mu(aw)_{p,q} = \pi(paw)/\pi(p)$. If $r = 1$, then $pa \in X$ and $\mu(aw)_{p,q} = \pi(pa)\pi(w)/\pi(p)$. Since $\pi(pa)\pi(w) = \pi(paw)$ by Proposition 3.7.6, the formula holds also in this case. It follows that

$$(|\mathcal{A}|, w) = I\mu(w)T = \sum_{q \in Q} \mu(w)_{1,q} = \mu(w)_{1, 1 \cdot w} = \pi(w). \qquad \square$$

Example 3.7.7 (*continued*) The probability distribution π is defined by the matrices

$$\mu(a) = \begin{bmatrix} p & q \\ 1 & 0 \end{bmatrix}, \quad \mu(b) = \begin{bmatrix} 0 & q \\ 0 & 0 \end{bmatrix}.$$

Let (X, π) be a recurrent event on the alphabet A. Recall from Chapter 1 that $F_X(t) = \sum_{n \geq 0} \pi(X \cap A^n)t^n$ and $F_{X^*}(t) = \sum_{n \geq 0} \pi(X^* \cap A^n)t^n$ are the probability generating series of X and of X^*. The next result has been proved for arbitrary codes in Chapter 2 (Proposition 2.4.3) in the case of Bernoulli distributions.

Proposition 3.7.9 *For any recurrent event (X, π), one has*

$$F_{X^*}(t) = \frac{1}{1 - F_X(t)}.$$

Proof. Since the sets X^k for $k \geq 0$ are pairwise disjoint, $F_{X^*}(t) = \sum_{n \geq 0} \pi(X^* \cap A^n)t^n = \sum_{n \geq 0} \sum_{k \geq 0} \pi(X^k \cap A^n)t^n$. It follows that $F_{X^*}(t) = \sum_{k \geq 0} \sum_{n \geq 0} \pi(X^k \cap A^n)t^n = \sum_{k \geq 0} F_{X^k}(t)$. Since π is multiplicative on X^*, one has $\pi(X^n) = \pi(X)^n$, and it follows that $F_{X^n}(t) = F_X(t)^n$, by the same argument as in the proof of Proposition 2.4.3. Thus $F_{X^*}(t) = \sum_{n \geq 0} F_X(t)^n$. This implies the formula. $\qquad \square$

Given a set K of words and a probability distribution π such that $\pi(K) = 1$, the *average length* of K with respect to π is defined by

$$\lambda(K) = \sum_{x \in K} |x| \pi(x).$$

It is a nonnegative real number or infinite. The context always indicates which is the underlying probability distribution. We therefore omit the reference to it in the notation.

The quantity $\lambda(K)$ is in fact the mean of the random variable assigning to each $x \in K$ its length $|x|$.

Since $\lambda(K) = \sum_{n \geq 0} n\pi(K \cap A^n)$ we have the following useful formula for persistent events.

Proposition 3.7.10 *Let* (X, π) *be a persistent recurrent event. Then*

$$\lambda(X) = F_X'(1).$$

$\square$

Proposition 3.7.11 *Let* (X, π) *be a persistent recurrent event and let* $P = XA^-$ *be the set of proper prefixes of elements of* X. *Then* $\lambda(X) = \pi(P)$.

Proof. By Proposition 3.7.6, for each $p \in P$ we have $\pi(pA^* \cap X) = \pi(p)$. Then we have

$$\pi(P) = \sum_{p \in P} \pi(pA^* \cap X) = \sum_{x \in X} \sum_{p < x} \pi(x) = \sum_{x \in X} \pi(x)|x|,$$

the last equality resulting from the fact that each term $\pi(x)$ appears exactly $|x|$ times in the sum. $\square$

Corollary 3.7.12 *Let* X *be a finite maximal prefix code and* $P = XA^-$. *For any probability distribution* π *on* A^*, *one has* $\lambda(X) = \pi(P)$.

Proof. This follows from the preceding proposition and Proposition 3.7.2. $\square$

For a Bernoulli distribution, the finiteness condition can be replaced by the condition to be thin.

Corollary 3.7.13 *Let* X *be a thin maximal prefix code, and* $P = XA^-$. *For any positive Bernoulli distribution* π *on* A^*, *the recurrent event* (X, π) *is persistent and one has* $\lambda(X) = \pi(P)$. *Further, the average length* $\lambda(X)$ *is finite.*

Proof. The code X being maximal, Theorem 2.5.16 shows that $\pi(X) = 1$. Thus, (X, π) is persistent and the equality $\lambda(X) = \pi(P)$ follows from Proposition 3.7.11. Moreover, P is thin since each factor of a word in P is also a factor of a word in X. By Proposition 2.5.12, $\pi(P)$ is finite. $\square$

We shall see in Chapter 13 that the average length is still finite in the more general case of thin maximal codes.

Example 3.7.14 Let $A = \{a, b\}$ and $X = a^*b$. Let π be a positive Bernoulli distribution. Then $\lambda(X) = \pi(a^*) = 1/\pi(b)$.

Example 3.7.15 Let D be the Dyck code over $A = \{a, b\}$ (see Example 2.4.10). We have seen that for a uniform Bernoulli distribution, one has

$$F_D(t) = 1 - \sqrt{1 - t^2}.$$

We have

$$F'_D(t) = \frac{2t}{\sqrt{1 - t^2}}.$$

Thus, for a uniform Bernoulli distribution, the Dyck code defines a persistent recurrent event but the average length is infinite.

Example 3.7.16 Recall from Example 3.4.4 that the Golomb–Rice code of order k is given by the regular expression

$$GR_k = 1^*0(0 + 1)^k. \tag{3.28}$$

For the Bernoulli distribution π with $\pi(0) = p$ and $\pi(1) = q$, the corresponding probability generating series is $F_{GR_k}(t) = \sum_{n \geq 0} \frac{pt^{k+1}}{1 - qt}$. Thus $\pi(GR_k) = F_{GR_k}(1) = 1$. The average length can be computed directly as $F'_{GR_k}(1) = k + 1/p$. One may also obtain this value by computing $\pi(P)$, where P is the set of proper prefixes of GR_k. One has $P = 1^* \cup 1^*0\left(\bigcup_{0 \leq i < k}\{0, 1\}^i\right)$. Since $\pi(1^*) = 1/p$ and $\pi(1^*0) = 1$, one has $\pi(P) = 1/p + \sum_{0 \leq i < k} \pi(1^*0)\pi(\{0, 1\}^i) = 1/p + k$.

We now consider the computation of the average length of semaphore codes. We start with an interesting identity.

Proposition 3.7.17 *Let $X \subset A^+$ be a semaphore code, $P = XA^-$ and let S be the minimal set for which $X = A^*S \setminus A^*SA^+$. For $s, t \in S$, let*

$$X_s = X \cap A^*s, \qquad R_{s,t} = \{w \in A^* \mid sw \in A^*t \text{ and } |w| < |t|\}.$$

Then, for all $t \in S$,

$$\underline{P}t = \sum_{s \in S} \underline{X_s R_{s,t}}. \tag{3.29}$$

Proof. First, we observe that each product $X_s R_{s,t}$ is unambiguous, since X_s is prefix. Further any two terms of the sum are disjoint, since $X = \bigcup X_s$ is prefix. Thus, it suffices to show that

$$Pt = \bigcup_{s \in S} X_s R_{s,t}.$$

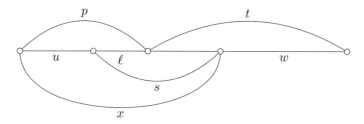

Figure 3.29 Factorizations of pt.

First let $p \in P$, and let x be the shortest prefix of pt which is in A^*S. Then $x \in X$ and

$$pt = xw$$

for some $w \in A^*$. Next $x \in X_s$ for some $s \in S$. Set $x = us$. The word p being in P we have $|p| < |x|$, whence $|w| < |t|$ (see Figure 3.29). Now p cannot be a proper prefix of u, since otherwise s would be a proper factor of t, contradicting Proposition 3.5.11 and the minimality of S. Thus, u is a prefix of p and $sw \in A^*t$, showing that $w \in R_{s,t}$.

Conversely, let $x \in X_s$ and $w \in R_{s,t}$ for some $s, t \in S$. Then $x = us$ and $sw = \ell t$ for a proper prefix ℓ of s. Then $u\ell$ is a proper prefix of $us = x$; thus, $u\ell \in P$ and $xw = u\ell t \in Pt$. □

Corollary 3.7.18 *With the notation of Proposition 3.7.17, we have for any Bernoulli distribution π, the following system of equations:*

$$\lambda(X)\pi(t) = \sum_{s \in S} \pi(X_s)\pi(R_{s,t}), \quad (t \in S), \tag{3.30}$$

$$\sum_{s \in S} \pi(X_s) = 1. \tag{3.31}$$

Proof. Equation (3.30) follows from Equation (3.29) by applying π to both sides and observing that $\lambda(X) = \pi(P)$. Equation (3.31) comes from the fact that X is a disjoint union of the codes X_s and is itself a thin maximal code. □

In the case of a finite set S, the system (3.30) and (3.31) is a set of $1 + \text{Card}(S)$ linear equations in the $1 + \text{Card}(S)$ unknown variables $\pi(X_s)$ and $\lambda(X)$. This gives a method to compute $\lambda(X)$. In the special case where S is a singleton, we get

Corollary 3.7.19 *Let $s \in A^+$, let $X = A^*s \setminus A^*sA^+$ and $R = \{w \in A^* \mid sw \in A^*s$ and $|w| < |s|\}$. Then for any positive Bernoulli distribution π, we have*

$$\lambda(X) = \pi(R)/\pi(s). \qquad \square$$

Example 3.7.20 Let $A = \{a, b\}$ and consider $s = aba$. The corresponding set R is $R = \{1, ba\}$. Setting $p = \pi(a)$ and $q = \pi(b) = 1 - p$, we get for $X = A^*aba \setminus$

$A^* aba A^+$

$$\lambda(X) = \frac{1 + pq}{p^2 q}.$$

Now, choose $s' = baa$. The corresponding R' is the set $R' = \{1\}$. Thus, for $X' = A^* baa \setminus A^* baa A^+$, we have

$$\lambda(X) = \frac{1}{qp^2}.$$

For $p = q = 1/2$, this gives $\lambda(X) = 10$, $\lambda(X') = 8$. This is an interesting paradox: we have to wait longer for the first appearance of aba than for the first appearance of baa!

3.8 Length distributions

Let X be a prefix code on the alphabet A with k letters. Let $f_X(z) = \sum_{n \geq 0} u_n z^n$ with $u_n = \mathrm{Card}(X \cap A^n)$. Recall that the sequence (u_n) is the *length distribution* of X and f_X is the *generating series* of X.

By Theorem 2.4.12, one has $f_X(1/k) = \sum_{n \geq 0} u_n k^{-n} \leq 1$. Conversely, if $u(z) = \sum_{n \geq 0} u_n z^n$ is a series with nonnegative coefficients then, in view of Theorem 2.4.12, if $u(1/k) \leq 1$, there exists a prefix code X on k letters such that $u(z) = f_X(z)$.

If X is a thin maximal prefix code, then by Theorem 2.5.16, $f_X(1/k) = 1$. Conversely, if $u(z) = \sum_{n \geq 0} u_n z^n$ is a series with nonnegative coefficients, and $u(1/k) = 1$, then there exists a prefix code X on k letters such that $f_X(z) = u(z)$. This code is clearly a maximal code, hence a maximal prefix code.

Example 3.8.1 It follows from Formula (3.9) that the generating series of the Golomb–Rice code of order k is

$$f_{GR_k}(z) = \frac{2^k z^{k+1}}{1 - z} = \sum_{i \geq k+1} 2^k z^i.$$

Let X be a rational prefix code. The generating series $f_X(z)$ is $\mathbb{N}$-rational by Proposition 1.10.11. The following statement proves the converse.

Theorem 3.8.2 *A series $u(z) = \sum_{n \geq 0} u_n z^n$ is the generating series of a rational prefix code on k letters if and only if it is $\mathbb{N}$-rational, $u_0 = 0$ and it satisfies the inequality $u(1/k) \leq 1$.*

The conditions are obviously necessary. To prove that they are sufficient, we prove several intermediary results. We assume from now on that u is an $\mathbb{N}$-rational series and that $u(1/k) \leq 1$. Since $u_0 = 0$, there is a normalized weighted automaton recognizing u by Proposition 1.10.10. We assume that u is not the null series.

The following lemma is the first step of the proof.

Lemma 3.8.3 *If $\mathcal{A} = (Q, i, t)$ is a normalized weighted automaton recognizing u, the adjacency matrix of $\mathcal{A}$ has a k-approximate eigenvector w which is positive and such that $w_i = w_t$.*

Proof. Let $\mathcal{A} = (Q, i, t)$ be a normalized weighted automaton recognizing u. Let $\overline{\mathcal{A}}$ be the weighted automaton on the set of states $\overline{Q} = Q \setminus t$ obtained by merging i and t. Let M be the adjacency matrix of $\mathcal{A}$ and let $\overline{M}$ be the adjacency matrix of $\overline{\mathcal{A}}$. Since $\mathcal{A}$ is trim, $\overline{M}$ is irreducible. By Proposition 1.10.12, $(\overline{Q}, i, i)$ recognizes $u^*(z) = 1/(1 - u(z))$. Since $u(1/k) \leq 1$, the radius of convergence ρ of u^* satisfies $\rho \geq 1/k$. By Proposition 1.10.14 the spectral radius λ of $\overline{M}$ is $1/\rho$. Thus $\lambda \leq k$ and by Proposition 1.9.6, there is a positive k-approximate eigenvector $\overline{w}$ of $\overline{M}$. Let w be the Q-vector defined by $w_q = \overline{w}_q$ for every $q \neq t$ and $w_t = \overline{w}_i$. By definition $w_i = \overline{w}_i = w_t$. Let us show that w is a positive k-approximate eigenvector of M. We have to prove that $\sum_{q \in Q} M_{pq} w_q \leq k w_p$ for all $p \in Q$. Since $\mathcal{A}$ is normalized, $M_{p,i} = 0$ for all $p \in Q$. Next, for $p \in \overline{Q}$, we have

$$\sum_{q \in Q} M_{pq} w_q = \sum_{q \in Q \setminus \{i, t\}} M_{pq} w_q + M_{pt} w_t = \sum_{q \in \overline{Q} \setminus i} \overline{M}_{pq} \overline{w}_q + \overline{M}_{pi} \overline{w}_i$$

$$= \sum_{q \in \overline{Q}} \overline{M}_{pq} \overline{w}_q \leq k \overline{w}_p = k w_p.$$

Moreover, since $M_{tq} = 0$ for all $q \in Q$, the inequality holds trivially for $p = t$ because $w_t \geq 0$. □

We will use the following two combinatorial lemmas of some independent interest. These will be used in the proof of Lemma 3.8.6. For a Q-vector $x = (x_q)_{q \in Q}$, we denote by $d(x)$ the sum of its coefficients $d(x) = \sum_{q \in Q} x_q$ and for two Q-vectors $x = (x_q)_{q \in Q}$ and $y = (y_q)_{q \in Q}$, we denote by $x \cdot y$ their scalar product defined by $x \cdot y = \sum_{q \in Q} x_q y_q$.

The first combinatorial lemma is a variant of the pigeon-hole principle.

Lemma 3.8.4 *For any integer $m \geq 1$ and any Q-vectors $z, w \in \mathbb{N}^Q$ such that $d(z) = m$, there is a Q-vector z' such that $0 < z' \leq z$ and $z' \cdot w \equiv 0 \mod m$.*

Proof. Since $d(z) = m$, there exists a sequence $x^{(1)}, x^{(2)}, \ldots, x^{(m)}$ of Q-vectors such that $0 < x^{(1)} < x^{(2)} < \cdots < x^{(m)} = z$. Indeed, this is clear if $m = 1$. Assume $m > 1$. There exists an index k such that $z_k > 0$. Define a Q-vector u by $u_i = z_i$ for $i \neq k$ and $u_k = z_k - 1$. Then $d(u) = m - 1 \geq 1$, and by induction there exists a sequence $x^{(1)}, x^{(2)}, \ldots, x^{(m-1)}$ of Q-vectors such that $0 < x^{(1)} < x^{(2)} < \cdots < x^{(m-1)} = u$. Setting $x^{(m)} = z$, we obtain the desired sequence because $u < z$.

Consider the sequence $x^{(1)}, x^{(2)}, \ldots, x^{(m)}$. If all residues $x^{(i)} \cdot w$ modulo m are distinct, then there is an index i with $1 \leq i \leq m$ such that $x^{(i)} \cdot w \equiv 0 \mod m$. In this case, we set $z' = x^{(i)}$. Otherwise, there exist indices i, j with $1 \leq i < j \leq m$

such that $x^{(i)} \cdot w \equiv x^{(j)} \cdot w \bmod m$. In this case, we set $z' = x^{(j)} - x^{(i)}$. Observe that $0 < z' < x^{(j)} \leq z$. Consequently, in both cases, $z \geq z' > 0$ and $z' \cdot w \equiv 0 \bmod m$.

$\square$

Lemma 3.8.5 *For any integer* $m \geq 1$ *and* $y, w \in \mathbb{N}^Q$, *there exist* $n \geq 0$ *and* $n + 1$ *vectors* $v^{(0)}, v^{(1)}, \ldots, v^{(n)} \in \mathbb{N}^Q$ *such that* $y = \sum_{j=0}^{n} v^{(j)}$, *with*

(i) $d(v^{(j)}) \leq m$ *for* $0 \leq j \leq n$, *and*
(ii) $v^{(j)} \cdot w \equiv 0 \bmod m$ *for* $1 \leq j \leq n$.

Proof. We proceed by induction on $d(y)$. If $d(y) \leq m$, then the properties hold with $n = 0$ and $v^{(0)} = y$. Indeed condition (ii) is vacuous for $n = 0$. Otherwise, we write $y = z + y'$ with $d(z) = m$. By Lemma 3.8.4, there is a Q-vector z' such that $0 < z' \leq z$ and $z' \cdot w \equiv 0 \bmod m$. We write $z = z' + s$. Then $y = z' + y''$ with $y'' = s + y'$. Since $z' > 0$, we have $d(y'') < d(y)$ and we can apply the induction hypothesis to y''. The set of vectors for y'' together with z' gives the desired result for y since $d(z') \leq d(z) \leq m$.

$\square$

Lemma 3.8.6 *There exists a normalized weighted automaton* $\mathcal{A} = (Q, i, t)$ *recognizing* u *such that the adjacency matrix of* $\mathcal{A}$ *has a positive* k-*approximate eigenvector* w *satisfying* $w_i = w_t = 1$.

Proof. We start with a normalized weighted automaton $\mathcal{A} = (Q, i, t)$ recognizing u. Let M be the adjacency matrix of $\mathcal{A}$. By Lemma 3.8.3, there is a positive k-approximate eigenvector w of M such that $w_i = w_t$. Set $m = w_i = w_t$. Let I be the characteristic Q-vector of i defined by $I_i = 1$ and $I_q = 0$ for $q \neq i$ and let T be the characteristic Q-vector T of t, defined similarly. Let $K = \{r \in \mathbb{N}^Q \mid d(r) \leq m, r_t = 0\}$, and let $R = K \cup \{T\}$. Since $i \neq t$, and $d(I) = 1$, the vector I is in K.

We define a weighted automaton $\mathcal{B} = (R, I, T)$ by defining its adjacency matrix N as follows.

Consider r in R and set $z = rM$ and $y = z - z_t T$. Thus $y_t = 0$. We apply Lemma 3.8.5 to the pair of vectors y, w, where w and $m = w_i = w_t$ are as defined above. The lemma gives a decomposition $y = \sum_{j=0}^{n} v^{(j)}$, where each $v^{(j)}$ is in K because $y_t = 0$. We set

$$N_{r,s} = \begin{cases} \text{Card}\{j \mid 0 \leq j \leq n \text{ and } v^{(j)} = s\} & \text{if } s \neq T, \\ z_t & \text{otherwise.} \end{cases}$$

Since $rM = y + z_t T$, we have

$$rM = \sum_{s \in R} N_{r,s} s. \tag{3.32}$$

Note that whenever $N_{r,s} \neq 0$ in the right-hand side, then $s \cdot w \equiv 0 \bmod m$ except possibly for one value of s for which $N_{r,s} = 1$, corresponding to the vector $v^{(0)}$. Indeed, this is true for $s \neq T$ by condition (ii) of Lemma 3.8.5, and it holds also for $s = T$ since $T \cdot w = w_t = m$.

We will verify that $\mathcal{B}$ recognizes u and that its adjacency matrix N has a positive k-eigenvector w' satisfying $w'_I = w'_T = 1$.

Let U be the $R \times Q$-matrix defined by $U_{r,q} = r_q$ for $q \in Q$. Thus the row of index r of U is the Q-vector r itself. It follows that for each Q-vector z, one has $(Uz)_r = \sum_{q \in Q} U_{r,q} z_q = r \cdot z$. Observe also that by construction $UM = NU$, since the row of index r in UM is rM, and $(NU)_{r,p} = \sum_{s \in R} N_{r,s} U_{s,p} = \sum_{s \in R} N_{r,s} s_p = (rM)_p$ by (3.32), showing that the row of index r in NU is rM.

Let I' (resp. T') be the characteristic R-vector of the state I (resp. of the state T). We obtain, considering I, I' as row vectors and T, T' as column vectors the equalities $I'U = I$ and $UT = T'$. Indeed, $(I'U)_p = \sum_{r \in R} I'_r U_{r,p} = I'_I U_{I,p} = U_{I,p} = I_p$, and for $r \in R$, $(UT)_r = \sum_{p \in Q} U_{r,p} T_p = U_{r,t} = r_t$. This shows that $UT = T'$ since $r_t = 0$ for all $r \in R$ except for $r = T$.

Since $UM^n = N^n U$ for all $n \geq 1$, we have

$$u_n = IM^nT = I'UM^nT = I'N^nUT = I'N^nT'.$$

This shows that u is recognized by $\mathcal{B}$. We also have $NUw = UMw \leq kUw$ and thus $w' = Uw$ is a k-approximate eigenvector of N. Note that $w'_I = w'_T = m$. Indeed,

$$w'_I = I' \cdot w' = I' \cdot Uw = I'U \cdot w = I \cdot w = w_i,$$

and, since the row of index T of U is the Q-vector T,

$$w'_T = (Uw)_T = T \cdot w = w_t.$$

For each $r \in R$, we have

$$\sum_{s \in R} N_{r,s} w'_s \leq kw'_r.$$

Since $w'_s = (Uw)_s = s \cdot w$, we have $w'_s \equiv 0 \mod m$ for all s except possibly for one index s_0 for which $N_{r,s_0} = 1$. We rewrite the inequality as

$$\sum_{s \neq s_0} N_{r,s} w'_s + N_{r,s_0} w'_{s_0} \leq kw'_r.$$

Dividing both sides by m gives

$$\sum_{s \neq s_0} N_{r,s} w'_s/m + N_{r,s_0} w'_{s_0}/m \leq kw'_r/m.$$

Taking the ceiling of both sides gives

$$\left\lceil \sum_{s \neq s_0} N_{r,s} w'_s/m + N_{r,s_0} w'_{s_0}/m \right\rceil \leq \lceil kw'_r/m \rceil.$$

Since on the left-hand side, all terms are integers except possibly the last one, and since $N_{r,s_0} = 1$, this implies

$$\sum_{s \neq s_0} N_{r,s} w'_s/m + N_{r,s_0} \lceil w'_{s_0}/m \rceil \leq \lceil kw'_r/m \rceil \leq k \lceil w'_r/m \rceil.$$

This shows that the vector w'' defined by $w''_r = \lceil w'_r/m \rceil$ is a positive k-approximate eigenvector such that $w''_i = w''_{t'} = 1$.   □

Proof of Theorem 3.8.2. We first show that there exists a normalized weighted automaton recognizing u such that each state has at most k outgoing edges.

According to Lemma 3.8.6, we start with a normalized weighted automaton $\mathcal{A} = (Q, i, t)$ recognizing u with state set Q such that the adjacency matrix M of $\mathcal{A}$ has a positive k-approximate eigenvector w with $w_i = w_t = 1$. We are going to define a weighted automaton $\mathcal{A}' = (R, i', t')$ by its adjacency matrix N. This matrix will have the property that there exists a nonnegative matrix U such that

$$MU = UN.$$

By construction, the sum of each row of the matrix N will be at most k.

The set R contains w_q copies of each state q in Q. Since $w_i = 1$, the set R contains only one copy of the initial state i. Formally, R is the set of pairs (q, j) for $q \in Q$ and $1 \leq j \leq w_q$. For given $p, q \in Q$, we define $N_{(p,i),(q,j)}$ for $1 \leq i \leq w_p$ and $1 \leq j \leq w_q$ in the following way.

For $p \in Q$, let $X(p) = \{(q, j, m) \mid q \in Q, 1 \leq j \leq w_q, 1 \leq m \leq M_{p,q}\}$. Thus $X(p)$ contains $M_{p,q}$ copies of each state $(q, j) \in R$. The set $X(p)$ has by definition $\sum_{q \in Q} M_{p,q} w_q$ elements. Since $\sum_{q \in Q} M_{p,q} w_q \leq kw_p$, we may partition the set $X(p)$ into w_p sets $X_{p,1}, \ldots, X_{p,w_p}$ having each at most k elements. We denote by $X_{p,\ell,q,j}$ the subset of the set $X_{p,\ell}$ composed of the elements of the form (q, j, m) for some m. We then define $N_{(p,\ell),(q,j)} = \mathrm{Card}(X_{p,\ell,q,j})$. Since N is the adjacency matrix of the automaton under construction, $N_{(p,\ell),(q,j)}$ is the weight of the edge from (p, ℓ) to (q, j). The sum of the weights of the edges going out of each state (p, ℓ) is the cardinality of $X_{p,\ell}$, and thus at most k. Note also that $\sum_{1 \leq \ell \leq w_p} N_{(p,\ell),(q,j)} = M_{p,q}$ since the sum is the number of elements of the set $X(p)$ of the form (q, j, m) for some m, that is precisely $M_{p,q}$.

Define the $Q \times R$-matrix U by $U_{q,(q,j)} = 1$ for $1 \leq j \leq w_q$, the other components being 0. Then we have $MU = UN$. Indeed, $MU_{p,(q,j)} = \sum_{s \in Q} M_{p,s} U_{s,(q,j)} = M_{p,q} U_{q,(q,j)} M_{p,q}$ and $UN_{p,(q,j)} = \sum_{r \in R} U_{p,r} N_{r,(q,j)} = \sum_{1 \leq \ell \leq w_p} U_{p,(p,\ell)} N_{(p,\ell),(q,j)} = M_{p,q}$.

Let $\mathcal{A}' = (R, i', t')$ be the weighted automaton with adjacency matrix N and with $i' = (i, 1)$ and $t' = (t, 1)$. By construction, this automaton is normalized. Then $\mathcal{A}'$ recognizes u. Indeed, let I (resp. T) be the characteristic Q-vector of i (resp. of t). Since the automaton $\mathcal{A}$ recognizes u, we have for $n \geq 0$, $u_n = IM^nT$. Let similarly I' (resp T') be the characteristic R-vector of i' (resp. of t'). By definition of i' and t', we have $IU = I'$ and $T = UT'$. Since $MU = UN$, we have also $M^nU = UN^n$ for all $n \geq 0$ and thus $I'N^nT' = IUN^nT' = IM^nUT' = IM^nT = u_n$.

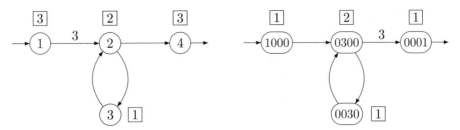

Figure 3.30 A trim normalized weighted automaton of u and the first transformation.

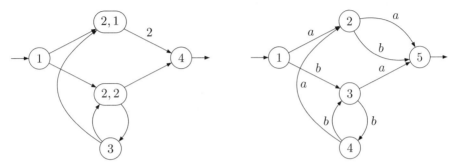

Figure 3.31 The second transformation and the final result.

By construction, the sum on each row of N is at most k and thus $\mathcal{A}'$ satisfies the required property.

We now label the edges going out of each state with different letters. Since there is only one initial state and no edge going out of the terminal state, the automaton obtained recognizes a prefix code with generating series u. □

Example 3.8.7 Let $u(z) = 3z^2/(1 - z^2)$. We have $u(1/2) = 1$. The series u is recognized by the trim normalized weighted automaton of the left of Figure 3.30. The result of the transformation realized in the proof of Lemma 3.8.6 is represented on the right. The coordinates of the 2-approximate eigenvector in both cases is indicated in a square.

We compute only the accessible part of the automaton $\mathcal{B}$. This gives the four vectors shown in the states of the automaton on the right of Figure 3.30. The matrices M, N, and U of the proof of Lemma 3.8.6 are

$$
M = \begin{bmatrix} 0 & 3 & 0 & 0 \\ 0 & 0 & 1 & 1 \\ 0 & 1 & 0 & 0 \\ 0 & 0 & 0 & 0 \end{bmatrix}, \quad
N = \begin{bmatrix} 0 & 1 & 0 & 0 \\ 0 & 0 & 1 & 3 \\ 0 & 1 & 0 & 0 \\ 0 & 0 & 0 & 0 \end{bmatrix}, \quad
U = \begin{bmatrix} 1 & 0 & 0 & 0 \\ 0 & 3 & 0 & 0 \\ 0 & 0 & 3 & 0 \\ 0 & 0 & 0 & 1 \end{bmatrix}.
$$

The second transformation (proof of the theorem) gives the weighted automaton of Figure 3.31 on the left. Note that the state with weight 2 is a split in two states $(2, 1)$ and $(2, 2)$ and that its output is distributed amongst them. The matrices M, N, and U

of the proof are

$$M = \begin{bmatrix} 0 & 1 & 0 & 0 \\ 0 & 0 & 1 & 3 \\ 0 & 1 & 0 & 0 \\ 0 & 0 & 0 & 0 \end{bmatrix}, \quad N = \begin{bmatrix} 0 & 1 & 1 & 0 & 0 \\ 0 & 0 & 0 & 0 & 2 \\ 0 & 0 & 0 & 1 & 1 \\ 0 & 1 & 1 & 0 & 0 \\ 0 & 0 & 0 & 0 & 0 \end{bmatrix}, \quad U = \begin{bmatrix} 1 & 0 & 0 & 0 & 0 \\ 0 & 1 & 1 & 0 & 0 \\ 0 & 0 & 0 & 1 & 0 \\ 0 & 0 & 0 & 0 & 1 \end{bmatrix}.$$

A deterministic labeling gives the automaton represented on the right. It recognizes the regular prefix code $X = (b^2)^*\{aa, ab, ba\}$. A final minimization would merge 1 and 4. The code X is maximal, which is not surprising because $u(1/2) = 1$.

3.9 Optimal prefix codes

Let X be a code over some alphabet A, and assume that each letter $a \in A$ has a *cost* $c(a)$ associated with it. The cost of a word w is by definition the sum of the costs of its letters.

Assume next that each codeword $x \in X$ has a *weight* $p(x)$ associated with it. The *weighted cost* of X is

$$C_X = \sum_{x \in X} p(x)c(x).$$

The *prefix coding problem* is to find a prefix code X with minimal weighted cost, for given weights. In what follows, weights and costs are positive numbers.

As usual, the code X can be viewed through a *coding morphism*, that is a bijection $\beta : B \to X$ for some alphabet B which extends into an injective morphism from B^* into A^*. With this in mind, the weight of a word $x \in C$ is in fact the weight of the letter $b \in B$ such that $x = \beta(b)$. So the weighted cost of X is also

$$C_X = \sum_{b \in B} p(b)c(\beta(b)).$$

In the case where all letters $a \in A$ have equal cost, the cost of a word over A is merely its length. In this case, the prefix coding problem reduces to the construction of a prefix code which minimizes

$$C_X = \sum_{x \in X} p(x)|x|.$$

In the case $\sum_x p(x) = 1$, the number C_X is just the average length of the words of X.

An encoding β which solves the optimal prefix problem for equal letter costs is called a *Huffman encoding*. The following greedy algorithm computes a solution in the binary case in time $O(n \log n)$, and in time $O(n)$ if the weights are available in increasing order. Let $A = \{0, 1\}$, and let $p : B \to \mathbb{R}$ be the weight function.

If B has just one element c, set $\beta(c) = 1$; otherwise, select two elements c_1, c_2 in B of minimal weight, that is such that $p(c_1), p(c_2) \le p(c)$ for all $c \in B \setminus \{c_1, c_2\}$.

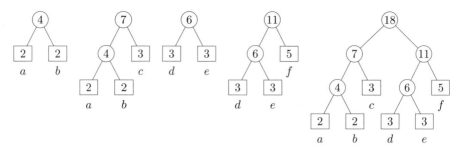

Figure 3.32 Computing an optimal Huffman encoding by combining trees.

Let

$$B' = (B \setminus \{c_1, c_2\}) \cup \{d\},$$

where d is a new symbol not in B, and define $p' : B' \to \mathbb{R}_+$ by $p'(c) = p(c)$ for all $c \neq d$ and $p'(d) = p(c_1) + p(c_2)$.

Let β' be a Huffman encoding of (B', p') and define $\beta : B \to A^*$ by

$$\beta(c) = \beta'(c) \text{ for } c \in B \setminus \{c_1, c_2\}, \quad \beta(c_1) = \beta'(d)0, \quad \beta(c_2) = \beta'(d)1.$$

Let us verify that β is a Huffman encoding of (B, p). For this, we show that there is an optimal encoding β such that $\beta(c_1), \beta(c_2)$ are words of maximal length differing only by the last letter. This will prove the claim.

Consider a prefix code $X = \beta(B)$ such that C_X is minimal. Let $c_1, c_2 \in B$ be letters with lowest weights $p(c_1), p(c_2)$. Let $x, y \in X$ be two words of maximal length which differ only by their last letter. Let $c, d \in B$ be such that $\beta(c) = x$, $\beta(d) = y$. Define the encoding β' derived from β by exchanging the values of c_1, c_2 with the values of c, d, and set $X' = \beta'(B)$. One gets $C_{X'} \leq C_X$ and thus $C_{X'} = C_X$.

Example 3.9.1 Consider the alphabets $B = \{a, b, c, d, e, f\}$ and $A = \{0, 1\}$, and the weights given in the table

	a	b	c	d	e	f
p	2	2	3	3	3	5

The steps of the algorithm are presented in the sequence of trees given in Figure 3.32.

In the case where the letters used for the encoding have *unequal costs*, less is known on the prefix coding problem. The problem is motivated by coding morphisms where different characters may have different transmission times. One example is the *telegraph channel*, in which the dash "–" has twice the cost of a dot ".". Another example is the family of binary *run-length limited* codes, where two consecutive symbols 1 must be separated by at least a and at most b adjacent 0's. In this model, each word $0^k 1$ with $a \leq k \leq b$ may be replaced by a single symbol in a new alphabet, and the cost of this symbol is $k + 1$.

The prefix coding problem with unequal letter costs has been considered mainly in the case where the costs are integers. A special case is known as the *Varn coding*

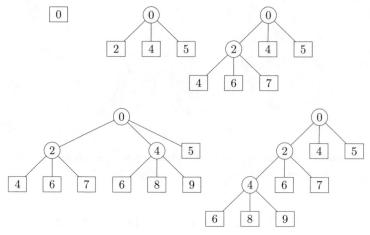

Figure 3.33 Varn's algorithm for 7 words and a 3-letter channel alphabet. At each step, a leaf of minimal cost is replaced by a node with 3 leaves. There are two choices for the last step. Both give an optimal tree.

problem. This is the prefix coding problem when all the weights of the codewords are equal. This problem has an amazingly simple $O(n \log n)$ time solution.

Assume that all n codewords have weight equal to 1. An optimal code minimizes the cost

$$C_X = \sum_{x \in X} c(x),$$

where the cost $c(x)$ is the sum of the costs of its letters, that is

$$c(x) = \sum_{a \in A} c(a)|x|_a.$$

We construct an optimal code over a k-letter alphabet A, assuming that $n = q(k - 1) + 1$ for some integer q. So the prefix code obtained is complete and its tree is complete with q internal nodes and n leaves. The algorithm starts with a tree composed solely of its root, and iteratively replaces the leaf of minimal cost by an internal node which has k leaves, one for each letter. The number of leaves increases by $k - 1$, so in q steps one gets a tree with n leaves.

Example 3.9.2 Assume we are looking for a code with seven words over the ternary alphabet $\{a, b, c\}$, and that the cost for letter a is 2, for letter b is 4, and for letter c is 5. We start with a tree composed of a single leaf, and then build the tree by applying the algorithm. There are two solutions, both of cost 45, given in Figure 3.33. The left tree defines the prefix code $\{aa, ab, ac, ba, bb, bc, c\}$, and the right tree gives the code $\{aaa, aab, aac, ab, ac, b, c\}$.

In order to get complexity $O(n \log n)$ for the construction, the leaves of the tree are managed through a priority queue: then insertion of a leaf is done in $O(\log n)$

operations, and the same time complexity holds for retrieval of a leaf with minimal cost. For a proof of correctness, see Exercise 3.9.2.

VARNCODING()
1   $T \leftarrow$ root
2   (By definition, the cost of the root is 0
3   $Q \leftarrow$ PRIORITYQUEUE()
4   ADD(Q, root)
5   **while** the number of leaves is $\neq n$ **do**
6       $f \leftarrow$ EXTRACTMIN(Q)
7   **for** each $a \in A$ **do**
8           $c \leftarrow$ MAKECHILD(f)
9   cost(c) $\leftarrow$ cost(f) + cost(a)
10   ADD(Q, c)
11 **return** T

A special case of prefix coding is a coding which is compatible with a given ordering of the input alphabet. Consider a coding morphism $\beta : B^* \to A^*$, where A and B are alphabets equipped with an order. Then β is an *ordered coding* or *alphabetic coding* if

$$b < b' \implies \beta(b) < \beta(b'),$$

where the order in A^* is the lexicographic order induced by the order on A. If β is a prefix coding, and if the prefix code $X = \beta(B)$ is viewed as a tree, this means that the leaves of the tree, read from left to right, correspond to the encoding of the input letters in B, read in alphabetic order. Such a tree is called *ordered* or *alphabetic*. The *ordered prefix code problem* is to find an ordered coding that with minimal weighted cost

$$C_X = \sum_{b \in B} p(b)|\beta(b)|,$$

where $p(b)$ is the weight of b.

Example 3.9.3 Consider the alphabet $B = \{a, b, c\}$, with weights $p(a) = p(c) = 1$ and $p(b) = 4$. Figure 3.34 shows on the left an optimal tree for these weights, and on the right an optimal ordered tree. This example shows that Huffman's algorithm does not give the optimal ordered tree.

Example 3.9.4 Consider the sequence of weights $(4, 3, 3, 4)$. An optimal tree is given in Figure 3.35. It shows that in an optimal ordered tree, leaves with minimal weight need not be adjacent.

Let $B = \{b_1, \ldots, b_n\}$ be an ordered alphabet with n letters, and let p_i be the weight of letter b_i. We present an algorithm for computing an optimal ordered tree due to

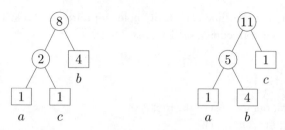

Figure 3.34 Two trees for the given weights. The left tree has weighted cost 8, it is optimal but not ordered. The right tree is ordered and has weighted cost 11.

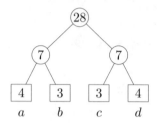

Figure 3.35 The optimal ordered tree for weights (4, 3, 3, 4).

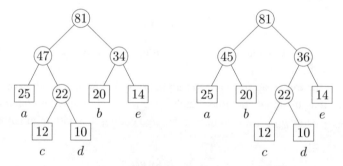

Figure 3.36 The two steps of the algorithm: On the left the unordered tree obtained in the combination phase, and on the right the ordered tree, obtained by recombination.

Garsia and Wachs (see Notes). The idea is to use a variant of Huffman's algorithm by grouping together pairs of elements with minimal weights which are consecutive in the ordering. The algorithm can be implemented to run in time $O(n \log n)$.

The algorithm is composed of three parts. In the first part, called the *combination* part, one starts with the sequence of weights $p = (p_1, \ldots, p_n)$ and constructs an optimal binary tree T' for a permutation $b_{\sigma(1)}, \ldots, b_{\sigma(n)}$ of the alphabet. The leaves, from left to right, have weights $p_{\sigma(1)}, \ldots, p_{\sigma(n)}$. In general, this permutation is not the identity, so the tree is not ordered, see Figure 3.36. Here the number in a node is its weight, that is the sum of the weights of the leaves of its subtree. In the second part, called the *level assignment*, one computes the levels of the leaves. In the last part, called the *recombination* part, one constructs a tree T which has the weights $p_1, \ldots, p_n$ associated to its leaves from left to right, and where each leaf with weight

p_i appears at the same level as in the tree T'. This tree is ordered by construction (see Figure 3.36). Since the leaves have the same level in T and in T', the corresponding codewords have the same length, and therefore the trees T and T' have the same cost. Thus T is an optimal ordered tree.

We now give the details of the algorithm. For ease of description, we introduce the following terminology. A sequence $(p_1, \ldots, p_k)$ of numbers is *2-descending* if $p_i > p_{i+2}$ for $1 \le i \le k - 2$. Clearly a sequence is 2-descending if and only if the sequence of "two-sums" $(p_1 + p_2, \ldots, p_{k-1} + p_k)$ is strictly decreasing.

Let $p = (p_1, \ldots, p_n)$ be a sequence of (positive) weights. We extend it by setting $p_0 = p_{n1} = \infty$. The *left minimal pair* or simply *minimal pair* of p is the pair (p_{k-1}, p_k), where $(p_1, \ldots, p_k)$ is the longest 2-descending chain that is a prefix of p. The index k is the *position* of the pair. In other words, k is the integer such that

$$p_{i-1} > p_{i+1} \ (1 < i < k) \quad \text{and} \quad p_{k-1} \le p_{k+1}.$$

Observe that the left minimal pair can be defined equivalently by the conditions

$$p_{i-1} + p_i > p_i + p_{i+1} \ (1 < i < k) \quad \text{and} \quad p_{k-1} + p_k \le p_k + p_{k+1}.$$

The *target* is the index j with $1 \le j < k$ such that

$$p_{j-1} \ge p_{k-1} + p_k > p_j, \ldots, p_k.$$

Example 3.9.5 For $(14, 15, 10, 11, 12, 6, 8, 4)$, the left minimal pair is $(10, 11)$ and the target is 1, whereas for the sequence $(28, 8, 15, 20, 7, 5)$, the left minimal pair is $(15, 20)$ and the target is 2.

The pair (j, k) composed of the position of the left minimal pair and of its target is called the *scope* of the sequence p. Observe that the sequence $(p_{j-1}, p_{k-1} + p_k, p_j, \ldots, p_{k-2})$ is 2-descending since $p_{j-1} \ge p_{k-1} + p_k > p_j, p_{j+1}$.

The three phases of the algorithm work as follows.

Combination Associate a singleton tree to each weight. Repeat the following steps as long as the sequence of weights has more than one element.

(i) Compute the *left minimal pair* (p_{k-1}, p_k).
(ii) Compute the *target* j.
(iii) Remove the weights p_{k-1} and p_k.
(iv) Insert $p_{k-1} + p_k$ between p_{j-1} and p_j.
(v) Associate to $p_{k-1} + p_k$ a new tree with weight $p_{k-1} + p_k$, and which has, as left and right subtrees, the tree for p_{k-1} and for p_k respectively.

Level assignment Compute, for each letter b in B, the level of its leaf in the tree T'.

Recombination Construct an ordered tree T in which the leaves of the letters have the levels computed by the level assignement.

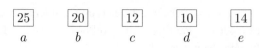

Figure 3.37 The initial sequence of trees.

Figure 3.38 The next two steps.

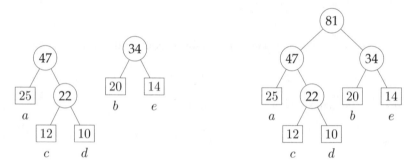

Figure 3.39 The two last steps of the combination part.

Example 3.9.6 Consider the following weights for an alphabet of five letters.

	a	b	c	d	e
p	25	20	12	10	14

The initial sequence of trees is given in Figure 3.37. The left minimal pair is 12, 10, its target is 2, so the leaves for c and d are combined into a tree which is inserted just to the right of the first tree. Now the minimal pair is (20, 14) (there is an infinite weight at the right end), so the leaves for letters b and e are combined, and inserted at the beginning. This gives the two sequences of Figure 3.38.

Next the two last trees are combined and inserted at the beginning as shown on the left of Figure 3.39, and finally, the two remaining trees are combined, as shown on the right.

The tree T' obtained at the end of the first phase is not ordered. The prescribed levels for the letters of the example are:

	a	b	c	d	e
level	2	2	3	3	2

The optimal ordered tree with these levels is given by recombination. It is the tree given on the right of Figure 3.36. The weighted cost of this tree is 184.

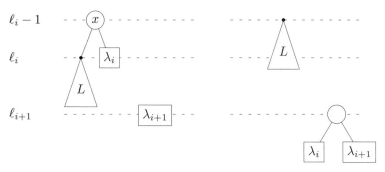

Figure 3.40 Reorganizing leaves in Lemma 3.9.8.

We now give a proof of the algorithm. Let T be some binary tree with n leaves labelled by the letters $b_1, \ldots, b_n$ of the alphabet B, with weights $p_1, \ldots, p_n$. We denote by ℓ_i^T (or simply ℓ_i) the level of the leaf of b_i in T, that is the length of the codeword coding the letter b_i. Each of the partial trees constructed in the algorithm will be identified with its root, considered as a leaf. The leaf corresponding to the letter b_i will be denoted by λ_i.

We first state two simple lemmas.

Lemma 3.9.7 *Let T be some binary tree. If $\ell_i > \ell_{i+1}$, then λ_i is a right leaf. Symmetrically, if $\ell_i < \ell_{i+1}$, then λ_i is a left leaf.*

Proof. Assume indeed that λ_i is a left leaf. Then its right sibling is a tree containing the leaf λ_{i+1}. Thus $\ell_i \le \ell_{i+1}$. ☐

The following statement is a first step to the proof of the correctness of the algorithm.

Lemma 3.9.8 *If $p_{i-1} > p_{i+1}$, then $\ell_i \le \ell_{i+1}$ in every optimal ordered tree. If $p_{i-1} = p_{i+1}$, then $\ell_i \le \ell_{i+1}$ in some optimal ordered tree.*

Proof. Suppose $p_{i-1} \ge p_{i+1}$, and consider a tree T with $\ell_i > \ell_{i+1}$. In this tree, the leaf λ_i is a right child by Lemma 3.9.8, and its left sibling is a tree L with weight $p(L) \ge p_{i-1}$, see Figure 3.40. Build a new tree T' as follows: replace the parent of L by L itself, replace the leaf of λ_{i+1} by a node having as children the leaves λ_i and λ_{i+1}. The difference of the costs is

$$C_{T'} - C_T = -p(L) + p_{i+1} - p_i(\ell_i - \ell_{i+1} - 1) \le p_{i+1} - p_{i-1}$$

because $\ell_i \ge \ell_{i+1} + 1$. If $p_{i-1} > p_{i+1}$, then this expression is < 0 and T is not optimal. If $p_{i-1} = p_{i+1}$ and if T is optimal, then T' is also optimal, and $\ell_i^{T'} = \ell_i^T$. ☐

Observe that the symmetric statement also holds.

Corollary 3.9.9 *If $p_{i-1} < p_{i+1}$, then $\ell_{i-1} \ge \ell_i$ in every optimal ordered tree. If $p_{i-1} = p_{i+1}$, then $\ell_{i-1} \ge \ell_i$ in some optimal ordered tree.*

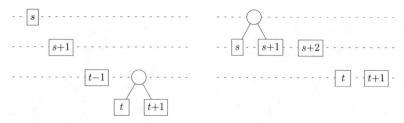

Figure 3.41 Proof of Proposition 3.9.11. On the left before the shift, on the right after the shift.

We use Lemma 3.9.8 in the following form.

Corollary 3.9.10 *If the subsequence* $(p_{j-1}, \ldots, p_k)$ *is 2-descending, then* $\ell_j \leq \cdots \leq \ell_k$ *in every optimal ordered tree.* $\square$

We now show that we always may assume that the minimal tree for a sequence p has some special form. Such a tree will be called *flat*.

Proposition 3.9.11 *Let* (j, k) *be the scope of the sequence* $p = (p_1, \ldots, p_n)$. *There exists a minimal tree for* p *satisfying* $\ell_{k-1} = \ell_k$ *and one of the two conditions*

(a) $\ell_k = \ell_j + 1$ *or*
(b) $\ell_k = \ell_j$ *and* λ_j *is a left leaf.*

Proof. Since the sequence $(p_1, \ldots, p_k)$ is 2-descending (and $p_0 = +\infty$), one has $\ell_1 \leq \ell_2 \leq \cdots \leq \ell_k$ in every minimal tree by Corollary 3.9.10. Next $p_{k-1} \leq p_{k+1}$. If $p_{k-1} < p_{k+1}$ then $\ell_{k-1} \geq \ell_k$ in every minimal tree, and if $p_{k-1} = p_{k+1}$ then $\ell_{k-1} \geq \ell_k$ in some minimal tree. Thus $\ell_{k-1} = \ell_k$ in some minimal tree.

Consider this tree. We prove that $\ell_j = \ell_k$ or $\ell_j = \ell_k - 1$. Assume the contrary. Then $\ell_j \leq \ell_k - 2$. Let s be the greatest index such that $\ell_s \leq \ell_k - 2$. Then $s < k - 1$ because $\ell_{k-1} = \ell_k$. Let t be the smallest index such that $\ell_t = \ell_k$. Then

$$\ell_j \leq \cdots \leq \ell_s < \ell_{s+1} \leq \cdots \leq \ell_{t-1} < \ell_t = \cdots = \ell_k.$$

It is quite possible that $s + 1 = t$. Observe that λ_{s+1} is left leaf by Lemma 3.9.8 because $\ell_s < \ell_{s+1}$. Similarly, λ_t is a left leaf, and λ_t and λ_{t+1} are siblings. We now make the following transformation, see Figure 3.41. Leaf λ_s is replaced by a node with the two siblings λ_s and λ_{s+1}. Each of the leaves $\lambda_{s+2}, \ldots, \lambda_{t-1}$ is shifted to the left. The leaf λ_t replaces λ_{t-1}, and the parent of λ_{t+1} is replaced by λ_{t+1} itself. The extra cost of this transformation is at most $p_s - p_t - p_{t+1}$ because the level of λ_s increases by 1, the level of λ_{s+1} does not increase, the levels of λ_t and λ_{t+1} decrease by 1. Now $p_s - p_t - p_{t+1} \leq p_s - p_{k-1} - p_k$ because $p_t + p_{t+1} \geq p_{k-1} + p_k$ (equality is possible because one might have $t = k - 1$, and the extra cost is < 0 because $j > s$ and therefore $p_s < p_{k-1} + p_k$). This gives a contradiction and shows that $\ell_j \geq \ell_k - 1$.

It remains to consider the case where $\ell_j = \ell_k$. Arguing by contradiction, assume that λ_j is a right leaf. Then, since $\ell_{j-1} \leq \ell_j$, the leaf λ_{j-1} is a left leaf and is the sibling

Figure 3.42 Second transformation in Proposition 3.9.11. Before the transformation on the left, and after the transformation on the right

Figure 3.43 The case $\ell_j = \ell_k$. Before and after the circular shift.

Figure 3.44 The case $\ell_j = \ell_k - 1$: A circular shift. Before and after the first shift.

of λ_j. Then make the following transformation, see Figure 3.42. Replace the common parent of λ_{j-1} and λ_j by λ_{j-1}, shift $\lambda_j, \ldots, \lambda_{k-2}$ one position to the right, and replace the leaf λ_k by a node with children λ_{k-1} and λ_k. Since the leaves $\lambda_{j-1}, \ldots, \lambda_k$ have the same level before the transformation, the extra cost is $-p_{j-1} + p_{k-1} + p_k$. This value is ≤ 0 by the definition of the target. Since the tree was minimal before the transformation, the tree after transformation has the same cost. In this new tree, one has indeed $\ell_k = 1 + \ell_j$. □

A tree T for p is *k-minimal* if it is minimal among all trees where the leaves for p_{k-1} and p_k are siblings.

A *level preserving permutation* σ of tree T is a tree T^σ that has the same leaves as T at the same levels. By definition, the cost of T^σ is equal to the cost of T.

Lemma 3.9.12 *Let $p = (p_1, \ldots, p_n)$ be a sequence of weights with scope (j, k) and let T be an optimal flat tree for p. Let*

$$p' = (p_1, \ldots, p_{j-1}, p_{k-1}, p_k, p_j, p_{j+1}, \ldots, p_{k-2}, p_{k+1}, \ldots, p_n).$$

There exists a level preserving permutation that transforms T into a tree T' for p' such that the leaves for p_{k-1} and p_k are siblings.

Proof. Since T is flat, $\ell_j = \ell_k$ or $\ell_j = \ell_k - 1$. If $\ell_j = \ell_k$, one makes a circular shift of the leaves $\lambda_j, \ldots, \lambda_k$ two positions to the right. Since λ_j was a left child before the shift, the leaves λ_{k-1} and λ_k are siblings after the shift, see Figure 3.43.

If $\ell_j = \ell_k - 1$, let s be such that $\ell_s = \ell_j, \ell_{s+1} = \ell_k$. Then one first makes a circular shift of the leaves $\lambda_{s+1}, \ldots, \lambda_k$ two positions to the right, as before, see Figure 3.44.

Figure 3.45 The case $\ell_j = \ell_k - 1$: Before and after the second shift.

Figure 3.46 The case $\ell_x = \ell_{k-2}$ in Theorem 3.9.13: Before and after the shift.

Then one applies a circular shift, one position to the right, of the sequence $\lambda_j, \ldots, \lambda - s, x$, where x is the parent node of λ_{k-1} and λ_k, see Figure 3.45. This is a transformation that preseves levels of leaves and therefore the resulting tree has the same cost as the tree T we started with.   □

Theorem 3.9.13 *Let $p = (p_1, \ldots, p_n)$ be a sequence of weights with scope (j, k) and let $\widehat{p} = (p_1, \ldots, p_{j-1}, p_{k-1} + p_k, p_j, p_{j+1}, \ldots, p_{k-2}, p_{k+1}, \ldots, p_n)$. Let $\widehat{T}$ be a minimal tree for $\widehat{p}$, and let T' be the tree obtained by substituting a tree with two leaves λ_{k-1} and λ_k to the leaf corresponding to $p_{k-1} + p_k$ in $\widehat{T}$. There exists a minimal tree T for p of cost $c(T) = c(T')$ which is obtained by a level preserving permutation of T'.*

Proof. Let $\widehat{T}$ be an optimal tree for $\widehat{p}$. Since $c(T') = c(\widehat{T}) + p_{k-1} + p_k$, the tree T' is k-minimal for

$$p' = (p_1, \ldots, p_{j-1}, p_{k-1}, p_k, p_j, p_{j+1}, \ldots, p_{k-2}, p_{k+1}, \ldots, p_n).$$

If $j - 1 = k - 2$, then $p' = p$ and there is nothing to prove. Otherwise, observe that sequence

$$p_{j-1}, p_{k-1} + p_k, p_j, p_{j+1}, \ldots, p_{k-2}$$

is a 2-descending factor of the sequence $\widehat{p}$ because $p_{j-1} \geq p_{k-1} + p_k > p_j$ and $p_{k-1} + p_k > p_{j+1}$. Therefore, denoting by x the leaf in $\widehat{T}$ with weight $p_{k-1} + p_k$, one has $\ell_x^{\widehat{T}} \leq \ell_j^{\widehat{T}} \leq \cdots \leq \ell_{k-2}^{\widehat{T}}$ by Corollary 3.9.10. The node x is also the parent node of the leaves for p_{k-1} and p_k in T', and since $\ell^{\widehat{T}} = \ell^{T'}$ for all nodes of $\widehat{T}$, one has $\ell_x \leq \ell_j \leq \cdots \leq \ell_{k-2}$ in T'.

We distinguish two cases. If $\ell_x = \ell_{k-2}$ then one makes the following transformation: the nodes $x, \lambda_j, \ldots, \lambda_{k-2}$ are cyclically permuted one position to the left, giving the nodes $\lambda_j, \ldots, \lambda_{k-2}, x$ and therefore the leaves $\lambda_j, \ldots, \lambda_{k-2}, \lambda_{k-1}, \lambda_k$, see Figure 3.46. The resulting tree S verifies $c(T) = c(T')$ and the permutation is level preserving.

If $\ell_x < \ell_{k-2}$, let s such that $\ell_x = \ell_s < \ell_{s+1}$. Then a first transformation (see Figure 3.47) similar to the previous one but on $x, \ldots, \lambda_s$ gives a tree where

Figure 3.47 The case $\ell_x < \ell_{k-2}$: first transformation. Before the first shift on the left, after this shift on the right.

Figure 3.48 The case $\ell_x < \ell_{k-2}$: second transformation. Before the first shift on the left, after this shift on the right.

the leaf sequence is $\lambda_j \ldots, \lambda_{s-1}, \lambda_{k-1}, \lambda_k, \lambda_{s+1}, \ldots, \lambda_{k-2}$. One has $\ell_{k-1} = \ell_k \le \ell_{s+1}\ell \cdots \le \ell_{k-2}$. A circular permutation by two positions to the left of the leaves $\lambda_{k-1}, \lambda_k, \lambda_{s+1}, \ldots, \lambda_{k-2}$ gives the sequence $\lambda_{s+1}, \ldots, \lambda_{k-2}, \lambda_{k-1}, \lambda_k$, see Figure 3.48.

By Lemma 3.9.14 below, the cost of the resulting tree S is less than the cost of T' unless $\ell_{k-2} = \ell_k$. But in view of Lemma 3.9.12, $c(S)$ cannot be strictly less than $c(T')$. □

Lemma 3.9.14 *Let $m \ge 3$, let $\ell_1 = \ell_2 \le \cdots \le \ell_m$ be integers and let $(p_1, p_2, \ldots, p_m)$ be a 2-descending chain. Set*

$$c = p_{m-1}\ell_1 + p_m\ell_2 + p_1\ell_3 + \cdots + p_{m-2}\ell_m,$$
$$c' = p_1\ell_1 + p_2\ell_2 + \cdots + p_m\ell_m.$$

Then $c' \le c$, and equality holds only if $\ell_m = \ell_1$.

Proof. If $m = 3$, then $c' - c = (p_1 - p_3)(\ell_1 - \ell_3) \le 0$ and indeed $c' = c$ only if $\ell_1 = \ell_3$.

If $m \ge 4$, then

$$c' - c = p_1(\ell_1 - \ell_3) + p_2(\ell_2 - \ell_4) + \cdots + p_{m-2}(\ell_{m-2} - \ell_m)$$
$$+ p_{m-1}(\ell_{m-1} - \ell_1) + p_m(\ell_m - \ell_2).$$

Since $(p_1, p_2, \ldots, p_m)$ is 2-descending, the $m - 2$ first terms of this sum may be grouped and bounded. If m is even

$$c' - c \le p_{m-3}(\ell_1 - \ell_{m-1}) + p_{m-2}(\ell_2 - \ell_m) + p_{m-1}(\ell_{m-1} - \ell_1) + p_m(\ell_m - \ell_2)$$
$$= (p_{m-3} - p_{m-1})(\ell_{m-1} - \ell_1) + (p_{m-2} - p_m)(\ell_m - \ell_2) \le 0$$

and equality holds only if $\ell_{m-1} = \ell_1$ and $\ell_m = \ell_2$, so only if $\ell_1 = \cdots = \ell_m$. Similarly, if m is odd, and because $\ell_1 = \ell_2$, one gets

$$c' - c \le p_{m-2}(\ell_1 - \ell_m) + p_{m-3}(\ell_2 - \ell_{m-1}) + p_{m-1}(\ell_{m-1} - \ell_1) + p_m(\ell_m - \ell_2)$$

$$= (p_{m-3} - p_{m-1})(\ell_1 - \ell_{m-1}) + (p_{m-2} - p_m)(\ell_1 - \ell_m) \le 0$$

Again, equality holds only if $\ell_1 = \cdots = \ell_m$. □

3.10 Exercises

Section 3.1

3.1.1 Let A be a finite alphabet, and let P be a prefix-closed subset of A^*. Show that P is infinite if and only if there exists an infinite sequence $(p_n)_{n \ge 1}$ of elements in P such that

$$p_1 < p_2 < p_3 < \cdots$$

3.1.2 Let A be a finite alphabet of k letters and let $X \subset A^+$ be a prefix code. For $n \ge 1$, let $\alpha_n = \text{Card}(X \cap A^n)$. Show that $\text{Card}(XA^* \cap A^n) = \sum_{i=1}^n \alpha_i k^{n-i}$ and

$$\sum_{n \ge 1} \alpha_n k^{-n} \le 1.$$

(This gives an elementary proof of Corollary 2.4.6 for prefix codes. See also Proposition 3.7.1)

Section 3.2

3.2.1 Let $X \subset A^+$ be a prefix code. Let $P = XA^-$ and let $\mathcal{A} = (P, 1, 1)$ be the literal automaton of X^*. Consider an automaton $\mathcal{B} = (Q, i, i)$ which is deterministic, trim, and such that $X^* = \text{Stab}(i)$. Show that there is a surjective function $\rho : P \to Q$ with $\rho(1) = i$ and such that for $a \in A$, $\rho(p \cdot a) = \rho(p) \cdot a$.

3.2.2 A prefix code X is a *chain* if there exist disjoint nonempty sets Y, Z such that $Y \cup Z$ is prefix and $X = Y^*Z$.

Let X be a nonempty prefix code over A, and let $\mathcal{A}(X) = (Q, i, t)$ be the minimal automaton of X. Show that the following conditions are equivalent:

(i) $\text{Stab}(i) \ne 1$.
(ii) X is a chain.
(iii) There exists a word $u \in A^+$ such that $u^{-1}X = X$.

Section 3.3

3.3.1 Let A be an alphabet, and let $M(A)$ be the monoid of prefix subsets of A^* equipped with the induced product. Show that $M(A)$ is a free monoid and that the set of maximal (resp. recognizable) prefix sets is a right unitary submonoid of $M(A)$. (*Hint*: Use Exercise 2.2.8 and set $\lambda(X) = \min_{x \in X} |x|$.)

Section 3.4

3.4.1 Show that the number of prefix-closed sets with n elements on a k-letter alphabet is

$$\frac{1}{kn+1}\binom{kn+1}{n} = \frac{1}{(k-1)n+1}\binom{kn}{n}.$$

For this, let L be the unique set of words on $\{a, b\}$ such that $L = aL^k \cup b$. Set $\|w\| = (k-1)|w|_a - |w|_b$. Prove that

(i) L is the set of words w such that $\|w\| = -1$ and $\|u\| \geq 0$ for any proper prefix u of w.

(ii) Any word w on $\{a, b\}$ such that $\|w\| = -1$ has exactly one conjugate in the set L.

(iii) There exists a bijection between prefix-closed sets on a k-letter alphabet and words of L.

3.4.2 Let X and Y be finite nonempty subsets of A^* such that the product XY is unambiguous. Show that if XY is a maximal prefix code, then X and Y are maximal prefix codes. (*Hint*: Use the fact that $\pi(X)\pi(Y) = 1$ for any positive Bernoulli distribution on A and use Proposition 2.5.29.)

3.4.3 Let X and Y be two prefix codes over A, and

$$P = A^* \setminus XA^*, \quad Q = A^* \setminus YA^*.$$

Set $R = P \cap Q$. Show that there exists a unique prefix code Z such that

$$Z = RA \setminus R.$$

Show that

$$Z = (X \cap Q) \cup (X \cap Y) \cup (P \cap Y).$$

Show that if X and Y are maximal prefix sets, then so is Z.

3.4.4 Let A be a finite alphabet. Show that the family of recognizable maximal prefix codes is the least family $\mathcal{F}$ of subset of A^* such that

(i) $A \in \mathcal{F}$,

(ii) if $X, Y \in \mathcal{F}$ and if $X = X_1 \cup X_2$ is a partition in recognizable sets X_1, X_2, then

$$Z = X_1 \cup X_2 Y \in \mathcal{F},$$

(iii) if $X \in \mathcal{F}$ and if $X = X_1 \cup X_2$ is a partition in recognizable sets, then

$$Z = X_1^* X_2 \in \mathcal{F}.$$

(*Hint*: Use an induction on the number of edges of the minimal deterministic automaton of an element of $\mathcal{F}$.)

Section 3.5

3.5.1 Let $X \subset A^*$ be a prefix code. Show that the following conditions are equivalent.

(i) $A^*X = X^+$.

(ii) X is a semaphore code, and the minimal set of semaphores $S = X \setminus A^+X$ satisfies
$SA^* \cap A^*S = SA^*S \cup S$.

Note that for a code $X = A^*w \setminus A^*wA^+$, the conditions are satisfied provided w is unbordered.

3.5.2 Let $J \subset A^+$ be a two-sided ideal. For each $x \in J$, denote by $\|x\|$ the greatest integer n such that $x \in J^n$, and set $\|x\| = 0$ for $x \notin J$. Show that, for all $x, y \in A^*$,

$$\|x\| + \|y\| \le \|xy\| \le \|x\| + \|y\| + 1.$$

Section 3.6

3.6.1 Let $X \subset A^+$ be a finite maximal prefix code. Show that if X contains a letter $a \in A$, then there is an integer $n \ge 1$ such that a^n is synchronizing.

3.6.2 Let $\mathcal{A}$ be a complete deterministic automaton with n states. Show that if $\mathcal{A}$ is synchronized, there exists a synchronizing word of length at most n^3 in $\mathcal{A}$.

3.6.3 Let $n \ge 1$ be an integer and let M be the monoid of mappings from $Q = \mathbb{Z}/n\mathbb{Z}$ into itself generated by the two maps a, b defined for $i \in Q$ by $ia = i + 1$ and

$$ib = \begin{cases} j > i + 1 & (0 \le i < n - t), \\ i + 1 & (n - t \le i < n) \end{cases}$$

for some integer t with $1 \le t \le n$. The aim of this exercise is to show that the minimal rank d of the elements of M divides n, and that $ib \equiv i + 1 \bmod d$ for all $i \in Q$.

For each e, f with $0 \le e < f \le n$, let $I_{e,f} = \{e, e + 1, \dots, f - 1\}$ and let $M_{e,f} = \{m \in M \mid Qm = I_{e,f}$ and $im = i$ for all $i \in I_{e,f}\}$.

(a) Show that for each $j \in Q$

$$I_{e,f}a^j = I_{e+j,f+j} \quad \text{and} \quad a^{-j}M_{e,f}a^j = M_{e+j,f+j}.$$

(b) Show that $M_{0,t}$ is not empty. (*Hint*: Show that ba^{-1} has a power in $M_{n-t,n}$.)

(c) Let d be the least integer such that $M_{0,d}$ is not empty. Show that $M_{0,d}$ is formed of one element m such that $im \equiv i \bmod d$ for all $i \in Q$. (*Hint*: Arguing by contradiction, let j be the least integer such that $jm \not\equiv j \bmod d$. Use $a^{j-d}m$ to show that one may reduce to the case $j = d$. Then show that some power of ma fixes an interval of less than d elements.)

(d) Show that d divides n. (*Hint*: Let $n = dq + r$ with $q \ge 1$ and $0 \le r < d$. Show that some power of $a^{n-r}m$ is in M_r.)

(e) Show that $ib \equiv i + 1 \bmod d$ for each $i \in Q$.

3.6.4 Let X be a maximal prefix code on the alphabet $A = \{a, b\}$. Let $a^n \in X$ and let $Y = X \cap a^*ba^*$. Set $Y = \{y_0, y_1, \ldots, y_{n-1}\}$ with $y_i = a^i ba^j$. Suppose that

(i) there is an integer $m \geq 1$ such that a^m is not a factor of a word in X,
(ii) for each i, we have $|y_i| \leq n$ with equality if and only if $n - t \leq i \leq n - 1$,
(iii) the lengths of the words of Y are relatively prime.

Show that the code X is synchronized. (*Hint*: Use Exercise 3.6.3.)

3.6.5 Let $X \subset A^+$ be a prefix code and let $X = Y \circ Z$ be its maximal decomposition. Show that if $X = Y' \circ Z'$ with Z' prefix and Y' maximal prefix, then $Z'^* \subset Z^*$.

Section 3.7

3.7.1 Let $X \subset A^+$ be a thin maximal code and let $\pi : X \to]0, 1]$ be a function such that

$$\sum_{x \in X} \pi(x) = 1.$$

Define the *entropy* of X (relative to π) by

$$H(X) = -\sum_{x \in X} \pi(x) \log_k \pi(x),$$

where $k = \text{Card}(A)$. Set $\lambda(X) = \sum_{x \in X} |x| \pi(x)$.
Show that $H(X) \leq \lambda(X)$ and that the equality holds if and only if $\pi(x) = k^{-|x|}$ for $x \in X$.
Show that if X is finite and has n elements, then $H(X) \leq \log_k n$.

Section 3.8

3.8.1 Show that $u(z) = \sum_n u_n z^n$ is the generating series of a thin maximal prefix code on k letters if and only if

(i) $\sum_{n \geq 1} u_n k^{-n} = 1$,

(ii) there is an integer $p \geq 1$ such that the series $v(z) = \sum_n v_n z^n$ defined by $u(z) - 1 = v(z)(kz - 1)$ satisfies $v_{n+p} \leq v_n(k^p - 1)$ for all $n \geq 1$.

(*Hint*: Show that if condition (ii) is satisfied, then u is the length distribution of a maximal prefix code X such that a^{2p} is not a factor of the words of X.)

3.8.2 Let X be a thin maximal prefix code such that the gcd of the length of the words in X is 1. Show that there exists a code with the same length distribution which is thin, maximal, and synchronized. (*Hint*: Use Exercise 3.6.4.)

Section 3.9

3.9.1 The aim of this exercise is to show that the Golomb codes of Example 3.4.3 are optimal prefix codes for a source of integers with the *geometric distribution* given by

$$\pi(n) = p^n q \tag{3.33}$$

for positive real numbers p, q with $p + q = 1$.

Show that there is a unique integer m such that

$$p^m + p^{m+1} \leq 1 < p^{m-1} + p^m. \tag{3.34}$$

Show that the application of the Huffman algorithm to a geometric distribution given by (3.33) produces a code with the same length distribution as the Golomb code of order m where m is defined by (3.34). This shows the optimality of the Golomb code. (*Hint:* Operate on a truncated, but growing source since Huffman's algorithm works only on finite alphabets.)

3.9.2 Prove that the code produced by Varn's algorithm is indeed optimal. (*Hint:* Consider a complete prefix code X_1 built by the algorithm and assume it is not optimal, and consider a complete prefix code X_2 which is optimal. Show that there is a word x_1 in X_1 which is in $X_2 A^-$, and there is a word x_2 in X_2 which is in $X_1 A^-$. Consider a word p in X_2 which has x_1 as a prefix and such that $pA \subset X_2$ are leaves, and build $X_3 = X_2 \setminus (pA \cup x_2) \cup p \cup x_2 A$. Show that X_3 has cost less or equal to the cost of X_2 and is closer to X_1 in the sense that $\mathrm{Card}(X_1 \cup X_1 A^-) \cap (X_3 \cup X_3 A^-)$ is greater than $\mathrm{Card}(X_1 \cup X_1 A^-) \cap (X_2 \cup X_2 A^-)$.)

3.11 Notes

The results of the first four sections belong to folklore, and they are known to readers familiar with automata theory or with trees. The Elias code (Example 3.1.1) is introduced in Elias (1975).

Some particular codes are used for compression purposes to encode numerical data subject to known probability distribution. They appear in particular in the context of digital audio and video coding. The data encoded are integers and thus these codes are infinite. Example 3.4.3 presents the Golomb codes introduced in Golomb (1966). Golomb–Rice codes were introduced in Rice (1979). Exponential Golomb–Rice codes are introduced in Teuhola (1978), see also Salomon (2007). Exponential Golomb codes are used in practice in digital transmissions. In particular, they are a part of the video compression standard technically known as H.264/MPEG-4 Advanced Video Coding (AVC), see for instance Richardson (2003).

The hypothesis of unambiguity is necessary in Proposition 3.4.13, as shown by Bruyère (1987).

Semaphore codes were introduced in Schützenberger (1964) under the name of $\mathcal{J}$ codes. All the results presented in Section 3.5 can be found in that paper which also contains Theorem 3.6.12 and Proposition 3.7.17.

The notion of synchronized prefix code has been extensively studied in the context of automata theory. The significance of synchronized prefix codes for error recovery has been emphasized in Capoceli *et al.* (1992). In Freiling *et al.* (2003), it is proved that almost all finite maximal binary prefix codes are synchronized. This means that if $\sigma(n)$ denotes the number of synchronized maximal binary prefix codes with n elements, and $\alpha(n)$ denotes the total number of maximal binary prefix codes with n elements, then $\sigma(n)/\alpha(n)$ tends to 1. Recall from (3.13) that $\alpha(n)$ is the nth Catalan number. In Biskup (2008), it is proved that a synchronized maximal binary prefix code with n elements has a synchronizing word of length at most $O(hn \log n)$ where h is the maximal length of the words of X.

Let us mention Černý's problem: given a complete deterministic automaton with n states which is synchronized, what is the least upper bound to the length of a synchronizing word as a function of n? *Černý's conjecture* asserts that any synchronized strongly connected deterministic automaton has a synchronizing word of length at most $(n-1)^2$. See Exercise 3.6.2, Moore (1956), Černý (1964), Pin (1978) and the section on research problems. Example 3.6.13 is obtained by a construction of Perrin (1977a) (see Exercise 14.1.9). Exercise 3.6.4 is due to Schützenberger (1967). The maximal decomposition of prefix codes and Propositions 3.6.14, is due to Perrot (1972).

The results of Section 3.7 are given in another terminology in Feller (1968).

Theorem 3.8.2 is from Bassino *et al.* (2000). The method of state splitting used in the proof of Lemma 3.8.6 is inspired from symbolic dynamics (see Marcus (1979) or Adler *et al.* (1983)). The transformations between the various weighted automata recognizing a given series used in the proof of the theorem have been systematically studied in Béal *et al.* (2005).

Huffman's algorithm, originally described in Huffman (1952), is presented in most textbooks on algorithms. It has numerous applications in data compression, and variations such as the adaptative Huffman algorithm have been developed, see Knuth (1985).

Run-length limited codes have applications in practical coding, see Lind and Marcus (1995).

The case of codewords with equal weights and unequal letter cost has been solved by Varn (1971). Another algorithm is in Perl *et al.* (1975).

Karp (1961) gave the first algorithm providing a solution of the general problem with integer costs. His algorithm reduces to a problem in integer programming.

Another approach by Golin and Rote (1998) uses dynamic programming. Their algorithm produces the solution in time $O(n^{\kappa+2})$, where n is the number of codewords and κ is the greatest of the costs of the letters of A. This algorithm has been improved to $O(n^{\kappa})$ in the case of a binary alphabet in Bradford *et al.* (2002).

Ordered prefix codes are usually called alphabetic trees. The use of dynamic programming technique for the construction of optimal alphabetic trees goes back to Gilbert and Moore (1959). Their algorithm is $O(n^3)$ in time and $O(n^2)$ in space. Knuth (1971) reduces time to $O(n^2)$.

We follow Knuth (1998) for the exposition and the proof of the Garsia–Wachs algorithm (see also Garsia and Wachs (1977); Kingston (1988)). The Garsia–Wachs algorithm is simpler than a previous algorithm given in Hu and Tucker (1971) which was also described in the first edition of Knuth's book. For a proof and a detailed description of the Hu–Tucker algorithm, and complements see Hu and Shing (2002); Hu and Tucker (1998).

There is no known polynomial time algorithm for the general problem, nor is the problem known to be NP-hard. A polynomial time approximation scheme, that is an algorithm that produces a solution which is optimal up to $1 + \epsilon$ in time $O(n \log n \exp(O(\frac{1}{\epsilon^2} \log \frac{1}{\epsilon^2})))$ is given by Golin *et al.* (2002).

An algorithm in cubic time for solving the optimal alphabetic prefix problem with unequal letter cost has been given in Itai (1976).

The results of Problems 3.8.1 and 3.8.2 are due to Schützenberger (1967). There is a strong relation with the road coloring theorem proved in Chapter 10.

The monoid of prefix subsets defined in Exercise 3.3.1 has been further studied by Lassez (1973). Exercise 3.4.1 is a well-known result in combinatorics, see Lothaire (1997). Exercises 9.5.3, 9.5.4 and 9.5.5 are from Bruyère *et al.* (1998). Exercise 3.9.1 follows Gallager and van Voorhis (1975). The geometric distribution of this exercise arises from *run-length encoding* where a sequence of $0^n 1$ is encoded by n. If the source produces 0's and 1's independently with probability p and q, the probability of $0^n 1$ is precisely $\pi(n)$. This is of practical interest if p is large since then long runs of 0's are expected and the run-length encoding realizes a logarithmic compression.

4

Automata

In this chapter, we study unambiguous automata. The main idea is to replace computations on words by computations on paths labeled by words. This is a technique which is well known in formal language theory. It will be used here in a special form related to the characteristic property of codes.

Within this frame, the main fact is the equivalence between codes and unambiguous automata. The uniqueness of paths in unambiguous automata corresponds to the uniqueness of factorizations for a code. Unambiguous automata appear to be a generalization of deterministic automata in the same manner as the notion of a code extends the notion of a prefix code.

We present devices for encoding and decoding, using transducers. A special class of transducers, called sequential transducers, is introduced. It will be shown in Chapter 5 to be related to the deciphering delay.

The chapter is organized as follows.

In the first section, we study unambiguous automata in connection with codes. In the next section, the flower automaton is defined. We show that it is a universal automaton in the sense that any unambiguous automaton associated with a code can be obtained by a reduction of the flower automaton of this code. We also show how to decompose the flower automaton of the composition of two codes.

In the last section, we use transducers. We introduce an algorithm to transform a transducer realizing a function into a sequential (possibly infinite) transducer.

4.1 Unambiguous automata

An automaton $\mathcal{A} = (Q, I, T)$ over A is *unambiguous* if for all $p, q \in Q$ and $w \in A^*$, there is at most one path from p to q with label w in $\mathcal{A}$.

Recall from Section 1.10 that $|\mathcal{A}|$ denotes the *behavior* of $\mathcal{A}$. For each word u, the coefficient $(|\mathcal{A}|, u)$ is the number of successful paths labeled by u in $\mathcal{A}$.

Proposition 4.1.1 *Let $\mathcal{A} = (Q, i, t)$ be a trim automaton with a unique initial and a unique final state. Then $\mathcal{A}$ is unambiguous if and only if $|\mathcal{A}|$ is a characteristic series.*

Proof. If $\mathcal{A}$ is unambiguous, then clearly $|\mathcal{A}|$ is a characteristic series. Conversely, if there are two distinct paths from p to q labeled with w for some $p, q \in Q$ and

$w \in A^*$, then choosing paths $i \xrightarrow{u} p$ and $q \xrightarrow{v} t$, we have

$$(|\mathcal{A}|, uwv) \geq 2. \qquad \square$$

Proposition 4.1.2 *Let $X \subset A^+$ and let $\mathcal{A}$ be an automaton such that $|\mathcal{A}| = \underline{X}$. Then X is a code if and only if the star $\mathcal{A}^*$ of $\mathcal{A}$ is an unambiguous automaton.*

Recall from Section 1.10 that the star $\mathcal{A}^*$ associated with an automaton $\mathcal{A}$ is such that $|\mathcal{A}^*| = |\mathcal{A}|^*$.

Proof. According to Proposition 1.10.5, we have $|\mathcal{A}^*| = (\underline{X})^*$. Since $\mathcal{A}^*$ is trim, Proposition 4.1.1 shows that $\mathcal{A}^*$ is unambiguous if and only if $|\mathcal{A}^*|$ is a characteristic series. Since $L(\mathcal{A}^*) = X^*$, this means that $\mathcal{A}^*$ is unambiguous if and only if $\underline{X^*} = (\underline{X})^*$. Thus we get the proposition from Proposition 2.6.1. $\qquad \square$

In view of Proposition 4.1.2, we can determine whether a set X given by an unambiguous automaton $\mathcal{A}$ is a code, by computing $\mathcal{A}^*$ and testing whether $\mathcal{A}^*$ is unambiguous. For doing this, we may use the following method.

Let $\mathcal{A} = (Q, I, T)$ be an automaton over A. The *square* $\mathcal{S}$ of $\mathcal{A}$ is the automaton

$$\mathcal{S}(\mathcal{A}) = (Q \times Q, I \times I, T \times T)$$

constructed by defining

$$(p_1, p_2) \xrightarrow{a} (q_1, q_2)$$

to be an edge of $\mathcal{S}(\mathcal{A})$ if and only if

$$p_1 \xrightarrow{a} q_1 \quad \text{and} \quad p_2 \xrightarrow{a} q_2$$

are edges of $\mathcal{A}$.

Proposition 4.1.3 *An automaton $\mathcal{A} = (Q, I, T)$ is unambiguous if and only if there is no path in $\mathcal{S}(\mathcal{A})$ of the form*

$$(p, p) \xrightarrow{u} (r, s) \xrightarrow{v} (q, q) \qquad (4.1)$$

with $r \neq s$.

Proof. The existence of a path of the form (4.1) in $\mathcal{S}(\mathcal{A})$ is equivalent to the existence of the pair of paths

$$p \xrightarrow{u} r \xrightarrow{v} q \quad \text{and} \quad p \xrightarrow{u} s \xrightarrow{v} q$$

with the same label uv in $\mathcal{A}$. $\qquad \square$

To decide whether a recognizable set X given by an unambiguous finite automaton $\mathcal{A}$ is a code, it suffices to compute $\mathcal{A}^*$ and to test whether $\mathcal{A}^*$ is unambiguous by inspecting the finite automaton $\mathcal{S}(\mathcal{A}^*)$, looking for paths of the form (4.1).

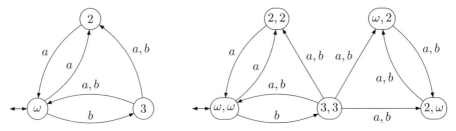

Figure 4.1 An unambiguous automaton, and part of the square of this automaton.

Example 4.1.4 Consider again the automaton $\mathcal{A}^*$ of Example 1.10.7 repeated here for convenience on the left of Figure 4.1. The automaton $S(\mathcal{A}^*)$ is given on the right of this figure, where only the part accessible from the states (q, q) is drawn. It shows that $\mathcal{A}^*$ is unambiguous.

The following proposition is a complement to Proposition 4.1.2.

Proposition 4.1.5 *Let* $\mathcal{A} = (Q, 1, 1)$ *be an unambiguous automaton over A with a single initial and final state. Then its behavior* $|\mathcal{A}|$ *is the characteristic series of some free submonoid of* A^*.

Proof. Let $M \subset A^*$ be such that $|\mathcal{A}| = \underline{M}$. Clearly the set M is a submonoid of A^*. We shall prove that M is a stable submonoid. For this, suppose that

$$u, wv, uw, v \in M.$$

Then there exist in $\mathcal{A}$ paths

$$1 \xrightarrow{u} 1, \quad 1 \xrightarrow{wv} 1, \quad 1 \xrightarrow{uw} 1, \quad 1 \xrightarrow{v} 1.$$

The two middle paths factorize as

$$1 \xrightarrow{w} p \xrightarrow{v} 1, \quad 1 \xrightarrow{u} q \xrightarrow{w} 1$$

for some $p, q \in Q$. Thus there exist two paths

$$1 \xrightarrow{u} 1 \xrightarrow{w} p \xrightarrow{v} 1$$

$$1 \xrightarrow{u} q \xrightarrow{w} 1 \xrightarrow{v} 1.$$

Since $\mathcal{A}$ is unambiguous, these paths coincide, whence $1 = p = q$. Consequently $w \in M$. Thus M is stable, and by Proposition 2.2.5, M is free. $\qquad \square$

The next result concerns the determinant of a matrix which is associated in a natural way with an automaton. It is of independent interest, and it will be useful later, in Chapter 7. Recall that we denote by $\alpha(w)$ the commutative image of a word $w \in A^*$ and by $\alpha(\sigma)$ the commutative image of the formal series σ. Formula (4.2) gives an expression of the polynomial $1 - \alpha(\underline{X})$ for a finite code X.

Proposition 4.1.6 *Let $X \subset A^+$ be a finite code and let $\mathcal{A} = (Q, 1, 1)$ be a unambiguous trim finite automaton recognizing X^*. Let M be the $Q \times Q$-matrix with elements in $\mathbb{Q}[A]$ such that $M_{p,q}$ is the sum of the elements of the set*

$$A_{pq} = \{a \in A \mid p \xrightarrow{a} q\}.$$

Then

$$1 - \alpha(\underline{X}) = \det(I - M). \tag{4.2}$$

Proof. Any path $q \xrightarrow{w} q$ with $q \neq 1$ and $w \in A^+$ passes through state 1. Otherwise $uw^*v \subset X$ for words u, v such that $1 \xrightarrow{u} q \xrightarrow{v} 1$, contradicting the finiteness of X. Thus we can set $Q = \{1, 2, \dots, n\}$ in such a way that whenever $i \xrightarrow{a} j$ for $a \in A$, $j \neq 1$, then $i < j$. Define for $i, j \in Q$, an element of $\mathbb{Q}\langle A \rangle$ by

$$r_{ij} = \delta_{ij} - \underline{A}_{ij} \tag{4.3}$$

where δ_{ij} is the Kronecker symbol. Let Δ be the polynomial

$$\Delta = \sum_{\sigma \in \mathfrak{S}_n} (-1)^{\epsilon(\sigma)} r_{1,1\sigma} r_{2,2\sigma} \cdots r_{n,n\sigma},$$

where $\epsilon(\sigma) = \pm 1$ denotes the *signature* of the permutation σ. By definition, $\epsilon(\sigma) = 1$ if σ is an even permutation, and $\epsilon(\sigma) = -1$ otherwise. According to the well-known formula for determinants we have

$$\det(I - M) = \alpha(\Delta).$$

Thus it suffices to show that

$$\Delta = 1 - \underline{X}. \tag{4.4}$$

For this, let

$$\Delta_\sigma = r_{1,1\sigma} r_{2,2\sigma} \cdots r_{n,n\sigma},$$

so that

$$\Delta = \sum_{\sigma \in \mathfrak{S}_n} (-1)^{\epsilon(\sigma)} \Delta_\sigma.$$

Consider a permutation $\sigma \in \mathfrak{S}_n$ such that $\Delta_\sigma \neq 0$. If $\sigma \neq 1$, then it has at least one cycle $(i_1, i_2, \dots, i_k)$ of length $k \geq 2$. Since $\Delta_\sigma \neq 0$, by (4.3) the sets $A_{i_1 i_2}, A_{i_2 i_3}, \dots, A_{i_k i_1}$ are nonempty. This implies that the cycle $(i_1, \dots, i_k)$ contains state 1. Consequently each permutation σ with $\Delta_\sigma \neq 0$ is composed of fixed points and of one cycle containing 1. If this cycle is $(i_1, i_2, \dots, i_k)$ with $i_1 = 1$, then

$$1 < i_2 < \cdots < i_k$$

by the choice of the ordering of states in $\mathcal{A}$. Set $X_\sigma = A_{1i_2} A_{i_2i_3} \cdots A_{i_k,1}$. Then $\Delta_\sigma = (-1)^k X_\sigma$ and also $(-1)^{\epsilon(\sigma)} = (-1)^{k+1}$ since a cycle of length k has the same parity as $k + 1$.

The set X_σ is composed of words $a_1 a_2 \cdots a_k$ with $a_i \in A$ and such that

$$1 \xrightarrow{a_1} i_2 \xrightarrow{a_3} i_3 \longrightarrow \cdots \longrightarrow i_k \xrightarrow{a_k} 1.$$

These words are in X. Denote by S the set of permutations $\sigma \in \mathfrak{S} \setminus 1$ having just one nontrivial cycle, namely, the cycle containing 1. Then $X = \sum_{\sigma \in S} X_\sigma$ since each word in X is the label of a unique path $(1, i_2, \ldots, i_k, 1)$ with $1 < i_2 < \cdots < i_k$. It follows that

$$\Delta = 1 + \sum_{\sigma \in S} (-1)^{\epsilon(\sigma)} \Delta_\sigma = 1 - \sum_{\sigma \in S} \underline{X}_\sigma = 1 - \underline{X}. \qquad \square$$

Example 4.1.7 Let $X = \{aa, ba, bb, baa, bba\}$. This is the code of Example 2.3.5. The unambiguous automaton given on the left of Figure 4.1 recognizes X^*. The matrix M is here

$$M = \begin{bmatrix} 0 & a & b \\ a & 0 & 0 \\ a+b & a+b & 0 \end{bmatrix}$$

and one easily checks that indeed $\det(I - M) = 1 - \alpha(\underline{X})$.

The *unambiguous rational operations* on sets of words are

(i) disjoint union,
(ii) unambiguous product,
(iii) star operation of a code.

Recall that the product XY is unambiguous if $xy = x'y'$ with $x, x' \in X$, $y, y' \in Y$ implies $x = x'$ and $y = y'$. The star of a code is of course a free submonoid.

The family of *unambiguous rational subsets* of A^* is the smallest family of subsets of A^* containing the finite sets and closed under unambiguous rational operations. A description of a rational set by unambiguous rational operations is called an *unambiguous rational expression* or an unambiguous regular expression.

Proposition 4.1.8 *Every rational set is unambiguous rational.*

Proof. By Proposition 1.4.1, every rational set is recognized by a finite deterministic automaton. In this case, Formulas (1.11)–(1.13) provide an unambiguous rational expression for this set. $\qquad \square$

Example 4.1.9 Let $A = \{a, b\}$. An unambiguous rational expression for the set $A^* b A^*$ is $a^* b A^*$ (or $A^* b a^*$).

Figure 4.2 The edges of $\mathcal{A}_D(X)$ for $x = a_1 a_2 \cdots a_n$.

4.2 Flower automaton

We describe in this section the construction of a "universal" automaton recognizing a submonoid of A^*.

Let X be an arbitrary subset of A^+. We define an automaton

$$\mathcal{A}_D(X) = (Q, I, T)$$

by

$$Q = \{(u, v) \in A^* \times A^* \mid uv \in X\}, \quad I = 1 \times X, \quad T = X \times 1,$$

with edges $(u, v) \xrightarrow{a} (u', v')$ if and only if $ua = u'$ and $v = av'$. In other words, the edges of $\mathcal{A}_D$ are

$$(u, av) \xrightarrow{a} (ua, v), \qquad uav \in X.$$

It is equivalent to say that the set of edges of the automaton $\mathcal{A}_D$ is the disjoint union of the sets of edges given by Figure 4.2 for each $x = a_1 a_2 \cdots a_n$ in X. The automaton $\mathcal{A}_D(X)$ is unambiguous and recognizes X, that is,

$$|\mathcal{A}_D(X)| = \underline{X}.$$

The *flower automaton* of X is by definition the star of the automaton $\mathcal{A}_D(X)$, as obtained by the construction described in Section 1.10. It is denoted by $\mathcal{A}_D^*(X)$ rather than $(\mathcal{A}_D(X))^*$. We denote by φ_D the associated representation. Thus, following the construction of Section 1.10, the automaton $\mathcal{A}_D^*(X)$ is obtained in two steps as follows. Starting with $\mathcal{A}_D(X)$, we add a new state ω, and the edges

$$\begin{aligned} \omega &\xrightarrow{a} (a, v) && \text{for } av \in X, \\ (u, a) &\xrightarrow{a} \omega && \text{for } ua \in X, \\ \omega &\xrightarrow{a} \omega && \text{for } a \in X. \end{aligned}$$

This automaton is now trimmed. The states in $1 \times X$ and $X \times 1$ are no longer accessible or coaccessible and consequently disappear. Usually, the state ω is denoted by $(1, 1)$. Then $\mathcal{A}_D^*(X)$ takes the form

$$\mathcal{A}_D^*(X) = (P, (1, 1), (1, 1)),$$

with

$$P = \{(u, v) \in A^+ \times A^+ \mid uv \in X\} \cup \{(1, 1)\},$$

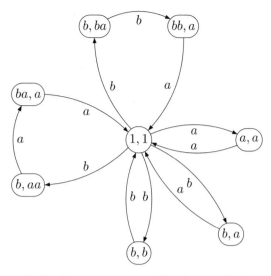

Figure 4.3 The flower automaton of $X = \{aa, ba, bb, baa, bba\}$.

and there are four types of edges

$$
\begin{aligned}
(u, av) &\xrightarrow{a} (ua, v) & \text{for } uav \in X, \ (u, v) \neq (1, 1), \\
(1, 1) &\xrightarrow{a} (a, v) & \text{for } av \in X, \quad v \neq 1, \\
(u, a) &\xrightarrow{a} (1, 1) & \text{for } ua \in X, \quad u \neq 1, \\
(1, 1) &\xrightarrow{a} (1, 1) & \text{for } a \in X.
\end{aligned}
$$

The terminology is inspired by the graphical representation of this automaton. Indeed each word $x \in X$ defines a simple path

$$(1, 1) \xrightarrow{x} (1, 1)$$

in $\mathcal{A}_D^*(X)$. If $x = a \in A$, it is the edge

$$(1, 1) \xrightarrow{a} (1, 1).$$

If $x = a_1 a_2 \cdots a_n$ with $n \geq 2$, it is the path

$$(1, 1) \xrightarrow{a_1} (a_1, a_2 \cdots a_n) \xrightarrow{a_2} (a_1 a_2, a_3 \cdots a_n) \to \cdots \to (a_1 a_2 \cdots a_{n-1}, a_n) \xrightarrow{a_n} (1, 1).$$

Example 4.2.1 Let $X = \{aa, ba, bb, baa, bba\}$. The flower automaton is given in Figure 4.3.

Theorem 4.2.2 *Let X be a subset of A^+. The following conditions are equivalent:*

(i) *X is a code.*
(ii) *For any unambiguous automaton $\mathcal{A}$ recognizing X, the automaton $\mathcal{A}^*$ is unambiguous.*
(iii) *The flower automaton $\mathcal{A}_D^*(X)$ is unambiguous.*

(iv) *There exists an unambiguous automaton* $\mathcal{A} = (Q, 1, 1)$ *recognizing* X^* *and* X
is the minimal set of generators of X^*.

Proof. (i) $\Longrightarrow$ (ii) is Proposition 4.1.2. The implication (ii) $\Longrightarrow$ (iii) is clear. To prove
(iii) $\Longrightarrow$ (iv), it suffices to show that X is the minimal generating set of X^*. Assume
the contrary, and let $x \in X$, $y, z \in X^+$ be words such that $x = yz$. Then there exists in
$\mathcal{A}_D^*(X)$ a simple path $(1, 1) \xrightarrow{x} (1, 1)$ and a path $(1, 1) \xrightarrow{y} (1, 1) \xrightarrow{z} (1, 1)$ which
is also labeled by x. These paths are distinct, so $\mathcal{A}_D^*(X)$ is ambiguous. Finally, for
(iv) $\Longrightarrow$ (i), observe that by Proposition 4.1.5, X^* is free. Thus X is a code. $\qquad\square$

We shall now describe explicitly the paths in the flower automaton of a code.

Proposition 4.2.3 *Let* $X \subset A^+$ *be a code. The following conditions are equivalent
for all words* $w \in A^*$ *and all states* (u, v), (u', v') *in the automaton* $\mathcal{A}_D^*(X)$:

(i) *There exists in* $\mathcal{A}_D^*(X)$ *a path* $c : (u, v) \xrightarrow{w} (u', v')$.
(ii) $w \in vX^*u'$ *or* $(uw = u'$ *and* $v = wv')$.
(iii) $uw \in X^*u'$ *and* $wv' \in vX^*$.

Proof. (i) $\Longrightarrow$ (ii). If c is a simple path, then it is a path in $\mathcal{A}_D$. Consequently, $uw = u'$
and $v = wv'$ (Figure 4.4(a)). Otherwise c decomposes into

$$c : (u, v) \xrightarrow{v} (1, 1) \xrightarrow{x} (1, 1) \xrightarrow{u'} (u', v')$$

with $w = vxu'$ and $x \in X^*$ (Figure 4.4(b)).

(ii) $\Longrightarrow$ (iii). If $w \in vX^*u'$, then $uw \in uvX^*u' \subset X^*u'$ and $w \in vX^*u'v' \subset vX^*$,
since $uv, u'v' \in X \cup 1$. If $uw = u'$ and $v = wv'$, then the formulas are clear.

(iii) $\Longrightarrow$ (i). By hypothesis, there exist $x, y \in X^*$ such that $uw = xu'$, $wv' = vy$.
Let $z = uwv'$. Then

$$z = uwv' = xu'v' = uvy \in X^*.$$

Each of these three factorizations determines a path in $\mathcal{A}_D^*(X)$ (see Figure 4.4):

$$
\begin{array}{lcccccccc}
c : (1, 1) & \xrightarrow{u} & (\bar{u}, \bar{v}) & \xrightarrow{w} & (\bar{u}', \bar{v}') & \xrightarrow{v'} & (1, 1), \\
c' : (1, 1) & \xrightarrow{x} & (1, 1) & \xrightarrow{u'} & (u', v') & \xrightarrow{v'} & (1, 1), \\
c'' : (1, 1) & \xrightarrow{u} & (u, v) & \xrightarrow{v} & (1, 1) & \xrightarrow{y} & (1, 1),
\end{array}
$$

(The paths $(1, 1) \xrightarrow{u} (u, v) \xrightarrow{v} (1, 1)$ and $(1, 1) \xrightarrow{u'} (u', v') \xrightarrow{v'} (1, 1)$ may have
length 0.) Since X is a code, the automaton $\mathcal{A}_D^*(X)$ is unambiguous and consequently
$c = c' = c''$. We obtain that $(u, v) = (\bar{u}, \bar{v})$ and $(u', v') = (\bar{u}', \bar{v}')$. Thus

$$(u, v) \xrightarrow{w} (u', v').$$
$\qquad\square$

The flower automaton of a code has "many" states. In particular, the flower automaton of an infinite code is infinite, even though there exist finite unambiguous automata

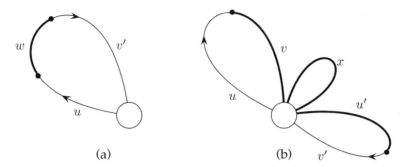

Figure 4.4 Paths in the flower automaton.

recognizing X^* when the code X is recognizable. We show that $\mathcal{A}_D^*(X)$ is universal among the automata recognizing X^*, in the following sense.

Consider two unambiguous automata

$$\mathcal{A} = (P, 1, 1) \quad \text{and} \quad \mathcal{B} = (Q, 1, 1),$$

and their associated representations $\varphi_\mathcal{A}$ and $\varphi_\mathcal{B}$. A function $\rho : P \to Q$ is a *reduction* of $\mathcal{A}$ *onto* $\mathcal{B}$ if it is surjective, $\rho(1) = 1$ and if, for all $w \in A^*$,

$$(q, \varphi_\mathcal{B}(w), q') = 1$$

if and only if there exist $p, p' \in P$ with

$$(p, \varphi_\mathcal{A}(w), p') = 1, \quad \rho(p) = q, \quad \rho(p') = q'.$$

The definition means that if $p \xrightarrow{w} p'$ is a path in $\mathcal{A}$, then $\rho(p) \xrightarrow{w} \rho(p')$ is a path in $\mathcal{B}$. Conversely, a path $q \xrightarrow{w} q'$ can be "lifted" in some path $p \xrightarrow{w} p'$ with $p \in \rho^{-1}(q)$, $p' \in \rho^{-1}(q')$.

Another way to see the definition is the following. The matrix $\varphi_\mathcal{B}(w)$ can be obtained from $\varphi_\mathcal{A}(w)$ by partitioning the latter into blocks indexed by a pair of classes of the equivalence defined by ρ, and then by replacing null blocks by 0, and nonnull blocks by 1.

Observe that if ρ is a reduction of $\mathcal{A}$ onto $\mathcal{B}$, then for all $w, w' \in A^*$, the following implication holds:

$$\varphi_\mathcal{A}(w) = \varphi_\mathcal{A}(w') \implies \varphi_\mathcal{B}(w) = \varphi_\mathcal{B}(w').$$

Thus there exists a unique surjective morphism

$$\widehat{\rho} : \varphi_\mathcal{A}(A^*) \to \varphi_\mathcal{B}(A^*)$$

such that $\varphi_\mathcal{B} = \widehat{\rho} \circ \varphi_\mathcal{A}$. The morphism $\widehat{\rho}$ is called the *morphism associated* with the reduction ρ.

Proposition 4.2.4 *Let $\mathcal{A} = (P, 1, 1)$ and $\mathcal{B} = (Q, 1, 1)$ be two unambiguous trim automata. Then there exists at most one reduction of $\mathcal{A}$ onto $\mathcal{B}$. If $\rho : P \rightarrow Q$ is a reduction, then*

1. $|\mathcal{A}| \subset |\mathcal{B}|$,
2. $|\mathcal{A}| = |\mathcal{B}|$ *if and only if* $\rho^{-1}(1) = 1$.

Proof. Let $\rho, \rho' : P \rightarrow Q$ be two reductions of $\mathcal{A}$ onto $\mathcal{B}$. Let $p \in P$, and let $q = \rho(p), q' = \rho'(p)$. Let $u, v \in A^*$ be words such that $1 \xrightarrow{u} p \xrightarrow{v} 1$ in the automaton $\mathcal{A}$. Then we have, in the automaton $\mathcal{B}$, the paths

$$1 \xrightarrow{u} q \xrightarrow{v} 1, \qquad 1 \xrightarrow{u} q' \xrightarrow{v} 1.$$

Since $\mathcal{B}$ is unambiguous, $q = q'$. Thus $\rho = \rho'$.

1. If $w \in |\mathcal{A}|$, there exists a path $1 \xrightarrow{w} 1$ in $\mathcal{A}$; thus there is a path $1 \xrightarrow{w} 1$ in $\mathcal{B}$. Consequently $w \in |\mathcal{B}|$.

2. Let $w \in |\mathcal{B}|$. Then there is a path $p \xrightarrow{w} p'$ in $\mathcal{A}$ with $\rho(p) = \rho(p') = 1$. If $1 = \rho^{-1}(1)$, then this is a successful path in $\mathcal{A}$ and $w \in |\mathcal{A}|$. Conversely, let $p \neq 1$. Let $1 \xrightarrow{u} p \xrightarrow{v} 1$ be a simple path in $\mathcal{A}$. Then $uv \in X$, where X is the base of $|\mathcal{A}|$. Now in $\mathcal{B}$, we have $1 \xrightarrow{u} \rho(p) \xrightarrow{v} 1$. Since $|\mathcal{A}| = |\mathcal{B}|$, we have $\rho(p) \neq 1$. Thus $\rho^{-1}(1) = 1$. ☐

Proposition 4.2.5 *Let $X \subset A^+$ be a code, and let $\mathcal{A}_D^*(X)$ be its flower automaton. For each unambiguous trim automaton $\mathcal{A} = (Q, 1, 1)$ recognizing X^*, there exists a reduction of $\mathcal{A}_D^*(X)$ onto $\mathcal{A}$.*

Proof. Let $\mathcal{A}_D^*(X) = (P, (1, 1), (1, 1))$. Define a function $\rho : P \rightarrow Q$ as follows. Let $p = (u, v) \in P$. If $p = (1, 1)$, then set $\rho(p) = 1$. Otherwise $uv \in X$, and there exists a unique path $c : 1 \xrightarrow{u} q \xrightarrow{v} 1$ in $\mathcal{A}$. Then set $\rho(p) = q$.

The function ρ is surjective. Let indeed $q \in Q, q \neq 1$. Let

$$c_1 : 1 \xrightarrow{u} q, \qquad c_2 : q \xrightarrow{v} 1$$

be two simple paths in $\mathcal{A}$. Then $uv \in X$, and $p = (u, v) \in P$ satisfies $\rho(p) = q$.

We now verify that ρ is a reduction. For this, assume first that for a word $w \in A^*$, and $q, q' \in Q$, there is a path in $\mathcal{A}$ from q to q' labeled by w. Consider two simple paths in $\mathcal{A}$, $e : 1 \xrightarrow{u} q, e' : q' \xrightarrow{v'} 1$. Then in $\mathcal{A}$, there is a path

$$1 \xrightarrow{u} q \xrightarrow{w} q' \xrightarrow{v'} 1.$$

Consequently $uwv' \in X^*$. Thus for some $x_i \in X$, $uwv' = x_1 x_2 \cdots x_n$. Since e is simple, u is a prefix of x_1, and similarly v' is a suffix of x_n. Setting $x_1 = uv$, $x_n = u'v'$, we have

$$uwv' = uvx_2 \cdots x_n = x_1 \cdots x_{n-1} u'v',$$

whence $uw \in X^*u'$, $wv' \in uX^*$. In view of Proposition 4.2.3, $((u, v), \varphi_D(w), (u', v')) = 1$.

Suppose now conversely that

$$(p, \varphi_D(w), p') = 1 \tag{4.5}$$

for some $p = (u, v)$, $p' = (u', v')$, and $w \in A^*$. Let $q = \rho(p)$, $q' = \rho(p')$. By construction, there are in $\mathcal{A}$ paths

$$1 \xrightarrow{u} q \xrightarrow{v} 1 \qquad \text{and} \qquad 1 \xrightarrow{u'} q' \xrightarrow{v'} 1. \tag{4.6}$$

In view of Proposition 4.2.3, Formula (4.5) is equivalent to

$$\{uw = u' \text{ and } v = wv'\} \text{ or } \{w = vxu' \text{ for some } x \in X^*\}.$$

In the first case, $uv = uwv' = u'v'$. Thus the two paths (4.6) coincide, giving the path in $\mathcal{A}$,

$$1 \xrightarrow{u} q \xrightarrow{w} q' \xrightarrow{v'} 1.$$

In the second case, there is in $\mathcal{A}$ a path

$$q \xrightarrow{v} 1 \xrightarrow{x} 1 \xrightarrow{u'} q',$$

Thus, $(q, \varphi_{\mathcal{A}}(w), q') = 1$ in both cases. □

Example 4.2.6 For the code $X = \{aa, ba, bb, baa, bba\}$, the flower automaton is given in Figure 4.5.

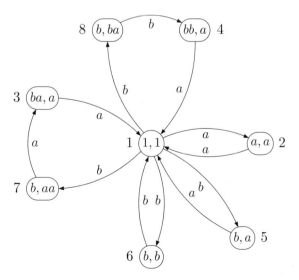

Figure 4.5 The flower automaton of X with its states renumbered.

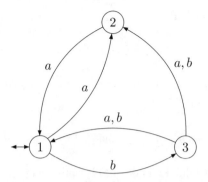

Figure 4.6 Another automaton recognizing X^*.

Consider the automaton given in Figure 4.6. The function $\rho : P \to \{1, 2, 3\}$ is given by

$$\rho((a, a)) = \rho((ba, a)) = \rho((bb, a)) = 2,$$

$$\rho((b, a)) = \rho((b, b)) = \rho((b, aa)) = \rho((b, ba)) = 3,$$

$$\rho((1, 1)) = 1.$$

The matrices of the associated representations (with the states numbered as indicated in Figures. 4.5 and 4.6) are

$$\varphi_D(a) = \begin{array}{c} 1 \\ 2 \\ 3 \\ 4 \\ 5 \\ 6 \\ 7 \\ 8 \end{array} \begin{bmatrix} 0 & 1 & 0 & 0 & 0 & 0 & 0 & 0 \\ 1 & 0 & 0 & 0 & 0 & 0 & 0 & 0 \\ 1 & 0 & 0 & 0 & 0 & 0 & 0 & 0 \\ 1 & 0 & 0 & 0 & 0 & 0 & 0 & 0 \\ 1 & 0 & 0 & 0 & 0 & 0 & 0 & 0 \\ 0 & 0 & 0 & 0 & 0 & 0 & 0 & 0 \\ 0 & 0 & 1 & 0 & 0 & 0 & 0 & 0 \\ 0 & 0 & 0 & 0 & 0 & 0 & 0 & 0 \end{bmatrix}, \qquad \varphi(a) = \begin{bmatrix} 0 & 1 & 0 \\ 1 & 0 & 0 \\ 1 & 1 & 0 \end{bmatrix},$$

$$\varphi_D(b) = \begin{array}{c} 1 \\ 2 \\ 3 \\ 4 \\ 5 \\ 6 \\ 7 \\ 8 \end{array} \begin{bmatrix} 0 & 0 & 0 & 0 & 1 & 1 & 1 & 1 \\ 0 & 0 & 0 & 0 & 0 & 0 & 0 & 0 \\ 0 & 0 & 0 & 0 & 0 & 0 & 0 & 0 \\ 0 & 0 & 0 & 0 & 0 & 0 & 0 & 0 \\ 0 & 0 & 0 & 0 & 0 & 0 & 0 & 0 \\ 1 & 0 & 0 & 0 & 0 & 0 & 0 & 0 \\ 0 & 0 & 0 & 0 & 0 & 0 & 0 & 0 \\ 0 & 0 & 0 & 1 & 0 & 0 & 0 & 0 \end{bmatrix}, \qquad \varphi(b) = \begin{bmatrix} 0 & 0 & 1 \\ 0 & 0 & 0 \\ 1 & 1 & 0 \end{bmatrix}.$$

The concept of a reduction makes it possible to indicate a relation between the flower automata of a composed code and those of its components.

Proposition 4.2.7 *Let $Y \subset B^+$, $Z \subset A^+$ be two composable codes and let $X = Y \circ_\beta Z$. If Y is complete, then there exists a reduction of $\mathcal{A}_D^*(X)$ onto $\mathcal{A}_D^*(Z)$.*

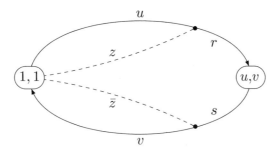

Figure 4.7 Decomposing a petal.

Moreover, $\mathcal{A}_D^*(Y)$ can be identified, through β with the restriction of $\mathcal{A}_D^*(X)$ to the states in $Z^* \times Z^*$.

Proof. Let P and S be the sets of states of $\mathcal{A}_D^*(X)$ and $\mathcal{A}_D^*(Z)$ respectively, and let φ_X and φ_Z be the representations associated to $\mathcal{A}_D^*(X)$ and $\mathcal{A}_D^*(Z)$.

We define the function $\rho : P \to S$ as follows. First, let $\rho((1, 1)) = (1, 1)$. Next, consider $(u, v) \in P \setminus (1, 1)$. Then $uv \in Z^+$. Consequently, there exist unique $z, \bar{z} \in Z^*$, and $(r, s) \in S$ such that

$$u = zr, \qquad v = s\bar{z}$$

(see Figure 4.7). Then let $\rho(u, v) = (r, s)$. The function ρ is surjective. Indeed, each word in Z appears in at least one word in X; thus each state in S is reached in a refinement of a state in P.

To show that ρ is a reduction, suppose that

$$((u, v), \varphi_X(w), (u', v')) = 1.$$

Let $(r, s) = \rho((u, v))$, $(r', s') = \rho((u', v'))$, and let $z, \bar{z}, z', \bar{z}' \in Z^*$ be such that

$$u = zr, \qquad v = s\bar{z}, \qquad u' = z'r', \qquad v' = s'\bar{z}'.$$

By Proposition 4.2.3, $uw \in X^*u'$, $wv' \in vX^*$. Thus $zrw \in Z^*r'$, $ws'\bar{z}' \in sZ^*$, implying that $zrws' \in Z^*$ and $rws'\bar{z} \in Z^*$. This in turn shows, in view of the stability of Z^*, that $rws' \in Z^*$. Set $zrw = \hat{z}r'$, with $\hat{z} \in Z^*$. Then

$$\hat{z}(r's') = z(rws'),$$

and each of the four factors in this equation is in Z^*. Thus Z being a code, either $\hat{z} = zt$ or $z = \hat{z}t$ for some $t \in Z^*$. In the first case, we get $tr's' = rws'$, whence $rw \in Z^*r'$. The second case implies $r's' = trws'$. Since $r's' \in 1 \cup Z$, this forces $t = 1$ or $rws' = 1$. In both cases, $rw \in Z^*r'$. Thus $rw \in Z^*r'$, and similarly $ws' \in sZ^*$. By Proposition 4.2.3,

$$((r, s), \varphi_Z(w), (r', s')) = 1.$$

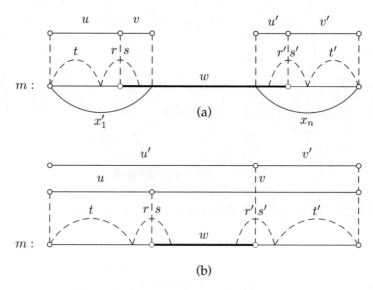

Figure 4.8 The cases of (a) $n > 1$ and (b) $n = 1$.

Assume conversely that

$$((r, s), \varphi_Z(w), (r', s')) = 1.$$

Then by Proposition 4.2.3

$$rw = zr', \qquad ws' = sz'$$

for some $z, z' \in Z^*$. Then $rws' \in Z^*$, and Y being complete, there exist $t, t' \in Z^*$ such that $m = trws't' \in X^*$. Let

$$m = trws't' = trsz't' = tzr's't' = x_1 \cdots x_n$$

with $n \geq 1, x_1, \ldots, x_n \in X$. We may assume that t and t' have been chosen of minimal length, so that t is a proper prefix of x_1 and t' is a proper suffix of x_n. But then, since $m \in Z^*$ and also $trs \in Z^*$, trs is a prefix of x_1 and $r's't'$ is a suffix of x_n (Figure 4.8). Define

$$
\begin{aligned}
x_1 &= uv && \text{with } u = tr, \ v \in sZ^*, \\
x_n &= u'v' && \text{with } u' = t'r', \ v' \in s'Z^*.
\end{aligned}
$$

Then (u, v) and (u', v') are states of $\mathcal{A}_D^*(X)$, and moreover

$$\rho((u, v)) = (r, s), \qquad \rho((u', v')) = (r', s'),$$

and

$$m = uwv' = uvx_2 \cdots x_n = x_1 \cdots x_{n-1}u'v'.$$

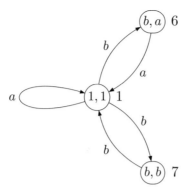

Figure 4.9 The flower automaton of Z.

Thus

$$uw \in X^*u' \quad \text{and} \quad wv' \in vX^*.$$

Finally, consider the set R of states of $\mathcal{A}_D^*(Y)$. Then R can be identified with

$$R' = \{(u, v) \in P \mid u, v \in Z^*\}.$$

The edges of $\mathcal{A}_D^*(Y)$ correspond to those paths $(u, v) \to (u', v')$ of $\mathcal{A}_D^*(X)$ with endpoints in R', and with label in Z. □

Example 4.2.8 Recall from Chapter 2 that the code $X = \{aa, ba, bb, baa, bba\}$ is a composition of $Y = \{cc, d, e, dc, ec\}$ and $Z = \{a, ba, bb\}$. The flower automaton $\mathcal{A}_D^*(X)$ is given in Figure 4.5. The flower automaton $\mathcal{A}_D^*(Z)$ is given in Figure 4.9. It is obtained from $\mathcal{A}_D^*(X)$ by the reduction

$$\rho(1) = \rho(2) = \rho(3) = \rho(4) = \bar{1},$$
$$\rho(6) = \rho(8) = \bar{6},$$
$$\rho(5) = \rho(7) = \bar{7}.$$

The flower automaton $\mathcal{A}_D^*(Y)$ is given in Figure 4.10.

4.3 Decoders

Let $X \subset A^+$ be a code and let $\beta : B^* \to A^*$ be a coding morphism for X. Since β is injective, there exists a partial function,

$$\gamma : A^* \to B^*$$

with domain X^* and such that $\gamma(\beta(u)) = u$ for all $u \in B^*$. We say that γ is a *decoding function* for X.

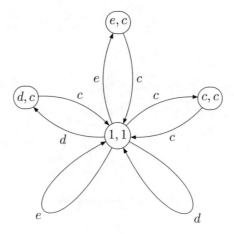

Figure 4.10 The flower automaton of Y.

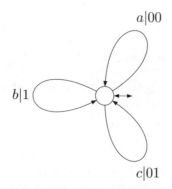

Figure 4.11 A simple encoder.

A coding morphism $\beta : B^* \to A^*$ can be realized by a one-state literal transducer, with the set of labels of edges being simply the pairs $(b, \beta(b))$ for b in B.

Example 4.3.1 Consider the encoding defined by $\gamma(a) = 00$, $\gamma(b) = 1$, and $\gamma(c) = 01$. The corresponding encoding transducer is given in Figure 4.11

Transducers for decoding are more interesting. For the purpose of coding and decoding, we are concerned with transducers which define single-valued mappings in both directions. We need two additional notions.

A literal transducer is called *deterministic* (resp. *unambiguous*) if its associated input automaton is deterministic (resp. unambiguous).

Clearly, the relation realized by a deterministic transducer is a function. Whenever there is a path $p \xrightarrow{u|w} q$ starting in p with input label u and output label w, we write $p \cdot u$ for q and $p * u$ for w. Observe that $p \cdot uv = p \cdot u \cdot v$. This is Equation (1.8). Also,

$$p * uv = (p * u)(p \cdot u * v). \qquad (4.7)$$

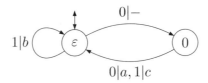

Figure 4.12 A deterministic decoder for $X = \{1, 00, 01\}$. A dash means no output. Here ε denotes the empty word.

Indeed, if there is a path starting in p with input label uv, then it is of the form $p \xrightarrow{u|w} q \xrightarrow{v|z} r$ for states $q = p \cdot u$ and $r = q \cdot v$ and output labels $w = p * u$ and $z = q * v$. It follows that $wz = (p * u)(p \cdot u * v)$ as claimed.

Let $\beta : B^* \rightarrow A^*$ be a coding morphism with finite alphabets A and B, and let $X = \beta(B)$. The *prefix transducer* T over B and A associated to β has as states the set of proper prefixes of words in X. The state corresponding to the empty word 1 is the initial and terminal state. There is an edge $p \xrightarrow{a|-} pa$, where the dash $(-)$ represents the empty word, for each prefix p and letter a such that pa is a prefix, and an edge $p \xrightarrow{a|b} 1$ for each p and letter a with $pa = \beta(b) \in X$. Note that for each edge $p \xrightarrow{a|v} q$ of the prefix transducer, one has

$$pa = \beta(v)q. \qquad (4.8)$$

Note also that the prefix transducer is finite when B is finite, and thus when the code X is finite.

Proposition 4.3.2 *For any coding morphism $\beta : B^* \rightarrow A^*$, the prefix transducer T associated to β is unambiguous and realizes the decoding function. When the code $\beta(B)$ is prefix, then the transducer T is deterministic.*

Proof. Let A be the input automaton of T. Then $A = B^*$, where B is the automaton whose states are the prefixes of the words in X. By Proposition 4.1.2, the automaton A is unambiguous. Moreover, each simple path $1 \rightarrow 1$ is labeled by construction with $(\beta(b), b)$ for some letter $b \in B$. Thus T realizes the associated decoding function. When the code is prefix, the decoder is deterministic.   □

Example 4.3.3 The decoder corresponding to the prefix code $X = \{1, 00, 01\}$ is represented in Figure 4.12.

Example 4.3.4 Consider the code $X = \{00, 10, 100\}$. The decoder given by the construction is represented in Figure 4.13.

Observe that the transducer constructed in the proof is finite (that is has a finite number of states) whenever the code is finite.

Assume now that the code X is finite. As a consequence of the proposition, decoding can always be realized in linear time with respect to the length of the encoded string

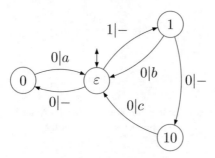

Figure 4.13 A unambiguous decoder for the code $X = \{00, 10, 100\}$ which is not prefix. Again ε denotes the empty word.

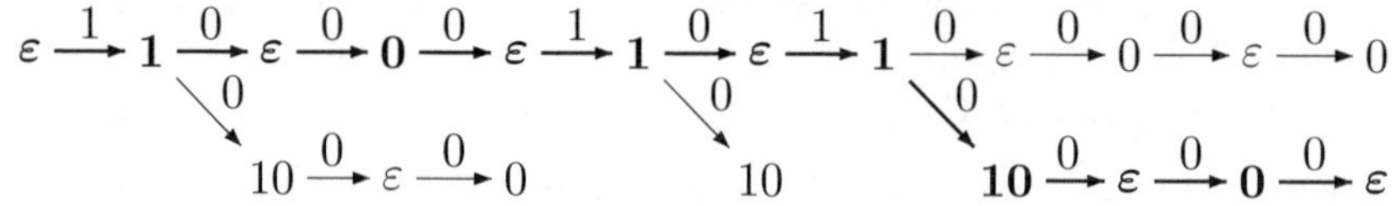

$$\varepsilon \xrightarrow{1} 1 \xrightarrow{0} \varepsilon \xrightarrow{0} 0 \xrightarrow{0} \varepsilon \xrightarrow{1} 1 \xrightarrow{0} \varepsilon \xrightarrow{1} 1 \xrightarrow{0} \varepsilon \xrightarrow{0} 0 \xrightarrow{0} \varepsilon \xrightarrow{0} 0$$

with branches $\searrow 0$ leading to $10 \xrightarrow{0} \varepsilon \xrightarrow{0} 0$, $\searrow 0$ leading to 10, and $\searrow 0$ leading to $10 \xrightarrow{0} \varepsilon \xrightarrow{0} 0 \xrightarrow{0} \varepsilon$.

Figure 4.14 The decoding of 10001010000. Here also ε denotes the empty word.

(considering the number of states of the transducer as a constant). Indeed, given a word $w = a_1 \cdots a_n$ of length n to be decoded, one computes the sequence of sets S_i of states accessible from the initial state for each prefix $a_1 \cdots a_i$ of length i of w, with the convention $S_0 = \{\varepsilon\}$. Of course the terminal state ε is in S_n. Working backwards, we set $q_n = \varepsilon$ and we identify in each set S_i the unique state q_i such that there is an edge $q_i \xrightarrow{a_i} q_{i+1}$ in the input automaton. The uniqueness comes from the unambiguity of the transducer. The corresponding sequence of output labels gives the decoding.

Example 4.3.5 Consider again the code $C = \{00, 10, 100\}$. The decoding of the sequence 10001010000 is represented in Figure 4.14. Working from left to right produces the tree of possible paths in the decoder of Figure 4.13. Working backwards from the state ε in the last column produces the successful path indicated in boldface.

The notion of deterministic transducer is too constrained for the purpose of coding and decoding because it does not allow a lookahead on the input or equivalently a delay on the output. The notion of sequential transducer to be introduced now fills this gap.

A *sequential transducer* over the input alphabet A and the output alphabet B is composed of a deterministic transducer over A and B and of an output function. This function maps the terminal states of the transducer into words on the output alphabet B. The function $f : A^* \to B^*$ realized by a sequential transducer is obtained by appending, to the value of the deterministic transducer, the image of the output function on the arrival state. Formally, the value on the input word $x \in A^*$ is

$$f(x) = g(x)\sigma(i \cdot x),$$

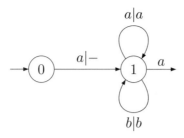

Figure 4.15 A sequential transducer realizing a cyclic shift on words starting with the letter a.

where $g(x) \in B^*$ is the value of the deterministic transducer on the input word x, $i \cdot x$ is the state reached from the input state i by the word x, and σ is the output function. This is defined only if the state $i \cdot x$ is a terminal state.

Deterministic transducers are a special case of sequential transducers. They are obtained when the output function takes always the value 1.

Example 4.3.6 The automaton given in Figure 4.15 computes, for each input word of the form aw, the output word wa. It is undefined on input words that do not start with the letter a. The initial state is 0 and the state 1 is terminal. The output function σ satisfies $\sigma(1) = a$ (the value of σ is indicated on the figure as the label of the outgoing edge).

Contrary to automata, it is not always true that a finite transducer is equivalent to a finite sequential transducer. Nonetheless, there is a procedure to compute a (possibly infinite) sequential transducer S that is equivalent to a given literal transducer T realizing a function.

Let $T = (Q, I, T)$ be a literal transducer realizing a function $A^* \to B^*$. We define a sequential transducer S as follows. The states of S are sets of pairs (u, p). Each pair (u, p) is composed of an output word $u \in B^*$ and a state $p \in Q$ of T.

The edges of S are the following. For a state s of S and an input letter $a \in A$, one first computes the set $\bar{s}$ of pairs (uv, q) such that there is a pair (u, p) in s and an edge $p \xrightarrow{a|v} q$ in T. In a second step, one chooses the longest common prefix z of all words uv, and one defines a set t by $t = \{(w, q) \mid (zw, q) \in \bar{s}\}$. The set t is a state of S. This defines an edge from state s to state t labeled with (a, z). The initial state is $\{(1, i) \mid i \in I\}$. The terminal states are the sets t containing a pair (u, q) with $q \in T$ terminal in T. Since T realizes a function, two pairs (u, q) and (u', q') in the same terminal state t with $q, q' \in T$ satisfy $u = u'$.

The output function σ of S is defined on the state t of S by $\sigma(t) = u$, where u is the unique word such that (u, q) is in t for some $q \in T$. The states of S are the sets of pairs which are accessible from the initial state of S. The words u appearing as first components in the pairs (u, p) will be called *remainders*.

The process of building new states of S will not halt if the lengths of the remainders is not bounded. There exist a priori bounds for the maximal length of the remainders whenever the determinization is possible. This makes the procedure effective in this case.

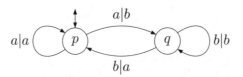

Figure 4.16 Another transducer realizing a cyclic shift on words starting with the letter a.

Example 4.3.7 Consider the transducer given in Figure 4.16. The result of the determinization algorithm is the transducer of Figure 4.15. State 0 is composed of the pair $(1, p)$, and state 1 is formed of the pairs (a, p) and (b, q).

Let $S = (P, I, S)$ be a literal transducer over the alphabets A, B and let $T = (Q, J, T)$ be a literal transducer over the alphabets B, C. We denote by $S \circ T$ the literal transducer U over the alphabets A, C given by $U = (P \times Q, I \times J, S \times T)$ with edges

$$(p, q) \xrightarrow{a|w} (r, s)$$

for all edges $p \xrightarrow{a|v} r$ in S and paths $q \xrightarrow{v|w} s$ in T. The transducer $U = S \circ T$ is the transducer *composed* of S and T.

Proposition 4.3.8 *The relation realized by the composed transducer $S \circ T$ is the composition of the relations realized by S and T.*

Proof. There is a path $(p, q) \xrightarrow{u|w} (r, s)$ in $U = S \circ T$ if and only if there is a path $p \xrightarrow{u|v} r$ in S and a path $q \xrightarrow{v|w} s$ in T. Thus $(u, w) \in A^* \times C^*$ is an element of the relation realized by U if and only if there exists $v \in B^*$ such that (u, v) is an element if the relation realized by S and (v, w) belongs to the relation realized by T. $\square$

Proposition 4.3.9 *If S and T are unambiguous, then $S \circ T$ is unambiguous.*

Proof. Let $u = a_1 a_2 \cdots a_n$ be a word with $a_i \in A$ and $n \geq 0$. Suppose that there are two paths in $U = S \circ T$ with the same input label u and the same starting and ending states. More precisely, assume that in U, there are paths

$$(p_0, q_0) \xrightarrow{a_1|w_1} (p_1, q_1) \cdots (p_{n-1}, q_{n-1}) \xrightarrow{a_n|w_n} (p_n, q_n),$$

$$(p_0', q_0') \xrightarrow{a_1|w_1'} (p_1', q_1') \cdots (p_{n-1}', q_{n-1}') \xrightarrow{a_n|w_n'} (p_n', q_n')$$

with $(p_0, q_0) = (p_0', q_0')$ and $(p_n, q_n) = (p_n', q_n')$. Then there exist in the transducer S two paths $p_0 \xrightarrow{a_1|v_1} p_1 \cdots p_{n-1} \xrightarrow{a_n|v_n} p_n$ and $p_0' \xrightarrow{a_1|v_1'} p_1' \cdots p_{n-1}' \xrightarrow{a_n|v_n'} p_n'$ for appropriate words $v_1, \ldots, v_n, v_1', \ldots, v_n'$ and, in the transducer T, two paths $q_0 \xrightarrow{v_1|w_1} q_1 \cdots q_{n-1} \xrightarrow{v_n|w_n} q_n$ and $q_0' \xrightarrow{v_1'|w_1'} q_1' \cdots q_{n-1}' \xrightarrow{v_n'|w_n'} q_n'$. Since S is unambiguous, the two paths coincide and thus $p_i = p_i'$ and $v_i = v_i'$. Since T is unambiguous and the two

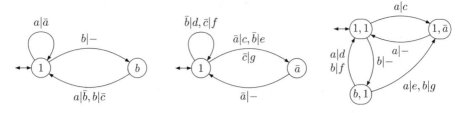

Figure 4.17 The transducers $\mathcal{T}$, $\mathcal{S}$ and $\mathcal{S} \circ \mathcal{T}$.

paths have the same input label, they coincide. Therefore $q_i = q_i'$ and $w_i = w_i'$. Thus the two paths in $\mathcal{U}$ coincide. □

Corollary 4.3.10 *Let $X = Y \circ Z$ be a code over A composed of the code Y over B and the code Z over A, and let $\gamma : B^* \to C^*$ and $\delta : A^* \to B^*$ be the decoding functions for Y and Z. If $\mathcal{S}$ and $\mathcal{T}$ are unambiguous transducers realizing γ and δ, then $\mathcal{T} \circ \mathcal{S}$ realizes the decoding function $\gamma \circ \delta : A^* \to C^*$.* □

Example 4.3.11 Let $X = \{aa, ba, baa, bb, bba\}$, $Y = \{\bar{a}\bar{a}, \bar{b}, \bar{b}\bar{a}, \bar{c}, \bar{c}\bar{a}\}$, and $Z = \{a, ba, bb\}$. Then $X = Y \circ_\beta Z$ with $B = \{\bar{a}, \bar{b}, \bar{c}\}$ and $\beta(\bar{a}) = a$, $\beta(\bar{b}) = ba$ and $\beta(\bar{c}) = bb$. The prefix transducer $\mathcal{S}$ of Z, the suffix transducer $\mathcal{T}$ of Y and their composition are shown in Figure 4.17, with $C = \{c, d, e, f, g\}$.

Proposition 4.3.12 *If $\mathcal{S}$ and $\mathcal{T}$ are deterministic, then $\mathcal{S} \circ \mathcal{T}$ is deterministic.*

Proof. Let $(p, q) \xrightarrow{a|w} (r, s)$ and $(p, q) \xrightarrow{a|w'} (r', s')$ be two edges of $\mathcal{U} = \mathcal{S} \circ \mathcal{T}$. Then there exist edges $p \xrightarrow{a|v} r$ and $p \xrightarrow{a|v'} r'$ in $\mathcal{S}$ and paths $q \xrightarrow{v|w} s$ and $q \xrightarrow{v'|w'} s'$ in $\mathcal{T}$. Since $\mathcal{S}$ is deterministic, $v = v'$ and $r = r'$. Since $\mathcal{T}$ is deterministic, this in turn implies that $w = w'$ and $s = s'$. Thus the two edges in $\mathcal{U}$ coincide. □

4.4 Exercises

Section 4.1

4.1.1 Show that a submonoid M of A^* is recognizable and free if and only if there exists an unambiguous trim finite automaton $\mathcal{A} = (Q, 1, 1)$ that recognizes M.

Section 4.2

4.2.1 Let X be a subset of A^+ and let $\mathcal{A}_D^*(X) = (P, (1, 1), (1, 1))$ be the flower automaton of X. Let φ be the associated representation. Show that for all $(p, q), (r, s) \in P$ and $w \in A^*$ we have

$$((p, q), \varphi(w), (r, s)) = (q(\underline{X})^* r, w) + (pw, r)(q, ws).$$

4.2.2 Let $\mathcal{A} = (P, i, T)$ and $\mathcal{B} = (Q, j, S)$ be two automata, and let $\rho : P \to Q$ be a reduction from $\mathcal{A}$ on $\mathcal{B}$ such that $i = \rho^{-1}(j)$. Show that if $\mathcal{A}$ is deterministic, then so is $\mathcal{B}$.

4.5 Notes

Unambiguous automata and their relation to codes appear in Schützenberger (1961d, 1965b). They appear also under the name of *information lossless machines* in Huffman (1959), see also Kohavi (1978).

Unambiguous automata are closely related to the notion of *finite-to-one maps* used in symbolic dynamics (see Lind and Marcus (1995)). The connection is the fact that in a finite unambiguous automaton, any word is the label of a finite number of paths. This number is bounded by the square of the number of states of the automaton. Indeed, for any pair p, q of states of $\mathcal{A}$ and any word w, there is at most one path $p \xrightarrow{w} q$.

Proposition 4.1.6 appears in Schützenberger (1965b). Formula (4.2) can be written in noncommutative variables using the notion of *quasideterminant* (see Gel'fand and Retakh (1991)).

For a comprehensive presentation of transducers, one may consult Eilenberg (1974) or Berstel (1979). For a recent exposition, see Sakarovitch (2009).

For the determinization algorithm of transducers, see Lothaire (2005). The decoding in linear time with the help of an unambiguous transducer is based on the *Schützenberger covering* of an unambiguous automaton, see Sakarovitch (2008).

Deciphering delay

This chapter is devoted to codes with finite deciphering delay. Intuitively, codes with finite deciphering delay can be decoded, from left to right, with a finite lookahead. There is an obvious practical interest in this condition. Codes with finite deciphering delay form a family intermediate between prefix codes and general codes. There are two ways to define the deciphering delay, counting either codewords or letters. The first one is called verbal delay, or simply delay for short, and the second one literal delay.

The first section is devoted to codes with finite verbal deciphering delay. We present first some preliminary material. In particular we prove a characterization of the deciphering delay in terms of simplifying words.

In the second section, we prove Schützenberger's theorem (Theorem 5.2.4) saying that a finite maximal code with finite deciphering delay is prefix. We prove that any rational code with finite deciphering delay is contained in a maximal rational code with the same delay (Theorem 5.2.9).

The next section considers the literal deciphering delay, that is the deciphering delay counted in terms of letters instead of words of the code. A code with finite literal deciphering delay is called weakly prefix. We introduce the notion of automata with finite delay, also called weakly deterministic. We prove the equivalence between weakly prefix codes and weakly deterministic automata (Proposition 5.3.4). We use this characterization to give yet another proof of Schützenberger's theorem. Next, we show that a rational completion with the same literal deciphering delay exists (Theorem 5.3.7).

5.1 Deciphering delay

A subset X of A^+ is said to have *finite verbal deciphering delay* if there exists an integer $d \geq 0$ such that the following condition holds: For $x, x' \in X$, $y \in X^d$, $y' \in X^*$,

$$xy \leq x'y' \quad \text{implies} \quad x = x'. \tag{5.1}$$

(Recall that we write $u \leq u'$ to express that u is a prefix of u'.) If this condition holds for an integer d, we say that X has verbal deciphering delay d. We omit the term verbal when possible.

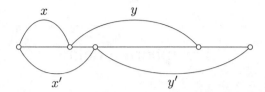

Figure 5.1 Forbidden configuration for finite deciphering delay.

The definition can be rephrased as follows. Let $w \in A^*$ be a word having two prefixes in X^+, and such that the shorter one is in X^{1+d}. Then the two prefixes start with the same word in X.

If X has deciphering delay d, it also has deciphering delay d' for $d' \geq d$. The smallest integer d satisfying (5.1) is called the *minimal deciphering delay* of X. If no such integer exists, the set X has *infinite deciphering delay*.

This notion of deciphering delay is clearly oriented from left to right. It is straightforward to define a dual notion (working from right to left). The terminology is justified by the following consideration: During a left-to-right parsing of an input word, the delay between the moment when a possible factor of an X-factorization is discovered, and the moment when these factors are definitively valid, is bounded by the deciphering delay.

If the deciphering delay of X is infinite, then there exist $x, x' \in X$ with $x \neq x'$ and $y_1, y_2, \ldots, y'_1, y'_2, \ldots \in X$ such that for all $n \geq 1$, $x y_1 y_2 \cdots y_n$ is a prefix of $x' y'_1 y'_2 \cdots y'_n$ or vice versa.

It follows from the definition that the sets with delay $d = 0$ are the prefix codes. This is the reason why prefix codes are also called instantaneous codes. In this sense, codes with finite delay are a natural generalization of prefix codes.

Proposition 5.1.1 *A subset X of A^+ which has finite deciphering delay is a code.*

Proof. Let X have deciphering delay d. We may suppose $X \neq \emptyset$. Assume there is an equality

$$w = x_1 x_2 \cdots x_n = y_1 y_2 \cdots y_m,$$

with $n, m \geq 1$, $x_1, \ldots, x_n, y_1, \ldots, y_m \in X$. Let $z \in X$. Then $w z^d \in y_1 X^*$. By (5.1), we have $x_1 = y_1$, $x_2 = y_2$ and so on. Thus, X is a code. □

Example 5.1.2 The suffix code $X = \{aa, ba, b\}$ has infinite deciphering delay. Indeed, for all $d \geq 0$, the word $b(aa)^d \in X^{1+d}$ is a prefix of $y(aa)^d$ with $y = ba \neq b$.

For a set $X \subset A^+$, define, as in Section 2.3, a sequence $(U_n)_{n \geq 0}$ of subsets of A^* by setting

$$U_1 = X^{-1} X \setminus 1 \qquad U_{n+1} = X^{-1} U_n \cup U_n^{-1} X, \quad n \geq 1.$$

Proposition 5.1.3 *The set X has finite deciphering delay if and only if the set U_n is empty for some n.*

Proof. By Lemma 2.3.3, for $n \geq 1$ one has $u \in U_n$ if and only if there are $x_1, \ldots, x_i$, $y_1, \ldots y_j \in X$ with $x_1 \neq y_1$, $i + j = n + 1$ and u suffix of y_j such that $x_1 \cdots x_i u = y_1 y_2 \cdots y_j$. We first verify that if X has deciphering delay d then $U_{2d+1} = \emptyset$. Suppose the contrary. Let $x_1, \ldots, x_i, y_1, \ldots y_j \in X$ be such that $x_1 \cdots x_i u = y_1 y_2 \cdots y_j$ with $i + j = 2d + 2$, u suffix of y_j and $x_1 \neq y_1$. Then $i - 1 \leq d - 1$ since otherwise $x_1 = y_1$. Similarly, $j - 2 \leq d - 1$ since otherwise, with $y_j = vu$, we have $y_1 y_2 \cdots y_{j-1} v = x_1 \cdots x_i$ and thus $x_1 = y_1$ again. Thus $i + j \leq 2d + 1$, a contradiction.

Conversely we show that if $U_n = \emptyset$, then X has deciphering delay $n - 1$. Let indeed $x, x' \in X$, $y \in X^{n-1}$, $y' \in X^j$ for $j \geq 0$ and $u \in A^*$ be such that $xyu = x'y'$. If $x \neq x'$, then $u \in U_m$ for some $m \geq n$, a contradiction. This forces $x = x'$ proving that X has deciphering delay $n - 1$. □

Example 5.1.4 The set $X = \{a, ab, bc, cd, de\}$ has deciphering delay 2. We obtain $U_1 = \{b\}$, $U_2 = \{c\}$, $U_3 = \{d\}$, $U_4 = \{e\}$, $U_5 = \emptyset$.

We reformulate the definition of deciphering delay as follows. Let X be a code. A word $s \in A^*$ is said to be *simplifying* for X if for all $x \in X^*$ and $v \in A^*$,

$$xsv \in X^* \Rightarrow sv \in X^*.$$

Proposition 5.1.5 *A code X has deciphering delay d if and only if all words of X^d are simplifying.*

Proof. Let us first suppose that X has delay d. Let $x \in X^d$, $x_1, \ldots, x_p \in X$ and $v \in A^*$ be such that $x_1 \cdots x_p x v \in X^*$. Thus

$$x_1 \cdots x_p x v = y_1 \cdots y_q$$

for some $y_1, \ldots, y_q \in X$. Since X has delay d, it follows that $x_1 = y_1, \ldots, x_p = y_p$, whence $q \geq p$ and $xv = y_{p+1} \cdots y_q$. Thus $xv \in X^*$. This shows that x is simplifying.

Conversely, suppose $y \in X^d$. Let $x, x' \in X$ and $u \in A^*$ be such that $xyu \in x'X^*$. Then $yu \in X^*$. Since X is a code, this implies $x = x'$. Thus X has deciphering delay d. □

The following statement characterizes the decoders of codes with finite deciphering delay in terms of sequential transducers introduced in Section 4.3.

Proposition 5.1.6 *Let $X \subset A^+$ be a finite code, and let $\beta : B^* \to A^*$ be a coding morphism for X. The corresponding decoding function $A^* \to B^*$ is realizable by a finite sequential transducer if and only if X has finite verbal deciphering delay.*

Proof. Suppose first that X has verbal deciphering delay d. By Proposition 4.3.2, the prefix transducer $\mathcal{T}$ associated with β realizes the corresponding decoding function γ from A^* to B^*. Let $\mathcal{S}$ be the sequential transducer obtained from $\mathcal{T} = (Q, 1, 1)$ by the determinization procedure described in Section 4.3. Let U be the set of remainders, that is of words $u \in B^*$ such that (u, p) belongs to a state of $\mathcal{S}$ for some state p of

Table 5.1 *States and output function*
for the sequential transducer S.

state	1	2	3
pairs	$(1, 1)$	$(\bar{a}, 1)$ $(1, a)$	$(\overline{ab}, 1)$ $(1, ab)$
output	1	$\bar{a}$	$\overline{ab}$

T. We show that any $u \in U$ has length at most d. This will prove that S is finite, and thus that the decoding function is realizable by a finite sequential transducer.

For this, we observe that if two pairs $(w, q), (w', q') \in B^* \times Q$ belong to the same state of S, then $\beta(w)q = \beta(w')q'$. This is true for the initial state $(1, 1) \in B^* \times Q$ (here the second 1 is the initial state of T). Next, if $(w, q), (w', q') \in t$ are two pairs belonging to some state $t \neq (1, 1)$ of S, then there is, by definition of S, and edge $s \xrightarrow{a,z} t$ in S for some $a \in A$, $z \in B^*$. Thus there are two pairs $(u, p), (u', p')$ in s and two edges $p \xrightarrow{a|v} q$ and $p' \xrightarrow{a|v'} q'$ in T such that $uv = zw$ and $u'v' = zw'$. We argue by induction on the length of the path from the initial state to t in S. Thus we may assume that $\beta(u)p = \beta(u')p'$. Since $p \xrightarrow{a|v} q$ and $p' \xrightarrow{a|v'} q'$ are edges in T, we have by (4.8), $pa = \beta(v)q$ and $p'a = \beta(v')q'$. This implies in turn $\beta(uv)q = \beta(u'v')q'$. Simplifying both sides by $\beta(z)$ gives $\beta(w)q = \beta(w')q'$.

Consider now a pair $(u, p) \in B^+ \times Q$ which belongs to a state of S. Since the word u is nonempty, by definition of S, there is another pair (u', p') in the same state of S such that u, u' have no nonempty common prefix. By the above observation, we have $\beta(u)p = \beta(u')p'$. Since p' is a prefix of some codewords, the word $\beta(u)$ is a prefix of a word $\beta(u'b)$ for some $b \in B$. Now set $\beta(u) = xy$, $\beta(u'b) = x'y'$ with $x, x' \in X$, $y, y' \in X^*$. Since u and u' start with distinct letters, one has $x \neq x'$. By the definition of the deciphering delay, this implies that $|u| \leq d$, completing the proof of the first implication.

Conversely, suppose that $S = (Q, i, \sigma)$ is a sequential transducer with output function σ realizing γ. Let d be the maximal length of the words $\sigma(p)$ for $p \in Q$. In view of applying again Equation (5.1), let $x, x' \in X$ and $y, y' \in X^*$ be such that $xy \leq x'y'$ with $x \neq x'$. We show that $y \in X^{d'}$ with $d' < d$. Let p be the state reached from the initial state i by reading x. There is no output along this reading because xy is a prefix of $x'y'$ and, since $x \neq x'$, it cannot be decided whether to output $\gamma(x)$ or $\gamma(x')$. Thus we have $i \xrightarrow{xy|1} p$. Moreover, if u is defined by $\beta(u) = xy$, then $\sigma(p) = u$. Since $|u| \leq d$ and $\beta(u) \in X^{1+d'}$, one has $1 + d' \leq d$, and thus $d' < d$. Thus X has verbal deciphering delay d.   □

Example 5.1.7 Consider the code $X = \{a, b, abc\}$ on the alphabet $A = \{a, b, c\}$, with $B = \{\bar{a}, \bar{b}, \bar{c}\}$ and coding morphism given by $\bar{a} \mapsto a, \bar{b} \mapsto b, \bar{c} \mapsto abc$. It has deciphering delay 2. The prefix transducer T and the sequential transducer S obtained by determinization are shown in Figure 5.2. The states of S are renumbered $1, 2, 3$, and the correspondence with the states obtained by the determinization procedure, and the output function σ are given in Table 5.1.

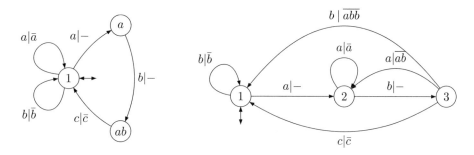

Figure 5.2 The transducers $\mathcal{T}$ and $\mathcal{S}$.

5.2 Maximal codes

We now study maximal codes with finite deciphering delay. The following result is similar to Proposition 2.2.5.6.

Proposition 5.2.1 *Let X be a subset of A^+ which has finite deciphering delay. If $y \in A^+$ is an unbordered word such that*

$$X^* y A^* \cap X^* = \emptyset,$$

then $Y = X \cup y$ has finite deciphering delay.

Proof. Consider the set $V = X^* y$. It is a prefix code. Indeed, assume that $v = xy$ and $v' = x'y$ with $x, x' \in X^*$, and $v < v'$. Then necessarily $v \le x'$ since y is unbordered. But then $x' \in X^* y A^*$, a contradiction. Note also that

$$V^+ A^* \cap X^* = \emptyset$$

since $V^+ A^* \subset V A^*$.

Let X have deciphering delay d and let $e = d + |y|$. We show that Y has deciphering delay e. For this, let us consider a relation

$$w = y_1 y_2 \cdots y_{e+1} u = y_1' y_2' \cdots y_n'$$

with $y_1, \dots, y_{e+1}, y_1', \dots y_n' \in Y$, $u \in A^*$ and, arguing by contradiction, assume that $y_1 \ne y_1'$.

First, let us verify that one of $y_1, \dots, y_{e+1}$ is equal to y. Assume the contrary. Then $y_1 \cdots y_{d+1} \in X^{d+1}$. Let q be the smallest integer such that (Figure 5.3)

$$y_1 \cdots y_{d+1} \le y_1' \cdots y_q'.$$

The delay of X being d, and $y_1 \ne y_1'$, one among $y_1', \dots, y_q'$ must be equal to y. We cannot have $y_i' = y$ for an index $i < q$, since otherwise $y_1 \cdots y_{d+1} \in V^+ A^* \cap X^*$. Thus $y_q' = y$ and $y_1' \cdots y_q' \in V$. Note that $y_1' \cdots y_{q-1}' \le y_1 \cdots y_{d+1}$. Next, $|y_{d+2} \cdots y_{e+1}| \ge e - d = |y|$. It follows that

$$y_1' \cdots y_q' \le y_1 \cdots y_{e+1}.$$

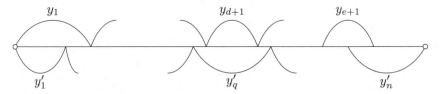

Figure 5.3 Two factorizations of the word w.

But then $y_1 \cdots y_{e+1} \in X^* \cap X^* y A^*$, which is impossible. This shows the claim, namely, that one of $y_1, \ldots, y_{e+1}$ is equal to y.

It follows that w has a prefix $y_1 y_2 \cdots y_p$ in V with $y_1, \ldots, y_{p-1} \in X$ and $y_p = y$. By the hypothesis, one of $y'_1, \ldots, y'_n$ must be equal to y. Thus w has also a prefix $y'_1 y'_2 \cdots y'_q$ in V with $y'_1, \ldots, y'_{q-1} \in X$ and $y'_q = y$. The code V being prefix, we have

$$y_1 y_2 \cdots y_{p-1} = y'_1 y'_2 \cdots y'_{q-1}.$$

Since X is a code, this and the assumption $y_1 \neq y'_1$ imply that $p = q = 1$. But then $y_p = y = y'_q$. This gives the final contradiction. □

Proposition 5.2.1 has the following interesting consequence.

Theorem 5.2.2 *Let X be a thin subset of A^+. If X has finite deciphering delay, then the following conditions are equivalent:*

(i) *X is a maximal code.*
(ii) *X is maximal in the family of codes with finite deciphering delay.*

Proof. The case where A has just one letter is clear. Thus, we suppose that Card(A) $\geq$ 2. It suffices to prove (ii) $\Longrightarrow$ (i). For this, it is enough to show that X is complete. Assume the contrary and consider a word u which is not a factor of a word in X^*. According to Proposition 1.1.3.6, there exists $v \in A^*$ such that $y = uv$ is unbordered. But then $A^* y A^* \cap X^* = \emptyset$ and by Proposition 5.2.1, $X \cup y$ has finite deciphering delay. This gives the contradiction. □

A word p is *strongly right completable* (for X) if, for all $u \in A^*$, there exists $v \in A^*$ such that $puv \in X^*$. Clearly, a strongly right completable word is right completable. The set of strongly right completable words is denoted by $E(X)$.

The following statement is the counterpart of Theorem 2.5.5 for codes with finite deciphering delay since it shows that maximal codes with finite deciphering delay satisfy a condition which is stronger than being complete.

Proposition 5.2.3 *Let $X \subset A^+$ be a maximal code with deciphering delay d. Then for any $x \in X^d$ and $u \in A^*$ there exists a word $v \in A^*$ such that $xuv \in X^*$. In other words $X^d \subset E(X)$.*

Proof. The case of a one letter alphabet is clear. Thus, assume that Card(A) $\geq$ 2. Let $x \in X^d$ and $u \in A^*$. By Proposition 1.1.3.6, there is a word $v \in A^*$ such that

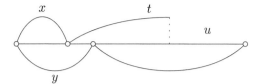

Figure 5.4 An element t of T.

$y = xuv$ is unbordered. This implies that

$$X^* y A^* \cap X^* \neq \emptyset.$$

Indeed, otherwise $X \cup y$ would be a code by Proposition 5.2.1 and Proposition 5.1.1, contradicting the maximality of X.

Consequently, there exist $z \in X^*$, $w \in A^*$ such that $zyw \in X^*$. By Proposition 5.1.5, x is simplifying. Thus, $zyw = zxuvw \in X^*$ implies $xuvw \in X^*$. This shows that x is strongly right completable. □

We now state and prove an important result.

Theorem 5.2.4 (Schützenberger) *A finite maximal code with finite deciphering delay is prefix.*

In an equivalent manner, a maximal finite code is either prefix or has infinite deciphering delay.

Proof. We argue by contradiction and suppose that X is not a prefix code. Denote by P the set of prefixes of the words in X^*. Define (see Figure 5.4)

$$T = \{t \in P \mid \exists x, y \in X, \ x \neq y \text{ and } xtA^* \cap yX^* \neq \emptyset\}.$$

We first observe that T contains the empty word. Indeed, since X is not a prefix code, there exist $x, y \in X$ with $y = xu$ for some $u \in A^+$. Thus $xA^* \cap \{y\}$ is nonempty. This shows that $1 \in T$. Thus T is not empty.

We next show that T is finite. Let L be the maximum length of the words in X. Suppose that there exists $t \in T$ of length $|t| \geq dL$, where X has deciphering delay d. Since $t \in T$, one has $t = x_1 \cdots x_d t'$ for some codewords $x_1, \ldots, x_d \in X$ and some $t' \in P$.

Let $x, y \in X, x \neq y$ be words such that $xtA^* \cap yX^*$ is nonempty. We have $xtu = yw$ for some word $w \in X^*$. Consequently $xx_1 \cdots x_d t'u = yw$, and since X has delay d, we obtain $x = y$, a contradiction. Therefore t cannot be in T. This shows that all words in T have length $< dL$, and thus T is finite.

We consider now some t in T of maximal length. We have, for some $x, y \in X, x \neq y$, that $xtA^* \cap yX^*$ is nonempty. Hence $xtu \in yX^*$ for some word u, and we may suppose that $u \in A^+$. Indeed, if $u = 1$, we replace u by any word of X. Set $u = au'$, where a is the first letter of u. We are going to show that $ta \in P$, which implies $ta \in T$, a contradiction.

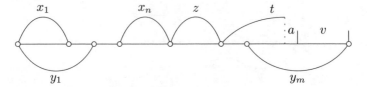

Figure 5.5 A completion of $w = zta$.

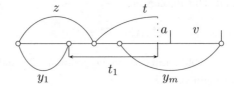

Figure 5.6 A consequence $y_1 \neq z$ is that $zt = y_1 t_1$.

Set $w = zta$, where z is a word of maximal length in the (finite) code X. By Proposition 2.5.6, $X^* w A^* \cap X^*$ is nonempty. Therefore there are $x_1, \ldots, x_n, y_1, \ldots, y_m$ in X and v in A^* such that (see Figure 5.5) $x_1 \cdots x_n z t a v = y_1 \cdots y_m$.

Take n minimal. If $n \geq 1$, we have $x_1 (x_2 \cdots x_n z t) a v = y_1 \cdots y_m$ and $t' = x_2 \cdots x_n z t \in P$, since $t \in P$. Thus $x_1 t' A^*$ intersects $y_1 X^*$, and since $t' \notin T$, we must have $x_1 = y_1$. Thus $x_2 \cdots x_n z t a v = y_2 \cdots y_m$ and this contradicts the minimality of n. Hence $n = 0$ and $z t a v = y_1 \cdots y_m$ (see Figure 5.6).

Note that, since z is of maximal length, y_1 is a prefix of z. Suppose by contradiction that $y_1 \neq z$. Then for some prefix t_1 of $y_2 \cdots y_m$, we have $y_1 t_1 = zt$. Since $t \in P$, the set $y_1 t_1 A^*$ intersects $z X^*$ and we conclude that $t_1 \in T$, a contradiction since $|y_1| < |z| \Rightarrow |t_1| > |t|$.

Thus $y_1 = z$ and $t a v = y_2 \cdots y_m$. Hence $ta \in P$, as claimed. This concludes the proof.   □

The following examples show that Theorem 5.2.4 is optimal in several directions.

Example 5.2.5 The suffix code $X = \{aa, ba, b\}$ is a finite maximal code and has infinite deciphering delay.

Example 5.2.6 The code $\{ab, abb, baab\}$ has minimal deciphering delay 1. It is neither prefix nor maximal : indeed, the word $bbab$, for instance, can be added to it.

Example 5.2.7 The code $X = ba^*$ is maximal and suffix. It has minimal deciphering delay 1. It is not prefix, but it is infinite.

The rest of this section is devoted to the proof of an analogue of Theorem 2.5.24 for codes with finite deciphering delay. The following example shows that the construction used in the proof of Theorem 2.5.24 does not apply in this context.

Example 5.2.8 Let $X = \{a, ab\}$, $A = \{a, b\}$ and $y = bba$ as in Example 2.5.26. The set $Y = X \cup y(Uy)^*$ with $U = A^* \setminus (X^* \cup A^* y A^*)$ constructed in the proof of

Theorem 2.5.24 is a maximal code but it has infinite deciphering delay. Indeed, the word $y' = ya^d bby$ is in Y for any $d \geq 0$, and has the proper prefix ya^d in Y^{d+1}.

Theorem 5.2.9 *Each rational code having deciphering delay d may be embedded into a maximal one with the same delay d.*

Let X be a nonempty code with deciphering delay d. If $d = 0$, X is prefix and the result is easy: let L be the set of proper prefixes of words in X, and let $\bar{L} = A^* \setminus L$ be its complement. Let $X' = \bar{L} \setminus \bar{L} A^+$. Then $Y = X \cup X'$ is easily seen to be a maximal prefix code containing X. If X is rational, then Y is rational.

We assume in what follows that $d \geq 1$. Let Q be the set of words having no prefix in X and which are not a factor of any word in X. Now, let P be the set of words in Q which are minimal for the prefix order: $P = Q \setminus QA^+$. Note that P is a prefix code. Moreover, words in P and X are incomparable for the prefix order.

We say that a pair $(w, p) \in X^* \times P$ is *good* if w is the longest prefix in X^* of wp. Note that if (w, p) is good, then this pair is completely determined by the word wp. Note also that any pair $(1, p)$ for $p \in P$ is good.

We say that the pair $(w, p) \in X^* \times P$ is *very good* if (uw, p) is good for any $u \in X^*$. Note that if (w, p) is very good, then so is (uw, p) for any $u \in X^*$.

We let S' be the set of words v of the form $v = wp$ with (w, p) good but not very good. Then we define $S = P \cup S'$. Note that $P \cap S'$ may be nonempty, and that any element in $S' \setminus P$ is of the form wp, with (w, p) good but not very good and $w \in X^+$. Moreover, let R be the set of words v of the form $v = xwp$ with $x \in X$, $w \in X^*$, (xw, p) very good, and $wp \in S$ with (w, p) good. Then we define

$$Y = X \cup RS^*. \tag{5.2}$$

Proposition 5.2.10 *Y is a code with deciphering delay d.*

The proof relies on a series of lemmas.

Lemma 5.2.11 *If (m, p) is good but not very good, there exists $x' \neq x''$ in X, a factorization $p = p_1 p_2$ with $p_1 \neq \epsilon$, and $w, v \in X^*$ such that $x'wmp = x''vp_2$.*

Proof. Since (m, p) is not very good, we may find $w', v' \in X^*$ and a factorization $p = p_1 p_2$ with $p_1 \neq \epsilon$ such that $w'mp = v'p_2$. Choose such a relation of shortest length. Then w' is nonempty, since (m, p) is good, and v' is nonempty because $|p| > |p_2|$. Thus (see Figure 5.7) $w' = x'w$, $v' = x''v$ with $w, v \in X^*$, $x', x'' \in X$. Necessarily, $x' \neq x''$ by minimality. $\square$

Lemma 5.2.12 *The set $S \cap X^d A^*$ is empty.*

Proof. Suppose that $s = ut$ with $s \in S$, $u \in X^d$, and $t \in A^*$. Note that s is not in P since it has prefix in X. Hence $s = mp$, with (m, p) good but not very good. We have $mp = ut$ and u cannot be longer than m, since (m, p) is good. Thus $m = um'$ with $m' \in A^*$. Next, we can find, by Lemma 5.2.11, two words x', x'' in X with $x' \neq x''$, a factorization $p = p_1 p_2$ with $p_1 \neq \epsilon$, and $w, v \in X^*$ such that $x'wmp = x''vp_2$.

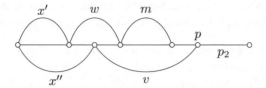

Figure 5.7 A good pair which is not a very good pair.

Thus $x''vp_2 = x'wum'p_1p_2$ and it follows that $x''v = x'wum'p_1$, which contradicts the fact that X has deciphering delay d, since $v \in X^*$ and $u \in X^d$.   □

Lemma 5.2.13 *Let $u, v \in X^*, r = mp \in R$ with (m, p) very good.*

  (i) *ur cannot be a prefix of v. In other words, $X^*RA^* \cap X^*$ is empty.*
 (ii) *If v is a prefix of ur, not shorter than um, then $v = um$.*
(iii) *If um is a prefix of v and if ur and v are comparable for the prefix order, then $um = v$.*

Proof. (i) Suppose that $urt = v$ for some $t \in A^*$. Then $umpt = v$. Since p is not a factor of any word in X, we find, by decoding $v \in X^*$, that $p = p_1p_2$ with $p_1, p_2 \neq \epsilon$ and $ump_1 \in X^*$, a contradiction with the fact that (m, p) is very good.
(ii) We have $ump = ur = vt$ with $t \in A^*$. Since $|um| \leq |v|$, v ends in p: there is a factorization $p = p_1p_2$ such that $ump_1 = v$. Since (m, p) is very good, we must have $p_1 = \epsilon$ and $v = um$.
(iii) Since ur and v are comparable, one of them is a prefix of the other. By (i), v is a prefix of ur. Since um is a prefix of v, (ii) applies, and we find $v = um$.   □

Lemma 5.2.14 *Let $v \in X^*$ and let $s = mp \in S$ with (m, p) good.*

  (i) *s cannot be a prefix of v. In other words, $SA^* \cap X^* = \emptyset$.*
 (ii) *If v is a prefix of s, not shorter than m, then $v = m$.*
(iii) *If m is a prefix of v and s, v are comparable for the prefix order, then $m = v$.*

Proof. (i) Suppose that $v = st$ for some $t \in A^*$. Then $v = mpt$. Since p is not a factor of any word in X, we have $p = p_1p_2$ with $p_1, p_2 \neq \epsilon$ and $v = mp_1$. This contradicts the fact that (m, p) is good.
(ii) Suppose that $mp = s = vt$ for some $t \in A^*$. Since $|m| \leq |v|$, we obtain $p = p_1p_2$ with $v = mp_1$. Since (m, p) is good, we must have $p_1 = 1$ and $v = m$.
(iii) One of s and v is a prefix of the other. By (i), it must be v which is a prefix of s. Since m is a prefix of v, (ii) applies and we find $m = v$.   □

Lemma 5.2.15 *The sets X^*R and S are prefix codes.*

Proof. We first consider X^*R. Suppose that $u, u' \in X^*, r, r' \in R$, and ur is a prefix of $u'r'$. We write $r = mp, r' = m'p'$, where $(m, p), (m', p')$ are very good. Then ump is a prefix of $u'm'p'$. Hence um is a prefix of $u'm'$ or conversely. Moreover, ur and $u'm'$ are comparable, and so are $u'r'$ and um (since all these four words are prefixes

$u'r'$). Hence, we find by Lemma 5.2.13 (iii) that $um = u'm'$. Thus p is a prefix of p'. Hence $p = p'$, since P is a prefix code. This shows that $ur = u'r'$ and thus X^*R is a prefix code.

We have $S = S' \cup P$. Since the words in P and X are incomparable for the prefix order, since $S' \setminus P$ is contained in X^+P, and since P is itself prefix, we are reduced to show that S' is prefix. Let u, u' be in S', and set $u = wp$, $u' = w'p'$, where (w, p), (w', p') are good pairs. Suppose that $wp \leq w'p'$. If $w = w'$, then $p = p'$ and the pairs are equal. We assume $w \neq w'$.

One has $w < w'$ because otherwise $w' < w$ and since w is a prefix of $w'p'$, the pair (w', p') would not be good. In fact, $wp \leq w'$ because otherwise $w < w' \leq wp$ and (w, p) would not be a good pair.

Thus, wp is a prefix of w'. Since p is not a factor of a word in X, there is a factorization $p = p_1p_2$, with $p_1, p_2 \neq 1$, such that wp_1 is in X^*, which contradicts the fact that (w, p) is a good pair. $\qquad\square$

Lemma 5.2.16 *We have*

(i) $SA^* \cap X^*RA^* = \emptyset$.
(ii) $SA^* \cap Y^* = \emptyset$.

Proof. Let $s \in S$, $r \in R$, and $v \in X^*$ be such that s and vr are comparable for the prefix order. We cannot have $s \in P$ since $vr \in X^+A^*$. Write $s = mp$, $r = m'p'$ where (m, p) is good but not very good and (m', p') is very good. Then m and vm' are comparable.

If vm' is a prefix of m, since vr and m are comparable, Lemma 5.2.13 (iii) shows that $vm' = m$. If, on the contrary, m is a prefix of vm', since s and vm' are comparable, Lemma 5.2.14 (iii) shows that $m = vm'$. So, we obtain that $m = vm'$ in both cases. Since $s = mp$, $vr = vm'p'$, we find that p, p' are comparable. Thus $p = p'$, since P is a prefix code. We conclude that $s = vr$.

Since $(vm', p) = (m, p)$ is not very good, we reach a contradiction with the fact that $(m', p') = (m', p)$ is very good.

(ii) By Lemma 5.2.14 (i), $SA^* \cap X^* = \emptyset$. Since $Y = X \cup RS^*$, we see that $Y^* \subset X^* \cup X^*RA^*$, so that (i) shows that $SA^* \cap Y^* = \emptyset$ $\qquad\square$

Proof of Proposition 5.2.10. We only have to show that Y has deciphering delay d, since it is then necessarily a code by Proposition 5.1.1. By contradiction, suppose that Y does not have deciphering delay d. We may find words $y_1, \ldots, y_{d+1}, z_1, \ldots, z_n$ in Y, $w \in A^*$ such that

$$y_1 y_2 \cdots y_{d+1} w = z_1 \cdots z_n \qquad (5.3)$$

with $y_1 \neq z_1$. Without loss of generality, we may assume that $|w| < |z_n|$ (otherwise, z_n is a suffix of w and we may shorten the relation by simplifying by z_n).

Since X has deciphering delay d, not all of $y_1, \ldots, y_{d+1}, z_1, \ldots, z_n$ are in X. Thus, if the z_j are all in X, then some y_i is in $Y \setminus X$, hence in RA^*. Then $y_1 \cdots y_{d+1}w \in X^*RA^*$ and $z_1 \cdots z_n \in X^*$. This contradicts Lemma 5.2.13 (i). We conclude that some z_j is in $Y \setminus X$.

Suppose now that all y_i are in X. By the length assumption on w, the word $y_1 \cdots y_{d+1}$ is in $z_1 \cdots z_{n-1} A^*$. If one of $z_1, \ldots, z_{n-1}$ is in $Y \setminus X$, then $y_1 \cdots y_{d+1} \in X^* \cap X^* R A^*$, which contradicts Lemma 5.2.13 (i). Thus $z_1, \ldots, z_{n-1} \in X$ and $z_n \in Y \setminus X$. Since $z_n \in RS^*$, we may write $z_n = xupm$, with $x \in X$, $u \in X^*$, $m \in S^*$, (xu, p) a very good pair, and $up \in S$, (u, p) good.

We have $y_1 \cdots y_{d+1} w = z_1 \cdots z_{n-1} xupm$. Therefore, $z_1 \cdots z_{n-1} xup$ and $y_1 \cdots y_{d+1}$ are comparable for the prefix order. If $z_1 \cdots z_{n-1} xu$ is a prefix of $y_1 \cdots y_{d+1}$, then by Lemma 5.2.13 (iii), they are equal. But $y_1 \cdots y_{d+1} = z_1 \cdots z_{n-1} xu$ implies $y_1 = z_1$ since X is a code, a contradiction.

Thus $y_1 \cdots y_{d+1}$ is a prefix of $z_1 \cdots z_{n-1} xu$. Since $y_1 \neq z_1$, and since X has deciphering delay d, we must have $n = 1$ and $y_1 = x$. Thus $y_1 \cdots y_{d+1}$ is a prefix of xu, hence $y_2 \cdots y_{d+1}$ is a prefix of u, hence $up \in S$, which contradicts Lemma 5.2.12.

All this shows that some y_i and some z_j are not in X, hence are in RS^*. Take i and j minimal. Then $y_i = ru$, $z_j = r'u'$ with $r, r' \in R$. Moreover $y_1 \cdots y_{i-1} r$ and $z_1 \cdots z_{j-1} r'$ are comparable by Equation (5.3). We deduce then from Lemma 5.2.15 that $y_1 \cdots y_{i-1} r = z_1 \cdots z_{j-1} r'$. We may write $r = xmp$, $r' = x'm'p'$, where (xm, p), $(x'm', p')$ are very good pairs and (m, p), (m', p') are good and $mp, m'p' \in S$. Then the equation $y_1 \cdots y_{i-1} xmp = z_1 \cdots z_{j-1} x'm'p'$ forces by the definition of a very good pair $p = p'$ since $y_1, \ldots, y_{i-1}, z_1, \ldots, z_{j-1}, x, x', m, m'$ are all in X^*. Thus $y_1 \cdots y_{i-1} xm = z_1 \cdots z_{j-1} x'm'$. If $i, j \geq 2$, then $y_1 = z_1$ since X is a code, a contradiction.

It follows from this that we must have $i = 1$ or $j = 1$, that is y_1 or z_1 is in RS^*. Suppose that $i = 1$ and $j > 1$. Then we obtain $xm = z_1 \cdots z_{j-1} x'm'$, which shows that $x = z_1$ and $m = z_2 \cdots z_{j-1} x'm'$. Note that $m \neq 1$. We know that the pair $(x'm', p)$ is very good. Hence $(z_2 \cdots z_{j-1} x'm', p)$ is also very good. Now this pair is equal to (m, p), which is not very good, a contradiction.

Thus, we cannot have $i = 1$ and $j > 1$. Similarly, we cannot have $i > 1$ and $j = 1$. Thus, we have $i = j = 1$, that is $y_1, z_1 \in RS^*$. Since R and S are prefix codes by Lemma 5.2.15, we have either $y_1 = rs_1 s_2$, $z_1 = rs_1$ or $y_1 = rs_1$, $z_1 = rs_1 s_2$ with $r \in R, s_1, s_2 \in S^*, s_2 \neq \epsilon$. In the first case, we have by Equation (5.3) and upon simplification by z_1, $z_2 \cdots z_n = s_2 y_2 \cdots y_{d+1} w$ which contradicts Lemma 5.2.16 (ii). Thus the second case holds. Again by Equation (5.3), we have $y_2 \cdots y_{d+1} w = s_2 z_1 \cdots z_n$. To avoid the same contradiction, we must have that $y_2 \cdots y_{d+1}$ is a proper prefix of s_2. We deduce from Lemma 5.2.16 (i) that $y_2, \ldots, y_{d+1}$ are all in X.

We may write $s_2 = ss_3$, where $s \in S$, $s_3 \in S^*$. Since y_2 is a prefix of s_2 (because $d \geq 1$), y_2 is a prefix of s or vice-versa. Hence $s \notin P$ and thus $s \in S'$. We deduce that we may write $s = mp$ for some good but not very good pair (m, p), and by Lemma 5.2.11, the existence of $f, n \in X^*$, $x, x' \in X$ with $x \neq x'$ such that $xnmp = x'fq$ with $|q| < |p|$.

We know that $y_2 \cdots y_{d+1}$ is a proper prefix of $s_2 = mps_3$. Now, m is not a prefix of $y_2 \cdots y_{d+1}$ (otherwise, by Lemma 5.2.14 (iii), we deduce $m = y_2 \cdots y_{d+1}$ and $mp \in S$ has a prefix in X^d, contradicting Lemma 5.2.12). Thus $y_2 \cdots y_{d+1}$ is a prefix of m. Let $m = y_2 \cdots y_{d+1} g$. Then $xny_2 \cdots y_{d+1} gp = x'fq$ and because $|q| < |p|$ and $n, f \in X^*$, this contradicts the fact that X has deciphering delay d. $\qquad\square$

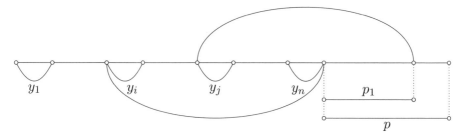

Figure 5.8 A factorization of $y_1 \cdots y_n p$ with $y_i, \ldots, y_n \in X$ and $y_j \cdots y_n p_1 \in X^*$.

Proposition 5.2.17 *The set Y is a complete code.*

If X is dense, then Y is dense and therefore is complete. So, we may assume that X is a thin code. The proof of Proposition 5.2.17 relies on the following lemma.

Lemma 5.2.18 *If X is a thin code, then the set $P \cup (X \setminus X A^+)$ is a maximal prefix code.*

Proof. Let $Z = P \cup (X \setminus X A^+)$. The two terms of this union are prefix codes. Moreover, any word in P is incomparable (for the prefix order) with any word of X. Hence Z is a prefix code (since $1 \notin Z$ because $X \neq \emptyset$ by assumption).

We show that Z is right complete. Let $w \in A^*$. Suppose that w is not comparable with X. Choose some word u which is factor of no word in X (such a word exists since X is thin). Then wu is not a factor of any word in X, and has no prefix in X. Therefore, wu has a prefix in P and we conclude that $wA^* \cap ZA^*$ is nonempty. □

Proof of Proposition 5.2.17. Choose some word $v \in X^d$. We show that for any word w, $vwA^* \cap Y^*$ is nonempty (this will imply that Y is complete). By contradiction, suppose that

$$vwA^* \cap Y^* = \emptyset. \tag{5.4}$$

We may write $vw = y_1 \cdots y_n u$ with $y_i \in Y$ and with u of minimal length among all such factorizations. Note that since v is in $X^d \subset Y^*$, the word v is necessarily a prefix of $y_1 \cdots y_n$. By Lemma 5.2.18, we find p in $P \cup (X \setminus X A^+)$ such that p and u are comparable.

We claim that if p_1 is a nonempty prefix of p, then $y_1 \cdots y_n p_1 \notin Y^*$. Indeed, if $y_1 \cdots y_n p_1 \in Y^*$, then since p and u are comparable, either p_1 is a prefix of u, contradicting the minimality of u, or u is a prefix of p_1, and this contradicts Equation (5.4).

By the claim, p is not in X, hence p is in P. Choose now $i \in \{1, \ldots, n+1\}$ minimal such that $y_i, y_{i+1}, \ldots, y_n$ are in X ($i = n+1$ means $y_n \notin X$). Then for any j with $i \leq j \leq n$, the pair $(y_j y_{j+1} \cdots y_n, p)$ is good: indeed, if not, then $p = p_1 p_2$ with $p_1 \neq 1$ and $y_j \cdots y_n p_1 \in X^*$, contradicting the claim (see Figure 5.8).

Take $n + 1 \geq j \geq i$ minimum such that $y_j y_{j+1} \cdots y_n p \in S$ (j exists since $p \in S$). If $j > i$, then $y_{j-1} y_j \cdots y_n p \in R$ (indeed $(y_{j-1} y_j \cdots y_n, p)$ is a very good pair). Since $R \subset Y$, this contradicts the claim.

Hence $j = i$. If $i > 1$, then y_{i-1} is not in X, hence is in RS^*. Then $y_{i-1}y_i \cdots y_n p \in RS^*$ (since $y_i \cdots y_n p \in S$), and we find a contradiction with the claim.

Thus we are reduced to $i = 1$ and $y_1 \cdots y_n p \in S$. This implies that $(y_1 \cdots y_n, p)$ is a good pair which is not very good because $y_1 \cdots y_n \neq 1$. Thus by Lemma 5.2.11, we find x, x' in X distinct, such that $xX^*y_1 \cdots y_n p \cap x'X^*p_2$ is not empty, for some factorization $p = p_1 p_2$, $p_1 \neq \epsilon$. Since v is a prefix of $y_1 \cdots y_n$, this contradicts the fact that X has delay d.   □

The above proof implies the following property: if a thin code $X \subset A^+$ with deciphering delay d is complete, then for any $x \in X^d$ and $u \in A^*$ there is a $v \in A^*$ such that $xuv \in X^*$. Indeed, a thin complete code is maximal by Theorem 2.5.13 and thus $X = Y$. Note that this property is also a consequence of Proposition 5.2.3.

Proposition 5.2.19 *If the code X is rational, then Y is a rational code.*

Proof. Since X is rational, the set $F(X)$ of its factors is rational. Consequently, $Q = A^* \setminus (F(X) \cup XA^*)$ is rational. Since, $P = Q \setminus QA^+$, the set P is also rational.

Let c be a new letter not in A and let $\pi : (A \cup c)^* \to A^*$ be the projection that erases c. For $u, p \in A^*$, we say that the word ucp is *good* (resp. *very good*) if so is the pair (u, p). We denote by S_0 (resp S_1) the sets of these words.

Let $L = (\pi^{-1}(X^*) \cap A^*cA^+)A^*$. Thus L is the set of words starting with a word $z = ucw$ with $w \neq \varepsilon$ and $uw \in X^*$. The set L is rational. We claim that $S_0 = X^*cP \setminus L$, which implies that S_0 is rational.

In order to prove the claim, let $ucp \in S_0$. Then evidently $u \in X^*$ and $p \in P$. Moreover, suppose $ucp \in L$, then there is a factorization $p = ww'$ such that $w \neq \varepsilon$ and $uw \in X^*$, contradicting the fact that (u, p) is good. Conversely, if $u \in X^*, p \in P$, and $ucp \notin L$, there is no prefix of up in X^* strictly longer than u. Thus (u, p) is good and $ucp \in S_0$.

Similarly $ucp \in S_1$ if and only if $u \in X^*, p \in P$, and $X^*ucp \cap L = \emptyset$. This implies that $S_1 = X^*cP \setminus (X^*)^{-1}L$ is rational.

Let R_0 be the set of words of the form $xucp$, with $x \in X$, $u \in X^*$, which are very good and such that $u = 1$ or ucp is good but not very good. In other words, $R_0 = S_1 \cap X(P \cup (S_0 \setminus S_1))$. This shows that R_0 is rational. Clearly $R = \pi(R_0)$. Recall that S' is the set of words of the form up with (u, p) good but not very good. Consequently $S' = \pi(S_0 \setminus S_1)$.

This shows that S' and R are rational. Thus $S = P \cup S'$ and $Y = X \cup RS^*$ are rational.   □

Proof of Theorem 5.2.9. Let X be a rational code with deciphering delay d. Then the code Y defined by Equation 5.2 has delay d by Proposition 5.2.10. By Propositions 5.2.17 and 5.2.19 it is a rational complete code. Since a rational code is thin by Proposition 2.5.20, and since a thin and complete code is maximal by Theorem 2.5.13, the conclusion follows.   □

Note that if X is thin, then Y also is thin (Exercise 5.1.10). Thus, any thin code with deciphering delay d is contained in a maximal one with the same delay.

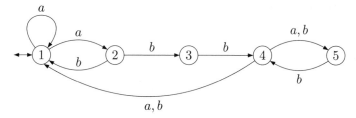

Figure 5.9 An automaton recognizing Y^*.

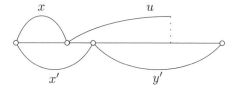

Figure 5.10 A forbidden configuration for weakly prefix codes.

Example 5.2.20 The finite code $X = \{a, ab\}$ has delay 1. We have $P = \{ba, bb\}$. The good pairs are those of the form (x, bb) and (x, ba) with $x \in X^*ab \cup 1$. They are also very good except when $x = 1$. Thus $S = P$ and $R = \{ab^3, ab^2a\}$. Finally $Y = \{a, ab\} \cup \{ab^3, ab^2a\}\{bb, ba\}^*$ is a complete code with deciphering delay 1 containing X. An automaton recognizing Y^* is represented on Figure 5.9.

Observe that there is a much simpler complete code with delay 1 containing X, namely the code ab^*. It would be interesting to have a completion procedure which gives this code directly. We will see in the next section a procedure which gives this code, but for a different definition of the delay (see Example 5.3.9).

5.3 Weakly prefix codes

There is another definition, close to the previous one where one counts the delay in letters instead of words of the code. A set $X \subset A^+$ is said to be *weakly prefix* if there exists an integer $d \geq 0$ such that the following condition holds: If xu is a prefix of $x'y'$ with $x, x' \in X$, u a prefix of a word in X^*, and $y' \in X^*$, then $|u| \geq d$ implies $x = x'$. If this holds, we also say that X has *literal deciphering delay d*.

The least integer d such that the implication above holds is called the *minimal literal deciphering delay*. If no such integer exists, the set has *infinite literal deciphering delay*.

Proposition 5.3.1 *Let X be a set with minimal verbal deciphering delay d and minimal literal deciphering delay e. Then*

$$d \leq e \leq d \max\{|x| \mid x \in X\}.$$

Proof. Indeed, assume that X has literal deciphering delay e, and consider $x, x' \in X$, $y \in X^e$, and $y' \in X^*$ such that $xy \leq x'y'$. Since $|y| \geq e$, one has $x = x'$, showing that X has verbal deciphering delay e.

Conversely, assume that X has verbal deciphering delay d. Let $x, x' \in X$ and u a prefix of a word in X^* and $y' \in X^*$ such that $xu \le x'y'$ with $|u| \ge d \max\{|x| \mid x \in X\}$. By the condition on the length, there is a word $y \in X^d$ which is a prefix of u. Thus $xy \le xu \le x'y'$. Since X has verbal deciphering delay d, we obtain $x = x'$.   □

Thus a finite set has simultaneously finite delay for both notions, but the example of $X = b \cup ba^*c \cup a^*d$ shows that the definitions differ when X is infinite. Indeed this set X has verbal deciphering delay 1, but has infinite literal deciphering delay since for all n, the condition of the definition is not satisfied with $x = b$, $u = a^n$, $x' = ba^nc$, $y' = 1$.

Proposition 5.3.2 *A weakly prefix set is a code.*

Proof. Let X have literal deciphering delay d. By Proposition 5.3.1, it has verbal deciphering delay d. By Proposition 5.1.1, the set X is a code.   □

An automaton $\mathcal{A}$ is said to have *delay* $d \ge 0$ if for any pair of paths

$$p \xrightarrow{a} q \xrightarrow{z} r, \quad p \xrightarrow{a} q' \xrightarrow{z} r',$$

if $|z| = d$ then $q = q'$. Thus a deterministic automaton has delay 0. An automaton with finite delay is also called *weakly deterministic*. Observe that if $\mathcal{A}$ has delay d, then for any word w, and for any pair of paths

$$p \xrightarrow{w} q \xrightarrow{z} r, \quad p \xrightarrow{w} q' \xrightarrow{z} r',$$

with $|z| = d$, the paths $p \xrightarrow{w} q$ and $p \xrightarrow{w} q'$ are equal.

Proposition 5.3.3 *A strongly connected weakly deterministic automaton is unambiguous.*

Proof. Indeed, let $c : p \xrightarrow{w} q$ and $c' : p \xrightarrow{w} q$ be two paths from p to q with the same label w. Since the automaton is strongly connected, there exists, for any $d \ge 0$, a path $q \xrightarrow{z} r$ with $|z| = d$. It follows that $c = c'$.   □

The following result proves that a code X is weakly prefix if and only if X^* is recognized by some weakly deterministic automaton $\mathcal{A} = (Q, 1, 1)$.

Proposition 5.3.4 *Let X be a code and $\mathcal{A} = (Q, 1, 1)$ be an automaton with delay d recognizing X^*. Then X has literal deciphering delay d. Conversely, if X has finite literal deciphering delay, the automaton can be chosen to have the same delay as X.*

Proof. Let us first suppose that X^* is recognized by $\mathcal{A} = (Q, 1, 1)$ with delay d. We show that X has delay d. Let $x, x' \in X$, let $u \in A^*$ be a prefix of a word in X^* with $|u| = d$ and $y' \in X^*$ such that $xu \le x'y'$. Since $\mathcal{A}$ recognizes X^*, there are paths $c : 1 \xrightarrow{x} 1 \xrightarrow{u} p$ and $c' : 1 \xrightarrow{x'} 1 \xrightarrow{y'} 1$. Since xu is a prefix of $x'y'$, the path c' has a decomposition $c' : 1 \xrightarrow{x} q \xrightarrow{u} p' \xrightarrow{w} 1$ for some states q, p' and some word w. Since $|u| = d$, the two paths c and c' have the same prefix of length $|x|$, and therefore $q = 1$.

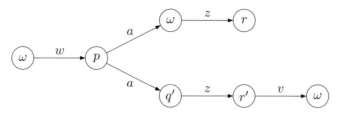

Figure 5.11 Two paths in the automaton $\mathcal{A}^*$.

Assume that x is a prefix of x'. Then $x' = xz$ for some $z \in A^*$, and the path $1 \xrightarrow{x'} 1$ decomposes into $1 \xrightarrow{x} 1 \xrightarrow{z} 1$. This shows that $z \in X^*$ and thus $z = 1$. Thus $x = x'$. The other case is handled symmetrically.

Conversely, let X have literal delay d and let $\mathcal{A} = (Q, i, T)$ be a trim deterministic automaton recognizing X and let $\mathcal{A}^* = (Q \cup \omega, \omega, \omega)$ be the star of the automaton $\mathcal{A}$. We show that $\mathcal{A}^*$ has delay d. Assume that

$$p \xrightarrow{a} q \xrightarrow{z} r, \quad p \xrightarrow{a} q' \xrightarrow{z} r'$$

with $|z| = d$. Then, by construction of $\mathcal{A}^*$ one of q, q' is ω. Let for example $q = \omega$. Since $\mathcal{A}^*$ is trim, there is a path $\omega \xrightarrow{w} p$ and we may suppose that this path does not pass by state ω inbetween. We also have a path $r' \xrightarrow{v} \omega$ (see Figure 5.11). Then $wa \in X$ and $wazv \in X^*$. Let $x = wa$ and let $wazv = x'y'$ with $x' \in X$ and $y' \in X^*$. Since X has literal deciphering delay d, we have $x = x'$. Consequently $y = zv$. Thus there are in $\mathcal{A}^*$ the paths $\omega \xrightarrow{x'} q' \xrightarrow{y'} \omega$ and $\omega \xrightarrow{x'} \omega \xrightarrow{y'} \omega$. Since $\mathcal{A}^*$ is unambiguous, this implies $q' = \omega$. Thus $\mathcal{A}^*$ has delay d. □

We may observe that the automaton $\mathcal{A}^*$ above can be used to check whether a code is weakly prefix, and to compute its minimal literal deciphering delay.

We now turn to maximal weakly prefix codes. The following result is the counterpart of Proposition 5.2.3.

Proposition 5.3.5 *Let X be a maximal code with literal deciphering delay d. Then any right completable word $u \in A^*$ of length d is strongly right completable.*

Proof. Let $v \in A^*$. By Proposition 1.3.6 there exists a word $w \in A^*$ such that uvw is unbordered. By Proposition 5.2.1, there exist $x \in X^*$ and $t \in A^*$ such that $xuvwt \in X^*$. Since X has literal deciphering delay d, and since the word u is right completable, this word is simplifying. Thus $uvwt \in X^*$, showing that uv is right completable. □

An automaton $\mathcal{A}$ is said to be *weakly complete* or *d-complete* if for any path $p \xrightarrow{w} q$ with $|w| = d$, there is a path $p \xrightarrow{wa} q'$ for each letter $a \in A$. Observe that this path is not required to start with the path $p \xrightarrow{w} q$.

If $\mathcal{A}$ is d-complete, then by induction for any path $p \xrightarrow{w} q$ with $|w| = d$, and for any word x, there is a path $p \xrightarrow{wx} q'$.

Proposition 5.3.6 *Let X be a thin code with literal deciphering delay d and let $\mathcal{A} = (Q, 1, 1)$ be a trim automaton with delay d recognizing X^*. The code X is complete if and only if $\mathcal{A}$ is d-complete.*

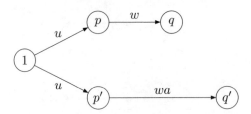

Figure 5.12 Showing that $\mathcal{A}$ is d-complete.

Proof. Suppose first that X is complete. Let $p \xrightarrow{w} q$ be a path in $\mathcal{A}$ with $|w| = d$ and let $a \in A$ be a letter. Since $\mathcal{A}$ is trim, there is a path $1 \xrightarrow{u} p$. Since X is thin and complete, it is a maximal code by Theorem 2.5.13. By Proposition 5.3.5, the word uwa is right completable. Thus there exists a path $1 \xrightarrow{u} p' \xrightarrow{wa} q'$. Since $\mathcal{A}$ has delay d and since $|w| = d$, we have $p = p'$ (see Figure 5.12). This shows that $\mathcal{A}$ is d-complete.

Conversely, let $x \in X^+$ be of length at least d. Then, for any $w \in A^*$, since $\mathcal{A}$ is d-complete, there is a path $1 \xrightarrow{xw} p$. This implies that X is complete since $\mathcal{A}$ is trim. □

We can use the previous result to give another proof of Theorem 5.2.4. Let X be a finite maximal code. We argue by contradiction and suppose that its verbal delay is strictly positive. Since X is finite, its literal delay d is also finite and strictly positive. By Proposition 5.3.4, there exists a finite d-complete automaton $\mathcal{A} = (Q, 1, 1)$ with minimal delay d recognizing X^*.

We first show that we may suppose the automaton *unfolded* in the sense that all states in $\mathcal{A}$ except the initial state 1 have indegree 1. This property can be obtained by applying the following state splitting method: Let $q \neq 1$ be a state with indegree $r > 1$. This state is split into r copies, each of which with indegree 1 and with the same outgoing edges. Since X is finite, all cycles in $\mathcal{A}$ contain state 1. Consequently, the state splitting can be repeated only a finite number of times. Clearly, state splitting preserves the delay and d-completeness.

Assume now that $\mathcal{A}$ is unfolded and has the minimal possible number of states. Since $\mathcal{A}$ has minimal delay d, there is a state q such that there are edges (q, a, r) and (q, a, r') with $r \neq r'$ and paths labeled $v \in A^{d-1}$ going out of r, r'. Let us prove that $r, r' \neq 1$. Arguing by contradiction, suppose that $r' = 1$. Let u be a word of maximal length such that there is a path $r \xrightarrow{vu} 1$, decomposing as $r \xrightarrow{v} s \xrightarrow{u} 1$ with a simple path $s \xrightarrow{u} 1$. Observe that vu is nonempty since otherwise $r = 1 = r'$. Let b be the first letter of uv. Note that no path exists labeled vb and going out of 1, since $\mathcal{A}$ has minimal delay d (otherwise, we would have two paths $q \xrightarrow{a} 1 \xrightarrow{vb}$ and $q \xrightarrow{a} r \xrightarrow{vb}$ labeled avb starting from q with different initial edges). Consider now the last letter c of vu and the state t such that $(t, c, 1)$ is the last edge of the path $r \xrightarrow{vu} 1$. Since $\mathcal{A}$ is d-complete, there exists a path labeled cvb going out of state t. Let (t, c, t') be the first edge of this path (see Figure 5.13 which corresponds to the case $u \neq 1$ and where $u = u'c$). We have $t' \neq 1$ since there is no path labeled vb going out of 1. Let

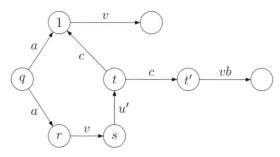

Figure 5.13 Showing that $r' \neq 1$.

$w \neq 1$ be a word such that there is a simple path $t' \xrightarrow{w} 1$. Then there is a simple path $s \xrightarrow{uw} 1$. This establishes the contradiction since uw is strictly longer than u.

Let $\mathcal{A}' = (Q', 1, 1)$ be the automaton obtained by merging r and r'. Since $r, r' \neq 1$ and since they both have indegree 1 and the same label on the incoming edge, the automaton $\mathcal{A}'$ also recognizes X^* and is unfolded. Since it has strictly less states than $\mathcal{A}$, we obtain the final contradiction.

We now prove the following result which is a variant of Theorem 5.2.9. The proof uses automata and it is illustrated in Example 5.3.10.

Theorem 5.3.7 *Each weakly prefix rational code can be embedded into a maximal one with the same delay.*

We shall use the following lemma. In the proof, we use the notation $q \xrightarrow{u}$ to denote some path starting in state q, and labeled with the word u.

Lemma 5.3.8 *Let $\mathcal{A} = (Q, 1, 1)$ be a trim automaton with delay d. One can obtain, by adding finitely many states and edges to $\mathcal{A}$, a trim automaton $\mathcal{B} = (Q', 1, 1)$ which has still delay d and which is d-complete.*

Proof. In the case $d = 0$ we simply add in $\mathcal{B}$ an edge $(q, a, 1)$ for all states q and letters $a \in A$, for which there is no edge leaving q and labeled a in $\mathcal{A}$. The proof for $d \geq 1$ consists in several steps.

1. We start with the definition of a new automaton $\mathcal{B}_0$. We add the set Q' of states denoted $q(w)$, for $w \in A^*$, with $1 \leq |w| \leq d$, and set $q(1) = 1$. We add the edges: $q(w) \xrightarrow{a} q(w')$, for $w = aw'$, $a \in A$.

Denote by $\mathcal{B}_0 = (Q \cup Q', 1, 1)$ this new automaton. Clearly, $\mathcal{B}_0$ also has delay d. Remark, for future use in the final step below, that each state of Q' is coaccessible, since for each $q(w)$, we have a path $q(w) \xrightarrow{w} 1$.

It will be convenient to call *future* of a state q the set of words w of length $\leq d$ such that there exists some path $q \xrightarrow{w}$. Note that in $\mathcal{B}_0$, the future of a state $q(w)$ with $|w| = d$ is the set of prefixes of w.

2. We construct now a sequence of automata $\mathcal{B}_1, \mathcal{B}_2, \ldots$ which all have the same states as $\mathcal{B}_0$. It will be clear that this sequence is finite. We will show that all $\mathcal{B}_i$ have delay d. Let $\mathcal{B}_n$ be its last element. This will be shown to be d-complete. If $\mathcal{B}_i$ is constructed and is not d-complete, then for some word $u \in A^d$, some letter b and

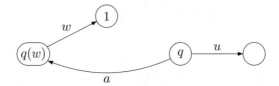

Figure 5.14 The new edge $(q, a, q(w))$ is added in $\mathcal{B}_{i+1}$ (with $ub = aw$, because there is no edge $q \xrightarrow{ub}$).

some state q of $\mathcal{B}_i$, a path $q \xrightarrow{u}$ exists, but no path $q \xrightarrow{ub}$. Then, writing $ub = aw$, with $a \in A$, we add to $\mathcal{B}_i$ the edge $q \xrightarrow{a} q(w)$, and this gives the automaton $\mathcal{B}_{i+1}$ (see Figure 5.14).

3. We now show a technical property: for each $i \geq 0$ and for each state p, the future of p in $\mathcal{B}_i$ is equal to the future of p in $\mathcal{B}_0$. This implies that for any word $m \in A^d$, the future in every $\mathcal{B}_i$ of $q(m)$ is the set of prefixes of m.

It suffices to prove that if there is a path $p \xrightarrow{v}$ in $\mathcal{B}_{i+1}$, with $|v| \leq d$, then there exists already a path $p \xrightarrow{v}$ in $\mathcal{B}_i$.

For this, we may suppose that the path $p \xrightarrow{v}$ in $\mathcal{B}_{i+1}$ involves the new edge $q \xrightarrow{a} q(w)$ created in step 2, where u is such that $ub = aw$, and $q \xrightarrow{u}$ in $\mathcal{B}_i$. Thus, we may suppose that this path has the form $p \xrightarrow{v_1} q \xrightarrow{a} q(w) \xrightarrow{v_2} p'$ with $v = v_1 a v_2$, where the last segment $q(w) \xrightarrow{v_2} p'$ is in $\mathcal{B}_i$. Now $|v_2| < d$, thus the induction hypothesis on the future of $q(w)$ implies that v_2 is a proper prefix of w. Thus, by construction of the new edge, there exists in $\mathcal{B}_i$ a path $q \xrightarrow{a v_2}$, since $a v_2$ is a prefix of u. Hence, we get in $\mathcal{B}_{i+1}$ a path $p \xrightarrow{v}$ with a smaller number of occurrences of the new edge. Consequently, a path $p \xrightarrow{v}$ exists in $\mathcal{B}_{i+1}$, with no occurrence of the new edge, and this path is therefore in $\mathcal{B}_i$, proving the induction step.

4. Suppose that $\mathcal{B}_i$ has delay d. We prove that $\mathcal{B}_{i+1}$ has the same delay. Suppose that for some states p, p_1, p_2, some letter c and some word $v \in A^d$, one has in $\mathcal{B}_{i+1}$ the two paths $p \xrightarrow{c} p_1 \xrightarrow{v}$ and $p \xrightarrow{c} p_2 \xrightarrow{v}$. Because of 3, some paths $p_1 \xrightarrow{v}$ and $p_2 \xrightarrow{v}$ exist in $\mathcal{B}_i$. If the edges $p \xrightarrow{c} p_1$ and $p \xrightarrow{c} p_2$ are in $\mathcal{B}_i$, then $p_1 = p_2$ because $\mathcal{B}_i$ has delay d. Otherwise, $p_1 \neq p_2$, and exactly one of the two edges $p \xrightarrow{c} p_1$ or $p \xrightarrow{c} p_2$, say $p \xrightarrow{c} p_1$, is the new edge $q \xrightarrow{a} q(w)$ and the other is in $\mathcal{B}_i$. Then $p = q$, $c = a$, $p_1 = q(w)$, so that $v = w$ by (2) because v has length d. Thus, considering the other edge (which is in $\mathcal{B}_i$), we see that there exists a path $q \xrightarrow{aw}$ in $\mathcal{B}_i$. This contradicts the assumption that led to the construction in step 2.

5. Let $\mathcal{B}' = (Q \cup Q'', 1, 1)$ be the trim part of $\mathcal{B} = (Q \cup Q', 1, 1)$. It has still delay d and we show that it is still d-complete. Assume there is a path $p \xrightarrow{u}$ in $\mathcal{B}'$, and let a be a letter. Since $\mathcal{B}$ is d-complete, there is a path $p \xrightarrow{ua}$ in $\mathcal{B}$. Since p is accessible, each state on this path is accessible. Since all states in Q' are coaccessible, all states on the path are both accessible and coaccessible. Thus this path is in $\mathcal{B}'$. This completes the proof.   □

Proof of Theorem 5.3.7. Let X be a nonempty rational code with literal deciphering delay d. By Proposition 5.3.4, there exists an unambiguous automaton $\mathcal{A} = (Q, 1, 1)$ with same delay d which recognizes X^*. We may suppose that $\mathcal{A}$ is trim. By

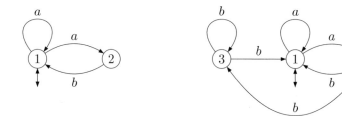

Figure 5.15 Completion of $X = \{a, ab\}$.

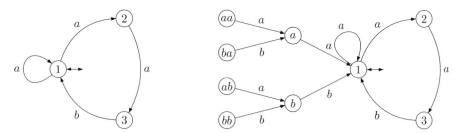

Figure 5.16 The automata $\mathcal{A}$ and $\mathcal{B}_0$.

Lemma 5.3.8, we may embed $\mathcal{A}$ into a trim automaton $\mathcal{B} = (Q', 1, 1)$ which has delay d and which is d-complete.

Since $\mathcal{B}$ is a strongly connected automaton with finite delay, it is unambiguous, as stated in Proposition 5.3.3. Thus the set recognized by $\mathcal{B}'$ is of the form Y^*, for some rational code Y containing X. Moreover, Y has deciphering delay d, by Proposition 5.3.4, and it is complete by Proposition 5.3.6. Thus Y is a maximal rational code with deciphering delay d containing X. $\qquad\square$

Example 5.3.9 Let $X = \{a, ab\}$ as in Example 5.2.20. Using Proposition 5.3.4, we obtain the automaton on the left of Figure 5.15. Applying the method of Theorem 5.3.7 to this automaton we obtain the automaton on the right of Figure 5.15. This gives the complete code $Y = ab^*$ containing X.

Example 5.3.10 Let $\mathcal{A}$ be the automaton represented in Figure 5.16 on the left. It has delay 2 and recognizes $\{a, aab\}^*$ which is a code with literal deciphering delay 2.

The automaton $\mathcal{B}_0$ is represented in Figure 5.16 on the right (we denote the new states w instead of $q(w)$ for simplicity). The final automaton $\mathcal{B}$ is represented in Figure 5.17 after removal of the states which are not accessible.

5.4 Exercises

Section 5.1

5.1.1 Show that the deciphering delay of a code X is infinite if and only if there is an infinite path in the graph G_X defined in 2.7 starting in a vertex in X. If X is finite,

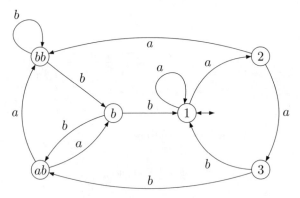

Figure 5.17 The automaton $\mathcal{B}$.

this happens if and only if there is a cycle in G_X that is accessible from some vertex in X.

5.1.2 (a) Show that a code X has deciphering delay d if any disjoint factorizations $x_1 \cdots x_n p = y_1 \cdots y_m$, where $x_1, \ldots, x_n, y_1, \ldots y_m$ are words in X and p is a prefix of a word in X, satisfy $n \le d$.

(b) Let $e_1 \cdots e_n$ be the sequence of edges of a path e from s to t in the prefix graph of a code X. The occurrence e_i is called *even* (*odd*) if the number of crossing edges among $e_1, \ldots, e_i$ is even (odd). Show that in the two factorizations

$$\text{(i)} \quad sy_1 \cdots y_\ell t = x_1 \cdots x_k \quad \text{or} \quad \text{(ii)} \quad sy_1 \cdots y_\ell = x_1 \cdots x_k t,$$

the number c of crossing edges is odd or even, according to (i) or (ii). Show next that ℓ is the number of even edges and k is the number of odd edges.

(c) Describe a linear time algorithm for computing the deciphering delay, assuming that there is no cycle in the prefix graph.

5.1.3 Let Y and Z be composable codes with finite deciphering delay $d(Y)$ and $d(Z)$. Show that $X = Y \circ Z$ has finite delay $d(X) \le d(Y) + d(Z)$. (*Hint*: Show that for $y \in X^{d(Y)}$, $z \in X^{d(Z)}$, the word yz is simplifying for X.)

5.1.4 Let $X = \{x, y\}$ be a two-element code. Show that X has finite deciphering delay. (*Hint*: Make use of an induction on $|x| + |y|$, and apply the result of Exercise 5.1.3.)

5.1.5 Let $X \subset A^*$ be a finite code.

(a) Show that there exists a smallest submonoid M containing X^* such that M is generated by a code with finite deciphering delay.

(b) Let $Y \subset A^*$ be the base of the submonoid whose existence is asserted in (a). Show by a proof analogous to that of Proposition 2.2.16 that

$$Y \subset X(Y^*)^{-1} \cap (Y^*)^{-1} X.$$

Deduce from this that if X does not have finite deciphering delay,

$$\text{Card}(Y) \leq \text{Card}(X) - 1.$$

5.1.6 Show that a code X has verbal deciphering delay d if and only if the code X^d has verbal deciphering delay 1.

5.1.7 Let $X \subset A^+$ be a code. Show that if both the sets $E(X)$ of strongly right completable words and $S(X)$ of simplifying words are nonempty, then they are equal.

5.1.8 Let $X \subset A^+$ be a code. Let $S(X)$ be the set of simplifying words and let $E(X)$ be the set of strongly right completable words. Let $U = S(X) \setminus S(X)A^+$. A *strict right context* of a word $w \in A^*$ is a word $v \in A^*$ such that there exist $x_1, \ldots, x_n \in X$ with $wv = x_1 x_2 \cdots x_n$ and v is a proper suffix of x_n. The set of strict right contexts of w is denoted by $C_r(w)$.

Show that if $S(X) = E(X) \neq \emptyset$ then, for all $w \in A^*$, we have

1. The set $C_r(w)U$ is prefix.
2. The product $C_r(w)U$ is unambiguous.
3. If $w \in S(X)$, then $C_r(w)U$ is maximal prefix.

5.1.9 Use Exercises 5.1.7, 5.1.8 and 3.4.2 to give a proof of Theorem 5.2.4.

5.1.10 Show that if X is a thin code with delay d, then the code Y defined by Equation (5.2) is thin. (*Hint*: Prove that if $p \in P$, $a \in A$, then $pa \notin P$. Then, prove successively that S, R, S^* are thin.)

Section 5.3

5.3.1 In this exercise, we call *right delay* of an automaton what is called delay in the text, and we call *left delay* the delay of the reversal of the automaton, obtained by reversing the edges. Similarly, we say that an automaton is is *right d-complete* if it is d-complete, and *left d-complete* if its reversal is d-complete.

We say that an automaton has *bidelay* (d, d') if it has left delay d and right delay d'. In the same way, we say that an automaton is (d, d')-complete if it is left d-complete and right d'-complete. We introduce a new notion to work with automata with finite bidelay.

An *extended automaton* with delay (d, d') is an automaton on a set of states Q where the set E of edges, in addition to ordinary edges, includes *boundary edges*. A *forward* boundary edge has an origin $q \in Q$ and a label $a \in A$ but no end. A *backward* boundary edge has a label $a \in A$ and an end $q \in Q$ but no origin. We extend the notion of a path by admitting that a path may possibly begin with a backward boundary edge and end with a forward boundary edge. We denote by $F(p)$ the set of edges starting at p and by $P(p)$ the set of edges ending at p. We denote by $\lambda(e)$ the label of the edge e.

Each state q of an extended automaton has attached to it a pair (U_q, V_q) where U_q is a set of words of length d and V_q is a set of words of length d'. Similarly, each edge e has such a pair $(U_e, V_e) \subset A^d \times A^{d'}$. These are subject to the following *compatibility conditions*.

1. For each state p the family of sets $\lambda(e)V_e$ for $e \in F(p)$ forms a partition of the set $V_p A$.
2. For each state p and each edge $e \in F(p)$, $U_p = U_e$.
3. For each state q, the family of sets $U_e\lambda(e)$ for $e \in P(q)$, forms a partition of the set AU_q.
4. For each state q and each edge $e \in P(q)$, $V_q = V_e$.

Show that the two following objects coincide:

 (i) An extended automaton with delay (d, d') without boundary edges.
(ii) A (d, d')-complete automaton with bidelay (d, d') with U_p (resp. V_p) equal for each state p to the set of labels of paths of length d (resp. d') ending at p (resp. starting at p).

(*Hint*: Show by induction on $k \geq 0$ that, in an extended automaton with delay (d, d') without boundary edges, for $0 \leq k \leq d' + 1$, the set of labels of paths of length $\leq k$ starting at p is the set of prefixes of $V_p A$ of length $\leq k$.)

5.3.2 Define, for a state p of an extended automaton, the noncommutative polynomial

$$\partial(p) = \underline{U_p V_p A} - \underline{A U_p V_p},$$

and for an edge e

$$\partial(e) = \varepsilon \underline{U_e}\lambda(e)\underline{V_e},$$

with $\varepsilon = 1$ if e is a forward boundary edge, $\varepsilon = -1$ if e is a backward boundary edge, and $\varepsilon = 0$ otherwise. Show that

$$\sum_{p \in Q} \partial(p) = \sum_{e \in E} \partial(e).$$

Derive that the sum of $\partial(e)$ for all boundary edges, called the *balance* of the automaton, belongs to the lattice $\mathcal{L}$ generated by the polynomials $f_w = w\underline{A} - \underline{A}w$ for $w \in A^{d+d'}$.

5.3.3 Show that the following labeled graphs satisfy the definition of an extended automaton.

1. The automaton $\mathcal{A}_0$ with set of states $Q = A^{d+d'}$, with $U_{uv} = u$ and $V_{uv} = v$ for $u \in A^d$, $v \in A^{d'}$. The set of edges is $A^{d+d'+1}$ with $U_{uav} = u$, $\lambda(uav) = a$ and $V_{uav} = v$. Moreover, $F(uv) = uvA$ and $P(uv) = Auv$.
2. The automaton $\mathcal{A}_{-x}$ obtained from $\mathcal{A}_0$ by deleting the single state x. Show that in $\mathcal{A}_{-x}$,

$$\sum_{e \in E} \partial(e) = -f_x.$$

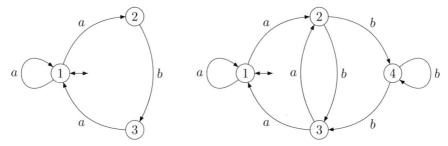

Figure 5.18 Automata with bidelay $(1, 1)$.

3. The automaton $\mathcal{A}_x$ obtained from $\mathcal{A}_0$ by deleting all edges except those incident
to state x. Show that in $\mathcal{A}_x$,

$$\sum_{e \in E} \partial(e) = f_x.$$

5.3.4 An edge e of an extended automaton is said to be *simple* if U_e and V_e have
just one element. Show that, by adding finitely many states and edges, any extended
automaton can be transformed in such a way that all boundary edges are simple.

5.3.5 Show that any extended automaton $\mathcal{A}$ can be embedded into an extended
automaton $\mathcal{B}$ having no boundary edge in the sense that every ordinary edge of $\mathcal{A}$ is
an edge of $\mathcal{B}$.
(*Hint*: First assume that all boundary edges are simple. Write $\sum_{e \in E} \partial(e) = \sum b_x f_x$
where the coefficients b_x are integers. If $b_x > 0$ add b_x copies of $\mathcal{A}_{-x}$, and if $b_x < 0$,
add b_x copies of $\mathcal{A}_x$. The resulting extended automaton is such that $\sum \partial(e) = 0$.
Finally merge each forward boundary edge e with a backward boundary edge e' such
that $\partial(e) + \partial(e') = 0$.)

5.3.6 The aim of this exercise is to show that any rational code with finite literal
delay in both directions is included in a maximal one.
    Let $\mathcal{A} = (Q, 1, 1)$ be an automaton with bidelay (d, d'). We use a series of steps
to transform $\mathcal{A}$ into an automaton with the same bidelay which is (d, d')-complete.
Show that if $\mathcal{A}$ is an automaton with bidelay (d, d'), one may first define the pairs
(U_q, V_q) and then add boundary edges to obtain an extended automaton.
    Conclude, using Exercise 5.3.5 that any code with literal bidelay (d, d') can be
embedded into a maximal one with the same literal bidelay.

5.3.7 Consider the automaton with bidelay $(1, 1)$ of Figure 5.18 on the left. Show that
the $(1, 1)$-complete automaton constructed as in Exercise 5.3.6 is the one represented
in Figure 5.18 on the right.

5.5 Notes

The notion of deciphering delay appears at the very beginning of the theory of
codes (Gilbert and Moore (1959); Levenshtein (1964)). Theorem 5.2.4 is due to

Schützenberger (1966). It was conjectured in Gilbert and Moore (1959). An incomplete proof appears in Markov (1962). A proof of a result which is more general than Theorem 5.2.4 has been given in Schützenberger (1966). The proof of Theorem 5.2.4 presented here is due to Véronique Bruyère (see Bruyère (1992) or Chapter 6 of Lothaire (2002)). The original proof of Schützenberger is given in Exercise 5.1.9. Proposition 5.1.6 is from Choffrut (1979).

Theorem 5.2.9 is due to Bruyère *et al.* (1990). We have followed their proof except for Proposition 5.2.19.

The notion of automaton with finite delay is known in early automata theory as *information lossless machines of finite order* (Kohavi (1978)). It is related with the notion of a *right closing map* in symbolic dynamics (see Lind and Marcus (1995)). The term was introduced by Kitchens (1981). Theorem 5.3.7 is due to Bruyère (1992).

The construction of Lemma 5.3.8 is from Ashley *et al.* (1993). We have followed the presentation of Bruyère and Latteux (1996), and Example 5.3.10 is also taken from here.

Exercise 5.1.5 is from Berstel *et al.* (1979). An analogous result is proved in Salomaa (1981). Exercises 5.1.6 is from Nivat (1966). Exercise 5.1.7 is from Schützenberger (1966). Exercises 5.3.1 to 5.3.7 are from Ashley *et al.* (1993), in which extended automata are introduced and called *molecules*. This name is used metaphorically and refers to the possibility to use the boundary edges as bindings.

Let us mention the following result which has not been reported here: For a three-element code $X = \{x, y, z\}$, there exists at most one right infinite word with two distinct X-factorizations (Karhumäki (1984)).

6

Bifix codes

The object of this chapter is to describe the structure of maximal bifix codes. This family of codes has quite remarkable properties and can be described in a rather satisfactory manner.

As in the rest of this book, we will work here within the family of *thin* codes. As we will see, this family contains all the usual examples, and most of the fundamental properties extend to this family when they hold in the simple (that is, finite or recognizable) case.

To each thin maximal bifix code, two basic parameters will be associated: its *degree* and its *kernel*. The degree is a positive integer which is, as we will see in Chapter 9, the degree of a permutation group associated with the code. The kernel is the set of codewords which are proper factors of some codeword. We shall prove that these two parameters characterize a thin maximal bifix code.

In the first section, we introduce the notion of a *parse* of a word with respect to a bifix code. It allows us to define an integer-valued function called the *indicator* of a bifix code. This function will be quite useful in this and later chapters.

In the second section, we give a series of equivalent conditions for a thin code to be maximal bifix. The fact that thin maximal bifix codes are extremal objects is reflected in the observation that a subset of their properties suffices to characterize them completely. We also give a transformation (called *internal transformation*) which preserves the family of maximal bifix codes.

Section 6.3 contains the definition of the degree of a thin maximal bifix code. It is defined as the number of *interpretations* of a word which is not a factor of a codeword. This number is independent of the word chosen. This fact will be used to prove most of the fundamental properties of bifix codes. We will prove that the degree is invariant under internal transformation.

In the fourth section, a construction of the thin maximal bifix code having a given degree and kernel is described. We also describe the *derived code* of a thin maximal bifix code. It is a code whose degree is one less than the degree of the original code. Both constructions are consequences of a fundamental result (Theorem 6.4.3) which characterizes those sets of words which can be completed in a finite maximal bifix code without modification of the kernel.

Section 6.5 is devoted to the study of *finite maximal* bifix codes. It is shown that for a fixed degree and a fixed size of the alphabet, there exists only a finite number

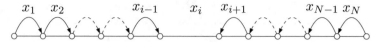

$$x_1 \quad x_2 \qquad\qquad x_{i-1} \qquad x_i \quad x_{i+1} \qquad\qquad x_{N-1} \ x_N$$

Figure 6.1 The decoding of a block of N codewords: $x_1 \cdots x_{i-1}$ is correctly decoded from left to right, the word $x_{i+1} \cdots x_N$ is correctly decoded from right to left. The error is located at x_i.

of such codes. Further it is proved that, on this finite set, the internal transformation acts transitively.

In the last section, we prove that any rational bifix code is contained in a maximal rational bifix code (Theorem 6.6.1).

6.1 Basic properties

A *bifix* code is a subset X of A^+ which is both prefix and suffix. In other words, we have

$$XA^+ \cap X = \emptyset, \qquad A^+X \cap X = \emptyset. \tag{6.1}$$

Example 6.1.1 Any code X composed of words of the same length is bifix.

Example 6.1.2 Let A be an alphabet containing two distinct letters a, b. Any set $X = a \cup bYb$ with $Y \subset (A \setminus b)^*$ is bifix.

Example 6.1.3 If X, Y are bifix codes, then XY is a bifix code.

Example 6.1.4 Let $A = \{a, b\}$. By inspection, the set

$$X = \{a^3, a^2ba, a^2b^2, ab, ba^2, baba, bab^2, b^2a, b^3\}$$

appears to be a bifix code. It will appear at several places later.

The use of bifix codes for transmissions is related to the possibility of limiting the consequences of errors occurring in the transmission using a bidirectional decoding scheme as follows. Assume that we use a binary bifix code to transmit data. Assume also that for the transmission, messages are grouped into blocks of N source symbols, encoded as N codewords.

Suppose that in a block $x_1 \cdots x_N$ of N codewords, an error has occurred during transmission that makes it impossible to decode x_i. The block $x_1 \cdots x_N$ is first decoded by using an ordinary left to right sequential decoding and the codewords x_1 up to x_{i-1} are correctly decoded. However, it is impossible to decode x_i. Then a new decoding process is started, this time from right to left. If at most one error has occurred, then again the codewords from x_N down to x_{i+1} are decoded correctly. Thus, in a block of N encoded source symbols, the incorrect codeword will be identified. These codes are used for the transmission of images, see Examples 6.2.5 and 6.2.6.

Let X be a subset of A^+. An *X-parse* (or simply a parse) of a word $w \in A^*$ is a triple (v, x, u) (see Figure 6.2) such that $w = vxu$ and

$$v \in A^* \setminus A^*X, \qquad x \in X^*, \qquad u \in A^* \setminus XA^*.$$

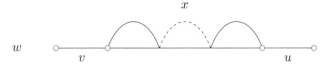

Figure 6.2 An X-parse (v, x, u) of w.

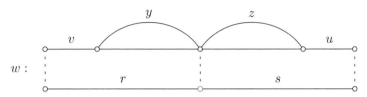

Figure 6.3 A parse of w passing through the point (r, s).

An *interpretation* of $w \in A^*$ is a triple (v, x, u) such that $w = vxu$ and

$$v \in A^- X, \qquad x \in X^*, \qquad u \in XA^-.$$

If X is a bifix code, then $A^- X \subset A^* \setminus A^*X$, and $XA^- \subset A^* \setminus XA^*$, thus any inter-pretation of w is also a parse of w.

A *point* in a word $w \in A^*$ is a pair $(r, s) \in A^* \times A^*$ such that $w = rs$. A word w thus has $|w| + 1$ points. A parse (v, x, u) of w is said to *pass* through the point (r, s) provided $x = yz$ for some $y, z \in X^*$ such that $r = vy$, $s = zu$ (see Figure 6.3).

Proposition 6.1.5 *Let $X \subset A^+$ be a bifix code. For each point of a word $w \in A^*$, there is one and only one parse passing through this point.*

Proof. Let (r, s) be a point of $w \in A^*$. The code X being prefix, there is a unique $z \in X^*$, and a unique $u \in A^* \setminus XA^*$ such that $s = zu$ (Theorem 3.1.6). Since X is suffix, we have $r = vy$ for a unique $v \in A^* \setminus A^*X$ and a unique $y \in X^*$. Clearly (v, yz, u) is a parse of w passing through (r, s). The uniqueness follows from the uniqueness of the factorizations of s and r. $\qquad\square$

Proposition 6.1.6 *Let $X \subset A^+$ be a bifix code. For any $w \in A^*$, there are bijections between the following sets:*

1. *the set of parses of w,*
2. *the set of prefixes of w which have no suffix in X,*
3. *the set of suffixes of w which have no prefix in X.*

Proof. Set $V = A^* \setminus A^*X$, $U = A^* \setminus XA^*$. For each parse (v, x, u) of w, the word v is in V and is a prefix of w. Thus v is in the set described in 2. Conversely, if $w = vw'$ and $v \in V$, set $w' = xu$ with $x \in X^*$ and $u \in U$ (this is possible since X is prefix). Then (v, x, u) is a parse. The uniqueness of the factorization $w' = xu$ shows that the mapping $(v, x, u) \mapsto v$ is a bijection from the set of parses on the set described in 2. $\qquad\square$

Let X be a subset of A^+. The *indicator* of X is the formal power series L_X (or simply L) which associates to any word w the number (L, w) of X-parses of w. Setting $U = A^* \setminus XA^*$, $V = A^* \setminus A^*X$, we have

$$L = \underline{V}\,\underline{X}^*\underline{U}. \tag{6.2}$$

Let X be a bifix code. We have $\underline{X}\,\underline{A}^* = \underline{XA}^*$ since X is prefix, and $\underline{A^*X} = \underline{A}^*\underline{X}$ since X is suffix. Thus $\underline{U} = \underline{A}^* - \underline{X}\,\underline{A}^* = (1 - \underline{X})\underline{A}^*$ and $\underline{V} = \underline{A}^*(1 - \underline{X})$. Substituting this in (6.2), we obtain

$$L = \underline{A}^*(1 - \underline{X})\underline{A}^*. \tag{6.3}$$

This can also be written as

$$L = \underline{V}\,\underline{A}^* = \underline{A}^*\underline{U}. \tag{6.4}$$

Note that this is an algebraic formulation of Proposition 6.1.6.

From Formula (6.3), we obtain a convenient expression for the number of parses of a word $w \in A^*$:

$$(L, w) = |w| + 1 - (\underline{A^*XA}^*, w). \tag{6.5}$$

The term $(\underline{A^*XA}^*, w)$ equals the number of occurrences of words in X as factors of w. Thus we see from (6.5) that for any bifix codes X, Y the following implication holds:

$$Y \subset X \Rightarrow L_X \le L_Y. \tag{6.6}$$

Recall that the notation $L_X \le L_Y$ means that $(L_X, w) \le (L_Y, w)$ for all w in A^*.

Proposition 6.1.7 *Let $X \subset A^+$ be a bifix code, let $U = A^* \setminus XA^*$, $V = A^* \setminus A^*X$, and let L be the indicator of X. Then*

$$\underline{V} = \underline{L}(1 - \underline{A}), \qquad \underline{U} = (1 - \underline{A})\underline{L}, \tag{6.7}$$

$$1 - \underline{X} = (1 - \underline{A})\underline{L}(1 - \underline{A}). \tag{6.8}$$

Proof. Formula (6.7) follows from (6.4), and (6.8) is an immediate consequence of (6.3). $\qquad \square$

Proposition 6.1.8 *Let $X \subset A^+$ be a bifix code and let L be its indicator. Then for all $w \in A^*$*

$$1 \le (L, w) \le |w| + 1. \tag{6.9}$$

In particular, $(L, 1) = 1$. Further, for all $u, v, w \in A^*$,

$$(L, v) \le (L, uvw). \tag{6.10}$$

Proof. For a given word w, there are at most $|w| + 1$ and at least one (namely, the empty word) prefixes of w which have no suffix in X. Thus (6.9) is a consequence of Proposition 6.1.6.

Next any parse of u can be extended to a parse of uvw. This parse of uvw is uniquely determined by the parse of v (Proposition 6.1.5). This shows (6.10). □

Example 6.1.9 The indicator L of the bifix code $X = \emptyset$ satisfies $(L, w) = |w| + 1$ for all $w \in A^*$.

Example 6.1.10 For the bifix code $X = A$, the indicator has value $(L, w) = 1$ for all $w \in A^*$.

The following proposition gives a characterization of formal power series which are indicators.

Proposition 6.1.11 *A formal power series $L \in \mathbb{Z}\langle\!\langle A \rangle\!\rangle$ is the indicator of a bifix code if and only if it satisfies the following conditions.*

(i) *For all $a \in A$, $w \in A^*$,*

$$0 \le (L, aw) - (L, w) \le 1, \tag{6.11}$$

$$0 \le (L, wa) - (L, w) \le 1. \tag{6.12}$$

(ii) *For all $a, b \in A$ and $w \in A^*$,*

$$(L, aw) + (L, wb) \ge (L, w) + (L, awb). \tag{6.13}$$

(iii) $(L, 1) = 1$.

Proof. Assume that L is the indicator of some bifix code X. It follows from Formula (6.7) that the coefficients of the series $L(1 - \underline{A})$ and $(1 - \underline{A})L$ are 0 or 1. For a word $w \in A^*$ and a letter $a \in A$, we have $(L(1 - \underline{A}), wa) = (L, wa) - (L, w)$. Thus, (6.12) holds and similarly for (6.11). Finally, Formula (6.8) gives for the empty word, the equality $(L, 1) = 1$, and for $a, b \in A$, $w \in A^*$,

$$-(\underline{X}, awb) = (L, awb) - (L, aw) - (L, wb) + (L, w),$$

showing (6.13).

Conversely, assume that L satisfies the three conditions. Set $S = (1 - \underline{A})L$. Then $(S, 1) = (L, 1) = 1$. Next for $a \in A$, $w \in A^*$, we have

$$(S, aw) = (L, aw) - (L, w).$$

By (6.11), $0 \le (S, aw) \le 1$, showing that S is the characteristic series of some set U containing the empty word 1. Next, if $a, b \in A$, $w \in A^*$, then by (6.13)

$$(S, aw) = (L, aw) - (L, w) \ge (L, awb) - (L, wb) = (S, awb).$$

Thus, $awb \in U$ implies $aw \in U$, showing that U is prefix-closed.

According to Theorem 3.1.6, the set $X = UA \setminus U$ is a prefix code and $1 - \underline{X} = \underline{U}(1 - \underline{A})$.

Symmetrically, the series $T = L(1 - \underline{A})$ is the characteristic series of some nonempty suffix-closed set V, the set $Y = AV - V$ is a suffix code and $1 - \underline{Y} = (1 - \underline{A})\underline{V}$.

Finally

$$1 - \underline{X} = \underline{U}(1 - \underline{A}) = (1 - \underline{A})L(1 - \underline{A}) = (1 - \underline{A})\underline{V} = 1 - \underline{Y}.$$

Thus, $X = Y$ and X is bifix with indicator L.   □

The following formulation is useful for the computation of the indicator.

Proposition 6.1.12 *Let $X \subset A^+$ be a bifix code, and L be its indicator. For any word $u \in A^*$, and any letter $a \in A$,*

$$(L, ua) = \begin{cases} (L, u) & \text{if } ua \in A^*X, \\ (L, u) + 1 & \text{otherwise.} \end{cases} \tag{6.14}$$

Proof. The formula results from Equation (6.7).   □

Example 6.1.13 Let $A = \{a, b\}$ and $X = \{a\}$. Then $L_X(w) = |w|_b + 1$. Indeed, this results directly from Equation (6.5). It can also be obtained from Equation (6.14): scanning the prefixes of w from left to right, the indicator remains constant whenever one meets an a.

The following result shows how the condition to be a bifix code can be expressed on a deterministic automaton recognizing X^*.

Proposition 6.1.14 *Let X be a prefix code over A and let $\mathcal{A} = (Q, 1, 1)$ be a trim deterministic automaton recognizing X^*. Then X is bifix if and only if for any $q \in Q$ and $w \in A^*$, $q \cdot w = 1 \cdot w$ implies $q = 1$.*

Proof. Assume first that the condition holds. We show that X^* is left unitary. Let u, v be words such that $u, vu \in X^*$. Set $q = 1 \cdot v$. Then $1 \cdot u = 1$ and $1 \cdot vu = (1 \cdot v) \cdot u = 1$. Set $q = 1 \cdot v$. Then $q \cdot u = 1$ and the condition implies $q = 1$. This shows that $1 \cdot v = 1$ and consequently $v \in X^*$.

Assume conversely that X^* is left unitary and let w be such that $1 \cdot w = q \cdot w$ for some $q \in Q$. Set $p = q \cdot w$ and let u, v be words such that $1 \cdot u = q$, $p \cdot v = 1$. Then $1 \cdot uwv = 1 \cdot wv = 1$, showing that $uwv, wv \in X^*$. Since X^* is left unitary, we obtain $u \in X^*$. This in turn implies that $q = 1$.   □

The above condition is satisfied by an automaton which is *bideterministic* in the sense that for any edges (p, a, q) and (r, a, s) with $p, q, r, s \in Q$ and $a \in A$, one has $p = r$ if and only if $q = s$. However, it is not always possible to recognize X^* by a bideterministic automaton for a bifix code X (see Exercise 6.1.2).

6.2 Maximal bifix codes

A bifix code $X \subset A^+$ is *maximal* if, for any bifix code $Y \subset A^+$, the inclusion $X \subset Y$ implies that $X = Y$. As in Chapter 3, it is convenient to note that the set $\{1\}$ is a maximal bifix set without being a code. We start by giving a series of equivalent conditions for a thin code to be maximal bifix.

Proposition 6.2.1 *Let X be a thin subset of A^+. The following conditions are equivalent.*

(i) *X is a maximal code and bifix.*
(ii) *X is a maximal bifix code.*
(iii) *X is a maximal prefix code and a maximal suffix code.*
(iv) *X is a left complete prefix code.*
(iv') *X is a right complete suffix code.*
(v) *X is a left complete and right complete code.*

Proof. (i) $\Rightarrow$ (ii) is clear. (ii) $\Rightarrow$ (iii). If X is maximal prefix, then by Theorem 3.3.8, X is a maximal code, therefore X is maximal suffix. Similarly, if X is maximal suffix, it is maximal prefix. Thus, assume that X is neither maximal prefix nor maximal suffix. Let $y, z \notin X$ be such that $X \cup y$ is prefix and $X \cup z$ is suffix. Since $X \cup yt$ is prefix for any word t, it follows that $X \cup yz$ is prefix, and so also bifix. Moreover, $yz \notin X$ (since otherwise $X \cup y$ would not be prefix). This contradicts (ii).

(iii) $\Rightarrow$ (iv') is a consequence of Proposition 3.3.3 stating that a maximal prefix code is right-complete (similarly for the implication (iii) $\Rightarrow$ (iv)).

(iv) $\Rightarrow$ (v) The code X is complete and thin. Thus, it is maximal. This shows that it is maximal prefix, which in turn implies that it is right complete.

(v) $\Rightarrow$ (i) A complete, thin code is maximal. By Theorem 3.3.8 a right-complete thin code is prefix. Similarly, X is suffix. $\qquad\square$

A code which is both maximal prefix and maximal suffix is always maximal bifix, and the converse holds, as we have seen, for thin codes. However, this may become false for codes that are not thin (see Example 6.2.4).

Example 6.2.2 A group code, as defined in Section 2.2, is bifix and is a maximal code.

Example 6.2.3 Let $A = \{a, b\}$ and

$$X = \{a^3, a^2ba, a^2b^2, ab, ba^2, baba, bab^2, b^2a, b^3\}.$$

By inspection of the literal representation (Figure 6.4), X is seen to be a maximal prefix code.

The reverse code $\tilde{X}$ represented on the right in Figure 6.4, is also maximal prefix. Thus X is a maximal bifix code. Observe that $\tilde{X}$ is equal to the set obtained from X by interchanging a and b (reflection with respect to the horizontal axis). This is an exceptional fact, which will be explained later (Example 6.5.3).

Bifix codes

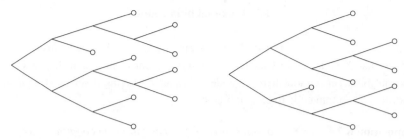

Figure 6.4 The literal representations of X on the left and of its reversal $\widetilde{X}$ on the right.

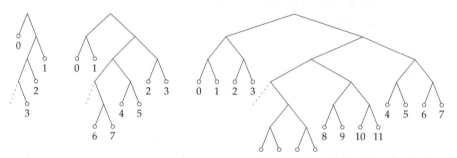

Figure 6.5 The reversible Golomb–Rice codes of orders 0, 1, 2.

Example 6.2.4 Let $A = \{a, b\}$ and $X = \{wab^{|w|} \mid w \in A^*\}$ (see Examples 2.4.11 and 3.3.9). It is a maximal, right-dense code which is suffix but not prefix. The set $Y = X \setminus XA^+$ is maximal prefix and suffix but not maximal suffix since $Y \neq X$. Thus, Y is also maximal bifix, satisfying condition (ii) in Proposition 6.2.1 without satisfying condition (iii).

Example 6.2.5 There is a reversible version of the Golomb–Rice codes described in Example 3.4.4. These are bifix codes having the same length distribution. The difference with the Golomb–Rice codes is that, in the base, the word $1^i 0$ is replaced by $10^{i-1}1$ for $i \geq 1$. Since the set of bases forms a bifix code, the set of all codewords is also a bifix code. The *reversible Golomb–Rice code* of order k, denoted RG_k is defined by the regular expression

$$RG_k = (0 + 10^*1)(0 + 1)^k.$$

Figure 6.5 represents the codes RG_k for $k = 0, 1, 2$.

Example 6.2.6 There is also a reversible version of the exponential Golomb codes (Example 3.4.5) which are bifix codes with the same length distribution. The code REG_0 is the bifix code

$$REG_0 = 0 + 1(00 + 10)^*(0 + 1)1,$$

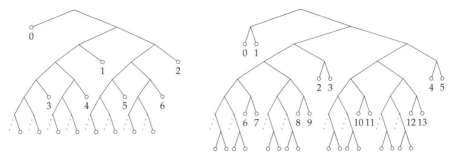

Figure 6.6 The reversible exponential Golomb codes of orders 0 and 1.

and the code of order k is

$$REG_k = REG_0(0 + 1)^k.$$

Note that REG_0 is equal to its reversal, that is $\widetilde{REG_0} = REG_0$. This shows that REG_0 is bifix. The other codes are also bifix because they are products of two bifix codes. The codes REG_k for $k = 0, 1, 2$ are represented on Figure 6.6.

The following result gives a different characterization of maximal bifix codes within the family of thin codes.

Proposition 6.2.7 *A thin code X is maximal bifix if and only if for all $w \in A^*$, there exists an integer $n \geq 1$ such that $w^n \in X^*$.*

Proof. Assume that for all $w \in A^*$, we have w^n in X^* for some $n \geq 1$. Then X clearly is right-complete and left-complete. Thus, X is maximal bifix by Proposition 6.2.1.

Conversely, let X be a maximal bifix code, and let $w \in A^*$. Consider a word $u \in \bar{F}(X)$, that is, which is not a factor of a word in X. The code X being right-complete, for all $i \geq 1$ there exists a word v_i such that

$$w^i u v_i \in X^*.$$

Since $u \in \bar{F}(X)$, there exists a prefix s_i of u such that $w^i s_i \in X^*$.

Let k, m with $k < m$ be two integers such that $s_k = s_m$. Then setting $n = m - k$, we have $w^k s_k \in X^*$, $w^m s_m = w^n w^k s_k \in X^*$. Since X^* is left-unitary, this implies that $w^n \in X^*$. $\qquad\square$

We now describe an operation which makes it possible to construct maximal bifix codes by successive transformations.

Proposition 6.2.8 *Let X be a code which is maximal prefix and maximal suffix, and let $w \in A^*$. Set*

$$G = Xw^{-1}, \qquad D = w^{-1}X,$$

$$G_0 = (wD)w^{-1}, \qquad D_0 = w^{-1}(Gw), \qquad (6.15)$$

$$G_1 = G \setminus G_0, \qquad D_1 = D \setminus D_0.$$

If $G_1 \neq \emptyset$ and $D_1 \neq \emptyset$, then the set

$$Y = (X \cup w \cup G_1(wD_0^*)D_1) \setminus (Gw \cup wD) \qquad (6.16)$$

is a maximal prefix and maximal suffix code. Further,

$$\underline{Y} = \underline{X} + (1 - \underline{G})w(1 - \underline{D_0}^*\underline{D_1}). \qquad (6.17)$$

Proof. By definition, Gw is the set of words in X ending with w. Similarly for wD. Next, G_0w is the set of words in X that start and end with w. Thus G_1w is the set of words in X which end with w and do not start with w.

Since $D_1 \neq \emptyset$, the set D is nonempty. Further $1 \notin D$, since otherwise $w \in X$, and X being bifix, this implies $G = D = \{1\}$, and $D_0 = \{1\}$ and finally $D_1 = \emptyset$, a contradiction. Thus, w is a proper prefix of a word in X, and by Proposition 3.4.9, the sets D and

$$Y_1 = (X \cup w) \setminus wD$$

are maximal prefix codes.

Next, $Gw = X \cap A^*w$ and $wD = X \cap wA^*$. Also $G_0w = wD \cap A^*w = X \cap wA^* \cap A^*w$. Similarly $wD_0 = Gw \cap wA^* = X \cap wA^* \cap A^*w$. Thus,

$$wA^* \cap A^*w \cap X = Gw \cap wD = wD_0 = G_0w. \qquad (6.18)$$

Now note that $G = G_0 \cup G_1$. From this and (6.18), we get

$$Gw \cup wD = G_0w \cup G_1w \cup wD = wD_0 \cup G_1w \cup wD = G_1w \cup wD,$$

since $D_0 \subset D$. Similarly

$$Gw \cup wD = Gw \cup wD_1.$$

Thus

$$Y = (Y_1 \cup G_1wD_0^*D_1) \setminus G_1w.$$

Note that $G_1w \subset Y_1$ because G_1w is the set of words in X which end with w and do not start with w, and thus $G_1w \subset X \setminus wD$. Since $D = D_1 \cup D_0$ is a maximal prefix code and $D_1 \neq \emptyset$, the set $D_0^*D_1$ is a maximal prefix code (Proposition 3.4.12). This and the fact that Y_1 is maximal prefix imply, according to Proposition 3.4.7, that Y is maximal prefix.

Symmetrically, it may be shown successively that $Y_2 = (X \cup w) \setminus wG$ and $Y' = (Y_2 \setminus wD_1) \cup G_1G_0^*wD_1$ are maximal suffix codes. From (6.18), we obtain by induction that $G_0^*w = wD_0^*$. Thus, $Y' = Y$ and consequently Y is also maximal suffix.

To prove (6.17), set

$$\sigma = \underline{X} + (1 - \underline{G})w(1 - \underline{D_0}^*\underline{D_1}).$$

Then

$$\sigma = \underline{X} + w - \underline{Gw} - w\underline{D_0}^*\underline{D_1} + \underline{GwD_0}^*\underline{D_1}$$
$$= \underline{X} + w - \underline{Gw} - w\underline{D_0}^*\underline{D_1} + \underline{G_0wD_0}^*\underline{D_1} + \underline{G_1wD_0}^*\underline{D_1}.$$

Since $\underline{G_0w} = w\underline{D_0}$, we obtain

$$\sigma = \underline{X} + w - \underline{Gw} - w\underline{D_0}^*\underline{D_1} + w\underline{D_0D_0}^*\underline{D_1} + \underline{G_1wD_0}^*\underline{D_1}$$
$$= \underline{X} + w - \underline{Gw} - w\underline{D_1} + \underline{G_1wD_0}^*\underline{D_1}.$$

The sets G_1w, D_0, D_1 are prefix, and $D_0 \neq 1$ (since otherwise $w \in X$). Thus, the products in the above expression are unambiguous. Next it follows from (6.18) that $G_1w \cap wD = \emptyset$. Consequently

$$\underline{Gw \cup wD} = \underline{G_1w} + w\underline{D}.$$

Thus

$$\sigma = \underline{X} + w + \underline{G_1wD_0^*D_1} - \underline{Gw \cup wD} = \underline{Y},$$

since $Gw \cup wD \subset X$. □

The code Y is said to be obtained from X by *internal transformation* (with respect to w).

Example 6.2.9 Let $A = \{a, b\}$, and consider the uniform code $X = A^2$. Let $w = a$. Then $G = D = A$ and $G_0 = D_0 = \{a\}$. Consequently, the code Y defined by Formula (6.16) is

$$Y = a \cup ba^*b.$$

Note that Y is a group code as is X.

From Formula (6.16), it is clear that for a finite code X, the code Y is finite if and only if $D_0 = \emptyset$. This case deserves particular attention.

Proposition 6.2.10 *Let X be a finite maximal bifix code and let $w \in A^*$. Set*

$$G = Xw^{-1}, \qquad D = w^{-1}X. \tag{6.19}$$

If $G \neq \emptyset$, $D \neq \emptyset$ and $Gw \cap wD = \emptyset$, then

$$Y = (X \cup w \cup GwD) \setminus (Gw \cup wD) \tag{6.20}$$

is a finite maximal bifix code, and

$$\underline{Y} = \underline{X} + (\underline{G} - 1)w(\underline{D} - 1). \tag{6.21}$$

Conversely, let Y be a finite maximal bifix code. Let $w \in Y$ be a word such that there exists a maximal prefix code D, and a maximal suffix code G with $GwD \subset Y$. Then

$$X = (Y \setminus (w \cup GwD)) \cup (Gw \cup wD) \tag{6.22}$$

is a finite maximal bifix code, and further Equations (6.19), (6.20), and (6.21) hold.

Proof. If $Gw \cap wD = \emptyset$, then we have, with the notations of Proposition 6.2.8, $G_0 = D_0 = \emptyset$ by Formula (6.18). Then (6.16) simplifies into (6.20). Formula (6.21) is a direct consequence of Formula (6.17).

Conversely, let us first show that X is a maximal prefix code. Set

$$Z = (Y \setminus w) \cup wD.$$

Since Y is maximal prefix by Proposition 6.2.1 and since D is maximal prefix and $w \in Y$, Corollary 3.4.8 implies that the set Z is a maximal prefix code. Next observe that

$$X = (Z \setminus GwD) \cup Gw.$$

The set Gw is contained in ZA^-, since $Gw \subset (Y \setminus w)A^-$. Next we show that Gw is prefix. Assume indeed that $gw = g'wt$ for some $g, g' \in G, t \in A^*$. Let d be a word in D of maximal length. The set D being maximal prefix, either td is a proper prefix of a word in D or td has a prefix in D. The first case is ruled out by the fact that d has maximal length. Thus, td has a prefix, say d' in D. The word $g'wd'$ is a prefix of $g'wtd = gwd$. Since both are in the prefix set Y, they are equal. Thus $d' = td$ and since d has maximal length, we get $t = 1$. This proves the claim.

Further, for all $g \in G$, we have $D = (gw)^{-1}Z$. Indeed, the inclusion $gwD \subset Z$ implies $D \subset (gw)^{-1}Z$, and D being a maximal prefix code, the equality follows.

In view of Proposition 3.4.10, the set X consequently is a maximal prefix code. Symmetrically, it may be shown that X is maximal suffix. Since X is finite, it is maximal bifix.

It remains to show that Y is obtained from X by internal transformation. First, the inclusion $Gw \subset X$ follows from (6.22), implying $G \subset Xw^{-1}$, and G being a maximal suffix code, this enforces the equality

$$G = Xw^{-1}.$$

Symmetrically $D = w^{-1}X$. Moreover, $G \neq \emptyset$, $D \neq \emptyset$, because they are maximal codes. Let us show that

$$Gw \cap wD = \emptyset.$$

If $gw = wd$ for some $g \in G, d \in D$, then $ggw = gwd \in GwD \subset Y$. Thus $w, ggw \in Y$; this is impossible, since Y is suffix.

From $w \in Y$ we get the result that $Gw \cap Y = \emptyset$; otherwise Y would not be suffix. Similarly $wD \cap Y = \emptyset$, because Y is prefix. Then as a result of (6.22), $X \setminus (Gw \cup wD) = Y \setminus (w \cup GwD)$, implying (6.20). $\square$

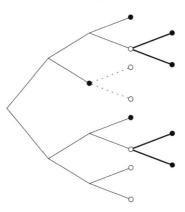

Figure 6.7 An internal transformation.

Example 6.2.11 Let $A = \{a, b\}$ and $X = A^3$. Consider the word $w = ab$. Then $G = D = A$ and $Gw \cap wD = \emptyset$. Thus Proposition 6.2.10 gives a finite code Y. This code is obtained by dropping in Figure 6.7 the dotted lines and by adjoining the heavy lines. The result is the maximal bifix code of Example 6.2.3.

6.3 Degree

In this section, we study the indicator of thin maximal bifix codes. For these bifix codes, some simplifications occur.

Let $X \subset A^+$ be a bifix code, set $U = A^* \setminus XA^*$, $V = A^* \setminus A^*X$, and let $L = \underline{V}\,\underline{X}^*\underline{U}$ be the indicator of X. If X is a maximal prefix code, then $U = P$ where $P = XA^-$ is the set of proper prefixes of words in X. In the same way, for a maximal suffix code, we have $V = S$ where $S = A^-X$ is the set of proper suffixes of words in X. It follows that if X is maximal prefix and maximal suffix, each parse of a word is an interpretation. Then we have

$$L = \underline{S}\,\underline{X}^*\underline{P} = \underline{S}\,\underline{A}^* = \underline{A}^*\underline{P}. \tag{6.23}$$

This basic formula will be used frequently. It means that the number of parses of a word is equal to the number of its suffixes which are in P, or equivalently the number of its prefixes which are in S. Let X be a subset of A^+. Denote by

$$H(X) = A^- X A^- = \{w \in A^* \mid A^+ w A^+ \cap X \neq \emptyset\}$$

the set of *internal factors* of words in X. Let

$$\bar{H}(X) = A^* \setminus H(X).$$

Clearly, each internal factor is a factor of a word in X. The converse may be false. The set $H(X)$ and the set

$$F(X) = \{w \in A^* \mid A^* w A^* \cap X \neq \emptyset\}$$

of factors of words in X are related by

$$F(X) = H(X) \cup XA^- \cup A^- X \cup X,$$

and for $\bar{F}(X) = A^* \setminus F(X)$,

$$A^+ \bar{H}(X) A^+ \subset \bar{F}(X) \subset \bar{H}(X).$$

These relations show that $\bar{H}(X)$ is nonempty if and only if $\bar{F}(X)$ is nonempty; thus X is thin if and only if $\bar{H}(X) \neq \emptyset$.

Theorem 6.3.1 *Let $X \subset A^+$ be a bifix code. Then X is a thin maximal code if and only if its indicator L is bounded. In this case,*

$$\bar{H}(X) = \{w \in A^* | (L, w) = d\}, \tag{6.24}$$

where d is defined as $d = \max\{(L, w) \mid w \in A^*\}$.

Proof. Let X be a thin maximal bifix code. Let $w \in \bar{H}(X)$ and $w' \in A^*$. According to Formula (6.23), $(L, ww') = (\underline{SA}^*, ww')$. Thus the number of parses of ww' is equal to the number of prefixes of ww' which are in $S = A^- X$. Since $w \in \bar{H}(X)$, it follows that no such prefix in S is strictly longer than w. Thus all these prefixes are prefixes of w. Again using Formula (6.23), this shows that $(L, ww') = (L, w)$. Now by Proposition 6.1.8, we have $(L, ww') \geq (L, w')$. Thus we get

$$(L, w') \leq (L, w),$$

showing that L is bounded on A^* by its value for a word in $\bar{H}(X)$. This shows also that L is constant on $\bar{H}(X)$. Thus

$$\bar{H}(X) \subset \{w \in A^* \mid (L, w) = d\}.$$

To show the converse inclusion, consider an internal factor $w \in H(X)$. Then there exist $p, s \in A^+$ such that $w' = pws \in X$. This implies that

$$(L, w') \geq (L, w) + 1.$$

Indeed, each parse of w can be extended in a parse of w', and w' has an additional parse, namely $(1, w', 1)$. This shows that for an internal factor w, the number (L, w) is strictly less than the maximal value d. Thus Formula (6.24) is proved.

Assume now conversely that X is a bifix code with bounded indicator L, let $d = \max\{(L, w) \mid w \in A^*\}$ and let $v \in A^*$ be a word such that $(L, v) = d$. We use Formula (6.3) which can be rewritten as

$$\underline{XA}^* = \underline{A}^* + (\underline{A} - 1)L.$$

Let $w \in A^+$ be any nonempty word, and set $w = au$, with $a \in A, u \in A^*$. Then

$$(\underline{XA}^*, wv) = (\underline{A}^* + (\underline{A} - 1)L, auv) = 1 + (L, uv) - (L, auv).$$

By Proposition 6.1.8, both (L, uv) and (L, auv) are greater than or equal to (L, v). By the choice of v, we have $(L, uv) = (L, auv) = d$.

Thus $(\underline{X}A^*, wv) = 1$. Thus we have proved that for all $w \in A^+$, $wv \in XA^*$. This shows that XA^* is right dense. This shows also that X is thin. Indeed, we have $v \in \bar{H}(X)$ since for all $g, d \in A^+$ we have $gv \in XA^*$ and therefore $gvd \notin X$. Thus X is a thin maximal prefix code. Symmetrically, it can be shown that X is maximal suffix. This gives the result by Proposition 6.2.1. □

Let X be a thin maximal bifix code, and let L be its indicator. The *degree* of X, denoted $d(X)$ or simply d, is the number

$$d(X) = \max\{(L, w) \mid w \in A^*\}.$$

According to Theorem 6.3.1, the degree d is the number of parses of any word which is not an internal factor of X. Before going on, let us illustrate the notion of degree with several examples.

Example 6.3.2 Let φ be a morphism from A^* onto a group G, and let G' be a subgroup of G. Let X be the group code for which $X^* = \varphi^{-1}(G')$. We have seen that X is a maximal bifix code, and that X is thin if and only if G' has finite index in G (Example 2.5.22).

The degree of X is equal to the index of G' in G. Indeed let $w \in \bar{H}(X)$ be a word which is not an internal factor of X, and consider the function ψ which associates, to each word $u \in A^*$, the unique word $p \in P = XA^-$ such that $uw \in X^*p$. Each p obtained in such a way is a suffix of w. The set $\psi(A^*)$ is the set of suffixes of w which are in P. Since $w \in \bar{H}(X)$, we have Card $\psi(A^*) = d(X)$. Next, we have for $u, v \in A^*$,

$$\psi(u) = \psi(v) \Leftrightarrow G'\varphi(u) = G'\varphi(v).$$

Indeed, if $\psi(u) = \psi(v) = p$, then $uw, vw \in X^*p$, and consequently $\varphi(u), \varphi(v) \in G'\varphi(p)\varphi(w)^{-1}$. Conversely, if $G'\varphi(u) = G'\varphi(v)$, let $r \in A^*$ be a word such that $uwr \in X^*$. Then $\varphi(vwr) \in G'\varphi(u)\varphi(wr) \subset G'$, whence $vwr \in X^*$. Since $\psi(u)$ and $\psi(v)$ are suffixes of w, one of the words $\psi(u)r$ and $\psi(v)r$ is a suffix of the other. Since X is a suffix code, it follows that $\psi(u) = \psi(v)$.

This shows that the index of G' in G is $d(X)$. By Proposition 1.13.1, $d(X)$ is also equal to the degree of the permutation group corresponding to the action of G on the cosets of G', as defined in Section 1.13.

Example 6.3.3 The only maximal bifix code with degree 1 over A is $X = A$.

Example 6.3.4 Any maximal bifix code of degree 2 over an alphabet A has the form

$$X = C \cup BC^*B, \tag{6.25}$$

where A is the disjoint union of B and C, with $B \neq \emptyset$.

Indeed, let $C = A \cap X$ and $B = A \setminus C$. Each $b \in B$ has two parses, namely $(1, 1, b)$ and $(b, 1, 1)$. Thus, a word which is an internal factor of a word $x \in X$

cannot contain a letter in B, since otherwise x would have at least three parses. Thus, the set H of internal factors of X satisfies $H \subset C^*$. Next consider a word x in X. Either it is a letter, and then it is in C, or otherwise it has the form $x = aub$ with $a, b \in A$ and $u \in H \subset C^*$. X being bifix, neither a nor b is in C. Thus $X \subset C \cup BC^*B$. The maximality of X implies the equality.

This shows that any maximal bifix code of degree 2 is a group code. Indeed, the code given by (6.25) is obtained by considering the morphism from A^* onto $\mathbb{Z}/2\mathbb{Z}$ defined by $\varphi(B) = \{1\}$, $\varphi(C) = \{0\}$. It shows also that any maximal bifix code of degree 2 is rational. This is false for degree 3 (see Example 6.4.8).

Example 6.3.5 Consider the set

$$Y = \{a^n b^n \mid n \geq 1\}.$$

It is a bifix code which is not maximal since $Y \cup ba$ is bifix. Also Y is thin since $ba \in \bar{F}(Y)$. The code Y is not contained in a thin maximal bifix code. Suppose indeed that X is a thin maximal bifix code of degree d containing Y. For any $n \geq 0$, the word a^n then has $n + 1$ parses, since it has $n + 1$ suffixes which all are proper prefixes of a word in Y, whence in X. Since $d \leq n$, this is impossible. In fact, Y is contained in the Dyck code over $\{a, b\}$ (see Example 2.2.11)

Example 6.3.6 Let $X, Y \subset A^+$ be two thin maximal bifix codes. Then XY is maximal bifix and thin and

$$d(XY) = d(X) + d(Y).$$

The first part of the claim follows indeed from Corollary 3.4.2. Next, let $w \in \bar{H}(XY)$ be a word which is not an internal factor of XY. Then, $w \in \bar{H}(X)$ and $w \in \bar{H}(Y)$. The prefixes of w which are also proper suffixes of XY are of two kinds. First, there are $d(Y)$ prefixes of w which are proper suffixes of words in Y. Next, there are $d(X)$ prefixes of w which are proper suffixes of words in X. For each such prefix u, set $w = uv$. The word v is not a proper prefix of a word in Y since otherwise w would be an internal factor of XY. Thus v has a prefix y in Y and uy is a prefix of w which is a proper suffix of a word in XY. These are the only prefixes of w which are in $A^-(XY)$. Since w has $d(XY)$ parses with respect to XY, this gives the formula.

We now define a formal power series associated to a code X and which plays a fundamental role in the following. Let X be a thin maximal bifix code over A. The *tower* over X is the formal power series T_X (also written T when no confusion is possible) defined by

$$(T_X, w) = d - (L_X, w). \tag{6.26}$$

The following proposition gives a simple way to compute the value of a tower.

Proposition 6.3.7 *Let $X \subset A^+$ be a thin maximal bifix code. For any word $u \in A^*$ and letter $a \in A$, one has*

$$(T_X, ua) = \begin{cases} (T_X, u) & \text{if } ua \in A^*X, \\ (T_X, u) - 1 & \text{otherwise.} \end{cases} \tag{6.27}$$

Proof. This results directly from Proposition 6.1.12.   □

The following proposition states some useful elementary facts about the series T.

Proposition 6.3.8 *Let X be a thin maximal bifix code of degree d over A, set $P = XA^-$, $S = A^-X$, and let T be the tower over X. Then*

$$(T, w) = 0 \iff w \in \bar{H}(X),$$

and for $w \in H(X)$,

$$1 \le (T, w) \le d - 1. \tag{6.28}$$

Further $(T, 1) = d - 1$ and

$$\underline{X} - 1 = (\underline{A} - 1)T(\underline{A} - 1) + d(\underline{A} - 1), \tag{6.29}$$

$$\underline{P} = (\underline{A} - 1)T + d, \tag{6.30}$$

$$\underline{S} = T(\underline{A} - 1) + d. \tag{6.31}$$

Proof. According to Theorem 6.3.1, $(T, w) = 0$ if and only if $w \in \bar{H}(X)$. For all other words, $1 \le (T, w)$. Also $(T, w) \le d - 1$ since all words have at least one parse, and $(T, 1) = d - 1$ since the empty word has exactly one parse.

Next, by definition of T, we have $T + L = dA^*$, whence

$$T(1 - \underline{A}) + L(1 - \underline{A}) = (1 - \underline{A})T + (1 - \underline{A})L = d.$$

The code X is maximal; consequently $P = A^* \setminus XA^*$ and $S = A^* \setminus A^*X$. Thus we can apply Proposition 6.1.7 with $P = U$, $S = V$. Together with the equation above, this gives Formulas (6.30), (6.31), and also (6.29) since

$$\underline{X} - 1 = \underline{P}(\underline{A} - 1) = ((\underline{A} - 1)\underline{T} + d)(\underline{A} - 1).$$ □

Proposition 6.3.8 shows that the support of the series T is contained in the set $H(X)$. Note that two thin maximal bifix codes X and X' having the same tower are equal. Indeed, by Proposition 6.3.8, they have the same degree since

$$(T, 1) = d(X) - 1 = d(X') - 1.$$

But then Equation (6.29) implies that $X = X'$.

Whenever a thin maximal bifix code of degree $d = d(X)$ satisfies the equation

$$\underline{X} - 1 = (\underline{A} - 1)T(\underline{A} - 1) + d(\underline{A} - 1),$$

for some T, then T must be the tower on X. The next result gives a sufficient condition to obtain the same conclusion without knowing that the integer d is equal to $d(X)$.

Proposition 6.3.9 *Let $T, T' \in \mathbb{Z}\langle\langle A \rangle\rangle$ and let $d, d' \geq 1$ be integers such that*

$$(\underline{A} - 1)T(\underline{A} - 1) + d(\underline{A} - 1) = (\underline{A} - 1)T'(\underline{A} - 1) + d'(\underline{A} - 1). \tag{6.32}$$

If there is a word $w \in A^*$ such that $(T, w) = (T', w)$, then $T = T'$ and $d = d'$.

Proof. After multiplication of both sides by $\underline{A}^* = (1 - \underline{A})^{-1}$, Equation (6.32) becomes

$$T - d\underline{A}^* = T' - d'\underline{A}^*.$$

If $(T, w) = (T', w)$, then $(d\underline{A}^*, w) = (d'\underline{A}^*, w)$. Thus, $d = d'$, which implies $T = T'$. $\square$

We now observe the effect of an internal transformation (Proposition 6.2.8) on the tower over a thin maximal bifix code X. Recall that, provided w is a word such that G_1, D_1 are both nonempty, where

$$G = Xw^{-1}, \quad D = w^{-1}X, \quad G_0 = (wD)w^{-1}, \quad D_0 = w^{-1}(Gw),$$

$$G_1 = G \setminus G_0, \quad D_1 = D \setminus D_0,$$

the code Y defined by

$$\underline{Y} = \underline{X} + (1 - \underline{G})w(1 - \underline{D_0}^*\underline{D_1})$$

is maximal bifix. By Proposition 3.4.9, the sets $G = Xw^{-1}$ and $D = w^{-1}X$, are maximal suffix and maximal prefix. Let U be the set of proper right factors of G, and let V be the set of proper prefixes of D. Then D_0^*V is the set of proper prefixes of words in $D_0^*D_1$, since $D = D_0 \cup D_1$. Consequently

$$\underline{G} - 1 = (\underline{A} - 1)\underline{U}, \quad \underline{D_0}^*\underline{D_1} - 1 = \underline{D_0}^*\underline{V}(\underline{A} - 1).$$

Going back to Y, we get

$$\underline{Y} - 1 = \underline{X} - 1 + (\underline{A} - 1)\underline{U}w\underline{D_0}^*\underline{V}(\underline{A} - 1).$$

Let T be the tower over X. Then using Equation (6.29), we get

$$\underline{Y} - 1 = (\underline{A} - 1)(T + \underline{U}w\underline{D_0}^*\underline{V})(\underline{A} - 1) + d(\underline{A} - 1).$$

Observe that since X is thin, both G and D are thin. Consequently also U and V are thin. Since $D_1 = D \setminus D_0 \neq \emptyset$, D_0 is not a maximal code. As a subset of D, the set D_0 is thin. By Theorem 2.5.13, D_0 is not complete. Thus D_0^* is thin. Thus UwD_0^*V, as a product of thin sets, is thin. Next $\text{supp}(T) \subset H(X)$ is thin. Thus $\text{supp}(T) \cup UwD_0^*V$ is thin.

Let u be a word which is not a factor of a word in this set. Then

$$(T + \underline{U}w\underline{D_0^*V}, u) = 0.$$

On the other hand, Formula (6.16) shows that since $G_1(w D_0^*)D_1$ is thin, the set Y is thin. Thus, the support of the tower T_Y over Y is thin. Let v be such that $(T_Y, v) = 0$, then

$$(T + \underline{U}w\underline{D_0^*V}, uv) = (T_Y, uv) = 0,$$

showing that Proposition 6.3.9 can be applied. Consequently,

$$d(X) = d(Y) \quad \text{and} \quad T_Y = T + \underline{U}w\underline{D_0^*V}.$$

Thus, the degree of a thin maximal bifix code remains invariant under internal transformations.

Example 6.3.10 The finite maximal bifix code $X = \{a^3, a^2ba, a^2b^2, ab, ba^2, baba, bab^2, b^2a, b^3\}$ over $A = \{a, b\}$ of Example 6.2.3 has degree 3. This can be seen by observing that no word has more than 3 parses, and the word a^3 has 3 parses, or also by the fact (Example 6.2.11) that X is obtained from the uniform code A^3 by internal transformation with respect to the word $w = ab$. Thus $d(X) = d(A^3) = 3$.

In this example, $D(= w^{-1}A^3) = G(= A^3 w^{-1}) = A$. Thus $T_X = T_{A^3} + w$. Clearly $T_{A^3} = 2 + a + b$. Consequently

$$T_X = 2 + a + b + ab.$$

We now give a characterization of the formal power series that are the tower over some thin maximal bifix code.

Proposition 6.3.11 *A formal power series $T \in \mathbb{N}\langle\!\langle A \rangle\!\rangle$ is the tower over some thin maximal bifix code if and only if it satisfies the following conditions.*

(i) *For all $a \in A$, $v \in A^*$,*

$$0 \leq (T, v) - (T, av) \leq 1, \tag{6.33}$$

$$0 \leq (T, v) - (T, va) \leq 1. \tag{6.34}$$

(ii) *For all $a, b \in A$, $v \in A^*$,*

$$(T, av) + (T, vb) \leq (T, v) + (T, avb). \tag{6.35}$$

(iii) *There exists a word $v \in A^*$ such that*

$$(T, v) = 0.$$

Proof. Let X be a thin maximal bifix code of degree d, let L be its indicator, and let $T = d\underline{A}^* - L$. Then Equations (6.33), (6.34), and (6.35) are direct consequences of

Equations (6.11), (6.12), and (6.13). Further (iii) holds for all $v \in \bar{H}(X)$, and this set is nonempty.

Conversely, assume that $T \in \mathbb{N}\langle\!\langle A \rangle\!\rangle$ satisfies the conditions of the proposition. Define

$$d = (T, 1) + 1, \qquad L = d\underline{A}^* - T.$$

Then by construction, L satisfies the conditions of Proposition 6.1.11, and therefore L is the indicator of some bifix code X. Next by assumption, T has nonnegative coefficients. Thus for all $w \in A^*$, we have $(T, w) = d - (L, w) \geq 0$. Thus, L is bounded. In view of Theorem 6.3.1, the code X is maximal and thin. Since $(T, v) = 0$ for at least one word v, we have $(L, v) = d$ and $d = \max\{(L, w) | w \in A^*\}$. Thus, d is the degree of X and $T = d\underline{A}^* - L$ is the tower over X. $\qquad\square$

The preceding result makes it possible to disassemble the tower over a bifix code.

Proposition 6.3.12 *Let T be the tower over a thin maximal bifix code X of degree $d \geq 2$. The series*

$$T' = T - \underline{H(X)}$$

is the tower over some thin maximal bifix code of degree $d - 1$.

Proof. First observe that T' has nonnegative coefficients. Indeed, by Proposition 6.3.8, $(T, w) \geq 1$ if and only if $w \in H(X)$. Consequently $(T', w) \geq 0$ for $w \in H(X)$, and $(T', w) = (T, w) = 0$ otherwise.

Next, we verify the three conditions of Proposition 6.3.11.

(i) Let $a \in A, v \in A^*$. If $av \in H(X)$, then $v \in H(X)$. Thus $(T', av) = (T, av) - 1$ and $(T', v) = (T', av) - 1$. Therefore the inequality (6.33) results from the corresponding inequality for T. Next, if $av \notin H(X)$, then $(T, av) = (T', av) = 0$. Consequently $(T, v) \leq 1$. If $(T, v) = 1$, then $v \in H(X)$ and thus $(T', v) = 0$. Otherwise, $v \in \bar{H}(X)$ and $(T', v) = 0$ as already observed above. In both cases, $(T', v) = 0$, and thus the inequality (6.33) holds for T'.

(ii) Let $a, b \in A$ and $v \in A^*$. If $avb \in H(X)$, then $(T', w) = (T, w) - 1$ for each of the four words $w = avb, av, vb$, and v. Thus, the inequality

$$(T', av) + (T', vb) \leq (T', v) + (T', avb)$$

results, in this case, from the corresponding inequality for T. On the other hand, if $avb \notin H(X)$, then as before $(T, av), (T, vb) \leq 1$ and $(T', av) = (T', vb) = 0$. Thus (6.35) holds for T'.

Condition (iii) of Proposition 6.3.11 is satisfied clearly for T' since $(T', w) = 0$ for $w \in \bar{H}(X)$. Thus T' is the tower over some thin maximal bifix code. Its degree is $1 + (T', 1)$. Since $1 \in H(X)$, we have $(T', 1) = d - 2$. This completes the proof. $\quad\square$

Let X be a thin maximal bifix code of degree $d \geq 2$, and let T be the tower over X. Let X' be the thin maximal bifix code with tower $T' = T - \underline{H(X)}$. Then X' has degree $d - 1$. The code X' is called the *code derived* from X. Since for the indicators L and L' of X and X', we have $L = d\underline{A}^* - T$ and $L' = (d-1)\underline{A}^* - T'$, it follows that $L - L' = \underline{A}^* - T + T' = \underline{A}^* - H(X) = \bar{H}(X)$, whence

$$L' = L - \bar{H}(X). \tag{6.36}$$

We denote by $X^{(n)}$ the code derived from $X^{(n-1)}$ for $d(X) \geq n + 1$, with $X^{(0)} = X$.

Proposition 6.3.13 *The tower over a thin maximal bifix code X of degree $d \geq 2$ satisfies*

$$T = \underline{H(X)} + \underline{H(X')} + \cdots + \underline{H(X^{(d-2)})}.$$

Proof. By induction, we have from Proposition 6.3.12

$$T = \underline{H(X)} + \underline{H(X')} + \cdots + \underline{H(X^{(d-2)})} + \widehat{T},$$

where $\widehat{T}$ is the tower over a code of degree 1. This code is the alphabet, and consequently $\widehat{T} = 0$. This proves the result. $\qquad\square$

We now describe the set of proper prefixes and the set of proper suffixes of words of the derived code of a thin maximal bifix code.

Proposition 6.3.14 *Let $X \subset A^+$ be a thin maximal bifix code of degree $d \geq 2$. Let $S = A^- X$, $P = X A^-$, and $H = A^* \setminus X A^-$, $\bar{H} = A^* \setminus H$.*

1. *The set $S \cap \bar{H}$ is a thin maximal prefix code. The set H is the set of its proper prefixes, that is, $S \cap \bar{H} = HA \setminus H$.*
2. *The set $P \cap \bar{H}$ is a thin maximal suffix code. The set H is the set of its proper suffixes, that is, $P \cap \bar{H} = AH \setminus H$.*
3. *The set $S \cap H$ is the set of proper suffixes of the derived code X'.*
4. *The set $P \cap H$ is the set of proper prefixes of the derived code X'.*

Proof. We first prove 1. Let T be the tower over X, and let T' be the tower over the derived code X'. By Proposition 6.3.12, $T = T' + \underline{H}$, and by Proposition 6.3.8

$$\underline{S} = T(\underline{A} - 1) + d.$$

Thus, $\underline{S} = T'(\underline{A} - 1) + d - 1 + \underline{H}(\underline{A} - 1) + 1$. The code X' has degree $d - 1$. Thus, the series $T'(\underline{A} - 1) + d - 1$ is, by Formula (6.31), the characteristic series of the set $S' = A^* \setminus X'$ of proper suffixes of words of X'. Thus,

$$\underline{S} = \underline{H}(\underline{A} - 1) + 1 + \underline{S'} \quad \text{and} \quad \underline{S'} = T'(\underline{A} - 1) + d - 1.$$

The set H is prefix-closed and nonempty. We show that H contains no right ideal. Indeed, the set $\bar{H}$ is not empty because X is thin, and thus it is an ideal. Thus, for each $h \in H$, and $k \in \bar{H}$, the word hk is not in H. By Proposition 3.3.3, the set

$Y = HA \setminus H$ is a maximal prefix code, and $H = YA^-$. Thus

$$\underline{Y} = \underline{H}(\underline{A} - 1) + 1.$$

Further, H being also suffix-closed, the set Y is in fact a semaphore code by Proposition 3.5.8. We now verify that $Y = S \cap \bar{H}$.

Assume that $y \in Y$. Then, from the equation $\underline{S} = \underline{Y} + \underline{S}'$, it follows that $y \in S$. Since $H = YA^-$, we have $y \notin H$. Thus $y \in S \cap \bar{H}$. Conversely, assume that $y \in S \cap \bar{H}$. Then $y \neq 1$, since $d \geq 2$ implies that $H \neq \emptyset$ and consequently $1 \in H$. Further, each proper prefix of y is in $SA^- = A^* \setminus XA^- = H$, thus is an internal factor of X. In particular, considering just the longest proper prefix, we have $y \in HA$. Consequently, $y \in HA \setminus H = Y$.

The second claim is proved in a symmetric way. To show 3, observe that by what we proved before, we have

$$\underline{S} = \underline{Y} + \underline{S}'. \tag{6.37}$$

Next $\underline{S} = (S \cap H) \cup (S \cap \bar{H}) = Y \cup (S \cap H)$, since $Y = S \cap \bar{H}$. Moreover, the union is disjoint, thus $\underline{S} = \underline{Y} + \underline{S \cap H}$. Consequently $S' = S \cap H$. In the same way, we get point 4. $\qquad\square$

Theorem 6.3.15 *Let X be a thin maximal bifix code of degree d. Then the set S of its proper suffixes is a disjoint union of d maximal prefix sets.*

Proof. If $d = 1$, then $X = A$ and the set $S = \{1\}$ is a maximal prefix set. If $d \geq 2$, then the set $Y = S \cap \bar{H}$, where $H = A^- X A^-$ and $\bar{H} = A^* \setminus H$, is maximal prefix by Proposition 6.3.14. Further, the set $S' = S \cap H$ is the set of proper suffixes of the code derived from X. Arguing by induction, the set S' is a disjoint union of $d - 1$ maximal prefix sets. Thus $S = Y \cup S'$ is a disjoint union of d maximal prefix sets. $\qquad\square$

It must be noted that the decomposition, in Theorem 6.3.15, of the set S into disjoint maximal prefix sets is not unique (see Exercise 6.3.1). The following corollary to Theorem 6.3.15 expresses the remarkable property that the average length of a thin maximal bifix code, with respect to a Bernoulli distribution, is an integer.

Corollary 6.3.16 *Let $X \subset A^+$ be a thin maximal bifix code. For any positive Bernoulli distribution π on A^*, the average length of X is equal to its degree.*

Proof. Set $d = d(X)$. Let π be a positive Bernoulli distribution on A^*, and let $\lambda(X)$ be the average length of X. By Corollary 3.7.13, the average length $\lambda(X)$ is finite and $\lambda(X) = \pi(S)$, where $S = A^- X$ is the set of proper suffixes of X. In view of Theorem 6.3.15, we have

$$\underline{S} = \underline{Y_1} + \underline{Y_2} + \cdots + \underline{Y_d},$$

where each Y_i is a maximal prefix code. As a set of factors of X, each Y_i also is thin. Thus $\pi(Y_i) = 1$ for $i = 1, \ldots, d$ by Theorem 2.5.16. Consequently,

$$\lambda(X) = \sum_{i=1}^{d} \pi(Y_i) = d. \qquad \square$$

Note that Corollary 6.3.16 can also be proved directly by starting with Formula 6.30. However, the proof we have given here is the most natural one.

We now prove a converse of Theorem 6.3.15.

Proposition 6.3.17 *Let X be a thin maximal suffix code. If the set of its proper suffixes is a disjoint union of d maximal prefix sets, then X is bifix, and has degree d.*

Proof. Let $S = A^- X$. By assumption $\underline{S} = \underline{Y_1} + \cdots + \underline{Y_d}$, where $Y_1, \ldots, Y_d$ are maximal prefix sets. Let U_i be the set of proper prefixes of Y_i. Then $\underline{A}^* = \underline{Y_i}^* \underline{U_i}$, and thus $(1 - \underline{Y_i})\underline{A}^* = \underline{U_i}$, whence

$$\underline{A}^* = \underline{U_i} + \underline{Y_i}\,\underline{A}^*.$$

Summing up these equalities gives

$$d\underline{A}^* = \sum_{i=1}^{d} \underline{U_i} + \underline{S}\,\underline{A}^*.$$

Multiply on the left by $\underline{A} - 1$. Then, since $(\underline{A} - 1)\underline{S} = \underline{X} - 1$,

$$-d = \sum_{i=1}^{d}(\underline{A} - 1)\underline{U_i} + (\underline{X} - 1)\underline{A}^*,$$

whence

$$\underline{X}\underline{A}^* = \underline{A}^* - \sum_{i=1}^{d}(\underline{A} - 1)\underline{U_i} - d.$$

From this formula, we derive the fact that XA^* is right dense. Indeed, let $w \in A^+$, and set $w = au$, with $a \in A$. Each of the sets Y_i is maximal prefix. Thus, each $Y_i A^*$ is right dense. We show that there exists a word v such that simultaneously $auv \in Y_i A^*$ for all $i \in \{1, \ldots, d\}$ and also $uv \in Y_i A^*$ for all $i \in \{1, \ldots, d\}$. Indeed, there exists a word v_1' such that $auv_1' \in Y_1 A^*$. There exists a word v_1'' such that $uv_1'v_1'' \in Y_1 A^*$. Set $v_1 = v_1'v_1''$. Then both $uv_1, auv_1 \in Y_1 A^*$. In the same way, there is a word v_2 such that both $uv_1 v_2$ and $auv_1 v_2$ are in $Y_1 A^*$ and in $Y_2 A^*$. Continuing in this way, there is a word v such that $uv, auv \in Y_i A^*$ for $i = 1, \ldots, d$. Thus for each $i \in \{1, \ldots, d\}$

$$((\underline{A} - 1)\underline{U_i}, wv) = (\underline{A}\,\underline{U_i}, wv) - (\underline{U_i}, wv)$$

$$= (\underline{U_i}, uv) - (\underline{U_i}, wv) = 0 - 0 = 0.$$

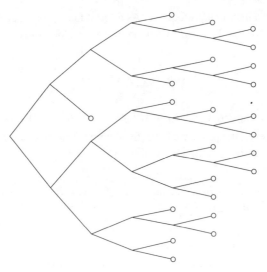

Figure 6.8 A maximal bifix code of degree 4.

Consequently

$$(\underline{X}\underline{A}^*, wv) = (\underline{A}^*, wv) = 1.$$

Thus, $wv \in XA^*$. Consequently XA^* is right dense or equivalently X is right complete. In view of Proposition 6.2.1, this means that X is maximal bifix.

Let $w \in \bar{H}(X)$ be a word which is not an internal factor of X. Then $w \notin U_i$ for $1 \leq i \leq d$. The set Y_i being maximal prefix, we have $w \in Y_i A^*$ for $1 \leq i \leq d$. Consequently, w has exactly d prefixes which are suffixes of words in X, one in each Y_i. Thus X has degree d. $\qquad\square$

Example 6.3.18 Let X be the finite maximal bifix code given in Figure 6.8. The tower T over X is given in Figure 6.9 (by its values on the set $H(X)$). The computation can be done by using Equation (6.27). The derived code X' is the maximal bifix code of degree 3 of Examples 6.2.3 and 6.3.10. The set S', or proper suffixes of X', is indicated in Figure 6.10. The set S of proper suffixes of X is indicated in Figure 6.11. The maximal prefix code $Y = S \cap \bar{H}$ is the set of words indicated in the figure by (⊙). It may be verified by inspection of Figures 6.9, 6.10, and 6.11 that $S' = S \cap H$.

6.4 Kernel

Let $X \subset A^+$, and let $H = A^- X A^-$ be the set of internal factors of X. The *kernel* of X, denoted $K(X)$, or K if no confusion is possible, is the set

$$K = X \cap H.$$

Thus a word is in the kernel if it is in X and is an internal factor of X. As we will see in this section, the kernel is one of the main characteristics of a maximal bifix code.

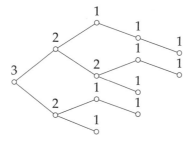

Figure 6.9 The tower T over X.

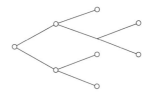

Figure 6.10 The set S' of proper suffixes of X'.

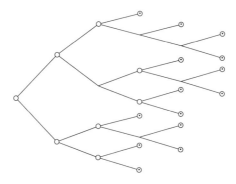

Figure 6.11 The set S of proper suffixes of X.

We start by showing how the kernel is related to the computation of the indicator.

Proposition 6.4.1 *Let $X \subset A^+$ be a thin maximal bifix code of degree d and let K be the kernel of X. Let Y be a set such that $K \subset Y \subset X$. Then for all $w \in H(X) \cup Y$,*

$$(L_Y, w) = (L_X, w). \tag{6.38}$$

For all $w \in A^*$,

$$(L_X, w) = \min\{d, (L_Y, w)\}. \tag{6.39}$$

Proof. By Formula (6.3), we have

$$L_X = \underline{A}^*(1 - \underline{X})\underline{A}^*, \quad L_Y = \underline{A}^*(1 - \underline{Y})\underline{A}^*.$$

Let $w \in A^*$, and let $F(w)$ be the set of its factors. For any word $x \in A^*$, the number $(A^* \underline{x} A^*, w)$ is the number of occurrences of x as a factor of w. It is nonzero only if $x \in F(w)$. Thus

$$(A^* \underline{X} A^*, w) = \sum_{x \in F(w) \cap X} (A^* x A^*, w),$$

showing that if $F(w) \cap X = F(w) \cap Y$, then $(L_X, w) = (L_Y, w)$. Thus, it suffices to show that $F(w) \cap X = F(w) \cap Y$ for all $w \in H(X) \cup Y$. From the inclusion $Y \subset X$, we get $F(w) \cap Y \subset F(w) \cap X$ for all $w \in A^*$. If $w \in H(X)$, then $F(w) \subset H(X)$ and $F(w) \cap X \subset K(X)$. Thus $F(w) \cap X \subset F(w) \cap Y$ in this case.

If $w \in Y$, then no proper prefix or suffix of w is in X, since X is bifix. Thus $F(w) \cap X = \{w\} \cup \{A^- w A^- \cap X\} \subset \{w\} \cup K(X) \subset Y$. Moreover $F(w) \cap X \subset F(w) \cap Y$ in this case also. This shows (6.38).

Now let $w \in H(X)$ be an internal factor of X. Then $(L_X, w) < d$ by Theorem 6.3.1. Consequently, $(L_X, w) = (L_Y, w)$ by Formula (6.38). Next let $w \in \bar{H}(X)$. Then $(L_X, w) = d$. By Formula (6.6), $(L_X, w) \le (L_Y, w)$. This proves (6.39). $\square$

Given two power series σ and τ, we denote by $\min\{\sigma, \tau\}$ the series defined by

$$(\min\{\sigma, \tau\}, w) = \min\{(\sigma, w), (\tau, w)\}.$$

Theorem 6.4.2 *Let X be a thin maximal bifix code with degree d, and let K be its kernel. Then*

$$L_X = \min\{d \underline{A}^*, L_K\}.$$

In particular, a thin maximal bifix code is determined by its degree and its kernel.

Proof. Take $Y = K(X)$ in the preceding proposition. Then the formula follows from (6.39). Assume that there are two codes X and X' of same degree d and same kernel. Since $K(X) = K(X')$, one has $L_{K(X)} = L_{K(X')}$ whence $L_X = L_{X'}$ which in turn implies $X = X'$ by Equation (6.8). This completes the proof. $\square$

Clearly, the kernel of a bifix code is itself a bifix code. We now give a characterization of those bifix codes which conversely are the kernel of some thin maximal bifix code. For this, it is convenient to introduce a notation: for a bifix code $Y \subset A^+$, let

$$\mu(Y) = \max\{(L_Y, y) \mid y \in Y\}. \tag{6.40}$$

It is a nonnegative integer or infinity. By convention, $\mu(\emptyset) = 0$.

Theorem 6.4.3 *A bifix code Y is the kernel of some thin maximal bifix code of degree d if and only if*

(i) *Y is not maximal bifix,*
(ii) *$\mu(Y) \le d - 1$.*

Proof. Let X be a thin maximal bifix code of degree d, and let $Y = K(X)$ be its kernel. Let us verify conditions (i) and (ii). To verify (i), consider a word $x \in X$ such that $(L_X, x) = \mu(X)$; we claim that $x \notin H(X)$. Thus, $x \notin K(X)$, showing that $Y \subsetneq X$. Assume the claim is wrong. Then $uxv \in X$ for some $u, v \in A^+$. Consequently, $(L_X, uxv) \geq 1 + (L_X, x)$ since the word uxv has the interpretation $(1, uxv, 1)$ which passes through no point of x. This contradicts the choice of x, and proves the claim. Next, for all $y \in Y$, we have $(L_X, y) = (L_Y, y)$ by Formula (6.38). Since $(L_X, y) \leq d - 1$ because $y \in H(X)$, condition (ii) is also satisfied.

Conversely, let Y be a bifix code satisfying conditions (i) and (ii). Let $L \in \mathbb{N}\langle\langle A \rangle\rangle$ be the formal power series defined for $w \in A^*$ by

$$(L, w) = \min\{d, (L_Y, w)\}.$$

Let us verify that L satisfies the three conditions of Proposition 6.1.11. First, let $a \in A$ and $w \in A^*$. By (6.11),

$$0 \leq (L_Y, aw) - (L_Y, w) \leq 1.$$

It follows that if $(L_Y, w) < d$, then $(L, w) = (L_Y, w)$. Since $(L_Y, aw) \leq (L_Y, w) + 1 \leq d$, one has $(L_Y, aw) = (L, aw)$. On the other hand, if $(L_Y, w) \geq d$, then $(L, aw) = (L, w) = d$. Thus in both cases

$$0 \leq (L, aw) - (L, w) \leq 1.$$

The symmetric inequality

$$0 \leq (L, wa) - (L, w) \leq 1$$

is shown in the same way. Thus the first of the conditions of Proposition 6.1.11 is satisfied.

Next, for $a, b \in A$, $w \in A^*$, $(L_Y, aw) + (L_Y, wb) \geq (L_Y, w) + (L_Y, awb)$. Consider first the case where $(L_Y, w) \geq d$. Then $(L, aw) = (L, wb) = (L, w) = (L, awb) = d$, and the inequality

$$(L, aw) + (L, wb) \geq (L, w) + (L, awb)$$

is clear. Assume now that $(L_Y, w) < d$. Then $(L_Y, aw) \leq d$ and $(L_Y, wb) \leq d$. Consequently

$$(L, aw) + (L, wb) = (L_Y, aw) + (L_Y, wb) \geq (L_Y, w) + (L_Y, awb)$$

$$\geq (L, w) + (L, awb)$$

since $L \leq L_Y$. This shows the second condition. Finally, we have $(L_Y, 1) = 1$, whence $(L, 1) = 1$.

Thus, according to Proposition 6.1.11, the series L is the indicator of some bifix code X. Further, L being bounded, the code X is thin and maximal bifix by Theorem 6.3.1. By the same argument, since the code Y is not maximal, the series L_Y is unbounded. Consequently, $\max\{(L, w) \mid w \in A^*\} = d$, showing that X has degree d.

We now prove that $Y = X \cap H(X)$, that is, Y is the kernel of X. First, we have the inclusion $Y \subset H(X)$. Indeed, if $y \in Y$, then $(L, y) \leq (L_Y, y) \leq \mu(Y) \leq d - 1$. Thus, by Theorem 6.3.1, $y \in H(X)$. Next, observe that it suffices to show that $X \cap H(X) = Y \cap H(X)$; this is equivalent to showing that $(\underline{X}, w) = (\underline{Y}, w)$ for all $w \in H(X)$. Let us prove this by induction on $|w|$. Clearly, the equality holds for $|w| = 0$. Next, let $w \in H(X) \setminus 1$. Then $(L, w) \leq d - 1$. Thus, $(L, w) = (L_Y, w)$. This in turn implies

$$(\underline{A}^* \underline{X} \underline{A}^*, w) = (\underline{A}^* \underline{Y} \underline{A}^*, w).$$

But $F(w) \subset H(X)$. Thus, by the induction hypothesis, $(\underline{X}, s) = (\underline{Y}, s)$ for all proper factors of w. Thus the equation reduces to $(\underline{X}, w) = (\underline{Y}, w)$. $\qquad\square$

We now describe the relation between the kernel and the operation of derivation.

Proposition 6.4.4 *Let X be a thin maximal bifix code of degree $d \geq 2$, and let $H = A^- X A^-$. Set*

$$K = X \cap H, \quad Y = HA \setminus H, \quad Z = AH \setminus H.$$

Then the code X' derived from X is

$$X' = K \cup (Y \cap Z). \tag{6.41}$$

Further,

$$K = X \cap X'. \tag{6.42}$$

Proof. Let $S = A^- X$ and $P = XA^-$ be the sets of proper right factors and of proper prefixes of words in X. Let $S' = S \cap H$ and $P' = P \cap H$. According to Proposition 6.3.14, S' is the set of proper suffixes of words in X' and similarly for P'. Thus,

$$\underline{X'} - 1 = (\underline{A} - 1)\underline{S'} = \underline{A}\,\underline{S'} - \underline{S'}.$$

From $S' = S \cap H$, we have $AS' = AS \cap AH$, and $\underline{A}\,\underline{S'} = \underline{A}\,\underline{S} \odot \underline{A}\,\underline{H}$, where $\odot$ denotes the Hadamard product (see Section 1.7). Thus,

$$\underline{X'} - 1 = (\underline{A}\underline{S} \odot \underline{A}\underline{H}) - \underline{S'}.$$

Now observe that, by Proposition 6.3.14, the set Z is a maximal suffix code with proper suffixes H. Thus, $\underline{Z} - 1 = (\underline{A} - 1)\underline{H}$ and $\underline{A}\underline{H} = \underline{Z} - 1 + \underline{H}$. Similarly, from $\underline{X} - 1 = (\underline{A} - 1)\underline{S}$ we get $\underline{A}\underline{S} = \underline{X} - 1 + \underline{S}$. Substitution gives

$$\underline{X'} - 1 = (\underline{X} - 1 + \underline{S}) \odot (\underline{Z} - 1 + \underline{H}) - \underline{S'}$$

$$= \underline{X \cap Z} + \underline{S \cap Z} + \underline{X \cap H} + \underline{S \cap H} + 1 - (1 \odot \underline{H}) - (\underline{S} \odot 1) - \underline{S'}.$$

Indeed, the other terms have the value 0 since neither X nor Z contains the empty word. Now $Z = P \cap \bar{H}$ (Proposition 6.3.14), whence $X \cap Z = X \cap P \cap \bar{H} = \emptyset$.

Also by definition $S' = S \cap H$ and $K = X \cap H$. Moreover $1 \odot \underline{H} = \underline{S} \odot 1 = 1$. Thus the equation becomes

$$\underline{X'} - 1 = \underline{S \cap Z} + \underline{K} - 1.$$

Finally, note that by Proposition 6.3.14, $Y = S \cap \bar{H}$. Thus, $S \cap Z = S \cap P \cap \bar{H} = Y \cap Z$ and

$$X' = K \cup (Y \cap Z),$$

showing (6.41). Next

$$X \cap X' = (K \cap X) \cup (X \cap Y \cap Z).$$

Now $X \cap Y \cap Z = X \cap P \cap S \cap \bar{H} = \emptyset$, and $K \cap X = K$. Thus, as claimed

$$X \cap X' = K. \qquad \square$$

Proposition 6.4.5 *Let X be a thin maximal bifix code of degree $d \geq 2$ and let X' be the derived code. Then*

$$K(X') \subset K(X) \subsetneq X'. \qquad (6.43)$$

Proof. First, we show that $H(X') \subset H(X)$. Indeed, let $w \in H(X')$. Then we have $(T_{X'}, w) \geq 1$, where $T_{X'}$ is the tower over X'. By Proposition 6.3.12, $(T_{X'}, w) = (T_X, w) - (H(X), w)$. Thus, $(T_X, w) \geq 1$. This in turn implies that $w \in H(X)$ by Proposition 6.3.8. By definition, $K(X') = X' \cap H(X')$. Thus, $K(X') \subset X' \cap H(X)$. By Proposition 6.4.4, $X' = K(X) \cup (Y \cap Z)$, where Y and Z are disjoint from $H(X)$. Thus $X' \cap H(X) = K(X)$. This shows that $K(X') \subset K(X)$. Next, Formula (6.42) also shows that $K(X) \subset X'$. Finally, we cannot have the equality $K(X) = X'$, since by Theorem 6.4.3, the set $K(X)$ is not a maximal bifix code. $\qquad \square$

The following theorem is a converse of Proposition 6.4.5.

Theorem 6.4.6 *Let X' be a thin maximal bifix code. For each set Y such that*

$$K(X') \subset Y \subsetneq X', \qquad (6.44)$$

there exists a unique thin maximal bifix code X such that $K(X) = Y$ and $d(X) = 1 + d(X')$. Moreover, the code X' is derived from X.

Proof. We first show that Y is the kernel of some bifix code. For this, we verify the conditions of Theorem 6.4.3. The strict inclusion $Y \subsetneq X'$ shows that Y is not a maximal code. Next, by Proposition 6.4.1, $(L_Y, y) = (L_{X'}, y)$ for $y \in Y$. Thus, setting $d = d(X') + 1$, we have $\mu(Y) \leq d(X') = d - 1$.

According to Theorem 6.4.3, there is a thin maximal bifix code X having degree d such that $K(X) = Y$. By Theorem 6.4.2, this code is unique. It remains to show that X' is the derived code of X. Let Z be the derived code of X. By Proposition 6.4.5,

$K(Z) \subset K(X) = Y \subsetneq Z$. Thus we may apply Proposition 6.4.1, showing that for all $w \in A^*$,

$$(L_Z, w) = \min\{d - 1, (L_Y, w)\}.$$

The inclusions of Formula 6.44 give, by Proposition 6.4.1,

$$(L_{X'}, w) = \min\{d - 1, (L_Y, w)\}$$

for all $w \in A^*$. Thus $L_{X'} = L_Z$ whence $Z = X'$.   □

Proposition 6.4.5 shows that the kernel of a code is located in some "interval" determined by the derived code. Theorem 6.4.6 shows that all of the "points" of this interval can be used effectively.

More precisely, Proposition 6.4.5 and Theorem 6.4.6 show that there is a bijection between the set of thin maximal bifix codes of degree $d \geq 2$, and the pairs (X', Y) composed of a thin maximal bifix code X' of degree $d - 1$ and a set Y satisfying (6.44). The bijection associates to a code X the pair $(X', K(X))$, where X' is the derived code of X.

Example 6.4.7 We have seen in Example 6.3.4 that any maximal bifix code of degree 2 has the form

$$X = C \cup BC^*B,$$

where the alphabet A is the disjoint union of B and C, and $B \neq \emptyset$. This observation can also be established by using Theorem 6.4.6. Indeed, the derived code of a maximal bifix code of degree 2 has degree 1 and therefore is A. Then for each proper subset C of A there is a unique maximal bifix code of degree 2 whose kernel is C. This code is clearly the code given by the above formula.

Example 6.4.8 The number of maximal bifix codes of degree 3 over a finite alphabet A having at least two letters is infinite. Indeed, consider an infinite thin maximal bifix code X' of degree 2. Its kernel $K(X')$ is a subset of A and consequently is finite. In view of Theorem 6.4.6, each set K containing $K(X')$ and strictly contained in X' is the kernel of some maximal bifix code of degree 3. Thus, there are infinitely many of them. Also, choosing a set $K(X)$ which is not rational gives a bifix code X of degree 3 which is not rational (Exercise 6.4.5).

6.5 Finite maximal bifix codes

Finite maximal bifix codes have quite remarkable properties which make them fascinating objects.

Proposition 6.5.1 *Let $X \subset A^+$ be a finite maximal bifix code of degree d. Then for each letter $a \in A$, $a^d \in X$.*

With the terminology introduced in Chapter 2, this is equivalent to say that the order of each letter is the degree of the code.

Proof. Let $a \in A$. According to Proposition 6.2.7, there is an integer $n \geq 1$ such that $a^n \in X$. Since X is finite, there is an integer k such that a^k is not an internal factor of X. The number of parses of a^k is equal to d. It is also the number of suffixes of a^k which are proper prefixes of words in X, that is n. Thus $n = d$. □

Note as a consequence of this result that it is, in general, impossible to complete a finite bifix code into a maximal bifix code which is finite. Consider, for example, $A = \{a, b\}$ and $X = \{a^2, b^3\}$. A finite maximal bifix code containing X would have simultaneously degree 2 and degree 3.

We now show the following result:

Theorem 6.5.2 *Let A be a finite set, and let $d \geq 1$. There are only a finite number of finite maximal bifix codes over A with degree d.*

Proof. The only maximal bifix code over A, having degree 1 is the alphabet A. Arguing by induction on d, assume that there are only finitely many finite maximal bifix codes of degree d. Each finite maximal bifix code of degree $d + 1$ is determined by its kernel which is a subset of X'. Since X' is a finite maximal bifix code of degree d there are only a finite number of kernels and we are finished. □

Denote by $\beta_k(d)$ the number of finite maximal bifix codes of degree d over a k letter alphabet A.

Clearly $\beta_k(1) = 1$. Also $\beta_k(2) = 1$; indeed $X = A^2$ is, in view of Example 6.2.4, the only finite maximal bifix code of degree 2. It is also clear that $\beta_1(d) = 1$ for all $d \geq 1$.

Example 6.5.3 Let us verify that

$$\beta_2(3) = 3. \tag{6.45}$$

Let indeed $A = \{a, b\}$, and let $X \subset A^+$ be a finite maximal bifix code of degree 3. The derived code X' is necessarily $X' = A^2$, since it is the only finite maximal bifix code of degree 2. Let $K = X \cap X'$ be the kernel of X. Thus $K \subset A^2$.

According to Proposition 6.5.1, both $a^3, b^3 \in X$. Thus K cannot contain a^2 or b^2. Consequently, $K \subset \{ab, ba\}$. We next rule out the case $K = \{ab, ba\}$. Suppose indeed that this equality holds. For each $k \geq 1$, the word $(ab)^k$ has exactly two X parses. But X being finite, there is an integer k such that $(ab)^k \in \bar{H}(X)$, and $(ab)^k$ should have three X parses. This is the contradiction.

Thus there remain three candidates for K: $K = \emptyset$ which correspond to $X = A^3$, then $K = \{ab\}$, which gives the code X of Example 6.2.3, and $K = \{ba\}$ which gives the reversal $\tilde{X}$ of the code X of Example 6.2.3. This shows (6.45). Note also that this explains why $\tilde{X}$ is obtained from X by exchanging the letters a and b: this property holds whenever it holds for the kernel.

We now show how to construct all finite maximal bifix codes by a sequence of internal transformations, starting with a uniform code.

Theorem 6.5.4 (Césari) *Let A be a finite alphabet and $d \geq 1$. For each finite maximal bifix code $X \subset A^+$ of degree d, there is a finite sequence of internal transformations which, starting with the uniform code A^d, gives X.*

Proof. Let K be the kernel of X. If $K = \emptyset$, then $X = A^d$ and there is nothing to prove. This holds also if $\mathrm{Card}(A) = 1$. Thus we assume $K \neq \emptyset$ and $\mathrm{Card}(A) \geq 2$. Let $x \in K$ be a word which is not a factor of another word in K. We show that there exist a maximal suffix code G and a maximal prefix code D such that

$$GxD \subset X. \tag{6.46}$$

Assume the contrary. Let $P = XA^-$. Since $x \in K$, x is an internal factor. Thus the set Px^{-1} is not empty. Then for all words $g \in Px^{-1}$, there exist two words d, d' such that

$$gxd, \ gxd' \in X \quad \text{and} \quad X(xd)^{-1} \neq X(xd')^{-1}.$$

Suppose the contrary. Then for some $g \in Px^{-1}$, all the sets $X(xd)^{-1}$, with d running over the words such that $gxd \in X$, are equal. Let $D = \{d \mid gxd \in X\}$ and let $G = X(xd)^{-1}$, where d is any element in D. Then $GxD \subset X$, contradicting our assumption. This shows the existence of d, d'.

Among all triples (g, d, d') such that

$$gxd, \ gxd' \in X \quad \text{and} \quad X(xd)^{-1} \neq X(xd')^{-1},$$

let us choose one with $|d| + |d'|$ minimal. For this fixed triple (g, d, d'), set

$$G = X(xd)^{-1} \quad \text{and} \quad G' = X(xd')^{-1}.$$

Then G and G' are distinct maximal suffix codes. Take any word $h \in G \setminus G'$. Then either h is a proper right factor of a word in G' or has a word in G' as a proper suffix. Thus, interchanging if necessary G and G', there exist words $u, g' \in A^+$ such that

$$g' \in G, \quad ug' \in G'.$$

Note that this implies

$$g'xd \in X, \quad ug'xd' \in X.$$

Now consider the word $ug'xd$. Of course, $ug'xd \notin X$. Next $ug'xd \notin P$, since otherwise $g'xd \in K$, and x would be a factor of another word in K, contrary to the assumption. Since $ug'xd \notin P \cup X$, it has a proper prefix in X. This prefix cannot be a prefix of $ug'x$, since $ug'xd' \in X$. Thus it has $ug'x$ as a proper prefix. Thus there is a factorization $d = d''v$ with $d'', v \in A^+$, and $ug'xd'' \in X$.

Now we observe that the triple (ug', d', d'') has the same properties as (g, d, d'). Indeed, both words $ug'xd'$ and $ug'xd''$ are in X. Also $X(xd')^{-1} \neq X(xd'')^{-1}$ since $gxd' \in X$, but $gxd'' \notin X$: this results from the fact that gxd'' is a proper prefix of

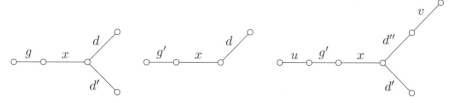

Figure 6.12 From triple (g, d, d') to triple (ug', d', d'').

$gxd \in X$ (Figure 6.12). Thus, (ug', d', d'') satisfies the same constraints as (g, d, d'): however, $|d'| + |d''| < |d'| + |d|$. This gives the contradiction and proves (6.46). Let

$$Y = (X \cup Gx \cup xD) \setminus (x \cup GxD). \tag{6.47}$$

In view of Proposition 6.2.10, the set Y is a finite maximal bifix code and, moreover, the internal transformation with respect to x transforms Y into X. Finally (6.47) shows that

$$\mathrm{Card}(Y) = \mathrm{Card}(X) + \mathrm{Card}(G) + \mathrm{Card}(D) - 1 - \mathrm{Card}(G)\,\mathrm{Card}(D)$$
$$= \mathrm{Card}(X) - (\mathrm{Card}(G) - 1)(\mathrm{Card}(D) - 1).$$

The code G being maximal suffix and $\mathrm{Card}(A) \geq 2$, we have $\mathrm{Card}(G) \geq 2$. For the same reason, $\mathrm{Card}(D) \geq 2$. Thus

$$\mathrm{Card}(Y) \leq \mathrm{Card}(X) - 1. \tag{6.48}$$

Arguing by induction on the number of elements, we can assume that Y is obtained from A^d by a finite number of internal transformations. This completes the proof. $\square$

Observe that by this theorem (and Formula (6.48)) each finite maximal bifix code $X \subset A^+$ of degree d satisfies

$$\mathrm{Card}(X) \geq \mathrm{Card}(A^d), \tag{6.49}$$

with an equality if and only if $X = A^d$. This result can be proved directly as follows (see also Exercise 3.7.1).

Let X be a finite maximal prefix code, and

$$\lambda = \sum_{x \in X} |x| k^{-|x|}$$

with $k = \mathrm{Card}(A)$. The number λ is the average length of X with respect to the uniform Bernoulli distribution on A^*. Let us show the inequality

$$\mathrm{Card}(X) \geq k^\lambda. \tag{6.50}$$

For a maximal bifix code X of degree d, we have $\lambda = d$ (Corollary 6.3.16), and thus (6.49) is a consequence of (6.50). To show (6.50), let $n = \text{Card}(X)$. Then

$$\lambda = \sum_{x \in X} k^{-|x|} \log_k k^{|x|},$$

$$\log_k n = \sum_{x \in X} k^{-|x|} \log_k n.$$

The last equality follows from $1 = \sum_{x \in X} k^{-|x|}$, which holds by the fact that X is a finite maximal prefix code. Thus,

$$\lambda - \log_k n = \sum_{x \in X} k^{-|x|} \log_k(k^{|x|}/n).$$

Since $\sum_{x \in X} k^{-|x|} = 1$ and since the function log is concave, we have

$$\sum_{x \in X} k^{-|x|} \log_k(k^{|x|}/n) \le \log\Big(\sum_{x \in X} k^{-|x|} \frac{k^{|x|}}{n}\Big),$$

and consequently

$$\lambda - \log_k n \le \log_k\Big(\sum_{x \in X} \frac{1}{n}\Big) = 0.$$

This shows (6.50).

Example 6.5.5 Let $A = \{a, b\}$ and let X be the finite maximal bifix code of degree 4 with literal representation given on the left of Figure 6.13. The kernel of X is $K = \{ab, a^2b^2\}$. There is no pair (G, D) composed of a maximal suffix code G and a maximal prefix code D such that $GabD \subset X$. On the other hand

$$Aa^2b^2A \subset X.$$

The code X is obtained from the code Y given on the right of Figure 6.13 by internal transformation relatively to a^2b^2. The code Y is obtained from A^4 by the sequence of internal transformations relatively to the words aba, ab^2, and ab.

We now describe the construction of a finite maximal bifix code from its derived code.

Let $Y \subset A^+$ be a bifix code. A word $w \in A^*$ is called *full* (with respect to Y) if there is an interpretation passing through any point of w. It is equivalent to say that w is full if any parse of w is an interpretation.

The bifix code Y is *insufficient* if the set of full words with respect to Y is finite.

Proposition 6.5.6 *A thin maximal bifix code over a finite alphabet A is finite if and only if its kernel is insufficient.*

Proof. Suppose first that X is finite. Let d be its degree, and let K be its kernel. Consider a word w in $\bar{H}(X)$. Then w has exactly d X-interpretations. These are not

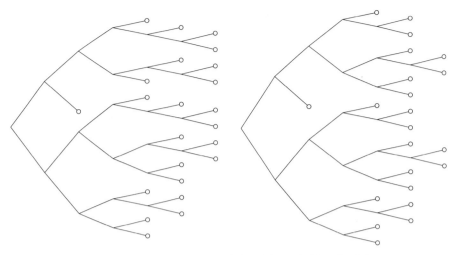

Figure 6.13 The code X on the left and the code Y on the right.

all K-interpretations, because K is a subset of the derived code of X, which has degree $d - 1$. Thus, there is a point of w through which no K-interpretation passes. Thus, w is not full (for K). This shows that the set of full words (with respect to K) is contained in $H(X)$. Since $H(X)$ is finite, the set K is insufficient.

Conversely, suppose that X is infinite. Since the alphabet A is finite, there is an infinite sequence $(a_n)_{n\geq 0}$ of letters such that, setting $P = XA^-$, we have for all $n \geq 0$,

$$p_n = a_0 a_1 \cdots a_n \in P.$$

We show there exists an integer k such that all words $a_k a_{k+1} \cdots a_{k+\ell}$ for $\ell \geq 1$ are full with respect to K. Note that there are at most $d(X)$ integers n for which p_n is a proper suffix of a word in X. Similarly, there exist at most $d(X)$ integers n such that for all $m \geq 1$,

$$a_{n+1} a_{n+2} \cdots a_{n+m} \in P.$$

Indeed, each such integer n defines an interpretation of each word $a_0 a_1 \cdots a_r, (r > n)$, which is distinct from the interpretations associated to the other integers.

These observations show that there exists an integer k such that for all $n \geq k$, the following hold: p_n has a suffix in X and $a_{n+1} a_{n+2} \cdots a_{n+m}$ is in X for some $m \geq 1$. The first property implies by induction that for all $n \geq k$, there is an integer $i \leq k$ such that $a_i \cdots a_n \in X^*$.

Let $w_\ell = a_k a_{k+1} \cdots a_{k+\ell}$ for $\ell \geq 1$. We show that through each point of w_ℓ passes a K-interpretation. Indeed, let

$$u = a_k a_{k+1} \cdots a_n, \quad v = a_{n+1} a_{n+2} \cdots a_{k+\ell},$$

for some $k \leq n \leq k + 1$. There exists an integer $i \leq k$ such that $a_i \cdots a_{k-1} u \in X^*$, and there is an integer $m \geq k + 1$ such that $v a_{k+1} \cdots a_m \in X^*$. In fact, these two words are in $H(X) \cap X^*$ and consequently they are in K^*. This shows that K is a sufficient set and completes the proof. □

The previous proposition gives the following result.

Theorem 6.5.7 *Let X' be a finite maximal bifix code of degree $d - 1$ and with kernel K'. For each insufficient subset K of X' containing K', there exists a unique finite maximal bifix code X of degree d, having kernel K. The derived code of X is X'.*

Proof. Since K is insufficient, K is not a maximal bifix code. Thus $K' \subset K \subsetneq X'$. In view of Theorem 6.4.6, there is a unique thin maximal bifix code X of degree d and kernel K. The derived code of X is X'. By Proposition 6.5.6, the code X is finite. □

The following corollary gives a method for the construction of all finite maximal bifix codes by increasing degrees.

Corollary 6.5.8 *For any integer $d \geq 2$, the function*

$$X \mapsto K(X)$$

is a bijection of the set of finite maximal bifix codes of degree d onto the set of all insufficient subsets K of finite maximal bifix codes X' of degree $d - 1$ such that

$$K(X') \subset K \subsetneq X'.$$

□

Example 6.5.9 Let $A = \{a, b\}$. For each integer $n \geq 0$, there exists a unique finite maximal bifix code $X_n \subset A^+$ of degree $n + 2$ with kernel

$$K_n = \{a^i b^i \mid 1 \leq i \leq n\}.$$

For $n = 0$, we have $K_0 = \emptyset$ and $X_0 = A^2$. Arguing by induction, assume X_n constructed. Then $K_n \subset X_n$ and also $a^{n+2}, b^{n+2} \in X_n$, since $d(X_n) = n + 2$. We show that $a^{n+1} b^{n+1} \in X_n$. Indeed, no proper prefix of $a^{n+1} b^{n+1}$ is in X_n since each has a suffix in X_n or is a proper suffix of a^{n+2}. Consider now a word $a^{n+1} b^{n+k}$ for a large enough integer k. Since X_n is finite, there is some prefix $a^{n+1} b^{n+r} \in X_n$ for some $r \geq 1$. If $r \geq 2$, then b^{n+2} is a suffix of this word. Thus $r = 1$, and $a^{n+1} b^{n+1} \in X_n$.

Clearly $K_n \subset K_{n+1}$. The set K_{n+1} is insufficient. In fact, a has no K_{n+1} interpretation passing through the point $(a, 1)$ and b has no interpretation passing through the point $(1, b)$. Therefore, the set of full words is $\{1\}$. Finally

$$K_n \subset K_{n+1} \subsetneq X_n.$$

This proves the existence and uniqueness of X_{n+1}, by using Theorem 6.5.7.

The code X_1 is the code of degree 3 given in Example 6.2.3. The code X_2 is the code of degree 4 of Example 6.5.5.

We end this section with some remarks on the length distribution of bifix codes. Contrary to the case of prefix codes, it is not true that any sequence $(u_n)_{n \geq 1}$ of integers such that $\sum_{n\geq 1} u_n k^{-n} \leq 1$ is the length distribution of a bifix code on k letters. For instance, there is no bifix code on the alphabet $\{a, b\}$ which has the same distribution as the prefix code $\{a, ba, bb\}$. Indeed, such a code must contain a letter, say a, and then the only possible word of length 2 is bb. We show that the following holds.

Proposition 6.5.10 *For any sequence* $(u_n)_{n\geq 1}$ *of integers such that*

$$\sum_{n\geq 1} u_n k^{-n} \leq \frac{1}{2} \tag{6.51}$$

there exists a bifix code on an alphabet of k letters with length distribution $(u_n)_{n\geq 1}$.

Proof. We show by induction on $n \geq 1$ that there exists a bifix code X_n of length distribution $(u_i)_{1\leq i\leq n}$ on an alphabet A of k symbols. It is true for $n = 1$ since $u_1 k^{-1} \leq 1/2$ and thus $u_1 < k$. Assume that the property is true for n. We have by (6.51)

$$\sum_{i=1}^{n+1} u_i k^{-i} \leq \frac{1}{2}$$

or equivalently, multiplying both sides by $2k^{n+1}$,

$$2(u_1 k^n + \cdots + u_n k + u_{n+1}) \leq k^{n+1}$$

whence

$$u_{n+1} \leq 2u_{n+1} \leq k^{n+1} - 2(u_1 k^n + \cdots + u_n k). \tag{6.52}$$

Since X_n is bifix by induction hypothesis, we have

$$\mathrm{Card}(X_n A^* \cap A^{n+1}) = \mathrm{Card}(A^* X_n \cap A^{n+1}) = u_1 k^n + \cdots + u_n k.$$

Thus, we have

$$\mathrm{Card}((X_n A^* \cup A^* X_n) \cap A^{n+1}) \leq \mathrm{Card}(X_n A^* \cap A^{n+1}) + \mathrm{Card}(A^* X_n \cap A^{n+1})$$

$$\leq 2(u_1 k^n + \cdots + u_n k)$$

It follows with Equation (6.52) that

$$u_{n+1} \leq k^{n+1} - 2(u_1 k^n + \cdots + u_n k)$$

$$\leq \mathrm{Card}(A^{n+1}) - \mathrm{Card}((X_n A^* \cup A^* X_n) \cap A^{n+1})$$

$$= \mathrm{Card}(A^{n+1} - (X_n A^* \cup A^* X_n))$$

This shows that we can choose a set Y of u_{n+1} words of length $n + 1$ on the alphabet A which do not have a prefix or a suffix in X_n. Then $X_{n+1} = Y \cup X_n$ is bifix, which ends the proof. $\square$

Table 6.1 *The list of maximal 2-realizable length distributions of length at most N ≤ 4.*

N	2			3				4				
	u_1	u_2	$u(1/2)$	u_1	u_2	u_3	$u(1/2)$	u_1	u_2	u_3	u_4	$u(1/2)$
	2	0	1.0000	2	0	0	1.0000	2	0	0	0	1.0000
	1	1	0.7500	1	1	1	0.8750	1	1	1	1	0.9375
				1	0	2	0.7500	1	0	2	1	0.8125
								1	0	1	3	0.8125
								1	0	0	4	0.7500
	0	4	1.0000	0	4	0	1.0000	0	4	0	0	1.0000
				0	3	1	0.8750	0	3	1	0	0.8750
								0	3	0	1	0.8125
				0	2	2	0.7500	0	2	2	2	0.8750
								0	2	1	3	0.8125
								0	2	0	4	0.7500
				0	1	5	0.8750	0	1	5	1	0.9375
								0	1	4	4	1.0000
								0	1	3	5	0.9375
								0	1	2	6	0.8750
								0	1	1	7	0.8125
								0	1	0	9	0.8125
				0	0	8	1.0000	0	0	8	0	1.0000
								0	0	7	1	0.9375
								0	0	6	2	0.8750
								0	0	5	4	0.8750
								0	0	4	6	0.8750
								0	0	3	8	0.8750
								0	0	2	10	0.8750
								0	0	1	13	0.9375
								0	0	0	16	1.0000

The bound $1/2$ in the statement of Proposition 6.5.10 is not the best possible. It is conjectured that the statement holds with $3/4$ instead of $1/2$. For convenience, we call a sequence (u_n) of integers k-*realizable* if there is a bifix code on k symbols with this length distribution.

We fix $N \geq 1$ and we order sequences $(u_n)_{1 \leq n \leq N}$ of integers by setting $(u_n) \leq (v_n)$ if and only if $u_n \leq v_n$ for $1 \leq n \leq N$. If $(u_n) \leq (v_n)$ and (v_n) is k-realizable then so is (u_n). We give in Table 6.1 the values of the maximal 2-realizable sequences for $N \leq 4$. We set $u(z) = \sum u_n z^n$. For each value of N, we list in decreasing lexicographic order the maximal realizable sequence with the corresponding value of the sum $u(1/2) = \sum u_n 2^{-n}$. The distributions with value 1 correspond to maximal bifix codes. For example, the distribution $(0, 1, 4, 4)$ corresponds to the maximal bifix code of Example 6.2.3.

It can be checked on this table that the minimal value of the sums $u(1/2)$ is $3/4$. Since the distributions listed are maximal for componentwise order, this shows that

Table 6.2 *The length distributions of binary finite maximal bifix codes of degree at most 4.*

d	1		2		3		4	
	2	1	0 4	1	0 0 8	1	0 0 0 16	1
							0 0 1 12 4	6
							0 0 2 8 8	6
							0 0 2 9 4 4	8
							0 0 3 5 8 4	6
							0 0 3 6 4 8	4
							0 0 3 6 5 4 4	4
							0 0 4 3 5 8 4	4
					0 1 4 4	2	0 1 0 5 12 4	2
							0 1 0 6 8 8	2
							0 1 0 6 9 4 4	4
							0 1 0 7 5 8 4	4
							0 1 0 7 6 5 4 4	2
							0 1 0 8 2 9 4 4	2
							0 1 1 3 9 8 4	4
							0 1 1 4 6 8 8	4
							0 1 1 4 6 9 4 4	4
							0 1 1 5 3 9 8 4	4
							0 1 2 2 4 9 12 4	2
		1				1		3			73

for any sequence $(u_n)_{1 \le n \le N}$ with $N \le 4$ such that $u(1/2) \le 3/4$, there exists a binary bifix code X such that $u_X = u$.

Since a thin maximal bifix code X is also maximal as a code (Proposition 6.2.1), its generating series satisfies $f_X(1/k) = 1$, where k is the size of the alphabet. Table 6.2 lists the length distributions of finite maximal bifix codes of degree $d \le 4$ over $\{a, b\}$. For each degree, the last column contains the number of bifix codes with this distribution, with a total number of 73 of degree 4. There are 39 of them with $\{a, b\}^3$ as derivative and 34 with one of the two other bifix codes of degree 3 (see the exercises).

6.6 Completion

For a finite bifix code X, a simple construction shows that it is contained in a maximal rational bifix code. Indeed, either X is already maximal, or it is, for each large enough integer d, the kernel of a maximal rational bifix code of degree d (Theorem 6.4.3 and Exercise 6.4.1).

For a rational bifix code X which is not maximal, it is not true in general that it is the kernel of a maximal rational bifix code. Instead of acting from the outside, adding words having the words of X as factors, one has to work from the inside, adding first words which are factors of words of X (and therefore are in the kernel of the result).

Theorem 6.6.1 *Any rational bifix code is contained in a maximal rational bifix code.*

Let $Y \subset A^*$ be a bifix code. Recall that its *indicator* is the formal series defined by

$$L_Y = \underline{A}^*(1 - \underline{Y})\underline{A}^*.$$

We shall need several properties of the indicator, grouped in the following lemma for convenience.

Lemma 6.6.2 *Let $Y \subset A^*$ be a bifix code and L its indicator. For any words u, v, w and any letter a, the following hold.*

(1) *For each i with $1 \le i \le (L, w)$, there is a prefix p of w such that $(L, p) = i$.*
(2) *If Y is a rational set and is not a maximal code, then for any word u, the set of values $\{(L, uv) \mid v \in A^*\}$ is unbounded.*
(3) *$(L, w) = (L, wa)$ if and only if wa has a suffix in Y.*
(4) *If $(L, v) = (L, uv)$, then uv has a prefix in Y.*
(5) *If $Y \subset Z$, then $L_Y \ge L_Z$.*

Proof. Property (1) is an easy consequence of Proposition 6.1.11, (6.12). For (2), we note that a rational code is thin (Proposition 2.5.20); if Y is rational and not maximal, L is unbounded (Theorem 6.3.1); hence, (L, v) is arbitrarily large, and so is $(L, uv) \ge (L, v)$ by Proposition 6.1.8.

By (6.5), (L, w) is equal to $|w| + 1 -$ the numbers of factors of w which are in Y. This number of factors is the same for wa, except if wa has a suffix in Y, in which case wa has exactly one more (since Y is a suffix code). This implies (3). For (4), assume $(L, v) = (L, uv)$. By Proposition 6.1.8, we have $(L, v) = (L, u'v)$ for each suffix u' of u; hence by the symmetric statement of (3), an easy induction on the length of u', starting with $|u'| = 1$, shows that $u'v$ has a prefix in Y. Thus uv has a prefix in Y. Property (5) is (6.6).   □

The idea of the construction for the proof of Theorem 6.6.1 is the following. Starting with a rational bifix code $X = X_0 \subset A^+$, we build an increasing sequence of sets $(X_n)_{n \ge 1}$ which all are shown to be rational bifix codes. It will then be proved that for some n, X_n is a maximal rational bifix code containing X, thereby proving the theorem.

For any set Y, we set $P(Y) = Y \setminus YA^+$. It is the set of words of Y which are minimal for the prefix order. Thus, $w \in P(Y)$ if and only if w is in Y and has no proper prefix in Y. The set $P(Y)$ is prefix. Next, $I(Y)$ denotes the set of words in A^* which are incomparable with Y for the prefix order. In other words, $w \in I(Y)$ if and only if w is not a prefix of a word in Y and has no prefix in Y. Sometimes the algebraic formulation $I(Y) = A^* \setminus (YA^- \cup YA^*)$ is useful. Finally, we denote by $\overline{Y}$ the set $P(I(Y))$. It is called the *companion* of Y. Thus $w \in \overline{Y}$ if and only if w is incomparable with Y, and each proper prefix of w is a prefix of a word in Y. Indeed, a proper prefix of w is a prefix of a word of Y or has a prefix in Y, but the second case is ruled out because it would imply that w itself has a prefix in Y and so is comparable with Y.

The companion of a set should not be confused with its complement. Recall also that A^-Y (resp. YA^-) denotes the set of proper suffixes (resp. prefixes) of words in Y.

Proposition 6.6.3 *Let $X = X_0$ be a bifix code. Define recursively, for $n \geq 0$:*

$$L_n = L_{X_n} \tag{6.53}$$

$$V_n = \{w \in A^* \mid (L_n, w) = n + 1\}, \tag{6.54}$$

$$Z_n = I(X_n) \cap P(V_n), \tag{6.55}$$

$$X_{n+1} = X_n \cup (Z_n \setminus A^- X). \tag{6.56}$$

For each $n \geq 1$, the set X_n is a bifix code and $(L_n, w) \leq n$ for all $w \in X_n \setminus X$.

Note that the union defining X_{n+1} is disjoint, since $Z_n \subset I(X_n)$ and $I(X_n)$ cannot intersect X_n.

Proof. Assume that X_n is a bifix code and satisfies the inequality in the statement. We show that the same hold for X_{n+1}. By Equation (6.55), Z_n is a prefix code which is incomparable with X_n for the prefix order. In view of Equation (6.56), X_{n+1} is the union of two prefix codes which are incomparable for the prefix order because the second is contained in $I(X_n)$. Thus X_{n+1} itself is a prefix code.

We show that X_{n+1} is a suffix code. By contradiction, suppose that for some $x, x' \in X_{n+1}$, x is a proper suffix of x'. By construction, we have two cases : either $x \in X_n$, or $x \in Z_n \setminus A^- X$.

In the first case, we have $x' \notin X_n$, since X_n is a suffix code by induction. Thus $x' \in Z_n \setminus A^- X$ and $x' \in P(V_n)$, hence x' is in V_n, and by definition of the latter, $(L_n, x') = n + 1$. Write $x' = wa$, $a \in A$. Since x' has a suffix in X_n (namely x itself), we have $(L_n, w) = (L_n, wa)$ by Lemma 6.6.2 (3). Thus $(L_n, w) = n + 1$, which implies that $w \in V_n$. This contradicts the fact that $x' \in P(V_n)$.

In the second case, $x \in Z_n$, hence $x \in V_n$ and $(L_n, x) = n + 1$. Moreover, $x' \notin X$ (otherwise $x \in A^- X$). Suppose that $x' \in X_n$. Then $x' \in X_n \setminus X$ and by the induction hypothesis, $(L_n, x') \leq n$. By Proposition 6.1.8, this gives a contradiction, since x is a factor of x'. Thus we have $x' \in Z_n \setminus A^- X$. This implies $x' \in V_n$ and consequently $(L_n, x') = n + 1 = (L_n, x)$. From Lemma 6.6.2 (4), we deduce that x' has a prefix in X_n, a contradiction, since $x' \in Z_n \subset I(X_n)$. We conclude that X_{n+1} is a bifix code. Observe that $L_{n+1} \geq L_n$ by Lemma 6.6.2 (5) because X_n is a subset of X_{n+1}.

It remains to prove that $(L_{n+1}, x) \leq n + 1$ for $x \in X_{n+1} \setminus X$. Let indeed $x \in X_{n+1} \setminus X$. Since $X_n \subset X_{n+1}$, we have by Lemma 6.6.2 (5), $(L_{n+1}, x) \leq (L_n, x)$. If $x \in X_n$, then $(L_n, x) \leq n$ by the induction hypothesis; if $x \notin X_n$, then $x \in Z_n \subset V_n$, and $(L_n, x) = n + 1$. In both case, we conclude that $(L_{n+1}, x) \leq n + 1$. □

Lemma 6.6.4 *Let $X = X_0$ be a rational bifix code. For each $n \geq 1$, the set X_n is a rational set.*

Proof. We prove the statement by induction on n. It is true for $n = 0$ by hypothesis. Suppose next that X_n is rational. Let $U_n = A^* \setminus X_n A^*$. This set is rational. According to (6.4), for any word z, (L_n, z) is the number of suffixes of z which are in U_n.

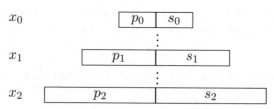

Figure 6.14 A chain for the factor order.

Let $\mathcal{A} = (Q, i, T)$ be a deterministic automaton recognizing U_n. Let $\mathcal{B} = (Q \cup \omega, \omega, T \cup \omega)$ with $\omega \notin Q$ be the automaton obtained as follows. The edges are those of $\mathcal{A}$ plus a loop (ω, a, ω) for each letter a in A and an edge (ω, a, q) for each edge (i, a, q) of $\mathcal{A}$.

Then, for any word z, the number of successful paths labeled by z starting in ω is equal to the number of suffixes of z which are in U_n. In other words, $(L_n, z) = (|\mathcal{B}|, z)$. Thus, by Proposition 1.10.4, the set V_n is rational. Since $I(X_n) = A^* \setminus (X_n A^- \cup X_n A^*)$, the set $I(X_n)$ is rational. Since $P(V_n) = V_n \setminus V_n A^+$, the set $P(V_n)$ is also rational. Thus Z_n is a rational set and so is X_{n+1}.   □

From now on, we assume that $X = X_0$ is a rational bifix code. In order to prove the theorem it is enough, in view of Lemma 6.6.3, to show that X_n is a maximal bifix code for some n. By Theorem 2.5.13 and Proposition 2.5.20, it is therefore enough to show that X_n is a right complete prefix code. This is the purpose of the following lemmas.

Given a partially ordered set S, the *height* of an element s of S, denoted $h(s)$, is the maximal length of the strictly increasing chains ending in s. The *height* of S is the maximal height of its elements, so it is simply the maximal length of a strictly increasing chain of elements in S. The height is finite or infinite. We denote by $S^{(i)}$ the set of elements of height i of S.

It follows from Proposition 3.2.9 that for a rational prefix code Y, the height of the set of suffixes of Y, ordered by the prefix order, is finite. A symmetric property holds for suffix codes. We denote by π the height of the set of prefixes of X for the suffix order.

Recall that $\overline{X} = P(I(X))$ denotes the companion of X. Thus, a word is in $\overline{X}$ if it is incomparable with the words of X for the prefix order and has no proper prefix with this property.

Lemma 6.6.5 *The height of $\overline{X}$ for the factor order is at most π.*

Proof. Assume, arguing by contradiction, that there is a strictly increasing chain for the factor order $x_0, x_1, x_2, \ldots, x_\pi$ of length $\pi + 1$ with $x_i \in \overline{X}$. Since $\overline{X}$ is a prefix code, x_i is not a prefix of x_{i+1}. We may write $x_i = p_i s_i$, in such a way that each p_i is a proper suffix of p_{i+1}, each s_i is a nonempty proper prefix of s_{i+1} (see Figure 6.14).

Note that $p_i \neq p_{i+1}$, since x_i is not a prefix of x_{i+1}. Hence $p_0, \ldots, p_\pi$ is a strictly increasing chain for the suffix order.

We prove that each p_i is a prefix of some word in X which gives a contradiction in view of the definition of π. Indeed, each p_i is a proper prefix of x_i. Since $x_i \in P(I(X))$, each proper prefix of x_i is a prefix of a word in X. Thus p_i is a prefix of a word in X. $\qquad\square$

Consider $\overline{X}$, the companion of X, ordered by the factor order. We set, for $i \geq 1$,

$$\overline{X}^{(i)} = \{w \in \overline{X} \mid h(w) \leq i\},$$

where $h(w)$ denotes the height of w in the set $\overline{X}$ for the factor order. In particular, $\overline{X}^{(1)}$ is the set of words in $\overline{X}$ which are minimal for the factor order. The previous lemma shows that $\overline{X}^{(\pi)} = \overline{X}$.

Let σ be equal to $1 +$ the height of the set of suffixes of X for the prefix order.

Lemma 6.6.6 *Let T be a set of words such that every proper suffix of a word of T is comparable for the prefix order with some word in X_n. Then L_n is bounded on T.*

Proof. Let $w \in T$. By Lemma 6.1.6, $(L_n, w) = 1 + \ell$, where ℓ is the number of proper suffixes of w which belong to $A^* \setminus X_n A^*$; now, since none of them is in $I(X_n)$, they all belong to $X_n A^-$.

Therefore ℓ is bounded by the maximal length of increasing chains of prefixes of X_n for the suffix order. This number is bounded, by the symmetric statement of Proposition 3.2.9, since X_n is rational. $\qquad\square$

Lemma 6.6.7 *There exists m such that L_m is bounded on the companion $\overline{X}$ of X.*

Proof. We prove by induction on $i \geq 1$ that there exists k such that L_k is bounded on $\overline{X}^{(i)}$.

For $i = 1$, we prove that L_0 is bounded on $\overline{X}^{(1)}$. For this, we show that we may apply Lemma 6.6.6 with $n = 0$ and $T = \overline{X}^{(1)}$. Indeed, assume on the contrary that some $v \in \overline{X}^{(1)}$ has a proper suffix s which is in $I(X)$. Then some prefix of s is in $P(I(X)) = \overline{X}$, and v has a proper factor in $\overline{X}$, which contradicts the definition of $\overline{X}^{(1)}$.

Suppose now that $i > 1$. By the induction hypothesis there are integers m and ℓ such that $L_m(w) \leq \ell$ for all $w \in \overline{X}^{(i-1)}$. We may suppose that $m \leq \ell$. Let $k = \ell + \sigma$ where σ was defined above. Since $m \leq \ell + \sigma$, we have $X_m \subset X_{\ell+\sigma}$ and $L_m \geq L_{\ell+\sigma}$ by Lemma 6.6.2 (5). Thus L_k is bounded on $\overline{X}^{(i-1)}$. It remains to show that L_k is bounded on $\overline{X}^{(i)}$.

Let $w \in \overline{X}^{(i)} \setminus \overline{X}^{(i-1)}$. We show that any proper suffix u of w is comparable with X_k for the prefix order.

Indeed, if u is comparable with X for the prefix order, then it is comparable with X_k (since $X \subset X_k$); if on the other hand, $u \in I(X)$, then u has a prefix v in $\overline{X}$. Then v is a proper factor of w, hence $v \in \overline{X}^{(i-1)}$ and $u \in \overline{X}^{(i-1)} A^*$ is comparable with X_k for the prefix order by Lemma 6.6.8 below with $T = \overline{X}^{(i-1)}$. Thus Lemma 6.6.6 applies with $T = \overline{X}^{(i)} \setminus \overline{X}^{(i-1)}$ and $n = k$, and we deduce that L_k is bounded on $\overline{X}^{(i)}$. $\qquad\square$

Lemma 6.6.8 *Let $T \subset \overline{X}$ and m, ℓ be two integers with $0 \leq m \leq \ell$. If $X_{\ell+\sigma}$ is not maximal and $(L_m, w) \leq \ell$ for any $w \in T$, then every word in $T A^*$ is comparable for the prefix order with a word in $X_{\ell+\sigma}$.*

Proof. Define $W_i = P(V_{\ell+i}) \cap T A^*$ for $i \geq 0$. The main step consists in showing that each word in W_σ has some prefix in $X_{\ell+\sigma}$.

For this, take a word $v \in W_\sigma$. Since $v \in V_{\ell+\sigma}$, we have $(L_{\ell+\sigma}, v) = \ell + \sigma + 1$. Let $i \in \{0, \ldots, \sigma\}$. Then $X_{\ell+i} \subset X_{\ell+\sigma}$ and thus we have by Lemma 6.6.2 (5) $(L_{\ell+i}, v) \geq (L_{\ell+\sigma}, v) = \ell + \sigma + 1 \geq \ell + i + 1$.

Thus by Lemma 6.6.2 (1), there exists a prefix p_i of v such that $(L_{\ell+i}, p_i) = \ell + i + 1$, and therefore $p_i \in V_{\ell+i}$. We may even assume, by choosing a shortest prefix, that $p_i \in P(V_{\ell+i})$. For $i < \sigma$, p_i is a proper prefix of p_{i+1}. Indeed, if on the contrary p_{i+1} is a prefix of p_i, then $\ell + i + 1 = (L_{\ell+i}, p_i) \geq (L_{\ell+i}, p_{i+1}) \geq (L_{\ell+i+1}, p_{i+1}) = \ell + i + 2$ by Proposition 6.1.8 and Lemma 6.6.2 (5), a contradiction.

Now, $v = tu$ for some $t \in T$ and $u \in A^*$. We have $\ell + i + 1 > \ell \geq (L_m, t)$ by the hypothesis in the Lemma and $(L_m, t) \geq (L_{\ell+i}, t)$ by Lemma 6.6.2 (5) because $X_m \subset X_{\ell+i}$. Since $(L_{\ell+i}, p_i) = \ell + i + 1$, the word t must be a prefix of p_i by Proposition 6.1.8. Thus $p_i \in TA^*$ and therefore $p_i \in W_i$.

Suppose, arguing by contradiction, that $v \in I(X_{\ell+\sigma})$. We first show that this implies that $p_i \in I(X_{\ell+i})$.

Indeed, p_i cannot have a prefix in $X_{\ell+i}$, since this word would be prefix of v, contradicting the assumption that v is not comparable with $X_{\ell+\sigma}$ which contains $X_{\ell+i}$. Next, suppose that p_i is a prefix of some $x \in X_{\ell+i}$. Then the word t which is a prefix of p_i is also a prefix of x. Since t is incomparable with X, the word x is not in X. Thus by Lemma 6.6.3, $(L_{\ell+i}, x) \leq \ell + i$, which implies by Proposition 6.1.8 that $(L_{\ell+i}, p_i) \leq (L_{\ell+i}, x) \leq \ell + i$. But $p_i \in W_i \subset V_{\ell+i}$, and this implies that $(L_{\ell+i}, p_i) = \ell + i + 1$, a contradiction.

We assume now $i < \sigma$. Since p_i is in $I(X_{\ell+i})$, it is in $Z_{\ell+i}$. Now, $p_i \notin X_{\ell+i+1}$, since otherwise v has a prefix in $X_{\ell+i+1} \subset X_{\ell+\sigma}$, which contradicts the assumption that $v \in I(X_{\ell+\sigma})$. Thus we must have $p_i \in A^-X$, since $Z_{\ell+i} \setminus A^-X \subset X_{\ell+i+1}$.

Since each p_i is a proper prefix of p_{i+1}, we obtain a chain of σ suffixes of X, a contradiction with the definition of σ.

We conclude that $v \notin I(X_{\ell+\sigma})$, and consequently there is some word $x \in X_{\ell+\sigma}$ which is comparable with v. If v is a prefix of x, then $x \notin X$, otherwise, t is comparable with X, contradicting the fact that $t \in T \subset \overline{X}$. Hence by Lemma 6.6.3, $(L_{\ell+\sigma}, x) \leq \ell + \sigma$. Now, $(L_{\ell+\sigma}, v) = \ell + \sigma + 1$, which is a contradiction by Proposition 6.1.8. Thus x is a prefix of v. Thus we have shown that each word in W_σ has a prefix in $X_{\ell+\sigma}$.

Let now $w = tu$ be any word in TA^* with $t \in T$. We have $(L_{\ell+\sigma}, t) \leq (L_m, t)$ (by Lemma 6.6.2 (5)) $\leq \ell < \ell + \sigma + 1$. Thus, by Proposition 6.1.8 and Lemma 6.6.2 (2), since $X_{\ell+\sigma}$ is not maximal, there is some word u', comparable with u for the prefix order, such that $L_{\ell+\sigma}(tu') = \ell + \sigma + 1$. Thus $v = tu' \in V_{\ell+\sigma}$, and one may even assume that $v \in P(V_{\ell+\sigma})$, hence $v \in W_\sigma$. By what we have already shown, v has a prefix in $X_{\ell+\sigma}$ and we conclude that w is comparable with a word in $X_{\ell+\sigma}$. □

Proof of Theorem 6.6.1. By Lemma 6.6.7, L_k is bounded on $\overline{X}$ for some k. Thus we may find ℓ such that $k \leq \ell$ and $(L_k, w) \leq \ell$ for any w in $\overline{X}$. Lemma 6.6.8 with $T = \overline{X}$ now implies that every word in $\overline{X}A^*$ is comparable for the prefix order with a word in $X_{\ell+\sigma}$. Let $w \in A^*$. If w is not comparable with a word in X, then it is in $\overline{X}A^*$, and therefore is comparable with a word in $X_{\ell+\sigma}$. Thus any word in A^* is comparable for the prefix order, with some word in $X_{\ell+\sigma}$. This shows that $X_{\ell+\sigma}$ is a maximal bifix code containing X. It is rational by Lemma 6.6.4. Hence the theorem is proved. □

Figure 6.15 The prefix codes $X = ba^*bb$ and $\overline{X} = a \cup ba^*ba$.

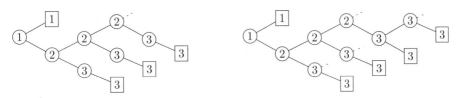

Figure 6.16 The bifix codes $X_2 = a \cup ba^*bb$ and $X_4 = a \cup ba^*ba^*b$.

We give now an example which may be illuminating. Let $X = X_0 = ba^*bb$. The tree representing X, viewed as prefix code, is in Figure 6.15 on the left where the values of the indicator on the prefixes are indicated. It follows that

$$I(X) = aA^* \cup b^2aA^* \cup babaA^* \cup ba^2baA^* \cdots = aA^* \cup ba^*baA^*.$$

Thus $\overline{X} = a \cup ba^*ba$. The prefix code $\overline{X}$ is indicated in Figure 6.15 on the right with the values of L_0 on its prefixes. It is easy to see that, by definition of L_0, $(L_0, a) = 2$ and $(L_0, ba^nba) = n + 4$, since a and ba^nba have no factor in X. Hence, by Proposition 6.1.8, $(L_0, w) \geq 2$ for any w in $I(X) = (a \cup ba^*ba)A^*$ and we deduce that $Z_0 = \emptyset$. Thus $X = X_1$ and $I(X) = I(X_1)$. Now the only possible word in $Z_1 = I(X_1) \cap P(V_1)$ is a; thus $Z_1 = \{a\}$ and $X_2 = X_1 \cup \{a\} = a \cup ba^*bb$, since $a \notin A^-X$ (see Figure 6.16).

Now, $I(X_2) = ba^*baA^*$. We have $(L_2, ba^nba) = n + 4 - (n + 1) = 3$, since the only factor of ba^nba in X_2 is a, with multiplicity $n + 1$. Moreover $(L_2, ba^nb) = 3$, hence $ba^nba \notin P(V_2)$ and likewise, no w in $I(X_2)$ is in $P(V_2)$. This implies that $Z_2 = \emptyset$ and $X_3 = X_2$.

We now have $Z_3 = P(V_3) \cap I(X_3) = ba^*ba^+b$. Indeed for $n, m \geq 0$ $(L_3, ba^nba^m) = 3$ and $(L_3, ba^nba^mb) = 4$. Thus $X_4 = a \cup ba^*bb \cup ba^*ba^+b = a \cup ba^*ba^*b$. It is easily checked that $I(X_4) = \emptyset$ and thus X_4 is right complete, hence maximal.

6.7 Exercises

Section 6.1

6.1.1 Let $X \subset A^+$ be a bifix code and $L = L_X$ its indicator. Show that if for $u, v \in A^*$ we have $(L, uvu) = (L, u)$, then for all $m \geq 0$, $(L, (uv)^m u) = (L, u)$.

6.1.2 Let $X \subset A^+$ be a bifix code and let H be the subgroup of the free group on A generated by X.

Show that the following conditions are equivalent:

(i) The minimal deterministic automaton of X^* is bideterministic.
(ii) For all $t, u, v, w \in A^*$, $tu, vu, vw \in X$ implies $tw \in X$.
(iii) $H \cap A^* = X^*$.

6.1.3 The aim of this exercise is to describe a method, which allows a decoding in both directions for any finite binary prefix code. Let X be a finite prefix code on the alphabet $\{0, 1\}$ and let ℓ be the maximal length of the words of X. Consider a sequence $x_1, x_2, \ldots, x_n$ of codewords. Let

$$w = x_1 x_2 \cdots x_n 0^\ell \oplus 0^\ell \tilde{x}_1 \tilde{x}_2 \cdots \tilde{x}_n \qquad (6.57)$$

where $\tilde{x}$ is the reversal of the word x and where $\oplus$ denotes the addition mod 2. Show that w can be decoded in both directions with finite delay.

Section 6.2

6.2.1 Let $X \subset A^+$ be a thin maximal prefix code. To each word $w = a_1 a_2 \cdots a_n \in \bar{F}(X)$ with $a_i \in A$, we will associate a function ρ_w from $\{1, 2, \ldots, n\}$ into itself.

(a) Show that for each integer i in $\{1, 2, \ldots, n\}$, there exists a unique integer $k \in \{1, 2, \ldots, n\}$ such that either $a_i a_{i+1} \cdots a_k$ or $a_i a_{i+1} \cdots a_n a_1 \cdots a_k$ is in X. Set $\rho_w(i) = k$. This defines, for each $w \in \bar{F}(X)$, a mapping ρ_w from $\{1, 2, \ldots, |w|\}$ into itself.
(b) Show that X is suffix if and only if the function ρ_w is injective for all $w \in \bar{F}(X)$.
(c) Show that X is left complete if and only if the function ρ_w is a surjection for all $w \in \bar{F}(X)$.
(d) Derive from this that a thin maximal prefix code is suffix if and only if it is left complete (see the proof of Proposition 6.2.1).

6.2.2 Let $P = \{w\tilde{w} \mid w \in A^*\}$ be the set of *palindrome* words of even length.

(a) Show that P^* is biunitary. Let X be the bifix code for which $X^* = P^*$. Then X is called the set of *palindrome primes*.
(b) Show that X is left complete and right complete.

6.2.3 Show that two maximal bifix codes which are obtained one from the other by internal transformation are either both recognizable or both not recognizable.

6.2.4 Show that a maximal bifix code $X \subset A^+$ is a group code if and only if for any $u, v, w, r \in A^*$,

$$uv, uw, rv \in X^* \Rightarrow rw \in X^*. \qquad (6.58)$$

(*Hint*: Use Exercise 6.1.2.)

Section 6.3

6.3.1 Let X be a thin maximal bifix code of degree d. Let $w \in \bar{H}(X)$ and let

$$1 = p_1, p_2, \ldots, p_d$$

be the sequence of the suffixes of w which are proper prefixes of X. Set $Y_1 = 1$ and $Y_i = p_i^{-1}X$ for $2 \le i \le d$. Show that each Y_i is a maximal prefix set, and that the set S of proper suffixes of X is the disjoint union of the Y_i's (see Theorem 6.3.15).

6.3.2 Let X be a thin maximal bifix code of degree d and let S be the set of its proper suffixes. Show that there exists a unique partition of S into a disjoint union of d prefix sets Y_i satisfying $Y_{i-1} \subset Y_i A^-$ for $2 \le i \le d$. (*Hint*: Set $Y_d = S \cap \bar{H}(X)$.)

Section 6.4

6.4.1 Let X be a finite bifix code. Show, using Theorem 6.4.3, that there exists a recognizable maximal bifix code containing X.

6.4.2 Show that if X is a recognizable maximal bifix code of degree $d \ge 2$, then the derived code is recognizable. (*Hint*: Use Proposition 6.3.14.)

6.4.3 Let X be a thin maximal bifix code of degree $d \ge 2$. Let $w \in \bar{H}(X)$, and let s be the longest prefix of w which is a proper suffix of X. Further, let x be the prefix of w which is in X. Show that the shorter one of s and x is in the derived code X'. (*Hint*: Prove that if $|x| \ge |s|$, then $s \in (HA \setminus H) \cap (AH \setminus H)$, with $H = A^- X A^-$.)

6.4.4 Let X_1 and X_2 be two thin maximal bifix codes having same kernel: $K(X_1) = K(X_2)$. Set

$$P_1 = A^* \setminus X_1 A^*, \quad P_2 = A^* \setminus X_2 A^*,$$

$$Z = (X_1 \cap P_2) \cup (X_1 \cap X_2) \cup (P_1 \cap X_2).$$

(see Exercise 3.4.3). Show that Z is thin, maximal, and bifix. Use this to prove directly that two thin maximal finite bifix codes with same kernel and same degree are equal. This is Theorem 6.4.2 for finite codes.

6.4.5 Show that there exists a maximal bifix code of degree 3 on $\{a, b\}$ which is not rational. (*Hint*: Choose a code with non rational kernel.)

Section 6.5

6.5.1 Let X be a finite maximal bifix code. Show that if a word $w \in A^+$ satisfies

$$pwq = rws \in X \tag{6.59}$$

for some $p, q, r, s \in A^+$, and $p \ne r$, then $w \in H(X')$, where X' is the derived code of X. (*Hint*: Start with a word of maximal length satisfying (6.59), consider the word rwq and use Proposition 6.3.14.)

6.5.2 For a finite code X, let $\ell(X) = \max\{|x| \mid x \in X\}$. Show, using Exercise 6.5.1, that if X is a finite maximal bifix code over a k letter alphabet, then

$$\ell(X) \le \ell(X') + k^{\ell(X')-1},$$

with X' denoting the derived code of X. Denote by $\lambda(k, d)$ the maximum of the lengths of the words of a finite maximal bifix code of degree d over a k letter alphabet. Show that for $d \ge 2$

$$\lambda(k, d) \le \lambda(k, d - 1) + k^{\lambda(k,d-1)-1}.$$

Compare with the bound given by Theorem 6.5.2.

6.5.3 Let $X \subset A^+$ be a finite maximal bifix code of degree d. Let $a, b \in A$, and define a function φ from $\{0, 1, \ldots, d - 1\}$ into itself by

$$a^i b^{d-\varphi(i)} \in X.$$

Show that φ is a bijection.

6.5.4 Show that for each $k \ge 2$, the number $\beta_k(d)$ of finite maximal bifix codes of degree d over a k letter alphabet is unbounded as a function of d.

6.5.5 A *quasipower* of order n is defined by induction as follows: a quasipower of order 0 is an unbordered word. A quasipower of order $n + 1$ is a word of the form uvu, where u is a quasipower of order n. Let k be an integer and let α_n be the sequence inductively defined by

$$\alpha_1 = k + 1, \qquad \alpha_{n+1} = \alpha_n(k^{\alpha_n} + 1) \quad (n \ge 1).$$

Show that any word over a k letter alphabet with length at least equal to α_n has a factor which is a quasipower of order n.

6.5.6 Let X be a finite maximal bifix code of degree $d \ge 2$ over a k letter alphabet. Show that

$$\max_{x \in X} |x| \le \alpha_{d-1} + 2,$$

where (α_n) is the sequence defined in Exercise 6.5.5. (*Hint*: Use Exercise 6.1.1.) Compare with the bound given by Exercise 6.5.2.

6.5.7 Show that the number of finite maximal bifix codes of degree 4 over a two-letter alphabet is $\beta_2(4) = 73$.

6.5.8 Let X be a thin maximal bifix code of degree d on k letters. Let S be the set of its suffixes and let $(U_i)_{1 \le i \le d}$ be disjoint maximal prefix codes such that S is their union. Let R_i be the set of prefixes of U_i. Define $t(z) = \sum_{i=1}^{d} f_{R_i}(z)$. Show that the generating series of X satisfies

$$f_X(z) - 1 = (kz - 1)d + (kz - 1)^2 t(z).$$

6.5.9 Let X be a thin maximal bifix code on k letters of degree d. We have $\frac{1}{k} f'_X(1/k) = d$, where the last expression can be viewed as the average length of the words of X with respect to the uniform Bernoulli distribution. Recall that the *variance* of the lengths of the words of X is the mean of the squares of the lengths minus the square of the mean of the lengths. Show that the variance is given by

$$v_X = 2t(1/k) + d - d^2,$$

where $t(z)$ is defined in Exercise 6.5.8.

Section 6.6

6.6.1 Show that if X is a prefix code, then $Y = X \cup \overline{X}$ is a maximal prefix code (where $\overline{X}$ denotes the companion of X). Show that if X is rational, so is Y.

6.8 Notes

The idea to study bifix codes goes back to Schützenberger (1956) and Gilbert and Moore (1959). These papers already contain significant results. The first systematic study is in Schützenberger (1961b), Schützenberger (1961c).

Propositions 6.2.1 and 6.2.7 are from Schützenberger (1961c). The internal transformation appears in Schützenberger (1961b). The fact that all finite maximal bifix codes can be obtained from the uniform codes by internal transformation (Theorem 6.5.4) is from Césari (1972). The fact that the average length of a thin maximal bifix code is an integer (Corollary 6.3.16) is already in Gilbert and Moore (1959). It is proved in Schützenberger (1961b) with the methods developed in Chapter 13. Theorem 6.3.15 and its converse (Proposition 6.3.17) appear in Perrin (1977a). The notion of derived code is due to Césari (1979).

The results of Section 6.4 are a generalization to thin codes of results in Césari (1979).

Theorem 6.5.2 appears already in Schützenberger (1961b) with a different proof (see Exercise 6.5.6). The rest of this section is due to Césari (1979). The enumeration of finite maximal bifix codes over a two-letter alphabet has been pursued by computer. A first program was written in 1975 by C. Precetti using internal transformations. It produced several thousands of them for $d = 5$. In 1984, a program written by M. Léonard using the method of Corollary 6.5.8 gave the exact number of finite maximal bifix codes of degree 5 over a two-letter alphabet. This number is 5 056 783. See Léonard (1988).

Bifix codes and their length distributions have been studied with a practical motivation, under the name of *reversible variable-length codes* (see Takishima *et al.* (1995); Gillman and Rivest (1995); Ye and Yeung (2001)). Proposition 6.5.10 is from Ahlswede *et al.* (1996).

It is conjectured (this is the so-called *3/4-conjecture*) that for any series $f(t) = \sum u_n t^n$ with integer nonnegative coefficients satisfying $f(1/k) \leq 3/4$ there exists a bifix code X on k letters such that $f_X = f$. Partial results are given in Yekhanin (2004) and Deppe and Schnettler (2006).

Theorem 6.6.1 is due to Zhang and Shen (1995). For the proof of the theorem, we have followed Bruyère and Perrin (1999).

A code X is *infix* if no word of X is a proper factor of another word of X. Thus infix codes are bifix. Infix codes have been studied for the first time in Ito *et al.* (1991). The problem of completing an infix code has·been solved by Ito and Thierrin (1994) for finite codes, and by Lam (2000) for rational codes.

Exercise 6.1.3 is due to Girod (1999); see also Salomon (2007). Exercise 6.2.4 appears in Long (1996). Exercises 6.3.2, 6.4.4, 6.5.1, and 6.5.2 are from Césari (1979). Exercise 6.4.5 is from Schützenberger (1961c).

7

Circular codes

In this chapter we study a particular family of codes called circular codes. The main feature of these codes is that they define a unique factorization of words written on a circle. The family of circular codes has numerous interesting properties. They appear in many problems of combinatorics on words, several of which will be mentioned here.

In Section 7.1 we give the definition of circular codes and we characterize the submonoid generated by a circular code. We also describe some elementary properties of circular codes. In particular we characterize maximal circular codes (Theorem 7.1.10).

In Section 7.2 we introduce successive refinements of the notion of a circular code. For this we define the notion of (p, q)-limitedness. We then proceed to a more detailed study of $(1, 0)$-limited codes. In particular, we show (Proposition 7.2.10) that $(1, 0)$-limited codes correspond to ordered automata. Comma-free codes are defined as circular codes satisfying the strongest possible condition.

Section 7.3 is concerned with length distributions of circular codes. Two important theorems are proved. The first gives a characterization of sequences of integers which are the length distribution of a circular code (Theorem 7.3.7). The second shows that for each odd integer n there exists a system of representatives of conjugacy classes of primitive words of length n which not only is circular but even comma-free (Theorem 7.3.11). The proofs of these results use similar combinatorial constructions. As a matter of fact they are based on the notion of factorization of free monoids studied in Chapter 8.

7.1 Circular codes

We define in this section a new family of codes which take into account, in a natural way, the operation of conjugacy.

By definition, a subset X of A^+ is a *circular code* if for all $n, m \geq 1$ and $x_1, x_2, \ldots,$ $x_n \in X$, $y_1, y_2, \ldots, y_m \in X$ and $p \in A^*$, $s \in A^+$, the equalities

$$sx_2x_3 \cdots x_n p = y_1 y_2 \cdots y_m, \tag{7.1}$$

$$x_1 = ps \tag{7.2}$$

imply $n = m$, $p = 1$, and $x_i = y_i$ for $1 \leq i \leq n$ (see Figure 7.1).

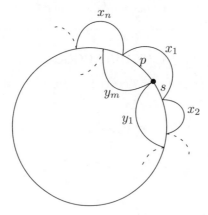

Figure 7.1 Two circular factorizations.

A circular code is clearly a code. The converse is false, as shown in Example 7.1.4. The asymmetry in the definition is only apparent, and comes from the choice of the cutting point on the circle in Figure 7.1. Clearly, any subset of a circular code is also a circular code.

Note that a circular code X cannot contain two distinct conjugate words. Indeed, if $ps, sp \in X$ with $s, p \in A^+$ then

$$s(ps)p = (sp)(sp).$$

Since X is circular, this implies $p = 1$ which gives a contradiction. Moreover, all words in X are primitive, since assuming $u^n \in X$ with $n \geq 2$, it follows that

$$u(u^n)u^{n-1} = u^n u^n.$$

This implies $u = 1$ and gives again a contradiction.

We now characterize in various ways the submonoids generated by circular codes. The first characterization facilitates the manipulation of circular codes. A submonoid M of A^* is called *pure* if for all $x \in A^*$ and $n \geq 1$,

$$x^n \in M \implies x \in M. \tag{7.3}$$

A submonoid M of A^* is *very pure* if for all $u, v \in A^*$,

$$uv, vu \in M \implies u, v \in M. \tag{7.4}$$

A very pure monoid is pure. The converse does not hold (see Example 7.1.4).

Proposition 7.1.1 *A submonoid of A^* is very pure if and only if its minimal set of generators is a circular code.*

Proof. Let M be a very pure submonoid. We show that M is stable. Let $m, m', xm, m'x \in M$. Then setting $u = x, v = mm'$, we have $uv, vu \in M$. This implies $x \in M$.

Thus M is stable, hence M is free. Let X be its base. Assume that (7.1) and (7.2) hold. Set $u = s$, $v = x_2x_3 \cdots x_n p$. Then uv, $vu \in M$. Consequently $s \in M$. Since ps, $x_2x_3 \cdots x_n p \in M$, the stability of M implies that $p \in M$. From $ps \in X$, it follows that $p = 1$. Since X is a code, this implies $n = m$ and $x_i = y_i$ for $i = 1, \ldots, n$.

Conversely, let X be a circular code and set $M = X^*$. To show that M is very pure, consider two nonempty words u, $v \in A^+$ such that uv, $vu \in M$. Set

$$uv = x_1x_2 \cdots x_n, \qquad vu = y_1y_2 \cdots y_m,$$

with x_i, $y_j \in X$. There exists an integer i with $1 \leq i \leq n$ such that

$$u = x_1x_2 \cdots x_{i-1}p, \qquad v = sx_{i+1} \cdots x_n,$$

with $x_i = ps$, $p \in A^*$, $s \in A^+$. Then vu may be written in two ways:

$$sx_{i+1} \cdots x_n x_1 x_2 \cdots x_{i-1}p = y_1y_2 \cdots y_m.$$

Since X is a circular code, this implies $p = 1$ and $s = y_1$. Thus u, $v \in M$, showing that M is very pure. $\qquad\qquad\qquad\qquad\qquad\qquad\qquad\qquad\qquad\qquad\qquad\quad\square$

Example 7.1.2 Let $A = \{a, b\}$ and $X = a^*b$. Then $X^* = A^*b \cup 1$. Thus if uv, $vu \in X^*$, the words u, v either are the empty word or end with the letter b; hence u, $v \in X^*$. Consequently X^* is very pure and X is circular.

Example 7.1.3 Let $A = \{a\}$ and $X = \{a^2\}$. The submonoid X^* clearly is not pure. Thus X is not a circular code.

Example 7.1.4 Let $A = \{a, b\}$ and $X = \{ab, ba\}$. The code X is not circular. However, X^* is pure (Exercise 7.1.1).

The following proposition characterizes the flower automaton of a circular code.

Proposition 7.1.5 *Let $X \subset A^+$ be a code and let φ be the representation associated with the flower automaton of X. The following conditions are equivalent:*

(i) *X is a circular code.*
(ii) *For all $w \in A^+$, the relation $\varphi(w)$ has at most one fixed point.*

Proof. For convenience, let 1 denote the state $(1, 1)$ of the flower automaton $\mathcal{A}_D^*(X)$.

(i) $\Longrightarrow$ (ii). Let $w \in A^+$, and let $p = (u, v)$, $p' = (u', v')$ be two states of $\mathcal{A}_D^*(X)$ which are fixed points of $\varphi(w)$, that is, such that $(p, \varphi(w), p) = (p', \varphi(w), p') = 1$.

Since $w \neq 1$, Proposition 4.2.3 shows that $w \in vX^*u$ and $w \in v'X^*u'$. Thus both paths $c : p \xrightarrow{w} p$ and $c' : p' \xrightarrow{w} p'$ pass through the state 1.

We may assume that $v \leq v'$. Let $z, t \in A^*$ be the words such that $v' = vz$ and $w = vzt$. Then the paths c, c' factorize as

$$c : p \xrightarrow{v} 1 \xrightarrow{z} r \xrightarrow{t} p, \qquad c' : p' \xrightarrow{v} s \xrightarrow{z} 1 \xrightarrow{t} p'.$$

Thus there are also paths

$$d : 1 \xrightarrow{z} r \xrightarrow{t} p \xrightarrow{v} 1, \qquad 1 \xrightarrow{t} p' \xrightarrow{v} s \xrightarrow{z} 1,$$

showing that $ztv, tvz \in X^*$. Since X^* is very pure, it follows that $z, tv \in X^*$. Consequently, there is a path $e : 1 \xrightarrow{z} 1 \xrightarrow{tv} 1$. By unambiguity, $d = e$, whence $r = 1$. Thus $1 \xrightarrow{t} p \xrightarrow{vz} 1$ which compared to d' gives $p = p'$. This proves that $\varphi(w)$ has at most one fixed point.

(ii) $\Longrightarrow$ (i). Let $u, v \in A^*$ be such that $uv, vu \in X^*$. Then there are two paths $1 \xrightarrow{u} p \xrightarrow{v} 1$ and $1 \xrightarrow{v} q \xrightarrow{u} 1$. Thus the relation $\varphi(uv)$ has two fixed points, namely 1 and q. This implies $q = 1$, and thus $u, v \in X^*$. $\qquad\square$

We now give a characterization of circular codes in terms of conjugacy. For this, the following terminology is used.

Let $X \subset A^+$ be a code. Two words $w, w' \in X^*$ are called *X-conjugate* if there exist $x, y \in X^*$ such that

$$w = xy, \qquad w' = yx.$$

The word $x \in X^*$ is called *X-primitive* if $x = y^n$ with $y \in X^*$ implies $n = 1$. The *X-exponent* of $x \in X^+$ is the unique integer $p \geq 1$ such that $x = y^p$ with y an X-primitive word. Let $\alpha : B \to A^*$ be a coding morphism for X. It is easily seen that $w, w' \in X^*$ are X-conjugate if and only if $\alpha^{-1}(w)$ and $\alpha^{-1}(w')$ are conjugate in B^*. Likewise, $x \in X^*$ is X-primitive if and only if $\alpha^{-1}(x)$ is a primitive word of B^*.

Thus, X-conjugacy is an equivalence relation on X^*. Of course, two words in X^* which are X-conjugate are conjugate. Likewise, a word in X^* which is primitive is also X-primitive. When $X = A$, we get the usual notions of conjugacy and primitivity.

Proposition 7.1.6 *Let $X \subset A^+$ be a code. The following conditions are equivalent:*

(i) *X is a circular code.*
(i) *X^* is pure, and any two words in X^* which are conjugate are also X-conjugate.*

Proof. (i) $\Longrightarrow$ (ii). Since X^* is very pure, it is pure. Next let $w, w' \in X^*$ be conjugate words. Then $w = uv, w' = vu$ for some $u, v \in A^*$. By (7.4), $u, v \in X^*$, showing that w and w' are X-conjugate.

(ii) $\Longrightarrow$ (i). Let $u, v \in A^*$ be such that $uv, vu \in X^*$. If $u = 1$ or $v = 1$, then $u, v \in X^*$. Otherwise, let x, y be the primitive words which are the roots of uv and vu: then $uv = x^n$, $vu = y^n$ for some $n \geq 1$. Since X^* is pure, we have $x, y \in X^*$. Next $uv = x^n$ gives a decomposition $x = rs, u = x^p r, v = sx^q$ for some $r \in A^*, s \in A^+$, and $p + q + 1 = n$. Substituting this in the equation $vu = y^n$ gives $y = sr$. Since x, y are conjugate, they are X-conjugate. But for primitive words x, y, there exists a unique pair $(r, s') \in A^* \times A^+$ such that $x = r's', y = s'r'$. Consequently $r, s \in X^*$. Thus $u, v \in X^*$, showing that X^* is very pure. $\qquad\square$

Proposition 7.1.7 *Let $X \subset A^+$ be a code and let $C \subset A^n$ be a conjugacy class that meets X^*. Then*

$$\sum_{m \geq 1} \frac{1}{m} \text{Card}(X^m \cap C) \geq \frac{1}{n} \text{Card}(C). \tag{7.5}$$

Moreover, equality holds if and only if the following two conditions are satisfied:

(i) *The exponent of the words in $C \cap X^*$ is equal to their X-exponent.*
(ii) *$C \cap X^*$ is a class of X-conjugacy.*

Proof. Let p be the exponent of the words in C. Then $\text{Card}(C) = n/p$. The set $C \cap X^*$ is a union of X-conjugacy classes. Let D be such a class, and set $C' = C \setminus D$. The words in D all belong to X^k for the same k, and all have the same X-exponent, say q. Then $\text{Card}(D) = k/q$. Since $C = C' \cup D$, the left side of (7.5) is

$$\sum_{m=1}^{n} \frac{1}{m} \text{Card}(X^m \cap C') + \sum_{m=1}^{n} \frac{1}{m} \text{Card}(X^m \cap D).$$

In the second sum, all terms vanish except for $m = k$. Thus this sum is equal to $(1/k) \text{Card}(X^k \cap D) = 1/q$. Thus

$$\sum_{m=1}^{n} \frac{1}{m} \text{Card}(X^m \cap C) = \frac{1}{q} + \sum_{m=1}^{n} \frac{1}{m} \text{Card}(X^m \cap C'). \tag{7.6}$$

Since $q \leq p$, we have $1/q \geq 1/p = (1/n) \text{Card}(C)$. This proves Formula (7.5).

Assume now that (i) and (ii) hold. Then $p = q$, and $D = C \cap X^*$. Thus $C' \cap X^* = \emptyset$. Thus the right side of (7.6) is equal to $1/p$, which shows that equality holds in (7.5). Conversely, assuming the equality sign in (7.5), it follows from (7.6) that

$$\frac{1}{p} = \frac{1}{q} + \sum_{m=1}^{n} \frac{1}{m} \text{Card}(X^m \cap C') \geq \frac{1}{q} \geq \frac{1}{p},$$

which implies $p = q$ and $C' \cap X^* = \emptyset$. $\qquad\square$

The proposition has the following consequence:

Proposition 7.1.8 *Let $X \subset A^+$ be a code. The following conditions are equivalent:*

(i) *X is a circular code.*
(ii) *For any integer $n \geq 1$ and for any conjugacy class $C \subset A^n$ that meets X^*, we have*

$$\sum_{m \geq 1} \frac{1}{m} \text{Card}(X^m \cap C) = \frac{1}{n} \text{Card}(C). \tag{7.7}$$

Proof. By Proposition 7.1.6, the code X is circular if and only if we have

(iii) X^* is pure.
(iv) Two conjugate words in X^* are X-conjugate.

Condition (iii) is equivalent to: the X-exponent of any word in X^* is equal to its exponent. Thus X is circular if and only if for any conjugacy class C meeting X^*, we have

(v) The exponent of words in $C \cap X^*$ equals their X-exponent.
(vi) $C \cap X^*$ is a class of X-conjugacy.

In view of Proposition 7.1.7, conditions (v) and (vi) are satisfied if and only if the conjugacy class $C \cap A^n$ satisfies the equality (7.7). This proves the proposition. $\square$

We now prove a result which is an analogue of Theorem 2.5.5.

Proposition 7.1.9 *Let $X \subset A^+$ be a circular code. If X is maximal as a circular code, then X is complete.*

Proof. If $A = \{a\}$, then $X = \{a\}$. Therefore, we assume $\mathrm{Card}(A) \geq 2$. Suppose that X is not complete. Then there is a word, say w, which is not a factor of a word in X^*. By Proposition 1.3.6, there is a word $v \in A^*$ such that $y = wv$ is unbordered.

Set $Y = X \cup y$. We prove that Y is a circular code. For this, let x_i $(1 \leq i \leq n)$ and y_i $(1 \leq i \leq m)$ be words in Y, let $p \in A^*$, $s \in A^+$ such that

$$s x_2 x_3 \cdots x_n p = y_1 y_2 \cdots y_m \qquad x_1 = ps.$$

If all x_i $(1 \leq i \leq n)$ are in X, then also all y_j are in X, because y is not a factor of a word in X^*. Since X is circular, this then implies that

$$n = m, \quad p = 1, \quad \text{and} \quad x_i = y_i \ (1 \leq i \leq n). \tag{7.8}$$

Suppose now that $x_i = y$ for some $i \in \{1, \ldots, n\}$ and suppose first that $i \neq 1$. Then x_i is a factor of $y_1 y_2 \cdots y_m$. Since $y \notin F(X^*)$, and since y is unbordered, this implies that there is a $j \in \{1, 2, \ldots, m\}$ such that $y_j = y$, and

$$s x_2 \cdots x_{i-1} = y_1 y_2 \cdots y_{j-1}, \qquad y_{i+1} \cdots x_n p = y_{j+1} \cdots y_m.$$

This in turn implies

$$s x_2 \cdots x_{i-1} x_{i+1} \cdots x_n p = y_1 y_2 \cdots y_{j-1} y_{j+1} \cdots y_m,$$

and (7.8) follows by induction on the length of the words.

Consider finally the case where $i = 1$, that is, $x_1 = y$. Since

$$x_1 x_2 \cdots x_n p = p y_1 y_2 \cdots y_m,$$

we have $yx_2 \cdots x_n p = p y_1 y_2 \cdots y_m$. Now p is a suffix of a word in Y^*; further $y \notin F(X^*)$ and y is unbordered. Thus $p = 1$ and $y_1 = y$. This again gives (7.8) by induction on the length of the words. Thus if X is not complete, then $Y = X \cup y$ is a circular code. Since $y \notin X$, X is not maximal as a circular code. $\qquad\square$

The preceding proposition and Theorem 2.5.13 imply

Theorem 7.1.10 *Let X be a thin circular code. The following three conditions are equivalent.*

(i) *X is complete.*
(ii) *X is a maximal code.*
(iii) *X is maximal as a circular code.* $\qquad\square$

Observe that a maximal circular code $X \subset A^+$ is necessarily infinite, except when $X = A$. Indeed, assume that X is a finite maximal circular code. Then by Theorem 7.1.10, it is a maximal code. According to Proposition 2.5.15, there is, for each letter $a \in A$, an integer $n \geq 1$ such that $a^n \in X$. Since X is circular, we must have $n = 1$, and consequently $a \in X$ for all $a \in A$. Thus $X = A$.

We shall need the following property which allows us to construct circular codes.

Proposition 7.1.11 *Let Y, Z be two composable codes, and let $X = Y \circ Z$. If Y and Z are circular, then X is circular.*

Proof. Let $\alpha : B^* \to A^*$ be a morphism such that $X = Y \circ_\alpha Z$. Let $u, v \in A^*$ be such that $uv, vu \in X^*$. Then $uv, vu \in Z^*$, whence $u, v \in Z^*$ because Z^* is very pure. Let $s = \alpha^{-1}(u), t = \alpha^{-1}(v)$. Then $st, ts \in Y^*$. Since Y^* is very pure, $s, t \in Y^*$, showing that $u, v \in X^*$. Thus X^* is very pure. $\qquad\square$

7.2 Limited codes

We introduce special families of circular codes which are defined by increasingly restrictive conditions concerning overlapping between words. The most special family is that of comma-free codes which is the object of an important theorem proved in the next section.

Let $p, q \geq 0$ be two integers. A submonoid M of A^* is said to satisfy condition $C(p, q)$ if for any sequence $u_0, u_1, \ldots, u_{p+q}$ of words in A^*, the assumptions

$$u_{i-1} u_i \in M \quad (1 \leq i \leq p + q) \tag{7.9}$$

imply

$$u_p \in M.$$

(see Figure 7.2). For example, the condition $C(1, 0)$ simply gives

$$uv \in M \implies v \in M,$$

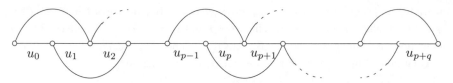

Figure 7.2 The condition $C(p, q)$ (for p odd and q even).

that is M is suffix-closed, and condition $C(1, 1)$ is

$$uv, vw \in M \implies v \in M.$$

It is easily verified that a submonoid M satisfying $C(p, q)$ also satisfies conditions $C(p', q')$ for $p' \geq p, q' \geq q$.

Proposition 7.2.1 *Let* $p, q \geq 0$ *and let* M *be a submonoid of* A^*. *If* M *satisfies condition* $C(p, q)$, *then* M *is very pure.*

Proof. Let $u, v \in A^*$ be such that $uv, vu \in M$. Define words $u_i (0 \leq i \leq p + q)$ to be equal to u (to v) for even (odd) i's. Then assumption (7.9) is satisfied and consequently either u or v is in M. Interchanging the roles of u and v, we get that both u and v are in M.   □

Let M be a submonoid satisfying a condition $C(p, q)$. By the preceding proposition, M is very pure. Thus M is free. Let X be its base. By definition, X is called a (p, q)-*limited* code. A code X is *limited* if there exist integers $p, q \geq 0$ such that X is (p, q)-limited.

Proposition 7.2.2 *Any limited code is circular.*   □

Example 7.2.3 The only $(0, 0)$-limited code over A is $X = A$.

Example 7.2.4 A $(p, 0)$-limited code X is prefix. Assume indeed X is $(p, 0)$-limited. If $p = 0$ then $X = A$. Otherwise take $u_0 = \cdots = u_{p-2} = 1$. Then for any u_{p-1}, u_p, we have

$$u_{p-1}, u_{p-1}u_p \in X^* \implies u_p \in X^*,$$

showing that X^* is right unitary. Likewise, a $(0, q)$-limited code is suffix. However, a prefix code is not always limited, since it is not even necessarily circular.

Example 7.2.5 The code $X = a^*b$ is $(1, 0)$-limited. It satisfies even the stronger condition

$$uv \in X \implies v \in X \cup 1.$$

Example 7.2.6 Let $A = \{a, b, c\}$ and $X = ab^*c \cup b$. The set X is a bifix code. It is neither $(1, 0)$-limited nor $(0, 1)$-limited. However, it is $(2, 0)$-limited and $(0, 2)$-limited.

Example 7.2.7 Let $A = \{a_i \mid i \geq 0\}$ and $X = \{a_i a_{i+1} \mid i \geq 0\}$. The code X is circular, as it is easily verified. However, it is not limited. Indeed, set $u_i = a_i$ for $0 \leq i \leq n$. Then $u_{i-1}u_i \in X$ for $i \in \{1, 2, \ldots, n\}$, but none of the u_i is in X^*.

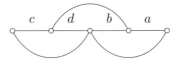

Figure 7.3 X is not (p, q)-limited for $p + q \leq 3$.

This example shows that the converse of Proposition 7.2.2 does not hold in general. However it holds for finite codes, as we shall see later (Theorem 10.2.7). It also holds for recognizable codes (Exercise 7.2.7).

One of the reasons which makes the use of (p, q)-limited codes convenient, is that they behave well with respect to composition. In the following statement, we do not use the notation $X = Y \circ Z$ because we do not assume that every word of Z appears in a word in X.

Proposition 7.2.8 *Let Z be a code over A, let $\beta : B^* \to A^*$ be a coding morphism for Z, and let Y be a code over B. If Y is (p, q)-limited and Z is (r, t)-limited, then $X = \beta(Y)$ is $(p + r, q + t)$-limited.*

Proof. Let $u_0, u_1, \ldots, u_{p+r+q+t} \in A^*$ be such that

$$u_{i-1}u_i \in X^* \quad (1 \leq i \leq p + r + q + t). \tag{7.10}$$

Since $X \subset Z^*$ and Z is (r, t)-limited, it follows from (7.10) that

$$u_r, u_{r+1}, \ldots, u_{r+p+q} \in Z^*. \tag{7.11}$$

Since Y is (p, q)-limited, (7.11) and (7.10) for $r + 1 \leq i \leq p + q + r$ show that $u_{r+p} \in X^*$. Thus X is $(p + r, q + t)$-limited. $\square$

Example 7.2.9 Let $A = \{a, b, c, d\}$ and $X = \{ba, cd, db, cdb, dba\}$. Then

$$X = Z_1 \circ Z_2 \circ Z_3 \circ Z_4,$$

with

$$Z_4 = \{b, c, d, ba\}$$

$$Z_3 \circ Z_4 = \{c, d, ba, db\}$$

$$Z_2 \circ Z_3 \circ Z_4 = \{d, ba, db, cd, cdb\}.$$

The codes Z_3 and Z_4 are $(0, 1)$-limited. The code $Z_3 \circ Z_4$ is not $(0, 1)$-limited, but it is $(0, 2)$-limited, in agreement with Proposition 7.2.8. The codes Z_1 and Z_2 are $(1, 0)$-limited. Thus X is $(2, 2)$-limited. It is not (p, q)-limited for any (p, q) such that $p + q \leq 3$, as shown by Figure 7.3.

We now give a characterization of $(1, 0)$-limited codes by means of automata. These codes occur in Section 8.2. For that, say that an automaton $\mathcal{A} = (Q, 1, 1)$ is *ordered* if it is deterministic and if the following conditions hold: Q is a partially

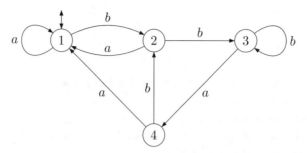

Figure 7.4 An ordered automaton.

ordered set, $q \leq 1$ for all $q \in Q$, and for all $p, q \in Q$, and $a \in A$, $p \leq q$ implies $p \cdot a \leq q \cdot a$.

Proposition 7.2.10 *Let $X \subset A^+$ be a prefix code. The set X^* is suffix-closed if and only if X^* is recognized by some ordered automaton.*

Proof. Assume first that X^* is suffix-closed. Let $\mathcal{A}(X^*) = (Q, 1, 1)$ be the minimal automaton of X^*. Define a partial order on Q by

$$p \leq q \quad \text{if and only if} \quad L_p \subset L_q,$$

where for each state p, $L_p = \{u \in A^* \mid p \cdot u = 1\}$. This defines an order on Q, since by the definition of a minimal automaton, $L_p = L_q \Leftrightarrow p = q$. Next let $q \in Q$, and let $u \in A^*$ be such that $1 \cdot u = q$. Then $v \in L_q$ if and only if $uv \in X^*$. Since X is $(1, 0)$-limited, $uv \in X^*$ implies $v \in X^*$, or also $v \in L_1$. Thus $L_q \subset L_1$, and therefore $q \leq 1$. Further, if $p, q \in Q$ with $p \leq q$, and $a \in A$, let $v \in L_{p \cdot a}$. Then $av \in L_p$, hence $av \in L_q$, and thus $v \in L_{q \cdot a}$. This proves that $\mathcal{A}(X^*)$ is indeed an ordered automaton for this order.

Conversely, let $\mathcal{A} = (Q, 1, 1)$ be an ordered automaton recognizing X^*. Assume that $uv \in X^*$, for some $u, v \in A^*$. Then $1 \cdot uv = 1$. Since $1 \cdot u \leq 1$, we have $1 \cdot uv \leq 1 \cdot v$. Thus $1 \leq 1 \cdot v \leq 1$, whence $1 \cdot v = 1$. Consequently $v \in X^*$. $\qquad \square$

Example 7.2.11 Consider the automaton $(Q, 1, 1)$ given in Figure 7.4. The set $Q = \{1, 2, 3, 4\}$ is equipped with the partial order given by $3 < 2 < 1$ and $4 < 1$. For this order, the automaton $(Q, 1, 1)$ is ordered. It recognizes the submonoid X^* generated by

$$X = (b^2 b^* a)^* \{a, ba\}.$$

Consequently, X is a $(1, 0)$-limited code.

The following proposition gives another characterization of $(1, 0)$-limited codes.

Proposition 7.2.12 *A prefix code $X \subset A^+$ is $(1, 0)$-limited if and only if the set $R = A^* \setminus XA^*$ of words having no prefix in X is a submonoid.*

Proof. By Theorem 3.1.6, $\underline{A}^* = \underline{X}^*\underline{R}$. Suppose first that X is $(1, 0)$-limited. Let $u, u' \in R$, and set $uu' = xr$ with $x \in X^*, r \in R$. Arguing by contradiction, suppose that $x \neq 1$. Then x is not a prefix of u. Consequently $x = uv$, $vr = u'$ for some $v \in A^*$. Since X is $(1, 0)$-limited, one has $v \in X^*$; this implies that $v = 1$, since v is a prefix of u'. Thus $x = u$, a contradiction. Consequently $x = 1$ and $uu' \in R$.

Conversely, suppose that R is a submonoid. Then, being prefix-closed, R is a left unitary submonoid. Thus $R = Y^*$ for some suffix code Y. From the power series equation, we get

$$\underline{A}^* = \underline{X}^*\underline{Y}^*.$$

Multiplication with $1 - \underline{Y}$ on the right gives $\underline{X}^* = \underline{A}^* - \underline{A}^*\underline{Y}$. Thus X^* is the complement of a left ideal. Consequently X^* is suffix-closed. Thus X is $(1, 0)$-limited. $\square$

Example 7.2.13 The code $X = (b^2 b^* a)^* \{a, ba\}$ of Example 7.2.11 gives, for $R = A^* \setminus XA^*$, the submonoid $R = \{b, b^2 a\}^*$.

We end this section with the definition of a family of codes which is the most restrictive of the families we have examined. A code $X \subset A^+$ is called *comma-free* if for all $x \in X^+, u, v \in A^*$,

$$uxv \in X^* \implies u, v \in X^*. \tag{7.12}$$

Comma-free codes are bifix. They are those with the easiest deciphering: if in a word $w \in X^*$, some factor can be identified to be in X, then this factor is one term of the unique X-factorization of w.

Proposition 7.2.14 *A code $X \subset A^+$ is comma-free if and only if it is (p, q)-limited for all p, q with $p + q = 3$, and if $A^+ X A^+ \cap X = \emptyset$. In particular, a comma-free code is circular.*

Proof. First suppose that X is comma-free. Let $u_0, u_1, u_2, u_3 \in A^*$ be such that $u_0 u_1, u_1 u_2, u_2 u_3 \in X^*$. If $u_1 = u_2 = 1$, then $u_0, u_3 \in X^*$. Otherwise $u_1 u_2 \in X^+$ and $u_0 u_1 u_2 u_3 \in X^+$. Thus by (7.12) $u_0, u_3 \in X^*$. Since X is prefix, $u_0, u_0 u_1 \in X^*$ implies that $u_1 \in X^*$, and X being suffix, $u_2 u_3, u_3 \in X$ implies that u_2 is in X^*. Thus $u_0, u_1, u_2, u_3 \in X^*$. Consequently, X is (p, q)-limited for all $p, q \geq 0$ with $p + q = 3$. Furthermore $A^+ X A^+ \cap X = \emptyset$. Indeed assume that $uxv, x \in X$. Then by (7.12) $u, v \in X^*$, whence $u = v = 1$.

Conversely, let $u, v \in A^*$ and $x \in X^+$ be such that $uxv \in X^*$. Since $A^+ x A^+ \cap X = \emptyset$, there exists a factorization $x = ps$, with $p, s \in A^*$, such that $up, sv \in X^*$. From $up, ps, sv \in X^*$ it follows, by the limitedness of X, that $u, p, s, r \in X^*$. Thus (7.12) holds. The last statement follows from Proposition 7.2.2. $\square$

Proposition 7.2.15 *Let X, Z be two composable codes and let $X = Y \circ Z$. If Y and Z are comma-free, then X is comma-free.*

Proof. Let $u, v \in A^*$ and $x \in X^+$ be such that $uxv \in X^*$. Since $X \subset Z^*$, we have $uxv \in Z^*$, $x \subset Z^+$. Since Z is comma-free, it follows that $u, v \in Z^*$. Since Y is comma-free, this implies that u, v are in X^*. Thus X is comma-free by (7.12). $\square$

Example 7.2.16 Let $A = \{a, b\}$ and $X = \{aab, bab\}$. The words aab and bab have a unique interpretation. This shows that X is comma-free.

7.3 Length distributions

We now study the length distributions of circular codes. Let X be a fixed circular code and let $(u_n)_{n \geq 1}$ be its length distribution. For each $n \geq 1$, let p_n be the number of words of length n which have a conjugate in X^*.

We set $u(z) = \sum_{n \geq 1} u_n z^n$ and $p(z) = \sum_{n \geq 1} p_n z^n$. Thus $u(z) = f_X(z)$ is the generating series of X.

Proposition 7.3.1 *The following relation holds between $u(z)$ and $p(z)$:*

$$\exp \sum_{n \geq 1} \frac{p_n}{n} z^n = \frac{1}{1 - u(z)}, \tag{7.13}$$

or equivalently

$$p(z) = \frac{zu'(z)}{1 - u(z)}, \tag{7.14}$$

where u' is the derivative of u.

Proof. We first assume that the code X is finite.

Let $\mathcal{A}$ be the flower automaton of X and let N be the adjacency matrix of the graph of $\mathcal{A}$, that is $N_{i,j}$ is the number of edges from i to j in $\mathcal{A}$. We have for each $n \geq 0$,

$$p_n = \text{Tr}(N^n).$$

Indeed, $\text{Tr}(N^n) = \sum N_{i,i}^n$ and $N_{i,i}^n$ is the number of paths of length n from i to i. In view of Proposition 7.1.5, each word w of length n which has a conjugate in X^* is the label of a unique closed path in $\mathcal{A}$. Conversely, each cycle contains the initial state, and thus its label has a conjugate in X^*. This shows the formula.

We now use Proposition 4.1.6. By assigning the same symbol z to all letters in Equation (4.2), the matrix M of (4.2) becomes Nz, and $\alpha(\underline{X})$ becomes $u(z)$. Thus

$$\det(I - Nz) = 1 - u(z).$$

Let $\lambda_1, \ldots, \lambda_k$ be the eigenvalues of the matrix N counted with their multiplicities. Then for each $n \geq 1$, $p_n = \text{Tr}(N^n) = \lambda_1^n + \cdots + \lambda_k^n$. Next, from elementary calculus, one has, for any complex number λ,

$$\exp \left(\sum_{n \geq 1} \frac{(\lambda z)^n}{n} \right) = \exp \left(\log \frac{1}{1 - \lambda z} \right) = \frac{1}{1 - \lambda z}.$$

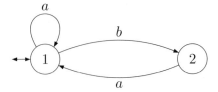

Figure 7.5 The flower automaton of the circular code $X = \{a, ba\}$.

Consequently

$$\exp \sum_{n\geq 1} \frac{p_n}{n} z^n = \exp \sum_{n\geq 1} \frac{\lambda_1^n + \cdots + \lambda_k^n}{n} z^n$$

$$= \exp \sum_{n\geq 1} \left(\frac{(\lambda_1 z)^n}{n} + \cdots + \frac{(\lambda_k z)^n}{n} \right)$$

$$= \frac{1}{1 - \lambda_1 z} \cdots \frac{1}{1 - \lambda_k z} = \frac{1}{\det(I - Nz)}.$$

This shows (7.13) for finite codes. In the general case, one considers, for each positive integer m, the set of words in X of length at most m. Since each p_n depends only on the first n terms of the sequence (u_n), (7.13) gives the relation up to m. Since this holds for each m, the formula is true also for infinite codes.

Formula (7.14) follows from (7.13) by logarithmic derivation, that is by taking the derivatives of the logarithms. Indeed, the equality $S = T$ of two series with constant term 1 is equivalent to the equality of their logarithmic derivatives. □

Example 7.3.2 Consider the circular code $X = \{a, ba\}$ on the alphabet $A = \{a, b\}$. We have $u(z) = z + z^2$ and thus by Formula 7.14

$$p(z) = \frac{z + 2z^2}{1 - z - z^2}.$$

The automaton $\mathcal{A}$ is represented on Figure 7.5. We have

$$M = \begin{bmatrix} 1 & 1 \\ 1 & 0 \end{bmatrix},$$

and thus $\det(I - Mz) = 1 - z - z^2$. The eigenvalues of M are the two roots $\varphi, \widehat{\varphi}$ of the polynomial $1 - z - z^2$ and $p_n = \varphi^n + \widehat{\varphi}^n$.

By Formula (7.14), we get $p(z) = zu'(z) + p(z)u(z)$, from which we obtain the following recurrence relation for p_n which is useful for numerical computations and which is known as *Newton's formula* (see the Notes):

$$p_n = nu_n + \sum_{i=1}^{n-1} p_i u_{n-i}. \tag{7.15}$$

There is also a closed formula for p_n. For each $i \geq 1$, let $u^{(i)} = (u_n^{(i)})_{n\geq 1}$ be the length distribution of X^i. Equivalently, $u_n^{(i)}$ is the coefficient of degree n of $u(z)^i$. Then

$$\sum_{n\geq 1} \frac{p_n}{n} z^n = \log \frac{1}{1 - u(z)} = \sum_{n\geq 1} \frac{u^{(i)}(z)}{i}.$$

Thus, for each $n \geq 1$, the explicit value of the numbers p_n in terms of the numbers $u_n^{(i)}$ is

$$p_n = \sum_{i=1}^{n} \frac{n}{i} u_n^{(i)}.$$

We now give a relation with primitive necklaces. Let ℓ_n be the number of primitive necklaces of length n which meet X^*. We start with a formula which is useful to compute the numbers ℓ_n.

Proposition 7.3.3 *For all $n \geq 1$,*

$$p_n = \sum_{d\mid n} d\ell_d. \tag{7.16}$$

Proof. Let u be a primitive word of length d which has a conjugate in X^*. Any power v of u has exactly d distinct conjugates and has a conjugate in X^*. Conversely, if v has a conjugate v' in X^*, let u be the unique primitive word such that v' is in u^+. Since X^* is pure, the word u is in X^*, and thus v itself is a power of a primitive word which has a conjugate in X^*. This shows the formula. $\square$

Using the Möbius inversion formula (Proposition 1.3.4), we obtain an explicit formula

$$\ell_n = \frac{1}{n} \sum_{d\mid n} \mu(n/d) p_d.$$

The following proposition establishes a direct relationship between the sequences (u_n) and (ℓ_n).

Proposition 7.3.4 *The following relation holds:*

$$\frac{1}{1 - u(z)} = \prod_{n\geq 1} \frac{1}{(1 - z^n)^{\ell_n}}. \tag{7.17}$$

Proof. Since, for each n,

$$\frac{p_n}{n} = \sum_{d\mid n} \frac{d\ell_d}{n},$$

we get

$$\sum_{n\geq 1} \frac{p_n}{n} z^n = \sum_{d,k\geq 1} \ell_d \frac{z^{dk}}{k} = \sum_{d\geq 1} \ell_d \log \frac{1}{1 - z^d} = \sum_{n\geq 1} \log \frac{1}{(1 - z^n)^{\ell_n}}.$$

Table 7.1 *The values of p_n and ℓ_n for $X = \{a, ab\}$.*

n	1	2	3	4	5	6	7
p_n	1	3	4	7	11	18	29
ℓ_n	1	1	1	1	2	2	4

Taking the exponential of both sides, we obtain

$$\exp \sum_{n \geq 1} \frac{p_n}{n} z^n = \prod_{n \geq 1} \frac{1}{(1 - z^n)^{\ell_n}}. \tag{7.18}$$

Putting together Formulas (7.13) and (7.18), we obtain Formula (7.17). $\qquad \square$

Given a series $u(z) = \sum u_n z^n$, Equation (7.14) defines directly the series $p(z)$, and Equation (7.16) allows us to compute the sequence (ℓ_n). These altogether are equivalent to Equation (7.17). To emphasize these dependencies, we write $\ell_n(u)$ and $p_n(u)$ for the sequences given by u.

In the special case of the series $u(z) = kz$, we write $\ell_n(k)$ instead of $\ell_n(u)$. This agrees with Chapter 1 where $\ell_n(k)$ denotes the number of primitive necklaces of length n on k symbols. It is clear that the sequence $(\ell_n(k))_{n \geq 1}$ corresponds to the code $X = A$ and in this case Identity (7.17) reads

$$\frac{1}{1 - kz} = \prod_{n \geq 1} \frac{1}{(1 - z^n)^{\ell_n(k)}}. \tag{7.19}$$

It can be shown that if $u_n \leq v_n$ for all n, then $\ell_n(u) \leq \ell_n(v)$ for all n (Exercise 7.3.4).

Example 7.3.5 Consider again the circular code $X = \{a, ab\}$ on the alphabet $A = \{a, b\}$. We have $u(z) = z + z^2$ and

$$p(z) = \frac{z + 2z^2}{1 - z - z^2}.$$

The first values of p_n and ℓ_n are given in Table 7.1.

We shall now characterize the length distributions of circular codes.

For this, we say that a finite or infinite sequence $(x_i)_{i \geq 1}$ of words in A^+ is a *Hall sequence* over A if it is obtained in the following way:

Let $X_1 = A$. Then x_1 is an arbitrary word in X_1. If x_i and X_i are defined, then the set X_{i+1} is defined by

$$X_{i+1} = x_i^*(X_i \setminus x_i),$$

and x_{i+1} is an arbitrary chosen element in X_{i+1} satisfying

$$|x_{i+1}| \geq |x_i|.$$

The sequence $(X_i)_{i \geq 1}$ is the sequence of codes *associated* with the sequence $(x_i)_{i \geq 1}$.

Proposition 7.3.6 *Let $(x_i)_{i \geq 1}$ be a Hall sequence over A and let $(X_i)_{i \geq 1}$ be the associated sequence of codes.*

1. *Each X_i, for $i \geq 1$, is a $(i-1, 0)$-limited code.*
2. *Each primitive word w such that $|w| > |x_i|$ has a conjugate in X_{i+1}^*.*

Proof. 1. $X_1 = A$ is $(0, 0)$-limited. Next

$$X_{i+1} = T \circ X_i,$$

where T is a code of the form $b^*(B \setminus b)$. Clearly T is $(1, 0)$-limited. Assuming by induction that X_i is $(i-1, 0)$-limited, the conclusion follows from Proposition 7.2.8.

2. Define $x_0 = 1$. We prove that the claim holds for all $i \geq 0$ by induction on i. For $i = 0$, the claim just states that any primitive word is in A^*. Thus assume $i \geq 1$, and let $w \in A^+$ be a primitive word of length $|w| > |x_i|$. Since $|x_i| \geq |x_{i-1}|$, one has $|w| > |x_{i-1}|$. By the induction hypothesis, there is a word w' conjugate of w which is in X_i^*. The word w' is not in x_i^* since w' is primitive and $|w'| > |x_i|$. Thus w' factorizes into $w' = uxv$ for some $u, v \in X_i^*$ and $x \in X_i \setminus x_i$. Then the conjugate $w'' = vux$ of w' is in $X_i^*(X_i \setminus x_i) \subset X_{i+1}^*$. Thus a conjugate of w is in X_{i+1}^*.   □

Theorem 7.3.7 *The sequence $u = (u_n)_{n \geq 1}$ is the length distribution of a circular code over k letters if and only if $\ell_n(u) \leq \ell_n(k)$, for all $n \geq 1$.*

Proof. Let A be an alphabet with k letters. Let X be a circular code with length distribution $u = (u_n)$. Since $\ell_n(u)$ is the number of primitive necklaces of length n which meet X^*, one has $\ell_n(u) \leq \ell_n(k)$.

For the converse, we build a Hall sequence. Arguing by induction on n, we suppose defined an integer $m = m(n)$ and a Hall sequence $x_1, \ldots, x_m$ of words of length at most n with the sequence $X_1, \ldots, X_m$ of associated codes and thus with $X_{i+1} = x_i^*(X_i \setminus x_i)$, such that the length distribution of X_m coincides with the sequence u on the n first terms. We set for convenience $Y_n = X_{m(n)}$. Thus, setting $v_i = \text{Card}(Y_n \cap A^i)$, one has $v_i = u_i$ for $1 \leq i \leq n$. We prove that

$$v_{n+1} - u_{n+1} = \ell_{n+1}(k) - \ell_{n+1}(u). \tag{7.20}$$

Take this equation for granted. Set $r = v_{n+1} - u_{n+1}$. Since $0 \leq r$ we may select r words $x_{m+1}, \ldots, x_{m+r}$ of length $n + 1$ in $Y_n = X_m$ to carry on the construction of the Hall sequence for r steps. In this way, the sequence $x_1, \ldots, x_m, x_{m+1}, \ldots, x_{m+r}$ forms altogether a Hall sequence. Setting $m(n + 1) = m + r$, the code $Y_{n+1} = X_{m(n+1)}$ satisfies $\text{Card}(Y_{n+1} \cap A^i) = u_i$ for $1 \leq i \leq n + 1$. This is clear for $i \leq n$. Next, $Y_{n+1} \cap A^{n+1}$ is obtained from $Y_n \cap A^{n+1}$ by removing r words of length $n + 1$. This finishes the induction, starting with $Y_0 = A$.

We now prove Equation (7.20). Since $u_i = v_i$ for $i = 1, \ldots, n$, one gets by Equation (7.15) that $p_i(u) = p_i(v)$ for $i = 1, \ldots, n$. Thus, again by Equation (7.15), one obtains that $p_{n+1}(v) - p_{n+1}(u) = (n + 1)(v_{n+1} - u_{n+1})$.

Table 7.2 *The list of componentwise maximal length distributions of binary circular codes of length at most 4.*

2	0	0	0	a, b
1	1	1	1	b, ab, a^2b, a^3b
1	1	0	2	b, ab, a^3b, a^2b^2
1	0	2	1	b, ab^2, a^2b, a^3b
1	0	1	2	b, a^2b, a^3b, ab^3
1	0	0	3	b, a^3b, ab^3, a^2b^2
0	1	2	3	$ab, a^2b, bab, a^3b, ba^2b, b^2ab$

Equation (7.16) and the equalities proved above show that $\ell_i(u) = \ell_i(v)$ for $i = 1, \ldots, n$. This implies $p_{n+1}(v) - p_{n+1}(u) = (n + 1)(\ell_{n+1}(v) - \ell_{n+1}(u))$ which in turn shows that $\ell_{n+1}(v) - \ell_{n+1}(u) = v_{n+1} - u_{n+1}$.

Since $|x_m| \leq n$, the property of Hall sequences stated in Proposition 7.3.6(2) shows that each primitive necklace of length $n + 1$ meets X_m^*. Thus $\ell_{n+1}(v) = \ell_{n+1}(k)$. This proves Equation (7.20). □

Example 7.3.8 Let $A = \{a, b\}$ and let $u = (0, 1, 1, 3, \ldots)$. The construction of the proof gives

$$X_1 = \{a, b\}$$
$$X_2 = \{b, ab, aab, aaab, \ldots\}$$
$$X_3 = \{ab, aab, abab, aaab, baab, bbab, \ldots\}$$
$$X_4 = \{ab, bab, aaab, baab, bbab, \ldots\}$$

corresponding to the Hall sequence $x_1 = a, x_2 = b, x_3 = aab$. One gets $Y_1 = X_3$ and $Y_2 = Y_3 = X_4$.

We have represented in Table 7.2 the componentwise maximal length distributions of binary circular codes of length at most 4. The list is presented in decreasing lexicographic order. The last column gives a circular code having the indicated distribution constructed using the method of the proof of Theorem 7.3.7.

Corollary 7.3.9 *Let A be an alphabet with $k \geq 1$ letters. For all $m \geq 1$, there exists a circular code $X \subset A^m$ such that $\mathrm{Card}(X) = \ell_m(k)$.*

Proof. Let $u = (u_n)_{n \geq 1}$ be the sequence with all terms zero except for u_m which is equal to $\ell_m(k)$. By (7.15) and (7.16), one has $\ell_n(u) = 0$ for $1 \leq n \leq m - 1$ and $\ell_m(u) = u_m$. Thus $\ell_n(u) \leq \ell_n(k)$ for $1 \leq n \leq m$. According to the proof of Theorem 7.3.7, this suffices to ensure the existence of a circular code X having u_m words of length m. Thus $X \cap A^m$ satisfies the claim. □

Corollary 7.3.9 can be formulated in the following way: It is possible to choose a system X of representatives of the primitive conjugacy classes of words of length m in such a manner that X is a circular code. The following example gives a more precise description of these codes for $m = 2$.

Example 7.3.10 Let X be a subset of $A^2 \setminus \{a^2 \mid a \in A\}$ and let θ be the relation over A defined by $a\theta b$ if and only if $ab \in X$. Then X is a circular code if and only if the reflexive and transitive closure θ^* of θ is an order relation.

Indeed, assume first that θ^* is not an order. Then

$$a_1 a_2, \ a_2 a_3, \ \ldots, \ a_{n-1} a_n, \ a_n a_1 \in X$$

for some $n \geq 1$, and $a_1, \ldots, a_n \in A$. If n is even, then setting $u = a_1$, $v = a_2 \cdots a_n$, one has $uv, vu \in X^*$ and $u \notin X^*$. If n is odd, then $(a_1 a_2 \cdots a_n)^2 \in X^*$ but not $a_1 a_2 \cdots a_n$. Thus X is not circular.

Assume conversely that θ^* is an order. Then A can be ordered in such a way that $A = \{a_1, a_2, \ldots, a_k\}$ and $a_i \theta a_j \implies i < j$. Then $X \subset \{a_i a_j \mid i < j\}$, and in view of Example 7.2.7, the set X is a circular code.

The codes $X \subset A^m$ in Corollary 7.3.9 are circular. The next theorem states that for m odd, X may even be chosen to be comma-free.

Theorem 7.3.11 *For any alphabet A with k letters and for any odd integer $m \geq 1$, there exists a comma-free code $X \subset A^m$ such that*

$$\mathrm{Card}(X) = \ell_m(k).$$

It follows from Example 7.3.10 that a circular code $X \subset A^2$ having $\ell_2(k) = k(k-1)/2$ elements has the form $X = \{a_i a_j \mid i < j\}$ for some numbering of the alphabet. For $k = 4$ and $A = \{a, b, c, d\}$, one gets the code $X = \{ab, ac, ad, bc, cd, bd\}$. It is not comma-free, since $abcd$ has the factorizations $(ab)(cd)$ and $a(bc)d$. Consequently, a result like Theorem 7.3.11 does not hold for even integers m.

To prove Theorem 7.3.11, we construct a Hall sequence $(x_i)_{i \geq 1}$ and the sequence $(X_i)_{i \geq 1}$ of associated codes by setting

$$X_1 = A, \quad X_{i+1} = x_i^*(X_i \setminus x_i), \quad (i \geq 1), \tag{7.21}$$

where x_i is an element of X_i of minimal odd length. By construction, $(x_i)_{i \geq 1}$ is indeed a Hall sequence. Set

$$U = \bigcup_{i \geq 1} X_i, \quad Y = U \cap (A^2)^*, \quad Z = U \cap A(A^2)^*.$$

Thus Y is the set of words of even length in U, and

$$Z = \{x_j \mid i \geq 1\}.$$

For any word $u \in U$, we define

$$v(u) = \min\{i \in \mathbb{N} \mid u \in X_i\} - 1,$$

$$\delta(u) = \sup\{i \in \mathbb{N} \mid u \in X_i\}.$$

Thus $v(u)$ denotes the last time before u appears in some X_i and $\delta(u)$ is the last time u appears in some X_i. Observe that $Y = \{u \in U \mid \delta(u) = +\infty\}$. Next, note that $\delta(x_i) = i$, and if $v(u) = q$ for some $u \in U \setminus A$, then $u \in X_{1+q}$ and $u \notin X_q$. Consequently $u = x_q v$ for some $v \in X_{q+1}$. Further, for all $u \in U$ and $n \geq 1$, we have

$$v(u) \leq n < \delta(u) \implies x_n u \in U. \tag{7.22}$$

We shall prove by a series of lemmas that, for any odd integer m, the code $Z \cap A^m$ satisfies the conclusion of Theorem 7.3.11.

Lemma 7.3.12 *For all odd integers m, we have* $\mathrm{Card}(Z \cap A^m) = \ell_m(k)$.

Proof. Let n be the smallest integer such that $|x_n| = m$. Let u be the length distribution of X_n. Then by construction of the Hall sequence (x_i), we have

$$Z \cap A^m = \{x_n, x_{n+1}, \ldots, x_{n+p}\}$$

for some integer p. Then $Z \cap A^m = X_n \cap A^m$, since for all $k \geq 1$, words in X_{n+k} which are not in X_n have length strictly greater than $|x_n|$. Thus $\mathrm{Card}(Z \cap A^m) = u_m$.

Next, by the definition of n, we have $m > |x_{n-1}|$. According to Proposition 7.3.6(2), each primitive word of length m has a conjugate in X_n^*. Thus $\ell_m(u) = \ell_m(k)$.

Let D be the set of odd integers d such that $1 \leq d \leq m - 2$. By construction of the Hall sequence, we have $u_d = 0$ for each d in D. We show by induction on d that $p_d(u) = 0$ for $d \in D$. It is true for $d = 1$ since $p_1 = u_1 = 0$. By Equation (7.15), we have $p_d = d u_d + \sum_{i=1}^{d-1} p_i u_{d-i}$. Each term of the right-hand side is zero since $u_d = 0$ and either $p_i = 0$ or $u_{d-i} = 0$ since i or $d - i$ is odd. Thus $p_d = 0$. Consequently, by Equation (7.16), we have $\ell_d(u) = 0$ for $d \in D$ and finally $p_m(u) = m u_m$ and $\ell_m(u) = u_m$.

We obtain in this way $\mathrm{Card}(Z \cap A^m) = \ell_m(k)$. $\qquad\square$

Lemma 7.3.13 *Each word $w \in A^*$ admits a unique factorization*

$$w = y z_1 z_2 \cdots z_n \tag{7.23}$$

with $y \in Y^*$, $z_i \in Z$, $n \geq 0$, and $\delta(z_1) \geq \delta(z_2) \geq \cdots \geq \delta(z_n)$.

Proof. First we show that for $n \geq 1$

$$X_n^* = X_{n+1}^* x_n^*.$$

Indeed, by definition $X_{n+1} = x_n^*(X_n \setminus x_n)$. The product of x_n^* with $X_n \setminus x_n$ is unambiguous since X_n is a code. Thus one has in terms of formal power series

$$\underline{X_{n+1}} = x_n^*(\underline{X_n} - x_n). \tag{7.24}$$

Consequently, $\underline{X_{n+1}} = \underline{x_n}^* \underline{X_n} - \underline{x_n}^+$ and $\underline{X_{n+1}} - 1 = \underline{x_n}^* \underline{X_n} - \underline{x_n}^* = \underline{x_n}^*(\underline{X_n} - 1)$. Formula (7.24) follows by inversion.

By successive substitutions in (7.24), starting with $A^* = X_1^*$, one gets for all $n \geq 1$

$$\underline{A}^* = \underline{X_{n+1}}^* x_n^* x_{n-1}^* \cdots x_1^*. \tag{7.25}$$

Now let $w \in A^*$ and set $p = |w|$. Let n be an integer such that X_{n+1} contains no word of odd length $\leq p$. By (7.25) there exists a factorization of w as

$$w = y z_1 z_2 \cdots z_k$$

with $\delta(z_1) \geq \delta(z_2) \geq \cdots \geq \delta(z_k)$, $z_i \in Z$ and $y \in X_{n+1}^*$. Since $|y| \leq p$, the choice of n implies that y is a product of words in X_{n+1} of even length. Consequently $y \in Y^*$. This proves the existence of one factorization (7.23). Assume that there is second factorization of the same type, say,

$$w = y' z_1' z_2' \cdots z_n'.$$

Let m be an integer greater than $\delta(z_1)$ and $\delta(z_1')$, and large enough to ensure $y, y' \in X_{m+1}^*$. Such a choice is possible since all even words of some code X_ℓ are also in the codes $X_{\ell'}$, for $\ell' \geq \ell$. Then according to (7.25), both factorizations of w are the same. $\qquad\square$

Now, we characterize successively the form of the factorization (7.23), for words which are prefixes and for words which are suffixes of words in U.

Lemma 7.3.14 *Each proper prefix w of a word in U admits a factorization (7.23) with $y = 1$.*

Proof. Each of the codes X_n is a maximal prefix code. This follows by iterated application of Proposition 3.4.13. Consequently for $n \geq 0$,

$$\underline{A}^* = \underline{X_{n+1}^* P_{n+1}}$$

where $P_{n+1} = X_{n+1} A^-$ is the set of proper prefixes of words of X_{n+1}. Comparing this equation with (7.25), we get

$$P_{n+1} = x_n^* x_{n-1}^* \cdots x_1^*. \tag{7.26}$$

Let now w be a proper prefix of some word u in U. Then $u \in X_{n+1}$ for some $n \geq 0$ and consequently $w \in P_{n+1}$. By Equation (7.26), w admits a factorization of the desired form. $\qquad\square$

Lemma 7.3.15 *For all n, $p \geq 1$, we have $x_n x_{n+p} \in Y^*$. Further for $z \in Z$ and $y \in Y$, we have $zy \in Y^* Z$.*

Proof. The first formula is shown by induction on p. For $p = 1$, we have $\nu(x_{n+1}) \leq n$ since $x_{n+1} \in X_{n+1}$. Thus according to Formula (7.22), we have $x_n x_{n+1} \in U$. Since $x_n x_{n+1}$ has even length, $x_n x_{n+1} \in Y$.

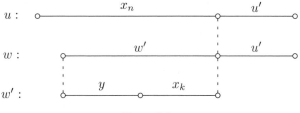

Figure 7.6

Assume that the property holds up to $p - 1$, and set $q = v(x_{n+p})$. We distinguish two cases. First assume $q \leq n$. Then by (7.22), with x_{n+p} playing the role of u, we have $x_n x_{n+p} \in U$. This word has even length. Thus $x_n x_{n+p} \in Y$.

Next suppose that $n \leq q$. Then $x_{n+p} \in U \setminus A$. Consequently $x_{n+p} = x_q u$ for some $u \in U$. Since $q \leq n + p = \delta(x_{n+p})$, we have $x_n x_q \in Y^*$ by the induction hypothesis. Next u has even length (because $|x_n|$, $|x_q|$ are both odd). Thus $u \in Y$, whence $x_n x_{n+p} \in Y^*$.

Let us prove the second formula. Set $n = \delta(Z)$ and $q = v(y)$. Then $z = x_n$ and $y = x_q x_t$ for some t. If $n \leq q$, then $x_n x_q \in Y^*$ by the preceding argument, and consequently $zy \in Y^* Z$. On the contrary, assume $q \leq n$. Then by (7.22) $x_n x_q x_t = x_n y \in U$. Since it has odd length, this word is in Z. □

Lemma 7.3.16 *Any suffix w of a word in U admits a factorization (7.23) with $n = 0$ or $n = 1$.*

Proof. Given a word $u \in U$, we prove that all its suffixes are in $Y^* Z \cup Y^*$, by induction on $|u|$. The case $|u| = 1$ is obvious, and clearly it suffices to prove the claim for proper suffixes of words in U.

Assume $|u| \geq 2$. Set $n = v(u)$. Since $u \in U \setminus A$, we have $u = x_n u'$ for some $u' \in U$.

Let w be a proper right factor of u. If w is a suffix of u', then by the induction hypothesis, w is in $Y^* Z \cup Y^*$. Thus we assume that $w = w' u'$, with w' a proper suffix of x_n. By induction, w' is in $Y^* Z \cup Y^*$. If $w' \in Y^*$, then $w' u' \in Y^*(Y \cup Z)$ and the claim is proved. Thus it remains the case where $w' \in Y^* Z$. In this case, set $w' = y x_k$ with $y \in Y^*$, $k \geq 1$. Observe that $k \leq n$ since $|x_k| \leq |w'| \leq |x_n|$ (see Figure 7.6).

We now distinguish two cases: First, assume $u' \in Y$. Then by Lemma 7.3.15, $x_k u' \in Y^* Z$. Consequently, $w = y x_k u' \in Y^* Z$. Second, suppose that $u' \in Z$. Then $u' = x_m$ for some m. We have $x_m \in X_{n+1}$, implying that $m > n$. Since $k \leq n$, we have $k \leq m$ and by Lemma 7.3.15, $x_k x_m \in Y^*$. Thus $w = y x_k x_m \in Y^*$. This concludes the proof. □

Proof of Theorem 7.3.11. Let m be an odd integer and let $X = Z \cap A^m$. Let $x, x', x'' \in X$. Assume that for some $u, v \in A^+$,

$$xx' = ux''v. \tag{7.27}$$

Then for some $w, t \in A^+$, we have $x = uw$, $x'' = wt$, $x' = tv$. Since x'' has odd length, one of the words w or t must have even length. Assume that the length of w is

Table 7.3 *A sequence satisfying the conditions of the construction.*

X_1	a, b				
X_2	b	ab	a^2b	a^3b	a^4b
X_3		ab	a^2b bab b^2ab	a^3b ba^2b b^2ab b^3ab	a^4b ba^3b b^2a^2b b^3ab
X_4		ab	bab	a^3b ba^2b b^2ab	a^4b ba^3b b^2a^2b b^3ab a^2bab
X_5		ab		a^3b ba^2b b^2ab	a^4b ba^3b b^2a^2b b^3ab a^2bab $babab$

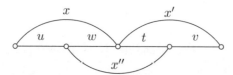

Figure 7.7 The case where w has even length.

even (see Figure 7.7). Since w is a proper prefix of $x'' \in Z$, we have by Lemma 7.3.14, a factorization $w = z_1z_2 \cdots z_n$ with $z_1, z_2, \ldots, z_n \in Z$ and $\delta(z_1) \geq \cdots \geq \delta(z_n)$ On the other hand, the word w is a suffix of $x \in Z$, and according to Lemma 7.3.16, we have $w \in Y^*Z \cup Y^*$. Since w has even length, $w \in Y^*$. Thus $n = 0$ and $w = 1$, showing that $u = x$, $x' = x''$, and $v = 1$. $\square$

Example 7.3.17 Let $A = \{a, b\}$. A sequence $(x_n)_{n \geq 1}$ satisfying the conditions of the construction given above is given in Table 7.3. We have represented only words of length at most five. Words of the same length are written in a column. Taking the words of length five in X_5, we obtain all words of length five in the code Z. Thus the following is a comma-free code $X \subset A^5$:

$$X = \{a^4b, ba^3b, b^2a^2b, b^3ab, a^2bab, babab\}.$$

It has $\mathrm{Card}(X) = \ell_2(5) = 6$ elements. The words of length three in X_3 give the comma-free code of Example 7.2.16.

7.4 Exercises

Section 7.1

7.1.1 Show that the submonoid $\{ab, ba\}^*$ is pure.

7.1.2 (Fine–Wilf theorem) Show that if two powers of words x and y have a common prefix of length $|x| + |y| - \gcd(|x|, |y|)$, then x and y are powers of a word z.

Section 7.2

7.2.1 A finite monoid is called *aperiodic* if it contains no nontrivial group. Let $X \subset A^+$ be a finite code and let $\mathcal{A} = (Q, 1, 1)$ be an unambiguous trim automaton recognizing X^*. Let φ be the associated representation. Show that X^* is pure if and only if the monoid $\varphi(A^*)$ is aperiodic.

7.2.2 A set $X \subset A^+$ is called (p, q)-*constrained* for some $p, q \geq 0$ if for each sequence $u_0, u_1, \ldots, u_{p+q}$ of words the condition $u_{i-1} u_i \in X$ for $1 \leq i \leq p + q$ implies $u_p \in X^*$.

(a) Show that, for $p + q \leq 2$, a set X is (p, q)-constrained if and only if it is (p, q)-limited.
(b) Let $A = \{a, b\}$ and $X = \{a, ab\}$. Show that X is $(3, 0)$-constrained but not $(3, 0)$-limited.

7.2.3 Show that a recognizable code is limited if and only if it is circular. (*Hint*: For a recognizable circular code X, let $\varphi : A^* \to M$ be the morphism on the syntactic monoid of X^*. Prove that X is (p, p)-limited for $p = \mathrm{Card}(M) + 1$.)

Section 7.3

7.3.1 Let A be a k letter alphabet and let $s \in A^+$ be a word of length p. Let R be the finite set

$$R = \{w \in A^* \mid sw \in A^*s, |w| < p\}.$$

Let X be the semaphore code $X = A^*s \setminus A^*sA^+$. Using Proposition 3.7.17, show that the generating series of X is

$$f_X(t) = \frac{t^p}{t^p + (1 - kt)f_R(t)}.$$

Now let $Z = (sA^+ \cap A^+s) \setminus A^+sA^+$. Show that $s + \underline{AX} = \underline{X} + \underline{Z}$. Let $U = Zs^{-1}$. Show that for all $n \geq p$, the code $U \cap A^n$ is comma-free and that the generating series of U is

$$f_U(t) = \frac{(kt - 1)}{t^p + (1 - kt)f_R(t)} + 1.$$

7.3.2 Show that for any sequence $(u_n)_{n \geq 1}$ of nonnegative integers, the sequence p_n defined by Formula (7.13) is formed of nonnegative integers.

7.3.3 Let $(u_n)_{n \geq 1}$ be a sequence of nonnegative integers. Let A be a *weighted alphabet* with u_n letters of weight n for each $n \geq 1$. The weight of a word is the sum of the weights of its letters. Show that $\ell_n(u)$ is the number of primitive necklaces on the alphabet A with weight n.

7.3.4 Let $(u_n)_{n \geq 1}$ and $(v_n)_{n \geq 1}$ be two sequences of integers such that $0 \leq u_n \leq v_n$ for each $n \geq 1$. Show that $\ell_n(u) \leq \ell_n(v)$ for all $n \geq 0$. (*Hint*: Use Exercise 7.3.3.)

7.3.5 For any sequence $(v_n)_{n \geq 1}$ of complex numbers, define the sequence (p_n) by

$$p_n = \sum_{d \mid n} d v_d^{n/d}.$$

Show that, in terms of generating series, one has

$$\exp \sum_{n \geq 1} \frac{p_n}{n} z^n = \prod_{n \geq 1} (1 - v_n z^n)^{-1}.$$

7.5 Notes

The definition of limited codes is from Schützenberger (1965c), where limited codes are defined by a condition denoted $\mathcal{U}_s(p, q)$ for $p \leq 0 \leq q$ which is our condition $C(-p, q)$. Theorem 7.1.10 is from de Luca and Restivo (1980). See also Lassez (1976) where the term "circular code" appears for the first time.

There is a close connection between the formulas concerning the length distributions of circular codes and symmetric functions. Actually, for a finite code, the numbers u_n are, up to the sign, the elementary symmetric functions of the roots of the polynomial $1 - u(z)$ and the p_n are the sums of powers. Formula (7.13) is well known in this context and Formula (7.15) is known as *Newton's formula* (see for instance Macdonald (1995)). Proposition 7.3.1 appears also in Stanley (1997).

The left side of Formula (7.13) is often called a *zeta function*. In the context of symbolic dynamics, the zeta function of a subshift S is defined as

$$\zeta_S(z) = \exp \sum_{n \geq 1} \frac{p_n}{n} z^n,$$

where p_n is the number of points of period n (see Lind and Marcus (1995)). This corresponds to our hypotheses, considering the subshift formed of all infinite words having a factorization in words of X. In this context, Formula (7.13) is a particular case of a result of Manning (1971) which is the following. Let S be the subshift formed of all two-sided infinite paths in a graph G. Let M be the adjacency matrix of G. Then

$$\zeta_S(z) = \frac{1}{\det(I - Mz)}.$$

The numbers $\ell_n(k)$ are called the *Witt numbers* and Identity (7.19) is called the *cyclotomic identity*. Other results on zeta functions and circular codes are given in

Keller (1991). The book by Stanley (1997) contains applications of these notions to enumerative combinatorics.

Theorem 7.3.7 is due to Schützenberger (1965c). The proof uses a method known in the context of free Lie algebras as *Lazard elimination method*.

The pair (v, p) defined as in Exercise 7.3.5 is called a *Witt vector* (see Lang (1965) or Metropolis and Rota (1983)). The link between Witt vectors and codes and the construction given in Exercise 7.3.5 is due to Luque and Thibon (2007).

The story of comma-free codes is interesting. They were introduced in Golomb *et al.* (1958). Some people thought at that time that the biological code is comma-free (Crick's hypothesis). The number of amino acids appearing in proteins is 20. They are coded by words of length three over the alphabet of bases A, C, G, U. Now, the number $\ell_3(4)$ which is the maximum number of elements in a comma-free (or circular) code composed of words of length three over a four-letter alphabet is precisely 20. Unfortunately for mathematics, it appeared several years later with the work of Niernberg that the biological code is not even a code in the sense of this book. Several triples of bases may encode the same acid (see Stryer (1975) or Lewin (1994)). This disappointment does not weaken the interest of circular codes, we believe.

Theorem 7.3.11 has been conjectured by Golomb *et al.* (1958) and proved by Eastman (1965). Another construction has been given by Scholtz (1969), on which the proof given here is based. Other constructions which are possible are described in Devitt and Jackson (1981). For even length, no formula is known giving the maximal number of elements of a comma-free code (see Jiggs (1963)). Further studies on comma-free codes include the corresponding completion problem, solved by Lam (2003), and the study of a more general family, called *solid* codes, by Shyr and Yu (1990), Lam (2001).

Exercise 7.1.2 is due to Fine and Wilf (see Lothaire (1997)). Exercise 7.2.1 is from Restivo (1974) (see also Hashiguchi and Honda (1976b)). These statements have a natural place within the framework of the theory of varieties of monoids (see Eilenberg (1976) or Pin (1986)).

Exercise 7.3.1 is from Guibas and Odlyzko (1978). The codes introduced in this exercise were defined by Gilbert (1960) and named *prefix-synchronized*. Gilbert has conjectured that $U \cap A^n$ has maximal size when the word s is chosen unbordered and of length $\log_k n$. This conjecture has been settled by Guibas and Odlyzko (1978). It holds for $k = 2, 3, 4$, but is false for $k \geq 5$.

8

Factorizations of free monoids

This chapter investigates in a systematic way the notion of factorization of free monoids already seen in particular cases in Chapter 7. The main result of Section 8.1 (Theorem 8.1.2) characterizes factorizations of free monoids. It shows in particular that the codes which appear in these factorizations are circular. The proof is based on an enumeration technique. For this, we define the logarithm in a ring of formal power series in noncommutative variables. The properties necessary for the proof are derived. We illustrate the factorization theorem by considering a very general family of factorizations obtained from sets called Lazard sets.

Section 8.2 is devoted to the study of factorizations into finitely many submonoids. We first consider factorizations into two submonoids called bisections. The main result (Theorem 8.2.4) gives a method to construct all bisections. We then study trisections, that is factorizations into three submonoids. We prove a difficult result (Theorem 8.2.6) showing that every trisection can be constructed by "pasting" together factorizations into four factors obtained by successive bisections.

8.1 Factorizations

Several times in the previous chapter, we have used special cases of the notion of factorization which will be defined here. We shall see in this section that these factorizations are closely related to circular codes. Let I be a totally ordered set and let $(X_i)_{i \in I}$ be a family of subsets of A^+ indexed by I. An *ordered factorization* of a word $w \in A^*$ is a factorization

$$w = x_1 x_2 \cdots x_n \qquad (8.1)$$

with $n \geq 0$, $x_i \in X_{j_i}$ such that $j_1 \geq j_2 \geq \cdots \geq j_n$.

A family $(X_i)_{i \in I}$ is a *factorization of the free monoid* A^* if each word $w \in A^*$ has exactly one ordered factorization.

If $(X_i)_{i \in I}$ is a factorization, then each X_i is a code, since otherwise the unique factorization would not hold for words in X_i^*. We shall see later (Theorem 8.1.2) that each X_i is in fact a circular code.

Let us give a formulation in terms of formal power series. Consider a family $(\sigma_i)_{i \in I}$ of formal power series over an alphabet A with coefficients in a semiring K, indexed by a totally ordered set I. Assume furthermore that the family $(\sigma_i)_{i \in I}$ is locally finite. Let $J = \{j_1, j_2, \ldots, j_n\}$ be a finite subset of I, with $j_1 \geq j_2 \geq \cdots \geq j_n$. Set

$$\tau_J = \sigma_{j_1} \sigma_{j_2} \cdots \sigma_{j_n}.$$

Then for all $w \in A^*$,

$$(\tau_J, w) = \sum_{x_1 x_2 \cdots x_n = w} (\sigma_{j_1}, x_1)(\sigma_{j_2}, x_2) \cdots (\sigma_{j_n}, x_n). \tag{8.2}$$

Let $\mathcal{S}$ be the set of all finite subsets of I. Then the family $(\tau_J)_{J \in \mathcal{S}}$ is locally finite. Indeed, for each word $w \in A^*$, the set $F(w)$ of factors of w is finite. For each $x \in F(w)$, the set I_x of indices $i \in I$ such that $(\sigma_i, x) \neq 0$ is finite. From (8.2), it follows that if $(\tau_J, w) \neq 0$, then $J \subset \bigcup_{x \in F(w)} I_x$. Consequently there are only finitely many sets J such that $(\tau_J, w) \neq 0$. These considerations allow us to define the product

$$\sigma = \prod_{i \in I} (1 + \sigma_i)$$

by the formula

$$\sigma = \sum_{J \in \mathcal{S}} \tau_J.$$

If I is finite, we obtain the usual notion of a product of a sequence of formal power series, and the latter expression is just the expanded form obtained by distributivity.

Consider a family $(X_i)_{i \in I}$ of subsets of A^+ indexed by a totally ordered set I. If the family is a factorization of A^*, then

$$\underline{A^*} = \prod_{i \in I} \underline{X_i^*}. \tag{8.3}$$

Conversely, if the sets X_i are codes and if the semigroups X_i^+ are pairwise disjoint, then the product $\prod_{i \in I} \underline{X_i^*}$ is defined and (8.3) implies that the family $(X_i)_{i \in I}$ is a factorization of A^*.

Example 8.1.1 Formula (7.25) states that the family $(X_{n+1}, x_n, \ldots, x_1)$ is a factorization of A^* for all $n \geq 1$. Lemma 7.3.13 says that the family of sets $(Y, \ldots, x_n, x_{n-1}, \ldots, x_1)$ is a factorization of A^*.

The main result of this section is the following theorem.

Theorem 8.1.2 (Schützenberger) *Let $(X_i)_{i \in I}$ be a family of subsets of A^+ indexed by a totally ordered set I. Two of the three following conditions imply the third.*

(i) *Each word $w \in A^*$ has at least one ordered factorization.*
(ii) *Each word $w \in A^*$ has at most one ordered factorization.*
(iii) *Each of the X_i ($i \in I$) is a circular code and each conjugacy class of nonempty words meets exactly one among the submonoids X_i^*.*

The proof is based on an enumeration technique. Before giving the proof, we need some results concerning the logarithm of a formal power series in commuting or noncommuting variables. For this, we shall consider a slightly more general situation, namely, the formal power series defined over monoids which are direct products of a finite number of free monoids. Let M be a monoid which is a direct product of finitely many free monoids. The set

$$S = \mathbb{Q}^M$$

of functions from M into the field $\mathbb{Q}$ of rational numbers is equipped with the structure of a semiring as was done for formal series over a free monoid. In particular if $\sigma, \tau \in S$, the product $\sigma\tau$ given by

$$(\sigma\tau, m) = \sum_{uv=m} (\sigma, u)(\tau, v)$$

is well defined since the set of pairs (u, v) with $uv = m$ is finite. As in the case of formal power series over a free monoid, a family $(\sigma_i)_{i \in I}$ of elements of S is locally finite if for all $m \in M$, the set $\{i \in I \mid (\sigma_i, m) \neq 0\}$ is finite. Define

$$S^{(1)} = \{\sigma \in S \mid (\sigma, 1) = 0\}.$$

For $\sigma \in S^{(1)}$, the family $(\sigma^n)_{n \geq 0}$ of powers of σ is locally finite. Indeed, for each $m \in M$, $(\sigma^n, m) = 0$ for all n greater than the sum of the lengths of the components of m. This allows us to define for all $\sigma \in S^{(1)}$,

$$\log(1 + \sigma) = \sigma - \sigma^2/2 + \sigma^3/3 - \cdots + (-1)^{n+1}\sigma^n/n + \cdots \tag{8.4}$$

$$\exp(\sigma) = 1 + \sigma + \frac{\sigma^2}{2!} + \cdots + \frac{\sigma^n}{n!} + \cdots \tag{8.5}$$

Let M and N be monoids which are finite direct products of free monoids. Let $S = \mathbb{Q}^M$ and $T = \mathbb{Q}^N$. A morphism

$$\gamma : M \to T$$

from the monoid M into the multiplicative monoid T is called *continuous* if and only if the family $(\gamma(m))_{m \in M}$ is locally finite. In this case, the morphism γ can be extended into a morphism, still denoted by γ, from the algebra S into the algebra T by the formula

$$\gamma(\sigma) = \sum_{m \in M} (\sigma, m)\gamma(m). \tag{8.6}$$

This sum is well defined since the family $(\gamma(m))_{m \in M}$ is locally finite. The extended morphism γ is also called a continuous morphism from S into T. For any locally finite family $(\sigma_i)_{i \in I}$ of elements of S, the family $\gamma(\sigma_i)_{i \in I}$ is also locally finite and

$$\sum_{i \in I} \gamma(\sigma_i) = \gamma\left(\sum_{i \in I} \sigma_i\right). \tag{8.7}$$

According to Formula (8.7), a continuous morphism $\gamma : S \to T$ is entirely determined by its definition on M, thus on a set X of generators for M. Furthermore, γ is continuous if and only if $\gamma(X \setminus \{1\}) \subset T^{(1)}$ and the family $(\gamma(x))_{x \in X}$ is locally finite. This is due to the fact that each $m \in M$ has only finitely many factorizations $m = x_1 x_2 \cdots x_k$ with $x_1, x_2, \ldots x_k \in X \setminus 1$. It follows from (8.6) that if $\sigma \in S^{(1)}$, then $\gamma(\sigma) \in T^{(1)}$. From (8.7), we obtain

$$\log(1 + \gamma(\sigma)) = \gamma(\log(1 + \sigma)), \tag{8.8}$$

$$\exp(\gamma(\sigma)) = \gamma(\exp(\sigma)). \tag{8.9}$$

According to classical results from elementary analysis, we have the following formulas in the algebra $\mathbb{Q}[[s]]$ of formal power series in the variable s:

$$\exp(\log(1 + s)) = 1 + s, \quad \log(\exp(s)) = s. \tag{8.10}$$

Furthermore, in the algebra $\mathbb{Q}[[s, t]]$ of formal power series in two commuting variables s, t, we have

$$\exp(s + t) = \exp(s)\exp(t), \quad \log((1 + s)(1 + t)) = \log(1 + s) + \log(1 + t). \tag{8.11}$$

Let M be a monoid which is a finite direct product of free monoids and let $S = \mathbb{Q}^M$. Let $\sigma \in S^{(1)}$ and let γ be the continuous morphism from the algebra $\mathbb{Q}[[s]]$ into S defined by $\gamma(s) = \sigma$. Then by formulas (8.8)–(8.10), we have

$$\exp(\log(1 + \sigma)) = 1 + \sigma, \quad \log(\exp(\sigma)) = \sigma \tag{8.12}$$

showing that exp and log are inverse bijections of each other from the set S onto the set

$$1 + S^{(1)} = \{1 + r \mid r \in S^{(1)}\}.$$

Now consider two series $\sigma, \tau \in S^{(1)}$ which commute, that is, such that $\sigma\tau = \tau\sigma$. Since the submonoid of S generated by σ and τ is commutative, the function γ from $s^* \times t^*$ into S defined by $\gamma(s^p t^q) = \sigma^p \tau^q$ is a continuous morphism from $\mathbb{Q}[[s, t]]$ into S and by (8.11),

$$\exp(\sigma + \tau) = \exp(\sigma)\exp(\tau),$$

$$\log((1 + \sigma)(1 + \tau)) = \log(1 + \sigma) + \log(1 + \tau). \tag{8.13}$$

These formulas do not hold when σ and τ do not commute. We shall give a property of the difference of the two sides of (8.13) in the general case. A series $\sigma \in \mathbb{Q}\langle\langle A \rangle\rangle$ is called *cyclically null* if for each conjugacy class $C \subset A^*$ one has

$$(\sigma, \underline{C}) = \sum_{w \in C}(\sigma, w) = 0.$$

Clearly any sum of cyclically null series still is cyclically null.

Proposition 8.1.3 *Let A be an alphabet and let* $S = \mathbb{Q}\langle\langle A \rangle\rangle$. *Let* $\gamma : S \to S$ *be a continuous morphism. For each cyclically null series* $\sigma \in S$, *the series* $\gamma(\sigma)$ *is cyclically null.*

Proof. Let $T \subset A^*$ be a set of representatives of the conjugacy classes of A^*. Denote by $C(t)$ the conjugacy class of $t \in T$. Let

$$\tau = \sum_{t \in T} \left(\sum_{w \in C(t)} (\sigma, w)(w - t) \right).$$

The family of polynomials $(\sum_{w \in C(t)}(\sigma, w)(w - t))_{t \in T}$ is locally finite. Thus the sum is well defined. Next

$$\tau = \sum_{t \in T} \sum_{w \in C(t)} (\sigma, w)w - \sum_{t \in T} \sum_{w \in C(t)} (\sigma, w)t = \sigma - \sum_{t \in T} (\sigma, \underline{C}(t))t.$$

Since σ is cyclically null, the second series vanishes and consequently $\tau = \sigma$. It follows that

$$\gamma(\sigma) = \sum_{t \in T} \left(\sum_{w \in C(t)} (\sigma, w)(\gamma(w) - \gamma(t)) \right).$$

In order to prove the claim, it suffices to show that each series $\gamma(w) - \gamma(t)$ for $w \in C(t)$ is cyclically null. For this, consider $w \in C(t)$. Then $t = uv$, $w = vu$ for some $u, v \in A^*$. Setting $\mu = \gamma(u)$, $v = \gamma(v)$, one has $\gamma(w) - \gamma(t) = v\mu - \mu v$. Next

$$v\mu = \sum_{x,y \in A^*} (v, x)(\mu, y)xy.$$

Thus

$$v\mu - \mu v = \sum_{x,y \in A^+} (v, x)(\mu, y)(xy - yx).$$

Since each polynomial $xy - yx$ clearly is cyclically null, the series $v\mu - \mu v$ and hence $\gamma(\sigma)$ is cyclically null. □

Proposition 8.1.4 *Let* $A = \{a, b\}$, *and let C be a conjugacy class of* A^*. *Then*

$$(\log((1 + a)(1 + b)), \underline{C}) = (\log(1 + a), \underline{C}) + (\log(1 + b), \underline{C}). \tag{8.14}$$

In other words, the series $\log((1 + a)(1 + b)) - \log(1 + a) - \log(1 + b)$ *is cyclically null.*

Proof. One has $(1 + a)(1 + b) = 1 + a + b + ab$ and

$$\log((1 + a)(1 + b)) = \sum_{m \geq 1} \frac{(-1)^{(m+1)}}{m}(a + b + ab)^m.$$

Let $w \in A^n$, and let d be the number of times ab occurs as a factor in w. Let us verify that

$$((a + b + ab)^m, w) = \binom{d}{n - m}. \tag{8.15}$$

Indeed, $((a + b + ab)^m, w)$ is the number of factorizations $w = x_1 x_2 \cdots x_m$ of w in m words, with $x_i \in \{a, b, ab\}$. Since w has length n and the x_i's have length 1 or 2, there are exactly $n - m$ x_i's which are equal to ab. Each factorization of w thus corresponds to a choice of $n - m$ factors of w equal to ab among the d occurrences of ab. Thus there are exactly $\binom{d}{n-m}$ factorizations. This proves (8.15).

Now let C be a conjugacy class, let n be the length of the words in C and let p be their exponent. Then $\mathrm{Card}(C) = n/p$. If $C \subset a^*$, then $C = \{a^n\}$. Then Formula (8.15) shows that $((a + b + ab)^m, a^n)$ equals 1 or 0 according to $n = m$ or not. Thus both sides of (8.14) in this case are equal to $(-1)^n/n$. The same holds if $C \subset b^*$. Thus we may assume that C is not contained in $a^* \cup b^*$. Then the right-hand side of (8.14) equals 0. Consider the left-hand side. Since each word in C contains at least one a, there is a word w in C whose first letter is a. Let d be the number of occurrences of ab as a factor in w. Among the n/p conjugates of w, there are d/p which start with the letter b and end with the letter a. Indeed, set $w = v^p$. Then the word v has d/p occurrences of the factor ab. Each of the d/p conjugates of w in bA^*a is obtained by "cutting" v in the middle of one occurrence of ab. Each of these d/p conjugates has only $d - 1$ occurrences of ab as a factor. The $(n - d)/p$ other conjugates of w have all d occurrences of the factor ab. According to Formula (8.15), we have for each conjugate u of w,

$$((a + b + ab)^m, u) = \begin{cases} \binom{d - 1}{n - m} & \text{if } u \in bA^*a, \\[2mm] \binom{d}{n - m} & \text{otherwise.} \end{cases}$$

Summation over the elements of C gives

$$((a + b + ab)^m, \underline{C}) = \frac{d}{p} \binom{d - 1}{n - m} + \frac{n - d}{p} \binom{d}{n - m}.$$

Since $\binom{d-1}{n-m} = \frac{d-n+m}{d} \binom{d}{n-m}$, we obtain $((a + b + ab)^m, \underline{C}) = (m/p)\binom{d}{n-m}$. Consequently

$$(\log(1 + a)(1 + b), \underline{C}) = \frac{1}{p} \sum_{m \geq 1} (-1)^{m+1} \binom{d}{n - m}. \tag{8.16}$$

Since $n > d$ and $d \neq 0$, this alternating sum of binomial coefficients equals 0. □

The following proposition is an extension of Proposition 8.1.4.

Proposition 8.1.5 *Let $(\sigma_i)_{i \in I}$ be a locally finite family of elements of $\mathbb{Q}\langle\langle A \rangle\rangle$ indexed by a totally ordered set I, such that $(\sigma_i, 1) = 0$ for all $i \in I$. The series*

$$\log\left(\prod_{i \in I}(1 + \sigma_i)\right) - \sum_{i \in I}\log(1 + \sigma_i) \tag{8.17}$$

is cyclically null.

Proof. Set $S = \mathbb{Q}\langle\langle A \rangle\rangle$, and $S^{(1)} = \{\sigma \in S \mid (\sigma, 1) = 0\}$. Let $\sigma, \tau \in S^{(1)}$. The series

$$\delta = \log((1 + \sigma)(1 + \tau)) - \log(1 + \sigma) - \log(1 + \tau)$$

is cyclically null. Indeed, either σ and τ commute and δ is null by (8.13), or the alphabet A has at least two letters a, b. Consider a continuous morphism γ such that $\gamma(a) = \sigma$, $\gamma(b) = \tau$. The series

$$d = \log((1 + a)(1 + b)) - \log(1 + a) - \log(1 + b)$$

is cyclically null by Proposition 8.1.4. Since $\delta = \gamma(d)$, Proposition 8.1.3 shows that δ is cyclically null. Now let $\tau_1, \tau_2, \ldots, \tau_n \in 1 + S^{(1)}$. Arguing by induction, assume that

$$\epsilon = \log(\tau_n \cdots \tau_2) - \sum_{i=2}^{n} \log \tau_i$$

is cyclically null. In view of the preceding discussion, the series

$$\epsilon' = \log(\tau_n \cdots \tau_2 \tau_1) - \log(\tau_n \cdots \tau_2) - \log \tau_1$$

is cyclically null. Consequently

$$\epsilon + \epsilon' = \log(\tau_n \cdots \tau_1) - \sum_{i=1}^{n} \log \tau_i$$

is cyclically null. This proves (8.17) for finite sets I. For the general case, we consider a fixed conjugacy class C. Let n be the length of words in C and let $B = \text{alph}(C)$. Then B is finite and $C \subset B^n$. Define an equivalence relation on S by $\sigma \sim \tau$ if and only if $(\sigma, w) = (\tau, w)$ for all $w \in B^{[n]}$. (Recall that $B^{[n]} = \{w \in B^* \mid |w| \le n\}$.) Observe first that $\sigma \sim \tau$ implies $\sigma^k \sim \tau^k$ for all $k \ge 1$. Consequently $\sigma \sim \tau$ and $\sigma, \tau \in S^{(1)}$ imply $\log(1 + \sigma) \sim \log(1 + \tau)$.

Consider the family $(\tau_i)_{i \in I}$ of the statement. Let

$$I_0 = \{i \in I \mid \sigma_i \sim 0\}, \quad I' = I \setminus I_0.$$

Then I' is finite. Indeed, for each $w \in B^{[n]}$ there are only finitely many indices i such that $(\sigma_i, w) \ne 0$. Since B is finite, the set $B^{[n]}$ is finite and therefore I' is finite.

Next observe that

$$\prod_{i \in I}(1 + \sigma_i) \sim \prod_{i \in I'}(1 + \sigma_i), \tag{8.18}$$

since in view of (8.2), we have $(\tau_J, w) = 0$ for $w \in B^{[n]}$ except when $J \subset I'$. It follows from (8.18) that

$$\log\Big(\prod_{i\in I}(1 + \sigma_i)\Big) \sim \log\Big(\prod_{i\in I'}(1 + \sigma_i)\Big).$$

Consequently

$$\Big(\log\Big(\prod_{i\in I}(1 + \sigma_i)\Big), \underline{C}\Big) = \Big(\log\Big(\prod_{i\in I'}(1 + \sigma_i)\Big), \underline{C}\Big).$$

Next, since $\sigma_i \sim 0$ for $i \in I_0$, we have $\log(1 + \sigma_i) \sim 0$ for $i \in I_0$. Thus

$$\Big(\sum_{i\in I}\log(1 + \sigma_i), \underline{C}\Big) = \Big(\sum_{i\in I'}\log(1 + \sigma_i), \underline{C}\Big).$$

From the finite case, one obtains

$$\Big(\log\Big(\prod_{i\in I'}(1 + \sigma_i)\Big), \underline{C}\Big) = \Big(\sum_{i\in I'}\log(1 + \sigma_i), \underline{C}\Big).$$

Putting all this together, we obtain

$$\Big(\log\Big(\prod_{i\in I}(1 + \sigma_i)\Big), \underline{C}\Big) = \Big(\sum_{i\in I'}\log(1 + \sigma_i), \underline{C}\Big) = \Big(\sum_{i\in I}\log(1 + \sigma_i), \underline{C}\Big).$$

Thus the proof is complete. $\qquad\square$

To prove Theorem 8.1.2, we need a final lemma which is a reformulation of Propositions 7.1.7 and 7.1.8.

Proposition 8.1.6 *Let $X \subset A^+$ be a code. For each conjugacy class C meeting X^*, we have $(\log \underline{X}^*, \underline{C}) \geq (\log \underline{A}^*, \underline{C})$, and equality holds if X is a circular code. Conversely if $(\log \underline{X}^*, \underline{C}) = (\log \underline{A}^*, \underline{C})$ for all conjugacy classes that meet X^*, then X is a circular code.*

Proof. We have $\underline{X}^* = (1 - \underline{X})^{-1}$. Thus $\log(\underline{X}^*(1 - \underline{X})) = 0$. Since the series $\underline{X}^*$ and $1 - \underline{X}$ commute, we have $0 = \log \underline{X}^* + \log(1 - \underline{X})$, showing that $\log \underline{X}^* = -\log(1 - \underline{X})$. Thus

$$\log \underline{X}^* = \sum_{m\geq 1}\frac{1}{m}\underline{X}^m.$$

In particular, if $C \subset A^m$ is a conjugacy class, then

$$(\log \underline{X}^*, \underline{C}) = \sum_{m\geq 1}\frac{1}{m}\,\mathrm{Card}(X^m \cap C).$$

For $X = A$, the formula becomes

$$(\log \underline{A}^*, \underline{C}) = \frac{1}{n}\,\mathrm{Card}(C).$$

The proposition is now a direct consequence of Propositions 7.1.6 and 7.1.7.   □

Proof of Theorem 8.1.2. Assume first that conditions (i) and (ii) are satisfied, that is, that the family $(X_i)_{i \in I}$ is a factorization of A^*. Then the sets X_i are codes and by Formula (8.3), we have

$$\underline{A}^* = \prod_{i \in I} \underline{X}_i^*. \tag{8.19}$$

Taking the logarithm on both sides, we obtain

$$\log \underline{A}^* = \log \left(\prod_{i \in I} \underline{X}_i^* \right). \tag{8.20}$$

By Proposition 8.1.5, the series

$$\delta = \log \underline{A}^* - \sum_{i \in I} \log \underline{X}_i^* \tag{8.21}$$

is cyclically null. Thus for each conjugacy class C

$$(\log \underline{A}^*, \underline{C}) = \sum_{i \in I} (\log \underline{X}_i^*, \underline{C}). \tag{8.22}$$

In view of Proposition 8.1.6, we have for each $i \in I$ and for each C that meets X_i^* the inequality

$$(\log \underline{A}^*, \underline{C}) \le (\log \underline{X}_i^*, \underline{C}). \tag{8.23}$$

Formulas (8.22) and (8.23) show that for each conjugacy class C, there exists a unique $j \in I$ such that C meets X_j^*. For this index j, we have

$$(\log \underline{A}^*, \underline{C}) = (\log \underline{X}_j^*, \underline{C}). \tag{8.24}$$

Thus if some X_j^* meets a conjugacy class, no other X_i^* ($i \in I \setminus j$) meets this conjugacy class. Since (8.24) holds, each of the codes X_i is a circular code by Proposition 8.1.6. This proves condition (iii).

Now assume that condition (iii) holds. Let C be a conjugacy class and let $i \in I$ be the unique index such that X_i^* meets C. Since X_i is circular, (8.24) holds by Proposition 8.1.6 and furthermore $(\log \underline{X}_j^*, \underline{C}) = 0$ for all $j \ne i$. Summing up all equalities (8.24), we obtain Equation (8.22). This proves that the series δ defined by (8.21) is cyclically null.

Let α be the canonical morphism from $\mathbb{Q}\langle\!\langle A \rangle\!\rangle$ onto the algebra $\mathbb{Q}[[A]]$ of formal power series in commutative variables in A. The set of words in A^* having the same image by α is a union of conjugacy classes, since $\alpha(uv) = \alpha(vu)$. Since the series δ is cyclically null, the series $\alpha(\delta)$ is null. Since α is a continuous morphism, we obtain, by applying α to both sides of (8.21),

$$0 = \log \alpha(\underline{A}^*) - \sum_{i \in I} \log \alpha(\underline{X}_i^*).$$

Hence

$$\log \alpha(\underline{A}^*) = \sum_{i \in I} \log \alpha(\underline{X_i^*}). \tag{8.25}$$

Next, condition (iii) ensures that the product $\prod_{i \in I} \underline{X_i^*}$ exists. By Proposition 8.1.5, the series

$$\log \left(\prod_{i \in I} \underline{X_i^*} \right) - \sum_{i \in I} \log \underline{X_i^*}$$

is cyclically null. Thus its image by α is null, whence

$$\log \alpha \left(\prod_{i \in I} \underline{X_i^*} \right) = \sum_{i \in I} \log \alpha(\underline{X_i^*}).$$

This together with (8.25) shows that

$$\log \alpha(\underline{A}^*) = \log \alpha \left(\prod_{i \in I} \underline{X_i^*} \right).$$

Since log is a bijection, this implies

$$\alpha(\underline{A}^*) = \alpha \left(\prod_{i \in I} \underline{X_i^*} \right).$$

This shows that $\alpha(\epsilon) = 0$, where

$$\epsilon = \underline{A}^* - \prod_{i \in I} \underline{X_i^*}.$$

Observe that condition (i) means that all the coefficients of ϵ are negative or zero. Condition (ii) says that all coefficients of ϵ are positive or zero. Thus, in both cases, all the coefficients of ϵ have the same sign. This together with the condition $\alpha(\epsilon) = 0$ implies that $\epsilon = 0$. This shows that if condition (iii) and either (i) or (ii) hold, then the other one of conditions (i) and (ii) also holds.   □

A factorization $(X_i)_{i \in I}$, is called *complete* if each X_i is reduced to a singleton x_i. The following result is a consequence of Theorem 8.1.2. Recall from Chapter 1 that $\ell_n(k)$ denotes the number of primitive necklaces of length n on a k-letter alphabet.

Corollary 8.1.7 *Let $(x_i)_{i \in I}$ be a complete factorization of A^*. Then the set $X = \{x_i \mid i \in I\}$ is a set of representatives of the primitive conjugacy classes. In particular, for all $n \geq 1$,*

$$\mathrm{Card}(X \cap A^n) = \ell_n(k) \tag{8.26}$$

with $k = \mathrm{Card}(A)$.

Proof. According to condition (iii) of Theorem 8.1.2, each conjugacy class intersects exactly one of the submonoids X_i^*. In view of the same condition, each code $\{x_i\}$ is circular and consequently each word x_i is primitive. This shows that X is a system of representatives of the primitive conjugacy classes. Formula (8.26) is an immediate consequence. $\qquad \square$

Now we describe a systematic procedure to obtain a large class of complete factorizations of free monoids. These include the construction used in Section 7.3.

A *Lazard set* is a totally ordered subset Z of A^+ satisfying the following property: For each $n \geq 1$, the set $Z \cap A^{[n]} = \{z_1, z_2, \ldots, z_k\}$ with $z_1 < z_2 < \cdots < z_k$ satisfies

$$z_i \in Z_i \quad \text{for} \quad 1 \leq i \leq k, \quad \text{and} \quad Z_{k+1} \cap A^{[n]} = \emptyset,$$

where the sets $Z_1, \ldots, Z_{k+1}$ are defined by

$$Z_1 = A, \quad Z_{i+1} = z_i^*(Z_i \setminus z_i) \quad (1 \leq i \leq k).$$

(Recall that $A^{[n]} = \{w \in A^* \mid |w| \leq n\}$.)

Example 8.1.8 Let $(x_n)_{n \geq 1}$ be a Hall sequence over A and let $(X_n)_{n \geq 1}$ be the associated sequence of codes. Assume that, for each n, the word x_n is a word of minimal length in X_n, and let $Z = \{x_n \mid n \geq 1\}$ be the subset of A^+ ordered by the indices. Then Z is a Lazard set.

Example 8.1.9 Let $(x_n)_{n \geq 1}$ be the sequence used in the proof of Theorem 7.3.11. Recall that we start with $X_1 = A$ and

$$X_{i+1} = x_i^*(X_i \setminus x_i) \quad i \geq 1,$$

where x_i is a word in X_i of minimal odd length. Denote by Y the set of even words in the set $\bigcup_{i \geq 1} X_i$. Now set $Y_1 = Y$ and for $i \geq 1$,

$$Y_{i+1} = y_i^*(Y_i \setminus y_i),$$

where $y_i \in Y_i$ is chosen with minimal length. Let $T = \{x_i, y_i \mid i \geq 1\}$ ordered by

$$x_1 < x_2 < \cdots < x_n < \cdots < y_1 < y_2 < \cdots .$$

The ordered set T is a Lazard set. Indeed, let $n \geq 1$ and

$$T \cap A^{[n]} = \{x_1, x_2, \ldots, x_r, y_1, y_2, \ldots, y_s\}.$$

Set

$$Z_i = X_i, \quad (1 \leq i \leq r+1), \quad Z_{r+i+1} = y_i^*(Z_{r+i} \setminus y_i), \quad (1 \leq i \leq s).$$

We show by induction on i that

$$Z_{r+i} \cap A^{[n]} = Y_i \cap A^{[n]} \quad (1 \leq i \leq s+1). \tag{8.27}$$

Indeed, words in $X_{r+1} = Z_{r+1}$ of length at most n all have even length (since the words with odd length are $x_1, x_2, \ldots, x_r$). Thus all these words are in $Y = Y_1$. Conversely, any word of even length $\leq n$ is already in X_{r+1}, since $|x_{r+1}| > n$.

Next, consider $y \in Y_{i+1} \cap A^{[n]}$. Then $y = y_i^p y'$ for some $y' \in Y_i \setminus y_i$. Since $|y'| \leq n$, we have by the induction hypothesis $y' \in Z_{r+i}$, whence $y \in Z_{r+i+1}$. The converse is proved in the same way.

Equation (8.27) shows that $y_i \in Z_{r+i}$ for $1 \leq i \leq s$ and that $Z_{r+s+1} \cap A^{[n]} = \emptyset$. Thus T is a Lazard set.

Proposition 8.1.10 *Let $Z \subset A^+$ be a Lazard set. Then the family $(z)_{z \in Z}$ is a complete factorization of A^*.*

Proof. Let $w \in A^*$ and $n = |w|$. Set $Z \cap A^{[n]} = \{z_1, z_2, \ldots, z_k\}$ with $z_1 < z_2 < \cdots < z_k$. Let $Z_1 = A$ and $Z_{i+1} = Z_i^*(Z_i \setminus z_i)$ for $i = 1, 2, \ldots, k$. Then $z_i \in Z_i$ for $i = 1, 2, \ldots, k$ and $Z_{k+1} \cap A^{[n]} = \emptyset$. As in the proof of Lemma 7.3.13, we have for $1 \leq i \leq k$,

$$Z_i^* = Z_{i+1}^* z_i^*,$$

whence by successive substitutions

$$A^* = Z_{k+1}^* z_k^* \cdots z_1^*. \tag{8.28}$$

Thus there is a factorization $w = y z_{i_1} z_{i_2} \cdots z_{i_n}$ with $y \in Z_{k+1}^*$ and $i_1 \geq i_2 \geq \cdots \geq i_n$. Since $Z_{k+1} \cap A^{[n]} = \emptyset$, we have $y = 1$. This proves the existence of an ordered factorization. Assume there is another factorization, say $w = t_1 t_2 \cdots t_m$, with $t_j \in Z$, $t_1 \geq t_2 \geq \cdots \geq t_m$. Then $t_i \in Z \cap A^{[n]}$ for each i. Thus by (8.28) both factorizations coincide. $\quad\square$

We conclude this section with an additional example of a complete factorization. Consider a totally ordered alphabet A. Recall that the *lexicographic* or *alphabetic order*, denoted $\prec$, on A^* is defined by setting $u \prec v$ if u is a proper prefix of v, or if $u = ras, v = rbt, a < b$ for $a, b \in A$ and $r, s, t \in A^*$. Recall also that the alphabetic order has the property

$$u \prec v \Leftrightarrow wu \prec wv.$$

By definition, a *Lyndon word* is a primitive word which is minimal in its conjugacy class. In an equivalent way, a word $w \in A^+$ is a Lyndon word if and only if $w = uv$ with $u, v \in A^+$ implies $w \prec vu$. Let L denote the set of Lyndon words. We shall show that $(\ell)_{\ell \in L}$ is a complete factorization of A^*. For this we establish propositions which are interesting on their own.

Proposition 8.1.11 *A word is a Lyndon word if and only if it is smaller than all its proper nonempty right factors.*

Proof. The condition is sufficient. Let $w = uv$, with $u, v \in A^+$. Since $w \prec v$ and $v \prec vu$, we have $w \prec vu$. Consequently $w \in L$. Conversely, let $w \in L$ and consider a factorization $w = uv$ with $u, v \in A^+$. First, let us show that v is not a prefix of w. Assume the contrary. Then $w = vt$ for some $t \in A^+$. Since $w \in L$, we have $w \prec tv$. But $w = uv$ implies $uv \prec tv$. This in turn implies $u \prec t$ whence, multiplying on the left by v,

$$vu \prec vt = w,$$

a contradiction. Suppose that $v \prec uv$. Since v is not a prefix of w, this implies that $vu \prec uv$ and $w \notin L$, a contradiction. Thus $uv \prec v$, and the proof is completed. $\square$

Proposition 8.1.12 *Let ℓ, m be two Lyndon words. If $\ell \prec m$, then ℓm is a Lyndon word.*

Proof. First we show that $\ell m \prec m$. If ℓ is a prefix of m, let $m = \ell m'$. Then $m \prec m'$ by Proposition 8.1.11. Thus $\ell m \prec \ell m' = m$. If ℓ is not a prefix of m, then the inequality $\ell \prec m$ directly implies $\ell m \prec m$. Let v be a nonempty proper suffix of ℓm. If v is a suffix of m, then by Proposition 8.1.11, $m \prec v$. Hence $\ell m \prec m \prec v$. Otherwise $v = v'm$ for some proper nonempty suffix v' of ℓ. Then $\ell \prec v'$ and consequently $\ell m \prec v'm$. Thus in all cases $\ell m \prec v$. By Proposition 8.1.11, this shows that $\ell m \in L$. $\square$

Theorem 8.1.13 *The family $(\ell)_{\ell \in L}$ is a complete factorization of A^*.*

Proof. We prove that conditions (i) and (iii) of Theorem 8.1.2 are satisfied. This is clear for condition (iii) since L is a system of representatives of primitive conjugacy classes. For condition (i), let $w \in A^+$. Then w has at least one factorization $w = \ell_1 \ell_2 \cdots \ell_n$ with $\ell_j \in L$. Indeed each letter is already a Lyndon word. Consider a factorization $w = \ell_1 \ell_2 \cdots \ell_n$ into Lyndon words with minimal n. Then this is an ordered factorization. Indeed, otherwise, these would be some index i such that $\ell_i \prec \ell_{i+1}$. But then $\ell_i \ell_{i+1} \in L$ and w would have a factorization into $n - 1$ Lyndon words. Thus condition (i) is satisfied. $\square$

It can be proved (see Exercises 8.1.3, 8.1.4) that the set L is a Lazard set.

8.2 Finite factorizations

In this section we consider factorizations $(X_i)_{i \in I}$ with I a finite set. These are families $X_n, X_{n-1}, \ldots, X_1$ of subsets of A^+ such that

$$A^* = \underline{X_n^*} \underline{X_{n-1}^*} \cdots \underline{X_1^*}. \tag{8.29}$$

According to Theorem 8.1.2, each X_i is a circular code and each conjugacy class meets exactly one of the X_i^*. The aim of this section is to refine these properties. We shall see that in some special cases the codes X_j are limited. The question whether all codes appearing in finite factorizations are limited is still open. We start with the study of *bisections*, that is, factorizations of the form (X, Y). Here X is called the *left* *factor* and Y is called the *right factor* of the bisection. Then

$$\underline{A}^* = \underline{X}^* \underline{Y}^*. \tag{8.30}$$

Example 8.2.1 Let $A = \{a, b\}$. The pair (a^*b, a) is a bisection of A^*. More generally, if $A = A_0 \cup A_1$ is a partition of A, the pair $(A_0^* A_1, A_0)$ is a bisection of A^*.

Formula (8.30) can be written as

$$\underline{Y}\underline{X} + \underline{A} = \underline{X} + \underline{Y}. \tag{8.31}$$

Indeed, (8.30) is equivalent to $1 - \underline{A} = (1 - \underline{Y})(1 - \underline{X})$ by taking the inverses. This gives (8.31). Equations (8.30) and (8.31) show that a pair (X, Y) of subsets of A^+ is a bisection if and only if the following are satisfied:

$$A \subset X \cup Y, \tag{8.32}$$

$$X \cap Y = \emptyset, \tag{8.33}$$

$$YX \subset X \cup Y, \tag{8.34}$$

each $z \in X \cup Y, z \notin A$ factorizes uniquely into $z = yx$ with $x \in X, y \in Y$. (8.35)

We shall see later (Theorem 8.2.6) that a subset of these conditions is already enough to ensure that a pair (X, Y) is a bisection.

Before doing that, we show that for a bisection (X, Y) the code X is $(1, 0)$-limited and the code Y is $(0, 1)$-limited. Recall that a $(1, 0)$-limited code is prefix and that by Proposition 7.2.12 a prefix code X is $(1, 0)$-limited if and only if the set $R = A^* \setminus XA^*$ is a submonoid. Symmetrically, a suffix code Y is $(0, 1)$-limited if and only if the set $S = A^* \setminus A^*Y$ is a submonoid.

Proposition 8.2.2 *Let X, Y be two subsets of A^+. The following conditions are equivalent:*

(i) (X, Y) *is a bisection of A^*.*
(ii) X, Y *are codes, X is $(1, 0)$-limited and $Y^* = A^* \setminus XA^*$.*
(iii) X, Y *are codes, Y is $(0, 1)$-limited and $X^* = A^* \setminus A^*Y$.*

Proof. (i) $\Rightarrow$ (ii). From $\underline{A}^* = \underline{X}^* \underline{Y}^*$ we obtain by multiplication on the left by $1 - \underline{X}$ the equation $(1 - \underline{X})\underline{A}^* = \underline{Y}^*$, showing that $\underline{Y}^* = \underline{A}^* - \underline{X}\underline{A}^*$. The number of prefixes in X of any word $w \in A^*$ is $(\underline{X}\underline{A}^*, w)$. The equation shows that this number is 0 or 1, according to $w \in Y^*$ or $w \notin Y^*$. This proves that X is a prefix code. This also gives the set relation $Y^* = A^* \setminus XA^*$. Thus $A^* \setminus XA^*$ is a submonoid and by Proposition 7.2.12, the code X is $(1, 0)$-limited.

(ii) $\Rightarrow$ (i). By Theorem 3.1.8, we have $\underline{A}^* = \underline{X}^* \underline{R}$ with $R = A^* \setminus XA^*$. Since $R = Y^*$ and Y is a code, we have $\underline{R} = \underline{Y}^*$. Thus $\underline{A}^* = \underline{X}^* \underline{Y}^*$.

Consequently (i) and (ii) are equivalent. The equivalence between (i) and (iii) is shown in the same manner. $\qquad\square$

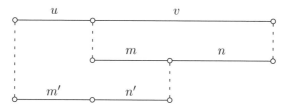

Figure 8.1 Factorizations.

Corollary 8.2.3 *The left factors of bisections are precisely the* $(1, 0)$-*limited codes.*

□

Observe that for a bisection (X, Y), either X is maximal prefix or Y is maximal suffix. Indeed, we have $Y^* = A^* \setminus XA^*$. If Y^* contains no right ideal, then XA^* is right dense and consequently X is maximal prefix. Otherwise, Y^* is left dense and thus Y is maximal suffix.

Proposition 8.2.4 *Let* M, N *be two submonoids of* A^* *such that* $\underline{A}^* = \underline{M}\,\underline{N}$. *Then* M *and* N *are free and the pair* (X, Y) *of their bases is a bisection of* A^*.

Proof. Let u, v be in A^* such that $uv \in M$. Set $v = mn$ with $m \in M, n \in N$. Similarly set $um = m'n'$ for some $m' \in M$, $n' \in N$ (see Figure 8.1). Then $uv = m'(n'n)$. Since $uv \in M$, the unique factorization property implies $n = n' = 1$, whence $v \in M$. This shows that M satisfies condition $C(1, 0)$. Thus M is generated by a $(1, 0)$-limited code X. Similarly N is generated by a $(0, 1)$-limited code Y. Clearly (X, Y) is a factorization. □

Example 8.2.5 Let M and N be two submonoids of A^* such that

$$M \cap N = \{1\}, \quad M \cup N = A^*.$$

We shall associate a special bisection of A^* with the pair (M, N). For this, let

$$R = \{r \in A^* \mid r = uv \Rightarrow v \in M\}$$

be the set of words in M having all its suffixes in M. Symmetrically, define

$$S = \{s \in A^* \mid s = uv \Rightarrow u \in N\}.$$

The set R is a submonoid of A^* because M is a submonoid. Moreover, R is suffix-closed. Consequently the base of R, say X, is a $(1, 0)$-limited code. Similarly S is a free submonoid and its base, say Y, is a $(0, 1)$-limited code. We prove that (X, Y) is a bisection. In view of Proposition 8.2.2, it suffices to show that $Y^* = A^* \setminus XA^*$. First, consider a word $y \in Y^* = S$. Then all its prefixes are in N. Thus no prefix of y is in X. This shows that $Y^* \subset A^* \setminus XA^*$. Conversely, let $w \in A^* \setminus XA^*$. We show that any prefix u of w is in N by induction on $|u|$. This holds clearly for $|u| = 0$. Next, if $|u| \geq 1$, then u cannot be in $R = X^*$ since otherwise w would have a prefix in X.

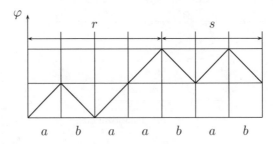

Figure 8.2 The path of values of φ for $\varphi(a) = 1$, $\varphi(b) = -1$ and $w = abaabab$.

Thus there exists a factorization $u = u'v'$ with $v' \notin M$. Hence $v' \in N$ and $v' \neq 1$. By the induction hypothesis, $u' \in N$. Since N is a submonoid, $u = u'v' \in N$. This proves that $w \in S = Y^*$.

A special case of this construction is obtained by considering a morphism $\varphi : A^* \to \mathbb{Z}$ into the additive monoid $\mathbb{Z}$ and by setting

$$M = \{m \in A^* \mid \varphi(m) > 0\} \cup \{1\}, \quad N = \{n \in A^* \mid \varphi(n) \leq 0\}.$$

Given a word $w \in A^*$, we obtain a factorization $w = rs$ with $r \in R$, $s \in S$ as follows. The word r is the shortest prefix of w such that the value $\varphi(r)$ is maximal in the set of values of φ on the prefixes of w (see Figure 8.2).

The construction of Example 8.2.5 can be considered as a special case of the very general following result.

Theorem 8.2.6 *Let* (P, Q) *be a partition of* A^+. *There exists a unique bisection* (X, Y) *of* A^* *such that* $X \subset P$ *and* $Y \subset Q$. *This bisection is obtained as follows.*
Let

$$X_1 = P \cap A, \quad Y_1 = Q \cap A, \tag{8.36}$$

and for $n \geq 2$,

$$Z_n = \bigcup_{i=1}^{n} Y_i X_{n-i}, \tag{8.37}$$

$$X_n = Z_n \cap P, \quad Y_n = Z_n \cap Q. \tag{8.38}$$

Then

$$X = \bigcup_{n \geq 1} X_n \quad and \quad Y = \bigcup_{n \geq 1} Y_n. \tag{8.39}$$

Proof. We first prove uniqueness. Consider a bisection (X, Y) of A^* such that $X \subset P$ and $Y \subset Q$. We show that for $n \geq 1$, we have $X \cap A^n = X_n$, $Y \cap A^n = Y_n$, with X_n and Y_n given by (8.36) and (8.38). Arguing by induction, we consider $n = 1$. Then $X \cap A \subset P \cap A = X_1$. Conversely we have $A \subset X \cup Y$ by (8.32) and $P \cap Y = \emptyset$.

Consequently $P \cap A \subset X$ and therefore $X \cap A = X_1$. For $n \geq 2$, we have $Z_n \subset YX \cap A^n$ by the induction hypothesis. Thus by (8.34), $Z_n \subset (X \cup Y) \subset A^n$. This implies that $Z_n \cap P \subset X \cap A^n$ and $Z_n \cap Q \subset Y \cap A^n$. Conversely, let $z \in (X \cup Y) \cap A^n$. Then by (8.35) $z = yx$ for some $y \in Y$, $x \in X$. By the induction hypothesis, $y \in Y_i$ and $x \in X_{n-i}$ for $i = |y|$. In view of (8.37), we have $z \in Z_n$. This shows that $(X \cup Y) \cap A^n \subset Z_n$. Hence $X \cap A^n \subset Z_n \cap P$ and $Y \cap A^n \subset Z_n \cap Q$.

To prove the existence of a bisection, we consider the pair (X, Y) given in (8.39). We proceed in several steps. Define $Z_1 = A$ and set $Z = \bigcup_{n \geq 1} Z_n$. In view of (8.36) and (8.38) we have $Z = X \cup Y$. Observe first that by Formula (8.37)

$$YX \cup A = X \cup Y. \tag{8.40}$$

Clearly (8.40) implies $YX \subset X \cup Y$. By induction, we obtain

$$Y^*X^* \subset X^* \cup Y^*. \tag{8.41}$$

Next, we have

$$A^* = X^*Y^*. \tag{8.42}$$

Indeed, let $w \in A^*$. Since $A \subset Z$, the word w has at least one factorization $w = z_1 z_2 \cdots z_n$ with $z_j \in Z$. Choose such a factorization with n minimal. Then we cannot have $z_i \in Y$, $z_{i+1} \in X$ for some $1 \leq i \leq n - 1$, since this would imply that $z_j z_{j+1} \in Z$ by (8.40) contradicting the minimality of n. Consequently there is some $j \in \{1, \ldots, n\}$ such that $z_1, \ldots, z_j \in X$ and $z_{j+1}, \ldots, z_n \in Y$, showing that $w \in X^*Y^*$.

Now we prove that X^* is suffix-closed. For this, it suffices to show that

$$uv \in X \Rightarrow v \in X^*. \tag{8.43}$$

Indeed, assuming (8.43), consider a word $w = rs \in X^*$. Then $r = r'u$, $s = vs'$ for some $r', s' \in X^*$, and $uv \in X \cup 1$. By (8.43), v is in X^*, and consequently, $s \in X^*$, showing that X^* is suffix-closed. We prove (8.43) by induction on the length of $x = uv$. Clearly the formula holds for $|x| = 1$. Assume $|x| \geq 2$. Then by (8.40) $x = y_1 x_1$ for some $y_1 \in Y$, $x_1 \in X$. If y_1 is not a letter, then again by (8.40), $y_1 = y_2 x_2$ for some $y_2 \in Y$, $x_2 \in X$. Iterating this operation, we obtain a factorization $x = y_k x_k \cdots x_2 x_1$ with $y_k \in Y_n \cap A$ and $x_1, \ldots, x_k \in X$.

Each proper suffix v of x has the form $v = v_p x_{p-1} \cdots x_1$ for some suffix v_p of x_p and $1 \leq p \leq k$. By the induction hypothesis, $v_p \in X^*$. Consequently $v \in X^*$. This proves (8.43). An analogous proof shows that Y^* is prefix-closed.

Next we claim that

$$X^* \cap Y^* = \{1\}, \tag{8.44}$$

and prove this claim by induction, showing that $X^* \cap Y^*$ contains no word of length $n \geq 1$. This holds for $n = 1$ because $X \cap Y = \emptyset$. Assume that for some $w \in A^n$, there are two factorizations $x = x_1 x_2 \cdots x_p = y_1 y_2 \cdots y_q$ with $x_i \in X$, $y_j \in Y$. Since Y^* is prefix-closed, we have $x_1 \in Y^*$. Since X^* is suffix-closed $y_q \in X^*$. Thus $x_1 \in$

$X \cap Y^*$ and $y_q \in X^* \cap Y$. By the induction hypothesis this is impossible if x_1 and y_q are shorter than w. Therefore we have $p = q = 1$. But then $w \in X \cap Y = \emptyset$, a contradiction. This proves (8.44). Now we prove that X is prefix. For this, we show by induction on $n \geq 1$ that no word in X of length n has a proper prefix in X. This clearly holds for $n = 1$.

Consider $uv \in X \cap A^n$ with $n \geq 2$ and suppose that $u \in X$. In view of (8.40), we have $uv = yx$ for some $y \in Y, x \in X$. The word u cannot be a prefix of y, since otherwise u would be in $X \cap Y^*$ because Y^* is prefix-closed and this is impossible by (8.44). Thus there is a word $u' \in A^+$ such that $u = yu'$, $u'v = x$.

By (8.43), $u' \in X^*$. Moreover $|x| \leq n$. By the induction hypothesis, the equation $x = u'v$ implies $v = 1$. Thus $u = uv$, showing the claim for n. Consequently X is prefix. A similar proof shows that Y is suffix.

We now are able to show that (X, Y) is a bisection. Equation (8.42) shows that any word in A^* admits a factorization. To show uniqueness, assume that $xy = x'y'$ for $x, x' \in X^*$ and $y, y' \in Y^*$. Suppose $|x| \geq |x'|$. Then $x = x'u$ and $uy = y'$ for some word u. Since X^* is suffix-closed and Y^* is prefix-closed, we have $u \in X^* \cap Y^*$. Thus $u = 1$ by (8.44). Consequently $x = x'$ and $y = y'$. Since X and Y are codes, this completes the proof.   □

Theorem 8.2.6 shows that the following method allows us to construct all bisections.

(i) Partition the alphabet A into two subsets X_1 and Y_1.
(ii) For each $n \geq 2$, partition the set $Z_n = \bigcup_{i=1}^{n-1} Y_i X_{n-i}$ into two subsets X_n and Y_n.
(iii) Set $X = \bigcup_{n \geq 1} X_n$ and $Y = \bigcup_{n \geq 1} Y_n$.

In other words, it is possible to construct the components of the partition (P, Q) progressively during the computation. A convenient way to represent the computations is to display the words in X and Y in two columns when they are obtained. This is illustrated by the following example.

Example 8.2.7 Let $A = \{a, b\}$. We construct a bisection of A^* by distributing iteratively the products yx $(x \in X, y \in Y)$ into two columns as shown in Figure 8.3. All the remaining products are put into the set R. This gives a defining equation for R, since from $A \cup YX = X \cup Y$ and $X = \{a, ba\} \cup R$ we obtain $R = \{b, b^2a\}R \cup b^2a\{a, ba\}$. Thus $R = \{b, b^2a\}^*b^2a\{a, ba\}$ or also $R = (b^2b^*a)^*b^2b^*a\{a, ba\}$. Consequently $X = (b^2b^*a)^*\{a, ba\}$, which is the code of Example 7.2.11.

The following convention will be used for the rest of this section. Given a code X over A, a pair (U, V) of subsets of A^* will be called a *bisection* of X^* if

$$\underline{X^* = U^*V^*}.$$

To fit into the ordinary definition of bisection, it suffices to consider a coding morphism for X.

	X	Y
1	a	b
2	ba	
3		bba
≥4	R	

Figure 8.3 A bisection of A^*.

A *trisection* of A^* is a triple (X, Y, Z) of subsets of A^+, which form a factorization of A^*, that is

$$\underline{A}^* = \underline{X}^* \underline{Y}^* \underline{Z}^*. \tag{8.45}$$

We shall prove the following result which gives a relationship between bisections and trisections.

Theorem 8.2.8 *Let (X, Y, Z) be a trisection of A^*. There exist a bisection (U, V) of Y^* and a bisection (X', Z') of A^* such that (X, U) is a bisection of X'^* and (V, Z) is a bisection of Z'^*,*

$$\underline{A}^* = \underline{X}^* \underline{Y}^* \underline{Z}^* = (\underline{X}^* \underline{U}^*)(\underline{V}^* \underline{Z}^*) = \underline{X'}^* \underline{Z'}^*.$$

Before giving the proof we establish some useful formulas.

Proposition 8.2.9 *Let (X, Y, Z) be a trisection of A^*.*

1. *The set $X^* Y^*$ is suffix-closed and the set $Y^* Z^*$ is prefix-closed.*
2. *One has the inclusions*

$$Y^* X^* \subset X^* \cup Y^* Z^*, \tag{8.46}$$

$$Z^* Y^* \subset Z^* \cup X^* Y^*. \tag{8.47}$$

3. *The codes X, Y and Z are $(2, 0)$-, $(1, 1)$-, and $(0, 2)$-limited, respectively.*

Proof. We first prove 1. Let $w \in X^* Y^*$, and let v be a suffix of w (see Figure 8.4). Then $w = uv$ for some u. Set $v = xyz$ with $x \in X^*$, $y \in Y^*$, and $z \in Z^*$. Set also $uxy = x'y'z'$ with $x' \in X^*$, $y' \in Y^*$, and $z' \in Z^*$. Then

$$w = uv = uxyz = x'y'(z'z).$$

Uniqueness of factorization implies $z' = z = 1$. This shows that $v \in X^* Y^*$ and proves that $X^* Y^*$ is suffix-closed. Likewise $Y^* Z^*$ is prefix-closed. We now verify (8.46). Let $x \in X^*$ and $y \in Y^*$. Set $yx = x'y'z'$ with $x' \in X^*$, $y' \in Y^*$, and $z' \in Z^*$. If $x' = 1$, then $yx \in Y^* Z^*$. Thus assume that $x' \neq 1$. The word x' cannot be a prefix of

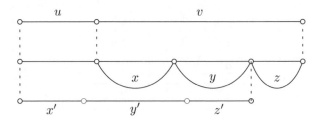

Figure 8.4 The set X^*Y^* is suffix-closed.

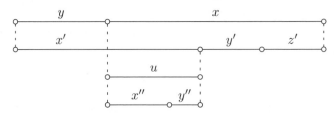

Figure 8.5 $Y^*X^* \subset X^* \cup Y^*Z^*$.

y since $y \in Y^*Z^*$ and Y^*Z^* is prefix-closed and $X^* \cap Y^*Z^* = \{1\}$. Therefore there is a word u such that $x' = yu$ and $x = uy'z'$ (see Figure 8.5). Since u is a suffix of $x' \in X^*Y^*$, it is itself in X^*Y^*. Consequently $u = x''y''$ for some $x'' \in X^*$ and $y'' \in Y^*$. This shows that $x = x''y''y'z'$. Uniqueness of factorization implies $y'' = y' = z' = 1$. Consequently $yx = x'y'z' = x' \in X^*$. This proves (8.46). Formula (8.47) is proved symmetrically.

The code X is $(2, 0)$-limited. Indeed, let $u, v, w \in A^+$ be words such that $uv, vw \in X^*$. Since v and w are suffixes of words in X^* and since X^*Y^* is suffix-closed, both v and w are in X^*Y^*. Thus we have

$$v = x'y', \quad w = xy$$

for some $x, x' \in X^*$, $y, y' \in Y^*$ (see Figure 8.6). The word $y'x$ is a suffix of $uvx \in X^*$. By the same argument, $y'x$ is in X^*Y^* and consequently $y'x = x''y''$ for some $x'' \in X^*$ and $y'' \in Y^*$, whence $vw = x'x''y''y$. Since by assumption $vw \in X^*$, uniqueness of factorization implies that $y'' = y = 1$. Thus $w = x \in X^*$. This proves that X is $(2, 0)$-limited. Likewise Z is $(0, 2)$-limited.

To show that Y is $(1, 1)$-limited, consider words $u, v, w \in A^*$ such that $uv, vw \in Y^*$. Then $v \in X^*Y^*$ because v is a suffix of the word uv in X^*Y^* and also $v \in Y^*Z^*$ as a left factor of the word vw in Y^*Z^*. Thus $v \in X^*Y^* \cap Y^*Z^*$. Uniqueness of factorization implies that $v \in Y^*$. This completes the proof. □

Proof of Theorem 8.2.8. Set $S = \{s \in Y^* \mid sX^* \subset X^*Y^*\}$. First, we observe that

$$S = \{s \in Y^* \mid sX^* \subset X^* \cup Y^*\}. \tag{8.48}$$

Indeed, consider a word $s \in Y^*$. If $sX^* \subset X^* \cup Y^*$, then clearly, $sX^* \subset X^*Y^*$. Assume conversely that $sX^* \subset X^*Y^*$. Since $s \in Y^*$ we have $sX^* \subset Y^*X^*$ and

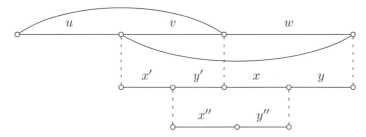

Figure 8.6 The code X is $(2, 0)$-limited.

it follows by (8.46) that $sX^* \subset X^* \cup Y^*Z^*$. Thus $sX^* \subset X^*Y^* \cap (X^* \cup Y^*Z^*) = (X^*Y^* \cap X^*) \cup (X^*Y^* \cap Y^*Z^*) = X^* \cup Y^*$ by uniqueness of factorization. This proves (8.48). Next, S is a submonoid. Indeed, $1 \in S$ and if $s, t \in S$, then $stX^* \subset sX^*Y^* \subset X^*Y^*$. We show that the monoid S, considered as a monoid on the alphabet Y, satisfies condition $C(1, 0)$. In other words, $s, t \in Y^*$ and $st \in S$ imply $t \in S$. Indeed, consider $x \in X^*$. Since tx is a suffix of $stx \in X^*Y^*$ and since X^*Y^* is suffix-closed, $tx \in X^*Y^*$. Thus $t \in S$. This shows that S is a free submonoid of Y^* generated by some $(1, 0)$-limited code $U \subset Y^+$. Note that U is $(1, 0)$-limited as a code over Y. According to Proposition 8.2.2, the code U is the left factor of some bisection (U, V) of Y^*, with $V^* = Y^* \setminus UY^*$. We shall give another definition of V. For this, set

$$R = \{r \in Y^* \mid rX^* \cap Z^* \neq \emptyset\}.$$

Clearly $R \cap S = 1$. We prove that

$$R^* = V^*. \tag{8.49}$$

First, we show that $R \subset V^*$. Let $r \in R \setminus 1$. Set $r = st$ with $s \in Y^+$, $t \in Y^*$. Since $r \in R$, we have $stx \in Z^*$ for some $x \in X^*$. By (8.46), we have $tx \in X^* \cup Y^*Z^*$. If $tx \in Y^*Z^*$, then $st \in Y^+Z^*$ which is impossible since $stx \in Z^*$. Consequently $tx \in X^*$. Thus $s \in R$. Since $R \cap S = \{1\}$, it follows that $s \notin S$. This shows that no prefix $s \in Y^+$ of r is in S. In other words, no prefix of r is in the code U. This proves that r is in V^*.

Second, we prove that $V^* \subset R^*$. We proceed by induction on the length of words in V, the case of the empty word being trivial. Let $v \in V^+$. Since (U, V) is a factorization, we have $U^* \cap V^* = \{1\}$. Consequently $v \in U^* = S$. Thus by (8.48), there is some $x \in X^*$ such that $vx \in X^* \cup Y^*$. Since $v \in Y^*$, we have by (8.46) $vx \in Y^*Z^*$, and by a previous remark even $vx \in Y^*Z^+$. Set $vx = yz$ with $y \in Y^*$, $z \in Z^+$. Then z cannot be a suffix of x, since otherwise z would be in $X^*Y^* \cap Z^+$, which is impossible. Thus there is some word $w \in A^+$ such that $v = yw$ and $wx = z$. Since w is a suffix of $v \in X^*Y^*$, we have $w \in X^*Y^*$. Similarly w is a prefix of $z \in Y^*Z^*$. This implies that $w \in Y^*Z^*$. Uniqueness of factorization implies $w \in Y^*$. The word y is in V^*. Indeed, $y \in Y^*$ is a prefix of v, and since V^* is prefix-closed as a subset of Y^*, $y \in V^*$. Since $|y| \leq |v|$, we have $y \in R^*$ by the induction hypothesis. On the other hand, $w \in Y^*$ and $wx = z \in Z^*$ imply $w \in R$. Thus $v = yw \in R^*$. This completes the proof of (8.49). Up to now, we have proved that

$$\underline{A}^* = \underline{X}^* \underline{U}^* \underline{V}^* \underline{Z}^*, \tag{8.50}$$

with $\underline{Y}^* = \underline{U}^* \underline{V}^*$, $S = U^*$, and $R^* = V^*$. To finish the proof, it suffices to show that
the products $M = X^* U^*$ and $N = V^* Z^*$ are submonoids. Indeed, since the product
(8.50) is unambiguous, we have $\underline{M} = \underline{X}^* \underline{U}^*$ and $\underline{N} = \underline{V}^* \underline{Z}^*$ whence $\underline{A}^* = \underline{M} \underline{N}$. By
Proposition 8.2.4, the monoids M and N then are free and their bases constitute the
desired bisection (X', Y'). To show that $X^* U^*$ is a submonoid it suffices to show that
$U^* X^* \subset X^* \cup U^*$. Thus, let us consider words $x \in X^*$ and $s \in U^* = S$. Then by
(8.48) $sx \in X^* \cup Y^*$. But $sx \in Y^*$ implies $sx \in S$ because $sx X^* \subset s X^* \subset X^* \cup Y^*$.
Consequently $sx \subset X^* \cup S$, showing that $X^* U^*$ is a submonoid. Finally we show
that $V^* Z^*$ is a submonoid. For this, we show that

$$Z^* R \subset R \cup Z^*. \tag{8.51}$$

This will imply that $Z^* R^* \subset R^* \cup Z^*$ which in turn proves the claim in view of
(8.49) To show (8.51), let $z \in Z^*$ and $r \in R$. Since $r \in Y^*$, Formula (8.47) implies
that $zr \in Z^* \cup X^* Y^*$. Next, by definition of R, $rx \in Z^*$ for some $x \in X^*$, showing
that $zrx \in Z^*$. Since $Y^* Z^*$ is prefix-closed, we have $z \in Y^* Z^*$. By the uniqueness
of factorization, $zr \in Z^* \cup Y^*$. If $zr \in Y^*$, then $zr \in R$, since $zrx \in Z^*$. Thus $zr \in$
$Z^* \cup R$ and this proves (8.51). □

Theorem 8.2.8 shows that all trisections can be built by "pasting" together quadrisections obtained by a sequence of bisections. The following example shows that, on
the contrary, a trisection cannot always be obtained by two bisections.

Example 8.2.10 Let $A = \{a, b\}$. The suffix code $Z' = \{b, ba, ba^2\}$ is $(0, 1)$-limited.
Thus Z' is the right factor of the bisection (X', Z') of A^* with $X'^* = A^* \setminus A^* Z'$. The
equation

$$\underline{Z}' \underline{X}' + \underline{A} = \underline{Z}' + \underline{X}'$$

derived from (8.31) gives $\underline{A} - \underline{Z}' = (1 - \underline{Z}') \underline{X}'$ whence $\underline{X}' = \underline{Z}'^* (\underline{A} - \underline{Z}')$. It follows
that

$$\underline{X}' = \underline{Z}'^* (a - ba - ba^2)$$
$$= (\underline{Z}'^* - \underline{Z}'^* b - \underline{Z}'^* ba) a$$
$$= (1 + \underline{Z}'^* (b + ba + ba^2) - \underline{Z}'^* b - \underline{Z}'^* ba) a$$
$$= (1 + \underline{Z}'^* ba^2) a.$$

Thus

$$X' = Z'^* ba^3 \cup \{a\}.$$

Next define

$$U = (ba)^*ba^3, \quad V = ba, \quad Z = \{b, ba^2\}(ba)^*.$$

The pair (V, Z) is clearly a bisection of Z'^*. Moreover, by inspection $U \subset X'$. This inclusion shows that, over the alphabet X', the set U is the right factor of the bisection (X, U) of X'^* with $X = U^*(X' \setminus U)$. Moreover, $U^*V^* = \{ba, ba^3\}^*$. Then setting

$$Y = \{ba, ba^3\},$$

(U, V) is a bisection of Y^*. Thus we have obtained

$$\underline{A}^* = \underline{X}'^* \underline{Z}'^* = \underline{X}^* \underline{U}^* \underline{V}^* \underline{Z}^* = \underline{X}^* \underline{Y}^* \underline{Z}^*,$$

and (X, Y, Z) is a trisection of A^*. Neither X^*Y^* nor Y^*Z^* is a submonoid. Indeed, $ba \in Y$ and $a \in X$ (since $a \in X' \setminus U$). However, $ba^2 \in Z$ and consequently $ba^2 \notin X^*Y^*$. Similarly $b \in Z$ and $ba^3 \in Y$ but $b^2a^3 \in X$ whence $b^2a^3 \notin Y^*Z^*$. This means that the trisection (X, Y, Z) cannot be obtained by two bisections.

8.3 Exercises

Section 8.1

8.1.1 Let $A = \{1, 2, \ldots, n\}$ and for $j \in A$, let $X_j = j\{j+1, \ldots, n\}^*$. Show that the family $(X_j)_{1 \le j \le n}$ is a factorization of A^*.

8.1.2 Let $\varphi : A^* \to \mathbb{R}$ be a morphism into the additive monoid. For $r \in \mathbb{R}$, let

$$C_r = \{v \in A^+ \mid \varphi(v) = r|v|\}, \quad B_r = C_r \setminus \left(\bigcup_{s \ge r} C_s\right) A^+.$$

Show that the family $(B_r)_{r \in \mathbb{R}}$ (with the usual order on $\mathbb{R}$) is a factorization of A^*.

8.1.3 The (left) *standard factorization* of a Lyndon word $w \in L \setminus A$ is defined as the pair

$$\pi(w) = (\ell, m)$$

of words in A^+ such that $w = \ell m$ and ℓ is the longest proper prefix of w that is in L. Show that $m \in L$ and $\ell \prec m$. (*Hint*: Consider the factorization of m as a nonincreasing product of Lyndon words.)

Show that if $\pi(w) = (\ell, m)$ and $\pi(m) = (p, q)$, then $p \preceq \ell \prec m$.

8.1.4 Show that the set L of Lyndon words over A is a Lazard set. (*Hint*: Set $L \cap A^n = \{z_1, z_2, \ldots, z_k\}$ with $z_1 \le z_2 \le \cdots \le z_k$. Show that $z_i \in Z_i$ for $1 \le i \le k$ where

$$Z_1 = A,$$

$$Z_{i+1} = Z_i^*(Z_i \setminus z_i) \quad (1 \le i \le k).$$

Show that Z_i contains all z_r such that $\pi(z_r) = (z_s, z_t)$ with $s \le i \le r$.)

8.1.5 Show that the set L_n of Lyndon words of length n over a k letter alphabet is a circular code. Show that L_n is comma-free if and only if $n = 1$ or ($n = 2, k \le 3$) or ($n = 3, 4$ and $k \le 2$).

8.1.6 (Lyndon–Schützenberger theorem) Show that if three words x, y, z satisfy the equation $x^m y^n = z^p$ with $m, n, p \ge 2$, then the three words x, y, z belong to the same cyclic submonoid t^*. (*Hint*: First prove that the conclusion holds if $p \ge 3$ considering the conjugate z' of z which is a Lyndon word. Then solve the case $p = 2$ using the fact that for some conjugate x' of x, the equality $x'^m = u^2 y^n$ holds for some u.)

8.1.7 Let $X = \{x, y\}$ be a code with two elements. Show that if X^* is not pure, then the set $x^*y \cup y^*x$ contains a word which is not primitive. (*Hint*: Consider the least integer $i \ge 1$ such that $w^2 \in X^*xy^ixX^*$. Replacing w by an X-conjugate, suppose that y^ix is a prefix of w and x a suffix of w. Let w' be an X-conjugate of w such that $wh = hw'$ and with h shorter than the word $z \in X$ such that $w' \in X^*z$. Distinguish three cases: (1) $w' \in yX^*x$, (2) $w' \in xX^*x$, (3) $w' \in X^*y$ and $|hx| > |y^i|$. Discuss cases (2) and (3) according to $|hx| > y^i$ or not.)

8.1.8 Deduce from Exercise 8.1.7 that if $x = uv$ and $y = vu$ are conjugate primitive words, then $X^* = \{x, y\}^*$ is pure.

8.1.9 Show that the coefficient of z^n in the series of Equation (7.13) is equal to the number of multisets of primitive necklaces meeting X^* whose total degree (that is, the sum of the lengths of the necklaces) is n. Give two proofs, one using Equation (7.17), the other by applying to the free monoid X^* the property of complete factorizations given in Corollary 8.1.7, using the fact that X^* is a very pure submonoid.

8.1.10 Take the notations of Exercise 7.3.5, with p_n as at the beginning of Section 7.3. Show that the v_n are nonnegative integers. (*Hint*: They are already integers using Equation (7.13). By iteration of the fundamental bisection of Example (8.2.2), show the existence of codes X_n and C_n, defined by: $X_1 = X$, $C_n = \{x \in X_n : |x| = n\}$, $X_{n+1} = (X_n \setminus C_n)C_n^*$ such that the free monoid X^* has the factorization $X^* = C_1^* C_2^* \cdots C_n^* X_{n+1}^*$. Show that v_n is the cardinality of C_n.)

8.1.11 A set $L \subset A^*$ is called *cyclic* if (i) for any words $u, v \in A^*$, one has $uv \in L$ if and only if $vu \in L$, and (ii) for any word $w \in A^*$ and any positive integer n, $w \in L$ if and only if $w^n \in L$. The *zeta function* of a set is given by the left-hand side of Equation (7.13), where p_n is the number of words of length n in L.

Show that if X is a circular code, then the closure under conjugacy of X^* is a cyclic set. Show that the latter is rational if the former is. Show that its zeta function is equal to the generating function of X^*. Show that more generally, the zeta function of a cyclic set L has the expansion given in the right-hand side of Equation (7.17), where

ℓ_n denotes the number of primitive necklaces of length n contained in L. Deduce that it has therefore natural integer coefficients.

Section 8.2

8.2.1 Show that if a factorization $A^* = X_n^* X_{n-1}^* \cdots X_1^*$ is obtained by a composition of bisections, then X_i is a $(i-1, n-i)$-limited code. (*Hint*: Use induction on n.)

8.2.2 Let X be a $(2, 0)$-limited code over A. Let $M \subset A^*$ be the submonoid generated by the suffixes of words in X. Show that M is right unitary. Let U be the prefix code generating M. Show that there exists a bisection of A^* of the form (U, Z). Show that X, considered as a code over U is $(1, 0)$-limited. Derive from this a trisection (X, Y, Z) of A^*. This shows that any $(2, 0)$-limited code is a left factor of some trisection.

8.2.3 Let $A = \{a, b, c, d, e, f, g\}$ and let $Y = \{d, eb, fa, ged, dac\}$. Show that Y is $(1, 1)$-limited. Show that there is no trisection of A^* of the form (X, Y, Z). (*Hint*: Use Proposition 8.2.9.)

8.2.4 Let $y \in A^+$ be an unbordered word. Show that there exists a trisection of A^* of the form (X, y, Z). Show that a prefix (resp. a suffix) of y is in Z^* (resp. X^*). (*Hint*: First construct a bisection (X', Z) of A^* such that X'^* is the submonoid generated by the suffixes of y.)

8.4 Notes

The notion of a factorization has been introduced by Schützenberger (1965a) in the paper where he proves Theorem 8.1.2. The factorizations of free monoids are very closely related with decompositions in direct sums of free Lie algebras. A complete treatment of this subject can be found in Viennot (1978) and in Lothaire (1997). Proposition 8.1.4 is a special case of a statement known as the Baker–Campbell–Hausdorff formula (see, e.g., Lothaire (1997)). The notion of a Lazard set is due to Viennot (1978). A series of examples of other factorizations and a bibliography on this field can be found in Lothaire (1997). Finite factorizations were studied by Schützenberger and Viennot. Theorem 8.2.4 is from Schützenberger (1965a). Theorem 8.2.6 is due to Viennot (1974). Viennot (1974) contains other results on finite factorizations. Among them, there is a necessary and sufficient condition in terms of the construction of Theorem 8.2.4 for the factors of a bisection to be recognizable. He also gives a construction of trisections analogous to that of bisections given in Theorem 8.2.4. Quadrisections have been studied by Krob (1987).

The factorization of Example 8.1.8 is due to Spitzer (see Lothaire (1997)). Exercise 8.1.6 is a theorem of Lyndon and Schützenberger (1962). The proof given in the Solutions follows Harju and Nowotka (2004). Exercises 8.1.7 and 8.1.8 are from Lentin and Schützenberger (1969). The proof given in the Solutions follows Barbin-Le Rest and Le Rest (1985).

Zeta functions of cyclic sets were introduced in Berstel and Reutenauer (1990). It is shown there that the zeta function of a rational cyclic set is a rational function (see also Béal *et al.* (1996)). Exercise 8.1.11 shows that this is true if the cyclic set is the closure under conjugacy of a rational circular code. In Reutenauer (1997), it is shown that each rational cyclic set is the disjoint union of the closure under conjugacy of rational very pure monoids. This implies that the zeta function is $\mathbb{N}$-rational.

Exercises 8.2.2 and 8.2.3 are from Viennot (1974).

9

Unambiguous monoids of relations

To each unambiguous automaton corresponds a monoid of relations which is also called unambiguous. A relation in this monoid corresponds to each word and the computations on words are replaced by computations on relations.

The principal result of this chapter (Theorem 9.4.1) shows that very thin codes are exactly the codes for which the associated monoid satisfies a finiteness condition: it contains relations of finite positive rank. This result explains why thin codes constitute a natural family containing the recognizable codes. It makes it possible to prove properties of thin codes by reasoning in finite structures. As a consequence, we shall give, for example, an alternative proof of the maximality of thin complete codes which does not use probabilities.

The main result also allows us to define, for each thin code, some important parameters: the degree and the group of the code. The group of a thin code is a finite permutation group. The degree of the code is the number of elements on which this group acts. These parameters reflect properties of words by means of "interpretations". For example, the synchronized codes in the sense of Chapter 3 are those having degree 1.

This chapter is organized in the following manner. In Section 9.1, basic properties of unambiguous monoids of relations are proved. These monoids constantly appear in what follows, since each unambiguous automaton gives rise to an unambiguous monoid of relations. In Section 9.2, we define two representations of unambiguous monoids of relations, called the $\mathcal{R}$ and $\mathcal{L}$-representations or Schützenberger representations. These representations are relative to a fixed idempotent chosen in the monoid, and they describe the way the elements of the monoid act by right or left multiplication on the $\mathcal{R}$-class and the $\mathcal{L}$-class of the idempotent.

The notion of rank of a relation is defined in Section 9.3. The most important result in this section states that the minimal ideal of an unambiguous monoid of relations is formed of the relations having minimal rank, provided that rank is finite (Theorem 9.3.10). Moreover, in this case the minimal ideal has a well-organized structure.

In Section 9.4 we return to codes. We define the notion of a very thin code which is a refinement of the notion of thin code. The two notions coincide for a complete code. Then we prove the fundamental theorem: A code X is very thin if and only if the associated unambiguous monoid of relations contains elements of finite positive rank (Theorem 9.4.1). Several consequences of this result on the structure of codes are given.

Section 9.5 contains the definition of the group and the degree of a code. The definition is given through the flower automaton, and then it is shown that it is independent of the automaton considered. We also show how the degree may be expressed in terms of interpretations of words.

9.1 Unambiguous monoids of relations

A *relation* m over P and Q is a subset of $P \times Q$. If $P = Q$, we say that m is a relation over P. If $(p, q) \in m$, we write equivalently

$$(p, q) \in m \iff (p, m, q) = 1 \iff pmq \iff p \xrightarrow{m} q \iff m_{p,q} = 1.$$
(9.1)

Each of these notations refers to a specific view of a relation. The fourth allows us to consider a relation as a graph, the third mimics order relations, the last one refers to the view of a relation as a matrix. Of course, one has the negations

$$(p, q) \notin m \iff (p, m, q) = 0 \iff m_{p,q} = 0.$$
(9.2)

In these expressions, 0 and 1 refer to the elements of the Boolean semiring. In particular, viewed as matrices, relations are Boolean matrices. Since 0 and 1 are elements of every semiring, every relation can also be viewed as a matrix with entries in this semiring. Similarly, a *row* or a *column* of a relation is a row or a column of the corresponding matrix. Thus $m_{p*} = \{q \in Q \mid m_{pq} = 1\}$ and $m_{*q} = \{p \in Q \mid m_{pq} = 1\}$.

Each partial function from P to Q is a particular relation over P and Q. In particular, a permutation of Q is a relation over Q.

The *product* of a relation m over P and Q and a relation n over Q and R is the relation mn defined by

$$(p, r) \in mn \iff \exists q \in Q : (p, q) \in m \text{ and } (q, r) \in n.$$

The set $\mathfrak{P}(Q \times Q)$ of relations over a set Q is a monoid for this product. The product is *unambiguous* if for each (p, r), there exists at most one $q \in Q$ such that $(p, q) \in m$ and $(q, r) \in n$.

If the relations are viewed as graphs, this amounts to the uniqueness of paths of length 2, that is $p \xrightarrow{m} q \xrightarrow{n} r, p \xrightarrow{m} q' \xrightarrow{n} r$ imply $q = q'$. Viewed as matrices, the definition is equivalent to the property that the value of the product of m and n has the same value in any semiring. In particular, viewed as matrices with entries in $\mathbb{N}$, the sums $\sum_{q \in Q} m_{p,q} n_{q,r}$ take only the values 0 or 1. Another way to view this is to observe that if r is a row of m, and ℓ is a column of n, there is at most one $q \in Q$ such that $r_q = \ell_q = 1$.

Example 9.1.1 Let m and n be the relations given in matrix form by

$$m = \begin{bmatrix} 0 & 1 & 0 \\ 1 & 0 & 0 \\ 1 & 1 & 0 \end{bmatrix}, \quad n = \begin{bmatrix} 0 & 0 & 1 \\ 0 & 0 & 0 \\ 1 & 1 & 0 \end{bmatrix}.$$

One checks that the product over the integers gives

$$mn = \begin{bmatrix} 0 & 0 & 0 \\ 0 & 0 & 1 \\ 0 & 0 & 1 \end{bmatrix},$$

and therefore the product of the relations is unambiguous.

A monoid of relations over Q is *unambiguous* if for each $m, n \in M$, the product mn is unambiguous. As a submonoid of $\mathfrak{P}(Q \times Q)$ it contains the identity id_Q.

Example 9.1.2 Every monoid of relation over a set Q which is composed of partial functions is unambiguous.

Example 9.1.3 The reader may check that the monoid generated by the matrices of Example 9.1.1 is unambiguous and has nine elements.

Recall that a monoid M of relations over Q is said to be *transitive* if for all $p, q \in Q$, there exists $m \in M$ such that $(p, q) \in m$.

Let $\mathcal{A} = (Q, I, T)$ be an automaton over A. Recall that, for each word w, we denote by $\varphi_\mathcal{A}(w)$ the relation over Q defined by

$$(p, q) \in \varphi_\mathcal{A}(w) \iff p \xrightarrow{w} q.$$

It follows from the definition that $\varphi_\mathcal{A}$ is a morphism from A^* into the monoid of relations over Q.

The next statement relates unambiguous monoids of relations and unambiguous automata.

Proposition 9.1.4 *Let $\mathcal{A}$ be an automaton over A. Then $\mathcal{A}$ is unambiguous if and only if the monoid $\varphi_\mathcal{A}(A^*)$ is unambiguous. Moreover, if $\mathcal{A} = (Q, 1, 1)$, then $\mathcal{A}$ is trim if and only if the monoid $\varphi_\mathcal{A}(A^*)$ is transitive.*

Proof. Assume there are paths $p \xrightarrow{u} r \xrightarrow{v} q$ and $p \xrightarrow{u} r' \xrightarrow{v} q$ in $\mathcal{A}$. If $r \neq r'$, the product of $\varphi_\mathcal{A}(u)$ and $\varphi_\mathcal{A}(v)$ is ambiguous, and conversely.

Next let $\mathcal{A} = (Q, 1, 1)$ be a trim automaton. Let $p, q \in Q$. Let $u, v \in A^*$ be such that $p \xrightarrow{u} 1$ and $1 \xrightarrow{v} q$ are paths. Then $p \xrightarrow{uv} q$ is a path and consequently $p \varphi_\mathcal{A}(uv) q$. The converse is clear. $\qquad\square$

A relation m over Q is *invertible* if there is a relation n over Q such that $mn = nm = I_Q$ where I_Q is the identity relation over Q.

Proposition 9.1.5 *A relation is invertible if and only if it is a permutation.*

Proof. Let m be an invertible relation, and let n be a relation such that $mn = nm = I_Q$. For all $p \in Q$, there exists $q \in Q$ such that pmq, since from $pmnp$ we get $pmqnp$ for some $q \in Q$. This element q is unique: if pmq', then $qnpmq' = qI_Qq'$, whence $q = q'$. This shows that m is a function. Now if pmq and $p'mq$, then $pmqnp$ and $p'mqnp$, implying $p' = p$. Thus m is injective. Since $nm = I_Q$, m is also surjective. Consequently m is a permutation. The converse is clear. $\qquad\square$

Let m be a relation over a set Q. A *fixed point* of m is an element $q \in Q$ such that qmq. In matrix form, the fixed points are the indices q such that $m_{q,q} = 1$, in other words those for which there is a 1 on the diagonal. We denote by $\mathrm{Fix}(m)$ the set of fixed points of m.

Proposition 9.1.6 *Let M be an unambiguous monoid of relations over Q. Let $e \in M$ and let $S = \mathrm{Fix}(e)$. The following conditions are equivalent:*

(i) *e is idempotent.*
(ii) *For all $p, q \in Q$, we have $p \xrightarrow{\ e\ } q$ if and only if there exists an $s \in S$ such that $p \xrightarrow{\ e\ } s$ and $s \xrightarrow{\ e\ } q$.*
(iii) *We have*

$$e = \ell r \quad \text{and} \quad r\ell = I_S, \tag{9.3}$$

where $\ell \subset Q \times S$ and $r \subset S \times Q$ are the restrictions of e to $Q \times S$ and $S \times Q$, respectively.

If e is idempotent, then moreover in matrix form

$$\ell = \begin{bmatrix} I_S \\ \ell' \end{bmatrix}, \quad r = \begin{bmatrix} I_S & r' \end{bmatrix}, \quad e = \begin{bmatrix} I_S & r' \\ \ell' & \ell'r' \end{bmatrix},$$

with $\ell' \subset (Q \setminus S) \times S$, $r' \subset S \times (Q \setminus S)$ and $r'\ell' = 0$. In particular, e is the identity on $\mathrm{Fix}(e)$.

The decomposition (9.3) of an idempotent relation is called the *column-row decomposition* of the relation. Note that

$$e\ell = \ell, \quad re = r, \tag{9.4}$$

since for instance $re = r\ell r = rI_S = r$.

Proof. (i) $\Rightarrow$ (ii). Let $p, q \in Q$ be such that peq. Then pe^3q. Consequently, there are $s, t \in Q$ such that $peseteq$. It follows that $peseq$ and $peteq$. Since M is unambiguous, we have $s = t$, whence ses and $s \in S$. The converse is clear.

(ii) $\Rightarrow$ (iii). Let ℓ and r be the restrictions of e to $Q \times S$ and $S \times Q$, respectively. If peq, then there exists $s \in S$ such that pes and seq. Then $p\ell s$ and srq. Conversely if $p\ell s$ and srq, then we have $peseq$, thus peq. Since this fixed point s is unique, we have $e = \ell r$.

Now let $r, s \in S$ with $rr\ell s$. Then $rrq\ell s$ for some $q \in Q$. Thus req and qes. Moreover, rer and ses, whence

$$rereqes, \quad reqeses.$$

The unambiguity implies that $r = q = s$. Conversely we have $sr\ell s$ for all $s \in S$. Thus $r\ell = \mathrm{id}_S$.

(iii) $\Rightarrow$ (i). We have $e^2 = \ell r\ell r = \ell(r\ell)r = \ell r = e$. Thus e is idempotent.

Assume now that e is idempotent. The restriction of e to $S \times S$ is the identity. Indeed ses holds for all $s \in S$, and if ser with $s, r \in S$, then $seser$ and $serer$, implying $s = r$ by unambiguity. This shows that ℓ and r have the indicated form. Finally, the product $r\ell$ is

$$r\ell = I_S + r'\ell'.$$

Since $r\ell = I_S$, this implies that $r'\ell' = 0$, which concludes the proof. □

Let M be an unambiguous monoid of relations over Q and let $e \in M$ be an idempotent. Then eMe is a monoid, and e is the neutral element of eMe, since for all $m \in eMe$, $em = me = eme = m$. It is the greatest monoid contained in M and having neutral element e. It is called the *monoid localized* at e (cf. Section 1.2). The $\mathcal{H}$-class $H(e)$ of e is the group of units of the monoid eMe (Proposition 1.12.4).

Proposition 9.1.7 *Let M be an unambiguous monoid of relations over Q, let e be an idempotent in M and let $S = \mathrm{Fix}(e)$ be the set of fixed points of e. The restriction γ of the elements of eMe to $S \times S$ is an isomorphism of eMe onto an unambiguous monoid of relations over S. If $e = \ell r$ is the column-row decomposition of e, this isomorphism is given by*

$$\gamma : m \mapsto rm\ell. \tag{9.5}$$

The set $\gamma(H(e))$ is a permutation group over S. Further, if M is transitive, then $\gamma(eMe)$ is transitive.

The unambiguous monoid of relations $\gamma(eMe)$ is denoted by M_e, and the permutation group $\gamma(H(e))$ is denoted by G_e.

Proof. Let γ be the function defined by (9.5). If $m \in eMe$, then for $s, t \in S$,

$$(s, \gamma(m), t) = (s, rm\ell, t) = (s, m, t),$$

because we have srs and $t\ell t$. Thus $\gamma(m)$ is the restriction of the elements in eMe to $S \times S$. Further, γ is a morphism since

$$\gamma(e) = re\ell = \mathrm{id}_S$$

and for $m, n \in eMe$,

$$\gamma(mn) = \gamma(men) = r(men)\ell = rm\ell rn\ell = \gamma(m)\gamma(n).$$

Finally γ is injective since if $\gamma(m) = \gamma(n)$ for some $m, n \in eMe$, then also $\ell\gamma(m)r = \ell\gamma(n)r$. But $\ell\gamma(m)r = \ell rm\ell r = eme = m$. Thus $m = n$. The monoid

$$M_e = \gamma(eMe)$$

is a monoid of relations over S since it contains the relation id_S. It is unambiguous as any restriction of an unambiguous monoid of relations.

Finally $G_e = \gamma(H(e))$ is composed of invertible relations. By Proposition 9.1.5, it is a permutation group over S.

If M is transitive, consider $s, t \in S$. There exists $m \in M$ such that smt. Then also $semet$. Taking the restriction to S, we have $s\gamma(eme)t$. Since $\gamma(eme) \in M_e$ this shows that M_e is transitive. $\qquad\square$

Example 9.1.8 Consider the relation m given in matrix form by

$$
m = \begin{bmatrix} 0 & 1 & 0 & 0 \\ 1 & 0 & 1 & 0 \\ 0 & 0 & 0 & 0 \\ 1 & 0 & 1 & 0 \end{bmatrix}.
$$

Then

$$
m^2 = \begin{bmatrix} 1 & 0 & 1 & 0 \\ 0 & 1 & 0 & 0 \\ 0 & 0 & 0 & 0 \\ 0 & 1 & 0 & 0 \end{bmatrix},
$$

and $m^3 = m$. Thus m^2 is an idempotent relation. The monoid $M = \{1, m, m^2\}$ is an unambiguous monoid of relations. The fixed points of the relation $e = m^2$ are 1 and 2, and its column-row decomposition is

$$
e = \begin{bmatrix} 1 & 0 \\ 0 & 1 \\ 0 & 0 \\ 0 & 1 \end{bmatrix} \begin{bmatrix} 1 & 0 & 1 & 0 \\ 0 & 1 & 0 & 0 \end{bmatrix} = \ell r.
$$

We have

$$
m = \ell \begin{bmatrix} 0 & 1 \\ 1 & 0 \end{bmatrix} r,
$$

and the restriction of m to the set $\{1, 2\}$ is the transposition (12). The monoid M_e is equal to the group G_e which is isomorphic to $\mathbb{Z}/2\mathbb{Z}$.

Let M be an arbitrary monoid. We compare now the localized monoids of two idempotents of a $\mathcal{D}$-class. Let e, e' be two $\mathcal{D}$-equivalent idempotents of M. Since, by definition, $\mathcal{D} = \mathcal{RL}$, there exists an element $d \in M$ such that $e\mathcal{R}d\mathcal{L}e'$. By definition of these relations, there exists a quadruple

$$
(a, a', b, b') \tag{9.6}
$$

of elements of M such that

$$
ea = d, \quad da' = e, \quad bd = e', \quad b'e' = d. \tag{9.7}
$$

(see Figure 9.1). The quadruple (9.6) is a *passing system* from e to e'.

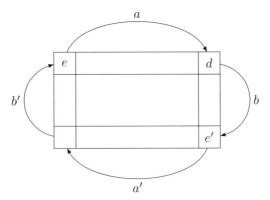

Figure 9.1 The passing system. Right multiplication by a or a' is represented by a horizontal arrow and left multiplication by b or b' is represented by a vertical arrow.

The following formulas are easily derived from (9.7) (Note that most of these identities appear in Section 1.12):

$$eaa' = e, \quad bea = e', \quad ea = b'e', \tag{9.8}$$

and

$$bb'e' = e', \quad b'e'a' = e, \quad be = e'a' \tag{9.9}$$

(the last formula is obtained by $be = bb'e'a' = e'a'$). Since e and e' are idempotents, the following hold also:

$$eabe = e, \quad e'a'b'e' = e'. \tag{9.10}$$

Indeed, we have by (9.8), $e' = e'e' = beabea$. Thus $b'e'a' = b'beabeaa'$. Since $be = e'a'$ by (9.9), one has $b'be = b'e'a' = e$ and since by (9.8), $eaa' = e$, we obtain $b'e'a' = e = eabe$. This proves the first equality. The second one is proved in the same way.

Two monoids of relations M over Q and M' over Q' are *equivalent* if there exists a relation $\theta \in \mathfrak{P}(Q \times Q')$ which is a bijection from Q onto Q' such that the function

$$m \mapsto \theta^t m \theta$$

is an isomorphism from M onto M' (θ^t is the transposed of θ). Since θ is a bijection, we have $\theta^t = \theta^{-1}$. Therefore, in the case where M and M' are permutation groups, this definition coincides with the one given in Section 1.13.

Proposition 9.1.9 *Let M be an unambiguous monoid of relations over Q, and let $e, e' \in M$ be two $\mathcal{D}$-equivalent idempotents. Then the monoids eMe and $e'Me'$ are isomorphic, the monoids M_e and $M_{e'}$ are equivalent, and the groups G_e and $G_{e'}$ are equivalent permutation groups. More precisely, let $S = \mathrm{Fix}(e)$, $S' = \mathrm{Fix}(e')$, let $e = \ell r$, $e' = \ell' r'$ be their column-row decompositions, let γ and γ' be the restrictions to $S \times S$ and $S' \times S'$ and let (a, a', b, b') be a passing system from e to e'. Then*

1. *The function $\tau : m \mapsto bma$ is an isomorphism from eMe onto $e'Me'$.*
2. *The relation $\theta = ra\ell' = rb'\ell' \in \mathfrak{P}(S \times S')$ is a bijection from S onto S'.*
3. *The function $\tau' : n \mapsto \theta'n\theta$ is an isomorphism from M_e onto $M_{e'}$.*
4. *The following diagram is commutative*

$$
\begin{array}{ccc}
eMe & \xrightarrow{\ \tau\ } & e'Me' \\
{\scriptstyle \gamma}\big\downarrow & & \big\downarrow{\scriptstyle \gamma'} \\
M_e & \xrightarrow{\ \tau'\ } & M_{e'}
\end{array}
$$

Proof. 1. Let $m \in eMe$. Then $\tau(m) = bma = bemea = e'a'mb'e'$, since by (9.8) and (9.9), $be = e'a'$ and $b'e' = ea$. This shows that $\tau(m)$ is in $e'Me'$. Next $\tau(e) = bea = e'$ by (9.8). For $m, m' \in eMe$, we have by (9.10)

$$\tau(m)\tau(m') = bmabm'a = bmeabem'a = bmem'a = bmm'a = \tau(mm').$$

Thus τ is a morphism. Finally, it is easily seen that $m' \mapsto b'm'a'$ is the inverse function of τ; thus τ is an isomorphism from eMe onto $e'Me'$.

2. We have $eae' = eb'e'$. Consequently $reae'\ell' = reb'e'\ell'$. Since by (9.4) $re = r$, $e'\ell' = \ell'$, we get that

$$\theta = ra\ell' = rb'\ell'.$$

The relation θ is left invertible since

$$(r'b\ell)\theta = r'b\ell ra\ell' = r'bea\ell' = r'e'\ell' = \mathrm{id}_{S'},$$

and it is right invertible, since we have

$$\theta(r'a'\ell) = rb'\ell'r'a'\ell = rb'e'a'\ell = re\ell = \mathrm{id}_S.$$

Thus θ is invertible and consequently is a bijection, and $\theta' = r'a'\ell = r'b\ell$.

4. For $m \in eMe$, we have

$$\tau'\gamma(m) = (r'b\ell)(rm\ell)(ra\ell') = r'bemea\ell' = r'(bma)\ell' = \gamma'\tau(m),$$

showing that the diagram is commutative.

3. Results from the commutativity of the diagram and from the fact that γ, τ, γ' are isomorphisms. $\qquad\square$

Example 9.1.10 Consider the matrices

$$
u = \begin{bmatrix} 0 & 0 & 0 & 1 \\ 1 & 0 & 1 & 0 \\ 0 & 0 & 0 & 0 \\ 0 & 1 & 0 & 0 \end{bmatrix}, \quad
v = \begin{bmatrix} 0 & 0 & 1 & 0 \\ 1 & 0 & 1 & 0 \\ 1 & 0 & 0 & 0 \\ 0 & 1 & 0 & 0 \end{bmatrix}.
$$

They generate an unambiguous monoid of relations (as we may verify by using, for instance, the method of Proposition 4.2.5). The matrix

$$uv = \begin{bmatrix} 0 & 1 & 0 & 0 \\ 1 & 0 & 1 & 0 \\ 0 & 0 & 0 & 0 \\ 1 & 0 & 1 & 0 \end{bmatrix}$$

is the matrix m of Example 9.1.8. The element

$$e = (uv)^2 = \begin{bmatrix} 1 & 0 & 1 & 0 \\ 0 & 1 & 0 & 0 \\ 0 & 0 & 0 & 0 \\ 0 & 1 & 0 & 0 \end{bmatrix}$$

is an idempotent. We have $\text{Fix}(e) = \{1, 2\}$, and the column-row decomposition is

$$e = \begin{bmatrix} 1 & 0 \\ 0 & 1 \\ 0 & 0 \\ 0 & 1 \end{bmatrix} \begin{bmatrix} 1 & 0 & 1 & 0 \\ 0 & 1 & 0 & 0 \end{bmatrix} = \boldsymbol{\ell r}.$$

The matrix

$$e' = (vu)^2 = \begin{bmatrix} 0 & 0 & 0 & 0 \\ 1 & 0 & 1 & 0 \\ 1 & 0 & 1 & 0 \\ 0 & 0 & 0 & 1 \end{bmatrix}$$

is also an idempotent. We have $\text{Fix}(e') = \{3, 4\}$, and e' has the column-row decomposition

$$e' = \begin{bmatrix} 0 & 0 \\ 1 & 0 \\ 1 & 0 \\ 0 & 1 \end{bmatrix} \begin{bmatrix} 1 & 0 & 1 & 0 \\ 0 & 0 & 0 & 1 \end{bmatrix} = \boldsymbol{\ell' r'}.$$

The idempotents e and e' lie in the same $\mathcal{D}$-class. Indeed, we may take as a passing system from e to e' the elements

$$a = b' = u, \qquad a' = b = vuv.$$

The bijection $\theta = \boldsymbol{r a \ell'}$ from the set $\text{Fix}(e) = \{1, 2\}$ onto the set $\text{Fix}(e') = \{3, 4\}$ is

$$\theta : 1 \mapsto 4, 2 \mapsto 3.$$

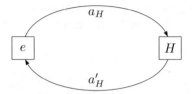

Figure 9.2 Two coordinates. The pair (a_H, a'_H) satisfies $ea_H \in H$ and $ea_H a'_H = e$.

9.2 The Schützenberger representations

We now describe a useful method for computing the permutation group G_e for an idempotent e in an unambiguous monoid of relations. This method requires us to make a choice between "left" and "right". We first present the right-hand case.

Let M be an unambiguous monoid of relations, and let e be an idempotent element in M. Let R be the $\mathcal{R}$-class of e, let Λ be the set of $\mathcal{H}$-classes of R and let $G = H(e)$ be the $\mathcal{H}$-class of e. For each $H \in \Lambda$, choose two elements $a_H, a'_H \in M$ such that

$$ea_H \in H, \quad ea_H a'_H = e,$$

with the convention that

$$a_G = a'_G = e.$$

(see Figure 9.2). Such a set of pairs $(a_H, a'_H)_{H \in \Lambda}$ is called a *system of coordinates* of R relative to the idempotent e. Then, by Proposition 1.12.2, $Ga_H = H$ and $Ha'_H = G$ since the elements a_H, a'_H realize by right multiplication two reciprocal bijections from G onto H.

Let $e = \boldsymbol{\ell r}$ be the column-row decomposition of e, and set

$$\boldsymbol{r}_H = \boldsymbol{r}a_H \quad \text{and} \quad \boldsymbol{\ell}_H = a'_H \boldsymbol{\ell} \quad \text{for } H \in \Lambda. \tag{9.11}$$

Note that the equality $\boldsymbol{r}_H = \boldsymbol{r}ea_H$ follows from $\boldsymbol{r} = \boldsymbol{r}e$, which is (9.4).

Each $m \in M$ defines a partial right action on the set Λ by setting, for all $H \in \Lambda$

$$H \cdot m = \begin{cases} Hm & \text{if } Hm \in \Lambda, \\ \emptyset & \text{otherwise.} \end{cases} \tag{9.12}$$

Now we define a partial function from $\Lambda \times M$ into G by setting

$$H * m = \begin{cases} \boldsymbol{r}_H m \boldsymbol{\ell}_{Hm} & \text{if } Hm \in \Lambda, \\ \emptyset & \text{otherwise.} \end{cases} \tag{9.13}$$

First, observe that $H \cdot m \neq \emptyset$ implies $H * m \in G_e$. Indeed, set $H' = Hm$. From $ea_H \in H$ we get $ea_H m \in H'$, showing that

$$ea_H m a'_{H'} \in G.$$

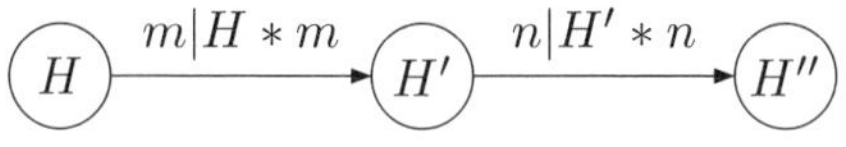

Figure 9.3 Composition of outputs. The label of an edge from H to $H' = H \cdot m$ is the pair $(m, H * m)$, denoted $m|H * m$.

It follows that

$$H * m = r_H m \ell_{H'} = (rea_H)m(a'_{H'}\ell)$$

$$= r(ea_H m a'_{H'})\ell \in G_e.$$

Observe also that for all $H \in \Lambda$,

$$H \cdot 1 = H \quad \text{and} \quad H * 1 = e. \tag{9.14}$$

Next, for all $m, n \in M$,

$$(H * m)(H \cdot m * n) = H * mn. \tag{9.15}$$

This formula shows that the functions $(H, m) \mapsto H \cdot m$ and $(H, m) \mapsto H * m$ are similar to those associated to a deterministic transducer, as defined in Chapter 4.

To verify Formula (9.15), let $H' = Hm$, $H'' = Hmn$ (the cases where $H \cdot m = \emptyset$ or $H \cdot mn = \emptyset$ are straightforward). See Figure 9.3. We have

$$(H * m)(H' * n) = r_H m \ell_{H'} r_H n \ell_{H''} = r_H m a'_{H'} \ell r a_{H'} n \ell_{H''}$$

$$= r a_H m a'_{H'} e a_{H'} n a'_{H''} \ell$$

$$= r((ea_H m a'_{H'})e)a_{H'} n a'_{H''}\ell.$$

(We have used (9.4).) Since $ea_H m a'_{H'} \in G$, we have $ea_H m a'_{H'} e = ea_H m a'_{H'}$. Thus

$$(H * m)(H' * n) = r((ea_H m)a'_{H'} a_{H'})na'_{H''}\ell.$$

Since $ea_H m \in H'$, and because the multiplication on the right by $a'_{H'} a_{H'}$ is the identity on H', we get

$$(H * m)(H' * n) = rea_H m n a'_{H''}\ell = r_H m n \ell_{H''} = H * mn.$$

This proves Formula (9.15). As a consequence, we have the following result.

Proposition 9.2.1 *Let M be an unambiguous monoid of relations generated by a set T. Let e be an idempotent of M, let R be its $\mathcal{R}$-class, let Λ be the set of $\mathcal{H}$-classes of R and let $(a_H, a'_H)_{H \in \Lambda}$ be a system of coordinates of R relative to e. Then the permutation group G_e is generated by the elements of the form $H * t$, for $H \in \Lambda$, $t \in T$, and $H * t \neq \emptyset$.*

Proof. The elements $H * t$, for $H \in \Lambda$ and $t \in T$ either are $\emptyset$ or are in G_e. Now let g be an element of $H(e)$. Then there are $t_1, \ldots, t_n \in T$ with

$$g = t_1 t_2 \cdots t_n,$$

because T generates M. Let $G = H(e)$ and let

$$H_i = G t_1 t_2 \cdots t_i$$

for $1 \leq i \leq n$. From $Gg = G$ it follows that $H_i t_{i+1} \cdots t_n = G$. Thus $H_i \in \Lambda$ and $G \cdot t_1 \cdots t_i = H_i$. By (9.15),

$$G * g = (G * t_1)(H_1 * t_2) \cdots (H_{n-1} * t_n).$$

But $G * g = r g \ell$. This shows the result. □

The pair of partial functions from $\Lambda \times M$ to Λ and to G_e defined by (9.12) and (9.13) is called the *right Schützenberger representation* or $\mathcal{R}$-*representation* of M relative to e and to the coordinate system $(a_H, a'_H)_{H \in \Lambda}$.

Let 0 be a new element such that $0g = g0 = 00 = 0$ for all $g \in G_e$. The function

$$\mu : M \to (G_e \cup 0)^{\Lambda \times \Lambda},$$

which associates to each $m \in M$ the $\Lambda \times \Lambda$-matrix defined by

$$(\mu m)_{H,H'} = \begin{cases} H * m & \text{if } Hm = H', \\ 0 & \text{otherwise,} \end{cases}$$

is a morphism from M into the monoid of row-monomial $\Lambda \times \Lambda$-matrices with elements in $G_e \cup 0$. This is indeed an equivalent formulation of Formula (9.15).

Symmetrically, we define the *left Schützenberger representation* or $\mathcal{L}$-*representation* of M relative to e as follows. Let L be the $\mathcal{L}$-class of e, and let Γ be the set of its $\mathcal{H}$-classes. For each $H \in \Gamma$, choose two elements $b_H, b'_H \in M$ such that

$$b_H e \in H, \quad b'_H b_H e = e,$$

with $b_G = b'_G = e$. Such a set of pairs $(b_H, b'_H)_{H \in \Gamma}$ is called a *system of coordinates* of L with respect to e. As in (9.11), we set $\ell^H = b_H c$, $r^H = r b'_H$ for $H \in \Gamma$.

For each $m \in M$, we define a partial left action on Γ by setting, for $H \in \Gamma$,

$$m \cdot H = \begin{cases} mH & \text{if } mH \in \Gamma, \\ \emptyset & \text{otherwise,} \end{cases} \tag{9.16}$$

and a partial function from $M \times \Gamma$ into G_e by setting

$$m * H = \begin{cases} r^{mH} m \ell^H & \text{if } mH \in \Gamma, \\ \emptyset & \text{otherwise.} \end{cases} \tag{9.17}$$

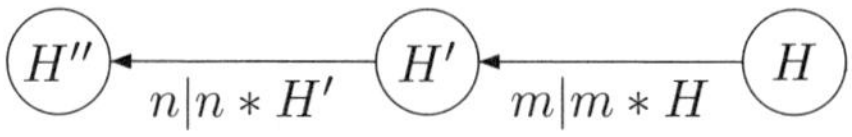

Figure 9.4 Composition of outputs. The label of an edge from H to $H' = m \cdot H$ is the pair $(m, m * H)$, denoted $m|m * H$. Note that the input is read from right to left and that the output is written from right to left.

Then Formula (9.15) becomes

$$(n * m \cdot H)(m * H) = nm * H \tag{9.18}$$

and Proposition 9.2.1 holds mutatis mutandis.

Note that for the computation of the $\mathcal{L}$-classes and the $\mathcal{R}$-classes of an unambiguous monoid of relations, we can use the following observation, whose verification is straightforward: If $m\mathcal{L}n$ (resp. if $m\mathcal{R}n$), then each row (resp. column) of m is a sum of rows (resp. columns) of n and vice versa. This yields an easy test to conclude that two elements are in *distinct* $\mathcal{L}$-classes (resp. $\mathcal{R}$-classes).

Example 9.2.2 Let us consider again the unambiguous monoid of Example 9.1.10, generated by the matrices

$$u = \begin{bmatrix} 0 & 0 & 0 & 1 \\ 1 & 0 & 1 & 0 \\ 0 & 0 & 0 & 0 \\ 0 & 1 & 0 & 0 \end{bmatrix}, \quad v = \begin{bmatrix} 0 & 0 & 1 & 0 \\ 1 & 0 & 1 & 0 \\ 1 & 0 & 0 & 0 \\ 0 & 1 & 0 & 0 \end{bmatrix}.$$

We consider the idempotent

$$e = (uv)^2 - \begin{bmatrix} 1 & 0 & 1 & 0 \\ 0 & 1 & 0 & 0 \\ 0 & 0 & 0 & 0 \\ 0 & 1 & 0 & 0 \end{bmatrix}.$$

Its $\mathcal{R}$-class R is formed of three $\mathcal{H}$-classes, numbered 0, 1, 2. In Figure 9.5 a representative is given for each of these $\mathcal{H}$-classes. The fact that the $\mathcal{L}$-classes are distinct is verified by inspecting the rows of e, eu, eu^2. Next, we note that $eu^3 = eu^2v = e$, showing that these elements are $\mathcal{R}$-equivalent. Further, $euv = (uv)^3 \mathcal{H} e$. Finally

$$ev = \begin{bmatrix} 1 & 0 & 1 & 0 \\ 1 & 0 & 1 & 0 \\ 0 & 0 & 0 & 0 \\ 1 & 0 & 1 & 0 \end{bmatrix}$$

has only one nonnull row (column) and consequently cannot be in the $\mathcal{D}$-class of e. We have reported in Figure 9.5 the effect of the right multiplication by u and v.

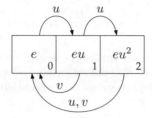

Figure 9.5 The $\mathcal{R}$-class of the idempotent e.

We choose a system of coordinates of R by setting

$$a_0 = a_0' = e,$$

$$a_1 = u, \quad a_1' = vuv,$$

$$a_2 = u^2, \quad a_2' = u.$$

Then

$$r_0 = \begin{bmatrix} 1 & 0 & 1 & 0 \\ 0 & 1 & 0 & 0 \end{bmatrix}, \quad \ell_0 = \begin{bmatrix} 1 & 0 \\ 0 & 1 \\ 0 & 0 \\ 0 & 1 \end{bmatrix},$$

$$r_1 = \begin{bmatrix} 0 & 0 & 0 & 1 \\ 1 & 0 & 1 & 0 \end{bmatrix}, \quad \ell_1 = \begin{bmatrix} 0 & 0 \\ 0 & 1 \\ 0 & 1 \\ 1 & 0 \end{bmatrix},$$

$$r_2 = \begin{bmatrix} 0 & 1 & 0 & 0 \\ 0 & 0 & 0 & 1 \end{bmatrix}, \quad \ell_2 = \begin{bmatrix} 0 & 1 \\ 1 & 0 \\ 0 & 0 \\ 0 & 1 \end{bmatrix}.$$

Let us denote by $H \xrightarrow{t|g} H'$ the fact that $H \cdot t = H'$ and $H * t = g$. Then the $\mathcal{R}$-representation of M relative to e and to this system of coordinates is obtained by completing Figure 9.5 and is given in Figure 9.6 with

$$i = \begin{bmatrix} 1 & 0 \\ 0 & 1 \end{bmatrix}, \quad j = \begin{bmatrix} 0 & 1 \\ 1 & 0 \end{bmatrix}.$$

The group G_e is of course $\mathfrak{S}_2$.

The concepts introduced in this paragraph are greatly simplified when we consider the case of a monoid of (total) *functions* from Q into itself, instead of an unambiguous monoid of relations.

For $a \in M$, write $pa = q$ instead of $(p, a, q) = 1$.

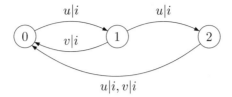

Figure 9.6 The $\mathcal{R}$-representation of M.

The *image* of a, denoted $\mathrm{Im}(a)$, is the set of $q \in Q$ such that $pa = q$ for some $p \in Q$. The *nuclear equivalence* of a, denoted $\mathrm{Ker}(a)$, is the equivalence relation on Q defined by $p \equiv q \bmod \mathrm{Ker}(a)$ if and only if $pa = qa$. If $b \in Ma$, then $\mathrm{Im}(b) \subset \mathrm{Im}(a)$. If $b \in aM$, then $\mathrm{Ker}(a) \subset \mathrm{Ker}(b)$ (note the inversion of inclusions).

A function $e \in M$ is idempotent if and only if its restriction to its image is the identity. Thus, its image is in this case equal to its set of fixed points: $\mathrm{Im}(e) = \mathrm{Fix}(e)$.

As a result of what precedes, if $a\mathcal{L}b$, then $\mathrm{Im}(a) = \mathrm{Im}(b)$ and if $a\mathcal{R}b$, then $\mathrm{Ker}(a) = \mathrm{Ker}(b)$. This gives a sufficient condition to ensure that two elements are in different $\mathcal{L}$-classes (resp. $\mathcal{R}$-classes).

To compute the $\mathcal{R}$-class of an idempotent function e over a finite set, we may use the following observation, where $S = \mathrm{Fix}(e)$. If the restriction of m to S is a permutation on S, then $e\mathcal{H}em$. Indeed, the restriction of m to S is a permutation on S, thus $em^p = e$ for some p, therefore $emm^{p-1} = e$ and thus $em\mathcal{H}e$.

Example 9.2.3 Let M be the monoid of functions from the set

$$Q = \{1, 2, \ldots, 8\}$$

into itself generated by the two functions u and v given in the following array

	1	2	3	4	5	6	7	8
u	4	5	4	5	8	1	8	1
v	2	3	4	5	6	7	8	1

where each column contains the images by u and v of the element of Q placed on the top of the column. The function $e = u^4$ is idempotent and has the set of fixed points $S = \{1, 4, 5, 8\}$,

	1	2	3	4	5	6	7	8
u^4	1	4	1	4	5	8	5	8

We get the pattern of Figure 9.7 for the $\mathcal{R}$-class R of e. These four $\mathcal{H}$-classes are distinct because the images of e, ev, ev^2, ev^3 are distinct. For the edges going back to the $\mathcal{H}$-class of e, we use the observation stated above; it suffices to verify that the restrictions to S of the functions u, vu, v^2u, v^3u, v are permutations. Choose a

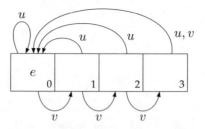

Figure 9.7 The $\mathcal{R}$-class of the idempotent e.

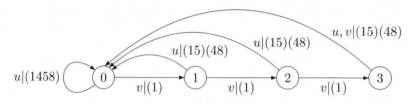

Figure 9.8 The $\mathcal{R}$-representation.

system of coordinates of R by taking

$$a_0 = a_0' = e,$$
$$a_1 = v, \quad a_1' = v^7,$$
$$a_2 = v^2, \quad a_2' = v^6,$$
$$a_3 = v^3, \quad a_3' = v^5.$$

For the computation of the $\mathcal{R}$-representation of M relative to e, we proceed as follows: if $H \cdot m = H'$, then the permutation $H * m$ on S is not computed by computing the matrix product $H * m = r_H m \ell_{H'}$ of Formula (9.13), but, observing that $H * m$ is the restriction to S of $e a_H m a'_{H'} e$, by evaluating this function on S. Thus we avoid unnecessary matrix computations when dealing with functions. Figure 9.8 shows the $\mathcal{R}$-representation obtained.

According to Proposition 9.2.1, the group G_e is generated by the permutations

$$(1458), \quad (15)(48), \quad (14)(58).$$

It is the *dihedral* group D_4 which is the group of all symmetries of the square.

$$
\begin{array}{ccc}
1 & \!\!\!-\!\!\! & 4 \\
| & & | \\
8 & \!\!\!-\!\!\! & 5
\end{array}
$$

It contains eight elements.

9.3 Rank and minimal ideal

Let m be a relation between two sets P and Q. The *rank* of m is the minimum of the cardinalities of the sets R such that there exist two relations $\ell \in \mathfrak{P}(P \times R)$ and $r \in \mathfrak{P}(R \times Q)$ with

$$m = \ell r, \tag{9.19}$$

and such that the product ℓr is unambiguous. The rank is denoted by $\text{rank}(m)$. It is a nonnegative integer or $+\infty$. A pair (ℓ, r) satisfying (9.19) is a *minimal decomposition* if there exists no unambiguous factorization $m = \ell' r'$ with $\ell' \in \mathfrak{P}(P \times R')$, $r' \in \mathfrak{P}(R' \times Q)$ and $R' \subsetneq R$. If $\text{rank}(m)$ is finite, this is the equivalent of saying that $\text{Card}(R)$ is minimal.

Example 9.3.1 The relation

$$m = \begin{bmatrix} 0 & 1 & 0 \\ 1 & 0 & 0 \\ 1 & 1 & 0 \end{bmatrix} = \begin{bmatrix} 0 & 1 \\ 1 & 0 \\ 1 & 1 \end{bmatrix} \begin{bmatrix} 1 & 0 & 0 \\ 0 & 1 & 0 \end{bmatrix}$$

has rank at most 2 in view of the above decomposition. It does not have rank 1 because it has two distinct nonzero columns. Thus, m has rank 2.

The following properties are used frequently. First, if the product nmn' is unambiguous, then

$$\text{rank}(nmn') \leq \text{rank}(m). \tag{9.20}$$

Indeed, each decomposition (ℓ, r) of m induces a decomposition $(n\ell, rn')$ of nmn'. If $p \xrightarrow{n} s \xrightarrow{\ell} t \xrightarrow{r} u \xrightarrow{n} q$ and $p \xrightarrow{n} s' \xrightarrow{\ell} t' \xrightarrow{r} u' \xrightarrow{n} q$, then $s = s'$ and $u = u'$ by the unambiguity of the product nmn'. The unambiguity of the product ℓr forces $t = t'$. Second

$$\text{rank}(m) \leq \min\{\text{Card}(P), \text{Card}(Q)\}.$$

If (ℓ, r) is a minimal decomposition of m, then

$$\text{rank}(m) = \text{rank}(\ell) = \text{rank}(r).$$

Further

$$\text{rank}(m) = 0 \Leftrightarrow m = 0.$$

If $P' \subset P$, $Q' \subset Q$, and if m' is the restriction of m to $P' \times Q'$, then

$$\text{rank}(m') \leq \text{rank}(m). \tag{9.21}$$

We get from the first inequality that two $\mathcal{J}$-equivalent elements of an unambiguous monoid of relations have the same rank. Thus, the rank is constant on a $\mathcal{D}$-class.

Consider two relations $m \in \mathfrak{P}(P \times S)$ and $n \in \mathfrak{P}(S \times Q)$. The pair (m, n) is called *trim* if no column of m is null and no row of n is null. This is equivalent to saying that for all $s \in S$, there exists at least one pair $(p, q) \in P \times Q$ such that $p \xrightarrow{m} s$ and $s \xrightarrow{n} q$.

Proposition 9.3.2 *Any minimal decomposition of a relation is trim.*

Proof. Let ℓr be a minimal decomposition of a relation m. Assume that ℓ contains a column which is null. Then we can delete this column and the row of same index of r without changing the value of the product. But this implies that (ℓ, r) is not a minimal decomposition. Thus no column of ℓ is null, and symmetrically no row of r is null. Consequently (ℓ, r) is trim. $\qquad\square$

Proposition 9.3.3 *For each set Q, $\mathrm{rank}(\mathrm{id}_Q) = \mathrm{Card}(Q)$.*

Proof. Let $\mathrm{id} = \ell r$ be a minimal decomposition of id_Q, with $\ell \in \mathfrak{P}(Q \times P)$ and $r \in \mathfrak{P}(P \times Q)$. Let $p \in P$. Since the pair (ℓ, r) is trim, there exist $q, q' \in Q$ such that $q \xrightarrow{\ell} p \xrightarrow{r} q'$. Since $\ell r = \mathrm{id}_Q$, one has $q = q'$, and there is no $q'' \neq q$ such that $p \xrightarrow{r} q''$. Thus r defines a mapping from P into Q. This mapping is surjective since id_Q is surjective. This implies that $\mathrm{Card}(P) = \mathrm{Card}(Q)$. $\qquad\square$

Proposition 9.3.4 *A permutation on Q has rank $\mathrm{Card}(Q)$.*

Proof. Let m be a permutation on Q and let n be its inverse. Then by Proposition 9.3.3 and Equation (9.20),

$$\mathrm{Card}(Q) = \mathrm{rank}(\mathrm{id}_Q) = \mathrm{rank}(mn) \le rank(m).$$

Thus $\mathrm{rank}(m) = \mathrm{Card}(Q)$. $\qquad\square$

Example 9.3.5 The rank of a partial function m from P to Q is

$$\mathrm{rank}(m) = \mathrm{Card}(\mathrm{Im}(m)).$$

Let m' be the restriction of m to $P \times \mathrm{Im}(m)$. Then $m = m'r$, where r is the restriction of id_Q to $\mathrm{Im}(m)$. This shows that $\mathrm{rank}(m) \le \mathrm{Card}(\mathrm{Im}(m))$. The partial function m' contains a bijection n of a cross-section of m onto $\mathrm{Im}(m)$ obtained by choosing one element in P for each set $m^{-1}(q)$, with $q \in \mathrm{Im}(m)$. By Proposition 9.3.4 and Equation 9.21, $\mathrm{rank}(m) \ge \mathrm{rank}(n) = \mathrm{Card}(\mathrm{Im}(m))$.

Thus the notion of rank that we defined in Section 3.6 coincides with the notion defined here.

Let us observe that the rank of a relation m over a finite set Q has strong connections with the usual notion of rank as defined in linear algebra. Let K be a field containing $\mathbb{N}$. The *rank* of a matrix m with coefficients in K, denoted by $\mathrm{rank}_K(m)$, is the maximal number of rows (or columns) which are linearly independent over K. We can observe (Exercise 9.3.2) that this number may be defined in a manner analogous to the definition of the rank of a relation. In particular,

$$\text{rank}_K(m) \le \text{rank}(m).$$

It is easy to see (Exercise 9.3.3) that usually the inequality is strict. However, in the case of relations which are functions, the two notions coincide.

The following proposition gives an easy method for computing the rank of an idempotent relation.

Proposition 9.3.6 *Let e be an idempotent element of an unambiguous monoid of relations. Then*

$$\text{rank}(e) = \text{Card}(\text{Fix}(e)).$$

Proof. Set $S = \text{Fix}(e)$. The column-row decomposition of e shows that $\text{rank}(e) \le \text{Card}(S)$. Moreover, in view of Proposition 9.1.6, the matrix e contains the identity matrix I_S. Thus $\text{Card}(S) = \text{rank}(I_S) \le \text{rank}(e)$ by Equation (9.21). $\qquad\square$

The following statement gives a characterization of relations of finite rank.

Proposition 9.3.7 *For any relation m, the following conditions are equivalent:*

(i) *m has finite rank.*
(ii) *The set of rows of m is finite.*
(iii) *The set of columns of m is finite.*

Proof. (i) $\Rightarrow$ (ii). Let $m = \ell r$, with $\ell \in \mathfrak{P}(P \times S)$ and $r \in \mathfrak{P}(S \times Q)$ be a minimal decomposition of m. If two rows of ℓ, say with indices p and q, are equal, then the corresponding rows m_{p*} and m_{q*} of m also are equal. Since S is finite, the matrix ℓ has at most $2^{\text{Card}(S)}$ distinct rows. Thus the set of rows of m is finite.

(ii) $\Rightarrow$ (i). Let $(m_{s*})_{r \in S}$ be a set of representatives of the rows of m. Then $m = \ell r$, where r is the restriction of m to $S \times Q$, and $\ell \in \mathfrak{P}(Q \times S)$ is defined by

$$\ell_{qr} = \begin{cases} 1 & \text{if } m_{q*} = m_{s*}, \\ 0 & \text{otherwise.} \end{cases}$$

This shows (i) $\Leftrightarrow$ (ii). The proof of (i) $\Leftrightarrow$ (iii) is identical. $\qquad\square$

Proposition 9.3.8 *Let m be a relation over a set Q of finite rank. Then the semigroup generated by m is finite.*

Proof. Let $m = \ell r$ be a minimal decomposition of m, with $\ell \in \mathfrak{P}(Q \times R)$ and $r \in \mathfrak{P}(R \times Q)$. Let u be the relation over R defined by $u = r\ell$. Then for all $n \ge 0$,

$$m^{n+1} = \ell(r\ell)^n r = \ell u^n r.$$

Since R is finite, the set of relations u^n is finite and the semigroup $\{m^n \mid n \ge 1\}$ is finite. $\qquad\square$

In particular it follows from this proposition that for any relation of finite rank, a convenient power is an idempotent relation.

Let M be an unambiguous monoid of relations over Q. The *minimal rank* of M, denoted by $r(M)$, is the minimum of the ranks of the elements of M other than the null relation,

$$r(M) = \min\{\text{rank}(m) \mid m \in M \setminus 0\}.$$

If M does not contain the null relation over Q, this is of course the minimum of the ranks of the elements of M. One has $r(M) > 0$ if $Q \neq \emptyset$ and $r(M) < \infty$ if and only if M contains a relation of finite positive rank.

We now study the monoids having finite minimal rank and we shall see that they have a regular structure. We must distinguish two cases: the case where the monoid contains the null relation, and the easier case where it does not.

Note that the null relation plays the role of a zero in view of the following, more precise statement.

Proposition 9.3.9 *If a transitive unambiguous monoid of relations over a nonempty set Q contains a zero, then the zero is the null relation.*

Proof. The null relation always is a zero. Conversely, if M has a zero z, let us prove that z is the null relation. If $\text{Card}(Q) = 1$, then $z = 0$. Thus we assume $\text{Card}(Q) \geq 2$, and $z \neq 0$. Let $p, q \in Q$ such that $z_{p,q} = 1$. Let $r, s \in Q$. By transitivity of M, there exist $m, n \in M$ such that

$$m_{rp} = n_{qs} = 1.$$

From $mzn = z$, it follows that $z_{rs} = 1$. Thus $z_{rs} = 1$ for all $r, s \in Q$, which contradicts the unambiguity of M. $\qquad\square$

Let M be an unambiguous monoid of relations over Q. For each $q \in Q$, the *stabilizer* of q is the submonoid

$$\text{Stab}(q) = \{m \in M \mid q \xrightarrow{m} q\}.$$

Theorem 9.3.10 *Let M be a transitive unambiguous monoid of relations over Q, containing the relation 0, and having finite minimal rank. Let K be the set of elements of M of minimal rank $r(M)$.*

1. *M contains a unique 0-minimal ideal J, which is $K \cup \{0\}$.*
2. *The set K is a regular $\mathcal{D}$-class whose $\mathcal{H}$-classes are finite.*
3. *Each $q \in Q$ is a fixed point of at least one idempotent e in K that is, $e \in K \cap \text{Stab}(q)$.*
4. *For each idempotent $e \in K$, the group G_e is a transitive group of degree $r(M)$.*
5. *The groups G_e, for e idempotent in K, are equivalent.*

Before we proceed to the proof, we establish several preliminary results.

Proposition 9.3.11 *Let M be an unambiguous monoid of relations over Q, and let $e \in M$ be an idempotent. If e has finite rank, then the localized monoid eMe is finite.*

Proof. Let S be the set of fixed points of e. By Proposition 9.3.6, the set S is finite. Thus the monoid M_e which is an unambiguous monoid of relations over S, is finite. Since, by Proposition 9.1.9, the monoid eMe is isomorphic to M_e, it is finite. □

We now verify a technical lemma which is useful to "avoid" the null relation.

Lemma 9.3.12 *Let M be a transitive unambiguous monoid of relations over Q.*

1. *For all $m \in M \setminus 0$, there exist $n \in M$ and $q \in Q$ such that $mn \in \mathrm{Stab}(q)$ (resp. $nm \in \mathrm{Stab}(q)$). Thus in particular $mn \neq 0$ (resp. $nm \neq 0$).*
2. *For all $m \in M \setminus 0$ and $q \in Q$, there exist $n, n' \in M$ such that $nmn' \in \mathrm{Stab}(q)$.*
3. *For all $m, n \in M \setminus 0$, there exists $u \in M$ such that $mun \neq 0$. In other terms, the monoid M is prime.*

Proof. 1. Let $q, r \in Q$ be such that $(q, m, r) = 1$. Since M is transitive, there exists $n \in M$ such that $(r, n, q) = 1$. Thus $(q, mn, q) = 1$.

2. There exist $p, r \in Q$ such that $(p, m, r) = 1$. Let $n, n' \in M$ be such that $(q, n, p) = 1$, $(r, n', q) = 1$. Then $(q, nmn', q) = 1$.

3. There exist $p, r, s, q \in Q$ such that $(p, m, r) = (s, n, q) = 1$. Take $u \in M$ with $(r, u, s) = 1$. Then $(p, mun, q) = 1$. □

Proposition 9.3.13 *Let M be a transitive unambiguous monoid of relations over Q, having finite minimal rank. Each right ideal $R \neq 0$ (resp. each left ideal $L \neq 0$) of M contains a nonnull idempotent.*

Proof. Let $r \in R \setminus 0$. By Lemma 9.3.12, there exist $n \in M$ and $q \in Q$ such that $rn \in \mathrm{Stab}(q)$. Let $m \in M$ be an element such that $\mathrm{rank}(m) = r(M)$. Again by Lemma 9.3.12, there exist $u, v \in M$ such that $umv \in \mathrm{Stab}(q)$. Consider the element $m' = rnumv$. Then $m' \in R$ and $m' \in \mathrm{Stab}(q)$.

Since $\mathrm{rank}(m') \leq \mathrm{rank}(m)$, the rank of m' is finite. According to Proposition 9.3.8, the semigroup generated by m' is finite. Thus there exists $k \geq 1$ such that $e = (m')^k$ is idempotent. Then $e \in R$ and $e \neq 0$ since $e \in \mathrm{Stab}(q)$. □

Proposition 9.3.14 *Let M be a transitive unambiguous monoid of relations over Q, having finite minimal rank and containing the null relation. For all $m \in M$, the following conditions are equivalent:*

(i) $\mathrm{rank}(m) = r(M)$.
(ii) *The right ideal mM is 0-minimal.*
(iii) *The left ideal Mm is 0-minimal.*

Proof. (i) $\Rightarrow$ (ii). Let $R \neq \{0\}$ be a right ideal contained in mM. We show that $R = mM$. According to Proposition 9.3.13, R contains an idempotent $e \neq 0$. Since $e \in R \subset mM$, there exists $n \in M$ such that $e = mn$. Since $\mathrm{rank}(e) \leq \mathrm{rank}(m)$ and rank(m) is minimal, we have rank(e) = rank(m). Let $m = \ell r$ be a minimal decomposition of m, with $\ell \in \mathfrak{P}(Q \times S)$, $r \in \mathfrak{P}(S \times Q)$. Then $e = (\ell r)n = \ell(rn)$. The product $\ell(rn)$ is easily checked to be unambiguous. Since rank(e) = $r(M)$ = Card(S), the pair (ℓ, rn) is a minimal decomposition of e. For all $k \geq 0$,

$$e = e^{k+1} = \ell(rn\ell)^k rn$$

with all products unambiguous. Since S is finite, there exists an integer $i \geq 1$ such that $(rn\ell)^i$ is an idempotent element of the unambiguous monoid of relations on S composed of the powers of $rn\ell$. Since rank($(rn\ell)^i$) = Card(S), each element in S is a fixed point of $(rn\ell)^i$. Consequently $(rn\ell)^i = \mathrm{id}_S$. Thus

$$em = e^i m = (\ell rn)^i m = (\ell rn)^i \ell r = \ell(rn\ell)^i r = \ell r = m.$$

The equality $em = m$ shows that $m \in R$, whence $R = mM$. Thus mM is a 0-minimal right ideal.

(ii) $\Rightarrow$ (i). Let $n \in M$ be such that rank(n) = $r(M)$. By Lemma 9.3.12, there exists $u \in M$ such that $mun \neq 0$. From $munM \subset mM$, we get $munM = mM$, whence $m \in munM$. Thus rank(m) $\leq$ rank(n), showing that rank(m) = rank(n).

(i) $\Leftrightarrow$ (iii) is shown in the same way. $\qquad\square$

Proof of Theorem 9.3.10.

1. By Lemma 9.3.12, the monoid M is prime. According to Proposition 9.3.14, the monoid M contains 0-minimal left and right ideals. In view of Corollary 1.12.10, the monoid M contains a unique 0-minimal ideal J which is the union of the 0-minimal right ideals (resp. left ideals). Once more by Proposition 9.3.14, J is the union of 0 and of the set K of elements of minimal positive rank. This proves claim 1.

2. In view of Corollary 1.12.10, the set K is a regular $\mathcal{D}$-class. All the $\mathcal{H}$-classes of K have same cardinality by Proposition 1.12.3. The finiteness of these classes will result from claim 4.

3. Let $q \in Q$ and $k \in K$. By Lemma 9.3.12, $nkn' \in \mathrm{Stab}(q)$ for some $n, n' \in M$. Since the semigroup generated by $m = nkn'$ is finite (Proposition 9.3.8), it contains an idempotent e. Then $e \in K \cap \mathrm{Stab}(q)$.

4. Let e be idempotent in K. Then the $\mathcal{H}$-class of e is $H \cup 0 = eM \cap Me = eMe = H(e) \cup 0$. The first equality is a result of the fact that the $\mathcal{R}$-class of e is $eM \setminus 0$. Next $eMe \subset eM \cap Me$, and conversely, if $n \in eM \cap Me$, then $en = ne = n$ whence $n = ene \in eMe$. This shows the second equality. Finally, $H(e) = H$ since H is a group.

According to Proposition 9.1.7, we have $M_e = G_e \cup 0$ and M_e is transitive. Thus G_e is a transitive permutation group. Its degree is $r(M)$.

5. is a direct consequence of Proposition 9.1.9. $\qquad\square$

Now let M be an unambiguous monoid of relations that does not contain the null relation. Theorem 9.3.10 admits a formulation which is completely analogous, and which goes as follows.

Theorem 9.3.15 *Let M be a transitive unambiguous monoid of relations over Q which does not contain the null relation and which has finite minimal rank. Let K be the set of elements of minimal rank $r(M)$.*

1. *The set K is the minimal ideal of M.*
2. *The set K is a regular $\mathcal{D}$-class and is a union of finite groups.*
3. *Each $q \in Q$ is the fixed point of at least one idempotent e in K that is $e \in K \cap \text{Stab}(q)$.*
4. *For each idempotent $e \in K$, the group G_e is a transitive group of degree $r(M)$, and these groups are equivalent.*

Proof. Let M_0 be the unambiguous monoid of relations

$$M_0 = M \cup 0.$$

We have $r(M) = r(M_0)$. Thus Theorem 9.3.10 applies to M_0. For all m in M, we have $mM_0 = mM \cup 0$. It follows easily that mM is a minimal right ideal of M if and only if mM_0 is a 0-minimal right ideal of M_0. The same holds for left ideals and for two-sided ideals. In particular, the 0-minimal ideal J of M_0 is the union of 0 and of the minimal ideal K of M. This proves 1. Next K is a $\mathcal{D}$-class of M_0 thus also of M. Since the product of two elements of M is never 0, each $\mathcal{H}$-class of K is a group. This proves 2. The other claims require no proof. □

Let M be a transitive unambiguous monoid of relations over Q, of finite minimal rank, and let

$$K = \{m \in M \mid \text{rank}(m) = r(M)\}.$$

The groups G_e, for each idempotent e in K, are equivalent transitive permutation groups. The *Suschkewitch group* of M is, by definition, any one of them.

9.4 Very thin codes

A code $X \subset A^+$ is called *very thin* if there exists a word x in X^* which is not a factor of a word in X. Recall that $F(X)$ is the set of factors of words in X, and that $\overline{F}(X) = A^* \setminus F(X)$. With these notations, X is very thin if and only if

$$X^* \cap \overline{F}(X) \neq \emptyset.$$

Any very thin code is thin (that is, satisfies $\overline{F}(X) \neq \emptyset$). Conversely, a thin code is not always very thin (see Example 9.4.13). However, a thin complete code X is very thin. Consider indeed a word $w \in \overline{F}(X)$. Since X is complete, there exist $u, v \in A^*$ such that $uwv \in X^*$. Then $uwv \in X^* \cap \overline{F}(X)$.

The aim of this section is to prove the following result. It shows, in particular, that a recognizable code is very thin. This is more precise than Proposition 2.5.20, which only asserts that a recognizable code is thin.

For ease of description, we use the following shorthand. Given an automaton $\mathcal{A}$, the *rank* of a word w in $\mathcal{A}$ is the rank of the relation $\varphi_{\mathcal{A}}(w)$. This agrees with the definition of rank given in Section 3.6 for deterministic automata, as shown in Example 4.2.6.

Theorem 9.4.1 *Let $X \subset A^+$ be a code and let $\mathcal{A} = (Q, 1, 1)$ be an unambiguous trim automaton recognizing X^*. The following conditions are equivalent.*

  (i) *X is very thin.*
 (ii) *The monoid $\varphi_{\mathcal{A}}(A^*)$ has finite minimal rank.*

The proof of this result is in several steps. We start with the following property used to prove that condition (i) implies condition (ii).

Proposition 9.4.2 *Let $X \subset A^+$ be a code and let $\mathcal{A} = (Q, 1, 1)$ be an unambiguous trim automaton recognizing X^*. For all $w \in \overline{F}(X)$, the rank of w in $\mathcal{A}$ is finite.*

Proof. Let us write φ instead of $\varphi_{\mathcal{A}}$. For each $p \in Q$, let $\Phi(p)$ be the set of prefixes of w which are labels of paths from p to 1:

$$\Phi(p) = \{u \in A^* \mid u \le w \text{ and } p\varphi(u)1\}.$$

We now show that if $\Phi(p) = \Phi(p')$ for some $p, p' \in Q$, then the rows of index p and p' in $\varphi(w)$ are equal. Consider a $q \in Q$ such that

$$p\varphi(w)q.$$

Since the automaton is trim, there exist $v, v' \in A^*$ such that $1\varphi(v)p$ and $q\varphi(v')1$. Thus $1\varphi(vwv')1$ and consequently $vwv' \in X^*$. Since $w \in \overline{F}(X)$, the path $p \overset{w}{\longrightarrow} q$ is not simple; therefore there exist $u, u' \in A^*$ such that $w = uu'$ and $vu, u'v' \in X^*$. Consequently there is, in $\mathcal{A}$, the path

$$1 \overset{v}{\longrightarrow} p \overset{u}{\longrightarrow} 1 \overset{u'}{\longrightarrow} q \overset{v'}{\longrightarrow} 1.$$

By definition, $u \in \Phi(p)$, whence $u \in \Phi(p')$. It follows that $p'\varphi(u)1\varphi(u')q$, and consequently $p'\varphi(w)q$. This proves the claim.

The number of sets $\Phi(p)$, for $p \in Q$, is finite. According to the claim just proved, the set of rows of $\varphi(w)$ also is finite. By Proposition 9.3.7, this implies that w has finite rank. $\square$

Example 9.4.3 Let X be the code $X = \{a^n b a^n \mid n \ge 0\}$. This is a very thin code since $b^2 \in X^* \cap \overline{F}(X)$. An automaton recognizing X^* is given in Figure 9.9. The image e of b^2 in the associated monoid of relations M is idempotent of rank 1. The finiteness of the rank also follows from Proposition 9.4.2 since b^2 is not factor of a word in X. The localized monoid eMe is reduced to e and 0 (which is the image of $b^2 a b^2$, for example). The monoid M has elements of infinite rank: this holds for the image of a. Indeed, clearly no power of this element can be idempotent; hence by Proposition 9.3.8, it has infinite rank. Moreover, M has elements of finite rank n for each integer $n \ge 0$: the word $ba^n ba^n b$ has rank $n + 1$, as the reader may verify.

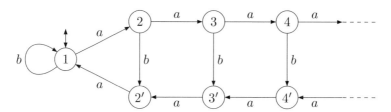

Figure 9.9 An automaton for X^*.

Proposition 9.4.4 *Let X be a code over A, let $\mathcal{A} = (Q, 1, 1)$ be an unambiguous trim automaton recognizing X^*, let φ be the associated representation and $M = \varphi(A^*)$.*

For each idempotent e in $\varphi(X^*)$ with finite rank such that the group G_e is transitive, the following assertions hold.

1. *There exist $v_1, v_2, \ldots, v_{n+1} \in \varphi^{-1}(H(e))$ with the following property: for all $y, z \in A^*$ such that*

$$y v_1 v_2 \cdots v_{n+1} z \in X^*$$

there is an integer i, $(1 \leq i \leq n)$ such that:

$$y v_1 v_2 \cdots v_i, \quad v_{i+1} \cdots v_{n+1} z \in X^*.$$

2. *The set $\varphi^{-1}(e) \cap \overline{F}(X)$ is nonempty.*

Proof. Let $e = \boldsymbol{\ell r}$ be the column-row decomposition of e, let S be the set of its fixed points and let $G = H(e)$. By Proposition 9.1.9, the restriction $\gamma : eMe \to M_e$ is the isomorphism $m \mapsto \boldsymbol{rm\ell}$, and its inverse is the function $n \mapsto \boldsymbol{\ell nr}$.

The set S contains the element 1, since $e \in \varphi(X^*)$. Set $S = \{1, 2, \ldots, n\}$. We first rule out the case where $\varphi^{-1}(e) = \{1\}$. Then e is the neutral element of M, and $S = Q$. Since $H(e) = \{1\}$ and G_e is assumed to be transitive, this forces $A = X$. Thus the result holds trivially.

We now assume that $\varphi^{-1}(e) \neq \{1\}$. Choose elements $g_2, g_3, \ldots, g_n \in G_e$ such that

$$2g_2 = 1, \quad 3g_2g_3 = 1, \ldots, \quad ng_2g_3 \cdots g_n = 1.$$

These elements exist because G_e is a transitive permutation group. The permutations $g_2, g_3, \ldots, g_n$ are the restrictions to S of elements $h_2, h_3, \ldots, h_n$ of $H(e)$ and one has $h_i = \boldsymbol{\ell g_i r}$. Thus $g_i = \boldsymbol{rh_i\ell} = \gamma(h_i)$. Let $v_1, v_2, \ldots, v_{n+1} \in A^+$ be such that

$$\varphi(v_1) = \varphi(v_{n+1}) = e, \quad \varphi(v_2) = h_2, \ldots, \quad \varphi(v_n) = h_n.$$

Set $w = v_1 v_2 \cdots v_{n+1}$. Consider words $y, z \in A^*$ such that $ywz \in X^*$. Then there exist $p, q \in Q$ such that

$$1 \xrightarrow{y} p \xrightarrow{w} q \xrightarrow{z} 1.$$

Note that

$$\varphi(w) = \boldsymbol{\ell}rh_2\cdots h_n\boldsymbol{\ell}r = \boldsymbol{\ell}\gamma(h_2\cdots h_n)r = \boldsymbol{\ell}g_2\cdots g_nr.$$

Since $p\varphi(w)q$, there exist $r, s \in S$ such that $p \xrightarrow{\ell} r$, $rg_2\cdots g_n = s$, and $s \xrightarrow{r} q$. Then $rg_2\cdots g_r = 1$ (with $g_2\cdots g_r = \mathrm{id}_S$ when $r = 1$). Since the g_i's are permutations, this implies

$$1g_{r+1}\cdots g_n = s.$$

Consequently $r \xrightarrow{h_2\cdots h_r} 1, 1 \xrightarrow{h_{r+1}\cdots h_n} s$, and since $\boldsymbol{\ell}_{p,r} = e_{p,r}, r_{s,q} = e_{s,q}$, we have

$$p \xrightarrow{eh_2\cdots h_r} 1, \quad 1 \xrightarrow{h_{r+1}\cdots h_n e} q.$$

This implies that

$$yv_1v_2\cdots v_r, v_{r+1}\cdots v_{n+1}z \in X^*.$$

Thus the words $v_1, \ldots, v_{n+1}$ satisfy the first statement.

To show the second part, we verify first that the word $w = v_1v_2\cdots v_{n+1}$ is in $\overline{F}(X)$. Assume indeed that $ywz \in X$ for some $y, z \in A^*$. Then there exists an integer i ($1 \leq i \leq n$) such that $yv_1\cdots v_i, v_{i+1}\cdots v_{n+1}z \in X^*$. Since $v_1, \ldots, v_{n+1} \in A^+$, these two words are in fact in X^+, contradicting the fact that X is a code. Thus $w \in \overline{F}(X)$.

Let h' be the inverse of $h = \varphi(w)$ in $H(e)$, and let w' be such that $\varphi(w') = h'$. Then $ww' \in \varphi^{-1}(e)$, and also $ww' \in \overline{F}(X)$. This concludes the proof. □

Proof of Theorem 9.4.1.

(i) $\Longrightarrow$ (ii). Let $x \in X^* \cap \overline{F}(X)$. According to Proposition 9.4.2, the rank of $\varphi(x)$ is finite. Since $x \in X^*$, we have $(1, \varphi_A(X), 1) = 1$ and thus $\varphi_A(x) \neq 0$. This shows that $\varphi_A(A^*)$ has finite minimal rank.

(ii) $\Longrightarrow$ (i). The monoid $M = \varphi_A(A^*)$ is a transitive unambiguous monoid of relations having finite minimal rank $r(M)$. Let

$$K = \{m \in M \mid \mathrm{rank}(m) = r(M)\}.$$

By Theorems 9.3.10 and 9.3.15, there exists an idempotent e in $K \cap \mathrm{Stab}(1)$, and the permutation group G_e is transitive of degree $r(M)$. By Proposition 9.4.4, the set $\varphi_A^{-1}(e) \cap \overline{F}(X)$ is not empty. Since $\varphi_A^{-1}(e) \subset X^*$, the code X is very thin. □

We now give a series of consequences of Theorem 9.4.1.

Corollary 9.4.5 *Let X be a complete code, and let $\mathcal{A} = (Q, 1, 1)$ be an unambiguous trim automaton recognizing X^*. The following conditions are equivalent.*

(i) *X is thin.*

(ii) *The monoid $\varphi_A(A^*)$ contains elements of finite rank.*

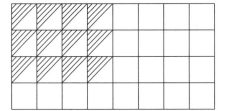

Figure 9.10 The minimal ideal.

Proof. Since X is complete, the monoid $\varphi_A(A^*)$ does not contain the null relation (Proposition 2.5.28). Thus the result follows directly from Theorem 9.4.1. $\quad\square$

Another consequence of Theorem 9.4.1 is an algebraic proof, independent of measures, of Theorem 2.5.13.

Corollary 9.4.6 *If X is a thin complete code, then X is a maximal code.*

Proof. Let $\mathcal{A} = (Q, 1, 1)$ be an unambiguous trim automaton recognizing X^* and let φ be the associated representation. Let $x \in X^*$ such that $e = \varphi(x)$ is an idempotent of the minimal ideal J of the monoid $\varphi(A^*)$. (Such an idempotent exists by Theorem 9.3.15, claim 3).

Let $y \notin X$. Then $e\varphi(y)e = \varphi(xyx)$ is in the $\mathcal{H}$-class of e. This $\mathcal{H}$-class is a finite group. Thus there exists an integer $n \geq 1$ such that $(\varphi(xyx))^n = e$. Consequently $(xyx)^n \in X^*$. This shows that $X \cup y$ is not a code. $\quad\square$

Let $X \subset A^+$ be a code and let $\mathcal{A} = (Q, 1, 1)$ be an unambiguous trim automaton recognizing X^*. We have shown that X is very thin if and only if the monoid $M = \varphi_A(A^*)$ has elements of finite, positive rank. Let r be the minimum of these nonzero ranks, and let K be the set of elements in M of rank r. Set $\varphi = \varphi_A$. It is useful to keep in mind the following facts.

1. $\varphi(X^*)$ meets K. Indeed $\varphi(X^*) = \mathrm{Stab}(1)$ and according to Theorems 9.3.10 and 9.3.15, K meets $\mathrm{Stab}(1)$.

2. Every $\mathcal{H}$-class H contained in K that meets $\varphi(X^*)$ is a group. Moreover, $\varphi(X^*) \cap H$ is a subgroup of H. These $\mathcal{H}$-classes are those which contain an idempotent having 1 as a fixed point.

Indeed, let H be an $\mathcal{H}$-class meeting $\varphi(X^*)$. Let $h \in H \cap \varphi(X^*)$. Then h^2 is not the null relation since $h^2 \in \mathrm{Stab}(1)$. Thus $h^2 \in H$ and consequently H is a group (Proposition 1.12.8). Let $N = H \cap \varphi(X^*)$. Since $\varphi(X^*)$ is a stable submonoid of M, N is a stable submonoid of H, hence a subgroup (Example 2.2.3).

Figure 9.10 represents, with slashed triangles, the intersection $K \cap \varphi(X^*)$. It expresses that the $\mathcal{H}$-classes of K meeting $\varphi(X^*)$ "form a rectangle" in K (see Exercise 9.3.4). Collecting together these facts, we have proved the following theorem.

Theorem 9.4.7 *Let $X \subset A^+$ be a very thin code. Let $\mathcal{A} = (Q, 1, 1)$ be an unambiguous trim automaton recognizing X^*. Let K be the set of elements of minimal nonzero rank in the monoid $M = \varphi_A(A^*)$.*

1. $\varphi_A(X^*)$ meets K.
2. Any $\mathcal{H}$-class H in K that meets $\varphi_A(X^*)$ is a group. Moreover, $H \cap \varphi_A(X^*)$ is a subgroup of H.
3. The $\mathcal{H}$-classes of K meeting $\varphi_A(X^*)$ are those whose idempotent has the state 1 as a fixed point.

Another consequence of the results of this section is the proof of the following lemma which was stated without proof in Chapter 2 (Lemma 2.6.5).

Lemma 9.4.8 *Let X be a complete thin code. For any word $u \in X^*$ there exists a word $w \in X^*uX^*$ satisfying the following property: if $ywz \in X^*$, then there exists a factorization $w = fug$ such that $yf, gz \in X^*$.*

Proof. Let φ be the representation associated with some unambiguous trim automaton recognizing X^*. Since X is thin, the monoid $M = \varphi(A^*)$ has a minimal ideal J. Since X is complete, M has no zero and thus $\varphi(X^+)$ meets J. Let e be an idempotent in $\varphi(X^*) \cap J$. The group G_e is transitive by Theorem 9.3.10 and, according to Proposition 9.4.4, there exist words $v_1, v_2, \ldots, v_{n+1} \in \varphi^{-1}(H(e))$ such that the word $v = v_1 v_2 \cdots v_{n+1}$ has the following property: if $yvz \in X^*$ for some $y, z \in A^*$, then there exists an integer i such that $yv_1 \cdots v_i, v_{i+1} \cdots v_{n+1}z \in X^*$.

We have $e\varphi(u)e \in eMe = H(e)$, and $e\varphi(u)e \in \varphi(X^*)$. Since $H(e) \cap \varphi(X^*)$ is a subgroup of $H(e)$, there exists $h \in H(e) \cap \varphi(X^*)$ such that $e\varphi(u)eh = e$. Since $h = eh$, we have $e\varphi(u)h = e$. Consider words $r \in \varphi^{-1}(e)$, $s \in \varphi^{-1}(h)$, set $u' = rus$ and consider the word

$$w = u'v_1u'v_2 \cdots u'v_{n+1}u'.$$

Let $y, z \in A^*$ be words such that $ywz \in X^*$. Since $\varphi(u') = e$, we have $\varphi(w) = \varphi(v)$. Consequently also yvz is in X^*. It follows that for some integer i,

$$yv_1v_2 \cdots v_i, v_{i+1} \cdots v_{n+1}z \in X^*.$$

Observe that

$$\varphi(v_1v_2 \cdots v_i) = \varphi(u'v_1u'v_2 \cdots u'v_i)$$

and

$$\varphi(v_{i+1} \cdots v_{n+1}) = \varphi(v_{i+1}u' \cdots u'v_{n+1}u').$$

Thus also $yu'v_1u'v_2 \cdots u'v_i$ and $v_{i+1}u' \cdots v_{n+1}u'z$ are in X^*.

Let

$$f = u'v_1u'v_2 \cdots u'v_ir, \quad g = sv_{i+1}u' \cdots v_{n+1}u'.$$

Since $r, s \in X^*$, we have $yf, gz \in X^*$ and this shows that the word $w = fug$ satisfies the property of the statement. $\square$

Finally, we note that for complete thin codes, some of the information concerning the minimal ideal are characteristic of prefix, suffix, or bifix codes.

Proposition 9.4.9 *Let X be a thin complete code over A, let φ be the representation associated with an unambiguous trim automaton $\mathcal{A} = (Q, 1, 1)$ recognizing X^*, let $M = \varphi(A^*)$ and J its minimal ideal. Let H_0, R_0, L_0 be an $\mathcal{H}, \mathcal{R}, \mathcal{L}$-class of J such that $H_0 = R_0 \cap L_0$ and $\varphi(X^*) \cap H_0 \neq \emptyset$.*

1. *X is prefix if and only if $\varphi(X^*)$ meets every $\mathcal{H}$-class in L_0.*
2. *X is suffix if and only if $\varphi(X^*)$ meets every $\mathcal{H}$-class in R_0.*
3. *X is bifix if and only if $\varphi(X^*)$ meets all $\mathcal{H}$-classes in J.*

Proof. 1. Let H be an $\mathcal{H}$–class in L_0, let e_0 be the idempotent of H_0 and let e be the idempotent of H (each $\mathcal{H}$-class in J is a group). We have $e_0e = e_0$ since $e \in L_0$ (for some m, we have $me = e_0$; consequently $e_0 = me = mee = e_0e$).

If X is prefix, then $\varphi(X^*)$ is right unitary. Since $e_0 \in \varphi(X^*)$ and $e_0 = e_0e$, it follows that $e \in \varphi(X^*)$. Thus $H \cap \varphi(X^*) \neq \emptyset$.

Conversely, let us show that $\varphi(X^*)$ is right complete. Let $m \in M$. Then $me_0 \in L_0$, and therefore $me_0 \in H$ for some $\mathcal{H}$-class $H \subset L_0$. If n is the inverse of me_0 in the group H, then $me_0n \in \varphi(X^*)$. Thus $\varphi(X^*)$ is right complete and X is prefix.

The proof of 2 is symmetric, and 3 results from the preceding arguments. $\qquad\square$

Proposition 9.4.9 can be generalized to codes which are not maximal (see Exercise 9.4.3).

Let $X \subset A^*$ be a thin, maximal prefix code, and let $\mathcal{A} = (Q, 1, 1)$ be a complete deterministic automaton recognizing X^*. The monoid $M = \varphi_\mathcal{A}(A^*)$ then is a monoid of (total) functions and we use the notation already introduced in Section 9.1. We will write, for $m \in M$, $qm = q'$ instead of $(q, m, q') = 1$. Let $m \in M$, and $w \in A^*$ with $m = \varphi(w)$. The *image* of m is

$$\text{Im}(m) = Qm = Q \cdot w,$$

and the *nuclear equivalence* of m, denoted by $\text{Ker}(m)$, is defined by

$$q \equiv q' \ (\text{Ker}(m)) \iff qm = q'm.$$

The number of classes of the equivalence relation $\text{Ker}(m)$ is equal to $\text{Card}(\text{Im}(m))$; both are equal to $\text{rank}(m)$, in view of Example 9.3.5.

A nuclear equivalence is *maximal* if it is maximal among the nuclear equivalences of elements in M. It is an equivalence relation with a number of classes equal to $r(M)$. Similarly, an image is *minimal* if it is an image of cardinality $r(M)$, that is, an image which does not strictly contain any other image.

Proposition 9.4.10 *Let $X \subset A^+$ be a thin maximal prefix code, let $\mathcal{A} = (Q, 1, 1)$ be a complete deterministic automaton recognizing X^*, let $M = \varphi_\mathcal{A}(A^*)$ and let K be the $\mathcal{D}$-class of the elements of M of rank $r(M)$. Then*

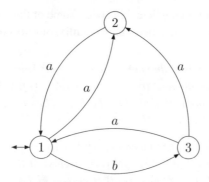

Figure 9.11 An automaton for X^*.

1. *There is a bijection between the minimal images and the $\mathcal{L}$-classes of K.*
2. *There is a bijection between the maximal nuclear equivalences and the $\mathcal{R}$-classes of K.*

Proof. 1. Let $n, m \in M$ be two $\mathcal{L}$-equivalent elements. We prove that $\mathrm{Im}(m) = \mathrm{Im}(n)$. There exist $u, v \in M$ such that $m = un, n = vm$. Thus $Qm = Qun \subset Qn$, and also $Qn \subset Qm$. This shows that $\mathrm{Im}(m) = \mathrm{Im}(n)$.

Conversely let $m, n \in K$ be such that $\mathrm{Im}(m) = \mathrm{Im}(n)$. K being a regular $\mathcal{D}$-class (Theorem 9.3.10), the $\mathcal{L}$-class of m contains an idempotent, say e, and the $\mathcal{L}$-class of n contains an idempotent f (Proposition 1.12.7). Then $\mathrm{Im}(e) = \mathrm{Im}(m)$ and $\mathrm{Im}(f) = \mathrm{Im}(n)$, in view of the first part. Thus $\mathrm{Im}(e) = \mathrm{Im}(f)$. We shall see that $ef = e$ and $fe = f$.

Let indeed $q \in Q$, and $q' = qe$. Then $q' \in \mathrm{Im}(e) = \mathrm{Im}(f)$, and $q' = q'f$ since f is idempotent. Consequently $qe = qef$. This shows that $e = ef$. The equality $fe = f$ is shown by interchanging e and f. These relations imply $e\mathcal{L}f$. Thus $m\mathcal{L}n$.

2. The proof is entirely analogous. □

Note also that in the situation described above, every state appears in some minimal image. This is indeed the translation of Theorem 9.3.15(4). This description of the minimal ideal of a monoid of functions, by means of minimal images and maximal equivalences, appears to be particularly convenient.

Example 9.4.11 Let $X = \{aa, ba, baa\}$. We consider the automaton given in Figure 9.11. The 0-minimal ideal of the corresponding monoid is the following: it is formed of elements of rank 1.

	001	110
011^t	* $\alpha\beta$	* $\alpha\beta\alpha$
100^t	β	* $\beta\alpha$
101^t	* $\alpha\alpha\beta$	* $\alpha\alpha\beta\alpha$

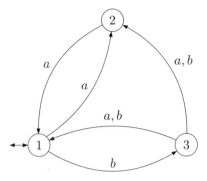

Figure 9.12 An automaton for X^*.

with

$$\alpha = \varphi(a) = \begin{bmatrix} 0 & 1 & 0 \\ 1 & 0 & 0 \\ 1 & 1 & 0 \end{bmatrix}, \quad \beta = \varphi(b) = \begin{bmatrix} 0 & 0 & 1 \\ 0 & 0 & 0 \\ 0 & 0 & 0 \end{bmatrix}.$$

For each element we indicate, on the top, its unique nonnull row, and, on its left, its unique nonnull column (with the convention $a_1 \cdots a_n{}^t = \begin{smallmatrix} a_1 \\ a_2 \\ \vdots \\ a_n \end{smallmatrix}$). The existence of an idempotent is indicated by an asterisk in the $\mathcal{H}$-class. The column-row decomposition of an idempotent is simply given by the vectors in the rows and columns of the array. For example, the column-row decomposition of $\alpha\beta$ is

$$\alpha\beta = \begin{bmatrix} 0 \\ 1 \\ 1 \end{bmatrix} \begin{bmatrix} 0 & 0 & 1 \end{bmatrix}.$$

The following array gives the fixed point of each idempotent

3	2
	1
3	1

Example 9.4.12 Let $X = \{aa, ba, baa, bb, bba\}$. We consider the automaton given in Figure 9.12. The corresponding monoid has no 0 (the code is complete).

The minimal ideal, formed of elements of rank 1, is represented by

	001	110
011^t	$^*\alpha\beta$	$^*\alpha\beta\alpha$
101^t	$^*\beta\alpha\beta$	$^*\beta\alpha$

$$, \quad \alpha = \varphi(a) = \begin{bmatrix} 0 & 1 & 0 \\ 1 & 0 & 0 \\ 1 & 1 & 0 \end{bmatrix}, \quad \beta = \varphi(b) = \begin{bmatrix} 0 & 0 & 1 \\ 0 & 0 & 0 \\ 1 & 1 & 0 \end{bmatrix}.$$

The fixed points of the idempotents are:

3	2
3	1

Example 9.4.13 Let $A = \{a, \bar{a}, b, \bar{b}\}$. Denote by θ the congruence on A^* generated by the relations

$$a\bar{a} \sim 1, \quad b\bar{b} \sim 1.$$

The class of 1 for the congruence θ is a biunitary submonoid. We denote by D'_2 the code generating this submonoid. This code is a *one-sided Dyck code*. The set $D'_2{}^*$ can be considered to be the set of "systems of parentheses" with two types of parentheses: a, b represent left parentheses, and $\bar{a}, \bar{b}$ the corresponding right parentheses.

The code D'_2 is thin since D'_2 is not complete. Indeed, for instance, $ab \notin F(D'_2)$ since $ab \notin F(D'_2{}^*)$. However, D'_2 is not very thin. Indeed, for all $w \in D'_2{}^*$, we have $aw\bar{a} \in D'_2$. The code D'_2 is bifix. Let $\mathcal{A}(D'_2{}^*) = (Q, 1, 1)$, let $\varphi = \varphi_A$ and let $M = \varphi(A^*)$. By Proposition 1.4.5, the monoid M is isomorphic with the syntactic monoid of $D'_2{}^*$. We have $D'_2{}^* = \varphi^{-1}(1)$ since $D'_2{}^*$ is the class of 1 for a congruence.

The monoid M contains a 0 and

$$\varphi^{-1}(0) = \overline{F}(D'_2{}^*).$$

The only two-sided ideals of M are M and 0. Indeed, if $m \in M \setminus 0$ and $w \in \varphi^{-1}(m)$, then $w \in F(D'_2{}^*)$. Therefore, there exist $u, v \in A^*$ such that $uwv \in D'_2{}^*$. Hence $\varphi(u)m\varphi(v) = 1$ whence $1 \in MmM$ and $MmM = M$.

This shows that M itself is a 0-minimal ideal. Nonetheless, M does not contain any 0-minimal right ideal. Suppose the contrary. By Proposition 1.12.9, M would be the union of all 0-minimal right ideals. Thus any element of $M \setminus 0$ would generate a 0-minimal right ideal. This is false as we shall see now.

For all $n \geq 1$, $\varphi(\bar{a}^n)M \supset \varphi(\bar{a}^{n+1})M$. This inclusion is strict, since if $\varphi(\bar{a}^n) = \varphi(\bar{a}^{n+1}w)$ for some $w \in A^*$, then $a^n\bar{a}^n \in D'_2{}^*$ would imply $a^n\bar{a}^{n+1}w \in D'_2{}^*$, whence $\bar{a}w \in D'_2{}^*$ which is clearly impossible.

This example illustrates the fact that for a code X which is not very thin, no automaton recognizing X^* has elements of finite positive rank (Theorem 9.4.1).

9.5 Group and degree of a code

Let $X \subset A^+$ be a very thin code, let $\mathcal{A}^*_D(X)$ be the flower automaton of X and let φ_D be the associated representation. By Theorem 9.4.1, the monoid $\varphi_D(A^*)$ has elements of finite, positive rank.

The *group of the code* X is, by definition, the Suschkewitch group of the monoid $\varphi_D(A^*)$ defined at the end of Section 9.3. It is a transitive permutation group of finite degree. Its degree is equal to the minimal rank $r(\varphi_D(A^*))$ of the monoid $\varphi_D(A^*)$.

We denote by $G(X)$ the group of X. Its degree is, by definition, the *degree of the code X* and is denoted by $d(X)$. Thus one has

$$d(X) = r(\varphi_D(A^*)).$$

We already met a notion of degree in the case of thin maximal bifix codes. We shall see below that the present and previous notions of degree coincide.

The definition of $G(X)$ and $d(X)$ rely on the flower automaton of X. In fact, these concepts are independent of the automaton which is considered. In order to show this, we first establish a result which is interesting in its own.

Proposition 9.5.1 *Let $X \subset A^+$ be a thin code. Let $\mathcal{A} = (P, 1, 1)$ and $\mathcal{B} = (Q, 1, 1)$ be two unambiguous trim automata recognizing X^*, and let φ and ψ be the associated representations. Let $M = \varphi(A^*)$, $N = \psi(A^*)$, $\Phi = \varphi(\overline{F}(X))$, $\Psi = \psi(\overline{F}(X))$, let E be the set of idempotents in Φ, and E' the set of idempotents in Ψ.*

Let $\rho : P \to Q$ be a reduction of $\mathcal{A}$ onto $\mathcal{B}$ and let $\widehat{\rho} : M \to N$ be the surjective morphism associated with ρ. Then

1. $\widehat{\rho}(E) = E'$.
2. *Let $e \in E$, $e' = \widehat{\rho}(e)$. The restriction of ρ to $\mathrm{Fix}(e)$ is a bijection from $\mathrm{Fix}(e)$ onto $Fix(e')$, and the monoids M_e and $N_{e'}$ are equivalent.*

Proof. Since $\mathcal{A}$ and $\mathcal{B}$ recognize the same set, we have $\rho^{-1}(1) = 1$ (Proposition 4.2.4). The morphism $\widehat{\rho} : M \to N$ defined by ρ satisfies $\psi = \widehat{\rho} \circ \varphi$.

1. Let $e \in E$. Then $\widehat{\rho}(e) = \widehat{\rho}(e^2) = \widehat{\rho}(e)^2$. Thus $\widehat{\rho}(e)$ is an idempotent. If $e = \varphi(w)$ for some $w \in \overline{F}(X)$, then $\widehat{\rho}(e) = \psi(w)$, whence $\widehat{\rho}(e) \in \Psi$. This shows that $\widehat{\rho}(E) \subset E'$.

Conversely, let $e' \in E'$, and let $w \in \overline{F}(X)$ with $e' = \psi(w)$. Then $\varphi(w)$ has finite rank by Proposition 9.4.2, and by Proposition 9.3.8, there is an integer $n \geq 1$ such that $(\varphi(w))^n$ is an idempotent. Set $e = (\varphi(w))^n$; then $e = \varphi(w^n)$ and $w^n \in \overline{F}(X)$. Thus $e \in E$. Next $\widehat{\rho}(e) = \psi(w^n) = e'^n = e'$. This shows that $\widehat{\rho}(E) = E'$.

2. Let $S = \mathrm{Fix}(e)$, $S' = \mathrm{Fix}(e')$. Consider $s \in S$ and let $s' = \rho(s)$. From ses, we get $s'e's'$ and consequently $\rho(S) \subset S'$. Conversely, if $s'e's'$, then peq for some $p, q \in \rho^{-1}(s')$. By Proposition 9.1.9(2), there exists $s \in S$ such that $peseq$. This implies that $s'e'\rho(s)e's'$ and, by unambiguity, $\rho(s) = s'$. It follows that $\rho(S) = S'$.

Now let $s, t \in S$ be such that $\rho(s) = \rho(t) = s'$. If $s = 1$ then $t = 1$, since $\rho^{-1}(1) = 1$. Thus we may assume that $s, t \neq 1$. Since $e \in \Phi$, there exist $w \in \overline{F}(X)$ with $e = \varphi(w)$ and factorizations $w = uv = u'v'$ such that $\varphi(uv) = \varphi(u'v') = e$ and

$$s \xrightarrow{u} 1 \xrightarrow{v} s, \quad t \xrightarrow{u'} 1 \xrightarrow{v'} t.$$

This implies that

$$s' \xrightarrow{u} 1 \xrightarrow{v} s', \quad s' \xrightarrow{u'} 1 \xrightarrow{v'} s',$$

whence in particular in $\mathcal{B}$

$$1 \xrightarrow{vu'} 1.$$

Since $\rho^{-1}(1) = 1$, this implies that there is also a path $1 \xrightarrow{vu'} 1$ in $\mathcal{A}$. This in turn implies that

$$s \xrightarrow{u} 1 \xrightarrow{vu'} 1 \xrightarrow{v'} t$$

or, equivalently, $(s, e, t) = 1$. Since e is an idempotent and $s, t \in S$, this implies that $s = t$. Thus the restriction of ρ to S is a bijection from S onto S'.

Since $\widehat{\rho}(eMe) = e'Ne'$, the restriction of ρ to S defines an equivalence between M_e and $N_{e'}$. $\qquad\square$

Proposition 9.5.2 *Let X be a very thin code over A. Let $\mathcal{A} = (Q, 1, 1)$ be an unambiguous trim automaton recognizing X^*, and let φ be the associated representation. Then the Suschkewitch group of $\varphi(A^*)$ is equivalent to $G(X)$.*

Proof. According to Proposition 4.2.7, there exists a reduction from $\mathcal{A}_D^*(X)$ onto $\mathcal{A}$. Let e be a nonnull idempotent in the 0-minimal ideal of $M = \varphi_D(A^*)$. The image of e by the reduction is a nonnull idempotent e' in the 0-minimal ideal of $N = \varphi(A^*)$. Both $\varphi_D(\overline{F}(X))$ and $\varphi(\overline{F}(X))$ are ideals which are nonnull because they meet $\varphi_D(X^*)$ and $\varphi(X^*)$ repectively. Thus $e \in \varphi_D(\overline{F}(X))$ and $e' \in \varphi(\overline{F}(X))$. By the preceding proposition, $M_e \simeq N_{e'}$. Thus $G(X) \simeq N_{e'} \setminus 0$ which is the Suschkewitch group of $\varphi(A^*)$. $\qquad\square$

Example 9.5.3 Let G be a transitive permutation group on a finite set Q, and let H be the subgroup of G stabilizing an element q of Q. Let φ be a morphism from A^* onto G, and let X be the (group) code generating $X^* = \varphi^{-1}(H)$. The group $G(X)$ then is equivalent to G and $d(X)$ is the number of elements in Q.

In particular, we have for all $n \geq 1$, $G(A^n) = \mathbb{Z}/n\mathbb{Z}$ and $d(A^n) = n$.

9.6 Interpretations

Proposition 9.5.2 shows that the group of a very thin code and consequently also its degree, are independent of the automaton chosen. Thus we may expect that the degree reflects some combinatorial property of the code. This is indeed the fact, as we will see now.

Let X be a very thin code over A. An *interpretation* of a word $w \in A^*$ (with respect to X) is a triple

$$(d, x, g)$$

with $d \in A^- X$, $x \in X^*$, $g \in XA^-$, and $w = dxg$. We denote by $I(w)$ the set of interpretations of w. Two interpretations (d, x, g) and (d', x', g') of w are *adjacent* or *meet* if there exist $y, z, y', z' \in X^*$ such that

$$x = yz, \quad x' = y'z', \quad dy = d'y', \quad zg = z'g'.$$

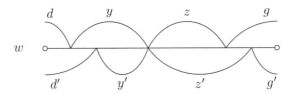

Figure 9.13 Two adjacent interpretations.

(see Figure 9.13). Two interpretations which do not meet are called *disjoint*. A set $\Delta \subset I(w)$ is *disjoint* if its elements are pairwise disjoint.

Let $w \in A^*$. The *degree* of w with respect to X is the nonnegative number $\delta_X(w)$ defined by

$$\delta_X(w) = \max\{\mathrm{Card}(\Delta) \mid \Delta \subset I(w), \Delta \text{ disjoint}\}.$$

Thus $\delta_X(w)$ is the maximal number of pairwise disjoint interpretations of w. Note that for $w \in \overline{F}(X)$,

$$\delta_X(uwv) \leq \delta_X(w).$$

Indeed, since w is not a factor of a word in X, every interpretation of uwv gives rise to an interpretation of w, and disjoint interpretations of uwv have their restriction to w also disjoint. Observe also that this inequality does not hold in general if $w \in F(X)$. In particular, a word in $F(X)$ may have no interpretation at all, whereas $\delta_X(w)$ is always at least equal to 1, for $w \in \overline{F}(X) \cap X^*$.

Proposition 9.6.1 *Let X be a very thin code. Then*

$$d(X) = \min\{\delta_X(w) \mid w \in X^* \cap \overline{F}(X)\}.$$

Proof. Let $\mathcal{A}_D^*(X) = (P, 1, 1)$ be the flower automaton of X, with the shorthand notation 1 instead of $(1, 1)$ for the initial and final state. Let $M = \varphi_D(A^*)$, let J be the 0-minimal ideal of M, let e be an idempotent in $\varphi_D(X^*) \cap J$ and let $S = \mathrm{Fix}(e)$. Then by definition $d(X) = \mathrm{Card}(S)$.

According to Proposition 9.4.4, we have $\varphi_D^{-1}(e) \cap \overline{F}(X) \neq \emptyset$. Take a fixed word $x \in \varphi_D^{-1}(e) \cap \overline{F}(X)$. Then $x \in X^* \cap \overline{F}(X)$, since $e \in \varphi_D(X^*)$.

Let $w \in X^* \cap \overline{F}(X)$ and let us verify that $d(X) \leq \delta_X(w)$. For this, it suffices to show that $d(X) \leq \delta_X(xwx)$, because of the inequality $\delta_X(xwx) \leq \delta_X(w)$. Now $\varphi_D(xwx) \in H(e)$, and consequently its restriction to S is a permutation on S. Thus for each $s \in S$, there exists one and only one $s' \in S$ such that $(s, \varphi_D(xwx), s') = 1$, or equivalently such that

$$s \xrightarrow{xwx} s'.$$

Since $w \in \overline{F}(X)$, this path is not simple. Setting $s = (u, d)$, $s' = (g, v)$ it factorizes into

$$s \xrightarrow{d} 1 \xrightarrow{y} 1 \xrightarrow{g} s'$$

and (d, y, g) is an interpretation of xwx. Thus each path from a state in S to another state in S, labeled by xwx, gives an interpretation of xwx. Two such interpretations are disjoint. Assume indeed the contrary. Then there are two interpretations (d_1, y_1, g_1) and (d_2, y_2, g_2) derived from paths $s_1 \xrightarrow{xwx} s_1'$ and $s_2 \xrightarrow{xwx} s_2'$ that are adjacent. This means that the paths factorize into

$$s_1 \xrightarrow{d_1} 1 \xrightarrow{z_1} 1 \xrightarrow{z_1'} 1 \xrightarrow{g_1} s_1',$$
$$s_2 \xrightarrow{d_2} 1 \xrightarrow{z_2} 1 \xrightarrow{z_2'} 1 \xrightarrow{g_2} s_2'$$

with $d_1 z_1 = d_2 z_2$ and also $z_1' g_1 = z_2' g_2$. Then there is also, in $\mathcal{A}_D^*(X)$, a path

$$s_1 \xrightarrow{d_1} 1 \xrightarrow{z_1} 1 \xrightarrow{z_2'} 1 \xrightarrow{g_2} s_2'$$

labeled xwx. This implies $(s_1, \varphi_D(xwx), s_2') = 1$; since $s_2' \in S$, one has $s_2' = s_1'$, whence $s_2 = s_1$.

Thus the mapping which associates, to each fixed point, an interpretation produces a set of pairwise disjoint interpretations. Consequently $\mathrm{Card}(S) \le \delta_X(xwx)$.

We now show that

$$\delta_X(x^3) \le d(X),$$

where x is the word in $\varphi_D^{-1}(e) \cap \bar{F}(X)$ fixed above. This will imply the proposition.

Let (d, y, g) be an interpretation of x^3. Let $p = (u, d), q = (g, v) \in P$. Then there is a unique path

$$p \xrightarrow{d} 1 \xrightarrow{y} 1 \xrightarrow{g} q, \qquad (9.22)$$

and moreover the paths $p \xrightarrow{d} 1, 1 \xrightarrow{g} q$ are simple or null. Since $\varphi_D(x) = e$, there exists a unique $s \in S$ such that the path (9.22) also factorizes into

$$p \xrightarrow{x} s \xrightarrow{x} s \xrightarrow{x} q.$$

Since $x \in \bar{F}(X)$, the word d is a prefix of x and g is a suffix of X.

Thus there exist words $z, \bar{z} \in A^*$ such that

$$y = zx\bar{z}, \qquad dz = x = \bar{z}g.$$

Observe that the fixed point $s \in S$ associated to the interpretation is independent of the endpoints of the path (9.22). Consider indeed another path

$$p' \xrightarrow{d} 1 \xrightarrow{y} 1 \xrightarrow{g} q'$$

associated to the interpretation (d, y, g), and a fixed point $s' \in S$ such that $p' \xrightarrow{x} s' \xrightarrow{x} s' \xrightarrow{x} q'$. Since $x = dz = \bar{z}g$, the above path factorizes in $p' \xrightarrow{d} 1 \xrightarrow{z} s' \xrightarrow{x} s' \xrightarrow{\bar{z}} 1 \xrightarrow{g} q'$. The uniqueness of the path $1 \xrightarrow{y} 1$ forces $s = s'$.

Thus we have associated, to each interpretation (d, y, g), a fixed point $s \in S$, which in turn determines two words $z, \bar{z}$ such that $y = zx\bar{z}$, and

$$1 \xrightarrow{z} s \xrightarrow{x} s \xrightarrow{\bar{z}} 1.$$

We now show that the fixed points associated to distinct interpretations are distinct. This will imply that $\delta_X(x^3) \leq \text{Card}(S) = d(X)$ and will complete the proof.

Let (d', y', g') be another interpretation of x^3, let $p' = (u', d')$, $q' = (g', v') \in P$, and assume that the path

$$p' \xrightarrow{d'} 1 \xrightarrow{y'} 1 \xrightarrow{g'} q'$$

decomposes into

$$p' \xrightarrow{d'} 1 \xrightarrow{z'} s \xrightarrow{x} s \xrightarrow{\bar{z}'} 1 \xrightarrow{g'} q'. \tag{9.23}$$

Since $x \in \overline{F}(X)$, the path $s \xrightarrow{x} s$ is not simple. Therefore there exist $h, \bar{h} \in A^*$ such that $x = h\bar{h}$ and

$$s \xrightarrow{h} 1 \xrightarrow{\bar{h}} s.$$

The paths (9.22) and (9.23) become

$$p \xrightarrow{d} 1 \xrightarrow{z} s \xrightarrow{h} 1 \xrightarrow{\bar{h}} s \xrightarrow{\bar{z}} 1 \xrightarrow{g} q$$
$$p' \xrightarrow{d'} 1 \xrightarrow{z'} s \xrightarrow{h} 1 \xrightarrow{\bar{h}} s \xrightarrow{\bar{z}'} 1 \xrightarrow{g'} q'.$$

This shows that $zh, \bar{h}\bar{z}, z'h, \bar{h}\bar{z}' \in X^*$. Next $dz = d'z' = x$. Thus $dzh = d'z'h$, showing that the interpretations (d, y, g) and (d', y', g') are adjacent. The proof is complete. $\qquad\square$

Now we are able to make the connection with the concept of degree of bifix codes introduced in the previous chapter. If $X \subset A^+$ is a thin maximal bifix code, then two adjacent interpretations of a word $w \in A^*$ are equal. This shows that $\delta_X(w)$ is the number of interpretations of w. As we have seen in Chapter 6, this number is constant on $\bar{H}(X)$, whence on $\overline{F}(X)$. By Proposition 9.6.1, the two notions of degree we have defined are identical.

9.7 Exercises

Section 9.1

9.1.1 Let e be an idempotent element of an unambiguous monoid of relations over a set Q. Show that if $p \xrightarrow{e} q \xrightarrow{e} r$ for $p, q, r \in Q$, then q is in $\text{Fix}(e)$.

9.1.2 The aim of this problem is to prove that for any stable submonoid N of a monoid M, there exists a morphism φ from M onto an unambiguous monoid of relations over some set Q and an element $1 \in Q$ such that $N = \mathrm{Stab}(1)$. For this let

$$D = \{(u, v) \in M \times M \mid uv \in N\}.$$

Let ρ be the relation over D defined by

$$(u, v)\rho(u', v') \iff Nu \cap Nu' \neq \emptyset \text{ and } vN \cap v'N \neq \emptyset.$$

Show that the equivalence classes of the transitive closure ρ^* of ρ are Cartesian products of subsets of M. (*Hint*: Prove that for any $(u, v), (u'v') \in D$ such that $(u, v)\rho(u', v')$, one has also $(u, v'), (u', v) \in D$ and $(u, v)\rho(u, v')\rho(u', v)$.)

Show that $N \times N$ is a class of ρ^*. Let Q be the set of classes of ρ^* and let 1 denote the class $N \times N$. Let φ be the function from M into $\mathfrak{P}(Q \times Q)$ defined by

$$(U \times V)\varphi(m)(U' \times V') \Leftrightarrow Um \subset U' \text{ and } mV' \subset V.$$

Show that φ is a morphism and that $N = \mathrm{Stab}(1)$. Show that in the case where $M = A^*$, the construction above coincides with the construction of the flower automaton.

9.1.3 Let K be a field and let m be an $n \times n$ matrix with elements in K. Show that $m = m^2$ if and only if there exist $\ell \in K^{n \times p}$ and $r \in K^{p \times n}$ such that

$$m = \ell r \quad \text{and} \quad r\ell = I_p,$$

where I_p denotes the identity matrix.

9.1.4 Let $\mathcal{A} = (P, 1, 1)$ and $\mathcal{B} = (Q, 1, 1)$ be two unambiguous trim automata. A reduction ρ from $\mathcal{A}$ to $\mathcal{B}$ is said to be *unambiguous* if there is a pair (λ, μ) of partial functions from P to Q which are restrictions of ρ and such that for each path $q \xrightarrow{w} q'$ in $\mathcal{B}$ there exists a unique pair $p \in \lambda^{-1}(q)$ and $p' \in \mu^{-1}(q')$ such that $p \xrightarrow{w} p'$ is a path in $\mathcal{A}$. Such a pair (λ, μ) is called an *unambiguous realization* of ρ.

(a) Verify that the functions λ, μ given below form an unambiguous realization of the reduction ρ of Example 4.2.6.

	1	2	3	4	5	6	7	8
ρ	1	2	2	2	3	3	3	3
λ	1	2	–	–	3	3	3	3
μ	1	2	2	2	3	–	–	–

(*Hint*: Show that there exists an invertible matrix R such that

$$R = \begin{bmatrix} U \\ L \\ V \end{bmatrix}, \quad R^{-1} = \begin{bmatrix} W & M & X \end{bmatrix}$$

where L is the matrix of the relation λ^{-1} and M is the matrix of the relation μ with

$$R\varphi_D(c)R^{-1} = \begin{bmatrix} 0 & 0 & 0 \\ * & \varphi(c) & 0 \\ * & * & 0 \end{bmatrix}$$

for each letter $c = a, b$.)

(b) Show that if the monoid $\varphi_A(A^*)$ has finite minimal rank, and if the automaton $\mathcal{B}$ is transitive, then any reduction from $\mathcal{A}$ to $\mathcal{B}$ is unambiguous. (*Hint*: Use Claim 2 of Proposition 9.1.9.)

Section 9.2

9.2.1 Let M be an unambiguous monoid of relations over a set Q. Let D be a $\mathcal{D}$-class of M containing an idempotent e. Let R (resp. L) be the $\mathcal{R}$-class (resp. the $\mathcal{L}$-class) of e and let Λ (resp. Γ) be the set of its $\mathcal{H}$-classes. Let $(a_H, a'_H)_{H \in \Lambda}$ be a system of coordinates of R, and let $(b_K, b'_K)_{K \in \Gamma}$ be a system of coordinates of L. Let $e = \ell r$ be the column-row decomposition of e and set $r_H = r a_H$, $\ell_K = b_K \ell$.

The *sandwich matrix* of D (with respect to these systems of coordinates) is defined as the $\Lambda \times \Gamma$ matrix with elements in $G_e \cup 0$ given by

$$S_{HK} = \begin{cases} r_H \ell_K & \text{if } e a_H b_K e \in H(e), \\ 0 & \text{otherwise.} \end{cases}$$

Show that for all $m \in M$, $H \in \Lambda$, $K \in \Gamma$,

$$(H * m)S_{H'K} = S_{HK'}(m * K),$$

with $H' = H \cdot m$, $K' = m \cdot K$.

Show that D is isomorphic with the semigroup formed by the triples $(H, g, K) \in \Gamma \times G_e \times \Lambda$ with the product defined by

$$(K, g, H)(K', g', H') = (K, g S_{HK'} g', H'). \tag{9.24}$$

Section 9.3

9.3.1 Let e be an idempotent element of an unambiguous monoid of relations over a set Q. Let $e = uv$ be a decomposition of e into an unambiguous product of relations $u : Q \to T$, $v : T \to Q$, where $\mathrm{Card}(T)$ is the rank of e. Show that there exists a bijection $\varphi : S \to T$, where S is the set of fixed points of e, such that $e = (u\varphi^{-1})(\varphi v)$ is the column-row decomposition of e.

9.3.2 Let K be a semiring and let m be a K-relation between P and Q. The *rank over K* of m is the minimum of the cardinalities of the sets R such that $m = \ell r$ for some K-relations $\ell \in K^{P \times R}$, $r \in K^{R \times Q}$. Denote it by $\mathrm{rank}_K(m)$. The rank of a relation, as defined in Section 9.3, is therefore also its rank when considered as an $\mathbb{N}$-relation.

Show that if K is a field and Q is finite, the rank over K coincides with the usual notion of rank in linear algebra.

9.3.3 Let

$$m = \begin{bmatrix} 1 & 0 & 0 & 1 \\ 1 & 1 & 0 & 0 \\ 0 & 1 & 1 & 0 \\ 0 & 0 & 1 & 1 \end{bmatrix}.$$

Show that $\text{rank}_{\mathbb{N}}(m) = 4$, but that $\text{rank}_{\mathbb{Z}}(m) = 3$.

9.3.4 Let M be an unambiguous monoid of relations over Q which is transitive and has finite minimal rank. Let $1 \in Q$ and $N = \text{Stab}(1)$. Let Λ (resp. Γ) be the set of 0-minimal or minimal left (resp. right) ideals of M, according to M contains or does not contain a zero. Let $R, R' \in \Gamma$, $L, L' \in \Lambda$. Show that if

$$R \cap L \cap N \neq \emptyset \quad \text{and} \quad R' \cap L' \cap N \neq \emptyset,$$

then also

$$R \cap L' \cap N \neq \emptyset \quad \text{and} \quad R' \cap L \cap N \neq \emptyset.$$

In other words, the set of pairs $(R, L) \in \Gamma \times \Lambda$ such that $R \cap L \cap N \neq \emptyset$ is a Cartesian product.

9.3.5 Let M be a transitive unambiguous monoid of relations on Q which has finite minimal rank and which does not contain the null relation. Let U be the set of nonzero rows of the elements of M. Show that the following conditions are equivalent for $v \in U$.

(i) v is a row of an element of M of minimal rank.
(ii) $0 \notin vM$.
(iii) v is maximal among the rows of the elements of M.
(iv) v is a row of an element of M with a minimal number of distinct nonzero rows.

9.3.6 Let X be a thin maximal code and let $\mathcal{A} = (Q, 1, 1)$ be a trim unambiguous automaton recognizing X^*. Let φ be the associated representation and let $M = \varphi(A^*)$.
(a) Show that a word w is strongly right completable if and only if $0 \notin \varphi(w)_{1*}M$.
(b) Let K be the minimal ideal of M. Show that any right completable word $w \in \varphi^{-1}(K)$ is simplifying and is strongly right completable. (*Hint*: Use Exercise 9.3.5.)

9.3.7 Let M be an unambiguous monoid of relations on a finite set Q. Let R (resp. L) be the set of rows (resp. columns) of the elements of M. Show that for each $r \in R$, $m \in M$, and $\ell \in L$, one has $rm\ell \leq 1$. Conversely, let R and L be sets of row and column vectors in $\mathfrak{P}(Q)$ such that

$$R = \{r \in \mathfrak{P}(Q) \mid r\ell \le 1 \text{ for all } \ell \in L\},$$
$$L = \{\ell \in \mathfrak{P}(Q) \mid r\ell \le 1 \text{ for all } r \in R\}. \tag{9.25}$$

Let $M = \{m \in \mathfrak{P}(Q \times Q) \mid rm\ell \le 1 \text{ for all } r \in R \text{ and } \ell \in L\}$.

(a) Show that M is a transitive unambiguous monoid of relations on Q which contains all products ℓr for $r \in R$ and $\ell \in L$.

(b) Show that any transitive unambiguous monoid of relations is a submonoid of one obtained in this way.

9.3.8 Let M be a transitive unambiguous monoid of relations on a finite set Q not containing the relation 0. Let R (resp. L) be the set of rows (resp. columns) of the elements of M which are maximal. Let U be the set of sums of the distinct rows of the elements of minimal rank of M and let $V = L$.

Show that for each $u \in U$, $m \in M$, and $v \in V$, one has $umv = 1$. Conversely, let U and V be sets of row and column vectors such that

$$U = \{u \in \mathfrak{P}(Q) \mid uv = 1 \text{ for all } v \in V\},$$
$$V = \{v \in \mathfrak{P}(Q) \mid uv = 1 \text{ for all } u \in U\}, \tag{9.26}$$

and such that for all $p \in Q$ there is a $u \in U$ (resp. $v \in V$) such that $u_p = 1$ (resp. $v_p = 1$). Let $M = \{m \in \mathfrak{P}(Q \times Q) \mid umv = 1 \text{ for all } u \in U \text{ and } v \in V\}$.

(a) Show that M is a transitive unambiguous monoid of relations on Q not containing 0.

(b) Show that any transitive unambiguous monoid of relations not containing 0 is a submonoid of one obtained in this way.

9.3.9 An unambiguous monoid of relations on a finite set Q with n elements is said to be *very transitive* if it contains a transitive group G of permutations on Q. The aim of this exercise is to show that all elements of a very transitive unambiguous monoid of relations have the same number n of elements (as subsets of $Q \times Q$).

Let e be an idempotent of minimal rank. Let u be the sum of the distinct rows of e and let v be a column of e. Let $r = \operatorname{Card}(u)$ and $s = \operatorname{Card}(v)$. Let $U = uG$ be the orbit of u under the right action of G and let $V = Gv$ be the orbit of v under the left action of G. Let $p = \operatorname{Card}(U)$ and $q = \operatorname{Card}(V)$.

(a) Show that for each $q \in Q$, the number of elements of U containing q is independent of q. Let h be this integer. In the same way, let k be the number of elements of V containing a given $q \in Q$.

(b) Show that $rp = hn$, $sq = kn$ and $rk = p$, $sh = q$.

(c) Show that for each $m \in M$, $pq = thk$ where t is the cardinality of m (as a subset of $Q \times Q$). Conclude that $t = n$.

9.3.10 Show that for any transitive unambiguous monoid of relations M on a finite set Q, there is a finite set R containing Q and a transitive unambiguous monoid of relations N on R not containing 0 such that the elements of M are subsets of the restriction to $Q \times Q$ of elements of N.

9.3.11 Let G be a graph. A *clique* in G is a set of vertices such that there is an edge between all pairs of vertices. A set of vertices is *stable* if no pair of vertices is connected by an edge of G. Consider the set L of cliques in G and the set R of stable sets. Show that the pair (L, R) satisfies the equalities (9.25) of Exercise 9.3.7 when identifying an element of L with its column characteristic vector and an element of R with its row characteristic vector.

Let U (resp. V) be the set of maximal cliques (resp. stable sets). Show that if the graph G has the property that any maximal clique intersects any maximal stable set, then (U, V) satisfies the relations (9.26) of Exercise 9.3.8.

9.3.12 Let M be a transitive unambiguous monoid of relations not containing zero. Show that for two elements m, m' of M, if $m \le m'$ then $m = m'$. (*Hint*: Use Exercise 9.3.5.)

9.3.13 Let $\mathcal{A}$ be an n-state strongly connected unambiguous automaton. Assume that the minimal rank of the words in $\mathcal{A}$ is 1. Show that there is a word of length at most $(n^2 - n + 2)(n - 1)/2$ that has rank one. (*Hint*: Prove first the following claim: For a state $p \in Q$ and a word $u \in A^*$, if $\varphi(u)_{p*}$ is not a maximal row, there is a state q and a word v of length at most $n(n - 1)/2$ such that $\varphi(u)_{p*} < \varphi(vu)_{q*}$.)

Section 9.4

9.4.1 Let $X \subset A^+$ be a very thin code. Let M be the syntactic monoid of X^* and let φ be the canonical morphism from A^* onto M. Show that M has a unique 0-minimal or minimal ideal J, according to M contains a zero or not. Show that $\varphi(X^*)$ meets J, that J is a $\mathcal{D}$-class, and that each $\mathcal{H}$-class contained in J and which meets $\varphi(X^*)$ is a finite group.

9.4.2 Let $X \subset A^+$ be a very thin code, let $\mathcal{A} = (Q, 1, 1)$ be an unambiguous trim automaton recognizing X^*. Let φ be the associated morphism and $M = \varphi(A^*)$. Let J be the minimal or 0-minimal ideal of M and $K = J \setminus 0$. Let $e \in M$ be an idempotent of minimal rank, let R be its $\mathcal{R}$-class and L be its $\mathcal{L}$-class. Let Λ (resp. Γ) be the set of $\mathcal{H}$-classes contained in R (resp. L), and choose two systems of coordinates

$$(a_H, a'_H)_{H \in \Lambda}, \quad (b_K, b'_K)_{K \in \Gamma}$$

of R and L, respectively. Let

$$\mu : M \to (G_e \cup 0)^{\Lambda \times \Lambda}$$

be the morphism of M into the monoid of row-monomial $\Lambda \times \Lambda$-matrices with elements in $G_e \cup 0$ defined by the $\mathcal{R}$-representation with respect to e. Similarly, let

$$\nu : M \to (G_e \cup 0)^{\Gamma \times \Gamma}$$

be the morphism associated with the $\mathcal{L}$-representation with respect to e. Let S be the sandwich matrix of J relative to the systems of coordinates introduced (see Exercise 9.2.1). Show that for all $m \in M$,

$$\mu(m)S = Sv(m).$$

Show that for all $m, n \in M$,

$$\mu(m) = \mu(n) \Leftrightarrow (\forall H \in \Lambda, r_H m = r_H n),$$

$$v(m) = v(n) \Leftrightarrow (\forall K \in \Gamma, m\ell_K = n\ell_K),$$

where $r_H = \ell a_H$, $\ell_K = b_K \ell$, and ℓr is the column-row decomposition of e.

Show, using these relations, that the function

$$m \mapsto (\mu(m), v(m))$$

is injective.

9.4.3 Let $X \subset A^+$ be a very thin code. Let φ be the representation associated with an unambiguous trim automaton $\mathcal{A}$ recognizing X^*, let $M = \varphi(A^*)$ and let J be its minimal ideal.

Show that X is prefix if and only if, for any idempotent e in J not in $\varphi(X^*)$, one has $Me \cap \varphi(X^*) = \emptyset$.

9.4.4 Let $\mathcal{A} = (Q, 1, 1)$ be a strongly connected complete deterministic automaton. Let M be the adjacency matrix of $\mathcal{A}$. Let w be a positive left eigenvector of M for the eigenvalue $\text{Card}(A)$. For any subset P of Q, set $w(P) = \sum_{q \in P} w_q$.

A *maximal class* is any class of some maximal nuclear equivalence of the transition monoid of $\mathcal{A}$. Show that w is constant on the set of maximal classes, that is $w(P) = w(P')$ for any pair P, P' of maximal classes. Assume that w has integer coefficients. Show that the minimal rank of $\mathcal{A}$ divides $w(Q)$.

Section 9.5

9.5.1 Let X be a very thin code. Let M be the syntactic monoid of X^*, and let J be the 0-minimal or minimal ideal of M (see Exercise 9.4.1). Let G be an $\mathcal{H}$-class in J that meets $\varphi(X^*)$, and let $H = G \cap \varphi(X^*)$.

Show that the representation of G over the right cosets of H is injective, and that the permutation group obtained is equivalent to $G(X)$.

9.5.2 Let $X \subset A^+$ be a very thin code. Let φ be the representation associated with an unambiguous trim automaton $\mathcal{A} = (Q, 1, 1)$ recognizing X^*. Let $M = \varphi(A^*)$ and let D be a nonzero regular $\mathcal{D}$-class of M. Show that if D meets $\varphi(\overline{F}(X))$, then $D \cap \varphi(X^*)$ contains an idempotent.

Conclude that when X is finite, $\varphi(X^*)$ meets all regular nonzero $\mathcal{D}$-classes.

9.5.3 Let X be a thin maximal code. Show that if $z \in A^*$ is both strongly right and strongly left completable, then some power of z is in X^* (a word x is *strongly left completable* if for any $u \in A^*$ the word ux is left completable).

9.5.4 Let X, Y be two codes. We define the *meet* of X and Y, denoted $X \wedge Y$ as the basis of the submonoid $X^* \cap Y^*$. Show that the meet of two thin codes $X, Y \subset A^+$ is thin maximal over A if and only if there is a word $x \in X^*$ strongly left completable in Y^* and a word $y \in Y^*$ strongly right completable in X^*. (*Hint*: Use Exercise 9.5.3.)

9.5.5 Show that for any rational (resp. thin) code Z, there exist two rational (resp. thin) maximal codes X, Y such that $Z = X \wedge Y$. (*Hint*: Use Theorem 2.5.24 and Exercise 2.5.4 for embedding Z into a rational (resp. thin) code T.)

9.8 Notes

There are only a few research papers devoted to unambiguous monoids of relations, and this chapter is a systematic presentation of the topic. The study of the structure of the $\mathcal{D}$-classes in unambiguous monoids of relations is very close to the standard development for abstract semigroups presented in the usual textbooks. This holds in particular for the Schützenberger representations, see Clifford and Preston (1961) or Lallement (1979). The generalization of the results of Section 9.1 to arbitrary monoids of relations is partly possible. See, for instance, Lerest and Lerest (1980). The notion of rank and the corresponding results appear in Lallement (1979) for the particular case of monoids of functions. A significant step in the study of unambiguous monoids of relations using such tools as the column-row decomposition appears in Césari (1974). The degree of a very thin code, as defined in Section 9.5 is closely related to the degree of a finite-to-one map as defined in Lind and Marcus (1995). Actually, let $\mathcal{A}$ be an unambiguous automaton. As explained in the Notes of Chapter 4, there is a finite-to-one map λ corresponding to $\mathcal{A}$, associating to a path its label. Let M be the transition monoid of $\mathcal{A}$. Then the minimal rank of M is the degree of the map λ.

Theorem 9.4.1 is due to Schützenberger. An extension to sets which are not codes appears in Schützenberger (1979a). Problem 9.1.2 is a theorem due to Boë *et al.* (1979). Extensions may be found in Boë (1976). The notion of sandwich matrix (Exercise 9.2.1) is standard, see Clifford and Preston (1961).

Exercise 9.1.4 is from Carpi (1987). The notion of unambiguous reduction has some connections with the reduction of linear representations of rational series (see Berstel and Reutenauer (1988)).

Exercise 9.3.5 is due to Césari (1974). Exercises 9.3.7 to 9.3.11 are due to Boë (1991). Exercise 9.3.9 gives an alternative proof of a result of Perrin and Schützenberger (1977) (see Proposition 12.2.4). Exercise 9.3.10 is related to the embedding of codes into maximal ones, although it does not provide an alternative to prove that every rational code is included in a maximal one (the relations corresponding to the letters may generate a monoid which is not transitive). The graphs having the property that any maximal clique meets any maximal stable set have been characterized in Deng *et al.* (2004, 2005).

Exercise 9.3.12 is from Béal *et al.* (2008). Exercise 9.3.13 is from Carpi (1988). A simplified proof appears in Béal *et al.* (2008). It shows that for strongly connected unambiguous automata such that the minimal rank of words in the automaton is 1, there is a cubic upper bound for the length of a word of rank 1, as it is the case for synchronized deterministic automata (see Exercise 3.6.2). As for deterministic automata, the optimal upper bound is not known.

Exercise 9.4.3 is from Reutenauer (1981). Exercises 9.5.3, 9.5.4, and 9.5.5 are from Bruyère *et al.* (1998). Exercise 9.4.4 is from Friedman (1990).

10

Synchronization

The notion of synchronization for codes and automata refers to the ability of parsing an input into codewords with a limited amount of information. It addresses a more general situation than deciphering which is left-to-right oriented. The interest of synchronization lies in the possibility of recovering from errors by the specific nature of the involved decoders.

The chapter starts with the definition of synchronizing pairs, synchronizing words, and absorbing words. These notions have already been considered in Chapter 3 for prefix codes. Next, as for the deciphering delay, two notions of synchronization delay are introduced, the first related to the number of words involved, the second connected to local automata. We describe the connection between synchronization delay and the notions of circular codes and limited codes. Important results are the completion of rational uniformly synchronized codes and of locally parsable codes (Theorem 10.3.13 andTheorem 10.2.11).

In the final section, we give a necessary and sufficient condition to guarantee that a deterministic automaton can be transformed into a synchronizing one by modifying the labels of its edges (Theorem 10.4.2). This theorem has been conjectured over many years as the *road coloring problem*.

10.1 Synchronizing pairs

The section starts with the definition of synchronizing pairs, synchronizing words, and constants. Relations among these objects are described. Constants are characterized by their rank. Next, synchronized codes are defined, and shown to coincide with codes of degree 1. Finally, absorbing words are introduced.

The following definitions will be used later for the submonoid $S = X^*$ generated by a code $X \subset A^+$. Since the nature of S does not play a role, we choose the more general formulation.

A pair (x, y) of words of A^* is *synchronizing* for $S \subset A^*$ if for any words $u, v \in A^*$, one has

$$uxyv \in S \implies ux, yv \in S.$$

If (x, y) is a synchronizing pair for S, then any pair $(x'x, yy')$ is a synchronizing pair for S. Thus the components of a synchronizing pair can be assumed to be nonempty words.

A word $x \in A^*$ is *synchronizing* for S if

$$uxv \in S \implies ux, xv \in S.$$

This definition was already given in Chapter 3 for $S = X^*$ where X is a prefix code.

Proposition 10.1.1 *If $x, y \in A^*$ are synchronizing words for S, then the pair (x, y) is synchronizing for S.*

Proof. Let x, y be synchronizing words. If $uxyv \in S$, then $ux \in S$ because x is synchronizing, and $yv \in S$ because y is synchronizing. Thus (x, y) is a synchronizing pair. $\square$

Example 10.1.2 Let $A = \{a, b\}$ and $S = \{ab, ba\}^*$. The pair (b, b) is synchronizing for S, the word bb is not synchronizing but $abba$ is synchronizing.

Let $S \subset A^*$ be a set. Recall that $\Gamma_S(w)$, or simply $\Gamma(w)$ when S is understood, denotes the set of contexts of a word w in S, that is

$$\Gamma_S(w) = \{(u, v) \in A^* \times A^* \mid uwv \in S\}.$$

A word $w \in A^*$ is said to be a *constant* for S if for any $(u, v), (u', v') \in \Gamma_S(w)$ one has also $(u, v'), (u', v) \in \Gamma_S(w)$. This means that $\Gamma_S(w)$ is a direct product. More precisely, $\Gamma_S(w) = \Gamma_S^{(\ell)}(w) \times \Gamma_S^{(r)}(w)$, where $\Gamma_S^{(\ell)}(w) = \{u \in A^* \mid \exists v \in A^*, (u, v) \in \Gamma_S(w)\}$ and $\Gamma_S^{(r)}(w)$ is defined symmetrically.

Example 10.1.3 Let $A = \{a, b\}$ and $S = \{ab, ba\}^*$. The word bb is a constant for S. Indeed, the contexts of bb in S are the pairs (xa, ay) for $x, y \in S$.

The following statement shows that the set of constants for a set S forms a two-sided ideal.

Proposition 10.1.4 *If $w \in A^*$ is a constant for a set S, then for all $u, v \in A^*$, the word uwv is a constant for S.*

Proof. Let $p, p', s, s' \in A^*$ be words such that $(p, s), (p', s') \in \Gamma(uwv)$. Then (pu, vs) and $(p'u, vs')$ are in $\Gamma(w)$. Since w is a constant, we have also (pu, vs'), $(p'u, vs) \in \Gamma(w)$. Thus $(p, s'), (p', s) \in \Gamma(uwv)$. This shows that uwv is a constant. $\square$

Proposition 10.1.5 *If a word of S is a constant for S, then it is synchronizing for S.*

Proof. Let $x \in S$ be a constant for S. Let $u, v \in A^*$ be words such that uxv is in S. Then $(u, v) \in \Gamma_S(x)$. Since $(1, 1)$ also is in $\Gamma_S(x)$, it follows that $ux, xv \in S$. Thus x is synchronizing. $\square$

Proposition 10.1.6 *Let $S \subset A^*$ be a submonoid. If (x, y) is a synchronizing pair for S, then xy is a constant.*

Proof. Let $(x, y) \in A^* \times A^*$ be a synchronizing pair. Let $(u, v), (u', v') \in \Gamma_S(xy)$. Considering the words $uxyv$ and $u'xyv'$, one gets that $ux, yv, u'x, yv'$ are in S. Since S is a submonoid, it follows that $uxyv', u'xyv \in S$. Consequently, $(u, v'), (u', v) \in \Gamma_S(xy)$, showing that xy is a constant. $\square$

The next statement summarizes the relations between the notions introduced so far in the case of the submonoid generated by a code.

Proposition 10.1.7 *Let $X \subset A^+$ be a code. The following conditions are equivalent.*

(i) *There exists a synchronizing pair $(x, y) \in X^* \times X^*$ for X^*.*
(ii) *There exists a word in X^* that is a synchronizing word for X^*.*
(iii) *There exists a word in X^* that is a constant for X^*.*

Proof. (i) implies (iii) by Proposition 10.1.6, (iii) implies (ii) by Proposition 10.1.5 and (ii) implies (i) by Proposition 10.1.1. $\square$

A code X is called *synchronized* if there exist pairs of words in X^* which are synchronizing for X^*. In view of the preceding proposition, this terminology is compatible with that introduced in Chapter 3.

A synchronized code X is very thin. Indeed, let $(x, y) \in X^+ \times X^+$ be a synchronizing pair of nonempty words. Then xy is not a factor of a word of X, since $uxyv \in X$ implies $ux, yv \in X^+$.

The existence of a synchronizing pair (x, y) has the following meaning. When we try to decode a word $w \in A^*$, the occurrence of a factor xy in w implies that the factorization of w into words in X, whenever it exists, must pass between x and y: if $w = uxyv$, it suffices to decode separately ux and yv.

The next proposition gives a method to check whether a word is a constant. Recall that the rank of a word w in a deterministic automaton $\mathcal{A} = (Q, i, T)$ is simply $\text{Card}(Q \cdot w)$.

Proposition 10.1.8 *Let $\mathcal{A}$ be the minimal deterministic automaton recognizing a set $S \subset A^*$. A word $w \in A^*$ is a constant for S if and only if it has rank at most 1 in $\mathcal{A}$.*

Proof. Set $\mathcal{A} = (Q, i, T)$. Suppose first that w is a constant. Assume that $\text{rank}(w) \geq 1$. Let $p, p' \in Q \cdot w$. Let u, u', v, v' be such that $i \cdot uw = p, i \cdot u'w = p'$, and $p \cdot v$, $p' \cdot v' \in T$. Thus $uwv, u'wv' \in S$. Then, for any $r \in A^*$, $p \cdot r \in T$ implies $uwr \in S$ and therefore $u'wr \in S$, whence $p' \cdot r \in T$. Similarly, $p' \cdot r \in T$ implies $p \cdot r \in T$. This shows that $p = p'$. This shows that $\text{rank}(w) = 1$.

Conversely, if $\text{rank}(w) = 0$, the set of contexts of w in S is empty and w is a constant. Assume that $\text{rank}(w) = 1$. Suppose that $uwv, u'wv' \in S$. Since $i \cdot uw$ and $i \cdot u'w$ are defined, they are equal. Then $i \cdot uwv = i \cdot u'wv$ implies that $u'wv \in S$. Similarly, $uwv' \in S$. Thus w is a constant. $\square$

The following result shows that part of the previous proposition holds for nondeterministic automata.

Proposition 10.1.9 *Let* $A = (Q, I, T)$ *be an automaton recognizing a set* $S \subset A^*$. *A word* $w \in A^*$ *that has rank 1 in the automaton* A *is a constant for* S.

Proof. Suppose that $uwv, u'wv' \in S$. There are paths $i \xrightarrow{u} p \xrightarrow{w} q \xrightarrow{v} t$ and $i' \xrightarrow{u'} p' \xrightarrow{w} q' \xrightarrow{v'} t'$ with $i, i' \in I, t, t' \in T$. Since $\varphi_A(w)$ has rank 1, $\varphi_A(w) = \ell r$, with $\ell \subset Q \times \{s\}$ and $r \subset \{s\} \times Q$, for some state s. Thus $(p, s), (p', s) \in \ell$ and $(s, q), (s, q') \in r$. It follows that $(p, q'), (p', q) \in \varphi_A(w)$. This implies that w is a constant. $\square$

Proposition 10.1.10 *Let* $X \subset A^+$ *be a code and let* $A = (Q, 1, 1)$ *be a trim unambiguous automaton such that* $X^* = \mathrm{Stab}(1)$. *If* $x, y \in A^*$ *form a synchronizing pair, then* $\mathrm{rank}(\varphi_A(xy)) \leq 1$.

Proof. Let ℓ be the column of $\varphi_A(x)$ of index 1 and let r be the row of $\varphi_A(x)$ of index 1. We verify that $\varphi_A(xy) = \ell r$. Suppose first that $p \xrightarrow{xy} q$ for some $p, q \in Q$. Since A is trim, there exist $u, v \in A^*$ such that $1 \xrightarrow{u} p$ and $q \xrightarrow{v} 1$. Then $uxyv$ is in X^*. This implies $ux, yv \in X^*$. This shows that $\ell_p = r_q = 1$. Thus $\varphi_A(xy) \subset \ell r$. The converse inclusion is clear. $\square$

The following is a characterization of synchronized codes in terms of the degree introduced in Chapter 9.

Proposition 10.1.11 *A code is synchronized if and only if it has degree 1.*

Proof. Let $A = (Q, 1, 1)$ be an unambiguous trim automaton recognizing X^* and let φ be the associated representation. If X is synchronized, there is a synchronizing pair (x, y) with $x, y \in X^*$. By Proposition 10.1.10, the rank of $\varphi(xy)$ is at most 1. Since $xy \in X^*$, the rank is not 0 and thus $\varphi(xy)$ has rank 1. This shows that $d(X) = 1$. Conversely, let $w \in A^*$ be such that $\mathrm{rank}(\varphi(w)) = 1$. Since $\mathrm{rank}(\varphi(w)) \neq 0$, there exist $u, v \in A^*$ such that $uwv \in X^*$. Set $x = uwv$. By Proposition 10.1.9, x is a constant for X^*. This shows that X is synchronized. $\square$

A pair (x, y) of words of X^* is *absorbing* if $A^*x \cap yA^* \subset X^*$. A code X which has an absorbing pair is complete since for any word w, one has $ywx \in X^*$.

Example 10.1.12 Consider the suffix code $X = ab^*$ over $A = \{a, b\}$. Observe that $X^+ = aA^*$. Every word in X is synchronizing. Indeed, if $x \in X$ and $uxv \in X^*$, then ux and xv start with the letter a, and therefore are in X^+. Every pair of words of X is absorbing. Indeed, if a word w has a prefix in X, then it starts with the letter a and therefore is in X^+.

Proposition 10.1.13 *Let* $X \subset A^+$ *be a code. Any absorbing pair is synchronizing. Conversely, if* X *is complete, then any synchronizing pair of words of* X^* *is absorbing.*

Proof. Let (x, y) be an absorbing pair. Let $u, v \in A^*$ be such that $uxyv \in X^*$. Then $w = yuxyvx$ is in X^*. Since $w = (yux)(yvx) = y(uxyvx)$, and $y, yux, uxyvx, yvx$ are in X^*, it follows by stability that $ux \in X^*$. Similarly $yv \in X^*$.

Conversely, let (x, y) be a synchronizing pair and let $w \in A^*x \cap yA^*$. Thus $w = ux = yv$ for some words $u, v \in A^*$. Since X is complete, there exist words $u', v' \in A^*$ such that $u'xwyv' \in X^*$. Since (x, y) is synchronizing, we have $u'x, u'xw, wyv', yv' \in X^*$ by synchronization. By stability, this implies $w \in X^*$. □

As a consequence, we have the following characterization of complete synchronized codes.

Proposition 10.1.14 *Let $X \subset A^+$ be a code. Then X is complete and synchronized if and only if there exist absorbing pairs.* □

Example 10.1.15 The code $X = \{aa, ba, baa, bb, bba\}$ is synchronized. Indeed, the pair (aa, ba) is an example of a synchronizing pair: assume that $uaabav \in X^*$ for some $u, v \in A^*$. Since $ab \notin F(X)$, we have $uaa, bav \in X^*$. Since X is also a complete code, it follows by Proposition 10.1.13 that (aa, ba) is absorbing. Thus $baA^*aa \subset X^*$.

10.2 Uniformly synchronized codes

Let s be an integer. A code $X \subset A^+$ has *verbal synchronization delay s* if any $x \in X^s$ is a synchronizing word. For simplicity we talk of the synchronization delay, when no confusion arises. Thus a code $X \subset A^+$ has synchronization delay s if

$$x \in X^s, \ u, v \in A^*, \ uxv \in X^* \implies ux, xv \in X^*. \tag{10.1}$$

A code X is said to be *uniformly synchronized* if it has synchronization delay s for some s. The least s of this kind is called the *minimal synchronization delay* of X. It is denoted by $\sigma(X)$.

Example 10.2.1 Consider over $A = \{a, b\}$ the code $X = \{a, ab\}$. Every word in X is synchronizing. Therefore X has synchronizing delay 1. Consequently, every pair of words of X is synchronizing.

The following result shows that a code with finite synchronization delay has also finite deciphering delay. More precisely

Proposition 10.2.2 *The minimal deciphering delay of a code is less than or equal to its minimal synchronization delay.*

Proof. Let s be the minimal synchronization delay of X. Let $x \in X^*$, $y \in X^s$ and $u \in A^*$ be such that $xyu \in X^*$. Since X has synchronization delay s, we have $xy, yu \in X^*$. Thus y is simplifying. In view of Proposition 5.1.5, this shows that X has deciphering delay s. □

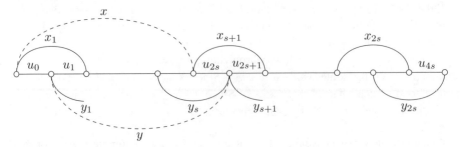

Figure 10.1 An X-factorization with $u_{i-1}u_i \in X^*$ for $1 \le i \le 4s$.

The following example shows that the minimal deciphering delay may be finite but not the synchronization delay.

Example 10.2.3 Let $X = \{ab, ba\}$. Since X is prefix, it has deciphering delay 0. It has infinite synchronization delay since for each $n \ge 1$, the word $x = (ab)^n$ satisfies $bxa \in X^*$ although $bx, xa \notin X^*$.

The following statements relate uniformly synchronized codes to limited codes as introduced in Chapter 7.

Proposition 10.2.4 *A uniformly synchronized code is limited.*

Proof. Let $X \subset A^+$ be a uniformly synchronized code, and let s be its minimal synchronization delay. We show that X is $(2s, 2s)$-limited (see Figure 10.1). Consider indeed words

$$u_0, u_1, \ldots, u_{4s} \in A^*,$$

and assume that $u_{i-1}u_i \in X^*$ for $1 \le i \le 4s$. Set, for $1 \le i \le 2s$,

$$x_i = u_{2i-2}u_{2i-1}, \qquad y_i = u_{2i-1}u_{2i}.$$

Let $y = y_1 y_2 \cdots y_s$ and $x = x_1 x_2 \cdots x_s$.

Assume first that $y_i \ne 1$ for all $i = 1, \ldots, s$. Then $y \in X^s X^*$. Since $u_0 y u_{2s+1} \in X^*$, the uniform synchronization shows that $u_0 y \in X^*$. Since $u_0 y = x u_{2s}$, this is equivalent to

$$x u_{2s} \in X^*. \tag{10.2}$$

Next, consider the case that $y_i = 1$ for some $i \in \{1, 2, \ldots, s\}$. Then $u_{2i-1} = u_{2i} = 1$. It follows that

$$y_{i+1} \cdots y_s = u_{2i+1} \cdots u_{2s} = x_{i+1} \cdots x_t u_{2s}.$$

Thus, in this case also $x u_{2s}$ is in X^*.

Setting $y' = y_{s+1} \cdots y_{2s}$, we prove in the same manner that

$$u_{2s} y' \in X^*. \tag{10.3}$$

Since X^* is stable, (10.2) and (10.3) imply that $u_{2s} \in X^*$. This shows that X is $(2s, 2s)$-limited. $\qquad\square$

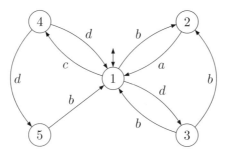

Figure 10.2 An unambiguous automaton recognizing X^*.

Example 10.2.5 Consider the $(2, 2)$-limited code $X = \{ba, cd, db, cdb, dba\}$ given in Example 7.2.7. We have $\sigma(X) = 1$. The words of X have rank 1 in the automaton of Figure 10.2. Indeed, a and c have rank 1 since $\varphi(a) = \{(2, 1)\}$ and $\varphi(c) = \{(1, 4)\}$. Further, we have $\varphi(db) = \{4, 1\} \times \{1, 2\}$ and thus db also has rank 1. Consequently each $x \in X$ is a constant, and therefore $\sigma(X) = 1$.

Example 10.2.6 Let $X = ab^*c \cup b$ be the limited code of Example 7.2.6. It is not uniformly synchronized. Indeed, for all $s \geq 0$, one has $b^s \in X^s$ and $ab^s c \in X$. However $ab^s, b^s c \notin X$. This example shows that the converse of Proposition 10.2.4 does not hold.

We now prove that in the case of finite codes, the concepts introduced coincide.

Theorem 10.2.7 *Let X be a finite code. The following conditions are equivalent.*

(i) *X is circular.*
(ii) *X is limited.*
(iii) *X is uniformly synchronized.*

For the proof of the theorem, we use a result about finite semigroups.

Proposition 10.2.8 *Let S be a finite semigroup and let J be an ideal of S. The following conditions are equivalent.*

(i) *There exists an integer $n \geq 1$ such that $S^n \subset J$.*
(ii) *All idempotents of S are in the ideal J.*

Proof. (i) $\Rightarrow$ (ii). For any idempotent e in S, we have $e = e^n \in J$.

(ii) $\Rightarrow$ (i). Set $n = 1 + \mathrm{Card}(S)$. We show the inclusion $S^n \subset J$. Indeed let $s \in S^n$. Then $s = s_1 s_2 \cdots s_n$, with $s_i \in S$. Let $t_i = s_1 s_2 \cdots s_i$, for $1 \leq i \leq n$. Then there exist indices i, j with $1 \leq i < j \leq n$ and $t_i = t_j$. Setting $r = s_{i+1} \cdots s_j$, we have $t_i r = t_i$, hence also $t_i r^k = t_i$ for all $k \geq 1$. Since S is finite, there exists an integer k such that $e = r^k$ is an idempotent. Then $e \in J$, and consequently

$$s = t_i s_{i+1} \cdots s_n = t_i e s_{i+1} \cdots s_n \in J.$$

This proves that (i) holds. □

Proof of Theorem 10.2.7. We have already proved the implications (iii) $\Longrightarrow$ (ii) $\Longrightarrow$ (i) without the finiteness assumption. Indeed, the first implication is Proposition 10.2.4, and the second is Proposition 7.2.2. Thus it remains to prove (i) $\Longrightarrow$ (iii).

Let $X \subset A^+$ be a finite circular code, and let $\mathcal{A}_D^*(X) = (P, 1, 1)$ be the flower automaton of X with the shorthand notation 1 for the state $(1,1)$. Let $M = \varphi_D(A^*)$, and let J be its 0-minimal ideal. Let $S = \varphi_D(A^+)$. By Proposition 7.1.5, each element in S has at most one fixed point. In particular, every nonzero idempotent in S has rank 1 and therefore is in J. By Proposition 10.2.8, there is an integer $n \geq 1$ such that $S^n \subset J$. Let $x \in X^n$. Then $\varphi_D(x) \in J$ and consequently x has rank 1. Thus x is a constant by Proposition 10.1.9, and therefore synchronizing by Proposition 10.1.5. It follows that each word of X^n is synchronizing, showing that X has synchronizing delay n. This shows that X is uniformly synchronized. $\qquad\square$

Example 10.2.9 Let $A = \{a_1, a_2, \ldots, a_{2k}\}$ and

$$X = \{a_i a_j \mid 1 \leq i < j \leq 2k\}.$$

We show that X is uniformly synchronized and $\sigma(X) = k$. First, $\sigma(X) \geq k$ since $(a_2 a_3)(a_4 a_5) \cdots (a_{2k-2} a_{2k-1}) \in X^{k-1}$ and also $(a_1 a_2) \cdots (a_{2k-1} a_{2k}) \in X^*$; however $a_1 a_2 \cdots a_{2k-1} \notin X^*$. Next, suppose that $x \in X^k$, and $uxv \in X^*$. If u and v have even length, then they are in X^*. Therefore we assume the contrary. Then $u = u' a_j$, $v = a_\ell v'$ with $a_j, a_\ell \in A$ and $u', v' \in X^*$. Moreover $a_j x a_\ell \in X^*$. Set $x = a_{i_1} \cdots a_{i_{2k}}$. Since $x \in X^*$, we have $i_1 < i_2, i_3 < i_4, \ldots, i_{2k-1} < i_{2k}$, and since $a_j x a_\ell \in X^*$, we have $j < i_1, i_2 < i_3, \ldots, i_{2k} < \ell$. Thus $1 \leq j < i_1 < i_2 < \cdots < i_{2k-1} < i_{2k} < \ell \leq 2k$, which is clearly impossible. Consequently u and v have even length, showing that $\sigma(X) \leq k$. This proves the equality.

Compare this example with Example 7.2.7, which is merely Example 10.2.9 with $k = \infty$. The infinite code is circular but not limited, hence not uniformly synchronized.

We prove now an analogue of Theorem 2.5.24 for uniformly synchronized codes. The construction of the proof of Theorem 2.5.24 cannot be used since it does not even preserve the finiteness of the deciphering delay (see Example 5.2.8).

The following example shows that the construction of the proof of Theorem 5.2.9 neither applies.

Example 10.2.10 Consider again the code $X = \{a, ab\}$ over $A = \{a, b\}$ which has synchronizing delay 1. We have seen in Example 5.2.20 that the construction used in Theorem 5.2.9 gives the code $Y = \{a, ab\} \cup \{ab^3, ab^2 a\}\{bb, ba\}^*$ which has deciphering delay 1. However, Y has infinite synchronization delay since every $(ab)^n$ is a factor of $ab(ba)^{n+1}$ which is in Y, and thus no pair $(ab)^k, (ab)^\ell$ is synchronizing.

Theorem 10.2.11 *Any rational uniformly synchronized code is contained in a complete rational code with the same minimal synchronization delay.*

Proof. Consider a nonempty code $X \subset A^+$ with synchronization delay s and consider

$$M = (X^s A^* \cap A^* X^s) \cup X^*. \tag{10.4}$$

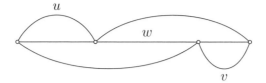

Figure 10.3 Proving that M is stable.

Observe that M is a submonoid of A^*. Let Y be the minimal generating set of M. We show that Y is a code having the desired properties. The proof is in several steps.

Let us first prove that Y is a code. For this, we prove that M is stable. Let $u, w, v \in A^*$ be such that $u, uw, wv, v \in Y^*$ (see Figure 10.3). We prove by induction on $|uwv|$ that $w \in Y^*$. It is true for $|uwv| = 0$. Suppose that it is true for any such triple u', w', v' with $|u'w'v'| < |uwv|$. We consider several cases.

Case 1. Suppose that $u \notin X^*$ (the case $v \notin X^*$ is symmetric). Then in particular $u \in A^+ X^s$ and thus $u = tz$ with $t \in A^+$ and $z \in X^s$. We distinguish two cases.

(i) If $uw \in X^*$, then, since $uw = tzw$, we have $tzw \in X^*$. Since z is synchronizing, we have $u = tz \in X^*$, a contradiction.

(ii) If $uw \notin X^*$, then in particular $uw \in A^+ X^s$. Thus $uw = t'z'$ with $t' \in A^+$ and $z' \in X^s$. Suppose first that $|zw| \geq |z'|$. Then $zw \in zA^* \cap A^*z'$ and $zw \in Y^*$. Therefore we may apply the induction hypothesis to the triple (z, w, v). Otherwise, we have $|zw| \leq |z'|$ and $z' = rzw$ for some $r \in A^*$. Then $rzw \in X^*$ implies that $rz \in X^*$. Consequently, we may apply the induction hypothesis to the triple (rz, w, v).

Case 2. We have now $u, v \in X^*$. Suppose that $wv \notin X^*$ (the case $uw \notin X^*$ is symmetric). Then $wv = zt$ with $z \in X^s$ and $t \in A^+$. But uwv is in X^* and $uwv = uzt$ implies $zt \in X^*$, a contradiction.

Case 3. Finally, if $u, uw, wv, v \in X^*$, then $w \in X^*$ since X is a code.

This proves that Y is a code.

We now prove that $X \subset Y$. Let indeed $x \in X$. Suppose that $x = yy'$ for two nonempty words of M. Then y or y' is not in X^*. We may suppose for instance that $y' \notin X^*$. Then $y' \in X^s A^*$ and thus $y' = zu$ with $z \in X^s$ and $u \in A^*$. Since z is synchronizing and $yzu \in X$, we have $y' = zu \in X^*$, a contradiction. Consequently $x \notin (Y^* \setminus 1)^2$, showing that $x \in Y$.

Next we show that Y is complete and has synchronization delay s. For this, we first prove that

$$Y^s \subset X^s A^* \cap A^* X^s. \tag{10.5}$$

Let indeed $y = y_1 y_2 \cdots y_s$ with $y_1, y_2, \ldots, y_s \in Y$. If all y_i are in X, the conclusion is true. Otherwise let i be the least index such that $y_i \notin X$. Then $y_i \in X^s A^*$ and since $y_1, \ldots, y_{i-1} \in X$, we obtain $y \in X^s A^*$. The proof of $y \in A^* X^s$ is symmetric.

Consider now $y \in Y^s$. Then by (10.5) for any $u \in A^*$, the word yuy starts and ends with a word in X^s, and thus is in Y^*. This shows that Y is complete.

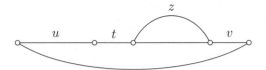

Figure 10.4 Proving that $y = tz$ is synchronizing.

To show that Y has synchronization delay s, suppose that $uyv \in Y^*$ for some $u, v \in A^*$ and $y \in Y^s$. Let us prove that $uy, yv \in Y^*$. We only prove that $uy \in Y^*$, the same reasoning holds for yv.

By (10.5), y has a suffix in X^s. Thus uy has a suffix in X^s. Let $y = tz$ with $t \in A^*$ and $z \in X^s$.

Since $uyv \in Y^*$, either $uyv \in X^*$ or uyv has a prefix in X^s. If $uyv \in X^*$, then since z is synchronizing, we have $utz = uy \in X^*$ and hence also $uy \in Y^*$ (see Figure 10.4).

Otherwise, uyv has a prefix x in X^s. If x is a prefix of uy, then $uy \in X^s A^* \cap A^* X^s$ and $uy \in Y^*$. Otherwise, uy is a prefix of x. Since z is synchronizing, $utz = uy \in X^*$. Thus again $uy \in Y^*$.   □

Example 10.2.12 Consider again the code $X = \{a, ab\}$ with synchronization delay 1 on the alphabet $A = \{a, b\}$. The set M defined by (10.4) is $M = aA^* \cap A^*X$ and the base of M is

$$Y = (abb^+)^* X.$$

Indeed, the words of Y are exactly the words starting with a, ending with a or ab and such that the number of occurrences of b between two a is at least 2.

10.3 Locally parsable codes and local automata

A code has *literal synchronization delay s* if any word of A^s is a constant for X^*. A code is *locally parsable* if there is an integer s such that it has literal synchronization delay s.

We use here constants instead of synchronizing words (as is done in the definition of uniformly synchronized codes). We could have used constants in the definition of the verbal synchronizing delay without changing the notion of uniformly synchronized code. Indeed, if every word in X^s is synchronizing, then every pair in $X^s \times X^s$ is synchronizing by Proposition 10.1.1 and thus every word in X^{2s} is a constant by Proposition 10.1.6. Conversely, if every word in X^s is a constant, then every word in X^s is synchronizing by Proposition 10.1.5.

Example 10.3.1 The code $X = \{a, aab\}$ has literal synchronization delay 2. Indeed $\Gamma(aa) = X^* \times \{1, b\}X^*$ and $\Gamma(b) = X^*aa \times X^*$.

Example 10.3.2 The prefix code $X = \{ba, ca, aba, cba, aca, acba, aaca\}$ is the *Franaszek code*. It has synchronization delay 4. Indeed, the minimal automaton of X^* is represented in Figure 10.5. One may verify that any word of length 4 is a

Table 10.1 *Transitions of the automaton*
of Figure 10.5.

	a	b	c	aa	ac	ca	aac
1	2	3	4	5	4	1	3
2	5	3	4	–	3	1	–
3	1	–	–	2	4	–	4
4	1	3	–	2	4	–	4
5	–	–	3	–	–	1	–

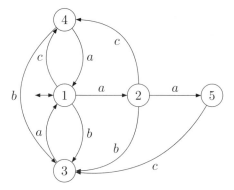

Figure 10.5 The minimal automaton of the Franaszek code.

constant. There is actually a unique word of length 3 which is not a constant, namely
aac. The two-sided ideal of constants is generated by the finite set $\{aaa, b, ca, cc\}$.
Some transitions of the automaton are represented in Table 10.1. They show in
particular the transitions of the words which are not constant.

Let X be a code with literal synchronization delay s. Let $P = X^*A^-$ and $S = A^-X^*$. It is a consequence of the definition that for any $u, v, w \in A^*$ such that
$uvw \in X^*$ and $|v| \geq s$, we have

$$v \in P \implies vw \in X^*. \tag{10.6}$$

Indeed, since $v \in P$, there is a $z \in A^*$ such that $vz \in X^*$. Then $(1, z), (u, w) \in \Gamma_{X^*}(v)$
implies $(1, w) \in \Gamma_{X^*}(v)$. Similarly

$$v \in S \implies uv \in X^*. \tag{10.7}$$

The following statement is the counterpart for the literal delay of Proposition 10.2.2.

Proposition 10.3.3 *The minimal literal deciphering delay of a code is at most equal
to its literal synchronization delay.*

Proof. Let $x \in X^*$, let y be a right completable word of length s and let $u \in A^*$ be
such that $xyu \in X^*$. By (10.6) we have $yu \in X^*$. Thus y is simplifying. This shows
that X has literal deciphering delay s. □

Proposition 10.3.4 *A locally parsable code is uniformly synchronized. The converse is true if the code is finite.*

Proof. Let $X \subset A^+$ be a code with literal synchronization delay s. Then any word of X^s is of length at least s and is therefore a constant and thus is synchronizing. It follows that X has verbal synchronization delay s.

Conversely, suppose that $X \subset A^+$ is a finite code with verbal synchronization delay s. Let ℓ be the maximal length of the words of X. Let w be a word of length $2\ell(s+1)$. If w is not completable, then it is a constant. Otherwise, there are words $x_1, x_2, \ldots, x_n$ in X such that w is a factor of $x_1 x_2 \cdots x_n$. We may suppose that $x_2 \cdots x_{n-1}$ is a factor of w. Then $|w| \le n\ell$ implies $2(s+1) \le n$ or $n-2 \ge 2s$.

Set $x_2 \cdots x_{n-1} = xy$ with $x \in X^s$ and $y \in X^{n-2-s}$. Then x and y are synchronizing words and thus xy is a constant by Propositions 10.1.1 and 10.1.6.

This implies that w is a constant. Consequently X has literal synchronization delay $2\ell(s+1)$.   □

A set $Y \subset A^*$ is said to be *strictly locally testable* if it is of the form

$$Y = T \cup (UA^* \cap A^*V) \setminus A^*WA^*, \tag{10.8}$$

where T, U, V, W are finite subsets of A^*.

Proposition 10.3.5 *A code X is locally parsable if and only if X^* is strictly locally testable.*

Proof. Suppose first that X has literal synchronization delay s. We may suppose $s \ge 1$. Let T be the set of words in X^* of length less than s. Let $U = X^*A^- \cap A^s$ and $V = A^-X^* \cap A^s$. Finally, let W be the set of words w of length $s+1$ which are not in the set $F(X^*)$ of factors of X^*. Let us verify that $X^* = T \cup (UA^* \cap A^*V) \setminus A^*WA^*$. The inclusion from left to right is clear.

Conversely, let x be in the set defined by the right-hand side. If $|x| < s$, then $x \in T$ and therefore $x \in X^*$. Otherwise, let us first show by contradiction that $x \in F(X^*)$. Suppose that x is not in $F(X^*)$. Let v be a factor of x of minimal length which is not in $F(X^*)$. Since x has no factor in W, we have $|v| > s+1$. Let $v = ahb$ with $a, b \in A$. Then $ah, hb \in F(X^*)$ imply that there exist $u_1, u_2, u_3, u_4 \in A^*$ such that $u_1 ahu_2, u_3 hbu_4 \in X^*$. But since $|ahb| > s+1$, h is a constant. Thus $u_1 ahbu_4 \in X^*$, a contradiction with the hypothesis $v = ahb \notin F(X^*)$. Finally, let $u, v \in A^*$ be such that $uxv \in X^*$. Since $x \in UA^*$, we have $xv \in X^*$. And since $x \in A^*V$, this implies in turn that $x \in X^*$. This shows that X^* is strictly locally testable.

Suppose conversely that X^* is strictly locally testable. Let T, U, V, W be finite sets of words such that (10.8) holds. Let s be the maximal length of the words of T, U, V, W. Let w be a word of length $s+1$ and let $(u, v), (u', v')$ be in $\Gamma(w)$. Since $|uwv|, |u'wv'| \ge s+1$, we cannot have $uwv \in T$ or $u'wv' \in T$. Thus $uwv, u'wv' \in UA^* \cap A^*V \setminus A^*WA^*$. Since $|uw|, |u'w| \ge s+1$, we have $uw, u'w \in UA^*$ and $wv, wv' \in A^*V$. For the same reason $uwv', u'wv \notin A^*WA^*$. It follows that (u, v') and (u', v) are in $\Gamma(w)$, showing that w is a constant. This implies that X has literal synchronization delay s.   □

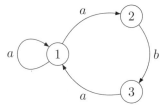

Figure 10.6 A local automaton.

Observe that, as a consequence of the above result, any locally parsable code is rational.

Example 10.3.6 Let $X = \{a, aab\}$ be the code with literal synchronization delay 2 of Example 10.3.1. Then

$$X^* = aa\,A^* \setminus A^*\{bb, bab\}A^*.$$

Example 10.3.7 Let $A = \{a, b, c\}$ and let X be the Franaszek code of Example 10.3.2. The sets U, V, W of (10.8) can be chosen as

$$U = \{aaca, ab, aca, acb, b, ca, cb\},$$
$$V = \{ba, ca\},$$
$$W = \{aaaa, aaab, bb, bc, cc\},$$

with $T = \emptyset$.

An automaton is called (ℓ, r)-*local* if for any paths $p \xrightarrow{u} q \xrightarrow{v} r$ and $p' \xrightarrow{u} q' \xrightarrow{v} r'$ with $|u| = \ell$ and $|v| = r$, one has $q = q'$. The integers ℓ, r are called the *memory* and the *anticipation*. The automaton is called *local* if it is (ℓ, r)-local for some $\ell, r \geq 0$.

Example 10.3.8 Let $\mathcal{A}$ be the automaton given in Figure 10.6. It is $(1, 1)$-local. Indeed, any path labeled aa uses state 1 in the middle and there is only one edge labeled b.

Let $\ell, r \geq 0$ and let $n = \ell + r + 1$. The *free* (ℓ, r)-*local* automaton is the automaton which has, for set of states, the words of length $\ell + r$, and for edges the triples (x, a, y) such that for some $w = a_1 \cdots a_n \in A^n$

$$x = a_1 \cdots a_{n-1}, \quad a = a_{\ell+1}, \quad y = a_2 \cdots a_n.$$

It is clear that this automaton is (ℓ, r)-local.

The free $(n, 0)$-local automaton is usually known as the *de Bruijn automaton* of order n.

Example 10.3.9 The free $(1, 1)$-local automaton on the alphabet $\{a, b\}$ is represented on Figure 10.7. The label of an edge is the second letter of its origin and the first letter of its end.

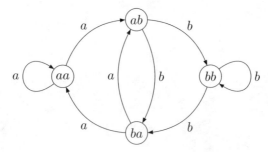

Figure 10.7 The free $(1, 1)$-local automaton.

The following result shows in particular that a strongly connected local automaton is unambiguous.

Proposition 10.3.10 *Let $\mathcal{A}$ be a strongly connected finite automaton on the alphabet A. The following conditions are equivalent.*

  (i) *$\mathcal{A}$ is local.*
 (ii) *$\mathcal{A}$ is unambiguous and there exists an integer s such that any word of length s has rank at most 1 in $\mathcal{A}$.*
(iii) *Distinct cycles in $\mathcal{A}$ have distinct labels.*

Proof. Suppose first that $\mathcal{A}$ is (ℓ, r)-local and let $s = \ell + r$. Let $u \in A^\ell$ and $v \in A^r$. Then there is at most one state q such that $p \xrightarrow{u} q \xrightarrow{v} r$ for some states p, r. If the rank of $\varphi_A(uv)$ is positive, such a unique q exists and $(p, r) \in \varphi_A(uv)$ if and only if $p \xrightarrow{u} q$ and $q \xrightarrow{v} r$. This shows that $\mathcal{A}$ is unambiguous and $\varphi_A(uv) = 1$. Thus (ii) holds.

If (ii) is true, for any word w of length s, the relation $\varphi(w)$ has at most one fixed point. This implies that (iii) is true.

Suppose finally that (iii) holds. First observe that $\mathcal{A}$ is unambiguous. Indeed, since $\mathcal{A}$ is strongly connected, any path is part of a cycle and thus there can be at most one path with given origin, end, and label. Let n be the number of states in $\mathcal{A}$. Consider paths $p \xrightarrow{v} q \xrightarrow{u} r$ and $p' \xrightarrow{v} q' \xrightarrow{u} r'$ such that $|u|, |v| \geq n^2$. Since $|u| \geq n^2$, there exists a pair s, s' which is repeated, that is such that $p \xrightarrow{h} s \xrightarrow{k} s \xrightarrow{k'} q$ and $p' \xrightarrow{h} s' \xrightarrow{k} s' \xrightarrow{k'} q'$ with $u = hkk'$. By condition (iii), we have $s = s'$. Thus, we have paths $p \xrightarrow{h} s \xrightarrow{u'} q$ and $p' \xrightarrow{h} s \xrightarrow{u'} q'$ with $u' = kk'$. In the same way there exist paths $q \xrightarrow{v'} t \xrightarrow{w} r$ and $q' \xrightarrow{v'} t \xrightarrow{w} r'$ for some state t with $v = v'w$. Since $\mathcal{A}$ is unambiguous, the uniqueness of the path from s to t with label $u'v'$ forces $q = q'$. This shows that $\mathcal{A}$ is (n^2, n^2)-local. $\qquad\square$

Let $\mathcal{A}$ be a local automaton. The least integer s such that any word of length s has rank 1 in $\mathcal{A}$ is called the *order* of the automaton.

Proposition 10.3.11 *An (ℓ, r)-local automaton has order at most $\ell + r$.*

Proof. Let $\mathcal{A}$ be a (ℓ, r)-local automaton. Let u and v be words of length ℓ and r respectively. We may assume that the rank of $\varphi_{\mathcal{A}}(uv)$ is not zero. Let $p \xrightarrow{uv} q$ and $p' \xrightarrow{uv} q'$ be two paths in the automaton. There exist states r, r' such that the paths factorize into $p \xrightarrow{u} s \xrightarrow{v} q$ and $p' \xrightarrow{u} s' \xrightarrow{v} q'$. Since the automaton is (ℓ, r)-local, one has $s = s'$. Consequently there are also paths $p \xrightarrow{u} s \xrightarrow{v} q'$ and $p' \xrightarrow{u} s \xrightarrow{v} q$. This shows that the relation $\varphi_{\mathcal{A}}(uv)$ is the product of the column of index s of $\varphi_{\mathcal{A}}(u)$ and the row of index s of $\varphi_{\mathcal{A}}(v)$. Thus $\varphi_{\mathcal{A}}(uv)$ has rank 1. We conclude that any word of length $\ell + r$ has rank at most 1 in $_{\mathcal{A}}$. ☐

The following result gives a characterization of locally parsable codes in terms of automata. It shows in particular that a code X is locally parsable if and only if X^* is the stabilizer of a state in a local automaton.

Proposition 10.3.12 *Let $\mathcal{A} = (Q, 1, 1)$ be a finite unambiguous automaton and let X be the code such that $\mathcal{A}$ recognizes X^*. If $\mathcal{A}$ is local, then X is locally parsable. Conversely, for any locally parsable code X, there exists a local automaton $\mathcal{A} = (Q, 1, 1)$ recognizing X^*.*

Proof. Suppose first that $\mathcal{A}$ is (ℓ, r)-local. Let w be a word of length $s = \ell + r$. By Proposition 10.3.11, w has rank at most 1 in $\mathcal{A}$. By Proposition 10.1.9, it is a constant for X^*. Thus X has literal synchronization delay s.

Conversely, let $\mathcal{A} = (Q, i, T)$ be the minimal deterministic automaton of X. Let $\mathcal{A}^* = (Q \cup \omega, \omega, \omega)$ be the star of the automaton $\mathcal{A}$. Let us show that $\mathcal{A}^*$ is local. For this consider two cycles $p \xrightarrow{w} p$ and $p' \xrightarrow{w} p'$ with the same label w. We will prove that $p = p'$. Since every long enough word is a constant, replacing w by some power, we may suppose that all words of the same length as w are constants.

Suppose first that state ω does not appear on these cycles. Then these paths are paths in $\mathcal{A}$. Let u, v, u', v' be such that $i \xrightarrow{u} p \xrightarrow{v} t$ and $i \xrightarrow{u'} p' \xrightarrow{v'} t$ are paths in $\mathcal{A}$. Since $uwv, u'wv' \in X$, we have $uwv', u'wv \in X^*$. Suppose that $v' = v_1' v_2'$ with $uwv_1' \in X$ and $v_2' \in X^*$. Then, since w is a constant, $u'wv_1'$ is also in X^*. Since X is a code, $u'wv'$ cannot have a second factorization in words of X and thus v_2' is empty. This shows that uwv' is in X. In the same way, we can show that $u'wv \in X$. Since $\mathcal{A}$ is the minimal automaton of X, this implies that $p = p'$.

Let us now suppose that ω appears in one of the cycles, say $p \xrightarrow{w} p$. We have $w = uv$ with $p \xrightarrow{u} \omega \xrightarrow{v} p$. Let u', v' be such that $\omega \xrightarrow{u'} p' \xrightarrow{v'} \omega$ is a path in $\mathcal{A}^*$. Then, since $|vu| = |w|$, vu is a constant. Since $vu, u'uvuvv' \in X^*$, we have also $u'uvu, vuvv' \in X^*$. Then $u'w^3v' = (u'uvu)(vuvv')$ is in X^* and thus we have a path $p' \xrightarrow{u} \omega \xrightarrow{v} p'$, which implies $p = p'$. ☐

We now prove the following result, which is the counterpart, for locally parsable codes of Theorem 10.2.11. The proof is similar to that of Theorem 10.2.11.

Theorem 10.3.13 *Any rational locally parsable code is contained in a complete rational code with the same delay.*

Proof. Let X be a nonempty rational code with literal synchronization delay s. Let P_s be the set of prefixes of length s of the words of X^* and let S_s be the set of suffixes of length s of the words of X^*. Let

$$M = (P_s A^* \cap A^* S_s) \cup X^*.$$

Then M is a submonoid. Let Y be the minimal generating set of M. We show that Y is a code with the desired properties. Let us first prove that M is stable. For this, let u, w, v be such that $u, wv, uw, v \in M$. We distinguish two cases.

Case 1. Suppose $|w| \geq s$. Then $uw \in M$ implies that w has a suffix in S_s and $wv \in M$ implies that w has a prefix in P_s. Thus $w \in M$.

Case 2. Suppose $|w| < s$. We first show that there exists $u' \in X^*$ such that $u'w \in A^* S_s$. If $u \in X^*$, then, since $uw \in A^* S_s$, we can take $u' = u$. Otherwise, we have $u = tr$ with $t \in A^*$ and $r \in S_s$. There exists $k \in A^*$ such that $u' = kr$ is in X^*. Since $|r| = s$, the suffix of uw which is in S_s is a suffix of rw and we have $u'w \in A^* S_s$. Symmetrically, one can prove that there exists a $v' \in X^*$ such that $wv' \in P_s A^*$. Let $u'w = zt$ and $wv' = pq$ with $z, q \in A^*$, $t \in S_s$, and $p \in P_s$. Let $h \in A^*$ be such that $ph \in X^*$. Since w is a prefix of p, $u'w = zt$ is a prefix of $u'ph$. Then, from $u'ph \in zt A^*$, we deduce by (10.7) that $u'w = zt \in X^*$. Similarly, we have $wv' \in X^*$. Since X^* is stable, this implies $w \in X^*$. Thus M is stable.

Let us prove that $X \subset Y$. Let $x \in X$ and suppose that $x = yy'$ with $y, y' \in M \setminus 1$. Since X is a code, we cannot have $y, y' \in X^*$. Let us suppose that $y' \notin X^*$. Then $y' \in P_s A^*$ and $yy' \in X$ imply by (10.6) that y' is in X^*, a contradiction.

Let $y \in P_s A^* \cap A^* S_s$. Then for any $u \in A^*$, we have $yuy \in P_s A^* \cap A^* S_s$. Thus Y is complete.

Finally, let us prove that Y has literal synchronization delay s. Let w be a word of length s. Let u, u', v, v' be such that $uwv, u'wv' \in M$. Then $uw, u'w \in P_s A^*$ and $wv, wv' \in A^* S_s$. Thus $uwv', u'wv$ are both in M, showing that w is a constant for M.   □

Example 10.3.14 Let $A = \{a, b\}$ and $X = \{a, ab\}$. Then X is a code with literal synchronization delay 1. The construction of the proof of Theorem 10.3.13 gives $Y = ab^*$.

10.4 Road coloring

All automata considered in this section are finite, complete, strongly connected, and deterministic.

The *road coloring problem* is the problem of the existence of a synchronizing word in an automaton, up to a relabeling of the edges. The name comes from the interpretation of the labels as colors. More details are given in the Notes. The aim of this section is to prove Theorem 10.4.2 below which states that this coloring of edges is indeed possible under mild and natural assumptions.

Recall from Chapter 3 that a word w is a synchronizing word for an automaton if $p \cdot w = q \cdot w$ for all pairs of states p, q. An automaton is synchronized if it has a synchronizing word.

The *period* of an automaton is the gcd of the lengths of the cycles in its underlying graph. We start by showing that a synchronized automaton must have period 1.

Proposition 10.4.1 *A synchronized automaton has period* 1.

Proof. Let p be the period of $\mathcal{A}$, and let ρ be the relation on the set of states defined by $r \equiv s \bmod \rho$ if there is a path of length multiple of p from r to s. Since the automaton is strongly connected, there is a path c from s to r. The length of the cycle resulting from the composition of the two paths is a multiple of p, so the length of the path c is a multiple of p. This show that $s \equiv r \bmod \rho$. Thus ρ is an equivalence relation.

We now show that any two states r and s are equivalent. Let w be a synchronizing word in $\mathcal{A}$, and let $q = r \cdot w = s \cdot w$. There is a path from q to s of length n such that $n + |w|$ is a multiple of p. This shows that there is a path form r to s of the same length.

This in turn implies that $p = 1$. Indeed, let r be a state and a a letter. Since $s = r \cdot a$ and r are equivalent, there exists a path from s to r of length n where n is a multiple of p. This path, together with the edge from r to s gives a cycle of length $n + 1$, and this number is also a multiple of p. Therefore $p = 1$. $\qquad\square$

We define the following equivalence relation between automata. Given an automaton $\mathcal{A}$, the automata *equivalent* to $\mathcal{A}$ are obtained from $\mathcal{A}$ by permuting the labels of the outgoing edges of the states, independently for each state. This implies that two equivalent automata have isomorphic underlying graphs, and conversely. Clearly, two equivalent automata have the same period.

We prove the following result, called the *road coloring theorem*, which shows that there are "many" synchronized automata.

Theorem 10.4.2 *An automaton which has period* 1 *is equivalent to a synchronized one.*

A set P of states of an automaton is said to be *synchronizable* if there exists a word u in A^* such that for all p, q in P, one has $p \cdot u = q \cdot u$. We also say that the word u synchronizes the states in P.

A pair p, q of states is said to be *strongly synchronizable* if for any word $u \in A^*$, the states $p \cdot u$ and $q \cdot u$ are synchronizable. We say that a deterministic automaton is *reducible* if it has two distinct strongly synchronizable states. Let ρ be the equivalence on the states of an automaton $\mathcal{A}$ defined by $p \equiv q \bmod \rho$ if p and q are strongly synchronizable. Then ρ is a congruence of $\mathcal{A}$ called the *synchronizability congruence*. We verify that ρ is transitive. Let indeed p, q, r be states such that $p \equiv q \bmod \rho$ and $q \equiv r \bmod \rho$. Let $u \in A^*$. There is a word v such that $p \cdot uv = q \cdot uv$. There exists w such that $q \cdot uvw = r \cdot uvw$. This shows that p and r are strongly synchronizable.

Lemma 10.4.3 *Let $\mathcal{A}$ be an automaton and let ρ be the synchronizability congruence. If the quotient $\mathcal{A}/\rho$ is equivalent to a synchronized automaton, then $\mathcal{A}$ itself is equivalent to a synchronized automaton.*

Proof. Let E be the set of edges of $\mathcal{A}$ and let F be the set of edges of $\mathcal{B} = \mathcal{A}/\rho$. Let φ be the map from E to F induced by ρ. Thus $\varphi(e) = f$ if $e = (p, a, q)$ and $f = (\bar{p}, a, \bar{q})$ where $\bar{p}, \bar{q}$ are the classes modulo ρ of p and q. Let $\mathcal{B}'$ be a synchronized automaton equivalent to $\mathcal{B}$. We define an automaton $\mathcal{A}'$ equivalent to $\mathcal{A}$ by changing the labels of its edges. The new label of an edge e is the label of $\varphi(e)$ in $\mathcal{B}'$. Let us show that $\mathcal{A}'$ is synchronized.

Consider first two states p, q in $\mathcal{A}$ which are strongly synchronizable. Let us prove that they are still synchronizable in $\mathcal{A}'$. We prove this by induction on the length of a shortest word w synchronizing such a pair p, q. For $|w| = 0$ we have $p = q$ and the property is true. For $|w| \geq 1$, set $w = au$ with a a letter. Let $e = (p, a, r)$ and $f = (q, a, s)$ be the edges of $\mathcal{A}$ labeled a going out of p and q. Since $p \equiv q$ modulo ρ and ρ is a congruence, we have $r = p \cdot a \equiv q \cdot a = s$. Since ρ is a congruence, r and s are strongly synchronizable in $\mathcal{A}$ and, by induction, r and s are synchronizable in $\mathcal{A}'$. Now $\varphi(e) = \varphi(f)$, hence the labels of e and f in $\mathcal{A}'$ are equal. This shows p and q are synchronizable in $\mathcal{A}'$.

Suppose now that p and q are not equivalent modulo ρ. Since $\mathcal{B}'$ is synchronized, the classes of p and q are synchronizable in $\mathcal{B}'$. Let w be a word synchronizing p and q. Then, in $\mathcal{A}'$, the states $p \cdot w$ and $q \cdot w$ are in the same class modulo ρ. The conclusion follows by the argument above.

Thus any pair of states of $\mathcal{A}'$ is synchronizable, which shows that $\mathcal{A}'$ is synchronized. $\qquad\square$

In the following lemma, we use the notion of *minimal image* of an automaton $\mathcal{A}$ (see Section 9.3). Recall that a set P of states of an automaton $\mathcal{A} = (Q, i, T)$ is a *minimal image* if it is of the form $P = Q \cdot w$ for some word w, and of minimal size with this property. Recall also that two minimal images have the same cardinality. This cardinality is the minimal rank of the elements of the transition monoid of $\mathcal{A}$. Also, if I is a minimal image and u is a word, then $I \cdot u$ is again a minimal image and $p \mapsto p \cdot u$ is one-to-one from I onto $I \cdot u$.

Lemma 10.4.4 *Let $\mathcal{A}$ be an automaton. If there exist two minimal images that differ by only one element, then $\mathcal{A}$ is reducible.*

Proof. Let I, J be minimal images such that $I = K \cup \{p\}$ and $J = K \cup \{q\}$ with $p, q \notin K$. For any $u \in A^*$, the sets $I \cdot u = K \cdot u \cup p \cdot u$ and $J \cdot u = K \cdot u \cup q \cdot u$ are minimal images. For any word v in A^* of minimal rank, the set $(I \cup J) \cdot uv$ is a minimal image. Indeed, $I \cdot uv \subset (I \cup J) \cdot uv \subset \mathrm{Im}(uv)$, hence all three are equal. But $(I \cup J) \cdot uv = K \cdot uv \cup p \cdot uv \cup q \cdot uv$. This forces $p \cdot uv = q \cdot uv$ since $p \cdot uv \notin K \cdot uv$, since otherwise $I \cdot uv$ would have less elements than I. Thus p, q are strongly synchronizable. $\qquad\square$

A state p is a *bunch* if all states $p \cdot a$ for a in A are equal. In this case, the state $p \cdot a$ is called the *target* of the bunch p.

Lemma 10.4.5 *If an automaton $\mathcal{A}$ has two distinct bunch states with the same target, then $\mathcal{A}$ is reducible.*

Proof. Let p, p' be such that all edges going out of p, p' end at q. The states p and p' are strongly synchronizable since for any letter a, one has $p \cdot a = p' \cdot a$. $\qquad\square$

$\mathcal{A}$

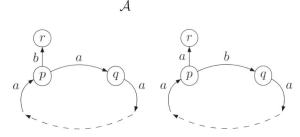

Figure 10.8 Case 1. All states have a-index 0.

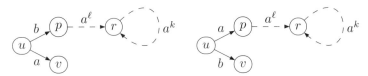

Figure 10.9 Case 2. u is not on C.

Let $\mathcal{A}$ be an automaton. The *a-index* of a state p with respect to a letter $a \in A$ is the least integer ℓ such that $p \cdot a^{\ell+k} = p \cdot a^{\ell}$ for some integer $k \geq 1$. An *a-cycle* is a cycle formed of edges all labeled by a. Thus, the a-index of a state is the least integer ℓ such that $p \cdot a^{\ell}$ is on an a-cycle. The state $p \cdot a^{\ell}$ is called the *a-basis* of p. If there is a path formed of edges all labeled by a from p to q, the state p is called an *a-ascendant* of a state q and q is said to be an *a-descendant* of p. Note that the set of states with given a-basis r forms a tree with root r. (In such a tree, the orientation is the reverse of the usual one.)

The following lemma is the key of the proof of Theorem 10.4.2.

Lemma 10.4.6 *Any automaton with period* 1 *is equivalent either to a reducible automaton, or to an automaton such that all states of maximal a-index for some letter a have the same a-basis.*

Proof. We assume that $\mathcal{A}$ is not equivalent to a reducible automaton, we fix a letter a and we assume that the automaton is chosen within its equivalence class in such a way that the number of states of a-index 0 is maximal. We distinguish a number of cases. Let ℓ be the maximal a-index of states.

Case 1. Suppose first that $\ell = 0$. If all states are bunches, the automaton consists of just one cycle and since the period of $\mathcal{A}$ is 1, the automaton has a single state.

Let p be a state which is not a bunch, let $q = p \cdot a$ and let $b \neq a$ be such that $r = p \cdot b$ satisfies $r \neq q$. Let us exchange the labels of these edges. The resulting automaton is equivalent to $\mathcal{A}$ and has just one state of maximal index, namely q (see Figure 10.8). Thus the conclusion holds in this case.

Assume now $\ell \geq 1$. Let p be a state of a-index ℓ. Since $\mathcal{A}$ is strongly connected, there is an edge $u \xrightarrow{b} p$ ending in p and one may suppose $u \neq p$. Since p has maximal a-index, the label of this edge is $b \neq a$. Let $v = u \cdot a$. One has $v \neq p$. Let $r = p \cdot a^{\ell}$ and let C be the a-cycle to which r belongs.

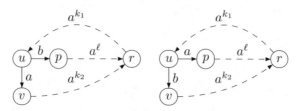

Figure 10.10 Case 3. $k_2 > \ell$.

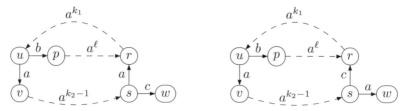

Figure 10.11 Case 4. The state s is not a bunch.

Case 2. Suppose first that u is not on C. We exchange the labels of $u \xrightarrow{b} p$ and $u \xrightarrow{a} v$ (see Figure 10.9). We have not destroyed the a-path from p to r. Indeed, this would mean that u was on this path and the exchange would have created a new cycle to which u and p belong, increasing the number of vertices with a-index 0. Since u is not on C, the exchange did not either modify the cycle C. In this new automaton, there are vertices of a-index at least $\ell + 1$. All vertices of a-index at least $\ell + 1$ have been created by this exchange, and are a-ascendants of u. Thus the vertices with maximal a-index are a-ascendants of u. Their basis is the same as the basis r of p. This proves the property.

Suppose now that u is on C. Let k_1 be the least integer such that $r \cdot a^{k_1} = u$. Since $u \cdot a = v$, the state v is also on C. Let k_2 be the least integer such that $v \cdot a^{k_2} = r$ in such a way that C has length $k_1 + k_2 + 1$ (see Figure 10.10).

Case 3. Suppose first that $k_2 > \ell$. We exchange as before the labels of $u \xrightarrow{b} p$ and $u \xrightarrow{a} v$. The a-index of v becomes k_2 and since $k_2 > \ell$, the states of maximal a-index are a-ascendants of v. Thus they all have a-basis equal to r and the property holds.

Suppose now that $k_2 \le \ell$. We have actually $k_2 = \ell$. Otherwise, exchange the labels of $u \xrightarrow{b} p$ and $u \xrightarrow{a} v$. This creates an a-cycle of length $k_1 + \ell + 1$ which replaces one of length $k_2 + k_1 + 1$. But the automaton obtained then has more states of a-index 0, contrary to the assumption made previously. Let s be the state of C such that $s \cdot a = r$. Observe that $k_2 = \ell \ge 1$ and therefore $v \ne r$.

Case 4. Suppose first that the state s is not a bunch (see Figure 10.11). Let $w = s \cdot c$ be such that $w \ne r$ with c a letter distinct of a. We exchange the labels of the edges $s \xrightarrow{a} r$ and $s \xrightarrow{c} w$. Then r is not anymore on an a-cycle. Indeed, otherwise, this cycle would begin with the path $r \xrightarrow{a^{k_1}} u \xrightarrow{a} v \xrightarrow{a^{k_2-1}} s \xrightarrow{a} w$ and would be longer than C. This would increase the number of states with a-index 0, contradicting the assumption made on $\mathcal{A}$. Thus, the a-index of r is positive and it is maximal among the states

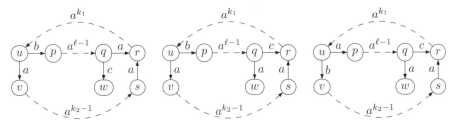

Figure 10.12 Case 5. The state s is a bunch.

which were before on the cycle C. The states with maximal a-index obtained in this way are a-ascendants of r and thus all have the same a-basis.

Case 5. Suppose now that s is a bunch. Let $q = p \cdot a^{\ell-1}$, which is the predecessor of r on the a-path from p to r. By Lemma 10.4.5 the state q is not a bunch since otherwise r would be the target of the bunches s and q. Thus there exists a letter c such that $r = q \cdot a \neq q \cdot c = w$. We exchange the labels of $q \xrightarrow{a} r$ and $q \xrightarrow{c} w$ (see Figure 10.12 middle). The state q cannot belong to $w \cdot a^*$ since otherwise we obtain an additional cycle $w \xrightarrow{a^{k_3}} q \xrightarrow{a} w$ and more states with a-index 0. In particular $w \neq p$.

Case 5(a) If the a-index of w is positive, then the maximal index becomes at least $\ell + 1$ and all states of maximal index are a-ascendants of w.

Case 5(b) Suppose now that the a-index of w is 0. If w is on the cycle C, the index of p remains ℓ and the only thing that has changed is the basis of p which becomes w instead of r. We proceed as in Case 3 and consider the least integer k_3 such that $v \cdot a^{k_3} = w$. We treat the case $k_3 > \ell$ in the same way and we are left with the case $k_3 = \ell$ (Case 4). But then $k_3 = \ell$ and $k_2 = \ell$ imply $k_2 = k_3$ which is impossible since $r \neq w$.

Case 5(c) Suppose finally that w is on a cycle distinct from C. We additionally exchange the labels of the edges $u \xrightarrow{a} v$ and $u \xrightarrow{b} p$ (see Figure 10.12 right). The maximal a-index has increased and the states of maximal index are all a-ascendants of u.

This concludes the proof of the lemma.   □

Proof of Theorem 10.4.2. We use an induction on the number n of states of the automaton. The property holds for $n = 1$. Let us suppose that it holds for automata with less than n states and consider an admissible automaton $\mathcal{A}$ with n states.

If $\mathcal{A}$ is reducible, we consider the quotient of $\mathcal{A}$ by the synchronizability congruence ρ. By induction hypothesis, the automaton $\mathcal{A}/\rho$ is equivalent to a synchronized automaton. Thus, by Lemma 10.4.3, $\mathcal{A}$ is equivalent to a synchronized automaton.

Suppose now that $\mathcal{A}$ is not equivalent to a reducible automaton. By Lemma 10.4.6, $\mathcal{A}$ is equivalent to an automaton in which, for some letter a, the states of maximal a-index have the same a-basis. Let ℓ be the maximal a-index and let r be the common a-basis. The states of a-index ℓ form a synchronizable set since the word a^ℓ maps all of them to r. Up to an automaton equivalence, we may assume that this property holds for $\mathcal{A}$. Let I be a minimal image containing a state p of maximal a-index ℓ. Then, since the other states of a-index ℓ are synchronizable with p, the a-index of the other elements of I is strictly less than ℓ (because I is an image of minimal cardinality). Let $J = I \cdot a^{\ell-1}$. Then all elements of J except $q = p \cdot a^{\ell-1}$ are on a cycle labeled by a. Let k be a multiple of the lengths of the cycles labeled a. Then $s \cdot a^k = s$ for each state s of J distinct of q and thus J and $J \cdot a^k$ are two distinct minimal images which differ by only one element. By Lemma 10.4.4, this is not possible. $\qquad\square$

The road coloring theorem has the following consequence for prefix codes. Say that two prefix codes are *flipping equivalent* if they have isomorphic associated (unlabeled) trees. The period of a prefix code is the gcd of the lengths of its words.

Theorem 10.4.7 *Any rational maximal prefix code with period* 1 *is flipping equivalent to a synchronized one.*

Proof. Let X be a rational maximal prefix code with period 1. Let $\mathcal{A} = (Q, 1, 1)$ be the minimal deterministic automaton of X^*. By Theorem 10.4.2, there is a synchronized automaton $\mathcal{A}'$ equivalent to $\mathcal{A}$. Let X' be the prefix code generating the stabilizer of state 1 in $\mathcal{A}'$. Then X and X' are flipping equivalent because the corresponding trees are obtained by unfolding the graph underlying $\mathcal{A}$ and $\mathcal{A}'$, duplicating the state 1 into two states one having all the input edges of 1 and the other all the output edges. Since $\mathcal{A}'$ is synchronizing, X' is synchronized. $\qquad\square$

The above result shows in particular that one may always find a synchronized prefix code among the prefix codes having a given length distribution provided the period is 1. In particular the code having an optimal length distribution for a given set of frequencies obtained by the Huffman algorithm can be chosen synchronized provided it is of period 1.

For another proof, see Exercise 3.8.2.

10.5 Exercises

Section 10.2

10.2.1 Let X be a code with (verbal) synchronization delay s. Show that

$$X^* = 1 \cup X \cup \cdots \cup X^{s-1} \cup (X^s A^* \cap A^* X^s) \setminus W \qquad (10.9)$$

with $W = \{w \in A^* \mid A^* w A^* \cap X^* = \emptyset\}$. Show that W has also the expression $W = A^* V A^*$ with

$$V = (A^* \setminus A^* X^{s+1} A^*) \setminus (A^* \setminus F(X^{s+2})) \qquad (10.10)$$

10.2.2 Show that a nonempty code X is complete and has finite synchronization delay if and only if there is an integer s such that

$$X^s A^* \cap A^* X^s \subset X^*.$$

10.2.3 Show that the code Y of the proof of Theorem 10.2.11 admits the expression

$$Y = X \cup (T \setminus W) \qquad (10.11)$$

where $T = (X^s A^* \setminus X^{s+1} A^*) \cap (A^* X^s \setminus A^* X^{s+1})$ and $W = A^* X^{2s} A^* \cup X^*$.

10.2.4 Show that a thin circular code is synchronized.

10.2.5 Let $X \subset A^+$ be a maximal prefix code. Show that the following conditions are equivalent.

(i) X has synchronization delay 1.
(ii) $A^* X \subset X^*$.
(iii) X is a semaphore code such that $S = X \setminus A^+ X$ satisfies $SA^* \cap A^* S = S \cup SA^* S$ (that is S is "non overlapping").

Section 10.3

10.3.1 Let $s \geq 1$ be an integer and let $\sim_s$ denote the equivalence on words of length at least s defined by $y \sim_s z$ if y and z have the same prefix of length s, the same suffix of length s and the same set of factors of length s. A set $Y \subset A^*$ is said to be *locally testable* of order s if there is an integer s such that for two words $y, z \in A^s A^*$ with $y \sim_s z$ one has $y \in Y$ if and only if $z \in Y$. Show that a set X is locally testable if and only if it is a finite Boolean combination of strictly locally testable sets.

10.3.2 The *syntactic semigroup* of a set $Y \subset A^+$ is the quotient of A^+ by the syntactic congruence. Show that a set $Y \subset A^*$ is strictly locally testable if and only if all idempotents of its syntactic semigroup are constants (where a constant in the syntactic semigroup is the image of a constant in A^+).

10.3.3 Show that if Y is locally testable, then for each idempotent e in the syntactic semigroup of Y, the semigroup eSe is idempotent and commutative.

10.3.4 Show that a code X is locally parsable if and only if X^* is locally testable. (*Hint*: Use Proposition 10.3.5, and Exercises 10.3.2, 10.3.3.)

10.6 Notes

The notion of synchronization delay was introduced in Golomb and Gordon (1965). It was proved in Bruyère (1998) that any rational code with finite synchronization delay is contained in a complete rational code with finite synchronization delay. However, the definition of synchronization delay used in Bruyère (1998) differs from ours. Her construction is basically the same, but does not preserve the delay. Exercise 10.2.3 is also from Bruyère (1998). Theorem 10.2.7 is in Restivo (1975). Exercise 10.2.1 is from Schützenberger (1975) (see also Perrin and Pin (2004)). The completion problem for rational synchronized codes has been solved by Guesnet (2003).

A set $X \subset A^*$ is called *star-free* if it can be obtained from the subsets of the alphabet by a finite number of set products and Boolean operations (including the complement). Thus star-free sets are those regular sets which can be obtained without using the star operation. Examples of star-free sets are $\emptyset$, A^* (the complement of $\emptyset$), the singletons $\{a\}$ for $a \in A$ and the ideals aA^* or $A^* a A^*$. Formulas (10.9) and

(10.10) are parts of a proof showing that if a code X with finite synchronization delay is star-free, then X^* is also star-free. Formula (10.4) shows that, if X is star-free, then Y^* and thus also Y are star-free. There is a deep link between codes with finite synchronization delay and star-free sets which has been investigated in Schützenberger (1975) (see Perrin and Pin (2004) for a connection with first-order logic).

The term "locally parsable" is from McNaughton and Papert (1971). Exercise 10.3.4 is from de Luca and Restivo (1980). Exercise 10.3.3 has a converse which is a difficult theorem due to McNaughton, Zalcstein, Bzrozowski and Simon (see Eilenberg (1976)).

The Franaszek code of Example 10.3.2 is used to encode arbitrary binary sequences into constrained sequences, see Lind and Marcus (1995).

The origin of the name "road coloring problem" is the following. Imagine a map with roads which are colored in such a way that a fixed sequence of colors, called a *homing sequence*, leads the traveler to a fixed place irrespective of its starting point. If the colors are replaced by letters, a homing sequence corresponds to a synchronizing word. The road coloring problem originates in Adler and Weiss (1970) and was explicitly formulated in Adler *et al.* (1977). It was proved in Trahtman (2008). The notion of strongly synchronizable states appears in Culik *et al.* (2002). Several partial solutions have appeared earlier (see O'Brien (1981) or Friedman (1990) in particular). Theorem 10.4.7 is proved in Perrin and Schützenberger (1992) for finite maximal prefix codes. The same result is also established in Perrin and Schützenberger (1992) with essentially the same proof for the commutative equivalence instead of the flipping equivalence (Theorem 14.6.10). Lemma 10.4.3 appears already in Culik *et al.* (2002).

Groups of codes

We have seen in Chapter 9 that there is a transitive permutation group $G(X)$ of degree $d(X)$ associated with every thin maximal code X which we called the group and the degree of the code. We have seen that a code has a trivial group if and only if it is synchronized.

In this chapter we study the relations between a code and its group. As an example, we will see that an indecomposable prefix code X has a permutation group $G(X)$ which is primitive (Proposition 11.1.6). We will also see that a thin maximal prefix code X has a regular group if and only if $X = U \circ V \circ W$ with U, W synchronized and V a regular group code (Proposition 11.2.3). This result is used to prove that any semaphore code is a power of a synchronized semaphore code (Theorem 11.2.1 already announced in Chapter 3). A direct combinatorial proof of this result would certainly be extremely difficult.

We study in more detail the groups of bifix codes. We start with the simplest class, namely the group codes in Section 11.3. We show in particular (Theorem 11.3.1) that a finite group code is uniform.

In the next two sections, we again examine the techniques introduced in Chapter 9 and particularize them to bifix codes. Specifically, we shall see that bifix codes are characterized by the algebraic property of their syntactic monoids being nil-simple (Theorem 11.5.2). The proof makes use of Schützenberger's theorem 5.2.4 concerning codes with finite deciphering delay. Section 11.6 is devoted to groups of finite maximal bifix codes. The main result is Theorem 11.6.8 stating that the group of a finite, indecomposable, nonuniform maximal bifix code is doubly transitive. For the proof of this theorem, we use difficult results from the theory of permutation groups without proof. The last section contains a series of examples of finite maximal bifix codes with special groups.

11.1 Groups and composition

We now examine the behavior of the group of a code under composition. Let G be a transitive permutation group over a set Q. Recall (see Section 1.13) that an imprimitivity equivalence of G is an equivalence relation θ on Q stable with respect to the action of G, that is, such that for all $p, q \in Q$ and $g \in G$,

$$p \equiv q \bmod \theta \Rightarrow pg \equiv qg \bmod \theta.$$

The action of G on the classes of θ defines a transitive permutation group denoted by G_θ and called the *imprimitivity quotient* of G for θ.

For any $q \in Q$, the restriction to the class mod θ of q of the subgroup

$$K = \{k \in G \mid qk \equiv q \bmod \theta\}$$

formed of the elements globally stabilizing the class of q mod θ is a transitive permutation group. The groups induced by G on the equivalence classes mod θ are all equivalent (see Section 1.13). Any one of these groups is called the *group induced* by G. It is denoted by G^θ.

Let $d = \text{Card}(Q)$ be the degree of G, let e be the cardinality of a class of θ (thus e is the degree of G^θ), and let f be the number of classes of θ (that is, the degree of G_θ). Then we have the formula $d = ef$.

Example 11.1.1 The permutation group over the set $\{1, 2, 3, 4, 5, 6\}$ generated by the two permutations

$$\alpha = (123456), \quad \beta = (26)(35)$$

is the group of symmetries of the hexagon,

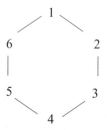

It is known under the name of *dihedral group* D_6, and has of course degree 6. It admits the imprimitivity partition $\{\{1, 4\}, \{2, 5\}, \{3, 6\}\}$ corresponding to the diagonals of the hexagon. The groups G_θ and G^θ are, respectively, equivalent to $\mathfrak{S}_3$ and $\mathbb{Z}/2\mathbb{Z}$.

Proposition 11.1.2 *Let X be a very thin code which decomposes into $X = Y \circ Z$ with Y a complete code. There exists an imprimitivity equivalence θ of $G = G(X)$ such that*

$$G^\theta = G(Y), \quad G_\theta = G(Z).$$

In particular, $d(X) = d(Y)d(Z)$.

Proof. Set $X = Y \circ_\beta Z$ with $B = \text{alph}(Y)$ and β a bijection from B onto Z. Let P and S be the sets of states of the flower automata $\mathcal{A}_D^*(X)$, $\mathcal{A}_D^*(Z)$, respectively. Let φ (resp. ψ) be the morphism associated to $\mathcal{A}_D^*(X)$ (resp. $\mathcal{A}_D^*(Z)$).

In view of Proposition 4.2.7, and since Y is complete, there exists a reduction $\rho : P \to S$. Actually, for $(u, v) \in P \setminus (1, 1)$ we have $\rho(u, v) = (r, s)$ where $u = zr$ and $v = s\bar{z}$ with $z, \bar{z} \in Z^*$ and $(r, s) \in S$.

Table 11.1 *The next state function of* $\mathcal{A}(X^*)$.

	1	2	3	4	5	6	7	8
a	4	5	4	5	8	1	8	1
b	2	3	4	5	6	7	8	1

Moreover, $\mathcal{A}_D^*(Y)$ can be identified through β with the restriction of $\mathcal{A}_D^*(X)$ to the states which are in $Z^* \times Z^*$. As usual, we denote by $\widehat{\rho}$ the morphism from $M = \varphi(A^*)$ onto $M' = \psi(A^*)$ induced by ρ. Thus $\psi = \widehat{\rho} \circ \varphi$.

Let J (resp. K) be the 0-minimal ideal of M (resp. of M'). Then $J \subset \widehat{\rho}^{-1}(K)$, since $\widehat{\rho}^{-1}(K)$ is a nonnull ideal. Thus $\widehat{\rho}(J) \subset K$. Since $\widehat{\rho}(J) \neq 0$, we have

$$\widehat{\rho}(J) = K.$$

Let e be an idempotent in $J \cap \varphi(X^*)$, let $R = \mathrm{Fix}(e) \subset P$ and let $G = G_e$. Let us verify that ρ is a surjective function from R onto $\mathrm{Fix}(\widehat{\rho}(e))$. Let indeed s be a fixed point of $f = \widehat{\rho}(e)$. By definition of a reduction, there exist $p, q \in P$ such that $\rho(p) = \rho(q) = s$ and $(p, e, q) = 1$. Since e is idempotent, there exists a fixed point r of e such that $(p, e, r) = (r, e, q) = 1$. Then $\rho(r) = s$ by unambiguity, proving the assertion.

Further, the nuclear equivalence of the restriction of ρ to R defines an equivalence relation θ on R which is an imprimitivity equivalence of G. Indeed, let $r, r' \in R$ be such that $\rho(r) = \rho(r')$. Let $g \in G$ and set $s = rg$, $s' = r'g$. By definition of G there is an $m \in M$ such that g is the restriction to R of eme. Then, since $\widehat{\rho}(eme)$ is a permutation on $\rho(R)$, we have $\rho(s) = \rho(s')$, proving the assertion. The group $G_{\widehat{\rho}(e)}$ is the corresponding imprimitivity quotient G_θ. This shows that $G(Z)$ is equivalent to G_θ.

Let $T = \{(u, v) \in P \mid u, v \in Z^*\}$. Then T can be identified with the states of the flower automaton of Y and moreover $T = \rho^{-1}(1, 1)$. Let L be the restriction to T of the submonoid $N = \varphi(Z^*)$ of M. Then

$$eNe = H(e) \cap N.$$

Indeed, one has $eNe \subset H(e) \cap N$ since $e \in \varphi(X^*)$ and $X^* \subset Z^*$. Conversely, if $n \in H(e) \cap N$, then $n = ene$ and thus $n \in eNe$. Since $H(e) \cap N$ is a group, this shows that eN is a minimal right ideal and Ne is a minimal left ideal. Thus e is in the minimal ideal of the monoid N. Moreover the restriction to $R \cap T$ of $H(e) \cap N$ is the Suschkewitch group of L.

Thus the restriction to $R \cap T$ of the group $H(e) \cap L$ is equivalent to the group $G(Y)$. On the other hand, since $T = \rho^{-1}(1, 1)$, this group is also the group G^θ induced by G on the classes of θ. $\square$

Example 11.1.3 Let $X = Z^n$ where Z is a very thin code and $n \geq 1$. Then $d(X) = nd(Z)$.

Example 11.1.4 Consider the maximal prefix code Z over $A = \{a, b\}$ given by

$$Z = (A^2 \setminus b^2) \cup b^2 A^2$$

and set $X = Z^2$. The automaton $\mathcal{A}(X^*)$ is given in Table 11.1. Let φ be the corresponding representation. The monoid $\varphi(A^*)$ is the monoid of functions of Example 9.2.3, when setting $\varphi(a) = u$, $\varphi(b) = v$.

The idempotent $e = \varphi(a^4)$ has minimal rank since the action of A on the $\mathcal{R}$-class of e given in Figure 9.7 is complete. Consequently, the group $G(X)$ is the dihedral group D_4. This group admits an imprimitivity partition with a quotient and an induced group both equivalent to $\mathbb{Z}/2\mathbb{Z}$. This corresponds to the fact that

$$G(Z) = \mathbb{Z}/2\mathbb{Z},$$

since

$$Z = T \circ A^2,$$

where T is a synchronized code.

In the case of prefix codes, we can continue the study of the influence of the decompositions of the prefix code on the structure of its group. We use the maximal decomposition of prefix codes defined in Proposition 3.6.14.

Proposition 11.1.5 *Let X be a very thin prefix code, and let*

$$X = Y \circ Z$$

be its maximal decomposition. Then Z is synchronized, and thus $G(X) = G(Y)$.

Proof. Let $D = X^*(A^*)^{-1}$, $U = \{u \in A^* \mid u^{-1}D = D\}$. Then $Z^* = U$. Let φ be the morphism associated with the automaton $\mathcal{A}(X^*)$. Let J be the 0-minimal ideal of the monoid $\varphi(A^*)$.

Consider $x \in X^*$ such that $\varphi(x) \in J$. First we show that

$$D = \{w \in A^* \mid \varphi(xw) \neq 0\}. \tag{11.1}$$

Indeed, if $w \in D$, then $xw \in D$ and thus $\varphi(xw) \neq 0$. Conversely, if $\varphi(xw) \neq 0$ for some $w \in A^*$, then the fact that the right ideal generated by $\varphi(x)$ is 0-minimal implies that there exists a word $w' \in A^*$ such that $\varphi(x) = \varphi(xww')$. Thus $xww' \in X^*$. By right unitarity, we have $ww' \in X^*$, whence $w \in D$. This proves (11.1).

Next $Dx^{-1} = Ux^{-1}$. Indeed $D \supset U$ implies $Dx^{-1} \supset Ux^{-1}$. Conversely, consider $w \in Dx^{-1}$. Then $wx \in D$. By (11.1), $\varphi(xwx) \neq 0$. Using now the 0-minimality of the *left* ideal generated by $\varphi(x)$, there exists a word $w' \in A^*$ such that $\varphi(w'xwx) = \varphi(x)$. Using again (11.1), we have, for all $w'' \in D$, $0 \neq \varphi(xw'') = \varphi(w'xwxw'')$. Then also $\varphi(xwxw'') \neq 0$ and, again by (11.1), $wxw'' \in D$. This shows that $D \subset (wx)^{-1}D$. For the reverse inclusion, let $w'' \in (wx)^{-1}D$. Then $wxw'' \in D$. Thus $\varphi(xwxw'') \neq 0$. This implies that $\varphi(xw'') \neq 0$, whence $w'' \in D$. Consequently $D = (wx)^{-1}D$, showing that $wx \in U$, hence $w \in Ux^{-1}$.

Now we prove that (x, x) is a synchronizing pair for Z. Let $w, w' \in A^*$ be such that $wxxw' \in Z^* = U$. Since $U \subset D$, we have $wxxw' \in D$ and thus $wx \in D$. By the equality $Dx^{-1} = Ux^{-1}$, this implies $wx \in U$. Since U is right unitary, xw' also is in U. Consequently Z is synchronized. In view of Proposition 11.1.2, this concludes the proof. □

We now prove a converse of Proposition 11.1.2 in the case of prefix codes. It is not known if it holds for arbitrary thin maximal codes.

Proposition 11.1.6 *Let X be a thin maximal prefix code. If the group $G = G(X)$ admits an imprimitivity equivalence θ, then there exists a decomposition of X into*

$$X = Y \circ Z$$

such that $G(Y) = G^\theta$ and $G(Z) = G_\theta$.

Proof. Let φ be the representation associated with the minimal automaton $\mathcal{A}(X^*) = (Q, 1, 1)$, and set $M = \varphi(A^*)$. Let J be the minimal ideal of M, let $e \in J \cap \varphi(X^*)$ be an idempotent, let L be the $\mathcal{L}$-class of e and Γ be the set of $\mathcal{H}$-classes of L. We have $G(X) = G_e$.

Since X is complete, each $H \in \Gamma$ is a group and therefore has an idempotent e_H with $\text{Im}(e) = \text{Im}(e_H)$ and thus $\text{Fix}(e_H) = \text{Fix}(e)$. The code X being prefix, e_H is in $\varphi(X^*)$ for all $H \in \Gamma$, by Proposition 9.4.9.

Set $S = \text{Fix}(e)$. By assumption, there exists an equivalence relation θ on S that is an imprimitivity equivalence of the group G_e. Consider the equivalence relation $\widehat{\theta}$ on the set Q of states of $\mathcal{A}(X^*)$ defined by $p \equiv q \bmod \widehat{\theta}$ if and only if, for all $H \in \Gamma$,

$$pe_H \equiv qe_H \bmod \theta.$$

Let us verify that $\widehat{\theta}$ is stable, that is, that

$$p \equiv q \bmod \widehat{\theta} \Rightarrow p \cdot w \equiv q \cdot w \bmod \widehat{\theta}$$

for $w \in A^*$. Indeed, let $m = \varphi(w)$. Note that for $H \in \Gamma$,

$$me_H = e_{mH}me_H = e_{mH}eme_H \tag{11.2}$$

since $e_{mH}e = e_{mH}$. Observe also that $eme_H \in H(e)$ since $en \in H(e)$ for all $n \in L$ and since $me_H \in L$ by (11.2). Assume now that $p \equiv q \bmod \widehat{\theta}$. Then by definition $pe_{mH} \equiv qe_{mH} \bmod \theta$ and θ being an imprimitivity equivalence, this implies

$$pe_{mH}eme_H \equiv qe_{mH}eme_H \bmod \theta.$$

By (11.2), it follows that $pme_H \equiv qme_H \bmod \theta$ for all $H \in \Gamma$. Thus $p \cdot w \equiv q \cdot w \bmod \widehat{\theta}$.

Moreover, the restriction of $\widehat{\theta}$ to the set $S = \text{Fix}(e)$ is equal to θ. Assume indeed that $p \equiv q \bmod \widehat{\theta}$ for some $p, q \in S$. Then $pe \equiv qe \bmod \theta$. Since $p = pe$ and $q = qe$, it

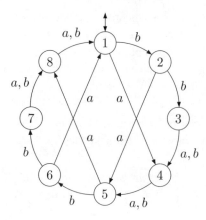

Figure 11.1 The minimal automaton of X^*.

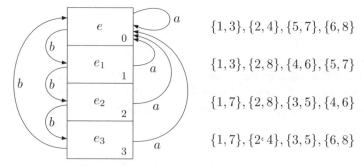

Figure 11.2 The $\mathcal{L}$-class of $e = \varphi(a^4)$.

follows that $p \equiv q \bmod \theta$. Conversely, if $p \equiv q \bmod \theta$, then for all $H \in \Gamma$, $pe_H = p$ and $qe_H = q$, because of the equality $\mathrm{Fix}(e_H) = S$. Consequently $p \equiv q \bmod \widehat{\theta}$.

Consider the prefix code Z defined by the right unitary submonoid

$$Z^* = \{z \in A^* \mid 1 \cdot z \equiv 1 \bmod \widehat{\theta}\}.$$

Then clearly $X \subset Z^*$, and the automaton $\mathcal{A}(X^*)$ being trim, $\mathrm{alph}_Z(X) = Z$. Thus, by Proposition 2.6.6, X decomposes over Z: $X = Y \circ Z$. The automaton $\mathcal{A}_{\widehat{\theta}}$ defined by the action of A^* on the classes of $\widehat{\theta}$ recognizes Z^* since Z^* is the stabilizer of the class of 1 modulo $\widehat{\theta}$. The group $G(Z)$ is the group G_θ. The automaton obtained by considering the action of Z on the class of 1 mod $\widehat{\theta}$ can be identified with an automaton recognizing Y^*, and its group is G^θ. $\qquad\square$

Corollary 11.1.7 *Let X be a thin maximal prefix code. If X is indecomposable, then the group $G(X)$ is primitive.* $\qquad\square$

Example 11.1.8 We consider once more the finite maximal prefix code $X = ((A^2 \setminus b^2) \cup b^2 A^2)^2$ of Example 11.1.4, with the minimal automaton of X^* given in Figure 11.1. Let φ be the associated representation. We have seen that $e = \varphi(a^4)$

Table 11.2 *The automaton of* $((a \cup bA^2)^4)^*$.

	1	2	3	4	5	6	7	8	9	10	11	12
a	4	3	4	7	6	7	10	9	10	1	12	1
b	2	3	4	5	6	7	8	9	10	11	12	1

Table 11.3 *The idempotent* $e = \varphi(a^4)$.

	1	2	3	4	5	6	7	8	9	10	11	12
a^4	1	10	1	4	1	4	7	4	7	10	7	10

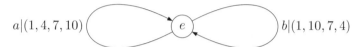

$$a|(1,4,7,10) \qquad e \qquad b|(1,10,7,4)$$

Figure 11.3 The $\mathcal{L}$-representation with respect to e.

is an idempotent of minimal rank. The group $G_e = G(X)$ is the dihedral group D_4. The partition $\theta = \{\{1, 5\}, \{4, 8\}\}$ is an imprimitivity partition of G_e.

The $\mathcal{L}$-class of e is composed of four $\mathcal{H}$-classes. They are represented in Figure 11.2 together with the associated nuclear equivalences.

The equivalence $\widehat{\theta}$ is

$$\widehat{\theta} = \{\{1, 3, 5, 7\}, \{2, 4, 6, 8\}\}.$$

The stabilizer of the class of 1 mod $\widehat{\theta}$ is the uniform code $Z = A^2$ with group $\mathbb{Z}/2\mathbb{Z}$. We have already seen that

$$X = (T \circ A^2)^2 = T^2 \circ A^2$$

for some synchronized code T. The decomposition of X into $X = T^2 \circ Z$ is that obtained by applying to X the method used in the proof of Proposition 11.1.6.

Example 11.1.9 Let Z be the finite complete prefix code over $A = \{a, b\}$ given by $Z = a \cup bA^2$, and consider $X = Z^4$. The automaton $\mathcal{A}(X^*)$ is given in Table 11.2. Let φ be the representation associated with $\mathcal{A}(X^*)$. The element $e = \varphi(a^4)$ is easily seen to be an idempotent of minimal rank 4, with $\text{Fix}(e) = \{1, 4, 7, 10\}$. It is given in Table 11.3. The minimal ideal of $\varphi(A^*)$ reduces to the $\mathcal{R}$-class of e, and we have $G(X) = \mathbb{Z}/4\mathbb{Z}$, as a result of computing the $\mathcal{L}$-representation (see Section 9.2) with respect to e given in Figure 11.3. The partition $\theta = \{\{1, 7\}, \{4, 10\}\}$ is an imprimitivity partition of G_e. The corresponding equivalence $\widehat{\theta}$ is

$$\widehat{\theta} = \{\{1, 3, 5, 7, 9, 11\}, \{2, 4, 6, 8, 10, 12\}\}.$$

The stabilizer of the class of 1 mod $\widehat{\theta}$ is the uniform code A^2, and we have $X \subset (A^2)^*$. Observe that we started with $X = Z^4$. In fact, the words in Z all have odd length, and consequently $Z^2 = Y \circ A^2$ for some Y. Thus X has the two decompositions

$$X = Z^4 = Y^2 \circ A^2.$$

11.2 Synchronization of semaphore codes

In this section, we prove the result announced in Chapter 3, namely the following theorem.

Theorem 11.2.1 *Let X be a semaphore code. There exist a synchronized semaphore code Z and an integer $d \geq 1$ such that $X = Z^d$.*

In view of Proposition 11.1.2, the integer d is of course the degree $d(X)$ of the code X. Observe that, by Proposition 3.5.9 and Corollary 3.5.10 a semaphore code is a thin maximal code and thus its degree $d(X)$ and its group $G(X)$ are well defined.

The proof of the theorem is in several parts. We first consider the group of a semaphore code. The following lemma is an intermediate step, since the theorem implies a stronger property, namely that the group is cyclic.

We recall that a transitive permutation group over a set is called *regular* if its elements, with the exception of the identity, have no fixed point (See Section 1.13).

Lemma 11.2.2 *The group of a semaphore code is regular.*

Proof. Let $X \subset A^+$ be a semaphore code, let $P = XA^-$ be the set of proper prefixes of words in X, and let $\mathcal{A} = (P, 1, 1)$ be the literal automaton of X^*. Let φ be the representation associated with $\mathcal{A}$, and set $M = \varphi(A^*)$.

A semaphore code is thin (by Proposition 3.5.9) and complete. Thus $0 \notin M$ and M has a minimal ideal denoted K. The ideal $\varphi(\bar{F}(X))$ of images of words which are not factors of words in X contains K. By Proposition 9.5.2, the Suschkewitch group of $\varphi(A^*)$ is equivalent to $G(X)$.

Let e be an idempotent in $\varphi(X^*) \cap K$, and let $R = \text{Fix}(e)$. These fixed points are words in P. They are totally ordered by their length. Indeed let w be in $\varphi^{-1}(e) \cap \bar{F}(X)$. Then we have $r \cdot w = r$ for all $r \in R$. Since w is not a factor of a word in X, no rw is in P. This implies that each word $r \in R$ is a suffix of w. Thus, for two fixed points of e, one is a suffix of the other.

Next, we recall that, by Corollary 3.5.7, $PX \subset X(P \cup X)$. By induction, this implies that for $n \geq 1$,

$$PX^n \subset X^n(P \cup X). \tag{11.3}$$

To show that G_e is regular, we verify that each $g \in H(e) \cap \varphi(X^*)$ increases the length, that is, for $r, s \in R$,

$$|r| < |s| \Rightarrow |rg| < |sg|. \tag{11.4}$$

This implies that g is the identity on R since the above property cannot be satisfied if g has a nontrivial cycle. Since $H(e) \cap \varphi(X^*)$ is composed of the elements of $H(e)$

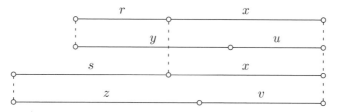

Figure 11.4 Comparison of rx and sx.

fixing 1, this means that only the identity of G_e fixes 1. Since G_e is transitive, this implies that G_e is regular.

For the proof of (11.4), let $g \in H(e) \cap \varphi(X^*)$, and let $x \in \varphi^{-1}(g)$. Then $x \in X^n$ for some $n \geq 0$. Let $r, s \in R$ with $|r| < |s|$. Then by (11.3)

$$rx = yu \text{ and } sx = zv \quad \text{with } y, z \in X^n, \ u, v \in P \cup X.$$

The word u is a suffix of v since otherwise $z \in A^* y A^+$ (see Figure 11.4) which implies $X^n \cap A^* X^n A^+ \neq \emptyset$, contradicting the fact that X^n is a semaphore code. Further, we have in $\mathcal{A}$

$$rg = u \text{ or } 1 \quad \text{according to } u \in P \text{ or } u \in X,$$

$$sg = v \text{ or } 1 \quad \text{according to } v \in P \text{ or } v \in X.$$

Since g is a permutation on R and $1g = 1$ and $s \neq 1$, we have $sg \neq 1$. Thus $sg = v$. Since $r \neq s$, we have $rg \neq sg$. Since u is a suffix of v, we have $|rg| < |sg|$ both in the two cases $rg = u$ and $rg = 1$. □

Now let $X \subset A^+$ be a group code. Then by definition,

$$X^* = \alpha^{-1}(H),$$

where $\alpha : A^* \to G$ is a surjective morphism onto a group G and H is a subgroup of G. The code X is called a *regular group code* if $H = \{1\}$. Then the permutation group $G(X)$ is the representation of G by multiplication on the right over itself. It is a regular group.

The following proposition is useful for the proof of Theorem 11.2.1. However, it is interesting in itself, because it describes the prefix codes having a regular group.

Proposition 11.2.3 *Let X be a thin maximal prefix code. Then the group $G(X)$ is regular if and only if*

$$X = U \circ V \circ W,$$

where V is a regular group code and U, W are synchronized codes.

Proof. The condition is sufficient. Indeed, if $X = U \circ V \circ W$, then by Proposition 11.1.2, we have $G(X) = G(V)$.

Conversely, let $\mathcal{A} = (Q, 1, 1)$ be an unambiguous trim automaton recognizing X^*, let φ be the associated representation and $M = \varphi(A^*)$. Since X is thin and complete, the minimal ideal J of M is a union of groups.

Consider an idempotent $e \in \varphi(X^*) \cap J$, let $G = H(e)$ be its $\mathcal{H}$-class, L its $\mathcal{L}$-class, and let Γ be the set of $\mathcal{H}$-classes contained in L. Each of them is a group, and the idempotent of H will be denoted by e_H. The set of pairs

$$\{(e_H, e) \mid H \in \Gamma\}$$

is a system of coordinates of L relative to e. Indeed $e_H e \in H$. Moreover, since $e \in Me_H$, $ee_H = e$ and thus $ee_He = e$. Let us consider the corresponding $\mathcal{L}$-representation of M. For this choice of coordinates, by (9.17), we have for $m \in M$ and $H \in \Gamma$,

$$m * H = rme_H\ell \tag{11.5}$$

where $e = \ell r$ is the column-row decomposition of e. Indeed, we have in this case $r_{mH} = re = r$ and $\ell_H = e_H\ell$.

Set

$$N = \{n \in M \mid n * H = n * G \text{ for all } H \in \Gamma\}.$$

The set N is composed of those elements $n \in M$ for which the mapping

$$H \in \Gamma \mapsto n * H \in G_e$$

is constant. It is a right-unitary submonoid of M. Indeed, first $1 \in N$ by (9.14). Next, if $n, n' \in N$, then

$$nn' * H = (n * n'H)(n' * H) \tag{11.6}$$
$$= (n * G)(n' * G)$$

which is independent of H. Thus $nn' \in N$. Assume now that $n, nn' \in N$. Then by (11.6), and since $n * n'H$ and $n * G$ have an inverse in G_e

$$n' * H = (n * n'H)^{-1}(nn' * H) = (n * G)^{-1}(nn' * G)$$

which is independent of H, showing that $n' \in N$. Therefore

$$\varphi^{-1}(N) = W^*$$

for some prefix code W.

The hypothesis that $G(X)$ is regular implies that $X^* \subset W^*$. Indeed, let $m \in \varphi(X^*)$. Then by (11.5) we have for $H \in \Gamma$,

$$m * H = rme_H\ell.$$

Since X is prefix, $e_H \in \varphi(X^*)$ by Proposition 9.4.9. Consequently $m * H$ fixes the state $1 \in Q$ (since r, m, e_H and ℓ do). Since $G(X)$ is regular, $m * H$ is the identity for all $H \in \Gamma$. This shows that $m \in N$.

We now consider the function

$$\theta : W^* \to G_e$$

which associates to each $w \in W^*$ the permutation $\varphi(w) * G$. By (11.6), θ is a morphism. Moreover, θ is surjective: if $g \in G$, then

$$g * G = rge\ell = rg\ell$$

which is the element of G_e associated to g. From $g * H = rge_H\ell = r(ge)e_H(e\ell) = rge\ell = rg\ell$, it follows that $g \in N$.

For all $x \in X^*$, since $\varphi(x) * G = 1$, we have $\theta(x) = 1$.

Since $X^* \subset W^*$ and X is a maximal code, we have by Proposition 2.6.14

$$X = Y \circ_\beta W,$$

where $\beta : B^* \to A^*$ is some injective morphism, $\beta(B) = W$ and $\beta(Y) = X$. Set

$$\alpha = \theta \circ \beta.$$

Then $\alpha : B^* \to G_e$ is a morphism and $Y^* \subset \alpha^{-1}(1)$ since for all $x \in X^*$, we have $\theta(x) = 1$. Let V be the regular group code defined by

$$V^* = \alpha^{-1}(1).$$

Then $Y = U \circ V$ and consequently

$$X = U \circ V \circ W.$$

By construction, $G(V) = G_e$. Thus $G(X) = G(V)$. The codes U and W are synchronized. Indeed $d(X) = d(V)$ and $d(X) = d(U)d(V)d(W)$ by Proposition 11.1.2 imply $d(U) = d(W) = 1$. This concludes the proof. $\qquad\square$

The following result is the final lemma needed for the proof of Theorem 11.2.1.

Lemma 11.2.4 *Let $Y \subset B^+$ be a semaphore code, and let $V \neq B$ be a regular group code. If $Y^* \subset V^*$, then $Y = (C^*D)^d$ for some integer d, where $C = B \cap V$ and $D = B \setminus C$. Moreover, C^*D is synchronized.*

Proof. Let $\alpha : B^* \to G$ be a morphism onto a group G such that $V^* = \alpha^{-1}(1)$. Since $V \neq B$, we have $G \neq \{1\}$. We have

$$C = \{b \in B \mid \alpha(b) = 1\}, \quad D = \{b \in B \mid \alpha(b) \neq 1\}.$$

The set D is nonempty. We claim that for $y \in Y$, $|y|_D > 0$. Assume the contrary, and let $y \in Y$ be such that $|y|_D = 0$. Let $b \in D$. Then $\alpha(bu) \neq 1$ for each prefix u of y since $\alpha(u) = 1$. Thus no prefix of by is in V, whence in Y. On the other hand, $B^*Y \subset YB^*$ because Y is a semaphore code (Proposition 3.5.4). This gives the contradiction and proves the claim.

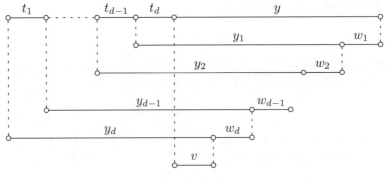

Figure 11.5

Set $T = C^*D$. Let d be the minimum value of $|y|_D$ for $y \in Y$. We will show that for any $t = t_1 t_2 \cdots t_d$, with $t_i \in T$ and $y \in Y$ such that $|y|_D = d$, there is a word v in Y such that $y = tv$ and v is a prefix of y.

Indeed, since Y is a semaphore code, $t_d y \in Y B^*$. Therefore

$$t_d y = y_1 w_1$$

for some $y_1 \in Y$, $w_1 \in B^*$. We have $|y_1|_D \geq d$ by the minimality of d and $|y_1|_D \leq d + 1$ since $|y_1|_D \leq |t_d y|_D = d + 1$. If $|y_1|_D = d + 1$, then $w_1 \in C^*$ and thus

$$\alpha(y_1) = \alpha(y_1 w_1) = \alpha(t_d) \neq 1,$$

a contradiction.

This implies that $|y_1|_D = d$, $|w_1|_D = 1$. In the same way, we get

$$t_{d-1} y_1 = y_2 w_2, \ldots, t_1 y_{d-1} = y_d w_d,$$

where each of the $y_2, \ldots, y_d$ satisfies $|y_i|_D = d$, and each $w_2, \ldots, w_d$ is in C^*DC^*. Composing these equalities, we obtain (see Figure 11.5)

$$ty = t_1 t_2 \cdots t_d y = y_d w_d w_{d-1} \cdots w_1. \tag{11.7}$$

Since $y_d \in (C^*D)^d C^*$ and $t \in (C^*D)^d$, we have

$$y_d = t_1 t_2 \cdots t_d v \in Y \tag{11.8}$$

for some $v \in C^*$ which is also a prefix of y. This proves the claim.

This property holds in particular if $t_1 \in D$, showing that Y contains a word x ($= y_d$) with d letters in D and starting with a letter in D, that is, $x \in (DC^*)^d$. Consequently x is one of the words in Y for which $|x|_D$ is minimal. Substitute x for y in (11.7). Then starting with any word $t = t_1 t_2 \cdots t_d \in T^d$, we obtain (11.8), with $v = 1$, since v is in C^* and is a prefix of x. This shows that $t \in Y$. Thus $T^d \subset Y$. Since T^d is a maximal code, we have $T^d = Y$. Since $B^*b \subset T^*$ for $b \in D$, the code T is synchronized.   □

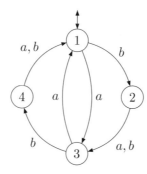

Figure 11.6 The automaton $\mathcal{A}(X^*)$.

Proof of Theorem 11.2.1. Let X be a semaphore code. By Lemma 11.2.2, the group $G(X)$ is regular. In view of Proposition 11.2.3, we have

$$X = U \circ V \circ W,$$

where V is a regular group code and U and W are synchronized. Set $Y = U \circ V$. If $d(V) = 1$, then X is synchronized and there is nothing to prove. Otherwise, according to Lemma 11.2.4, there exists a synchronized code T such that $Y = T^d$. Thus

$$X = T^d \circ W = (T \circ W)^d.$$

The code $Z = T \circ W$ is synchronized because T and W are. Finally, since $X = Z^d$ is a semaphore code, Z is a semaphore code by Corollary 3.5.12. This proves the theorem. □

Example 11.2.5 Let Z be the semaphore code $Z = \{a, ba, bb\}$ over $A = \{a, b\}$. This code is synchronized since $A^*a \subset Z^*$. Set $X = Z^2$. The minimal automaton $\mathcal{A}(X^*)$ is given by Figure 11.6.

Let φ be the associated representation and $M = \varphi(A^*)$. The element $e = \varphi(a^2)$ is an idempotent of minimal rank $2 = d(X)$. Its $\mathcal{L}$-class is composed of two groups $G_1 = H(e)$ and G_2. The $\mathcal{L}$-representation of M with respect to e is given in Figure 11.7, with the notation a instead of $\varphi(a)$ and the convention that the input is read from right to left and the output is written from right to left. The prefix code W of Proposition 11.2.3 is $W = Z$. Indeed, we have $a * 1 = a * 2 = (13)$; $ba * 1 = ba * 2 = (13)$; $bb * 1 = bb * 2 = (13)$. In this case, the code U is trivial.

Example 11.2.6 Consider, over $A = \{a, b\}$, the synchronized semaphore code $Z = a^*b$. Let $X = Z^2$. The automaton $\mathcal{A}(X^*)$ is given in Figure 11.8. Let φ be the associated representation. The element $e = \varphi(b^2)$ is an idempotent. Its set of fixed points is $\{1, 3\}$. The $\mathcal{L}$-class of e is reduced to the group $H(e)$, and the monoid N of the proof of Proposition 11.2.3 therefore is the whole monoid $\varphi(A^*)$. Thus $W = A$. The morphism α from A^* into G_e is given by

$$\alpha(a) = \mathrm{id}_{\{1,3\}}, \qquad \alpha(b) = (13).$$

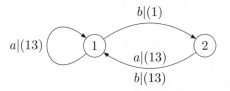

Figure 11.7 The $\mathcal{L}$-representation of M.

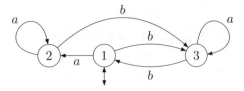

Figure 11.8 The automaton of $X^* = [(a^*b)^2]^*$.

We have $X = U \circ V$ with $V = a \cup ba^*b$. This example illustrates the fact that even when X is a semaphore code, the code U in the statement of Proposition 11.2.3 may be nontrivial and that Lemma 11.2.4 is needed to obtain the decomposition $X = Z^2$.

11.3 Group codes

Let us first recall the definition of a group code. Let G be a group, H a subgroup of G. Let $\varphi : A^* \rightarrow G$ be a surjective morphism. Then the submonoid $\varphi^{-1}(H)$ is biunitary. It is generated by a bifix code called a group code.

A group code is a maximal code (see Section 2.2). It is thin if and only if it is recognizable (Example 2.5.19), or equivalently, if the index of H in G is finite.

Rather than define a group code by an "abstract" group, it is frequently convenient to use a permutation group. This is always possible for a group code X by considering the minimal automaton of X^*. We give here the detailed description of the relation between the initial pair (G, H) and the minimal automaton of X^* (see also Section 1.13). Let G be a group and H a subgroup of G. Let Q be the set of the right cosets of H in G, that is, the set of subsets of the form Hg, for $g \in G$. To each element g in G, we associate a permutation $\pi(g)$ of Q as follows: for $p = Hk$, we define

$$p\pi(g) = Hkg.$$

It is easily verified that π is well defined and that it is a morphism from the group G into the symmetric group over Q. The subgroup H is composed of the elements of G whose image by π fixes the coset H. The index of H in G is equal to $\mathrm{Card}(Q)$. In particular H has finite index in G if and only if $\pi(G)$ is a finite group.

Now let $\varphi : A^* \rightarrow G$ be a surjective morphism. Let X be the code generating $X^* = \varphi^{-1}(H)$. For all $u, v \in A^*$,

$$H\varphi(u) = H\varphi(v) \Leftrightarrow u^{-1}X^* = v^{-1}X^*.$$

Indeed, set $g = \varphi(u)$, $k = \varphi(v)$. Then $Hg = Hk$ if and only if $g^{-1}H = k^{-1}H$ (since $(Hg)^{-1} = g^{-1}H$). Further $u^{-1}X^* = \varphi^{-1}(g^{-1}H)$, $v^{-1}X = \varphi^{-1}(k^{-1}H)$. This proves the formula.

According to Example 6.3.2, we have the equality

$$\text{Card}(Q) = d(X). \tag{11.9}$$

Theorem 11.3.1 *Let $X \subset A^+$ be a group code. If X is finite, then $X = A^d$ for some integer d.*

Proof. Let $\mathcal{A} = (Q, 1, 1)$ be the minimal automaton of X^*, and let φ be the associated representation. Let d be the degree of X. Then $d = \text{Card}(Q)$ by (11.9).

Consider the relation on Q defined as follows: for $p, q \in Q$, we have $p \leq q$ if and only if $p = q$ or $q \neq 1$ and there exists a simple path from p to q in $\mathcal{A}$. Thus $p \leq q$ if and only if $p = q$, or there exists a word $w \in A^*$ such that both $p \cdot w = q$ and $p \cdot u \neq 1$ for each left factor $u \neq 1$ of w. This relation is reflexive and transitive.

If X is finite, then the relation $\leq$ is an order on Q. Assume indeed that $p \leq q$ and $q \leq p$. Then either $p = 1$ and $q = 1$ or both $p \neq 1$, $q \neq 1$. In the second case, there exist simple paths $p \xrightarrow{w} q$ and $q \xrightarrow{w'} p$. There are also simple paths

$$1 \xrightarrow{u} p, \quad p \xrightarrow{v} 1.$$

This implies that, for all $i \geq 0$, the paths

$$1 \xrightarrow{u} p \xrightarrow{(ww')^i} p \xrightarrow{v} 1$$

are simple, showing that $u(ww')^*v \subset X$. Since X is finite, this implies $ww' = 1$, whence $p = q$. Thus $\leq$ is an order. Now let $a, b \in A$ be two letters. According to Proposition 6.5.1, we have

$$a^d, b^d \in X.$$

It follows that none of the states $1 \cdot a^i$, $1 \cdot b^i$ for $1 < i < d$ is equal to 1. Consequently,

$$1 < 1 \cdot a < 1 \cdot a^2 < \cdots < 1 \cdot a^i < \cdots < 1 \cdot a^{d-1}$$

and

$$1 < 1 \cdot b < 1b^2 < \cdots < 1 \cdot b^i < \cdots < 1 \cdot b^{d-1}.$$

Since Q has d states, this implies that $1 \cdot a^i = 1 \cdot b^i$ for all $i \geq 0$. Therefore $\varphi(a) = \varphi(b)$ for all $a, b \in A$. We get that for all $w \in A^*$ of length n, we have $w \in X^*$ if and only if $a^n \in X^*$, that is if and only if n is a multiple of d. This shows that $X = A^d$. □

The following theorem gives a sufficient condition, concerning the group $G(X)$, for a bifix code to be a group code. It will be useful later, in Section 11.6.

Theorem 11.3.2 *Let X be a thin maximal bifix code. If the group $G(X)$ is regular, then X is a group code.*

Proof. According to Proposition 11.2.3, there exist two synchronized codes U, W and a group code V such that

$$X = U \circ V \circ W.$$

Since X is thin maximal bifix, so are U and W (Proposition 2.6.13). Since U and W are synchronized, they are reduced to their alphabets (Example 3.6.6). Thus, $X = V$ and this gives the result.   □

Theorem 11.3.3 *Let $X \subset A^+$ be a code with $A = \mathrm{alph}(X)$. Then X is a regular group code if and only if X^* is closed under conjugacy.*

Proof. If X is a regular group code, the syntactic monoid of X^* is a group $G = \varphi(A^*)$ and $X^* = \varphi^{-1}(1)$. If $uv \in X^*$, then $\varphi(u)\varphi(v) = 1$, hence also $\varphi(v)\varphi(u) = 1$, showing that vu is in X^*.

To show the other implication, let us first show that X is bifix. Let u, $v \in A^*$ be such that u, $uv \in X^*$. Then also $vu \in X^*$. Since X^* is stable, it follows that $v \in X^*$. Thus, X^* is right unitary. The proof for left unitarity is analogous. Now let $M = \varphi(A^*)$ be the syntactic monoid of X^*. We verify that $\varphi(X^*) = 1$. For $x \in X^*$, we have the equivalences

$$uxv \in X^* \Leftrightarrow xvu \in X^* \Leftrightarrow vu \in X^* \Leftrightarrow uv \in X^*.$$

Therefore $\varphi(x) = \varphi(1)$. Since $\varphi(1) = 1$, it follows that $\varphi(X^*) = 1$.

Finally, we show that M is a group. From $A = \mathrm{alph}(X)$, for each letter $a \in A$, there exists $x \in X$ of the form $x = uav$. Then $avu \in X^*$, whence $\varphi(a)\varphi(vu) = 1$. This shows that all elements $\varphi(a)$, for $a \in A$, are invertible. This implies that M is a group.   □

Corollary 11.3.4 *Let $X \subset A^+$ be a finite code with $A = \mathrm{alph}(X)$. If X^* is closed under conjugacy, then $X = A^d$ for some $d \geq 1$.*

11.4 Automata of bifix codes

The general theory of unambiguous monoids of relations takes a nice form in the case of bifix codes, since the automata satisfy some additional properties. Thus, the property to be bifix can be "read" on the automaton.

Proposition 11.4.1 *Let X be a thin maximal prefix code over A, and let $\mathcal{A} = (Q, 1, 1)$ be a deterministic trim automaton recognizing X^*. The following conditions are equivalent.*

(i) *X is maximal bifix.*
(ii) *For all $w \in A^*$, we have $1 \in Q \cdot w$.*
(iii) *For all $w \in A^*$, $q \cdot w = 1 \cdot w$ implies $q = 1$.*

Proof. In a first step, we show that

$$\text{(ii)} \Leftrightarrow X \text{ is left complete.} \tag{11.10}$$

If (ii) is satisfied, consider a word w, and let $q \in Q$ be a state such that $q \cdot w = 1$. Choose $u \in A^*$ satisfying $1 \cdot u = q$. Then $1 \cdot uw = 1$, whence $uw \in X^*$. This shows that X is left complete. Conversely, assume X left complete. Let $w \in A^*$. Then there exists $u \in A^*$ such that $uw \in X^*$. Thus, $1 = 1 \cdot uw = (1 \cdot u) \cdot w$ shows that $1 \in Q \cdot w$.

Next, the equivalence

$$\text{(iii)} \Leftrightarrow X^* \text{is left unitary} \tag{11.11}$$

is precisely Proposition 6.1.14. In view of (11.10) and (11.11), the proposition is a direct consequence of Proposition 6.2.1. □

Let X be a thin maximal bifix code, and let $\mathcal{A} = (Q, 1, 1)$ be a trim deterministic automaton recognizing X^*. Then the automaton A is complete, and the monoid $M = \varphi_{\mathcal{A}}(A^*)$ is a monoid of (total) functions. The minimal ideal J is composed of the functions m such that $\text{Card}(\text{Im}(m)) = \text{rank}(m)$ equals the minimal rank $r(M)$ of M. The $\mathcal{H}$-classes of J are indexed by the minimal images and by the maximal nuclear equivalences (Proposition 9.4.10). Each state appears in at least one minimal image and the state 1 is in all minimal images. Each $\mathcal{H}$-class H meets $\varphi(X^*)$ and the intersection is a subgroup of H. Note the following important fact: If S is a minimal image and w is any word, then $T = S \cdot w$ is again a minimal image. Thus, $\text{Card}(S) = \text{Card}(T)$ and consequently w realizes a bijection from S onto T.

We will be interested in the minimal automaton $\mathcal{A}(X^*)$ of X^*. According to Proposition 3.3.11, this automaton is complete and has a unique final state coinciding with the initial state. This shows that $\mathcal{A}(X^*)$ is of the form considered above.

Let φ be the representation associated with the minimal automaton $\mathcal{A}(X^*) = (Q, 1, 1)$, and let $M = \varphi(A^*)$. Let J be the minimal ideal of M. We define

$$J(X) = \varphi^{-1}(J).$$

This is an ideal in A^*. Moreover, we have

$$w \in J(X) \Leftrightarrow S \cdot w = T \cdot w \text{ for all minimal images } S, T \text{ of } \mathcal{A}. \tag{11.12}$$

Indeed, let $w \in J(X)$. Then $U = Q \cdot w$ is a minimal image. For any minimal image T, we have $T \cdot w \subset Q \cdot w = U$, hence $T \cdot w = U$ since $T \cdot w$ is minimal. Thus, $T \cdot w = S \cdot w = Q \cdot w$. Conversely, assume that for $w \in A^*$, we have $S \cdot w = T \cdot w$ for all minimal images S, T. Set U equal to this common image. Since every state in Q appears in at least one minimal image, we have

$$Q \cdot w = \left(\bigcup_S S \right) \cdot w = \bigcup_S S \cdot w = U,$$

where the union is over the minimal images. This shows that $\varphi(w)$ has minimal rank, and consequently $w \in J(X)$. The equivalence (11.12) is proved.

Proposition 11.4.2 *Let X be a thin maximal bifix code and let $A(X^*) = (Q, 1, 1)$ be the minimal automaton of X^*. Let $p, q \in Q$ be two states. If $p \cdot h = q \cdot h$ for all $h \in J(X)$, then $p = q$.*

Proof. It suffices to prove that for all $w \in A^*$, $p \cdot w = 1$ if and only if $q \cdot w = 1$. The conclusion, namely that $p = q$, follows then by the definition of $A(X^*)$.

Let $h \in J(X) \cap X^*$. Let $w \in A^*$ be such that $p \cdot w = 1$. We must show that $q \cdot w = 1$. We have $p \cdot wh = (p \cdot w) \cdot h = 1 \cdot h = 1$, since $h \in X^*$. Now $wh \in J(X)$, hence by assumption $q \cdot wh = p \cdot wh = 1$. Thus, $(q \cdot w) \cdot h = 1$. By Proposition 11.4.1(iii), it follows that $q \cdot w = 1$. This proves the proposition.   □

For a transitive permutation group G of degree d it is customary to consider the number $k(G)$ which is the maximum number of fixed points of an element of G distinct from the identity. The *minimal degree* of G is the number $d - k(G)$. The group is regular if and only if $k(G) = 0$; it is a *Frobenius group* if $k(G) = 1$.

If X is a code of degree d and with group $G(X)$, we denote by $k(X)$ the integer $k(G(X))$. We will prove

Theorem 11.4.3 *Let $X \subset A^+$ be a thin maximal bifix code of degree d, and let $k = k(X)$. Then*

$$A^k \setminus A^* X A^* \subset J(X).$$

We use the following preliminary result.

Lemma 11.4.4 *With the above notation, let $A = (Q, 1, 1)$ be the minimal automaton recognizing X^*. For any two distinct minimal images S and T of A, we have*

$$\mathrm{Card}(S \cap T) \leq k.$$

Proof. Let $M = \varphi_A(A^*)$, and consider an idempotent $e \in M$ having image S, that is, such that $Qe = S$. Consider an element $t \in T \setminus S$, and set $s = te$. Then $s \in S$, and therefore, $s \neq t$. We will prove that there is an idempotent f separating s and t, that is, such that $sf \neq tf$.

According to Proposition 11.4.2, there exists $h \in J(X)$ such that $s \cdot h \neq t \cdot h$. Let $m = \varphi(h) \in J$, where J is the minimal ideal of M. Multiplying on the right by a convenient element $n \in M$, the element $mn \in J$ will be in the $\mathcal{L}$-class characterized by the minimal image T. Since n realizes a bijection from $\mathrm{Im}(m)$ onto $\mathrm{Im}(mn) = T$ we have $smn \neq tmn$. Let f be the idempotent of the $\mathcal{H}$-class of mn. Then f and mn have the same nuclear equivalence. Consequently $sf \neq tf$. Since $t \in T = \mathrm{Im}(mn) = \mathrm{Im}(f) = \mathrm{Fix}(f)$, we have $tf = t$.

Consider now the restriction to T of the mapping ef. For all $p \in S \cap T$, we obtain $pef = pf = p$. This shows that ef fixes the states in $S \cap T$. Further, since $s = te$, $t(ef) = sf \neq t$, showing that ef is not the identity on T. Thus, by definition of k, we have $\mathrm{Card}(S \cap T) \leq k$.   □

Proof of Theorem 11.4.3. Let $A = (Q, 1, 1)$ be the minimal automaton of X^*. Let $w \in A^* \setminus A^* X A^*$ and set $w = a_1 a_2 \cdots a_k$ with $a_i \in A$. Let S be a minimal image.

Table 11.4 *The automaton*
$\mathcal{A}(X^*)$.

	1	2	3	4	5
a	1	4	5	2	3
b	2	3	1	1	3

For each $i = 1, \ldots, k$, the word $a_1 a_2 \cdots a_i$ defines a bijection from S onto $S_i = S \cdot a_1 a_2 \cdots a_i$. Since S_i is a minimal image, it contains the state 1. Thus S_k contains all the $k + 1$ states

$$1 \cdot a_1 a_2 \cdots a_k, \ 1 \cdot a_2 \cdots a_k, \ldots, \ 1 \cdot a_k, 1.$$

These states are distinct. Indeed, assume that

$$1 \cdot a_i a_{i+1} \cdots a_k = 1 \cdot a_j \cdots a_k$$

for some $i < j$. Then setting $q = 1 \cdot a_i a_{i+1} \cdots a_{j-1}$, we get $q \cdot a_j \cdots a_k = 1 \cdot a_j \cdots a_k$. By Proposition 11.4.1, this implies $q = 1$. But then $w \in A^* X A^*$, contrary to the assumption.

This implies that $S \cdot w$ contains $k + 1$ states which are determined in a way independent from S. In other words, if T another minimal image, then $T \cdot w$ contains these same $k + 1$ states. This means that $\mathrm{Card}(T \cdot w \cap S \cdot w) \geq k + 1$, and by Lemma 11.4.4, we have $S \cdot w = T \cdot w$. Thus two arbitrary minimal images have the same image by w. This shows by (11.12) that w is in $J(X)$. ☐

Remark 11.4.5 Consider, in Theorem 11.4.3, the special case where $k = 0$, that is, where the group $G(X)$ is regular. Then $1 \in J(X)$. Now

$$1 \in J(X) \Leftrightarrow X \text{ is a group code.} \tag{11.13}$$

Indeed, if $1 \in J(X)$, then the syntactic monoid $M = \varphi_{\mathcal{A}(X^*)}(A^*)$ coincides with its minimal ideal. This minimal ideal is a single group since it contains the neutral element of M. The converse is clear. Thus we obtain, in another way, Theorem 11.3.2.

Example 11.4.6 If X is a thin maximal bifix code over A with degree $d(X) = 3$, then $k = 0$ (if $G(X) = \mathbb{Z}/3\mathbb{Z}$) or $k = 1$ (if $G(X) = \mathfrak{S}_3$). In the second case by Theorem 11.4.3, we have

$$A \setminus X \subset J(X).$$

The following example shows that the inclusion $A \subset J(X)$ does not always hold. Let X be the maximal prefix code over $A = \{a, b\}$ defined by the automaton $\mathcal{A}(X^*) = (Q, 1, 1)$ with $Q = \{1, 2, 3, 4, 5\}$ and transition function given in Table 11.4.

The set of images, together with the actions by a and b, is given in Figure 11.9. Each of the images contains the state 1. Consequently X is a bifix code. We have $d(X) = 3$

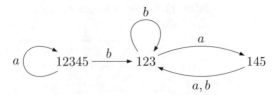

Figure 11.9 The diagram of images.

(which is the number of elements of the minimal images). We have $Q \cdot b = \{1, 2, 3\}$. Thus $\varphi_A(b)$ has minimal rank; consequently $b \in J(X)$. However, $a \notin J(X)$ since $Q \cdot a = Q$. In fact $a \in X$, in agreement with Theorem 11.4.3.

Theorem 11.4.7 *Let X be a thin maximal bifix code. Then the code X is indecomposable if and only if $G(X)$ is a primitive group.*

Proof. If $X = Y \circ Z$, then Y and Z are thin maximal bifix codes by Proposition 2.6.13. According to Proposition 11.1.2, there exists an imprimitivity partition θ of $G(X)$ such that $G^\theta = G(Y)$ and $G_\theta = G(Z)$. If $G(X)$ is primitive, then $G_\theta = 1$ or $G^\theta = 1$. In the first case, $d(Y) = 1$, implying $X = Z$. In the second case, $d(Z) = 1$, whence $Z = A$. Thus, the code X is indecomposable.

The converse implication follows directly from Corollary 11.1.7. □

11.5 Depth

Let S be a finite semigroup, and let J be its minimal (two-sided) ideal. We say that S is *nil-simple* if there exists an integer $n \geq 1$ such that

$$S^n \subset J. \tag{11.14}$$

The smallest integer $n \geq 1$ satisfying (11.14) is called the *depth* of S. Since S^n is, for all n, a two-sided ideal, (11.14) is equivalent to $S^n = J$, which in turn implies $S^n = S^{n+1}$.

We shall use nil-simple semigroups for a characterization of bifix codes. Before stating this result, we have to establish a property which is interesting in itself.

Proposition 11.5.1 *Let $X \subset A^+$ be a thin maximal bifix code, and let $\mathcal{A} = (Q, 1, 1)$ be an unambiguous trim automaton recognizing X^*. Let J be the minimal ideal of $\varphi_A(A^*)$. Then*

$$\varphi_A(\bar{H}(X)) \subset J.$$

Recall that $H(X) = A^- X A^-$ is the set of internal factors of X, and $\bar{H}(X) = A^* \setminus H(X)$.

Proof. Let φ_D be the representation associated with the flower automaton of X, set $M_D = \varphi_D(A^*)$ and let J_D be the minimal ideal of M_D. It suffices to prove the result for φ_D. Indeed, there exists by Proposition 4.2.5, a surjective morphism $\widehat{\rho} : M_D \to \varphi_A(A^*)$ such that $\varphi_A = \widehat{\rho} \circ \varphi_D$, we have $\widehat{\rho}(J_D) = J$.

Thus the inclusion $\varphi_D(\bar{H}(X)) \subset J_D$ implies $\varphi_A(\bar{H}(X)) \subset \hat{\rho}(J_D) = J$. It remains to prove the inclusion $\varphi_D(\bar{H}(X)) \subset J_D$.

Let $\mathcal{A}_D = (Q, (1, 1)(1, 1))$ be the flower automaton of X. Let $w \in \bar{H}(X)$. Then w has $d = d(X)$ interpretations. We prove that $\text{rank}(\varphi_D(w)) = d$. Since this is the minimal rank, it implies that $\varphi_D(w)$ is in J_D.

Clearly $\text{rank}(\varphi_D(w)) \geq d$. To prove the converse inequality, let I be the set composed of the d interpretations of w. We define two relations

$$\alpha \in \{0, 1\}^{Q \times I}, \quad \beta \in \{0, 1\}^{I \times Q}$$

as follows : if $(u, v) \in Q$, and $(s, x, p) \in I$, with $s \in A^- X, x \in X^*, p \in X A^-$, then

$$((u, v), \alpha, (s, x, p)) = \delta_{v,s}, \quad ((s, x, p,), \beta, (u, v)) = \delta_{p,u},$$

where δ is the Kronecker symbol. We claim that

$$\varphi_D(w) = \alpha\beta.$$

Assume first that $(u, v)\alpha\beta(u', v')$. Then there exists an interpretation $i = (v, x, u') \in I$ such that $(u, v)\alpha i \beta(u', v')$. Note that i is uniquely determined by v or by u', because X is bifix. Next $w \in vX^*u'$, showing that $((u, v), \varphi_D(w), (u', v')) = 1$.

Conversely, assume that $((u, v), \varphi_D(w), (u', v')) = 1$. Then either $uw = u'$ and $v = wv'$, or $w \in vX^*u'$. The first possibility implies the second one: Indeed, if $uw = u'$ and $v = wv'$, then $uwv' \in X$. Since $w \in \bar{H}(X)$ this implies $u = v' = 1 = u' = v$. It follows that $w \in vX^*u'$. Thus, $w = vxu'$ for some $x \in X^*$, showing that $i = (v, x, u')$ is an interpretation of w. Consequently, $(u, v)\alpha i$ and $i\beta(u', v')$. This proves (11.5). By (11.5), we have $\text{rank } \varphi_D(w) \leq \text{Card}(I) = d(X)$. □

The following result gives an algebraic characterization of finite maximal bifix codes. The proof uses Theorem 5.2.4 on codes with finite deciphering delay.

Theorem 11.5.2 *Let $X \subset A^+$ be a finite maximal code, and let $\mathcal{A} = (Q, 1, 1)$ be an unambiguous trim automaton recognizing X^*. The two following conditions are equivalent.*

(i) *X is bifix.*
(ii) *The semigroup $\varphi_A(A^+)$ is nil-simple.*

Proof. Set $\varphi = \varphi_A$, and set $S = \varphi(A^+)$. Let J be the minimal ideal of S.

(i) $\Rightarrow$(ii). Let n be the maximum of the lengths of words in X. A word in X of length n cannot be an internal factor of X, showing that $A^n A^* \subset \bar{H}(X)$. Observe that $A^n A^* = (A^+)^n$. This implies that $S^n = \varphi((A^+)^n) \subset \varphi(\bar{H}(X))$. By Proposition 11.5.1, we obtain $S^n \subset J$, showing that S is nil-simple.

(ii) $\Rightarrow$ (i). Let n be the depth of S. Then for all $y \in A^n A^* = (A^+)^n$, we have $\varphi(y) \in J$. We prove that for any $y \in X^n$, and for all $x \in X^*, u \in A^*$,

$$xyu \in X^* \Rightarrow yu \in X^*. \tag{11.15}$$

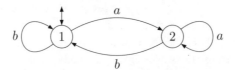

Figure 11.10 The minimal automaton of $(a^*b)^*$.

The semigroup S contains no zero. Further, the elements $\varphi(y)$ and $\varphi(yxy)$ of $\varphi(X^*)$ are in the same group, say G, of the minimal ideal, because $\varphi(yxy) = \varphi(yx)\varphi(y)$ and $\varphi(y) = [\varphi(yx)]^{-1}\varphi(yxy)$, showing that $\varphi(y)\mathcal{L}\varphi(yxy)$. The same argument holds for the other side. In fact, both $\varphi(yx)$ and $\varphi(yx)^{-1}$ are in the subgroup $G \cap \varphi(X^*)$. Thus there exists some $r \in X^*$ such that $\varphi(yx)^{-1} = \varphi(r)$, or also $\varphi(y) = \varphi(r)\varphi(yxy)$.

This gives

$$\varphi(yu) = \varphi(r)\varphi(y)\varphi(xyu) \in \varphi(X^*),$$

showing that $yu \in X^*$. This proves (11.15).

Formula (11.15) shows that every word in X^n is simplifying. In view of Proposition 5.1.5, the code X has deciphering delay n. According to Theorem 5.2.4, X is a prefix code. Symmetrically, X is suffix. Thus X is a bifix code.   □

Example 11.5.3 Consider again the maximal bifix code X of Example 11.4.6. The semigroup $\varphi_{A(X^*)}(A^+)$ is not nil-simple. Indeed, $\varphi(a)$ is a permutation of Q and thus $\varphi(a^n) \notin J$ for all $n \geq 1$. This shows that the implication (i) $\Rightarrow$ (ii) of Theorem 11.5.2 is in general false without the assumption of finiteness on the code.

Example 11.5.4 Let $A = \{a, b\}$ and $X = a^*b$. The code X is maximal prefix, but is not suffix. The automaton $A(X^*)$ is given in Figure 11.10.

The semigroup $\varphi(A^+)$ is nil-simple: it is composed of the two constant functions $\varphi(a)$ and $\varphi(b)$. This example shows in addition that the implication (ii) $\Rightarrow$ (i) of Theorem 11.5.2 may become false if the code is infinite.

11.6 Groups of finite bifix codes

In the case of a thin maximal bifix code X, the $\mathcal{L}$-representation, introduced in Chapter 9 (Section 9.2), of the minimal automaton of X^* takes a particular form which makes it easy to manipulate.

Consider a thin maximal bifix code $X \subset A^+$ of degree d, let $A(X^*) = (Q, 1, 1)$ be the minimal (deterministic) automaton of X^* and let $\varphi = \varphi_{A(X^*)}$ be the associated representation. Finally, let $M = \varphi(A^*)$ and let J be the minimal ideal of M. Each $\mathcal{H}$-class of J is a group. Fix an idempotent $e \in J$, let $S = \mathrm{Im}(e) = \mathrm{Fix}(e)$, and let Γ be the set of $\mathcal{H}$-classes of the $\mathcal{L}$-class of e. Denote by e_H the idempotent of the $\mathcal{H}$-class $H \in \Gamma$. The set of pairs $(e_H, e)_{H \in \Gamma}$ constitutes a system of coordinates. Indeed, for $H \in \Gamma$,

$$e_H e = e_H, \qquad e e_H = e.$$

If $e = \ell r$ is the column-row decomposition of e, then for $H \in \Gamma$, $e_H = \ell_H r$, with $\ell_H = e_H \ell$, is the column-row decomposition of e_H. The notations of Section 9.2 then simplify considerably. In particular, for $m \in M$ and $H \in \Gamma$,

$$m * H = rm\ell_H = r(eme_H)\ell.$$

Of course, $m * H \in G_e$. As we will see, this can be used to define a function

$$A^* \times J(X) \to G_e,$$

where $J(X) = \varphi^{-1}(J)$ as in the previous section. Let $u \in A^*$ and let $k \in J(X)$. Then $\varphi(k) \in J$, and corresponding to this element, there is an $\mathcal{H}$-class denoted $H^{(k)}$ in Γ which by definition is the intersection of the $\mathcal{R}$-class of $\varphi(k)$ and of the $\mathcal{L}$-class of e. In other words, $H^{(k)} = Me \cap \varphi(k)M$.

We define a function from $A^* \times J(X)$ into G_e by setting $u * k = \varphi(u) * H^{(k)}$. Then

$$u * k = r\varphi(u)\ell_{H^{(k)}} = re\varphi(u)e_{H^{(k)}}\ell.$$

Consequently $u * k \in G_e$. It is a permutation on the set $S = \mathrm{Fix}(e)$ obtained by restriction to S of the relation $e\varphi(u)e_{H^{(k)}}$.

The following explicit characterization of $u * k$ is the basic formula for the computations. For $u \in A^*$, $k \in J(X)$, we have for $s, t \in S$,

$$s(u * k) = t \iff s \cdot uk = t \cdot k. \tag{11.16}$$

In this formula, the computation of $s \cdot uk$ and $t \cdot k$ is of course done in the automaton $\mathcal{A}(X^*)$. Let us verify (11.16). If $s(u * k) = t$, then $se\varphi(u)e_{H^{(k)}} = t$. From $se = s$, it follows that $s\varphi(u)e_{H^{(k)}} = t$. Taking the image by $\varphi(k)$, we obtain

$$s\varphi(u)e_{H^{(k)}}\varphi(k) = t\varphi(k).$$

Since $e_{H^{(k)}}\varphi(k) = \varphi(k)$, we get that $s\varphi(uk) = t\varphi(k)$, or in other words, $s \cdot uk = t \cdot k$.

Conversely, assume that $s\varphi(uk) = t\varphi(k)$. Let $m \in M$ be such that $\varphi(k)m = e_{H^{(k)}}$. Then $s\varphi(u)\varphi(k)m = t\varphi(k)m$ implies $s\varphi(u)e_{H^{(k)}} = te_{H^{(k)}}$. Since $se = s$ and $te = t$, we get

$$se\varphi(u)e_{H^{(k)}} = tee_{H^{(k)}} = te = t,$$

showing that $s(u * k) = t$. This proves (11.16).

The function from $A^* \times J(X)$ into G_e defined above is called the *ergodic representation* of X (relative to e). We will manipulate it via the relation (11.16). Note the following formulas which are the translation of the corresponding relations given in Section 9.2, and which also can be simply proved directly using Formula (11.16). For $u \in A^*$, $k \in J(X)$, and $v \in A^*$,

$$u * kv = u * k, \tag{11.17}$$

$$uv * k = (u * vk)(v * k). \tag{11.18}$$

Proposition 11.6.1 *Let $X \subset A^+$ be a thin maximal bifix code, and let $R = J(X) \setminus J(X)A^+$ be the basis of the right ideal $J(X)$. Let e be an idempotent in the minimal ideal of $\varphi_{A(X^*)}(A^*)$ and let $S = \text{Fix}(e)$. The group $G(X)$ is equivalent to the permutation group over S generated by the permutations $a * r$, with $a \in A$, $r \in R$.*

Proof. It suffices to show that the permutations $a * r$ generate G_e, since G_e is equivalent to $G(X)$. Set $\varphi = \varphi_{A(X^*)}$. Every permutation $u * k$, for $u \in A^*$ and $k \in J(X)$, clearly is in G_e. Conversely, consider a permutation $\sigma \in G_e$. Let $g \in G(e)$ be the element giving σ by restriction to S, and let $u \in \varphi^{-1}(g)$, $k \in \varphi^{-1}(e)$. Then $u * k$ is the restriction to S of $e\varphi(u)e_{H(k)} = e\varphi(u)e = g$. Thus $u * k = \sigma$.

Consequently $G_e = \{u * k \mid u \in A^*, k \in J(X)\}$. For $u = a_1a_2 \cdots a_n$ with $a_i \in A$, and $k \in J(X)$, we get, by (11.18),

$$u * k = (a_1 * a_2a_3 \cdots a_nk)(a_2 * a_3 \cdots a_nk) \cdots (a_n * k).$$

This shows that G_e is generated by the permutations $a * k$, for a in A and k in $J(X)$. Now for each k in $J(X)$, there exists $r \in R$ such that $k \in rA^*$. By (11.17), we have $a * k = a * r$. This completes the proof.   □

Note that Proposition 11.6.1 can also be derived from Proposition 9.2.1.

Proposition 11.6.2 *Let X be a finite maximal bifix code over A of degree d and let $\varphi = \varphi_{A(X^*)}$. For each letter $a \in A$, we have $a^d \in J(X) \cap X$ and $\varphi(a^d)$ is an idempotent.*

Proof. Let $A(X^*) = (Q, 1, 1)$. By Proposition 6.5.1, we have $a^d \in X$ for $a \in A$. The states

$$1, \ 1 \cdot a, \ \ldots, \ 1 \cdot a^{d-1}$$

are distinct. Indeed, if $1 \cdot a^i = 1 \cdot a^j$ for some $0 \leq i < j \leq d - 1$, then setting $q = 1 \cdot a^j$, we would have $q \cdot a^{d-j} = 1$ and $1 \cdot a^{d-j+i} = 1$, whence $a^{d-j+i} \in X^*$. Since $d - j + i < d$, this contradicts the fact that X is prefix. Moreover, we have

$$\text{Im}(a^d) = Q \cdot a^d = \{1, 1 \cdot a, \ldots, 1 \cdot a^{d-1}\}.$$

Indeed, let $q \in Q, q \neq 1$, and let $w \in XA^-$ be a word such that $1 \cdot w = q$. Since X is right complete and finite there exists a power of a, say a^j, such that $wa^j \in X$. Then $j < d$ since X is suffix, and $j > 0$ since $w \notin X$. Thus $q \cdot a^j = 1$ and $q \cdot a^d = 1 \cdot a^{d-j} \in \{1, 1 \cdot a, \ldots, 1 \cdot a^{d-1}\}$. This proves that $\text{Im}(a^d) \subset \{1, 1 \cdot a, \ldots, 1 \cdot a^{d-1}\}$. The converse inclusion is a consequence of $(1 \cdot a^i) \cdot a^d = 1 \cdot a^{d+i} = 1 \cdot a^i$, for $i = 0, \ldots, d - 1$.

Thus $\varphi(a^d)$ has rank d, showing that $\varphi(a^d)$ is in the minimal ideal of $\varphi(A^*)$, which in turn implies that $a^d \in J(X)$. Next $(1 \cdot a^j) \cdot a^d = 1 \cdot a^j$ for $j = 0, \ldots, d - 1$. It follows that $\varphi(a^d)$ is the identity on its image. This proves that $\varphi(a^d)$ is an idempotent.   □

Proposition 11.6.2 shows that in the case of a finite maximal bifix code X, a particular ergodic representation can be chosen by taking, as basic idempotent for the

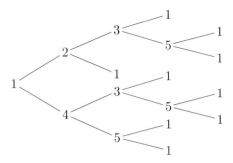

Figure 11.11 Transitions for a bifix code.

system of coordinates, the $d(X)$-th power of any of the letters a of the alphabet. More precisely, let $\mathcal{A}(X^*) = (Q, 1, 1)$ and let φ be the associated morphism, set $e = \varphi(a^d)$, and identify i with $1 \cdot a^{i-1}$, for $1 \le i \le d$. The ergodic representation relative to the idempotent $\varphi(a^d)$ is denoted by $*_a$. It is defined, for $u \in A^*$, $k \in J(X)$, and for $1 \le i, j \le d$, by

$$i(u *_a k) = j \Leftrightarrow i \cdot uk = j \cdot k \Leftrightarrow 1 \cdot a^{i-1}uk = 1 \cdot a^{j-1}k. \qquad (11.19)$$

Observe that for $u = a$ and for any $k \in J(X)$,

$$a *_a k = \alpha$$

with $\alpha = (1\ 2 \cdots d)$. Indeed, by (11.19) $i(a *_a k) = j$ if and only if $i \cdot ak = j \cdot k$, thus if and only if $(i+1) \cdot k = j \cdot k$. Since k induces a bijection from S onto $S \cdot k$, this implies $j = i + 1$, which is the claim.

Example 11.6.3 Let $A = \{a, b\}$, and consider the finite maximal bifix code $X \subset A^+$ of degree 3 with kernel $K(X) = \{ab\}$. The transitions of the minimal automaton of X^*, with states $\{1, 2, 3, 4, 5\}$, are given in Figure 11.11.

The letters a and b define mappings $\varphi(a)$ and $\varphi(b)$ of rank 3. Thus $a, b \in J(X)$. We consider the ergodic representation $*_a$, that is relative to the idempotent $e = \varphi(a^3)$. To compute it, it is sufficient (according to Proposition 11.6.1) to compute the four permutations $a *_a a$, $a *_a b$, $b *_a a$, $b *_a b$ by using (11.19). For instance, we have $i(a *_a a) = j \Leftrightarrow i \cdot a^2 = j \cdot a \Leftrightarrow i + 1 = j$ mod 3. The permutations are easily seen to be

$$a *_a a = a *_a b = (123), \quad b *_a a = (12), \quad b *_a b = (132).$$

The group $G(X)$ therefore is the symmetric group over S.

Proposition 11.6.4 *Let $X \subset A^+$ be a finite maximal bifix code of degree d, and let $a \in A$. Then $a *_a a^d$ is a cycle of length d.*

Proof. By (11.19),

$$i(a *_a a^d) = j \Leftrightarrow i \cdot a^{d+1} = j \cdot a^d.$$

This is equivalent to $i \cdot a = j$, or $i + 1 = j \mod d$. Thus $i(a *_a a^d) \equiv i + 1 \mod d$, proving the statement.   □

We are now ready to study the groups of finite bifix codes. We recall that a transitive permutation group G of degree $d \geq 2$ is called a *Frobenius group* if $k(G) = 1$.

Theorem 11.6.5 *Let X be a finite maximal bifix code of degree $d \geq 4$. Then $G(X)$ is not a Frobenius group.*

Proof. Let $\mathcal{A} = (Q, 1, 1)$ be the minimal automaton of X^*. Since $d \geq 4$, no letter is in X. Arguing by contradiction, we suppose that $G(X)$ is a Frobenius group. Thus $k(G(X)) = 1$. By Theorem 11.4.3, we have $A \subset J(X)$. This means that for all $a \in A$, $\mathrm{Im}(a)$ has d elements.

Let $a \in A$ be a letter, and set $S = \mathrm{Im}(a^d) = \{1, 2, \ldots, d\}$, where, for $1 \leq i \leq d$, $i = 1 \cdot a^{i-1}$. Consider the ergodic representation $*_a$, and set

$$\alpha = a *_a a, \qquad \beta = b *_a a,$$

where $b \in A$ is an arbitrary letter. We want to prove that $\beta = \alpha$.

Note that, by (11.19) and (11.6) we have for $i \in S$, $i \cdot ba = i\beta \cdot a$, and

$$i \cdot \alpha = \begin{cases} i + 1 & \text{if } i < d, \\ 1 & \text{if } i = d. \end{cases}$$

Since $S \cdot b$ is a minimal image, it contains the state 1. Thus there exists a (unique) state $q' \in S$ such that $q' \cdot b = 1$. For the same reason, there exists a unique state $q'' \in S$ such that $q'' \cdot ba = 1$. We claim that $q'\beta = 1, q''\beta = d$. Indeed, we have $1 \cdot a = q' \cdot ba = q'\beta \cdot a$. Next $q'' \cdot ba = q''\beta \cdot a = 1 = d \cdot a$. Since a defines a bijection from S onto itself, it follows that $1 = q'\beta$ and $q''\beta = d$. This proves the claim.

Now we verify that

$$q\beta \geq q \quad \text{for} \quad q \in S, \quad q \neq q'. \tag{11.20}$$

First, we observe that the inequality holds for q'', since $q''\beta = d$. Arguing by contradiction, suppose that $q\beta = p < q$ for some $q \in S, q \neq q', q''$. Then

$$q\beta \cdot a = q \cdot ba = p \cdot a = p + 1 \leq q.$$

Setting $n = q - (p + 1)$, it follows that $q \cdot ba^{n+1} = q$. Consider the path

$$q \xrightarrow{ba^{n+1}} q.$$

Since $q \neq q', q''$, we have $q \cdot b \neq 1, q \cdot ba \neq 1$. Also $q \cdot ba^i = p + i \neq 1$ for $i = 1, \ldots, n + 1$. Thus this path is simple. Consequently,

$$a^{q-1}(ba^{n+1})^* a^{d-q+1} \subset X$$

contradicting the finiteness of X. This proves (11.20).

It follows from this equality that there exists at most one state $q \in S$ such that $q\beta < q$, namely the state q'. This implies that the permutation β is composed of at most one cycle (of length > 1) and the remaining states are fixed points. Further, β cannot be the identity on S, since otherwise the relation $q'\beta = 1$ would imply $q' = 1$, hence $1 \cdot b = 1$ and $b \in X$ which is not true. Now by assumption, $G(X)$ is a Frobenius group. This shows that β has at most one fixed point. If β has no fixed point, then the inequalities in (11.20) are strict and this implies that

$$\beta = (123 \cdots d) = \alpha.$$

Assume now that β has just one fixed point i. Then $\beta = (123 \cdots i - 1i + 1 \cdots d)(i)$. This implies that

$$\beta^{-1}\alpha = \begin{cases} (i, i+1) & \text{if } i \neq d, \\ (d1) & \text{if } i = d. \end{cases}$$

Since $\beta^{-1}\alpha \in G(X)$ and $\beta^{-1}\alpha$ has $d - 2$ fixed points, $G(X)$ can be a Frobenius group only if $d \leq 3$. This gives a contradiction and proves that indeed $\alpha = \beta$.

It follows from (11.19) and from the equality $\alpha = \beta$ that $i \cdot ba = i \cdot a^2$ for $i \in S$. This shows that for $m \geq 0$,

$$1 \cdot a^m ba = 1 \cdot a^{m+2}. \tag{11.21}$$

Observe that this formula holds for arbitrary letters $a, b \in A$. This leads to another formula, namely, for $i \geq 0$ and $a, b \in A$,

$$a^i b = 1 \cdot b^{i+1}. \tag{11.22}$$

This formula holds indeed for $a, b \in A$ and $i = 0$. Arguing by induction, we suppose that (11.22) holds for some $i \geq 0$, and for all $a, b \in A$. Then we have, for $a, b \in A$, also $1 \cdot b^i a = 1 \cdot a^{i+1}$, whence $1 \cdot b^i ab = 1 \cdot a^{i+1}b$. Apply (11.21). We get

$$1 \cdot a^{i+1}b = 1 \cdot b^i ab = 1 \cdot b^{i+2}.$$

This proves (11.22).

Finally we show, by a descending induction on $i \in \{0, 1, \ldots, d\}$, that for all $a \in A$,

$$1 \cdot a^i A^{d-i} = \{1\}.$$

This holds for $i = d$, and for $i < d$ we have

$$1 \cdot a^i A^{d-i} = \bigcup_{b \in A} 1 \cdot a^i b A^{d-i-1} = \bigcup_{b \in A} 1 \cdot b^{i+1} A^{d-i-1} = 1$$

by using (11.22). This proves the formula. For $i = 0$, it becomes $1 \cdot A^d = \{1\}$, showing that $A^d \subset X$. This implies that $A^d = X$. Since $G(A^d)$ is a cyclic group, it is not a Frobenius group. This gives the contradiction and concludes the proof. $\quad\square$

Remark 11.6.6 Consider a finite maximal bifix code X of degree at most 3. If the degree is 1 or 2, then the code is uniform, and the group is a cyclic group. If $d(X) = 3$, then $G(X)$ is either the symmetric group $\mathfrak{S}_3$ or the cyclic group over 3 elements. The latter group is regular, and according to Theorems 11.3.2 and 11.3.1, the code X is uniform. Thus except for the uniform code, all finite maximal bifix codes of degree 3 have as a group $\mathfrak{S}_3$ which is a Frobenius group.

We now establish an interesting property of the groups of bifix codes. For this, we use a result from the theory of permutation groups which we formulate for convenience as stated in Theorem 11.6.7. References for proofs are given in the Notes. Recall that a permutation group G over a set Q is k-transitive if for all $(p_1, \ldots, p_k) \in Q^k$ and $(q_1, \ldots, q_k) \in Q^k$ composed of distinct elements, there exists $g \in G$ such that $p_1 g = q_1, \ldots, p_k g = q_k$. This shows that 1-transitive groups are precisely the transitive groups. A 2-transitive group is usually called *doubly transitive*.

Theorem 11.6.7 *Let G be a primitive permutation group of degree d containing a d-cycle. Then either G is a regular group or a Frobenius group or is doubly transitive.*

Theorem 11.6.8 *Let X be a finite maximal bifix code over A. If X is indecomposable and not uniform, then $G(X)$ is doubly transitive.*

Proof. According to Theorem 11.4.7, the group $G(X)$ is primitive. Let d be its degree. In view of Proposition 11.6.4, $G(X)$ contains a d-cycle. By Theorem 11.6.7, three cases may arise. Either $G(X)$ is regular and then, by Theorem 11.3.2, X is a group code and by Theorem 11.3.1 the code X is uniform. Or $G(X)$ is a Frobenius group. By Theorem 11.6.5, we have $d \le 3$. The only group of a nonuniform code then is $\mathfrak{S}_3$, as shown in the remark. This group is both a Frobenius group and doubly transitive. Thus in any case, the group is doubly transitive. $\square$

In Theorem 11.6.8, the condition on X to be indecomposable is necessary. Indeed, otherwise by Theorem 11.4.7, the group $G(X)$ would be imprimitive. But it is known that a doubly transitive group is primitive (Proposition 1.13.6).

There is an interesting combinatorial interpretation of the fact that the group of a bifix code is doubly transitive.

Proposition 11.6.9 *Let X be a thin maximal bifix code over A, and let $P = XA^-$. The group $G(X)$ is doubly transitive if and only if for all $p, q \in P \setminus \{1\}$, there exist $x, y \in X^*$ such that $px = yq$.*

Proof. Let φ be the representation associated with the literal automaton $\mathcal{A} = (P, 1, 1)$ of X^*. Let $d = d(X)$, and let e be an idempotent of rank d in $\varphi(X^*)$. Let $S = \text{Fix}(e)$. We have $1 \in S$, since $S = \text{Im}(e)$.

Let $p, q \in S \setminus \{1\}$, and assume that there exist $x, y \in X^*$ such that $px = yq$. We have $1 \cdot p = p$ and $1 \cdot q = q$, whence

$$p \cdot x = 1 \cdot px = 1 \cdot yq = 1 \cdot q = q.$$

This shows that for the element $e\varphi(x)e \in G(e)$, we have $pe\varphi(x)e = q$. Since $1e\varphi(x)e = 1$, this shows that the restriction to S of $e\varphi(x)e$, which is in the stabilizer of 1, maps p on q. Thus this stabilizer is transitive, and consequently the group $G_e = G(X)$ is doubly transitive. Assume now conversely that $G(X)$ is doubly transitive, and let $p, q \in P \setminus 1$. Let $i, j \in S$ be such that $pe = i$, $qe = j$. Then $i, j \neq 1$. Consider indeed a word $w \in \varphi^{-1}(e)$. Then $1 \cdot w = 1$; the assumption $i = 1$ would imply that $p \cdot w = pe = i = 1$, and since $1 \cdot w = 1$, Proposition 11.4.1 gives $p = 1$, a contradiction. Since $G(X)$ is doubly transitive, and $G(X)$ is equivalent to G_e there exists $g \in G(e)$ such that $ig = j$ and $1g = 1$.

Let $m \in \varphi(A^*)$ be such that $jm = q$, and let f be the idempotent of the group $G(em)$. Since e and f are in the same $\mathcal{R}$-class, they have the same nuclear equivalence. Therefore the equalities $qe = j = je$ imply $qf = jf$. Further $\mathrm{Im}(f) = \mathrm{Im}(em)$. Since $qem = jm = q$, we have $q \in \mathrm{Im}(f)$. Consequently q is a fixed point of f, and $jf = qf = q$. Consider the function egf. Then

$$1egf = 1gf = 1f = 1, \quad pegf = igf = jf = q.$$

Let x be in $\varphi^{-1}(egf)$. Then $x \in X^*$ and $p \cdot x = q$. This holds in the literal automaton. Thus there exists $y \in X^*$ such that $px = yq$. $\qquad \square$

11.7 Examples

The results of Section 11.6 show that the groups of finite maximal bifix codes are particular ones. This of course holds only for finite codes since every transitive group appears as the group of some group code. We describe, in this section, examples of finite maximal bifix codes with particular groups.

Call a permutation group G *realizable* if there exists a finite maximal bifix code X such that $G(X) = G$. We start with an elementary property of permutation groups.

Lemma 11.7.1 *For any integer $d \geq 1$, the group generated by $\alpha = (12 \cdots d)$ and one transposition of adjacent elements modulo d is the whole symmetric group $\mathfrak{S}_d$.*

Proof. Let $\beta = (1d)$. Then for $j \in \{1, 2, \ldots, d-1\}$,

$$\alpha^{-j}\beta\alpha^j = (j, j+1). \tag{11.23}$$

Next for $1 \leq i < j \leq d$, $(i, j) = \tau(j-1, j)\tau^{-1}$, where $\tau = (i, i+1)(i+1, i+2) \cdots (j-2, j-1)$. This shows that the group generated by α and β contains all transpositions. Thus it is the symmetric group $\mathfrak{S}_d$. Formula (11.23) shows that the same conclusion holds if β is replaced by any transposition of adjacent elements. $\qquad \square$

Proposition 11.7.2 *For all $d \geq 1$, the symmetric group $\mathfrak{S}_d$ is realizable by a finite maximal bifix code.*

Proof. Let $A = \{a, b\}$. For $d = 1$ or 2, the code $X = A^d$ can be used. Assume $d \geq 3$. By Theorems 6.4.2 and 6.4.3, there exists a unique maximal bifix code X of degree d with kernel $K = \{ba\}$. Indeed, $\mu(K) = (L_K, ba) = 2$. Recall that μ is defined in

Chapter 6 by (6.40). No word has more than one K-interpretation. Consequently K is insufficient as defined in Section 6.5 and by Proposition 6.5.6, the code X is finite. Let us verify that

$$X \cap a^* b a^* = ba \cup \{a^i b a^{d-i} \mid 1 \le i \le d - 2\} \cup a^{d-1} b. \tag{11.24}$$

For each integer $j \in \{0, 1, \ldots, d - 1\}$, there is a unique integer $i \in \{0, 1, \ldots, d - 1\}$ such that $a^i b a^j \in X$. It suffices to verify that the integer i is determined by Formula (11.24). Let $i, j \in \{0, 1, \ldots, d - 1\}$ be such that $a^i b a^j \in X$. By Formula (6.5) in Chapter 6, the number of X-interpretations of $a^i b a^j$ is

$$(L_X, a^i b a^j) = 1 + |a^i b a^j| - (\underline{A}^* \underline{X} \underline{A}^*, a^i b a^j)$$
$$= i + j + 2 - (\underline{A}^* \underline{X} \underline{A}^*, a^i b a^j).$$

The number $(\underline{A}^* \underline{X} \underline{A}^*, a^i b a^j)$ of occurrences of words of X in $a^i b a^j$ is equal to 1 plus the number of occurrences of words of K in $a^i b a^j$, except when $j = 1$ which implies $i = 0$ since $ba \in X$. Thus

$$(L_X, a^i b a^j) = \begin{cases} i + j & \text{if } i \in \{1, 2, \ldots, d - 1\}, \\ i + j + 1 & \text{if } i = 0 \text{ or } j = 0. \end{cases}$$

On the other hand, the word $a^i b a^j$ must have d interpretations since it is not in $K = K(X)$. This proves Formula (11.24). Now consider the automaton $\mathcal{A}(X^*) = (Q, 1, 1)$ and consider the ergodic representation $*_a$ associated to the idempotent $\varphi(a^d)$ defined in Section 11.6. Setting $i = 1 \cdot a^{i-1}$ for $i \in \{1, 2, \ldots d\}$, we have

$$a *_a a^d = (12 \cdots d).$$

Set $\beta = b *_a a^d$ and observe that $\beta = (1d)$. Indeed by Formula (11.19),

$$i\beta = j \iff 1 \cdot a^{i-1} b a^d = 1 \cdot a^{j-1} a^d \iff 1 \cdot a^{i-1} b a^d = 1 \cdot a^{j-1}.$$

Thus $i\beta = 1 \cdot a^{i-1} b a^d$. For $i = 1$, this gives $1\beta = 1 \cdot b a a^{d-1}$, whence $1\beta = 1 \cdot a^{d-1} = d$. Next, by (11.24), for $i = d$, we have $d\beta = 1 \cdot a^{d-1} b a^d = 1 \cdot (a^{d-1} b) a^d = 1$. Finally, if $1 < i < d$, then $i\beta = 1 \cdot a^{i-1} b a^{d-(i-1)} a^{i-1} = 1 \cdot a^{i-1} = i$. This shows that the group $G(X)$ contains the cycle

$$\alpha = (12 \cdots d)$$

and the transposition $\beta = (1d)$. In view of Lemma 11.7.1, $G(X) = \mathfrak{S}_d$. $\qquad \square$

For the next result, we prove again an elementary property of permutations.

Lemma 11.7.3 *Let d be an odd integer. The group generated by the two permutations*

$$\alpha = (1, 2, \ldots, d) \quad \text{and} \quad \gamma = \delta \alpha \delta,$$

where δ is a transposition of adjacent elements modulo d, is the whole alternating group $\mathfrak{A}_d$.

Proof. The group $\mathfrak{A}_d$ consists of all permutations $\sigma \in \mathfrak{S}_d$ which are a product of an even number of transpositions. A cycle of length k is in $\mathfrak{A}_d$ if and only if k is odd. Since d is odd, $\alpha, \gamma \in \mathfrak{A}_d$.

By Lemma 11.7.1, the symmetric group is generated by α and δ. Each permutation $\sigma \in \mathfrak{S}_d$ can be written as

$$\sigma = \alpha^{k_1} \delta \alpha^{k_2} \delta \cdots \alpha^{k_{n-1}} \delta \alpha^{k_n}$$

and $\sigma \in \mathfrak{A}_d$ if n is odd. In this case, setting $n = 2m + 1$,

$$\sigma = \alpha^{k_1} \beta_2 \alpha^{k_3} \beta_4 \cdots \beta_{2m} \alpha^{k_{2m+1}}$$

with $\beta_{2i} = \delta \alpha^{k_{2i}} \delta$ for $1 \le i \le m$. Since $\beta_{2i} = (\delta \alpha \delta)^{k_{2i}}$, this formula shows that $\mathfrak{A}_d$ is generated by α and $\delta \alpha \delta = \gamma$. □

Proposition 11.7.4 *For each odd integer d, the alternating group $\mathfrak{A}_d$ is realizable by a finite maximal bifix code.*

Proof. Let $A = \{a, b\}$. For $d = 1$ or 3, the code $X = A^d$ can be used. Assume $d \ge 5$. Let

$$I = \{1, 2, \ldots, d\}, \qquad J = \{1, 2, \ldots, d - 3, \overline{d - 2}, \overline{d - 1}, \overline{d}\},$$

and $Q = I \cup J$. Consider the deterministic automaton $\mathcal{A} = (Q, 1, 1)$ with transitions given by

$$i \cdot a = i + 1 \quad (1 \le i \le d - 1), \qquad d \cdot a = 1,$$
$$\overline{d - 2} \cdot a = d - 1, \quad \overline{d - 1} \cdot a = 1, \qquad \overline{d} \cdot a = d,$$

and

$$i \cdot b = i + 1 \quad (1 \le i \le d - 3),$$
$$(d - 2) \cdot b = \overline{d}, \qquad (d - 1) \cdot b = \overline{d - 1}, \qquad d \cdot b = 1,$$
$$\overline{d - 2} \cdot b = \overline{d - 1}, \qquad \overline{d - 1} \cdot b = \overline{d}, \qquad \overline{d} \cdot b = 1.$$

Let X be the prefix code such that $\mathcal{A}$ recognizes X^*. Since

$$I \cdot a = J \cdot a = I, \quad I \cdot b = J \cdot b = J,$$

the functions $\varphi(a)$ and $\varphi(b)$, of rank d, have minimal rank. Since I and J are the only minimal images, and since they contain the state 1, Proposition 11.4.1(ii) shows that X is maximal bifix code. It has degree d.

Let us show that X is finite. For this, consider the following order on Q :

$$1 < 2 < \cdots < d - 1 \quad \text{and} \quad d - 2 < \overline{d - 2} < d - 1 < \overline{d - 1} < \overline{d} < d.$$

For all $c \in \{a, b\}$ and $q \in Q$, either $q \cdot c = 1$ or $q \cdot c > q$. Thus, there are only finitely many simple paths in $\mathcal{A}$. Consequently, X is finite.

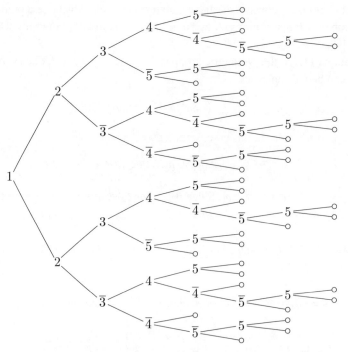

Figure 11.12   A finite maximal bifix code X with $G(X) = \mathfrak{A}_5$.

Now let us compute $G(X)$. Since $\varphi(a)$, $\varphi(b)$ have minimal rank, both $a, b \in J(X)$. According to Proposition 11.6.1, the group $G(X)$ is equivalent to the group generated by the four permutations

$$a *_a a, \quad a *_a b, \quad b *_a a, \quad b *_a b.$$

By Formula (11.6) we have $a *_a a = a *_a b = \alpha$, with $\alpha = (1, 2, \ldots, d)$. Next, by Formula (11.19)

$$b *_a a = \alpha, \quad b *_a b = \gamma$$

with $\gamma = (1, 2, \ldots, d - 3, d - 1, d - 2, d)$. In view of Lemma 11.7.3, $G(X) = \mathfrak{A}_d$.
□

Observe that for an even d, the group $\mathfrak{A}_d$ is not realizable. More generally, no subgroup of $\mathfrak{A}_d$ is realizable when d is even. Indeed, by Proposition 11.6.4, the group $G(X)$ of a finite maximal bifix code X contains a cycle of length d which is not in $\mathfrak{A}_d$ since d is even.

Example 11.7.5   We give, for $d = 5$, the figures of the automaton and of the code of the previous proof (Figures 11.12 and 11.13).

Example 11.7.6   For degree 5, the only realizable groups are $\mathbb{Z}/5\mathbb{Z}$, $\mathfrak{S}_5$, and $\mathfrak{A}_5$. It is known indeed that with the exception of these three groups, all transitive permutations groups of degree 5 are Frobenius groups. By Theorem 11.6.5, they are not realizable.

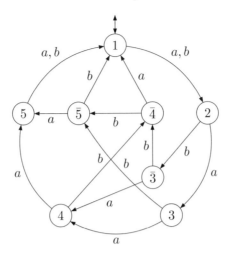

Figure 11.13 The automaton $\mathcal{A}(X^*)$.

Example 11.7.7 For degree 6, we already know, by the preceding propositions, that $\mathbb{Z}/6\mathbb{Z}$ and $\mathfrak{S}_6$ are realizable. We also know that no subgroup of $\mathfrak{A}_6$ is realizable. There exists, in addition to these two groups, another primitive group which is realizable. This group is denoted by $PGL_2(5)$ and is defined as follows. Let $P = \mathbb{Z}/5\mathbb{Z} \cup \infty$. The group $PGL_2(5)$ is the group of all homographies from P into P

$$p \mapsto \frac{xp + y}{zp + t}$$

for $x, y, z, t \in \mathbb{Z}/5\mathbb{Z}$ satisfying $xt - yz \neq 0$. Consider, for later use, the permutations

$$h = (\infty 01423), \quad k = (\infty 10243).$$

We have $h, k \in PGL_2(5)$. Indeed h and k are the homographies

$$h : p \mapsto \frac{2}{p+2}, \quad k : p \mapsto \frac{p-1}{p+2}$$

respectively. We verify now that h and k generate all $PGL_2(5)$. A straightforward computation gives

$$k^2 hk = (\infty 0421)(3), \quad k^2 hkh^{-1} = (\infty)(4)(0132).$$

The permutation h together with these two permutations show that the group G generated by h and k is 3-transitive. Now each element σ in $PGL_2(5)$ is characterized, as any homography, by its values on three points. Since G is 3-transitive, there exists an element $g \in G$ which takes the same three values on the points considered. Thus $\sigma = g$, whence $\sigma \in G$. This proves that $G = PGL_2(5)$.

To show that $PGL_2(5)$ is realizable, we consider the automaton $\mathcal{A} = (Q, 1, 1)$ given in Table 11.5. This automaton is minimal. Let X be the maximal prefix code

Table 11.5 *The transitions of the automaton* $\mathcal{A}$.

	1	2	3	4	5	6	$\bar{2}$	$\bar{3}$	$\bar{4}$	$\bar{5}$	$\bar{6}$
a	2	3	4	5	6	1	3	5	4	1	6
b	$\bar{2}$	$\bar{4}$	$\bar{3}$	$\bar{6}$	$\bar{5}$	1	3	$\bar{4}$	5	$\bar{6}$	1

Table 11.6 *The bijection* ρ.

1	2	3	4	5	6
∞	0	1	4	2	3

such that $\mathcal{A} = \mathcal{A}(X^*)$. Then X is a finite maximal bifix code. Indeed, the images

$$\mathrm{Im}(a) = \{1, 2, 3, 4, 5, 6\}, \quad \mathrm{Im}(b) = \{1, \bar{2}, \bar{3}, \bar{4}, \bar{5}, \bar{6}\}$$

are minimal images, containing both the state 1. By Proposition 11.4.1(ii), X is maximal bifix with degree 6. The code X is finite because if Q is ordered by

$$1 < 2 < \bar{2} < 3 < \bar{3} < \bar{4} < 4 < 5 < \bar{5} < \bar{6} < 6,$$

then the vertices on simple paths from 1 to 1 are met in strictly increasing order, with the exception of the last one. Next $a, b \in J(X)$, because of the minimality of the images $\mathrm{Im}(a)$, $\mathrm{Im}(b)$. Thus the group $G(X)$ is generated by the permutations

$$\alpha = a *_a a = a *_a b = (123456), \quad \beta = b *_a a, \quad \gamma = b *_a b.$$

Formula (11.19) shows that $\beta = \alpha$, $\gamma = (132546)$. This shows that $G(X)$ is generated by α and β. Let ρ be the bijection from $1, 2, 3, 4, 5, 6$ onto $P = \mathbb{Z}/5\mathbb{Z} \cup \infty$ given in Table 11.6. Then $h = \rho^{-1}\alpha\rho$ and $k = \rho^{-1}\gamma\rho$ where h, k are the generators of $PGL_2(5)$ defined previously. Consequently, the groups $G(X)$ and $PGL_2(5)$ are equivalent.

11.8 Exercises

Section 11.1

11.1.1 Let $X \subset A^+$ be a maximal prefix code. Let

$$R = \{r \in A^* \mid \forall x \in X^*, \exists y \in X^* : rxy \in X^*\}.$$

(a) Show that R is a right unitary submonoid containing X^*.

(b) Let Z be the maximal prefix code such that $R = Z^*$ and set $X = Y \circ Z$. Show that if X is thin, then Y is synchronized.

(c) Show that if $X = Y' \circ Z'$ with Y' synchronized, then $Z'^* \subset Z^*$.

(d) Suppose that X is thin. Let $\mathcal{A} = (Q, 1, 1)$ be a deterministic trim automaton recognizing X^* and let φ be the associated representation. Show that a word $r \in A^*$

is in R if and only if for all $m \in \varphi(A^*)$ with minimal rank, $1 \cdot r \equiv 1 \bmod \mathrm{Ker}(m)$. (*Hint*: Restrict to the case where $m \in \varphi(X^*)$.)

Section 11.3

11.3.1 Let $X \subset A^+$ be a finite code and let $\mathcal{A} = (Q, 1, 1)$ be an unambiguous trim automaton recognizing X^*. Show that the group of invertible elements of the monoid $\varphi_A(A^*)$ is a cyclic group.

11.3.2 Show that for every finite transitive permutation group G, there exists a finite bifix code X such that G is equivalent to G_e for some idempotent e in the transition monoid of the minimal automaton of X^*.

(*Hint*: Let G be a transitive group of permutations on a set and let H be the subgroup fixing some point of the set. Let $\psi : A^* \to G$ be a surjective morphism and let Z be the group code defined by $Z^* = \psi^{-1}(H)$. Since Z is recognizable, it is thin by Proposition 2.5.20. Let Y be a finite set of words in $\bar{F}(X)$ such that $\psi(Y)$ generates G. Show that the set $X = Z \cap F(Y^*)$ is a finite bifix code with the required property.)

Section 11.4

11.4.1 Let $X \subset A^+$ be a bifix code and let $\mathcal{A} = (Q, 1, 1)$ be a trim deterministic automaton recognizing X^*. Let $\varphi = \varphi_A$ be the associated representation and $M = \varphi(A^*)$. Show that for any idempotent $e \in \varphi(\bar{F}(X))$, the monoid of partial functions M_e is composed of injective functions.

Section 11.5

11.5.1 Let $X \subset A^+$ be a thin maximal bifix code, and let J_D be the minimal ideal of $\varphi_D(A^*)$. Show that for $\mathrm{Card}(A) \geq 2$,

$$\bar{H}(X) = \varphi_D^{-1}(J_D),$$

where $H(X) = A^- X A^-$ and $\bar{H}(X)$ is the complement of $H(X)$. (*Hint*: Use Exercise 9.3.5.)

11.5.2 Let $X \subset A^+$ be a finite maximal prefix code, let $\mathcal{A}(X^*) = (Q, 1, 1)$ be the minimal automaton of X^*, set $\varphi = \varphi_{A(X^*)}$. Let $a \in A$ and let n be the order of a in X ($a^n \in X$).

(a) Show that the idempotent in $\varphi(a^+)$ has rank n.
(b) Show, without using Theorem 11.5.2, that $\varphi(A^+)$ is not nil-simple when $n \geq 1 + d(X)$.

11.5.3 Let $X \subset A^+$ be a thin complete code. Then X is called *elementary* if there exists an unambiguous trim automaton $\mathcal{A} = (Q, 1, 1)$ recognizing X^* such that the semigroup $S = \varphi_A(A^+)$ has depth 1. Show that if X is elementary, then $X = Y \circ Z$, where Y is an elementary bifix code and $G(X) = G(Y)$. (*Hint*: Choose for Z the code generating the set of words which have a power in X^*.)

11.5.4 Let $A = (Q, 1, 1)$ be a complete, deterministic trim automaton and let φ be the associated representation. Suppose that $\varphi(A^+)$ has finite depth, and that $\varphi(A^+)$ has minimal rank 1. Show that the depth of $\varphi(A^+)$ is at most $\mathrm{Card}(Q) - 1$. (*Hint:* Consider the sequence θ_i of equivalence relations over Q defined by $p \equiv q \bmod \theta_i$ if and only if $p \cdot w = q \cdot w$ for all $w \in A^i$.)

11.5.5 Let $X \subset A^+$ be a finite bifix code. Let φ be the representation associated with $A(X^*)$ and $M = \varphi(A^*)$. Let J be the minimal ideal of M and let Λ be the set of its $\mathcal{L}$-classes. Let L_0 be a distinguished $\mathcal{L}$-class in Λ. Define a deterministic automaton $B = (\Lambda, L_0, L_0)$ by setting $L \cdot w = L\varphi(w)$. Let ψ be the representation associated with B, and let I be the minimal ideal of $\psi(A^*)$.

(a) Show that $\psi(A^*)$ has minimal rank 1, and that $\psi^{-1}(I) = \varphi^{-1}(J)$.

(b) Use Exercise 11.5.4 to show that $\varphi(A^+)$ has depth at most $\mathrm{Card}(\Lambda) - 1$.

Section 11.6

11.6.1 Let X be a finite maximal bifix code of degree d. Let $a \in A$ and $k \geq 0$ such that $a^k \in J(X)$. Show that for each integer $n \leq d - k$ and each word $u \in A^n$, there exist at least $d - k - n$ integers i, with $1 \leq i \leq d$ such that

$$i(u *_a a^k) \geq i - k + 1.$$

11.6.2 Derive directly Theorem 11.3.1 from Exercise 11.6.1 (take $k = 0$).

11.6.3 Derive the inequalities (11.20) from Exercise 11.6.1 (take $k = 1$, $u = b$).

11.6.4 Let G be a permutation group of degree d, let $k = k(G)$ and suppose that G contains the cycle $\alpha = (12 \cdots d)$. Show that if $d \geq 4k^2 + 8k + 2$, then every $\pi \in G$ which is not a power of α has at most $d - 2k - 2$ excedances (an *excedance* of a permutation π of $\{1, 2, \ldots, d\}$ is a value i such that $i\pi > i$).

11.6.5 Let X be a finite maximal bifix code of degree d and let $k = k(X)$. Assume that $d \geq 4k^2 + 8k + 2$.

(a) Show that for each $a \in A$ and $w \in A^k$, the permutation $\pi = w *_a a^d$ is in the subgroup generated by $\alpha = a *_a a^d$. (*Hint:* Use Theorem 11.4.3 to show that the permutation $\pi\alpha^k$ has at least $d - 2k$ excedances, and use Exercise 11.6.4.)

(b) Show that X does not contain words of length less than or equal to k.

11.6.6 Derive from Exercises 11.6.1 and 11.6.5 that a finite maximal nonuniform bifix code X of degree d satisfies $k(X) \geq (\sqrt{d}/2) - 1$.

Section 11.7

11.7.1 Let X be an elementary finite maximal bifix code of degree d on the alphabet $A = \{a, b\}$. Let $\alpha = (1, 2, \ldots, d)$, $\beta = b *_a a$, $\gamma = b *_a b$ with the usual convention to write i for $1 \cdot a^{i-1}$ for $1 \leq i \leq d$. Show that β and γ are such that

(i) $1\beta^{-1} = 1\gamma^{-1}$,

(ii) $\beta = (i_1, \ldots, i_k)$ with $1 = i_1 < \cdots < i_k$,

Table 11.7 *A finite code with group* $GL_3(2)$.

	1	2	3	4	5	6	7	8	9	10	11	12	13	14	15
a	2	3	4	5	6	7	1	9	4	14	15	13	1	6	7
b	8	9	12	11	10	1	13	9	10	11	12	13	1	12	13

Table 11.8 *A finite code with group* M_{11}.

	1	2	3	4	5	6	7	8	9	10	11	12	13	14
a	2	3	4	5	6	7	8	9	10	11	1	22	23	24
b	12	13	16	17	14	15	20	19	18	1	21	13	14	15

15	16	17	18	19	20	21	22	23	24	25	26	27	28	29
25	26	27	28	29	21	1	4	5	6	7	8	9	10	11
16	17	18	19	20	21	1	14	15	16	17	18	19	20	21

(iii) $\gamma = \tau^{-1}\alpha\tau$ where τ is a product of cycles of the form $(k, k+1, \ldots, k+m)$ with $k\beta \geq k+m$ or $k\beta = 1$.

Show that conversely, any choice of β and γ satisfying the above conditions defines a finite code.

11.7.2 Use Exercise 11.7.1 to show that for $A = \{a, b\}$, there are exactly six elementary finite maximal bifix codes over A with group equivalent to $PGL_2(5)$.

11.7.3 Show that the automaton in Table 11.7 defines a finite maximal bifix code X of degree 7. Show that $G(X)$ is equivalent to the group $GL_3(2)$ of invertible 3×3 matrices with elements in $\mathbb{Z}/2\mathbb{Z}$, considered as a permutation group acting on $(\mathbb{Z}/2\mathbb{Z})^3 \setminus 0$. (*Hint*: Identify $(\mathbb{Z}/2\mathbb{Z})^3$ with $\{1, 2, 3, 4, 5, 6, 7\}$ using the remainders of x^i modulo $1 + x + x^3$.)

11.7.4 Show that the automaton in Table 11.8 defines a finite maximal bifix code X of degree 11. Show that $G(X)$ is equivalent to the Mathieu group M_{11}.

11.9 Notes

Proposition 11.1.5 is due to Perrot (1972). The theorem on the synchronization of semaphore codes (Theorem 11.2.1) is in Schützenberger (1964). This paper contains also a difficult combinatorial proof of this result.

Theorem 11.3.1 already appears in Schützenberger (1956). Theorem 11.3.3 and Corollary 11.3.4 are from Reis and Thierrin (1979). The ergodic representation of Section 11.6 is described in Perrin (1979). It is used in Lallement and Perrin (1981) to describe a construction of finite maximal bifix codes. Theorem 11.6.7 is a combination of a theorem of Schur and of a theorem of Burnside. Schur's theorem is the following: "Let G be a primitive permutation group of degree d. If G contains a d-cycle and if d is not a prime number, then G is doubly transitive." This result is proved in Wielandt (1964), pp. 52–66. It is the final development of what H. Wielandt calls the "method of Schur". Burnside's theorem is the following: "A transitive permutation group of prime degree is either doubly transitive or a Frobenius group." Burnside's proof uses the theory of characters. It is reproduced in Huppert (1967), p. 609. An elementary proof (that is, without characters) is in Huppert and Blackburn (1982), Vol. III, pp. 425–434.

The other results of this chapter are from Perrin (1975), Perrin (1977b), Perrin (1978). Perrin (1975) gives a more exhaustive catalog of examples than the list of Section 5. Exercise 11.3.2 is from Perrin (1981) (see also Rindone (1983) and Perrin and Rindone (2003)). Exercise 11.4.1 is due to Margolis (1982). Exercise 11.5.4 is a well-known property of "definite" automata (Perles *et al.* (1963)).

The exercises of Section 11.6 are from Perrin (1978) and those of Section 11.7 are from Perrin (1975). The definition of the Mathieu group M_{11} used in the solution of Exercise 11.7.4 is from Conway (1971). It is a sharply 4-transitive group of order $11 \times 10 \times 9 \times 8$. The set H is known as the ternary *Golay code*.

Excedances of permutations are a well-known notion in combinatorics (see Lothaire (1997)). The result of Exercise 11.6.4 has been improved by Mantaci (1991). He proved the following result. Let d, k, ℓ be integers such that $d > 2k\ell - k$ and let G be a permutation group of degree d and minimal degree $d - k$, containing the cycle $\alpha = (12 \cdots d)$. Every permutation in $G \setminus \langle \alpha \rangle$ has at most $d - \ell - 1$ excedances. He also shows that the bound is the best possible. His result implies the statement of Exercise 11.6.4 taking $\ell = 2k + 1$.

12

Factorizations of cyclic groups

In this chapter we describe the links between codes and factorizations of cyclic groups. It happens that for any finite maximal code X one can associate with each letter a several factorizations of the cyclic group $\mathbb{Z}/n\mathbb{Z}$ where n is the integer such that a^n is in the code X. These factorizations play a role in several places in the theory of codes. They have appeared several times previously in this book. This chapter gives a systematic presentation.

We begin with an introduction to the notion of factorizations of cyclic groups (Section 12.1). We then study how factorizations arise in connection with two special kinds of words: bayonets (Section 12.2) and hooks (Section 12.3). We will see that factorizations of cyclic groups give insight into several properties of codes, like being synchronized or being finitely completable.

12.1 Factorizations of cyclic groups

Let G be a group written additively. Given two subsets L, R of G, we write $L + R = \{\ell + r \mid \ell \in L, r \in R\}$. The sum $L + R$ is *direct* if for any element g in G, there exists at most one pair (ℓ, r) with $\ell \in L$ and $r \in R$ such that $g = \ell + r$. This means that for finite sets L, R, the sum is direct if and only if $\mathrm{Card}(L + R) = \mathrm{Card}(L)\,\mathrm{Card}(R)$. The pair (L, R) is called a *factorization* if $G = L + R$ and the sum is direct. We also say that $G = L + R$ is a factorization of G.

Example 12.1.1 Let $G = \mathbb{Z}/6\mathbb{Z}$. The pair (L, R) defined by $L = \{0, 5\}$ and $R = \{0, 2, 4\}$ is a factorization of G. More generally, if R is a subgroup of some Abelian group G and L is a set representatives of the quotient G/R, then (L, R) is a factorization.

The following example illustrates how the coset decomposition may be iterated to form more complex factorizations.

Example 12.1.2 The pair (L, R) defined by $L = \{0, 4, 8, 9, 13, 17\}$ and $R = \{0, 3, 6\}$ is a factorization of $\mathbb{Z}/18\mathbb{Z}$. We have actually $L = \{0, 9\} + \{0, 4, 8\}$. Thus $\{0, 4, 8\} + R$ is a system of representatives of the residues modulo 9. Accordingly, $\mathbb{Z}/9\mathbb{Z} = \{0, 4, 8\} + \{0, 3, 6\}$ is a factorization.

Example 12.1.3 Let p, q be positive integers and let $L = \{0, 1\}$ and $R = \{0, p, q\}$ with $p < q$. The sum $L + R$ is direct in $\mathbb{Z}$ if and only if $1 < p < q - 1$.

We shall be interested in factorizations of Abelian and, more specifically of cyclic groups. Let $G = \mathbb{Z}/n\mathbb{Z}$, let L, R be two subsets of G, and let $U, V \subset \mathbb{Z}$ be sets of representatives of L, R. Then $G = L + R$ is a factorization if and only if for each integer k there exists a unique pair i, j with $i \in U$ and $j \in V$ such that $k \equiv i + j$ mod n.

Example 12.1.4 Let $L = \{0, 3, 8, 11\}$ and $R = \{0, 1, 7, 13, 14\}$. Since the numbers $\ell + r$ are all distinct, the sum $L + R$ is direct in $\mathbb{Z}$ or in $\mathbb{Z}/n\mathbb{Z}$ for large enough n. The pair (L, R) is not a factorization of $\mathbb{Z}/20\mathbb{Z}$ because $8 + 13 \equiv 0 + 1 \equiv 1 \mod 20$ and so the sum is not direct. It is not known whether there exists an integer n and sets L', R' such that $\mathbb{Z}/n\mathbb{Z} = L' + R'$ is a factorization with $R \subset R'$ and $L \subset L'$. See also Example 12.3.5.

The following statement gives a useful method to handle factorizations.

Proposition 12.1.5 *Let $G = L + R$ be a factorization of a finite Abelian group G. For any integer $q \in \mathbb{Z}$ prime to* $\mathrm{Card}(L)$, *$G = qL + R$ is a factorization.*

Proof. We may assume that $0 \in L$, since otherwise we replace L by $L' = L - \ell$ for some $\ell \in L$. If $G = qL' + R$ is a factorization, then so is $(qL' + q\ell) + R = qL + R$.

Consider first the case where $q = -1$. We clearly have $\mathrm{Card}(qL) = \mathrm{Card}(L)$ and we only need to prove that the sum $G = (-L) + R$ is direct. Suppose that $-\ell + r = -\ell' + r'$ with $\ell, \ell' \in L$ and $r, r' \in R$. Then $\ell' + r = \ell + r'$ and thus $r = r', \ell = \ell'$. This proves the result in this case.

Suppose next that $q \geq 1$ is prime. For $g = \ell + r$ with $\ell \in L$ and $r \in R$, we denote $\lambda(g) = \ell$ and $\rho(g) = r$.

As a first step, let us prove that for any $g \in G$, the map $\ell \mapsto \lambda(g + \ell)$ is a permutation of L. For this, let $\ell, \ell' \in L$ and assume $\lambda(g + \ell) = \lambda(g + \ell')$. Set $g + \ell = u + v$ and $g + \ell' = u + v'$ with $u \in L$ and $v, v' \in R$. Then $v - \ell = v' - \ell'$ and thus $\ell = \ell'$ since we have just shown that $R - L$ is a factorization.

We claim that for $g \in G$, there is an $x \in L$ such that $g = -qx + r$ for some $r \in R$. To prove this claim, consider the set T of q-tuples $(x_1, \ldots, x_q)$ of elements in L such that $\lambda(g + x_1 + \cdots + x_q) = 0$. For each choice of $x_1, \ldots, x_{q-1}$ in L the map $\ell \mapsto \lambda(g + x_1 + \cdots + x_{q-1} + \ell)$ is a permutation of L. Thus there is a unique $x_q \in L$ such that $(x_1, \ldots, x_q) \in T$. Consequently T has $\mathrm{Card}(L)^{q-1}$ elements. Since q is prime, and q does not divide $\mathrm{Card}(L)$ we obtain that $\mathrm{Card}(T) = \mathrm{Card}(L)^{q-1} \equiv 1$ mod q. The set T contains all cyclic shifts of its elements. Since q is prime, the number of distinct cyclic shifts of an element of T is either q or 1. Since $\mathrm{Card}(T) \equiv 1$ mod q there is at least one $t \in T$ such that all its cyclic shifts are equal, that is such that $t = (x, x, \ldots, x)$ for some $x \in L$. Since $\lambda(g + qx) = 0$, we have $g + qx = \rho(g + qx)$ and therefore $g = -qx + \rho(g + qx)$.

This shows that $G = (-qL) + R$. Since $\mathrm{Card}(-qL) \leq \mathrm{Card}(L)$, the sum is direct and thus $(-qL, R)$ is a factorization. By what we have seen above, this implies that $G = qL + R$ is also a factorization.

Table 12.1 *A non periodic factorization of* $\mathbb{Z}/72\mathbb{Z}$.

0	1	5	6	12	25	29	36	42	48	49	53
8	9	13	14	20	33	37	44	50	56	57	61
16	17	21	22	28	41	45	52	58	64	65	69
18	19	23	24	30	43	47	54	60	66	67	71
26	27	31	32	38	51	55	62	68	**2**	**3**	**7**
34	35	39	40	46	59	63	70	**4**	**10**	**11**	**15**

Finally, when $q \geq 1$ is prime to $\mathrm{Card}(L)$, we write q as a product of primes and apply iteratively the above argument. □

Example 12.1.6 When we start with the factorization $L = \{0, 4, 8, 9, 13, 17\}$ and $R = \{0, 3, 6\}$ of $\mathbb{Z}/18\mathbb{Z}$ given in Example 12.1.2, we obtain, for $q = 5$, the new factorization given by $5L = \{0, 2, 4, 9, 11, 13\}$ and R.

A subset H of a group G is said to be *periodic* if there is an element $g \in G \setminus \{e\}$ such that $g + H = H$. We refer to such elements g as *periods* of H. A factorization (L, R) of a group G is called periodic if L or R is periodic.

Example 12.1.7 The pair (M, S) defined by the two sets $M = \{0, 4, 8, 9, 13, 17\}$ and $S = \{0, 3, 6, 18, 21, 24\}$ is a periodic factorization of $\mathbb{Z}/36\mathbb{Z}$. Indeed, 18 is a period of the set S.

A group G is said to have the *Hajós property* if any factorization of G is periodic. The integer n is said to be a *Hajós number* if the group $\mathbb{Z}/n\mathbb{Z}$ has the Hajós property. If n is a Hajós number, then any divisor of n is (see Exercise 12.1.1). The following example shows that 72 is not a Hajós number.

Example 12.1.8 The pair (L, R) defined by $L = \{0, 8, 16, 18, 26, 34\}$ and $R = \{0, 1, 5, 6, 12, 25, 29, 36, 42, 48, 49, 53\}$ is a factorization of $\mathbb{Z}/72\mathbb{Z}$ which is not periodic.

One may verify that it is indeed a factorization by inspection of Table 12.1 in which R is the first row, L the first column and each entry is the sum of the elements in the first row and column (the elements appearing in boldface are those for which the sum exceeds 72). Alternatively, we may proceed as follows. Let $R_0 = \{0, 6, 12, 36, 42, 48\}$ and $R_1 = \{1, 5, 25, 29, 49, 53\}$ be the sets of even and odd elements of R. Let $M = \{0, 4, 8, 9, 13, 17\}$, $S = \{0, 3, 6, 18, 21, 24\}$ and $T = \{0, 2, 12, 14, 24, 26\}$. Then $L = 2M$, $R_0 = 2S$ and $R_1 = 2T + 1$. The pairs (M, S) and (M, T) are periodic factorizations of $\mathbb{Z}/36\mathbb{Z}$ (actually, (M, S) is the factorization of Example 12.1.7). Then $L + R = 2M + (2S \cup (2T + 1)) = 2(M + S) \cup (2(M + T) + 1)$ and thus (L, R) is a factorization.

See the Notes for a characterization of the Hajós integers. A group G is said to have the *Rédei property* if for any factorization $G = L + R$, either $\langle L \rangle \neq G$ or $\langle R \rangle \neq G$.

Table 12.2 *The sets L, H and R.*

		L		
0	36	72	108	144
100	136	172	208	244
200	236	272	308	344
225	261	297	333	369
325	361	397	433	469
425	461	497	533	569

		H		
0	180	360	540	{720}
(150)	330	510	690	870
300	480	660	840	120
(450)	630	810	90	{270}
[600]	[780]	[60]	[240]	[420]
(750)	30	210	390	570

		R		
0	180	360	540	{45}
(250)	330	510	690	870
300	480	660	840	120
(550)	630	810	90	{495}
[636]	[816]	[96]	[276]	[456]
(850)	30	210	390	570

(We denote by $\langle H \rangle$ the subgroup of G generated by H.) An integer n is called a *Rédei number* if the group $\mathbb{Z}/n\mathbb{Z}$ has the Rédei property.

It can be shown that a Hajós number is a Rédei number (see Exercise 12.1.2).

Example 12.1.9 Let $\mathbb{Z}/72\mathbb{Z} = L + R$ be the factorization of Example 12.1.8. Since all elements of L are even, the group $\langle L \rangle$ is contained in the subgroup of index 2 formed by the even residues modulo 72. Actually, 72 is a Rédei number (see the Notes section).

The following example shows that 900 is not a Rédei number.

Example 12.1.10 Let $n = 900$ and let L, H, R be the subsets of $G = \mathbb{Z}/900\mathbb{Z}$ listed in Table 12.2. We will show that $G = L + R$ is a factorization and that $\langle L \rangle = \langle R \rangle = G$. Let $x_1 = 225$, $x_2 = 100$ and $x_3 = 36$, which are elements of G of order 4, 9 and 25 respectively. The orders of x_1, x_2, x_3 are pairwise relatively prime with a product equal to 900. Thus $G = \langle x_1 \rangle + \langle x_2 \rangle + \langle x_3 \rangle$.

Let $L_1 = \{0, x_1\}$, $L_2 = \{0, x_2, 2x_2\}$, $L_3 = \{0, x_3, \ldots, 4x_3\}$ and $H_1 = \langle 2x_1 \rangle$, $H_2 = \langle 3x_2 \rangle$, $H_3 = \langle 5x_3 \rangle$. We have

$$L = L_1 + L_2 + L_3, \quad H = H_1 + H_2 + H_3.$$

Indeed, the first row of the array giving L in Table 12.2 is L_3, the first three rows form $L_3 + L_2$ and the last three rows form $L_3 + L_2 + x_1$. The first row of the second array is H_3, the rows 1, 3 and 5 form $H_2 + H_3$ and the other ones are obtained by adding $2x_1 = 450$.

Clearly, $G = L + H$ is a factorization. We now modify the set H as follows to obtain the set R in such a way that $x_1, x_2, x_3 \in \langle R \rangle$. We first add $x_2 = 100$ to each element of $H_2 + 2x_1$ (the corresponding elements are marked by ( ) in H and R). In this way, the set H' obtained is still such that $G = L + H'$. Indeed, we have $L + H_2 + 2x_1 + x_2 = L_1 + L_3 + \langle x_2 \rangle + 2x_1 = L + H_2 + 2x_1$. In a second step, we add $x_3 = 36$ to each element of $H_3 + 6x_2$ (the corresponding elements are marked [ ]). The set H'' obtained still satisfies $G = L + H''$ for a similar reason as previously. Finally, the set R is obtained by adding $x_1 = 225$ to each element of $H_1 + 20x_3$ (the elements are marked with { }).

The factorization $G = L + R$ is such that $\langle L \rangle = G$ and $\langle R \rangle = G$. The first equality follows from the fact that $x_1, x_2, x_3 \in L$. The second one can be verified as follows. Since $5x_3, 3x_2$ are in R (they already belong to H and have not been modified), we have $20x_3, 6x_2 \in \langle R \rangle$. Since, by construction of R, $20x_3 + x_1 \in R$, we have $x_1 \in \langle R \rangle$. Similarly, since $6x_2 + x_3 \in R$, we have $x_3 \in \langle R \rangle$. Finally, since $2x_1 + x_2$ is in R by construction, we have also $x_2 \in \langle R \rangle$. Thus $x_1, x_2, x_3 \in \langle R \rangle$ and $\langle R \rangle = G$.

12.2 Bayonets

In this section, we will see that, under appropriate hypotheses, given a code $X \subset A^+$ and a letter $a \in A$, the integers i, j such that $a^i w a^j \in X^*$ for $a \in A$ and $w \in A^*$ give rise to some factorizations of cyclic groups. We begin with the case of $w = b \in A$. A *bayonet* is a word of the form $a^\ell b a^r$ for $a, b \in A$.

We say that a pair (L, R) of sets of integers is *direct* modulo n if $\ell + r \equiv \ell' + r'$ mod n, with $\ell, \ell' \in L, r, r' \in R$ implies $\ell = \ell'$ and $r = r'$. In other words, (L, R) is direct if for any integer m there is at most one pair $(\ell, r) \in L \times R$ such that $m \equiv \ell + r$ mod n. This is equivalent to saying that (L, R) is direct modulo n if and only if the sum $\bar{L} + \bar{R}$ formed with the sets of residues modulo n of L, R is direct.

Observe that if (L, R) is direct modulo n and L, R are both nonempty, then the elements of L (and of R) are distinct representatives of classes of integers modulo n.

Given a word w and a subset H of $\mathbb{N}$, we write w^H for the set $\{w^h \mid h \in H\}$.

Proposition 12.2.1 *For $L, R \subset \mathbb{N}$ and $n \geq 1$, the set $X = a^n \cup a^L b a^R$ is a code on the alphabet $A = \{a, b\}$ if and only if (L, R) is direct modulo n. Moreover, the code X is maximal if and only if $L + R = \{0, \ldots, n - 1\}$.*

Proof. If (L, R) is direct modulo n, then X is a code. Consider indeed a word w in X^*. We prove that w has a unique decomposition into words in X. Set $w = a^{m_0} b a^{m_1} b \cdots b a^{m_k}$ for nonnegative integers $m_0, \ldots, m_k$. If $k = 0$, the word w is a unique power of a^n. So assume $k \geq 1$. For each i with $0 < i < k$ there is a unique pair $(r_i, \ell_{i+1}) \in R \times L$ such that $m_i \equiv r_i + \ell_{i+1}$ mod n. Moreover, there is a unique $\ell_1 \in L$ and a unique $r_k \in R$ such that $\ell_1 \equiv m_0$ mod n and $r_k \equiv m_k$ mod n. Thus the unique factorization of w is of the form $w = y_0 x_1 y_1 \cdots x_k y_k$ with $x_i = a^{\ell_i} b a^{r_i}$, and $y_i \in (a^n)^*$.

Conversely, assume that X is a code. In order to show that (L, R) is direct modulo n, let $\ell, \ell' \in L, r, r' \in R$ such that $\ell + r \equiv \ell' + r'$ mod n. There exist an integer k such that $\ell + r = \ell' + r' + kn$. By symmetry, we may assume $k \geq 0$. Then $(a^\ell b a^r)(a^\ell b a^r)$ and $(a^\ell b a^{r'})(a^n)^k (a^{\ell'} b a^r)$ are two factorizations of the word $a^\ell b a^{r+\ell} b a^r$. Since X is a code, this implies $k = 0$, $\ell = \ell'$ and $r = r'$.

Finally, let π be a Bernoulli distribution on A^* and set $p = \pi(a)$. Then $\pi(X) = p^n + (1 - p)(\sum_{\ell \in L} p^\ell)(\sum_{r \in R} p^r)$. Thus $\pi(X) = 1$ if and only if $\sum_{\ell \in L} p^\ell \sum_{r \in R} p^r = 1 + p + \cdots + p^{n-1}$, and this holds if and only if $L + R = \{0, \ldots, n - 1\}$. $\qquad\square$

The pairs (L, R) such that (L, R) is direct modulo n and $L + R = \{0, \ldots, n - 1\}$ are precisely the pairs such that every integer in $\{0, \ldots, n - 1\}$ has exactly one decomposition of the form $\ell + r$ with $\ell \in L, r \in R$. These pairs define particularly simple factorizations which are described in Exercise 12.2.2.

Example 12.2.2 For $n = 6$, the pair composed of $L = \{0, 1\}$ and $R = \{0, 3, 5\}$ is direct modulo n. The set $X = a^n \cup a^L ba^R$ is $\{a^6, b, ab, ba^3, aba^3, ba^5, aba^5\}$.

If X is an arbitrary finite maximal code on $A = \{a, b\}$, the set of bayonets contained in X does not necessarily have the form described above since the set of pairs (ℓ, r) such that $a^\ell ba^r \in X$ for some $a, b \in A$ needs not even be a Cartesian product.

Let X be a code and a be a letter such that $a^n \in X$ for some integer $n \geq 1$. For a word w, we denote by $C_a(w)$ the pairs of residues modulo n of integers $i, j \geq 0$ such that $a^i wa^j \in X^*$. In what follows, we denote by $\bar{k}$ the residue of k modulo n.

Recall that, given a finite maximal code X, the *order* of a letter a is the integer $n \geq 1$ such that $a^n \in X$. The order exists for each letter.

We start with a useful observation.

Lemma 12.2.3 *Let X be a finite maximal code over A, and let $a \in A$ be a letter. For any $w \in A^*$, one has $a^* wa^* \cap X^* \neq \emptyset$.*

Proof. Since X is finite and maximal, it is complete. Let ℓ be the maximal length of a word in X. The word $a^\ell wa^\ell$ is completable, thus $ua^\ell wa^\ell v \in X^*$ for some words u, v. By the definition of ℓ, there exist integers i, i', j, j' such that $ua^{i'}, a^i wa^j, a^{j'} v \in X^*$. $\qquad\square$

Proposition 12.2.4 *Let X be a finite maximal code on the alphabet A. Let $a \in A$ be a letter and let n be the order of a. For each word $w \in A^*$, the set $C_a(w)$ has exactly n elements.*

Proof. Let ℓ be the maximal length of the words of X and let $kn \geq 2\ell$. For each r with $0 \leq r < n$, we show that there is a bijection from the set $C_a(wa^{r+kn}w)$ onto the set of pairs of elements in $C_a(w)$ of the form $(i, p), (q, j)$ with $p + q \equiv r$ modulo n.

In a first step, we show that for each $(\bar{\imath}, \bar{\jmath}) \in C_a(wa^{r+kn}w)$ there is a well-defined pair $(\bar{p}, \bar{q})$ of residues modulo n such that $(\bar{\imath}, \bar{p}), (\bar{q}, \bar{\jmath}) \in C_a(w)$ and $\bar{p} + \bar{q} = \bar{r}$.

Indeed, consider a pair (i, j) of representatives of $(\bar{\imath}, \bar{\jmath}) \in C_a(wa^{r+kn}w)$. Then one has $a^i wa^{r+kn} wa^j \in X^*$. By the choice of k, there exist integers p, q such that $a^i wa^p, a^q wa^j \in X^*$ and $p + q = r + kn$.

Observe that if p', q' are such that $a^i wa^{p'}, a^{q'} wa^j \in X^*$ and $p' + q' = r + kn$, then assuming for instance $p' \geq p$, one has $a^{p'-p} \in X^*$ since X^* is stable. Thus $p \equiv p' \mod n$ and also $q \equiv q' \mod n$. Consequently, the pair $(\bar{p}, \bar{q})$ is well defined by the pair (i, j).

Next, if $i' \equiv i \bmod n$ and $j' \equiv j \bmod n$ and let $(\bar{p}', \bar{q}')$ be the pair corresponding to (i', j'). If for instance $i' \geq i$ then $a^i w a^p = a^{i'-i} a^i w a^p$ is in X^* and consequently $\bar{p}' = \bar{p}$. This defines a mapping $(\bar{\imath}, \bar{\jmath}) \to (\bar{\imath}, \bar{p}), (\bar{q}, \bar{\jmath})$ with $\bar{p} + \bar{q} = \bar{r}$.

This mapping is clearly injective. We prove that it is surjective. Indeed, consider a pair $a^i w a^p, a^q w a^j \in X^*$ with $\bar{p} + \bar{q} = \bar{r}$. If $p > \ell$, then $a^i w a^{p-n} \in X^*$. Thus we may assume $p \leq \ell$ and also $q \leq \ell$. There is an integer t such that $p + q + tn = r + kn$, and actually $t \geq 0$ because $tn = r + kn - p - q \geq r + kn - 2\ell \geq r \geq 0$. Thus $(a^i w a^p) a^{tn} (a^q w a^j) = a^i w a^{r+kn} w a^j$ is in X^* and $(\bar{\imath}, \bar{\jmath})$ is in $C_a(w a^{r+kn} w)$.

Let $c(w) = \mathrm{Card}(C_a(w))$. By Lemma 12.2.3, we have $c(w) > 0$. From the bijection, it follows that

$$c(w)^2 = \sum_{r=0}^{n-1} c(w a^{r+kn} w).$$

Now we prove that $c(w) = n$ for all $w \in A^*$. Recall that $0 < c(w) \leq n^2$. Let w be such that $c(w)$ is minimal. Since $\sum_{r=0}^{n-1} c(w a^{r+kn} w) \geq n c(w)$, we obtain $c(w)^2 \geq n c(w)$ and consequently $c(w) \geq n$. Next, let w be such that $c(w)$ is maximal. We have $\sum_{r=0}^{n-1} c(w a^{r+kn} w) \leq n c(w)$ and therefore $c(w) \leq n$. $\qquad \square$

Example 12.2.5 Let $X = \{aa, ba, baa, bb, bba\}$. There are four distinct sets $C_a(w)$ with respect to the letter a, namely $C_a(a) = \{(0, 1), (1, 0)\}$, $C_a(a^2) = \{(0, 0), (1, 1)\}$, $C_a(b) = \{(0, 0), (0, 1)\}$ and $C_a(ab) = \{(1, 0), (1, 1)\}$.

Theorem 12.2.6 *Let X be a finite maximal code. Let $\varphi : A^* \to M$ be the morphism from A^* onto the syntactic monoid of X^* and let K be the minimal ideal of M. Let a be a letter and let n be its order. For $u, v \in A^*$, let*

$$R(u) = \{i \geq 0 \mid u a^i \in X^*\}, \quad L(v) = \{j \geq 0 \mid a^j v A^* \cap X^* \neq \emptyset\},$$

and let $\bar{R}(u), \bar{L}(v)$ denote the sets of residues mod n of $R(u), L(v)$. If $u, v \in \varphi^{-1}(K)$ and u is right completable in X^*, then $\mathbb{Z}/n\mathbb{Z} = \bar{R}(u) + \bar{L}(v)$ is a factorization. Moreover, $\mathrm{Card}(\bar{L}(v))$ is a multiple of the degree of X.

Recall that a word $u \in A^*$ is called *right completable* in X^* if there is a word w such that $uw \in X^*$. A word $u \in A^*$ is called *strongly right completable* (with respect to some code X) if any word in uA^* is right completable in X^*. A word u is called *simplifying* if for any $x \in X^*$ and $v \in A^*$, $x, xuv \in X^*$ implies $uv \in X^*$. Clearly, the sets of strongly right completable and of simplifying words both are right ideals.

Proposition 12.2.7 *Let $X \subset A^+$ be a thin maximal code. Let $\varphi : A^* \to M$ be the morphism onto the syntactic monoid of X^* and let K be the minimal ideal of M. Then any right completable word $u \in \varphi^{-1}(K)$ is both strongly right completable and simplifying.*

Proof. To show that u is strongly right completable, observe that the right ideal $\varphi(u)M$ is minimal and consequently, for every $m = \varphi(v) \in M$ there exists $m' = \varphi(w)$ such that $\varphi(u)mm' = \varphi(uvw) = \varphi(u)$. Since u is right completable, this shows that uvw is right completable. It follows that u is strongly right completable.

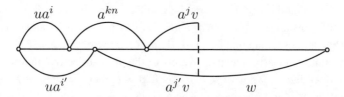

Figure 12.1 Proving that the sum is direct.

To show that u is simplifying, suppose first that $u \in X^*$. Let $x \in X^*$ and $v \in A^*$ be such that $xuv \in X^*$. Let $m = \varphi(u)$, $p = \varphi(x)$, and $q = \varphi(v)$. Then mpm belongs to the same group G as m. Let n be the inverse of mpm in G. Note that, since G is a finite group, n is a power of mpm and therefore $n \in \varphi(X^*)$. We have $mpmn = nmpm = e$ where e is the idempotent of G and thus $m(nmpm)q = meq = mq$. Hence $mq = mnmpmq = (m)(n)(m)(pmq)$ is in $\varphi(X^*)$, and $uv \in X^*$. This shows that u is simplifying in this case.

In the general case, since u is right completable, $uA^* \cap X^* \neq \emptyset$. Let $y \in uA^* \cap X^*$. Then $\varphi(y) \in K$, showing that the word y is simplifying by the preceding proof. Since the right ideal $\varphi(u)M$ is minimal, there exists $v \in A^*$ such that $\varphi(yv) = \varphi(u)$. To show that u is simplifying, consider $x \in X^*$ and $t \in A^*$ such that $xut \in X^*$. Since $\varphi(yv) = \varphi(u)$, one has $xyvt \in X^*$, and since y is simplifying, one gets $yvt \in X^*$. Since $\varphi(ut) = \varphi(yvt)$, this in turn shows that $ut \in X^*$. This proves that u is simplifying.

$\square$

For another proof of Proposition 12.2.7 see Exercise 9.3.6.

Proof of Theorem 12.2.6. Consider an integer $r \geq 0$ and let k be such that kn is larger than the maximum of the lengths of the words of X. Since u is strongly right completable, there is a word w such that $ua^{r+kn}vw \in X^*$. By the hypothesis on k, there exist i, j with $r + kn = i + j$ such that $ua^i, a^j vw \in X^*$. By definition $i \in R(u)$, $j \in L(v)$. This shows that $\bar{R}(u) + \bar{L}(v) = \mathbb{Z}/n\mathbb{Z}$.

Let us now show that the sum is direct.

Let $i, i' \in R(u)$ and $j, j' \in L(v)$ be such that $i + j \equiv i' + j' \mod n$. We may assume that $i + j \leq i' + j'$. Let $k \geq 0$ be such that $i + j + kn = i' + j'$. Then $ua^{i+j+kn}v = ua^{i'+j'}v$ (see Figure 12.1). Since $j' \in L(v)$, there is a word w such that $a^{j'}vw \in X^*$. Since $j \in L(v)$, the word $a^j v$ is right completable and there-fore is simplifying by Proposition 12.2.7. We have $ua^{i+kn} \in X^*$ and $ua^{i+kn}a^j vw = (ua^{i'})(a^{j'}vw) \in X^*$. Thus $a^j vw \in X^*$.

Since $ua^i, a^{kn+j}vw, ua^{i'}, a^{j'}vw \in X^*$ and X^* is stable, we have, assuming for instance that $i' \geq i$, $a^{i'-i} \in X^*$. This implies that $i \equiv i' \mod n$ and also $j \equiv j' \mod n$.

Finally, for $w \in A^*$, let

$$S(w) = \{j \geq 0 \mid a^j w \in X^*\} \tag{12.1}$$

and let $\bar{S}(w)$ denote the set of residues of the elements of $S(w)$. Let $e = \varphi(x)$ be an idempotent in $K \cap \varphi(X^*)$. Let $G = eMe$ be the group containing e and H be the subgroup $G \cap \varphi(X^*)$. Let $G = \bigcup_{i=1}^{d} Hg_i$ be the decomposition of G into right cosets of H and let $w_i \in \varphi^{-1}(g_i^{-1})$ for each $i = 1, \ldots, d$.

We claim that $L(v) = \bigcup_{i=1}^{d} S(vw_i)$ and moreover the sets $\bar{S}(vw_i)$ are disjoint. First, consider $j \in S(vw_i)$. By definition, $a^j vw_i \in X^*$ and thus $j \in L(v)$. Moreover, we have also $e\varphi(a^j v)g_i^{-1} \in H$ and consequently $e\varphi(a^j v)e \in Hg_i$, showing that the index i is uniquely determined by $\bar{j}$. Thus the sets $\bar{S}(vw_i)$ are disjoint.

Conversely, let $j \in L(v)$. Then since $e\varphi(a^j v)e \in G$, there is an index i such that $e\varphi(a^j v)e \in Hg_i$, which implies $e\varphi(a^j vw_i) \in \varphi(X^*)$. The word $a^j v$ is simplifying by Proposition 12.2.7. Hence $a^j vw_i \in X^*$, showing that $j \in S(vw_i)$.

Let

$$N(u) = \{i \geq 0 \mid A^* ua^i \cap X^* \neq \emptyset\} \tag{12.2}$$

and let $\bar{N}(u)$ denote the set of residues modulo n of the elements of $N(u)$. There is, by symmetry, an analogue factorization $\mathbb{Z}/n\mathbb{Z} = \bar{N}(u') + \bar{S}(v')$ for each $u', v' \in \varphi^{-1}(K)$ with v' left completable. Since for each w_i, $i = 1, \ldots, d$, the word vw_i is left completable, one gets d factorizations $\mathbb{Z}/n\mathbb{Z} = \bar{N}(u) + \bar{S}(vw_i)$. In particular all sets $\bar{S}(vw_i)$ have the same number s of elements. Thus $\mathrm{Card}(\bar{L}(v)) = \sum_{i=1}^{d} \mathrm{Card}(\bar{S}(vw_i)) = ds$ is a multiple of d. $\qquad\square$

Evidently, there is a symmetric statement for left completable words, using the sets $N(u)$ and $S(v)$ defined by (12.2) and (12.1), namely: if $u, v \in \varphi^{-1}(K)$ and v is left completable, then $\mathbb{Z}/n\mathbb{Z} = \bar{N}(u) + \bar{S}(v)$ is a factorization and $\mathrm{Card}(\bar{N}(u))$ is a multiple of d.

The previous theorem has a close connection with Theorem 14.2.4 and the factorization of the polynomial $1 - \underline{X}$ for a finite maximal code X. Actually, according to Lemma 14.4.3, there are polynomials P, Q, R with coefficients $0, 1$ such that

$$\underline{A}^* = P\underline{X}^*Q + R.$$

Taking $b = 0$ for all letters $b \neq a$, we obtain

$$a^* = U(a^n)^*V + W$$

for some polynomials U, V, W with coefficients $0, 1$. Multiplying both sides by $a^n - 1$, we obtain

$$1 + a + \cdots + a^{n-1} = UV + W(a^n - 1)$$

or

$$UV \equiv 1 + a + \cdots + a^{n-1} \pmod{(a^n - 1)},$$

which is equivalent to $U = a^L$, $V = a^R$ with (L, R) a factorization of $\mathbb{Z}/n\mathbb{Z}$.

We illustrate this statement in the following example.

Example 12.2.8 Let $A = \{a, b\}$ and let $X = (A^3 \setminus a^3) \cup a^3 A^3$ which is a finite maximal prefix code of degree 3 (the lengths of the words of X are multiples of 3). The transitions of the minimal automaton of X^* are represented in Table 12.3. Let $u = v = b$.

Table 12.3 *The minimal automaton of X^*.*

	0	1	2	3	4	5
a	1	2	3	4	5	0
b	4	5	0	4	5	0

The sets $S(b) = \{j \geq 0 \mid a^j b \in X^*\}$ and $N(b) = \{i \geq 0 \mid A^* ba^i \cap X^* \neq \emptyset\}$ satisfy $\bar{S}(b) = \{2, 5\}$ and $\bar{N}(b) = \{0, 1, 2\}$, giving a factorization of $\mathbb{Z}/6\mathbb{Z}$ such that $\text{Card}(\bar{N}(b)) = 3$.

Theorem 12.2.6 takes a simpler form when X is synchronized. We give here a direct proof, but the proposition follows from the theorem when the words x, y are taken in the inverse image of the minimal ideal of the syntactic monoid.

Proposition 12.2.9 *Let X be a finite maximal synchronized code. Let $a \in A$ and let $n \geq 1$ be its order. Let x, $y \in X^*$ be a synchronizing pair. Let $R(y) = \{r \geq 0 \mid ya^r \in X^*\}$ and $L(x) = \{\ell \geq 0 \mid a^\ell x \in X^*\}$. Let $\bar{L}$, $\bar{R}$ be the set of residues modulo n of the sets $L(x)$, $R(y)$. Then $(\bar{L}, \bar{R})$ is a factorization of $\mathbb{Z}/n\mathbb{Z}$.*

Proof. Recall that $yA^*x \subset X^*$. Let $w = ya^u x$ with u greater than the maximal length of the words in X. Then there is a pair r, ℓ of integers such that $ya^r, a^\ell x \in X^*$ and $u = r + \ell$. This proves that $\mathbb{N} = R(y) + L(x)$, and consequently that $\mathbb{Z}/n\mathbb{Z} = \bar{L} + \bar{R}$. The fact that the sum is direct is proved as in the proof of Proposition 12.2.1. $\square$

We illustrate the proposition in the example below.

Example 12.2.10 Let $A = \{a, b\}$. Consider the maximal prefix code $X = (A^2 \setminus b^2) \cup b^2 A$ and the maximal suffix code $Y = A^2 a \cup b$. Then $X^* \cap Y^*$ is generated by a finite maximal code Z which satisfies

$$\underline{Z} - 1 = (1 + a + b + b^2)((\underline{A} - 1)a(\underline{A} - 1) + \underline{A} - 1)(1 + a + a^2 + ba),$$

see Exercise 14.1.8. We have $a^6 \in Z$. The word $x = ab^2 a$ is synchronizing for X and the word $y = b^2$ is synchronizing for Y. Thus we have $yA^*x \subset yA^* \cap A^*x \subset Z^*$. We have $\bar{L}(x) = \{2, 5\}$, $\bar{R}(y) = \{1, 3, 5\}$. By a shift, we obtain the factorization $(\{0, 3\}, \{0, 2, 4\})$ of $\mathbb{Z}/6\mathbb{Z}$, in which both factors are periodic.

A consequence of Theorem 12.2.6 is the following statement (it appears also as Theorem 13.5.8 with a proof using probability distributions. It can also be obtained as a consequence of Theorem 14.2.1).

Proposition 12.2.11 *Let X be a finite maximal code on the alphabet A. The degree of X divides the greatest common divisor of the orders of the letters.*

Proof. Let a be a letter and let n be its order. According to Theorem 12.2.6, there exists a factorization $\mathbb{Z}/n\mathbb{Z} = R + L$ where $\text{Card}(L)$ is a multiple of the degree d of X. Since $\text{Card}(L)$ divides n, d divides n and the result follows. $\square$

In particular if the gcd of the orders of the letters is 1, then the code X is synchronized. This was proved for prefix codes, using factorizations implicitly, in Theorem 3.6.10.

12.3 Hooks

A *hook* is a word of the form $a^i b^j$ for some letters a, b and integers $i, j \geq 0$. In this section, we will show that, under adequate hypotheses, the hooks contained in a finite maximal code define factorizations of the cyclic groups $\mathbb{Z}/n\mathbb{Z}$ where n is the order of some letter.

Theorem 12.3.1 *Let X be a finite maximal code on the alphabet A and let $a, b \in A$ be such that $b \in X$. Let $n \geq 1$ be the order of a, and let*

$$L = \{\ell \geq 0 \mid a^\ell b^+ \cap X \neq \emptyset\}, \qquad R = \{r \geq 0 \mid b^+ a^r \cap X \neq \emptyset\}.$$

Let $\bar{L}, \bar{R}$ denote the sets of residues modulo n of L, R. Then $(\bar{L}, \bar{R})$ is a factorization of $\mathbb{Z}/n\mathbb{Z}$.

Proof. Let $k \geq 1$ be larger than the length of the words of X. Then, since $b \in X$, we have $b^k A^* b^k \subset X^*$. Thus, for any $i \geq 0$, the word $w = b^k a^{i+kn} b^k$ is in X^*. This implies that there exist integers p, q, r, ℓ such that $w \in b^*(b^p a^\ell)(a^n)^*(a^r b^q) b^*$ with $b^p a^\ell, a^r b^q \in X$. This shows that $i \equiv \ell + r \bmod n$.

The decomposition of i is unique. Suppose indeed that $r + \ell = r' + \ell' + tn$ for some integer t (with $t \geq 0$, the other case is symmetric) with $r, r' \in R$ and $\ell, \ell' \in L$. Let p', q' be such that $b^{p'} a^{\ell'}, a^{r'} b^{q'} \in X$. Then the word $b^k a^{\ell+r} b^k$ has the two factorizations

$$b^{k-p}(b^p a^\ell)(a^r b^q) b^{k-q} = b^{k-p'}(b^{p'} a^{\ell'}) a^{tn}(a^{r'} b^{q'}) b^{k-q'}.$$

Since X is a code, these factorizations are the same, and $p = p', \ell = \ell', r = r'$, and $q = q'$. $\qquad\square$

Example 12.3.2 Let $X = \{aaaa, ab, abaa, b, baa\}$. Then $n = 4$ and

$$L = \{0, 1\}, \quad R = \{0, 2\}.$$

It is possible to obtain Theorem 12.3.1 as a corollary of Theorem 12.2.6 (see Exercise 12.3.1). One may use Theorem 12.3.1 to prove that some codes are not contained in a finite maximal one.

Proposition 12.3.3 *Let $L, R \subset \mathbb{N}$ with $0 \in L \cap R$ and $n \geq 1$ be such that the pair (L, R) is direct modulo n and $\mathrm{Card}(L), \mathrm{Card}(R) \geq 2$. If n is a prime number, then $X = a^n \cup a^L b \cup ba^R$ is a code which is not contained in a finite maximal code.*

Proof. The fact that X is a code follows from Proposition 12.2.1. Let Y be a finite maximal code containing X. Then, by Theorem 12.3.1, the sets $\bar{R}, \bar{L}$ of residues modulo n of R, L are contained in sets $\bar{R}', \bar{L}'$ which form a factorization of $\mathbb{Z}/n\mathbb{Z}$.

Since (L, R) is direct, in particular $\mathrm{Card}(R) = \mathrm{Card}(\bar{R})$ and $\mathrm{Card}(L) = \mathrm{Card}(\bar{L})$. Thus $n = \mathrm{Card}(\bar{R}')\,\mathrm{Card}(\bar{L}')$ is a nontrivial factorization of n, a contradiction. $\qquad\square$

Example 12.3.4 The set $X = \{a^5, b, ab, ba^2\}$ is a code which is not contained in a finite maximal code.

Example 12.3.5 Let $X = ba^{R_1} \cup a^{\{3,8\}}ba^{R_2} \cup a^{11}ba^{R_3}$ with $R_1 = \{0, 1, 7, 13, 14\}$, $R_2 = \{0, 2, 4, 6\}$, $R_3 = \{0, 1, 2\}$. The set X is an example of a code which is not commutatively prefix (see Example 14.6.7).

It is not known whether X is contained in a finite maximal code. If it is the case, by Theorem 12.3.1 there exists an integer n and sets L, R such that $\mathbb{Z}/n\mathbb{Z} = L + R$ is a factorization with $\{0, 3, 8, 11\} \subset L$ and $\{0, 1, 7, 13, 14\} \subset R$ (see Example 12.1.4). This implies that n is not a Rédei number since $0, 1 \in R$ and $0, 3, 8 \in L$ and thus $\langle L \rangle = \langle R \rangle = \mathbb{Z}/n\mathbb{Z}$.

It is easy to see, using Proposition 12.1.5 that such an integer n is a multiple of $330 = 2 \times 3 \times 5 \times 11$. Indeed, if n were not divisible by 3, then $L + 3R$ would be a factorization, a contradiction with the fact that 3 is in L and in $3R$. The same argument shows that n is divisible by 2 and 11. Finally, if n is not divisible by 5, then $L + 5R$ is a factorization, a contradiction with the fact that $8 = 3 + 5 = 8 + 0$ has two decompositions.

A factorization L, R of $\mathbb{Z}/n\mathbb{Z}$ is a *Sands factorization* if there exist two relatively prime integers p, q which are not multiples of n such that $0, 1$ are in one of the factors L or R and $0, p, q$ are in the other factor. The hypothetical factorization discussed in the previous example would be a Sands factorization.

The following example shows that there exists a Sands factorization where, in addition, p is prime.

Example 12.3.6 We start with the factorization $G = L + R$ of Example 12.1.10 where $n = 900$ and the sets L and R are given in Table 12.2. Since 361 is an element of L prime to 900, it is invertible modulo 900. It is easily checked that $\ell = 541$ is its inverse. Since ℓ is prime to 30, setting $U = \ell L$, $G = U + R$ is still a factorization by Proposition 12.1.5, and $0, 1 \in U$. It remains to replace R by an appropriate factor. For this, consider the elements $r = 45$ and $s = 96$ of R. In the factorization $G = U + R$, the factor R can be replaced by $R - r$ to get the factorization $G = U + (R - r)$, and $0 \in R - r$. Next $96 - r = 51 = 3 \times 17$ is in $R - r$. Since 17 is relatively prime to 900, it is invertible and its inverse is 53. Since $m = 53$ is relatively prime to $\mathrm{Card}(R - r) = 30$, in the factorization $G = U + (R - r)$, we may replace the factor $R - r$ by $m(R - r)$ again by Proposition 12.1.5. We obtain the factorization $G = U + V$ with $V = m(R - r)$ which satisfy the conditions with $p = 3 \equiv m(96 - r)$ mod 900 and $q = 65 \equiv m(250 - r)$ mod 900. The sets U, V are represented in Table 12.4. This factorization is a Sands factorization since $0, 1 \in U$ and $0, p, q \in V$ with $p = 3$ and $q = 65$.

A *multiple factorization* is defined as follows. For an integer $d \geq 1$, a d-factorization of a group G is a pair (L, R) of subsets of G such that each $g \in G$

Table 12.4 *The sets U and V with 0, 1 ∈ U and 0, 3, 65 ∈ V.*

U					V				
0	576	252	828	504	315	855	495	135	**0**
100	676	352	28	604	**65**	705	345	885	525
200	776	452	128	704	15	555	195	735	375
225	801	477	153	729	665	405	45	585	450
325	**1**	577	253	829	723	363	**3**	543	183
425	101	677	353	29	365	105	645	285	825

can be written in d different ways $g = \ell + r$ with $\ell \in L$ and $r \in R$. Thus an ordinary factorization is a 1-factorization.

The concept of multiple factorization can be extended to the case of multisets (L, R). We say that (L, R) is an m-factorization of $\mathbb{Z}/n\mathbb{Z}$ if each element of $\mathbb{Z}/n\mathbb{Z}$ can be written in m different ways as the sum modulo n of an element of L and an element of R, with the multiplicity taken into account.

For example, $L = \{0, 0, 1, 5\}$, $R = \{0, 2, 4\}$ forms a 2-factorization of $\mathbb{Z}/6$.

A generalization of Theorem 12.3.1 is the following.

Proposition 12.3.7 *Let X be a finite maximal code on the alphabet A. Let $a, b \in A$ and let $n, m \geq 1$ be the integers such that $a^n, b^m \in X$. Let R, L be the multisets*

$$L = \{\ell \geq 0 \mid a^\ell b^+ \cap X \neq \emptyset\}, \quad R = \{r \geq 0 \mid b^+ a^r \cap X \neq \emptyset\}.$$

Let $\bar{L}, \bar{R}$ be the multisets of residues modulo n of L, R. Then the pair $(\bar{L}, \bar{R})$ is an m-factorization of $\mathbb{Z}/n\mathbb{Z}$.

Proof. We use Proposition 12.2.4. Let k be the maximal length of the words of X. Let $s \geq k$. By Proposition 12.2.4, there are m pairs of residues modulo m of integers $i, j \geq 0$ such that $b^i a^s b^j \in X^*$. Thus s is the sum in m ways of integers r, ℓ such that $b^i a^r, a^\ell b^j \in X^*$. □

Example 12.3.8 Let $X = \{aa, ba, baa, bb, bba\}$. Then $n = m = 2$ and $L = \{0\}$, $R = \{0, 1, 1, 2\}$. The statement is satisfied since 0 and 1 are obtained each in two ways as the residue modulo 2 of an element of R.

One may use Proposition 12.3.7, to prove that some codes are not contained in a finite maximal code (see Exercise 12.3.3).

12.4 Exercises

Section 12.1

12.1.1 Show that a divisor of a Hajós number is also a Hajós number.

12.1.2 Prove that a Hajós number is a Rédei number.

12.1.3 Show that if $\mathbb{Z} = L + R$ is a factorization of $\mathbb{Z}$ with L finite, then R is periodic. (*Hint*: Prove that if $L \subset \{0, 1, \ldots, d\}$, then R has period at most 2^d.)

Section 12.2

12.2.1 Let $L, R \subset \{0, 1, \ldots, n-1\}$ and consider the polynomials in the variable a

$$a^L = \sum_{\ell \in L} a^\ell, \quad a^R = \sum_{r \in R} a^r.$$

Show that if (L, R) is a factorization of $\mathbb{Z}/n\mathbb{Z}$, then $a^n - 1$ divides $a^L a^R (a - 1)$.

12.2.2 Let $n \geq 0$, and let P and Q be two sets of nonnegative integers such that any integer r in $\{0, 1, \ldots, n-1\}$ can be written in a unique way as a sum $r = p + q$ with $p \in P$ and $q \in Q$.

Show that there exist integers $n_1, n_2, \ldots, n_k$ with $n_1 | n_2 | \cdots | n_k$ and $n_k = n$ such that $\{0, 1, \ldots, n-1\} = \{0, 1, \ldots, n_1 - 1\} + \{0, n_1, 2n_1, \ldots, n_2 - 1\} + \cdots + \{0, n_{k-1}, \ldots, n_k - 1\}$ such that P and Q are obtained by grouping into two parts the terms of this sum. (*Hint*: Prove first the following remark: let $r < n-1$ and set $r = p + q$ with $p \in P$ and $q \in Q$. Show that $r + 1 = p' + q'$ where either p' is the successor of p in P and $q' \leq q$, or q' is the successor of q in Q and $p' \leq p$.)

Section 12.3

12.3.1 Deduce Theorem 12.3.1 from Proposition 12.2.9.

12.3.2 Let $m, n \geq 1$, and let H, K be subsets of $\mathbb{N}$ containing m. Let $\bar{H}, \bar{K}$ be the sets of residues modulo m of H and K and assume that the sum $\bar{H} + \bar{K}$ is direct. Similarly, let S, T be subsets of $\mathbb{N}$ containing n, and let $\bar{S}, \bar{T}$ be the sets of residues modulo n of S and T. Assume again that the sum $\bar{S} + \bar{T}$ is direct. Show that

$$X = \{a^n, b^m\} \cup b^H a^S \cup a^T b^K \setminus \{a^n b^m, b^m a^n\} \tag{12.3}$$

is a code.

12.3.3 Let $d, t > j > 0$ and let $m = dt + j$. Show that for any $n \geq 1$, when (S, T) is a factorization of $\mathbb{Z}/n\mathbb{Z}$ and $\mathrm{Card}(H) = d$, $\mathrm{Card}(K) = t$, the code defined by Equation (12.3) is not contained in a finite maximal code.

12.3.4 Use Exercise 12.3.3 to show that the code

$$Y = \{a^2, ba^2, b^2 a^2, b^{10}, a^2 b^3, a^2 b^6, ab^{10}, ab^3, ab^6\}$$

is not contained in a finite maximal code.

12.3.5 Show that if (L, R) is a factorization of $\mathbb{Z}/n\mathbb{Z}$ where n is a Hajós number, then the code $a^n \cup a^L b a^R$ is composed of prefix and suffix codes.

12.5 Notes

Factorizations of cyclic groups, or more generally of Abelian groups, form a subject with an interesting history, beginning with the proof by G. Hajós in 1941 of a conjecture of Minkovski. The books by Szabó (2004) and Szabó and Sands (2009) are recommended for an exposition of this subject. Two important results in this theory are the theorems of Hajós and Rédei. The first one asserts that if $G = A_1 + \cdots + A_n$ is a factorization of a finite Abelian group G where each A_i is a cyclic subset, then at least one of the factors must be a subgroup of G (a cyclic subset is of the form $0, a, 2a, \ldots, ra$ for some $a \in G$ and $r \geq 1$). The second one is a generalization of the Hajós theorem proved by L. Rédei (1965). The theorem says that if $G = A_1 + \cdots + A_n$ is a factorization of a finite Abelian group G such that each A_i has a prime number of elements and contains the neutral element, then at least one of the factors must be a subgroup of G.

The link between codes and factorizations of cyclic groups was first noted in Schützenberger (1979b).

Proposition 12.1.5 is due to Sands (2000). Example 12.1.8 is a counterexample to a conjecture of Hajós due to De Bruijn (1953). The Hajós numbers are known exactly. An integer n is a Hajós number if and only if it is a divisor of an integer of the form $p^a q$, $p^2 q^2$, $p^2 qr$ or $pqrs$ with $a \geq 1$ and p, q, r distinct prime numbers (see Szabó (2004)). The least integer n which is not a Hajós number is thus $n = 72$. Example 12.1.10 is due to Szabó (1985). The list of Rédei numbers is also known exactly. It is formed of the divisors of integers of the form $p^a q^b r$, $p^a qrs$, $pqrst$, where p, q, r, s, t are distinct primes and $a, b \geq 1$, Szabó (2006). Example 12.3.6 is a counterexample to a conjecture formulated in Restivo *et al.* (1989). The counterexample is due to Sands (2007).

Exercise 12.2.2 is a result of Krasner and Ranulac (1937).

Exercise 12.1.3 is a result due to Hajós (see Szabó (2004) p. 165 and also Newman (1977)). The optimal bound on the period of R is not known (see Szabó (2004) for an example where the period is quadratic in the size of R).

Proposition 12.2.4 is a result from Perrin and Schützenberger (1977). Theorem 12.3.1 is a result from Restivo *et al.* (1989) while Proposition 12.3.7 is due to Lam (1996). Proposition 12.3.3 is from Restivo (1977). It exhibits a class of codes which are not contained in any finite maximal code. Further results in this direction can be found in De Felice and Restivo (1985).

Exercise 12.3.5 is from Lam (1997). His result generalizes one of De Felice (1996) who proved the same result for a code X of the form $X = a^n \cup a^L b \cup ba^R$. For this smaller class De Felice also proved in De Felice (1996) that X is included in a finite maximal code with the additional property that for each word in X there are at most three occurrences of the letter b.

13

Densities

In this chapter we present a study of probabilistic aspects of codes. We have already seen in Chapters 2 and 3 that probability distributions play an important role in this theory.

In Section 13.1, we present some basics on probability measures, and we state and prove Kolmogorov's extension theorem. In Section 13.2, the notion of *density* of a subset L of A^* is introduced. It is the limit in mean, provided it exists, of the probability that a word of length n is in L. In Section 13.3, we introduce the topological entropy and we give a way to compute it for a free submonoid. We will see how it is related to the results of Chapter 2 on Bernoulli distributions.

In Section 13.4, we describe how to compute the density of a set of words by defining probabilities in abstract monoids. In Section 13.5, we use this study for the proof of a fundamental formula (Theorem 13.5.1) that relates the density of the submonoid generated by a thin complete code to that of its sets of prefixes and suffixes.

13.1 Probability

We start with a short description of probability spaces, random variables, infinite words, and a result on the average length of prefix codes. We then give a proof of Kolmogorov's extension theorem.

Let S be a set. A family $\mathcal{F}$ of subsets of S is a *Boolean algebra* of subsets of S if it contains S and is closed under finite unions and under complement. This means that for $E, F \in \mathcal{F}$, then $E \cup F \in \mathcal{F}$ and $\bar{E} \in \mathcal{F}$ where $\bar{E}$ denotes the complement of E. It is also closed under intersection since $E \cap F$ is the complement of $\bar{E} \cup \bar{F}$. A Boolean algebra is called a σ-*algebra* if it is closed under countable union. This means that if $(E_n)_{n \geq 0}$ is a sequence of elements of $\mathcal{F}$, then $\bigcup_{n \geq 0} E_n \in \mathcal{F}$.

Example 13.1.1 Let A be an alphabet. The family composed of A^*, the empty set, and the set of words of even (odd) length is a Boolean algebra of four elements.

Example 13.1.2 Let $\varphi : A^* \to M$ be a morphism of A^* onto a monoid. The family $\mathcal{F}$ of set $\varphi^{-1}(P)$, for $P \subset M$, is a σ-algebra. Indeed, the family of all subsets of M is σ-algebra, and so is $\mathcal{F}$.

A real valued function μ defined on a σ-algebra $\mathcal{F}$ is *additive* if for any disjoint sets E, $F \in \mathcal{F}$, one has $\mu(E \cup F) = \mu(E) + \mu(F)$. It is called *countably additive* if

$$\mu\left(\bigcup_{n \geq 0} E_n\right) = \sum_{n \geq 0} \mu(E_n)$$

for any sequence $(E_n)_{n \geq 0}$ of pairwise disjoint elements of $\mathcal{F}$. If μ is countably additive and takes nonnegative values, then it is *monotone* in the sense that if $E \subset F$ for E, $F \in \mathcal{F}$, then $\mu(E) \leq \mu(F)$ since indeed $\mu(F) = \mu(E \cup (F \setminus E)) - \mu(E) + \mu(F \setminus E) \geq \mu(E)$.

Proposition 13.1.3 *Let μ be a countably additive function on a σ-algebra $\mathcal{F}$ with nonnegative values. Then*

$$\mu\left(\bigcup_{n \geq 0} E_n\right) \leq \sum_{n \geq 0} \mu(E_n)$$

for any sequence of subsets $(E_n)_{n \geq 0}$ of elements of $\mathcal{F}$.

Proof. Indeed, let $F_n = E_n \setminus \bigcup_{i < n} E_i$ for $n \geq 0$. Then the sets F_n are pairwise disjoint subsets in $\mathcal{F}$ and $\bigcup_{n \geq 0} E_n = \bigcup_{n \geq 0} F_n$. Moreover $F_n \subset E_n$ for $n \geq 0$ and therefore $\mu(F_n) \leq \mu(E_n)$. Thus

$$\mu\left(\bigcup_{n \geq 0} E_n\right) = \mu\left(\bigcup_{n \geq 0} F_n\right) = \sum_{n \geq 0} \mu(F_n) \leq \sum_{n \geq 0} \mu(E_n). \qquad \square$$

Let $\mathcal{F}$ be a σ-algebra on a set S. A *probability measure* on $\mathcal{F}$ is a function μ from $\mathcal{F}$ into the interval $[0, 1]$ which is countably additive and such that $\mu(S) = 1$. The triple $(S, \mathcal{F}, \mu)$ is called a *probability space*. When the σ-algebra $\mathcal{F}$ is understood, we also say that μ is a *probability* on S.

Given a probability space $(S, \mathcal{F}, \mu)$, an integer valued *random variable* is a map V from S into $\mathcal{N} = \mathbb{N} \cup \infty$ such that $V^{-1}(n) \in \mathcal{F}$ for any $n \in \mathcal{N}$. The semirings $\mathcal{N}$ and $\mathcal{R}_+$ are defined in Section 1.6. In particular, $0\infty = 0$ in both semirings. We write $\mathrm{Prob}(V = n)$ for $\mu(V^{-1}(n))$. Note that $\sum_{n \in \mathcal{N}} \mathrm{Prob}(V = n) = 1$, since indeed one has $\sum_{n \in \mathcal{N}} \mathrm{Prob}(V = n) = \sum_{n \in \mathcal{N}} \mu(V^{-1}(n)) = \mu(\bigcup_{n \in \mathcal{N}} V^{-1}(n)) = \mu(S) = 1$. The *mean value* or *expectation* of V is is the finite or infinite sum

$$E(V) = \sum_{n \in \mathcal{N}} n\,\mathrm{Prob}(V = n) = \sum_{n \in \mathbb{N}} n\,\mathrm{Prob}(V = n) + \infty\,\mathrm{Prob}(V = \infty).$$

Thus $E(V)$ is infinite if $\mathrm{Prob}(V = \infty) > 0$, and it is equal to $\sum_{n \in \mathbb{N}} n\,\mathrm{Prob}(V = n)$ otherwise since $\infty 0 = 0$ in $\mathcal{R}_+$.

Proposition 13.1.4 *Let S be a countable set. Any function $\mu : S \to [0, 1]$ with $\sum_{s \in S} \mu(s) = 1$ defines a probability on the family of all subsets of S by $\mu(T) = \sum_{t \in T} \mu(t)$ for a subset T of S.*

Proof. It suffices to show that μ is countably additive. Consider a sequence $(E_n)_{n \geq 0}$ of pairwise disjoint subsets of S and let $T = \bigcup_{n \geq 0} E_n$. Then $\mu(\bigcup_{n \geq 0} E_n) = \sum_{s \in T} \mu(s) = \sum_{n \geq 0} \mu(E_n)$. $\qquad \square$

From now on, all alphabets considered in this chapter are assumed to be finite. Let A be an alphabet. We introduce the set of infinite words on an alphabet which appears to be the appropriate structure to define a probability measure on the set of all words.

An *infinite word* w on the alphabet A is a sequence $a_0, a_1, \ldots$ of elements of A. We write w as $w = a_0 a_1 \cdots$. The set of infinite words on A is denoted A^ω. For a word $u = a_0 a_1 \cdots a_n \in A^*$ and an infinite word $v = b_0 b_1 \cdots \in A^\omega$, we denote by uv the infinite word $a_0 a_1 \cdots a_n b_0 b_1 \cdots$ obtained by concatenating u and v. More generally, for a set $X \subset A^*$ of words, we denote $X A^\omega$ the set of infinite words xu for $x \in X$ and $u \in A^\omega$. In particular, if x is a word, the set $x A^\omega$ is the set of all infinite words starting with x. Thus the word x is a prefix of the word y if and only if $x A^\omega \supset y A^\omega$, and x and y are incomparable for the prefix order if and only if the sets $x A^\omega$ and $y A^\omega$ are disjoint.

The family of *Borel subsets* of A^ω is the smallest family of subsets of A^ω containing the sets of the form $x A^\omega$ for $x \in A^*$ and closed under countable union and complement. It is clear that it is a σ-algebra and that it is closed under countable intersections.

Example 13.1.5 Let $A = \{a, b\}$. The set reduced to the infinite word a^ω is a Borel subset of A^ω since it is the complement of $a^* b A^\omega$, and $a^* b A^\omega$ is the countable union of the sets $a^n b A^\omega$ for $n \geq 0$.

Example 13.1.6 For any set $X \subset A^*$, the set $X A^\omega$ of infinite words with a prefix in X is a Borel set since it is the countable union $X A^\omega = \bigcup_{x \in X} x A^\omega$.

Example 13.1.7 Let $X \subset A^+$ be a prefix code. Then the set X^ω of infinite words of the form $x_0 x_1 \cdots$ with $x_i \in X$ is

$$X^\omega = \bigcap_{n \geq 0} X^n A^\omega. \tag{13.1}$$

It is a Borel set. Indeed, let us show (13.1). The inclusion $X^\omega \subset \bigcap_{n \geq 0} X^n A^\omega$ is clear. Conversely, consider an infinite word $x = x_1 u_1 = \ldots = x_n u_n = \ldots$ for $x_n \in X^n$ and $u_n \in A^\omega$. Since X is prefix, we have for each $n \geq 2$, $x_n = x_{n-1} y_n$ with $y_n \in X$. Thus $x = y_1 y_2 \cdots$ is in X^ω. The Equation (13.1) shows that X^ω is a Borel set.

Let μ be a probability measure on the family of Borel subsets of A^ω and let π be the map from A^* into $[0, 1]$ defined for $u \in A^*$ by

$$\pi(u) = \mu(u A^\omega). \tag{13.2}$$

Then $\pi(1) = 1$ and moreover π satisfies the coherence condition

$$\sum_{a \in A} \pi(ua) = \pi(u)$$

for all $u \in A^*$. Indeed, the sets $ua A^\omega$ for $a \in A$ are disjoint, and consequently one has $\sum_{a \in A} \pi(ua) = \sum_{a \in A} \mu(ua A^\omega) = \mu(\bigcup_{a \in A} ua A^\omega) = \mu(u A^\omega) = \pi(u)$. This shows that π is a probability distribution, as defined in Section 1.11. The converse statement is the following theorem.

Theorem 13.1.8 (Kolmogorov's extension theorem) *For any probability distribution* π *on* A^*, *there is one and only one probability measure* μ *on the family of Borel subsets of* A^ω *such that* $\mu(xA^\omega) = \pi(x)$ *for all* $x \in A^*$.

We say that the probability distribution π on A^* defined by (13.2) and the probability distribution μ are *associated*. We postpone the proof of Theorem 13.1.8 to the end of this section.

Let π be the probability distribution on A^*, and let μ be the associated probability measure on A^ω. Let $X \subset A^*$ be a prefix code. Recall that by Proposition 3.7.1, we have $\pi(X) \le 1$. The proof now becomes obvious. Indeed, the sets xA^ω for $x \in X$ are pairwise disjoint. Consequently $\pi(X) = \sum_{x \in X} \mu(xA^\omega) = \mu(\bigcup_{x \in X} xA^\omega)$ and this number is at most 1 as for any subset of A^ω.

Suppose now that $\pi(X) = 1$. Observe that, since X is prefix, any infinite word $w \in A^\omega$ has at most one prefix of w in X. Let V be the random variable defined on A^ω by $V(w) = n$ if w has a prefix of length n in X and $V(w) = \infty$ if w has no prefix in X. Then $\mathrm{Prob}(V = \infty) = \mu(A^\omega \setminus XA^\omega) = 1 - \pi(X) = 0$. Next, for $n \ge 0$,

$$\mathrm{Prob}(V = n) = \mu((X \cap A^n)A^\omega) = \pi(X \cap A^n).$$

Recall that the average length of X is $\lambda(X) = \sum_{x \in X} |x| \pi(x)$. We show that the mean value of V is equal to $\lambda(X)$. Indeed,

$$E(V) = \sum_{n \ge 0} n \, \mathrm{Prob}(V = n) = \sum_{n \ge 0} n \pi(X \cap A^n) = \lambda(X).$$

Let π be a probability distribution on A^* and let μ be the associated probability measure on A^ω. The following statement shows that the quantity $\pi(T)$ for any set $T \subset A^*$ is the mean value of the random variable which assigns to an infinite word the number of its prefixes in T.

Proposition 13.1.9 *Let* T *be a subset of* A^*, *and let* V *be the random variable which assigns to an infinite word the number of its prefixes in* T. *Then* $\pi(T) = E(V)$.

Proof. For $n \ge 0$, let T_n be the set of words in T having n prefixes in T. Observe that the sets T_n are all prefix and that they are pairwise disjoint. Moreover $T = \bigcup_{n \ge 1} T_n$ and thus $\pi(T) = \sum_{n \ge 1} \pi(T_n)$. Let V be the random variable assigning to an infinite word the number of its prefixes in T. Let $p_n = \mathrm{Prob}(V = n)$ for $n \in \mathcal{N}$. For finite n, p_n is the probability that an infinite word has n prefixes in T and p_∞ is the probability that an infinite word has infinitely many prefixes in T.

We have $\pi(T_n) = \mu(T_n A^\omega)$. Since $T_n A^\omega$ is the set of infinite words having at least n prefixes in T, we have $\pi(T_n) = \sum_{m \ge n} p_m + p_\infty$ and thus

$$E(V) = \sum_{n \in \mathcal{N}} n p_n = \sum_{n \ge 1} \pi(T_n) = \pi(T). \qquad \square$$

Proposition 13.1.9 has the following interesting interpretation when one takes for the set T a code $X \subset A^+$. Then, by Theorem 2.4.5, one has $\pi(X) \le 1$ for any

Bernoulli distribution π on A^*. Thus the proposition shows that the average number of prefixes in X of an infinite word is at most one, as it is for a prefix code.

We give a second interpretation of Proposition 13.1.9. Let $X \subset A^+$ be a prefix code, and let π be a probability distribution on A^* such that $\pi(X) = 1$. Let P be the set of proper prefixes of X. We know by Proposition 3.7.11, that $\lambda(X) = \pi(P)$. This can be obtained as a consequence of Proposition 13.1.9 with T replaced by P. Indeed, the number of prefixes of an infinite word which are in P is equal to the length of its longest prefix in P plus 1. This number is equal to the length of the unique word in X which is a prefix of w, provided it exists. Now the probability of the set of infinite words having no prefix in X is zero because its complement has probability 1. So the average value is indeed $\lambda(X)$, showing that $\lambda(X) = \pi(P)$.

We will use the fact that

$$x A^\omega = \bigcup_{y \in A^n} xy A^\omega \tag{13.3}$$

for all $n \geq 0$ and $x \in A^*$. The formula indeed holds for $n = 0$, and since $A^\omega = \bigcup_{a \in A} a A^\omega$, one has by induction

$$x A^\omega = \bigcup_{y \in A^n} xy \left(\bigcup_{a \in A} a A^\omega \right) = \bigcup_{z \in A^{n+1}} xz A^\omega.$$

Let $\mathcal{F}$ be the family of sets of the form $X A^\omega$ where X is a *finite* subset of A^ω. Observe that there are countably many sets in $\mathcal{F}$. A set F in $\mathcal{F}$ has many different representations of the form $F = X A^\omega$, where X is a finite set. The following lemma describes some canonical representations.

Lemma 13.1.10 *For any set $F \in \mathcal{F}$, and for any sufficiently large integer n, there is a subset X of A^n such that $F = X A^\omega$.*

Proof. Let $F = Y A^\omega$ for some finite set $Y \subset A^*$, and let n be larger than the lengths of the words of Y. Let X be the set of words of length n which have a prefix in F. Then $X = \bigcup_{y \in Y} y A^{n - |y|}$. By Equation (13.3), one has $y A^\omega = y A^{n - |y|} A^\omega$ for all $y \in Y$, and consequently $X A^\omega = Y A^\omega = F$. $\quad\square$

Lemma 13.1.11 *For every sequence $(E_n)_{n \geq 0}$ of elements of $\mathcal{F}$ such that $E = \bigcup_{n \geq 0} E_n$ is in $\mathcal{F}$, there is an integer n such that $E = E_0 \cup \cdots \cup E_n$.*

Proof. Set $E = X A^\omega$ with $X \subset A^n$. For each $x \in X$ there is an integer $m = m(x)$ such that $x A^\omega \in E_{m(x)}$. Consequently $E = \bigcup_{x \in X} E_{m(x)}$. Let m be the maximal value of the integers $m(x)$ for $x \in X$. Then $E = E_0 \cup \cdots \cup E_m$. $\quad\square$

Lemma 13.1.12 *The family $\mathcal{F}$ is a Boolean algebra.*

Proof. The empty set and the set A^ω are in $\mathcal{F}$, by taking $X = \emptyset$ and $X = \{1\}$ in the definition. Since $X A^\omega \cup Y A^\omega = (X \cup Y) A^\omega$, the family $\mathcal{F}$ is clearly closed under union.

Let $F \in \mathcal{F}$. By Lemma 13.1.10, there are an integer $n \geq 0$ and a set $X \subset A^n$ such that $F = XA^\omega$. Set $Z = A^n \setminus Y$. Then ZA^ω is in $\mathcal{F}$, and it is the complement of XA^ω. This shows that $\mathcal{F}$ is closed under complementation. □

We now start the proof of Kolmogorov's extension theorem 13.1.8.

The proof is in several steps. First, one proves the existence of a function μ on the family $\mathcal{F}$ of sets of the form XA^ω where X is a *finite* subset of A^ω. Then, the definition is extended to the family of all subsets of A^ω. It is finally proved that the extended function is a probability measure on the Borel subsets of A^ω.

Let π be a probability distribution on A^*. We define a function μ from $\mathcal{F}$ into $[0, 1]$ by setting

$$\mu(XA^\omega) = \pi(X) \tag{13.4}$$

for $X \subset A^n$. This is indeed a map from $\mathcal{F}$ into $[0, 1]$ since by Lemma 13.1.10, each F in $\mathcal{F}$ can be written in this form. We first verify that the definition is consistent, that is that the value of μ is independent of the set X. Indeed, assume that $XA^\omega = YA^\omega$ for $Y \subset A^m$ with $n < m$. Then $Y = \bigcup_{x \in X} xA^{m-n}$ and thus $\pi(Y) = \sum_{x \in X} \pi(xA^{m-n}) = \pi(X)$ by the coherence condition for π.

Proposition 13.1.13 *The function μ is a probability measure on $\mathcal{F}$.*

Proof. Clearly $\mu(\emptyset) = 0$ and $\mu(A^\omega) = \pi(1) = 1$. We first prove that μ is additive. Let $E, F \in \mathcal{F}$ be disjoint. We may suppose, by Lemma 13.1.10 that $E = XA^\omega$ and $F = YA^\omega$ where X and Y are subsets of A^m for the same integer m. Since E and F are disjoint, one has $X \cap Y = \emptyset$ and $\mu(E \cup F) = \pi(X \cup Y) = \pi(X) + \pi(Y) = \mu(E) + \mu(F)$. This shows that μ is additive.

We now prove that μ is countably additive on $\mathcal{F}$. For this, let $(E_n)_{n \geq 0}$ be a sequence of pairwise disjoint elements in $\mathcal{F}$ such that $E = \bigcup_{n \geq 0} E_n \in \mathcal{F}$. By Lemma 13.1.11, there is an integer m such that $E = E_0 \cup \cdots \cup E_m$. Since the elements of the sequence $(E_n)_{n \geq 0}$ are pairwise disjoint, this implies that $E_n = \emptyset$ for $n > m$. Since μ is additive, one has $\mu(E) = \mu(E_0) + \cdots + \mu(E_m)$. Moreover, $\mu(E_n) = 0$ for $n > m$, and consequently $\mu(E) = \sum_{n \geq 0} \mu(E_n)$. Thus μ is countably additive on $\mathcal{F}$. □

The function μ is extended to a function μ^* defined on all subsets of A^ω as follows. Given any set $E \subset A^\omega$, we denote by $\mathcal{S}(E)$ the set of sequences $(E_n)_{n \geq 0}$ of elements $E_n \in \mathcal{F}$ such that $E \subset \bigcup_{n \geq 0} E_n$.

For an arbitrary set $E \subset A^\omega$, we define

$$\mu^*(E) = \inf\left\{\sum_{n \geq 0} \mu(E_n) \,\Big|\, (E_n)_{n \geq 0} \in \mathcal{S}(E)\right\}. \tag{13.5}$$

Observe that by definition, for any $E \subset A^\omega$ and any $\varepsilon > 0$, there exists a sequence $(E_n)_{n \geq 0} \in \mathcal{S}(E)$ such that $\mu^*(E) + \varepsilon \geq \sum_{n \geq 0} \mu(E_n)$.

Lemma 13.1.14 *The function μ^* is an extension of μ on $\mathcal{F}$, that is $\mu^*(E) = \mu(E)$ for $E \in \mathcal{F}$.*

Proof. Let $E \in \mathcal{F}$. Consider the sequence $(E_n)_{n \geq 0}$ defined by $E_0 = E$ and $E_n = \emptyset$ for $n \geq 0$. Then $(E_n)_{n \geq 0} \in \mathcal{S}(E)$ and $\sum_{n \geq 0} \mu(E_n) = \mu(E)$. Therefore $\mu^*(E) \leq \mu(E)$.

For the converse inequality, let $(E_n)_{n \geq 0}$ be a sequence in $\mathcal{S}(E)$. Let $F_n = E \cap E_n$ for $n \geq 0$. Then $(F_n)_{n \geq 0}$ is a sequence of elements of $\mathcal{F}$ and $\bigcup_{n \geq 0} F_n = E$. Thus $(F_n)_{n \geq 0}$ is in $\mathcal{S}(E)$. By Lemma 13.1.11, there is an integer m such that $E = F_0 \cup \cdots \cup F_m$. It follows that

$$\mu(E) = \mu\left(\bigcup_{0 \leq n \leq m} F_n \right) \leq \sum_{0 \leq n \leq m} \mu(F_n) \leq \sum_{n \geq 0} \mu(F_n) \leq \sum_{n \geq 0} \mu(E_n).$$

The last inequality holds because μ is monotone. This inequality is true for any sequence $(E_n)_{n \geq 0}$ in $\mathcal{S}(E)$. Consequently $\mu(E) \leq \mu^*(E)$. $\qquad\square$

A function ν defined on the subsets of a set U is *countably subadditive* if, for any sequence $(E_n)_{n \geq 0}$ of subsets of U, one has $\nu(\bigcup_{n \geq 0} E_n) \leq \sum_{n \geq 0} \nu(E_n)$.

Lemma 13.1.15 *The function μ^* is monotone and countably subadditive on the set of subsets of A^ω.*

Proof. We first prove that μ^* is monotone. Let $E \subset F \subset A^\omega$. A sequence $(F_n)_{n \geq 0}$ of subsets of $\mathcal{F}$ which is in $\mathcal{S}(F)$ is also in $\mathcal{S}(E)$, that is $\mathcal{S}(F) \subset \mathcal{S}(E)$. This shows that $\mu^*(E) \leq \mu^*(F)$. Thus μ^* is monotone.

We next show that μ^* is countably subadditive on the subsets of A^ω. Let $(E_n)_{n \geq 0}$ be a sequence of subsets of A^ω. For any $\varepsilon > 0$ and for each $n \geq 0$, there exists, by the definition of $\mu^*(E_n)$, a sequence $(E_{n,m})_{m \geq 0}$ of subsets of $\mathcal{F}$ such that $\sum_{m \geq 0} \mu(E_{n,m}) \leq \mu^*(E_n) + \varepsilon/2^{n+1}$. Set $E = \bigcup_{n \geq 0} E_n$. Since $\bigcup_{n,m \geq 0} E_{n,m} \supset \bigcup_{n \geq 0} E_n = E$, the family $(E_{n,m})_{n,m \geq 0}$ is in $\mathcal{S}(E)$. By definition of μ^*, one has

$$\mu^*(E) \leq \sum_{n \geq 0} \sum_{m \geq 0} \mu(E_{n,m}).$$

By the choice of the sequences $(E_{n,m})_{m \geq 0}$, it follows that

$$\sum_{n \geq 0} \sum_{m \geq 0} \mu(E_{n,m}) \leq \sum_{n \geq 0} \left(\mu^*(E_n) + \varepsilon/2^{n+1} \right) = \varepsilon + \sum_{n \geq 0} \mu^*(E_n).$$

This inequality holds for all ε. It follows that $\mu^*(E) \leq \sum_{n \geq 0} \mu^*(E_n)$. $\qquad\square$

In the next proposition, we denote by $\overline{E}$ the complement of E.

Proposition 13.1.16 *Let $\mathcal{U}$ be the family of subsets E of A^ω such that, for all $H \subset A^\omega$,*

$$\mu^*(H) = \mu^*(H \cap E) + \mu^*(H \cap \overline{E}).$$

The family $\mathcal{U}$ contains all Borel subsets of A^ω and μ^* is countably additive on $\mathcal{U}$.

Proof. The proof is in several steps.

1. We first show that $\mathcal{U}$ contains $\mathcal{F}$. Let $E \in \mathcal{F}$ and $H \subset A^\omega$. By the definition of $\mu^*(H)$, there exists, for any $\varepsilon > 0$ a sequence $(H_n)_{n \geq 0}$ in $\mathcal{S}(H)$ such that $\mu^*(H) +$

$\varepsilon \geq \sum_{n\geq 0} \mu(H_n)$. Next, $\mu(H_n) = \mu(H_n \cap E) + \mu(H_n \cap \overline{E})$ for all $n \geq 0$, and the sequence $(H_n \cap E)_{n\geq 0}$ is in $\mathcal{S}(H \cap E)$, and similarly $(H_n \cap \overline{E})_{n\geq 0}$ is in $\mathcal{S}(H \cap \overline{E})$. Consequently

$$\mu^*(H) + \varepsilon \geq \sum_{n\geq 0} \mu(H_n) = \sum_{n\geq 0}(\mu(H_n \cap E) + \mu(H_n \cap \overline{E}))$$

$$\geq \mu^*(H \cap E) + \mu^*(H \cap \overline{E}).$$

This inequality holds for any ε, whence $\mu^*(H) \geq \mu^*(H \cap E) + \mu^*(H \cap \overline{E})$. Moreover, since $H = (H \cap E) \cup (H \cap \overline{E})$, we have

$$\mu^*(H) = \mu((H \cap E) \cup (H \cap \overline{E})) \leq \mu^*(H \cap E) + \mu^*(H \cap \overline{E})$$

because μ^* is subadditive by Lemma 13.1.15. Thus $\mu^*(H) = \mu^*(H \cap E) + \mu^*(H \cap \overline{E})$ and this shows that $E \in \mathcal{U}$.

2. Next we prove that $\mathcal{U}$ is closed under union. Let indeed $E_1, E_2 \in \mathcal{U}$ and $H \subset A^\omega$. We have

$$\mu^*(H) = \mu^*(H \cap E_1) + \mu^*(H \cap \overline{E_1})$$

$$= \mu^*(H \cap E_1) + \mu^*(H \cap \overline{E_1} \cap E_2) + \mu^*(H \cap \overline{E_1} \cap \overline{E_2}).$$

The first two terms of the right-hand side sum to $\mu^*(H \cap (E_1 \cup E_2))$. Indeed, since $E_1 \in \mathcal{U}$, one has

$$\mu^*(H \cap (E_1 \cup E_2)) = \mu^*((H \cap (E_1 \cup E_2)) \cap E_1) + \mu^*((H \cap (E_1 \cup E_2)) \cap \overline{E_1})$$

and next $H \cap (E_1 \cup E_2) \cap E_1 = H \cap E_1$ and $H \cap (E_1 \cup E_2) \cap \overline{E_1} = H \cap \overline{E_1} \cap E_2$. Since $\overline{E_1} \cap \overline{E_2}$ is the complement of $E_1 \cup E_2$, it follows that $E_1 \cup E_2$ is in $\mathcal{U}$. Thus $\mathcal{U}$ is closed under union. It is clearly closed under complement and thus it is a Boolean algebra. If moreover E_1 and E_2 are disjoint, then

$$\mu^*(H \cap (E_1 \cup E_2)) = \mu^*(H \cap E_1) + \mu^*(H \cap E_2) \tag{13.6}$$

because then $H \cap (E_1 \cup E_2) \cap E_1 = H \cap E_1$ and $H \cap (E_1 \cup E_2) \cap \overline{E_1} = H \cap E_2$.

3. We show that $\mathcal{U}$ is closed under countable union and that μ^* is countably additive on $\mathcal{U}$. Consider first a sequence $(E_n)_{n\geq 0}$ of pairwise disjoint elements of $\mathcal{U}$. Set $E = \bigcup_{n\geq 0} E_n$.

Let $H \subset A^\omega$. Since the sets E_n are pairwise disjoint, it follows from (13.6) that for all $m \geq 0$, one has $\mu^*(H \cap \bigcup_{n\leq m} E_n) = \sum_{n\leq m} \mu^*(H \cap E_n)$. Set $F_m = \bigcup_{n\leq m} E_n$. The inclusion $F_m \subset E$ implies $\overline{F_m} \supset \overline{E}$ whence $H \cap \overline{F_m} \supset H \cap \overline{E}$.

Since $\mathcal{U}$ is a Boolean algebra, one has $F_m, \overline{F_m} \in \mathcal{U}$, and since μ^* is monotone, one gets $\mu^*(H \cap \overline{F_m}) \geq \mu^*(H \cap \overline{E})$. It follows that

$$\mu^*(H) = \mu^*(H \cap F_m) + \mu^*(H \cap \overline{F_m}) \geq \sum_{n\leq m} \mu^*(H \cap E_n) + \mu^*(H \cap \overline{E}).$$

This is true for every m, and consequently

$$\mu^*(H) \geq \sum_{n \geq 0} \mu^*(H \cap E_n) + \mu^*(H \cap \overline{E}) \geq \mu^*(H \cap E) + \mu^*(H \cap \overline{E}).$$

On the other hand, since μ^* is (countably) subadditive on all subsets of A^ω by Lemma 13.1.15, one has the inequality $\mu^*(H) = \mu^*((H \cap E) \cup (H \cap \overline{E})) \leq \mu^*(H \cap E) + \mu^*(H \cap \overline{E})$. This implies the equality

$$\mu^*(H) = \sum_{n \geq 0} \mu^*(H \cap E_n) + \mu^*(H \cap \overline{E}) = \mu^*(H \cap E) + \mu^*(H \cap \overline{E}).$$

This shows that $\mathcal{U}$ is closed under disjoint countable unions. To show that $\mathcal{U}$ is closed under all countable unions, consider any sequence $(E_n)_{n \geq 0}$ of elements in $\mathcal{U}$. Set $E = \bigcup_{n \geq 0} E_n$, and set $F_n = E_n \setminus (E_0 \cup \cdots \cup E_{n-1})$ for $n \geq 0$. The sets F_n are in $\mathcal{U}$ because $\mathcal{U}$ is a Boolean algebra. Moreover $\bigcup_{n \geq 0} F_n = E$. Thus E is a disjoint countable union and by the preceding proof, E is in $\mathcal{U}$.

Since the family $\mathcal{U}$ is a Boolean algebra containing $\mathcal{F}$ and closed under countable unions, it contains the family of Borel subsets of A^ω. It remains to show that μ^* is countably additive on $\mathcal{U}$. For this let $(E_n)_{n \geq 0}$ be a sequence of pairwise disjoint elements in $\mathcal{U}$ and set $E = \bigcup_{n \geq 0} E_n$. Then Equation (13.1) holds for any set H, and in particular for H replaced by E. This gives the equality

$$\mu^*(E) = \sum_{n \geq 0} \mu^*(E_n),$$

showing that μ^* is countably additive on $\mathcal{U}$. □

Proof of Theorem 13.1.8. Let π be a probability distribution on A^*, let μ be defined by Equation (13.4) and let μ^* be defined by Equation (13.5). By Proposition 13.1.16, μ^* is countably additive on the family of Borel subsets of A^ω, and therefore is a probability measure on this family.

To prove uniqueness, let μ' be another probability measure on the Borel subsets of A^ω such that $\mu'(x A^\omega) = \pi(x)$ for $x \in A^*$. Then $\mu' = \mu$ on $\mathcal{F}$ because μ' is additive. Next, let E be a subset of A^ω and let $(E_n)_{n \geq 0}$ be in $\mathcal{S}(E)$. Define $F_n = E_n \setminus (E_0 \cup \cdots \cup E_{n-1})$. Then $E \subset \bigcup_{n \geq 0} E_n = \bigcup_{n \geq 0} F_n$, and one has $\mu'(E) \leq \mu'(\bigcup_{n \geq 0} F_n) = \sum_{n \geq 0} \mu'(F_n) \leq \sum_{n \geq 0} \mu'(E_n)$.

Since $\mu' = \mu$ on $\mathcal{F}$ and $E_n \in \mathcal{F}$ for all $n \geq 0$, one has $\mu'(E) \leq \sum_{n \geq 0} \mu(E_n)$. This holds for all sequences $(E_n)_{n \geq 0}$ in $\mathcal{S}(E)$, and thus $\mu'(E) \leq \mu^*(E)$. By the same argument, $\mu'(\overline{E}) \leq \mu^*(\overline{E})$. Since $\mu^*(E) + \mu^*(\overline{E}) = \mu'(E) + \mu'(\overline{E}) = 1$ for a Borel subset, this forces $\mu'(E) = \mu^*(E)$. This shows the uniqueness. □

Example 13.1.17 Let $X \subset A^*$ be a prefix code. For any probability distribution π, with corresponding probability measure μ, one has

$$\mu(X^\omega) = \lim_{n \to \infty} \pi(X^n). \tag{13.7}$$

Indeed, we first observe that if $E = \bigcup_{n \geq 0} E_n$ for Borel subsets of A^ω, and $E_n \subset E_{n+1}$ for $n \geq 0$, then $\mu(E) = \lim_{n \to \infty} \mu(E_n)$. To see this, set $F_n = E_n \setminus (E_0 \cup \cdots \cup E_{n-1})$ for $n \geq 0$. Then the sets F_n are pairwise disjoint and since μ is countable additive, $\mu(E) = \sum_{n \geq 0} \mu(F_n)$. Next $\sum_{i \leq n} \mu(F_i) = \mu(E_n)$, which implies that $\sum_{n \geq 0} \mu(F_n) = \lim_{n \to \infty} \mu(E_n)$. By taking the complements, it follows that if $E = \bigcap_{n \geq 0} E_n$ and $E_n \supset E_{n+1}$ for $n \geq 0$, then again $\mu(E) = \lim_{n \to \infty} \mu(E_n)$. These conditions are satisfied for $E = X^\omega$ and $E_n = X^n A^\omega$ by Equation (13.1). Therefore $\mu(X^\omega) = \lim_{n \to \infty} \mu(X^n A^\omega) = \lim_{n \to \infty} \pi(X^n)$.

Example 13.1.18 Let D be the Dyck code on $A = \{a, b\}$. Let π be a Bernoulli distribution on A^* and set $p = \pi(a)$ and $q = \pi(b)$. By Example 2.4.10, we have $\pi(D) = 1 - |p - q|$. Let μ be the measure on A^ω corresponding to π. If $p \neq q$, then $\pi(D)^n \to 0$ for $n \to \infty$ and by (13.7) $\mu(D^\omega) = 0$. This means that with probability one, the event that the number of occurrences of a and b are equal will occur a finite number of times. If $p = q$, then $\pi(D)^n = 1$ for all n and $\mu(D^\omega) = 1$. This means that the same event will occur infinitely often with probability one.

Example 13.1.19 Consider the function π defined on $A^* = \{a, b\}^*$ as follows. For $x \notin a^* b^*$, one has $\pi(x) = 0$, and for $n \geq 0$, $j > 0$,

$$\pi(a^n) = 2^{-n}, \quad \pi(a^n b^j) = 2^{-n-1}.$$

Then $\pi(a) = \pi(b) = 1/2$, and $\pi(a^n) = \pi(a^{n+1}) + \pi(a^n b)$, $\pi(a^n b^j) = \pi(a^n b^{j+1})$. Thus π satisfies the coherence condition and therefore is a probability distribution on A^*. This corresponds to the following experiment: a and b are chosen at random with equal probability until the occurrence of the first b. Afterwards, the outcome is always b. The probability of no occurrence of a is $1/2$.

The probability measure μ corresponding to π is such that $\mu(b^\omega) = \mu(a A^\omega) = 1/2$. The maximal prefix code $X = b^* a$ is such that $\pi(X) = 1/2$ since $\pi(b^n a) = 0$ for $n > 0$. This is consistent with the fact that $A^\omega = X A^\omega \cup b^\omega$ and thus $1 = \mu(X A^\omega) + 1/2$.

13.2 Densities

We use the notation

$$A^{(n)} = \{1\} \cup A \cup \cdots \cup A^{n-1}.$$

In particular $A^{(0)} = \emptyset$, $A^{(1)} = \{1\}$.

Let π be a probability distribution on A^*. Let L be a subset of A^*. The set L is said to *have a density* with respect to π if the sequence of the $\pi(L \cap A^n)$ converges in mean, that is, if

$$\lim_{n \to \infty} \frac{1}{n} \sum_{k=0}^{n-1} \pi(L \cap A^k)$$

exists. If this is the case, the *density* of L (relative to π) denoted by $\delta(L)$, is this limit, which can also be written as

$$\delta(L) = \lim_{n\to\infty} (1/n)\pi(L \cap A^{(n)}).$$

An elementary result from analysis shows that if the sequence $\pi(L \cap A^n)$ has a limit, then its limit in mean also exists, and both are equal. This remark may sometimes simplify computations. Observe that $\delta(A^*) = 1$ and

$$0 \le \delta(L) \le 1$$

for any subset L of A^* having a density. If L and M are subsets of A^* having a density, then so has $L \cup M$, and

$$\delta(L \cup M) \le \delta(L) + \delta(M).$$

If $L \cap M = \emptyset$, and if two of the three sets L, M and $L \cup M$ have a density, then the third one also has a density and

$$\delta(L \cup M) = \delta(L) + \delta(M).$$

The function δ is a partial function from $\mathfrak{P}(A^*)$ into $[0, 1]$. Of course, $\delta(\{w\}) = 0$ for all $w \in A^*$. This shows that in general

$$\delta(L) \ne \sum_{w\in L} \delta(\{w\}).$$

Observe that if $\pi(L) < \infty$, then $\delta(L) = 0$ since $\pi(L \cap A^{(n)}) \le \pi(L)$, whence

$$\lim_{n\to\infty} \frac{1}{n}\pi(L \cap A^{(n)}) = 0.$$

Example 13.2.1 Let $L = (A^2)^*$ be the set of words of even length. Then

$$\pi(L \cap A^{(2k)}) = \pi(L \cap A^{(2k-1)}) = k.$$

Thus $\delta(L) = \frac{1}{2}$.

Example 13.2.2 Let $D^* = \{w \in A^* \mid |w|_a = |w|_b\}$ over $A = \{a, b\}$. The set D is the *Dyck code* (see Example 2.4.10). Let π be a Bernoulli distribution and set $p = \pi(a)$, $q = \pi(b)$. Then

$$\pi(D^* \cap A^{2n}) = \binom{2n}{n}p^n q^n, \quad \pi(D^* \cap A^{2n+1}) = 0.$$

Recall that *Stirling's formula* gives the following asymptotic equivalent for $n!$:

$$n! \sim \left(\frac{n}{e}\right)^n \sqrt{2\pi n}.$$

Using this formula, we get

$$\pi(D^* \cap A^{2n}) \sim \frac{1}{\sqrt{\pi n}} 4^n (pq)^n,$$

Since $pq \le 1/4$ for all values of p and q, this shows that $\lim_{n\to\infty} \pi(D^* \cap A^{2n}) = 0$. Thus $\delta(D^*) = 0$.

The definition of density clearly depends only on the values of the numbers $\pi(L \cap A^n)$. It appears to be useful to consider an analogous definition for power series. Let $f = \sum_{n \ge 0} f_n t^n$ be a power series. The *density* of f, denoted by $\delta(f)$ is the limit in mean, provided it exists, of the sequence f_n,

$$\delta(f) = \lim_{n\to\infty} \frac{1}{n} \sum_{i=0}^{n-1} f_i.$$

Recall from Section 1.11 that the probability generating series, denoted by $F_L(t)$, of a set $L \subset A^*$, is defined by

$$F_L(t) = \sum_{n \ge 0} \pi(L \cap A^n) t^n.$$

Clearly $F_L(t)$ has a density if and only if L has a density, and

$$\delta(L) = \delta(F_L).$$

We denote by ρ_L the *radius of convergence* of the series $F_L(t)$. Recall (see Section 1.8) that it is infinite if $F_L(z)$ converges for all real numbers, or it is the unique real positive number $\rho \in \mathbb{R}_+$, such that $F_L(z)$ converges for $|z| < \rho$ and diverges for $|z| > \rho$. For any set L, we have $\rho_L \ge 1$ since $\pi(L \cap A^n) \le 1$ for all $n \ge 0$.

The following proposition is a more precise formulation of Proposition 2.5.12. It implies Proposition 2.5.12, since if $\rho_L > 1$, then $\pi(L) = F_L(1)$ is finite.

Proposition 13.2.3 *Let L be a subset of A^* and let π be a positive Bernoulli distribution. If L is thin, then $\rho_L > 1$ and $\delta(L) = 0$.*

Proof. Let w be a word which is not a factor of a word of L and set $n = |w|$. Then we have, for $0 \le i < n$ and $k \ge 0$,

$$L \cap A^i (A^n)^k \subset A^i (A^n \setminus w)^k.$$

Hence

$$\pi(L \cap A^i (A^n)^k) \le (1 - \pi(w))^k.$$

Thus for any $\rho > 0$ satisfying $(1 - \pi(w))\rho^n < 1$, we have

$$F_L(\rho) \le \sum_{i=0}^{n-1} \sum_{k=0}^{\infty} (1 - \pi(w))^k \rho^{i+kn} = \sum_{i=0}^{n-1} \rho^i \left[\sum_{k=0}^{\infty} ((1 - \pi(w))\rho^n)^k \right] < +\infty.$$

This proves that

$$\rho_L \geq \left(\frac{1}{1-\pi(w)}\right)^{1/n} > 1.$$

This shows that $F_L(1)$ is finite, and consequently $\lim_{n\to\infty} \pi(L \cap A^n) = 0$. Therefore $\delta(L) = 0$. $\qquad\square$

For later use, we need an elementary result concerning the convergence of certain series. For the sake of completeness we include the proof.

Proposition 13.2.4 *Let* $f(t) = \sum_{n\geq 0} f_n t^n$, $g(t) = \sum_{n\geq 0} g_n t^n$ *be two power series satisfying*

(i) $0 < g(1) < \infty$,
(ii) $0 \leq f_n \leq 1$ *for all* $n \geq 0$.

Then $\delta(f)$ *exists if and only if* $\delta(fg)$ *exists and in this case, one has*

$$\delta(fg) = \delta(f)g(1). \tag{13.8}$$

Proof. Set

$$h = fg = \sum_{n=0}^{\infty} h_n t^n.$$

Then for $n \geq 1$,

$$\left(\sum_{i=0}^{n-1} f_i\right) g(1) = \left(\sum_{i=0}^{n-1} f_i\right)\left(\sum_{j=0}^{\infty} g_j\right) = \sum_{0 \leq i+j \leq n-1} f_i g_j + \sum_{i=0}^{n-1} f_i\left(\sum_{j=n-i}^{\infty} g_j\right)$$

$$= \sum_{k=0}^{n-1} h_k + \sum_{i=0}^{n-1} f_i r_{n-i},$$

where $r_i = \sum_{j=i}^{\infty} g_j$. Let $s_n = \sum_{i=0}^{n-1} f_i r_{n-i}$. Then for $n \geq 1$,

$$\left(\frac{1}{n}\sum_{i=0}^{n-1} f_i\right) g(1) = \left(\frac{1}{n}\sum_{k=0}^{n-1} h_k\right) + \frac{1}{n}s_n. \tag{13.9}$$

Furthermore

$$s_n = \sum_{i=0}^{n-1} f_i r_{n-i} \leq \sum_{i=0}^{n-1} r_{n-i} = \sum_{i=1}^{n} r_i. \tag{13.10}$$

Since $\sum g_n$ converges, we have $\lim_{i\to\infty} r_i = 0$. This shows that

$$\lim_{n\to\infty} \frac{1}{n}\sum_{i=1}^{n-1} r_i = 0,$$

and in view of (13.10),

$$\lim_{n\to\infty} \frac{1}{n} s_n = 0.$$

Since $g(1) \neq 0$, Equation (13.9) shows that $\delta(f)$ exists if and only if $\delta(h)$ exists and that $\delta(f)g(1) = \delta(h)$. This proves (13.8) and the proposition. $\qquad\square$

Proposition 13.2.5 *Let π be a positive Bernoulli distribution on A^*. Let L, M be subsets of A^* such that*

(i) $0 < \pi(M) < \infty$,
(ii) *the product LM is unambiguous.*

Then LM has a density if and only if L has a density, and if this is the case,

$$\delta(LM) = \delta(L)\pi(M). \qquad (13.11)$$

Proof. Since the product LM is unambiguous, we have

$$F_{LM} = F_L F_M.$$

In view of the preceding proposition

$$\delta(LM) = \delta(F_{LM}) = \delta(F_L)\sigma,$$

where $\sigma = \sum_{n\geq 0} \pi(M \cap A^n) = \pi(M)$. $\qquad\square$

This proposition will be useful below. Note that the symmetric version with LM replaced by ML also holds. As a first illustration of its use, we note the following corollary.

Corollary 13.2.6 *Each right (left) ideal I of A^* has a nonnull density. More precisely $\delta(I) = \pi(X)$, where $X = I \setminus IA^+$.*

Proof. Let I be a right ideal and let $X = I \setminus IA^+$. By Proposition 3.1.2, the set X is prefix and

$$I = XA^*.$$

The product XA^* is unambiguous because X is prefix. Further $\pi(X) \leq 1$ since X is a code, and $\pi(X) > 0$ since $I \neq \emptyset$ and consequently also $X \neq \emptyset$. Thus, applying the (symmetrical version of the) preceding proposition, we obtain

$$\delta(I) = \delta(XA^*) = \pi(X)\delta(A^*) = \pi(X) \neq 0. \qquad\square$$

Let X be a code over A. Then $\pi(X) \leq 1$ and $\pi(X) = 1$ if X is thin and complete. For a code X such that $\pi(X) = 1$ we define the *average length* of X (relative to π)

as the finite or infinite number $\lambda(X)$ defined by

$$\lambda(X) = \sum_{x \in X} |x| \pi(x) = \sum_{n \geq 0} n \, \pi(X \cap A^n). \tag{13.12}$$

The following fundamental theorem gives a link between the density and the average length.

Theorem 13.2.7 *Let $X \subset A^+$ be a code and let π be a positive Bernoulli distribution. If*

(i) $\pi(X) = 1$,

(ii) $\lambda(X) < \infty$,

then X^* has a density and $\delta(X^*) = 1/\lambda(X)$.

The theorem is a combinatorial interpretation of the following property of power series.

Proposition 13.2.8 *Let $f(t) = \sum_{n \geq 0} f_n t^n$ be a power series with real nonnegative coefficients, and with zero constant term. If $f(1) = 1$ and $f'(1) < \infty$, then*

$$\delta\left(\frac{1}{1 - f(t)}\right) = \frac{1}{f'(1)}.$$

Proof. Let $g(t) = \sum_{n=0}^{\infty} g_n t^n$ be defined by

$$g(t) = \frac{1 - f(t)}{1 - t}, \tag{13.13}$$

which can also be written as $f(t) = 1 + (t - 1)g(t)$. Identifying terms, we get $f_0 = 1 - g_0$ and $f_n = g_{n-1} - g_n$ for $n \geq 1$, whence for $n \geq 0$, $g_n = 1 - \sum_{i=0}^{n} f_i$. Since $f(1) = 1$, it follows that

$$g_n = \sum_{i=n+1}^{\infty} f_i.$$

By this equation, one has $g_n \geq 0$ for $n \geq 0$. Moreover

$$g(1) = \sum_{n=0}^{\infty} g_n = \sum_{n=0}^{\infty} \sum_{i=n+1}^{\infty} f_i = \sum_{i=0}^{\infty} i f_i = f'(1). \tag{13.14}$$

Since at least one f_i, for $i \geq 1$, is not null because $\sum_{i \geq 1} f_i = 1$, one has $f'(1) > 0$. Next

$$\frac{1}{1 - t} = \frac{1}{1 - f(t)} g(t). \tag{13.15}$$

Since $f'(1)$ is finite and not zero, we can apply Proposition 13.2.4 to (13.15), with f replaced by $1/(1 - f)$, provided we check that the coefficients of the series $1/(1 - f)$

are nonnegative and less than or equal to 1. This holds by (13.15), because $g(t)$ is not null.

Now $\delta(1/(1 - t)) = 1$, consequently in view of (13.8), Formula (13.15) gives

$$1 = \delta(\frac{1}{1 - f(t)})f'(1).$$

$\square$

Proof of Theorem 13.2.7. Set $f_n = \pi(X \cap A^n)$. Then $F_X(t) = \sum_{n=0}^{\infty} f_n t^n$. Since X is a code, $F_X(t)$ has zero constant term, and by assumption $F_X(1) = \pi(X) = 1$. We have as a consequence of Proposition 2.1.15,

$$F_{X^*}(t) = (1 - F_X(t))^{-1}. \tag{13.16}$$

Next $\lambda(X) = F_X'(1) < \infty$, so we can apply the previous proposition. This gives the formula.

$\square$

Note the following important special case of Theorem 13.2.7.

Theorem 13.2.9 *Let X be a thin complete code over A, and let π be a positive Bernoulli distribution. Then X^* has a density. Further $\delta(X^*) > 0$, $\lambda(X) < \infty$, and $\delta(X^*) = 1/\lambda(X)$.*

Proof. Since X is a thin and complete code, $\pi(X) = 1$. Next, since X is thin, $\rho_X > 1$ by Proposition 13.2.3. Thus the derivative of $F_X(t)$ which is the series

$$F_X'(t) = \sum_{n \geq 1} n \pi(X \cap A^n) t^{n-1},$$

also has a radius of convergence strictly greater than 1. Hence $F_X'(1)$ is finite. Now

$$F_X'(1) = \sum_{n \geq 1} n \pi(X \cap A^n) = \lambda(X).$$

Therefore $\lambda(X) < \infty$ and the hypotheses of Theorem 13.2.7 are satisfied.

$\square$

Example 13.2.10 Let X be a thin maximal bifix code. Then $\lambda(X) = d(X)$ by Corollary 6.3.16. Thus $\delta(X^*) = 1/d(X)$.

In the case of a prefix code, Theorem 13.2.7 holds for more general probability distributions. Recall from Section 3.7 that a *persistent recurrent event* on the alphabet A is a pair (X, π) composed of a prefix code X and a probability distribution π which is multiplicative on X^* and such that $\pi(X) = 1$.

Theorem 13.2.11 *Let (X, π) be a persistent recurrent event over an alphabet A. If $\lambda(X) < \infty$, then the density of X^* exists and $\delta(X^*) = 1/\lambda(X)$.*

Proof. We verify that the assumptions of Proposition 13.2.8 are satisfied for $f(t) = F_X(t)$. We have $F_X(1) = \pi(X) = 1$ since the recurrent event is persistent. Next, $F_X'(1) = \lambda(X)$ by Proposition 3.7.10. Thus $F_X'(1) < \infty$.

By Proposition 13.2.8, $\delta(1/(1 - F_X(t))) = 1/\lambda(X)$. Finally, $F_{X^*}(t) = 1/(1 - F_X(t))$ by Proposition 3.7.3. This shows that $\delta(X^*) = \delta(F_{X^*}(t)) = \delta(1/(1 - F_X(t))) = 1/\lambda(X)$. $\qquad\qquad\square$

13.3 Entropy

Given a set $X \subset A^*$, recall that the generating series of X is $f_X(t) = \sum_{n \geq 1} \text{Card}(X \cap A^n)t^n$. It is related to the probability generating series corresponding to the uniform Bernoulli distribution by $f_X(t) = F_X(kt)$ with $k = \text{Card}(A)$.

The *topological entropy* of a set $X \subset A^*$ is $h(X) = -\log r_X$ where r_X is the radius of convergence of the series $f_X(t)$. By convention, $h(X) = 0$ if $r_X = \infty$. In particular, $h(A^*) = \log k$ with $k = \text{Card}(A)$. Also $X \subset Y$ implies $h(X) \leq h(Y)$. Thus

$$0 \leq h(X) \leq \log k$$

with $k = \text{Card}(A)$.

Recall that $F(X)$ denotes the set of factors of words in X.

Proposition 13.3.1 *For any rational set $X \subset A^*$, one has $h(X) = h(F(X))$. In particular, if the set X is dense, then $h(X) = \log k$ with $k = \text{Card}(A)$.*

Given a probability distribution π on A^* and a set $X \subset A^*$, recall that ρ_X denotes the radius of convergence of the probability generating function $F_X(t)$ of X.

The proposition is a consequence of the following statement.

Proposition 13.3.2 *Let X be a rational set and let Y be the set of factors of the words of X. Then for any positive Bernoulli distribution π, one has $\rho_X = \rho_Y$.*

Proof. Let $F_X(t) = \sum_{n \geq 0} a_n t^n$ and $F_Y(t) = \sum_{n \geq 0} b_n t^n$. Let $\mathcal{A}$ be a trim finite automaton recognizing X with set of states Q. For each state q, there are words u_q and v_q, an initial state i_q, and a terminal state t_q such that $i_q \overset{u_q}{\to} q \overset{v_q}{\to} t_q$. For each word w of length n in Y, there exists a path $p \overset{w}{\to} q$ in $\mathcal{A}$ and, therefore, also words u_p and v_q such that $u_p w v_q \in X$ and conversely. Thus

$$Y = \bigcup_{p,q \in Q} u_p^{-1} X v_q^{-1}.$$

Let $w \in Y$, and u_p, v_q be words such that $u_p w v_q \in X$ and set $x = u_p w v_q$. Since π is a positive Bernoulli distribution, one has $\pi(w) = \frac{\pi(x)}{\pi(u_p)\pi(v_q)}$. Consequently, for each $n \geq 0$

$$\pi(u_p^{-1} X v_q^{-1}) = \frac{\pi(X \cap u_p A^n v_q)}{\pi(u_p)\pi(v_q)}.$$

Setting $m = \min_{p,q \in Q} \pi(u_p)\pi(v_q)$, one gets

$$\pi(Y \cap A^n) = \sum_{p,q \in Q} \frac{\pi(X \cap u_p A^n v_q)}{\pi(u_p)\pi(v_q)} \leq \frac{\pi(X \cap A^n) + \cdots + \pi(X \cap A^{n+k+\ell})}{m},$$

where k is the maximal length of the words u_p and ℓ is the maximal length of the words v_q. It follows that $a_n \leq b_n \leq \frac{1}{m}(a_n + a_{n+1} + \cdots + a_{n+k+\ell})$. This shows that the series $F_X(t)$ and $F_Y(t)$ have the same radius of convergence, because the operations of shift, addition, and multiplication by a nonzero scalar do not change the convergence radius. $\square$

Proof of Proposition 13.3.1. By definition, $h(X) = \log r_X$, where r_X is the radius of convergence of $f_X(t)$. Since $f_X(t) = F_X(kt)$ for the uniform Bernoulli distribution, with $k = \mathrm{Card}(A)$, one has $\rho_X = r_X/k$. Consequently $r_X = k\rho_X = k\rho_{F(X)} = r_{F(X)}$ by Proposition 13.3.2. $\square$

We will prove the following result.

Theorem 13.3.3 *Let X be a nonempty rational code. One has $h(X^*) = -\log r$, where r is the unique positive real number such that $f_X(r) = 1$.*

This is a consequence of the following more general statement.

Theorem 13.3.4 *Let X be a nonempty rational code and let π be a positive Bernoulli distribution. Then ρ_{X^*} is the unique positive real number r such that $F_X(r) = 1$.*

Theorem 13.3.4 implies that $\pi(X) = 1$ for a complete rational code (see Theorem 2.5.16). Indeed, we have $\rho_{X^*} = \rho_{F(X^*)}$ since X^* is rational by Proposition 13.3.2. Since X is complete, we have $F(X^*) = A^*$ and thus $\rho_{X^*} = \rho_{F(X^*)} = 1$. By Theorem 13.3.4 $F_X(1) = 1$. Since $\pi(X) = F_X(1)$, the claim follows.

For the proof of Theorem 13.3.4, we first prove the following statement.

Proposition 13.3.5 *Let $X \subset A^*$ be a nonempty code and let π be a positive Bernoulli distribution on A^*. If $\rho_X < \rho_{X^*}$, then ρ_{X^*} is the unique positive root of $F_X(r) = 1$.*

Proof. Since $F_{X^*}(t) = 1/(1 - F_X(t))$, the statement is a direct application of Proposition 1.8.4. $\square$

We will show that the hypothesis of Proposition 13.3.5 is satisfied for a rational code. We first prove the following result.

Proposition 13.3.6 *Let $X \subset A^+$ be a nonempty rational set. Then $F_X(\rho_X) = \infty$, that is $\rho_X = \infty$ or ρ_X is a pole of $F_X(t)$.*

Proof. We use induction on the number of operations in an unambiguous rational expression for X, see Section 4.1. The result holds if X is finite since then $\rho_X = \infty$. Next, the cases of a disjoint union and unambiguous product are straightforward. Finally, consider the case $X = Y^*$ with Y a code. Since $F_Y(\rho_Y) = \infty$ by induction hypothesis, and $F_Y(t)$ is continuous inside its interval of convergence, there exists $r > 0$ such that $F_Y(r) = 1$. Since Y is a code, one has $F_X(t) = \sum_{n \geq 0} F_Y(t)^n$. Since $F_Y(r) = 1$, one has $F_X(r) = \infty$. If $0 < s < r$, then $F_Y(s) < 1$ and thus $F_X(s)$ converges. This shows that r is the radius of convergence of $F_X(t)$. $\square$

The following example shows that Proposition 13.3.6 is not true without the hypothesis that X is rational.

Example 13.3.7 Let D be the Dyck code on the alphabet $A = \{a, b\}$. Let π be the uniform Bernoulli distribution on A. We have seen (Example 2.4.10) that $F_D(t) = 1 - \sqrt{1 - t^2}$. Thus $\rho_D = 1$. Since $\rho_{D^*} \le \rho_D$, this implies $\rho_{D^*} = 1$ although $F_D(1) = 1$.

Proof of Theorem 13.3.4. By Proposition 13.3.6, we have $F_X(\rho_X) = \infty$. Therefore, there is an $r > 0$ such that $F_X(r) = 1$. Since $F_{X^*}(t) = \sum_{n \ge 0} F_X(t)^n$, the series $F_{X^*}(t)$ converges for $t < r$ and diverges for $t = r$. This shows that $\rho_{X^*} = r$. □

The following example shows that Theorem 13.3.4 is not true for very thin codes.

Example 13.3.8 Let $A = \{a, b, c\}$ and let D be the Dyck code on $\{a, b\}$. Consider the prefix code $X = c^2 \cup D_a$ where $D_a = D \cap aA^*$. The code X is very thin since $c^4 \in X^*$ but $c^4 \notin F(X)$. Let π be the uniform Bernoulli distribution on A. We have $F_{D_a}(t) = f_{D_a}(t/3)$. On the other hand, $f_{D_a}(t) = 1/2 f_D(t)$, and $f_D(t) = F_D(2t)$, where $F_D(t)$ denotes the probability generating series for the uniform Bernoulli distribution on the alphabet $\{a, b\}$. Consequently $f_{D_a}(t) = (1 - \sqrt{1 - 4t^2})/2$ and thus $F_{D_a}(t) = (1 - \sqrt{1 - 4t^2/9})/2$. This shows that $\rho_{X^*} = \rho_{D_a} = 3/2$, although $F_X(3/2) = 1/4 + 1/2 = 3/4 < 1$.

Proof of Theorem 13.3.3. It is a direct consequence of Theorem 13.3.4 in the case of the uniform Bernoulli distribution. □

Example 13.3.9 Let $A = \{a, b\}$ and let $X = \{a, ba\}$. We have $f_X(t) = t + t^2$ and $h(X^*) = \log(1 + \sqrt{5})/2$.

The next example is an illustration of the use of Proposition 13.3.5 to compute the topological entropy of non rational codes.

Example 13.3.10 Let $A = \{a, b\}$ and let $X = \{a^n b^n \mid n \ge 1\}$. We have $f_X(t) = \sum_{n \ge 1} t^{2n} = t^2/(1 - t^2)$. Since $f_X(1/\sqrt{2}) = 1$, the topological entropy of X^* is $(\log 2)/2$.

The following result gives a useful relation between the entropy of X^* when X is a rational code and the spectral radius of the adjacency matrix of an unambiguous automaton recognizing X^*.

Proposition 13.3.11 *Let X be a rational code. Let $\mathcal{A} = (Q, 1, 1)$ be a trim unambiguous automaton recognizing X^*. The topological entropy of X^* is $h(X^*) = \log \lambda$, where λ is the spectral radius of the adjacency matrix of $\mathcal{A}$.*

Proof. Let M be the adjacency matrix of $\mathcal{A}$ and let $N_{p,q}(t)$ be the coefficient of index p, q of the matrix $N(t) = (I - Mt)^{-1}$. Since $I + N(t)Mt = N(t)$, we have $\delta_{p,q} + t \sum_{s \in Q} N_{p,s}(t) M_{s,q} = N_{p,q}(t)$. Thus if $N_{p,s}(t)$ diverges for $t = r$, all $N_{p,q}(r)$ also diverge for $q \in Q$. Similarly, the equality $I + MtN(t) = N(t)$ shows that if $N_{s,q}(t)$ diverges for $t = r$, then all $N_{p,q}(r)$ diverge for $p \in P$. This shows that all series $N_{p,q}(t)$ have the same radius of convergence as $N_{1,1}(t)$ which is ρ. Let λ be the spectral radius of M. We cannot have $\rho < 1/\lambda$ since otherwise $1/\rho$ would be

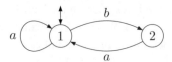

Figure 13.1 An automaton recognizing X^* for $X = \{a, ba\}$.

an eigenvalue of M larger than λ. We cannot have either $\rho > 1/\lambda$. Indeed, by the Perron–Frobenius theorem, λ is an eigenvalue of M and the matrix $M - \lambda I$ is not invertible. If $\rho > 1/\lambda$, then $N(t)$ converges for $t = 1/\lambda$ to a matrix which is the inverse of $I - \frac{1}{\lambda} M$, a contradiction.   □

Example 13.3.9 (*continued*) The automaton given in Figure 13.1 recognizes X^*. The matrix M is $\begin{bmatrix} 1 & 1 \\ 1 & 0 \end{bmatrix}$. Its spectral radius is $(1 + \sqrt{5})/2$.

13.4 Probabilities over a monoid

A detailed study of the density of a code, in relation to some of the fundamental parameters, will be presented in the next section. The aim of the present section is to prepare this investigation by the proof of some rather delicate results. We will show how certain monoids can be equipped with idempotent measures. This in turn allows us to determine the sets having a density, and to compute it.

We need the following lemma which is a generalization of Proposition 13.2.4.

Lemma 13.4.1 *Let I be a set, and for each $i \in I$, let*

$$f^{(i)}(t) = \sum_{n=0}^{\infty} f_n^{(i)} t^n, \qquad g^{(i)}(t) = \sum_{n=0}^{\infty} g_n^{(i)} t^n$$

be formal power series with nonnegative real coefficients satisfying

(i) $\sum_{i \in I} g^{(i)}(1) < \infty$,
(ii) $0 \le f_n^{(i)} \le 1$ *for all $i \in I$, $n \ge 0$,*
(iii) $\delta(f^{(i)})$ *exists for all $i \in I$.*

Then $\sum_{i \in I} f^{(i)} g^{(i)}$ admits a density and

$$\delta \left(\sum_{i \in I} f^{(i)} g^{(i)} \right) = \sum_{i \in I} \delta(f^{(i)}) g^{(i)}(1).$$

We first prove the following "dominated convergence" lemma. It gives a sufficient condition to allow one to extend the formula

$$\delta(f + g) = \delta(f) + \delta(g)$$

to an infinite sum.

Lemma 13.4.2 *Let I be a set and for each $i \in I$, let*

$$u^{(i)}(t) = \sum_{n=0}^{\infty} u_n^{(i)} t^n$$

be a formal power series with nonnegative real coefficients satisfying

(i) $\sum_{i \in I} u_n^{(i)} < \infty$ *for all $n \geq 0$,*
(ii) $\delta(u^{(i)})$ *exists for all $i \in I$,*
(iii) *there is a sequence $(v^{(i)})_{i \in I}$ of nonnegative real numbers such that $\sum_{i \in I} v^{(i)} < \infty$*
    and $u_n^{(i)} \leq v^{(i)}$ for all $i \in I$ and $n \geq 0$.

Then

$$\delta\left(\sum_{i \in I} u^{(i)}\right) = \sum_{i \in I} \delta(u^{(i)}).$$

Proof. Let $w_n = \sum_{i \in I} u_n^{(i)}$ and $w = \sum_{n \geq 0} w_n t^n$ in such a way that $w = \sum_{i \in I} u^{(i)}$. We show that

$$\left| \delta(w) - \sum_{i \in I} \delta(u^{(i)}) \right| < \epsilon$$

for arbitrary $\epsilon > 0$. Since the series $\sum_{i \in I} v^{(i)}$ is convergent, there is a finite set $F \subset I$ such that $\sum_{i \in I \setminus F} v^{(i)} < \epsilon$. Then $w_n - \sum_{i \in F} u_n^{(i)} < \epsilon$ and thus $\delta(w) - \sum_{i \in F} \delta(u^{(i)}) < \epsilon$. Since F is finite, $\delta(\sum_{i \in F} u^{(i)}) = \sum_{i \in F} \delta(u^{(i)})$ and the result follows. □

Proof of Lemma 13.4.1. Let $u^{(i)} = f^{(i)} g^{(i)}$ and $v^{(i)} = g^{(i)}(1)$. We verify that the conditions of Lemma 13.4.2 are satisfied.

Since $f_n^{(i)} \leq 1$ for all $i \in I$ and $n \geq 0$, we have $u_n^{(i)} \leq \sum_{\ell=0}^n g_\ell^{(i)}$. Thus $\sum_{i \in I} u_n^{(i)} \leq \sum_{i \in I} g^{(i)}(1) < \infty$. This shows that condition (i) is satisfied. Next, by Proposition 13.2.4, $\delta(u^{(i)})$ exists for all $i \in I$. Finally, $u_n^{(i)} \leq v^{(i)}$ and $\sum_{i \in I} v^{(i)} < \infty$, showing that condition (iii) is also satisfied. We can therefore apply Lemma 13.4.2 to obtain $\delta(\sum_{i \in I} f^{(i)} g^{(i)}) = \sum_{i \in I} \delta(f^{(i)} g^{(i)})$. We now apply Proposition 13.2.4 to obtain the desired result. □

Lemma 13.4.1 leads to the following proposition which extends Proposition 13.2.5.

Proposition 13.4.3 *Let I be a set and for each $i \in I$, let L_i and M_i be subsets of A^*. Let π be a Bernoulli distribution on A^* and suppose that*

(i) $\sum_{i \in I} \pi(M_i) < \infty$,
(ii) *the products $L_i M_i$ are unambiguous and the sets $L_i M_i$ are pairwise disjoint,*
(iii) *each L_i has a density $\delta(L_i)$.*

Then $\bigcup_{i \in I} L_i M_i$ has a density, and

$$\delta\left(\bigcup_{i \in I} L_i M_i\right) = \sum_{i \in I} \delta(L_i) \pi(M_i).$$

Proof. Set in Lemma 13.4.1,

$$f_n^{(i)} = \pi(L_i \cap A^n), \quad g_n^{(i)} = \pi(M_i \cap A^n).$$

Then $f^{(i)} = F_{L_i}$, $g^{(i)} = F_{M_i}$. Furthermore $\delta(f^{(i)}) = \delta(L_i)$, $g^{(i)}(1) = \pi(M_i)$, and in particular $\sum_{i \in I} \pi(M_i) < \infty$. According to Lemma 13.4.1, we have

$$\delta\left(\sum_{i \in I} f^{(i)} g^{(i)}\right) = \sum_{i \in I} \delta(L_i) \pi(M_i).$$

Since condition (ii) of the statement implies that

$$\sum_{i \in I} f^{(i)} g^{(i)} = \sum_{i \in I} F_{L_i} F_{M_i} = \sum_{i \in I} F_{L_i M_i} = F_{\bigcup_{i \in I} L_i M_i}$$

the proposition follows.   □

Let φ be a morphism from A^* onto a monoid M, and let π be a positive Bernoulli distribution on A^*. Provided M possesses certain properties which will be described below, each subset of A^* of the form $\varphi^{-1}(P)$, where $P \subset M$, has a density. The study of this phenomenon will lead us to give an explicit expression of the value of the densities of the sets $\varphi^{-1}(m)$ for $m \in M$, as a function of parameters related to M.

A monoid M is called *well founded* if it has a unique minimal ideal, if moreover this ideal is the union of the minimal left ideals of M, and also of the minimal right ideals, and if the intersection of a minimal right ideal and of a minimal left ideal is a finite group.

Any unambiguous monoid of relations of finite minimal rank is well founded by Proposition 9.3.14 and Theorem 9.3.15. It appears that the development given now does not depend on the fact that the elements of the monoid under concern are relations; therefore we present it in the more abstract frame of well-founded monoids.

Let $\varphi : A^* \to M$ be a morphism onto an arbitrary monoid, and let $m, n \in M$. We define

$$C_{m,n} = \{w \in A^* \mid m\varphi(w) = n\} = \varphi^{-1}(m^{-1}n).$$

Note that here $m^{-1}n$ is the left residual and m^{-1} is not the inverse of m. The set $C_{n,n}$ is a right-unitary submonoid of A^*: for $u, uv \in C_{n,n}$, we have $n\varphi(u) = n = n\varphi(uv) = n\varphi(u)\varphi(v) = n\varphi(v)$. Thus $C_{n,n}$ is free. Let X_n be its base. It is a prefix code. Let

$$Z_{m,n} = C_{m,n} \setminus C_{m,n} A^+$$

be the initial part of $C_{m,n}$. It is a prefix code. Next

$$C_{m,n} = Z_{m,n} X_n^*$$

and this product is unambiguous. Indeed, observe first that for all $m, n, p \in M$, one has $C_{m,n} C_{n,p} \subset C_{m,p}$ since if $w \in C_{m,n}$ and $w' \in C_{n,p}$, then $m\varphi(ww') = m\varphi(w)\varphi(w') = n\varphi(w') = p$. This shows in particular that $C_{m,n} \supset Z_{m,n} X_n^*$. Conversely, if $u \in C_{m,n}$, let $w \in Z_{m,n}$ and $t \in A^*$ be such that $u = wt$. Then $n =$

$m\varphi(wt) = m\varphi(w)\varphi(t) = n\varphi(t)$, showing that $t \in C_{n,n}$. The product is unambiguous because the code $Z_{m,n}$ is prefix. Note also that

$$C_{1,n} = \varphi^{-1}(n).$$

Proposition 13.4.4 *Let* $\varphi : A^* \to M$ *be a morphism onto a well-founded monoid* M, *and let* π *be a positive Bernoulli distribution on* A^*. *Let* K *be the minimal ideal of* M.

1. *For all* $m, n \in M$, *the set* $C_{m,n} = \varphi^{-1}(m^{-1}n)$ *has a density.*
2. *We have*

$$\delta(C_{m,n}) = \begin{cases} \pi(Z_{m,n})\delta(X_n^*) & \text{if } n \in K \text{ and } m^{-1}n \neq \emptyset, \\ 0 & \text{otherwise.} \end{cases}$$

3. *For* $m, n \in K$ *such that* $nM = mM$, *we have* $\pi(Z_{m,n}) = 1$ *and consequently*

$$\delta(C_{m,n}) = \delta(C_{n,n}) = \delta(X_n^*).$$

Proof. Let $n \in M$, with $n \notin K$. Then $m^{-1}n \cap K = \emptyset$. Indeed, assume that $p \in m^{-1}n \cap K$. Then $mp = n$ and since K is an ideal, $p \in K$ implies $n \in K$. Thus for an element $n \notin K$, the set $C_{m,n}$ does not meet the ideal $\varphi^{-1}(K)$. Consequently $C_{m,n}$ is thin, and by Proposition 13.2.3, $\delta(C_{m,n}) = 0$.

Consider now the case where $n \in K$. Let $R = nM$ be the minimal right ideal containing n. Consider the deterministic automaton over A, $\mathcal{A} = (R, n, n)$ with transition function defined by $r \cdot a = r\varphi(a)$ for $r \in R$, $a \in A$. We have $|\mathcal{A}| = X_n^*$. Since R is a minimal right ideal, the automaton is complete and trim and every state is recurrent. In particular, X_n is a complete code (Proposition 3.3.11).

Let us verify that the monoid $\varphi_\mathcal{A}(A^*)$ has finite minimal rank. For this, let $u \in A^*$ be a word such that $\varphi(u) = n$. Since $\mathcal{A}$ is deterministic, it suffices to compute $\text{rank}_\mathcal{A}(u)$. Now $\text{rank}(\varphi_\mathcal{A}(u)) = \text{rank}_\mathcal{A}(u) = \text{Card}(R \cdot u) = \text{Card}(Rn) = \text{Card}(nMn)$.

By assumption, nMn is a finite group. Thus $\text{rank}(\varphi_\mathcal{A}(u))$ is finite and the monoid $\varphi_\mathcal{A}(A^*)$ has finite minimal rank. By Corollary 9.4.5, the code X_n is complete and thin and according to Theorem 13.2.9, X_n^* has a positive density. Since $Z_{m,n}$ is a prefix set, we have $\pi(Z_{m,n}) \leq 1$. In view of Proposition 13.2.5, the set $C_{m,n}$ has a density and

$$\delta(C_{m,n}) = \pi(Z_{m,n})\delta(X_n^*).$$

Clearly

$$C_{m,n} = \emptyset \iff m^{-1}n = \emptyset \iff Z_{m,n} = \emptyset.$$

Moreover, π being positive, $\pi(Z_{m,n}) > 0$ if and only if $Z_{m,n} \neq \emptyset$. This shows that $\delta(C_{m,n}) \neq 0$ if $m^{-1}n \neq \emptyset$. This proves the claims (2) and (1).

To prove (3), let $u \in A^*$ be a word such that $n\varphi(u) = m$ and $n\varphi(u') \neq n$ for each proper nonempty prefix u' of u. Then

$$u Z_{m,n} \subset X_n.$$

Indeed, let $w \in Z_{m,n}$. We have $n\varphi(uw) = m\varphi(w) = n$, therefore $uw \in X_n^*$. We claim that $uw \in X_n$. Assume on the contrary that uw has a proper prefix u' which is in X_n. Then $n\varphi(u') = n$ and by the choice of u, the word u' is not a proper prefix of u. Thus u is a prefix of u'. If $u \neq u'$, then $u' = uu''$ and $n = n\varphi(u') = n\varphi(uu'') = m\varphi(u'')$, showing that u'' is in $Z_{m,n}$, contradicting the fact that $Z_{m,n}$ is prefix.

This shows that $Z_{m,n}$ is formed of suffixes of words in X_n, and in particular that $Z_{m,n}$ is thin. To show that $Z_{m,n}$ is right complete, let $w \in A^*$ and let $n' = m\varphi(w)$. Then $n' \in nM$, and since nM is a minimal right ideal, there exists $n'' \in M$ such that $n'n'' = n$. Let $v \in A^*$ be such that $\varphi(v) = n''$. Then $m\varphi(wv) = n$, and consequently $wv \in C_{m,n}$. This shows that $Z_{m,n} = \{1\}$ or $Z_{m,n}$ is a thin right complete prefix code, thus a maximal code. Therefore $\pi(Z_{m,n}) = 1$. Consequently $\delta(C_{m,n}) = \delta(X_n^*)$. $\qquad\square$

Let $\varphi : A^* \to M$ be a morphism onto a well-founded monoid, and let π be a positive Bernoulli distribution on A^*. We define a partial function v on the set of subsets of M as follows. The function v is defined for each subset F of M for which the density of the set $\varphi^{-1}(F)$ exists, and its value is this density

$$v(F) = \delta(\varphi^{-1}(F)).$$

It follows from Proposition 13.4.4 that $v(n)$ is defined for each $n \in M$ since $\varphi^{-1}(n) = C_{1,n}$. Note also that according to Corollary 13.2.6, every one-sided ideal R has a positive density. Thus v is defined for all ideals in M. We write $v = \delta\varphi^{-1}$ for short.

We shall see (Theorem 13.4.7 below) that v is defined for all subsets of M, so v is in fact a total function and, moreover, it is a probability measure on the set of subsets of M. We start with the following result

Theorem 13.4.5 *Let $\varphi : A^* \to M$ be a morphism onto a well-founded monoid, and let π be a positive Bernoulli distribution on A^*. Let K be the minimal ideal of M.*

1. *$v(n) \neq 0$ if and only if $n \in K$.*
2. *$v(K) = 1$.*
3. *For all $\mathcal{R}$-equivalent elements $m, n \in K$, one has $v(n) = v(m^{-1}n)v(nM)$.*
4. *For all $n \in K$,*

$$v(n) = \frac{v(nM)v(Mn)}{\mathrm{Card}(nM \cap Mn)}.$$

Proof. 1. One has $\varphi^{-1}(n) = C_{1,n}$. By Proposition 13.4.4, $\delta(C_{1,n}) \neq 0$ if and only if $n \in K$, since $C_{1,n}$ is never empty.

2. Let $Y = \varphi^{-1}(K) \setminus \varphi^{-1}(K)A^+$ be the initial part of the ideal $\varphi^{-1}(K)$. The set Y is prefix and $\varphi^{-1}(K) = YA^*$. Since the set $A^* \setminus \varphi^{-1}(K)$ is thin, we have $v(K) = 1$ by Proposition 13.2.3.

3. For each $\mathcal{R}$-class R of K, consider $Y_R = Y \cap \varphi^{-1}(R)$. Since the set Y is prefix, the set Y_R is prefix. We have $Y_R = \varphi^{-1}(R) \setminus \varphi^{-1}(R)A^+$. Indeed, consider first $y \in Y_R = Y \cap \varphi^{-1}(R)$. Then $y \in \varphi^{-1}(R)$ and $y \notin \varphi^{-1}(R)A^+$, since otherwise $y \in \varphi^{-1}(K)A^+$, in contradiction with the fact that $y \in Y$. Thus $Y_R \subset \varphi^{-1}(R) \setminus \varphi^{-1}(R)A^+$. Conversely, let $y \in \varphi^{-1}(R) \setminus \varphi^{-1}(R)A^+$. Then $y \in \varphi^{-1}(K)$ because $r \subset K$, and assuming $y \in \varphi^{-1}(K)A^+$, one has $y = uv$ with $u \in \varphi^{-1}(K)$, and since $y\mathcal{R}u$, one has $\varphi(u) \in R$. Consequently $u \in \varphi^{-1}(R)$ and $y \in \varphi^{-1}(R)A^+$, a contradiction. This implies that $y \notin \varphi^{-1}(K)A^+$, showing that $\varphi^{-1}(R) \setminus \varphi^{-1}(R)A^+ \subset Y_R$.

It follows that $\varphi^{-1}(R) = Y_R A^*$, and hence, $v(R) = \pi(Y_R)$ by the symmetric version of Corollary 13.2.6.

Let now $n \in R$. Then $R = nM$ and

$$\varphi^{-1}(n) = \bigcup_{r \in R}(Y_R \cap \varphi^{-1}(r))C_{r,n}. \tag{13.17}$$

Indeed, each word $w \in \varphi^{-1}(n)$ factorizes uniquely into $w = uv$, where u is the shortest prefix of w such that $\varphi(u) \in R$. Then $u \in Y_R \cap \varphi^{-1}(r)$ for some $r \in R$, and $v \in C_{r,n}$. The converse inclusion is clear. The union in (13.17) is disjoint, and the products are unambiguous because the sets $Y_R \cap \varphi^{-1}(r)$ are prefix. Indeed, they are subsets of the prefix code Y_R. Each $C_{r,n}$ has a density, and moreover

$$\sum_{r \in R} \pi(Y_R \cap \varphi^{-1}(r)) = \pi(Y_R) \leq 1.$$

We therefore can apply Proposition 13.4.3 to (13.17). This gives

$$v(n) = \sum_{r \in R} \pi(Y_R \cap \varphi^{-1}(r))\delta(C_{r,n}).$$

According to Proposition 13.4.4, all values $\delta(C_{r,n})$ for $r \in R$ are equal. Thus, for any $m \in R$,

$$v(n) = \delta(C_{m,n})\pi(Y_R) = v(m^{-1}n)\pi(Y_R) = v(m^{-1}n)v(R).$$

4. Set $R = nM$, $L = Mn$, and $H = R \cap L$. Then we claim that

$$L = \bigcup_{m \in H}(m^{-1}n \cap K)$$

and furthermore that the union is disjoint.

First consider an element $k \in m^{-1}n \cap K$ for some $m \in H$. Then $mk = n$. Thus $n \in Mk$, and since n is in the minimal ideal, $Mn = Mk$. Therefore, $k \in Mn = L$. This proves the first inclusion.

For the converse, let $k \in L = Mn$. The right multiplication by k, $m \mapsto mk$ is a bijection which exchanges the $\mathcal{L}$-classes in K and preserves $\mathcal{R}$-classes (Proposition 1.12.2). In particular, this function maps the $\mathcal{L}$-class L onto $Lk = L$ and thus onto itself. It follows that there exists $m \in L$ such that $mk = n$. The element m is $\mathcal{R}$-equivalent with n. Consequently $m \in H$ and therefore $k \in m^{-1}n$ for some $m \in H$.

Since the function $m \mapsto mk$ is a bijection, the sets $m^{-1}n$ are pairwise disjoint. Indeed, if $k \in m^{-1}n$ and $k \in m'^{-1}n$, then $mk = m'k$ and $m = m'$. This proves the formula.

For all $m, n \in K$,

$$v(m^{-1}n \cap K) = v(m^{-1}n)$$

since the set $\varphi^{-1}(m^{-1}n \cap (M \setminus K))$ is thin and therefore has density 0 by Proposition 13.2.3. The set H being finite, we have

$$v(L) = \sum_{m \in H} v(m^{-1}n).$$

Using the expression for $v(n)$ proved above, we obtain

$$v(L) = \sum_{m \in H} \frac{v(n)}{v(R)} = \mathrm{Card}(H) \frac{v(n)}{v(R)}.$$

This proves the last claim of the theorem. $\qquad \square$

The following elementary proposition is useful.

Proposition 13.4.6 *Let* $(\mu_n)_{n \geq 0}$ *and* μ *be probability measures on the family of subsets of a countable set* E, *and such that* $\mu(e) = \lim_{n \to \infty} \mu_n(e)$ *for every e in* E. *Then for all subsets* F *of* E,

$$\mu(F) = \lim_{n \to \infty} \mu_n(F).$$

Proof. The conclusion clearly holds when F is finite. In the general case, set

$$\sigma = \liminf \mu_n(F), \quad \tau = \limsup \mu_n(F),$$

and let $\bar{F} = E \setminus F$. Of course, $\sigma \leq \tau$ and

$$1 - \tau = \liminf \mu_n(\bar{F}).$$

Let F' be a finite subset of F. Then $\mu_n(F') \leq \mu_n(F)$ for all n, and taking the inferior limit, $\mu(F') \leq \sigma$. It follows that

$$\mu(F) = \sup_{\substack{F' \subset F \\ F' \text{ finite}}} \mu(F') \leq \sigma.$$

Similarly, $\mu(\bar{F}) \leq 1 - \tau$. Since $\mu(\bar{F}) + \mu(F) = \mu(E) = 1$, we obtain $1 \leq \sigma + (1 - \tau)$, whence $\sigma \geq \tau$. Thus $\sigma = \tau$. Since $\mu(F) \leq \sigma$ and $\mu(\bar{F}) \leq 1 - \sigma$, one has both $\mu(F) \leq \sigma$ and $\mu(F) \geq \sigma$, showing that $\mu(F) = \sigma$. $\qquad \square$

Theorem 13.4.7 *Let* $\varphi : A^* \to M$ *be a morphism onto a well-founded monoid, and let* π *be a positive Bernoulli distribution on* A^*. *For any subset* F *of* M, *the set* $\varphi^{-1}(F) \subset A^*$ *has a density. The function* $v = \delta \varphi^{-1}$ *is a probability measure on the family of subsets of* M.

Proof. Let K be the minimal ideal of M, let Γ be the set of its $\mathcal{R}$-classes and Λ the set of its $\mathcal{L}$-classes. By Theorem 13.4.5,

$$\nu(K) = 1.$$

Let Y (resp. Y_R) be the initial part of $\varphi^{-1}(K)$, (resp. of $\varphi^{-1}(R)$, with $R \in \Gamma$). Since K is the disjoint union of its $\mathcal{R}$-classes, we have

$$\pi(Y) = \sum_{R \in \Gamma} \pi(Y_R).$$

By Corollary 13.2.6, $\nu(K) = \pi(Y)$, $\nu(R) = \pi(Y_R)$. Thus

$$\nu(K) = \sum_{R \in \Gamma} \nu(R) = \sum_{L \in \Lambda} \nu(L) = 1,$$

where the intermediate assertion follows by symmetry.

Now consider a fixed $\mathcal{R}$-class $R \in \Gamma$. Then by Theorem 13.4.5,

$$\sum_{n \in R} \nu(n) = \sum_{n \in R} \frac{\nu(R)\nu(Mn)}{\mathrm{Card}(R \cap Mn)} = \nu(R) \sum_{L \in \Lambda} \sum_{n \in R \cap L} \frac{\nu(L)}{\mathrm{Card}(R \cap L)}$$

$$= \nu(R) \sum_{L \in \Lambda} \nu(L) = \nu(R)$$

and also

$$\sum_{n \in K} \nu(n) = \sum_{R \in \Gamma} \left(\sum_{n \in R} \nu(n) \right) = \sum_{R \in \Gamma} \nu(R) = 1.$$

Since $\nu(n) = 0$ for $n \notin K$, it follows that

$$\sum_{n \in M} \nu(n) = 1.$$

Define for any positive integer n and $F \subset M$

$$\nu_n(F) = \frac{1}{n} \pi(\varphi^{-1}(F) \cap A^{(n)}).$$

Then $\nu_n(m) = 0$ except for a finite number of elements of M. Since $\nu_n(M) = 1$, it follows that each ν_n is a probability measure on the family of all subsets of M.

Define for a subset F of M, $\mu(F) = \sum_{m \in F} \nu(m)$. Then, by Proposition 13.1.4, μ is a probability measure on the family of subsets of M. By Proposition 13.4.6 we have for any $F \subset M$, $\mu(F) = \lim_{n \to \infty} \nu_n(F)$. Since, on the other hand, the limit of $\nu_n(F)$ is by definition $\nu(F)$, it follows that $\nu(F)$ exists for any $F \subset M$ and is equal to $\mu(F)$. This concludes the proof. $\square$

The following result puts together the results obtained before.

Proposition 13.4.8 *Let $\varphi : A^* \to M$ be a morphism onto a well-founded monoid, and let π be a positive Bernoulli distribution on A^*. The function $\nu = \delta\varphi^{-1}$ is a* *probability measure on the set of subsets M. Let K be the minimal ideal of M. Then the following formulas hold:*

$$v(m) \neq 0 \qquad \text{if and only if } m \in K$$

$$v(m) = v(n^{-1}m)v(mM) \qquad \text{if } m, n \in K \text{ and } n\mathcal{R}m$$

$$v(m) = \frac{v(mM)v(Mm)}{\mathrm{Card}(mM \cap Mm)} \qquad \text{if } m \in K \tag{13.18}$$

$$v(M') = v(M' \cap K) \qquad \text{for } M' \subset M.$$

For each $\mathcal{H}$-class $H \subset K$, and $h \in H$,

$$v(h) = \frac{v(H)}{\mathrm{Card}(H)}. \tag{13.19}$$

Proof. The first assertion is Proposition 13.4.7. All the formulas with the exception of (13.19), are immediate consequences of the relations given in Theorem 13.4.5. For (13.19) observe that the value of v is the same for all $h \in H$ by Formula (13.18). Next $v(H) = \sum_{h \in H} v(h)$. This proves (13.19).   □

Example 13.4.9 Let $\varphi : A^* \to G$ be a morphism onto a finite group. Let π be a positive Bernoulli distribution. For $g \in G$,

$$v(g) = \frac{1}{\mathrm{Card}(G)} \tag{13.20}$$

in view of Formula (13.19) and observing that $H = K = G$. This gives another method for computing the density in Example 13.2.1. To that example corresponds a morphism $\varphi : A^* \to \mathbb{Z}/2\mathbb{Z}$ onto the additive group $\mathbb{Z}/2\mathbb{Z}$ with $\varphi(a) = 1$ for any letter a in A.

Example 13.4.10 Let $\varphi : A^* \to M$ be the morphism from A^* onto the unambiguous monoid of relations M over $Q = \{1, 2, 3\}$ defined by $\alpha = \varphi(a)$, $\beta = \varphi(b)$, with

$$\alpha = \begin{bmatrix} 0 & 1 & 0 \\ 1 & 0 & 0 \\ 1 & 1 & 0 \end{bmatrix}, \qquad \beta = \begin{bmatrix} 0 & 0 & 1 \\ 0 & 0 & 0 \\ 1 & 1 & 0 \end{bmatrix}.$$

This monoid has already been considered in Example 9.4.12. Its minimal ideal J is composed of elements of rank 1 and is represented in Figure 13.2.

Let π be a positive Bernoulli distribution and set $p = \pi(a)$, $q = \pi(b)$. Let us compute the probability measure $v = \delta\varphi^{-1}$ over M. With the notations of Figure 13.2, we have the equalities

$$\begin{array}{ll} L_1\alpha = L_2, & L_1\beta = L_2, \\ L_2\alpha = L_2, & L_2\beta = L_1. \end{array} \tag{13.21}$$

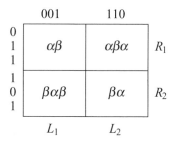

Figure 13.2 The minimal ideal of the monoid M.

Set $X_1 = \varphi^{-1}(L_1)$, $X_2 = \varphi^{-1}(L_2)$. By (13.21),

$$
\begin{aligned}
X_1 a^{-1} \cap \varphi^{-1}(J) &= \emptyset, & X_1 b^{-1} \cap \varphi^{-1}(J) &= X_2, \\
X_2 a^{-1} \cap \varphi^{-1}(J) &= X_1 \cup X_2, & X_2 b^{-1} \cap \varphi^{-1}(J) &= X_1.
\end{aligned} \qquad (13.22)
$$

Indeed consider, for instance, the last equation: if $w \in X_1$, then $\varphi(w) \in L_1$, hence $\varphi(wb) \in L_2$ by the fact that $L_1 \beta = L_2$. This implies that $wb \in X_2$, and $w \in X_2 b^{-1} \cap \varphi^{-1}(J)$. Conversely, let $w \in X_2 b^{-1} \cap \varphi^{-1}(J)$. Since $w \in \varphi^{-1}(J)$, $w \in X_1 \cup X_2$. But if $w \in X_2$, then $\varphi(wb) \in L_1$, showing that $wb \in X_1$, whence $w \notin X_2 b^{-1}$. Thus $w \in X_1$.

In view of (13.22),

$$
\begin{aligned}
X_1 a^{-1} &= T_1, & X_1 b^{-1} &= X_2 \cup T_1', \\
X_2 a^{-1} &= X_1 \cup X_2 \cup T_2, & X_2 b^{-1} &= X_1 \cup T_2',
\end{aligned}
$$

where T_1, T_1', T_2, T_2' are disjoint from $\varphi^{-1}(J)$. Multiplication by a and b on the right gives, since $X_i = (X_i a^{-1})a \cup (X_i b^{-1})b$ for $i = 1, 2$, by adding both sides on each row of the equations above,

$$
X_1 = X_2 b \cup (T_1 a \cup T_1' b),
$$

$$
X_2 = X_1 a \cup X_2 a \cup X_1 b \cup (T_2 a \cup T_2' b).
$$

Since T_1 is thin, $\delta(T_1 a) = \delta(T_1)\pi(a) = 0$, and similarly for the other T's. Therefore

$$
\delta(X_1) = \delta(X_2)q, \qquad \delta(X_2) = \delta(X_1) + \delta(X_2)p,
$$

which together with $\delta(X_1) + \delta(X_2) = 1$ gives

$$
\delta(X_1) = \frac{q}{1+q}, \qquad \delta(X_2) = \frac{1}{1+q}.
$$

Thus

$$
v(L_1) = \frac{q}{1+q}, \qquad v(L_2) = \frac{1}{1+q}.
$$

An analogous computation gives

$$v(R_1) = \frac{p}{1+p}, \qquad v(R_2) = \frac{1}{1+p}.$$

In particular, since $R_2 \cap L_2 = \{\beta\alpha\}$, we obtain

$$v(\beta\alpha) = \frac{v(L_2)v(R_2)}{\mathrm{Card}(L_2 \cap R_2)} = \frac{1}{(1+p)(1+q)}.$$

13.5 Strict contexts

Let $X \subset A^+$ be a thin complete code. We have seen that the *degree* $d(X)$ of X is the integer which is the minimal rank of the monoid of relations associated with any unambiguous trim automaton recognizing X^*. It is also the degree of the permutation group $G(X)$, and it is also the minimum of the number of disjoint interpretations in X (see Section 9.5). In this section, we shall see that $d(X)$ is related in a quite remarkable manner to the density $\delta(X^*)$. A word is *left (right) completable* in X^* if it is a suffix (prefix) of some word in X^*. The set of left completable (right completable) words is denoted by G_X (D_X).

Theorem 13.5.1 *Let $X \subset A^*$ be a thin complete code, and let π be a positive Bernoulli distribution on A^*. Then*

$$\delta(X^*) = \frac{1}{d(X)}\delta(G_X)\delta(D_X). \tag{13.23}$$

Proof. Let $\mathcal{A} = (Q, 1, 1)$ be an unambiguous trim automaton recognizing X^*, let φ be the associated morphism and $M = \varphi(A^*)$. In view of Corollary 9.4.5, the monoid M is well founded. Set $v = \delta\varphi^{-1}$. By Proposition 13.4.8, v is a probability measure over the set of subsets of M, and the values of v may be computed by the formulas of this proposition.

Let K be the minimal ideal of M. Since v vanishes outside of K, we have

$$\delta(X^*) = v(\varphi(X^*) \cap K).$$

Let $\widehat{R}$ be the union of the $\mathcal{R}$-classes in K meeting $\varphi(X^*)$, and similarly let $\widehat{L}$ be the union of those $\mathcal{L}$-classes in K that meet $\varphi(X^*)$. Then

$$v(\varphi(X^*) \cap K) = v(\varphi(X^*) \cap \widehat{R} \cap \widehat{L}) = \sum_H v(\varphi(X^*) \cap H),$$

where the sum is over all $\mathcal{H}$-classes H contained in $\widehat{R} \cap \widehat{L}$. For such an $\mathcal{H}$-class H, we have

$$v(\varphi(X^*) \cap H) = \sum_{m \in \varphi(X^*) \cap H} v(m) = \sum_{m \in \varphi(X^*) \cap H} \frac{v(R)v(L)}{\mathrm{Card}(H)},$$

where R and L are the $\mathcal{R}$-class and $\mathcal{L}$-class containing H. Therefore

$$v(\varphi(X^*) \cap H) = \frac{\mathrm{Card}(\varphi(X^*) \cap H)}{\mathrm{Card}(H)} v(R)v(L).$$

Now observe that for any $\mathcal{H}$-class $H \subset \widehat{R} \cap \widehat{L}$, since $\varphi(X^*) \cap H$ is a subgroup of index $d(X)$ of the group H,

$$\frac{\mathrm{Card}(\varphi(X^*) \cap H)}{\mathrm{Card}(H)} = \frac{1}{d(X)}.$$

Thus the formula becomes

$$\delta(X^*) = \sum_H \frac{1}{d(X)} v(R)v(L) = \frac{1}{d(X)} v(\widehat{R})v(\widehat{L}).$$

Next

$$\varphi^{-1}(\widehat{R}) = D_X \cap \varphi^{-1}(K). \tag{13.24}$$

Indeed, let $w \in D_X \cap \varphi^{-1}(K)$. Then $wu \in X^*$ for some word u. Consequently, $\varphi(wu) = \varphi(w)\varphi(u) \in \varphi(X^*) \cap K$, showing that the $\mathcal{R}$-class of $\varphi(w)$, which is the same as the $\mathcal{R}$-class of $\varphi(wu)$, meets $\varphi(X^*)$. This implies that $\varphi(w) \in \widehat{R}$. Conversely, let $w \in \varphi^{-1}(\widehat{R})$. Then $\varphi(w) \in \widehat{R}$ and there is some $m \in M$ such that $\varphi(w)m \in \varphi(X^*) \cap K$. Therefore $w\varphi^{-1}(m) \cap X^* \neq \emptyset$ and we derive that $w \in D_X$.

It follows from (13.24) that $v(\widehat{R}) = \delta(\varphi^{-1}(\widehat{R})) = \delta(D_X \cap \varphi^{-1}(K))$. Since $A^* \setminus \varphi^{-1}(K)$ is thin, we have

$$\delta(D_X) = \delta(D_X \cap \varphi^{-1}(K)).$$

Thus $\delta(D_X) = v(\widehat{R})$ and similarly $v(\widehat{L}) = \delta(G_X)$. This concludes the proof. $\square$

The following corollary is a consequence of Theorem 13.2.9.

Corollary 13.5.2 *Let $X \subset A^*$ be a thin complete code, and let π be a positive Bernoulli distribution on A^*. Then*

$$\lambda(X) = \frac{d(X)}{\delta(G_X)\delta(D_X)}. \tag{13.25}$$

$\square$

We observe that for a thin maximal bifix code $X \subset A^*$, we have $G_X = D_X = A^*$. Thus in this case, (13.25) becomes $\lambda(X) = d(X)$. This gives another proof of Corollary 6.3.16. Proposition 6.3.17 is also a consequence of (13.25).

Example 13.5.3 Let $A = \{a, b\}$ and consider our old friend $X = \{aa, ba, baa, bb, bba\}$ which is a finite complete code. In Figure 13.3 an automaton $\mathcal{A} = (Q, 1, 1)$ recognizing X^* is represented.

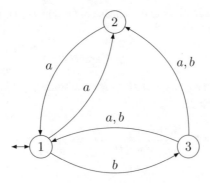

Figure 13.3 An unambiguous trim automaton recognizing X^*.

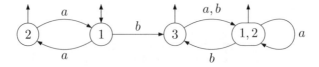

Figure 13.4 A deterministic automaton for D_X.

To derive more easily an expression for D_X, we compute the deterministic trim automaton associated to the automaton $\mathcal{A}$ by the subset construction and take all states as final states. This gives the automaton of Figure 13.4. We obtain

$$D_X = a^* \cup (a^2)^* b A^*.$$

A similar computation gives

$$G_X = b^* \cup A^* a (b^2)^*.$$

Let π be a positive Bernoulli distribution and set $p = \pi(a)$, $q = \pi(b)$. Then

$$\delta(D_X) = \delta(a^*) + \delta((a^2)^* b A^*) = \delta((a^2)^* b A^*)$$

since $\delta(a^*) = 0$. Since $(a^2)^* b$ is a prefix code, the product of $(a^2)^* b$ and A^* is unambiguous, and $\pi((a^2)^* b)$ is finite. We get

$$\delta(D_X) = \pi((a^2)^* b),$$

and

$$\delta(D_X) = \frac{q}{1 - p^2} = \frac{1}{1 + p}.$$

In a similar fashion, we obtain

$$\delta(G_X) = \frac{1}{1 + q}.$$

On the other hand, $d(X) = 1$ since the monoid $\varphi_A(A^*)$ has minimal rank 1. By Formula (13.25),

$$\lambda(X) = (1 + p)(1 + q).$$

This can also be verified by a direct computation of the average length of X. The computations made in this example are of course similar to those of Example 13.4.10.

Let $X \subset A^*$ be a code. A *strict context* of a nonempty word $w \in A^+$ is a pair (u, v) of words such that the following two conditions hold. There exist $n \geq 1$ and words $x_1, \ldots, x_n \in X$ with

$$uwv = x_1 x_2 \cdots x_n$$

and

$$|u| < |x_1|, \quad |v| < |x_n|.$$

The set of strict contexts of a word $w \in A^*$ (with respect to X) is denoted by $C(w)$. The set $C(1)$ is defined as $C(1) = \{(u, v) \in A^+ \times A^+ \mid uv \in X\} \cup \{(1, 1)\}$. The strict contexts of a word can be interpreted in terms of paths in the flower automaton $\mathcal{A}_D^*(X) = (P, (1, 1), (1, 1))$.

Lemma 13.5.4 *In the flower automaton* $\mathcal{A}_D^*(X) = (P, (1, 1), (1, 1))$, *the function that maps the path*

$$c : (u, u') \xrightarrow{w} (v', v)$$

onto the pair (u, v) *is a bijection between the set* $P(w)$ *of paths labeled* w *in the flower automaton and the set* $C(w)$ *of strict contexts of* w.

Proof. Let

$$c : (u, u') \xrightarrow{w} (v', v)$$

be a path labeled w in $\mathcal{A}_D^*(X)$. Then $uwv \in X^*$. Thus either $uwv = 1$, or

$$uwv = x_1 x_2 \cdots x_n$$

with $x_j \in X$ and $n > 0$. In that case, $|u| < |x_1|$ and $|v| < |x_n|$. This shows that, in both cases, (u, v) is a strict context. Consider another path

$$\bar{c} : (u, \bar{u}') \xrightarrow{w} (\bar{v}', v).$$

Then both paths

$$(1, 1) \xrightarrow{u} (u, u') \xrightarrow{w} (v', v) \xrightarrow{v} (1, 1),$$

$$(1, 1) \xrightarrow{u} (u, \bar{u}') \xrightarrow{w} (\bar{v}', v) \xrightarrow{v} (1, 1)$$

are labeled uwv. By unambiguity, $c = \bar{c}$. Conversely, if (u, v) is a strict context of w and $uwv = x_1 x_2 \cdots x_n$, define two words u', v' by

$$u' = \begin{cases} u^{-1} x_1 & \text{if } u \neq 1, \\ 1 & \text{otherwise,} \end{cases} \qquad v' = \begin{cases} x_n v^{-1} & \text{if } v \neq 1, \\ 1 & \text{otherwise.} \end{cases}$$

Then (u, u') and (v', v) are states in $\mathcal{A}_D^*(X)$, and there is a path $(u, u') \xrightarrow{w} (v', v)$. $\square$

The following result shows a strong relationship between all sets of strict contexts.

Theorem 13.5.5 *Let* $X \subset A^*$ *be a thin complete code, and let* π *be a positive Bernoulli distribution on* A^*. *For all* $w \in A^*$,

$$\lambda(X) = \sum_{(u,v)\in C(w)} \pi(uv).$$

Proof. Let $\mathcal{A}_D^*(X) = (P, (1, 1), (1, 1))$ be the flower automaton of X, let $M = \varphi_D(A^*)$ and set $v = \delta \varphi_D^{-1}$. Let $w \in A^*$, set $m = \varphi_D(w)$, and define a set $T(m)$ and a number $t(m)$ by

$$T(m) = \{(r, \ell) \in M \times M \mid rm\ell \in \varphi_D(X^*)\}, \quad t(m) = \sum_{(r,\ell)\in T(m)} v(r)v(\ell).$$

We compute $t(m)$ in two ways. First define, for each state $p \in P$,

$$R_p = \{r \in M \mid r_{1,p} = 1\}, \quad L_p = \{\ell \in M \mid \ell_{p,1} = 1\}.$$

Then $rm\ell \in \varphi_D(X^*)$ if and only if there exist $p, q \in P$ such that $r_{1,p} = 1, m_{p,q} = 1, \ell_{q,1} = 1$. Consequently,

$$T(m) = \bigcup_{\substack{(p,q) \\ m_{p,q}=1}} R_p \times L_q.$$

Thus

$$t(m) = \sum_{\substack{(p,q) \\ m_{p,q}=1}} v(R_p)v(L_q).$$

Set $p = (u, u')$ and $q = (v', v)$. Then $m_{p,q} = 1$ if and only if there is a path $c : p \to q$ labeled w. According to the bijection defined above, this holds if and only if $(u, v) \in C(w)$. Next,

$$\varphi_D^{-1}(R_p) = X^*u, \quad \varphi_D^{-1}(L_q) = vX^*,$$

hence

$$v(R_p) = \delta(X^*u) = \delta(X^*)\pi(u), \quad v(L_q) = \delta(vX^*) = \pi(v)\delta(X^*).$$

Consequently

$$t(m) = \sum_{(u,v)\in C(w)} \delta(X^*)\pi(u)\pi(v)\delta(X^*) = [\delta(X^*)]^2 \sum_{(u,v)\in C(w)} \pi(uv).$$

This is the first expression for $t(m)$.

Now we compute $t(m)$ in the monoid M. Let K be the minimal ideal of M. Since v vanishes for elements not in K, we have

$$t(m) = \sum_{\substack{(r,\ell)\in K \times K \\ rm\ell\in\varphi_D(X^*)}} v(r)v(\ell).$$

Let $N = \varphi_D(X^*) \cap K$. Then

$$t(m) = \sum_{n \in N} \sum_{\substack{(r,\ell) \in K \times K \\ rm\ell = n}} v(r)v(\ell) = \sum_{n \in N} \sum_{r \in K} v(r)v((rm)^{-1}n).$$

Let $r \in K$. Since $(rm)^{-1}n \neq \emptyset$ if and only if $rm\mathcal{R}n$, and since $r\mathcal{R}rm$, we have $(rm)^{-1}n \neq \emptyset$ if and only if $r \in nM$ and

$$t(m) = \sum_{n \in N} \sum_{r \in nM} v(r)v((rm)^{-1}n) = \sum_{n \in N} \sum_{r \in nM} v(r)\frac{v(n)}{v(nM)}$$

by Proposition 13.4.8. Further

$$t(m) = \sum_{n \in N} v(n) \sum_{r \in nM} \frac{v(r)}{v(nM)} = \sum_{n \in N} v(n) = v(N) = \delta(X^*).$$

Comparing both expressions for $t(m)$, we get

$$1 = \delta(X^*) \sum_{(u,v) \in C(w)} \pi(uv).$$

The result follows from the fact that $\delta(X^*) = 1/\lambda(X)$ by Theorem 13.2.9. $\square$

There is an interesting interpretation of the preceding result. With the notations of the theorem, set for any word $w \in A^*$,

$$\gamma(w) = \frac{1}{\lambda(X)} \sum_{(u,v) \in C(w)} \pi(uwv).$$

Call $\gamma(w)$ the *contextual probability* of w. Then Theorem 13.5.5 claims that if π is a Bernoulli distribution we have identically

$$\gamma(w) = \pi(w).$$

The fact that the distributions γ and π coincide is particular to Bernoulli distributions (see Exercise 13.5.3). We now study one-sided strict contexts. Let $X \subset A^+$ be a code, and let $w \in A^*$. The set of *strict right contexts* of w is

$$C_r(w) = \{v \in A^* \mid (1, v) \in C(w)\}.$$

Thus $v \in C_r(w)$ if and only if $wv = x_1 x_2 \cdots x_n$, $(x_i \in X)$ with $|v| < |x_n|$.
    Symmetrically, the set of *strict left contexts* of w is

$$C_\ell(w) = \{u \in A^* \mid (u, 1) \in C(w)\}.$$

We observe that

$$C_r(w)X^* = w^{-1}X^*. \tag{13.26}$$

The product $C_r(w)X^*$ is unambiguous, because X is a code.

Proposition 13.5.6 *Let $X \subset A^*$ be a thin complete code and let $\mathcal{A} = (Q, 1, 1)$ be an unambiguous trim automaton recognizing X^*. Let K be the minimal ideal of the monoid $M = \varphi_{\mathcal{A}}(A^*)$. Let π be a positive Bernoulli distribution. For all $w \in \varphi_{\mathcal{A}}^{-1}(K) \cap D_X$, we have*

$$\pi(C_r(w))\delta(D_X) = 1. \tag{13.27}$$

For all $w \in \varphi_{\mathcal{A}}^{-1}(K) \cap G_X$, we have

$$\pi(C_\ell(w))\delta(G_X) = 1. \tag{13.28}$$

Proof. Set $\varphi = \varphi_{\mathcal{A}}$, $\nu = \delta\varphi^{-1}$, and let $\widehat{R}$ (resp. $\widehat{L}$) be the union of the $\mathcal{R}$-classes (resp. $\mathcal{L}$-classes) in K that meet $\varphi(X^*)$. We have seen, in the proof of Theorem 13.5.1, that $\delta(D_X) = \nu(\widehat{R})$ and $\delta(G_X) = \nu(\widehat{L})$. According to Formula (13.26),

$$\delta(w^{-1}X^*) = \pi(C_r(w))\delta(X^*).$$

Set $n = \varphi(w)$ and $T = \{k \in K \mid nk \in \varphi(X^*)\}$. Then $T \subset \widehat{L}$ since for $k \in T$, we have $nk \in Mk \cap \varphi(X^*)$, showing that the left ideal Mk meets $\varphi(X^*)$. Let H be an $\mathcal{H}$-class contained in $\widehat{L}$. The function $h \mapsto nh$ is a bijection from H onto the $\mathcal{H}$-class nH. Since $n \in \widehat{R}$, we have $nH \subset \widehat{R}$; since $H \subset \widehat{L}$ we have $nH \subset \widehat{L}$. Thus $nH \subset \widehat{R} \cap \widehat{L}$. This implies that $nH \cap \varphi(X^*) \neq \emptyset$. Indeed let R and L denote the $\mathcal{R}$-class and $\mathcal{L}$-class containing nH, and take $m \in R \cap \varphi(X^*)$, $m' \in L \cap \varphi(X^*)$. Then $mm' \in R \cap L \cap \varphi(X^*) = nH \cap \varphi(X^*)$.

Setting $d = d(X)$, it follows that

$$\frac{\text{Card}(nH \cap \varphi(X^*))}{\text{Card}(nH)} = \frac{1}{d}.$$

Since $H \cap T = \{k \in H \mid nk \in \varphi(X^*)\}$ is in bijection with $nH \cap \varphi(X^*)$, we have

$$\text{Card}(H \cap T) = \text{Card}(nH \cap \varphi(X^*)) = \frac{1}{d}\text{Card}(H).$$

Therefore

$$\nu(T) = \sum_{H \subset \widehat{L}} \nu(H \cap T) = \sum_{H \subset \widehat{L}} \frac{\nu(H)}{\text{Card}(H)} \text{Card}(H \cap T)$$

$$= \sum_{H \subset \widehat{L}} \frac{\nu(H)}{d} = \frac{1}{d}\nu(\widehat{L}).$$

We observe that $\varphi^{-1}(T) = w^{-1}X^* \cap \varphi^{-1}(K)$. According to (13.18), we have $\nu(T) = \nu(T \cap K) = (1/d)\nu(\widehat{L})$. Since also $\nu(\widehat{L}) = \delta(G_X)$, we obtain

$$\pi(C_r(w))\delta(D_X) = \frac{\delta(w^{-1}X^*)}{\delta(X^*)}\delta(D_X) = \frac{1}{d}\frac{\delta(G_X)\delta(D_X)}{\delta(X^*)}. \tag{13.29}$$

By Theorem 13.5.1, the last expression is equal to 1. $\qquad\square$

Proposition 13.5.7 *Let $X \subset A^+$ be a thin complete code. Let π be a positive Bernoulli distribution on A^*. For all $w \in A^*$ the following conditions are equivalent.*

(i) *The set $C_r(w)$ is maximal among the sets $C_r(u)$, for $u \in A^*$.*
(ii) $\pi(C_r(w))\delta(D_X) = 1$.

Proof. With the notations of Proposition 13.5.6, consider a word $x \in \varphi^{-1}(K) \cap X^*$. Then $C_r(w) \subset C_r(xw)$, hence also $\pi(C_r(w)) \leq \pi(C_r(xw))$. On the other hand $xw \in \varphi^{-1}(K) \cap D_X$. Indeed the right ideal generated by x is minimal, and therefore there exists $v \in A^*$ such that $\varphi(xwv) = \varphi(x)$. Thus $xwv \in X^*$. By Proposition 13.5.6, we have $\pi(C_r(xw))\delta(D_X) = 1$ showing that

$$\pi(C_r(w)) \leq 1/\delta(D_X). \tag{13.30}$$

Now assume $C_r(w)$ maximal. Then $C_r(w) = C_r(xw)$, implying the equality sign in the formula. This proves (i) $\implies$ (ii). Conversely Formula (13.30) shows the implication (ii) $\implies$ (i). $\qquad\square$

In fact, the set of words $w \in A^*$ such that the set of strict right contexts is maximal is an old friend: in Chapter 5, Section 5.1, we defined the sets of strongly right completable and simplifying words by

$$E(X) = \{u \in A^* \mid \forall v \in A^*, \exists w \in A^* : uvw \in X^*\},$$

$$S(X) = \{u \in A^* \mid \forall x \in X^*, \forall v \in A^* : xuv \in X^* \implies uv \in X^*\}.$$

We have seen (Exercise 5.1.7) that these sets are equal provided they are both nonempty. It can be shown (Exercise 13.5.1) that, for a thin complete code X, the following three conditions are equivalent for all words $w \in A^*$:

(i) $w \in E(X)$.
(ii) $w \in S(X)$.
(iii) $C_r(w)$ is maximal.

This leads to a natural interpretation of Formula (13.27) (see Exercise 13.5.2). We now establish, as a corollary of Formula (13.27) a property of finite maximal codes which generalizes the property for prefix codes shown in Chapter 3 (Theorem 3.6.10).

Theorem 13.5.8 *Let $X \subset A^+$ be a finite maximal code. For any letter $a \in A$, the order of a is a multiple of $d(X)$.*

Recall that the order of a is the integer n such that $a^n \in X$.

Proof. Let π be a positive Bernoulli distribution on A^*. Let $\mathcal{A} = (Q, 1, 1)$ be a trim unambiguous automaton recognizing X^*. Let K be the minimal ideal of the monoid $M = \varphi_\mathcal{A}(A^*)$. Let $x \in X^* \cap \varphi_\mathcal{A}^{-1}(K)$. According to Proposition 13.5.6,

$$\pi(C_r(x))\delta(D_X) = 1, \quad \pi(C_\ell(x))\delta(G_X) = 1.$$

By Formula (13.25), the average length of X is

$$\lambda(X) = \frac{d(X)}{\delta(G_X)\delta(D_X)}.$$

Consequently

$$\lambda(X) = d(X)\pi(C_r(x))\pi(C_\ell(x)).$$

The proof would be complete if we could set $\pi(a) = 1$ and $\pi(b) = 0$ for $b \neq a$. Indeed, we have then $\lambda(X) = n$, and thus $d(X)$ divides n. However this distribution is not positive, and so Proposition 13.5.6 cannot be applied.

Let a be a fixed letter and let n be its order. Consider a sequence $(\pi_k)_{k\geq 0}$ of positive Bernoulli distributions such that $\lim_{k\to\infty} \pi_k(a) = 1$ and $\lim_{k\to\infty} \pi_k(b) = 0$ for any $b \in A \setminus a$. For any word $w \in A^*$, we have $\lim_{k\to\infty} \pi_k(w) = 1$ if $w \in a^*$, and $\lim_{k\to\infty} \pi_k(w) = 0$ otherwise. For any $k \geq 0$, denote by $\lambda_k(X)$ the average length of X with respect to π_k. Then

$$\lambda_k(X) = d(X)\pi_k(C_r(x))\pi_k(C_\ell(x)),$$

and also, by definition

$$\lambda_k(X) = \sum_{x\in X} |x|\pi_k(x).$$

Since X is finite, this sum is over a finite number of terms, and going to the limit, we get

$$\lim_{k\to\infty} \lambda_k(X) = \sum_{x\in X} |x| \lim_{k\to\infty} \pi_k(x).$$

Since $\lim_{k\to\infty} \pi_k(x) = 0$ unless $x \in a^*$, we have $\lim_{k\to\infty} \lambda_k(X) = n$, where n is the order of a. On the other hand,

$$\pi_k(C_r(x)) = \sum_{v\in C_r(x)} \pi_k(v).$$

The words in $C_r(x)$ are suffixes of words in X. Since X is finite, $C_r(x)$ is finite. Thus, going to the limit, we have

$$\lim_{k\to\infty} \pi_k(C_r(x)) = \sum_{v\in C_r(x)} \lim_{k\to\infty} \pi_k(v) = \mathrm{Card}(C_r(x) \cap a^*).$$

Similarly

$$\lim_{k\to\infty} \pi_k(C_\ell(x)) = \sum_{v\in C_\ell(x)} \lim_{k\to\infty} \pi_k(v) = \mathrm{Card}(C_\ell(x) \cap a^*).$$

Consequently

$$n = d(X)\,\mathrm{Card}(C_r(x) \cap a^*)\,\mathrm{Card}(C_\ell(x) \cap a^*).$$

This proves that $d(X)$ divides n. $\qquad\square$

13.6 Exercises

Section 13.1

13.1.1 A probability distribution π on A^* is said to be *invariant* if for any $w \in A^*$

$$\sum_{a \in A} \pi(aw) = \pi(w).$$

Let $\mathcal{A} = (Q, I, T)$ be a stochastic automaton with adjacency matrix P, and let π be the probability distribution defined by $\mathcal{A}$. Show that if $IP = I$, then π is an invariant distribution.

Section 13.2

13.2.1 Let $\mathcal{A} = (Q, i, t)$ be a complete deterministic strongly connected finite automaton and let π be a positive Bernoulli distribution on A^*. Let P be the $Q \times Q$-matrix defined by $P_{p,q} = \sum_{a \in A, p \cdot a = q} \pi(a)$.

A nonnegative Q-vector I with $\sum_{q \in Q} I_q = 1$ is said to be *stationary* for $\mathcal{A}$ if $IP = I$.

Show that $\mathcal{A}$ admits a unique stationary vector, given by $I_q = 1/\lambda(X_q)$ for any $q \in Q$, where X_q is the prefix code such that X_q^* is the stabilizer of the state q in $\mathcal{A}$.

Section 13.3

13.3.1 Let $X \subset A^+$ be a rational code. Show that if Y is a code such that

$$X \subset Y \quad \text{and} \quad Y^* \subset F(X^*),$$

then $X = Y$ (this generalizes the fact that a complete rational code is maximal).

Section 13.4

13.4.1 Let M be a monoid, and let μ, ν be two probability measures over M. The *convolution* of μ and ν is defined as the probability measure given by

$$\mu * \nu(m) = \sum_{uv=m} \mu(u)\nu(v).$$

(a) Show that

$$\left(\lim_{n \to \infty} \mu_n \right) * \nu = \lim_{n \to \infty} (\mu_n * \nu).$$

(b) Let π be a positive Bernoulli distribution on A^*. For $n \geq 0$, let $\pi^{(n)}$ be the probability measure on the subsets of A^* defined by

$$\pi^{(n)}(L) = \pi(L \cap A^n)$$

for $L \subset A^*$. Show that

$$\pi^{(n+1)} = \pi^{(n)} * \pi^{(1)}.$$

(c) Let $\varphi : A^* \to M$ be a morphism onto a well-founded monoid. Let π be as above and let $v = \delta\varphi^{-1}$ be the probability measure over M defined in Proposition 13.4.8. Show that v is *idempotent*, that is

$$v * v = v.$$

13.4.2 Let $\mathcal{A} = (Q, i, T)$ be a finite automaton over A. Assume moreover that $\mathcal{A}$ is complete, deterministic, and strongly connected. Let φ be the associated representation and let $M = \varphi(A^*)$. Let π be a positive Bernoulli distribution on A^*. Let d be the minimal rank of M. Let $\mathcal{E}$ be the set of minimal images of $\mathcal{A}$. Let $\mathcal{B}$ be the deterministic automaton with states $\mathcal{E}$ and with the action induced by $\mathcal{A}$. Show that the stationary vectors I of $\mathcal{A}$ and J of $\mathcal{B}$ are related, for $q \in Q$, by

$$I_q = \frac{1}{d} \sum_{E \in \mathcal{E}_q} J_E,$$

where $\mathcal{E}_q$ is the set of E in $\mathcal{E}$ such that $q \in E$.

Section 13.5

13.5.1 Let $X \subset A^+$ be a thin complete code. Let $S(X)$ and $E(X)$ be the sets of simplifying and strongly left completable words defined in Chapter 5. Show that for $w \in A^*$ the following conditions are equivalent:

(i) $w \in S(X)$.
(ii) $w \in E(X)$.
(ii) $C_r(w)$ is maximal among all $C_r(u)$, $u \in A^*$.

13.5.2 Use Exercise 5.1.8 to give another proof of Formula (13.27).

13.5.3 Let $X \subset A^+$ be a code and $\alpha : B^* \to A^*$ a coding morphism for X, that is, $\alpha(B) = X$. Let π be an invariant distribution on B^*. Show that the function π^α from A^* into $[0, 1]$ defined by

$$\pi^\alpha(w) = \frac{1}{\lambda(\alpha)} \sum_{(u,v) \in C(w)} \pi(\alpha^{-1}(uwv))$$

with $\lambda(\alpha) = \sum_{x \in X} |x| \pi(\alpha^{-1}(x))$ is an invariant distribution on A^*. Compare with the definition of the contextual probability.

13.7 Notes

The presentation of measure spaces follows Halmos (1950). We have followed this book for the proof of Kolmogorov's extension theorem. The term "process" is used in Shields (1996) where many additional properties of measures related to words are presented. Theorem 13.2.11 is due to Feller. A more precise statement is the following: Let (X, π) be a persistent recurrent event. Let p be the gcd of the lengths of the words in X. Then the sequence $\pi(X^* \cap A^{np})$ for $n \geq 0$ has a limit, which is

0 or $p/\lambda(X)$, according to $\lambda(X) = \infty$ or not (see Feller (1968), Theorem XIII.3.3). Theorem 13.2.7 is less precise on two points: (i) we only consider the case where $\lambda(X) < \infty$ and (ii) we only consider the limit in mean of the sequence $\pi(X^* \cap A^n)$.

The notion of topological entropy is well known in symbolic dynamics (Lind and Marcus (1995)). The word "topological" is used to distinguish this notion from probabilistic entropy, such as mentioned in Exercise 3.7.1. The results of Section 13.4 and related results can be found in Greenander (1963) and Martin-Löf (1965). Theorem 13.5.1 is due to Schützenberger (1965b). Theorem 13.5.5 is from Hansel and Perrin (1983).

A stationary vector, as introduced in Exercise 13.2.1, is usually called a stationary distribution in the theory of Markov chains.

The statement of Exercise 13.3.1 is a particular case of a result of Restivo (1990) who proved it under the more general hypothesis that X is a thin code.

Further developments of the results presented in this chapter may be found in Blanchard and Perrin (1980), Hansel and Perrin (1983), or Blanchard and Hansel (1986). In particular these papers discuss the relationship of the concepts developed in this chapter with ergodic theory.

14

Polynomials of finite codes

There is a noncommutative polynomial canonically associated with a finite code: it is the sum of the codewords, minus 1. When the code is maximal, this polynomial has some striking factorization properties, which reflect probabilistic and combinatorial properties of the code, such as the property of being prefix, suffix or synchronizing. When the code is prefix, the factorization is directly related to the tree representation of the code. When the code is bifix, one has even more combinatorial evidence for the factorization, as described in Chapter 6. In the general case, the factorization of the polynomial has no direct combinatorial interpretation, but is related via the *factorization conjecture* to a kind of coset decomposition of the free monoid with respect to the submonoid generated by the code. The factorization conjecture is the main open problem in the theory of codes.

The chapter is organized as follows. In Section 14.1 we define positive factorizations. In Section 14.2, we state the factorization theorem (Theorem 14.2.1), which is the main result of this chapter. Section 14.3 presents some results on noncommutative polynomials which are used in the proof of the factorization theorem. Section 14.4 contains the proof of the theorem. Section 14.5 presents some applications of the factorization theorem.

Section 14.6 introduces another equivalence, called the *commutative equivalence*. It is conjectured that any finite maximal code is commutatively equivalent to a prefix code. This is a consequence of the factorization conjecture. Indeed, it is shown that any positively factorizing maximal code is commutatively prefix (Corollary 14.6.6). Section 14.7 presents a specialized topic concerning the reducibility property of the linear representation associated to an automaton. We prove that the minimal representation associated with the submonoid generated by a maximal code is completely reducible if and only if the code is bifix (Theorems 14.7.5 and 14.7.7).

14.1 Positive factorizations

Let X be a subset of A^+. A pair (P, S) of subsets of A^* is called a *positive factorization* for the set X if each word $w \in A^*$ factorizes uniquely into

$$w = sxp \tag{14.1}$$

with $p \in P$, $s \in S$, $x \in X^*$. In terms of formal power series, (14.1) can be expressed as

$$\underline{A}^* = \underline{S}\,\underline{X}^*\underline{P}\,. \qquad (14.2)$$

Note the analogy with the coset decomposition of a group with respect to a subgroup. Observe that $1 \in P$ and $1 \in S$. Taking the inverses in (14.2), we obtain the equivalent formulation

$$1 - \underline{X} = \underline{P}\,(1 - \underline{A})\,\underline{S} \qquad (14.3)$$

or also

$$\underline{X} - 1 = \underline{P}\,\underline{A}\,\underline{S} - \underline{P}\,\underline{S}\,. \qquad (14.4)$$

This equation shows that each word in X can be written in at least one way as $x = pas$ with $p \in P$, $a \in A$, $s \in S$.

Proposition 14.1.1 *A set X for which there is a positive factorization (P, S) is a code.*

Proof. Indeed, (14.4) implies that $\underline{A}^* = \underline{S}(\underline{X})^*\underline{P}$ which in turn shows that $(\underline{X})^*$ has only coefficients 0 or 1. $\qquad \square$

A code X is *positively factorizing* if there exists a pair (P, S) of sets which is a positive factorization for X.

A prefix code X is positively factorizing. Indeed, let $P = A^* \setminus XA^*$ be the set of words having no prefixes in X. Then $\underline{A}^* = \underline{X}^*\underline{P}$ and thus $(P, \{1\})$ is a positive factorization for X. Conversely, if $(P, \{1\})$ is a positive factorization for X, then the code X is prefix. Indeed, if u, $uv \in X^*$, then setting $v = xp$ with $x \in X^*$ and $p \in P$, we obtain $(ux)p \in X^*$, which implies $p = 1$ by the uniqueness of factorization. Thus X^* is right unitary.

Symmetrically, for a suffix code X, one has $\underline{A}^* = \underline{S}\,\underline{X}^*$ with $S = A^* \setminus A^*X$. If X is a bifix code, then simultaneously

$$\underline{A}^* = \underline{X}^*\underline{P} \quad \text{and} \quad \underline{A}^* = \underline{S}\,\underline{X}^*$$

with $P = A^* \setminus XA^*$ and $S = A^* \setminus A^*X$. This shows in particular that there may exist several positive factorizations for a code (see also Exercise 14.1.8).

Recall that by Proposition 6.3.8, for a thin maximal bifix code X, we have

$$\underline{X} - 1 = d(\underline{A} - 1) + (\underline{A} - 1)T(\underline{A} - 1),$$

where T is the tower over X and d isthe degree of X. The series T has nonnegative coefficients. Hence $\underline{A}^* = \underline{X}^*\underline{P} = \underline{S}\,\underline{X}^*$ with

$$\underline{P} = d + (\underline{A} - 1)T, \quad \underline{S} = d + T(\underline{A} - 1). \qquad (14.5)$$

Let $X \subset A^+$ be a positively factorizing code and let (P, S) be a positive factorization for X. If P and S are thin, then X is a thin maximal code. Indeed, Equation (14.4) shows that $X \subset PAS$. Since P, A, S are thin, the product PAS is thin also and consequently X is thin. Furthermore, X is complete. Indeed, let $u \in \bar{F}(S)$ and $v \in \bar{F}(P)$. For each w in A^* the word uwv is in SX^*P. By the choice of u and v, it follows that w is in $F(X^*)$. Thus X is complete.

As a special case, note that if P and S are finite, then X is a finite maximal code. We shall see later that, conversely, if (P, S) is a positive factorization for a finite maximal code, then P and S are finite. There exist finite codes which are not positively factorizing. An example will be given in Section 14.6. However, no finite maximal code is known which is not positively factorizing. Whether any finite maximal code is positively factorizing is still unknown. This constitutes the *factorization conjecture*.

Proposition 14.1.2 *The composition of two positively factorizing codes is again a positively factorizing code.*

Proof. Let $X, Y \subset A^+$ and $Y \subset B^+$ be codes and let $\beta : B \to Z$ be a bijection such that $X = Y \circ_\beta Z$. By assumption, Y and Z are positively factorizing codes. Thus there are sets $S, P \subset A^*$ and $Q, R \subset B^*$ such that

$$\underline{A}^* = \underline{S} \, \underline{Z}^* \underline{P}, \quad \underline{B}^* = \underline{Q} \, \underline{Y}^* \underline{R}.$$

Set $U = \beta(Q)$ and $V = \beta(R)$. We extend β to series over B. Since β is bijective, we get $\underline{U} = \beta(\underline{Q})$, $\underline{V} = \beta(\underline{R})$, and also $\underline{Z} = \beta(\underline{B}^*)$, $\underline{X}^* = \beta(\underline{Y}^*)$. This shows that $\underline{Z}^* = \underline{U} \, \underline{X}^* \underline{V}$ and consequently

$$\underline{A}^* = \underline{S} \, \underline{U} \, \underline{X}^* \underline{V} \, \underline{P}.$$

Since the left-hand side of this equation is a characteristic series, the products of the right-hand side only give coefficients 0 and 1, and consequently

$$\underline{A}^* = \underline{S} \, \underline{U} \, \underline{X}^* \underline{V} \, \underline{P},$$

showing that X is positively factorizing. $\qquad\square$

Example 14.1.3 Let $A = \{a, b\}$, and let

$$X = \{a^4, ab, aba^6, aba^3b, aba^3ba^2, aba^2ba, aba^2ba^3, aba^2b^2, aba^2b^2a^2, b, ba^2\}.$$

The set X is a positively factorizing code. Indeed, an easy computation gives

$$1 - \underline{X} = (1 + a + aba^2(1 + a + b))(1 - a - b)(1 + a^2). \tag{14.6}$$

Thus this is a positive factorization (P, S) with

$$P = \{1, a, aba^2, aba^3, aba^2b\}, \quad S = \{1, a^2\}.$$

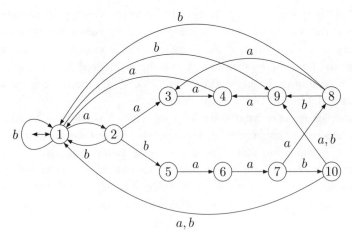

Figure 14.1 The automaton $\mathcal{A}$.

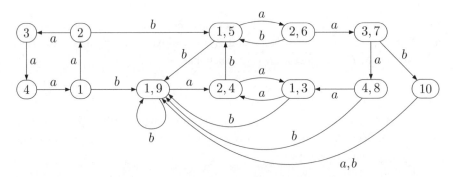

Figure 14.2 The result of the determinization.

Since P and S are finite, X is a maximal code. We may verify that X is indecomposable. This is the smallest known example of a finite maximal indecomposable code which is neither prefix nor suffix (see Example 2.6.11 and Exercise 14.1.7).

The remaining part of this example illustrates the relation between the positive factorization and the structure of the transition monoid of an unambiguous automaton. The computation allows us, in cases such as the present one, to recover the positive factorization directly from the monoid (see also Exercises 14.1.1 and 14.1.2).

An unambiguous automaton $\mathcal{A}$ recognizing X^* is represented in Figure 14.1. This automaton can be used as follows to recover the positive factorization for X given by (14.6). We first compute the deterministic automaton obtained by applying the determinization algorithm to the automaton $\mathcal{A}$ starting from $\{1\}$. The result is shown in Figure 14.2. This automaton has a unique minimal strongly connected component corresponding to the rows of the elements of the minimal ideal of the monoid $M = \varphi_{\mathcal{A}}(A^*)$.

We then apply the determinization algorithm backwards to the automaton $\mathcal{A}$ starting also from state $\{1\}$. The result is shown in Figure 14.3 (we represent only part of the result, containing the unique minimal strongly connected component). Let L be the

Table 14.1 *The minimal ideal of M.*

5	6	3	4	10	9	4	3
1	2	7	8	10	1	2	1
1	6	7	4	10	1	4	1
5	2	3	8	10	9	2	3

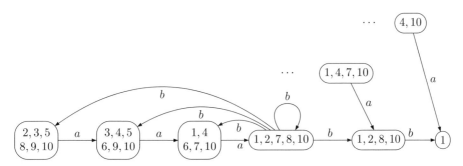

Figure 14.3 The result of the backwards determinization.

set of states of the strongly connected component of the automaton of Figure 14.2 and let C be the set of states of the strongly connected component of Figure 14.3. Any element of L intersects any element of C in exactly one element, as shown in Table 14.1 in which the elements of L appear as the columns and the elements of C as the rows (this is true for any thin maximal code, see Exercise 9.3.8).

We select the state $\ell = \{1, 9\}$ in L and the state $c = \{1, 2, 7, 8, 10\}$ in C. The set of labels of simple paths from ℓ to 1 is $S = \{1, aa\}$ and the sets of labels of simple paths from 1 to c is $P = \{1, a, abaa, abaaa, abaab\}$. Since all paths from $\{1, 9\}$ to $\{1, 2, 7, 8, 10\}$ pass through state 1, the pair (P, S) is a positive factorization for X.

14.2 The factorization theorem

Recall that the degree of a finite maximal code has been defined in Section 9.5. The following theorem is the main result of this chapter.

Theorem 14.2.1 *Let $X \subset A^*$ be a finite maximal code and d its degree. Then for some polynomials P, Q, S in $\mathbb{Z}\langle A \rangle$, one has*

$$\underline{X} - 1 = P(d(\underline{A} - 1) + (\underline{A} - 1)Q(\underline{A} - 1))S. \tag{14.7}$$

Moreover, if X is prefix (resp. suffix), one can choose $S = 1$ (resp. $P = 1$).

Note that in all known cases, the polynomial Q has nonnegative coefficients, and moreover P, S have coefficients 0, 1. Thus, P and S can be viewed as representing sets of prefixes and suffixes. The polynomial Q is not, in general, a characteristic polynomial.

Example 14.2.2 Let

$$X = \{a^3, a^2ba^2, a^2bab, a^2b^2, aba^3, aba^2ba^2, aba^2bab, aba^2b^2,$$
$$ababa^2, abababa^2, (ab)^4, ababab^2, abab^2, ab^2a, ab^3a^2, ab^3aba^2,$$
$$ab^3abab, ab^3ab^2, ab^4, ba, b^2a^2, b^2aba^2, b^2abab, b^2ab^2, b^3\}$$

be the maximal prefix code of degree 3 of Example 3.6.13. We have, in agreement with Theorem 14.2.1,

$$\underline{X} - 1 = (1 + ab)(3(\underline{A} - 1) + (\underline{A} - 1)Q(\underline{A} - 1)),$$

with $A = \{a, b\}$ and $Q = 2 + a + b + ba + (1 + b)ab(1 + a)$. This can be checked directly or by observing that one has

$$\underline{X} = (1 + ab)(a^3 + a^2b(a^2 + ab + b) + abab(a^2 + b)$$
$$+ ba + b^2a(a + b(a^2 + ab + b)) + b^3) + (ab)^4.$$

Corollary 14.2.3 *For any finite maximal code X over A, there exist polynomials P, S in $\mathbb{Z}\langle A \rangle$ such that*

$$\underline{X} - 1 = P(\underline{A} - 1)S. \tag{14.8}$$

□

Observe that the expression (14.8) with P, S having coefficients $0, 1$ defines a positive factorization for X, in the sense defined previously.

The previous result has the following converse. Thus finite maximal codes are completely characterized by Corollary 14.2.3.

Theorem 14.2.4 *Let W be a polynomial in $\mathbb{N}\langle A \rangle$ without constant term, and let P, S be polynomials in $\mathbb{C}\langle A \rangle$ such that*

$$W - 1 = P(\underline{A} - 1)S.$$

Then W is the characteristic polynomial of a finite maximal code X. If moreover S (resp. P) is constant, then X is a prefix (resp. suffix) code.

Proof. Since $W - 1 = P(\underline{A} - 1)S$ and since W has no constant term, P and S are invertible in $\mathbb{C}\langle\langle A \rangle\rangle$, and we obtain

$$\underline{A}^* = SW^*P. \tag{14.9}$$

Define $X = \text{supp}(W)$ (recall that $\text{supp}(T)$ denotes the support of the series T). Then X is finite. We show that X is complete. Indeed, let w be any word, and choose u of length $\geq \deg(S), \deg(P)$. Then uwu appears in the left-hand side of Equation (14.9), and we obtain $uwu = smp$, for some words $s \in \text{supp}(S)$, $m \in X^*$, $p \in \text{supp}(P)$. By the choice of u, it follows that w is a factor of m. Thus $X^* \cap A^*wA^*$ is not empty, and X is complete.

Now we show that $\pi(X) = 1$, where π is some Bernoulli distribution. This implies that X is a maximal code by Theorem 2.5.19.

Since X is complete and finite, we have $\pi(X) \geq 1$, by Proposition 2.5.11. On the other hand, we extend π naturally to a morphism from $\mathbb{C}\langle A \rangle$ to $\mathbb{C}$, and we obtain

$$\pi(W) - 1 = \pi(P)\pi(\underline{A} - 1)\pi(S) = 0$$

and therefore $\pi(W) = 1$. Next, since W has coefficients in $\mathbb{N}$, one has $\pi(X) \leq \pi(W) = 1$, and therefore $\pi(X) = 1$.

If S is a constant, we may suppose that $S = 1$ and Equation (14.9) becomes $\underline{A}^* = W^* P$. A similar argument as before shows that X is right complete. By Theorem 3.3.8, X is a prefix code. $\qquad \square$

14.3 Noncommutative polynomials

Let K be a commutative ring. We begin with a result on the division of polynomials which is a version of Euclidean division in several noncommutative variables. Given two polynomials X, Y in $K\langle A \rangle$, we say that Y is *weak left divisor* of X in $K\langle A \rangle$ if there exist polynomials Q, R in $K\langle A \rangle$ such that

$$X = YQ + R \quad \text{with} \quad \deg(R) < \deg(Y).$$

The polynomial R is called the *remainder*. Observe that in one variable, this relation is just Euclidean division. Weak left division is not always possible if A has more than one letter (for instance take $X = a$ and $Y = b$ for distinct letters a, b).

The next result gives a sufficient condition for weak divisibility (this condition is easily seen to be also necessary, see Exercise 14.3.1).

Theorem 14.3.1 *Let K be a field. Let X, Y, P, Q be polynomials in $K\langle A \rangle$ with $\deg(Q) \leq \deg(P)$ and $P \neq 0$. If Y is a weak left divisor of $XP + Q$, then Y is a weak left divisor of X.*

The following consequence is immediate.

Corollary 14.3.2 *If X, Y, X', Y' are nonzero polynomials such that $XY' = YX'$, then Y is a weak left divisor of X and X is a weak left divisor of Y.* $\qquad \square$

We fix an order on A and use the corresponding radix order on A^*. Given a nonzero polynomial P we denote by $\max(P)$ the *maximal word* (with respect to the radix order) appearing in the support of P. One checks easily that $\max(P + Q) = \max(P)$ if $\deg(Q) < \deg(P)$, and $\max(PQ) = \max(P)\max(Q)$.

Proof of Theorem 14.3.1. Let Q' and R' be polynomials such that

$$XP + Q = YQ' + R' \tag{14.10}$$

with $\deg(R') < \deg(Y)$. We have $Y \neq 0$ since $\deg(R') < \deg(Y)$. We may assume $\deg(Y) \geq 1$, since the case $\deg(Y) = 0$ is immediate. The case $\deg(X) < \deg(Y)$

is also easy. So we may assume $\deg(X) \geq \deg(Y) \geq 1$. Observe that $\deg(Q) \leq \deg(P) < \deg(XP)$ and $\deg(R') < \deg(Y) \leq \deg(X) \leq \deg(XP)$. This shows that Q' is nonzero. By (14.10), we have $\max(XP) = \max(XP + Q - R') = \max(YQ')$, and $\max(X)\max(P) = \max(Y)\max(Q')$. Thus the word $\max(Y)$ is a prefix of $\max(X)$ and we may write $\max(X) = \max(Y)u$ for some $u \in A^*$. Hence for some $\alpha \in K$, we have $X = X' + \alpha Yu$, with $\max(X') < \max(X)$. By (14.10), we obtain

$$X'P + Q = Y(Q - \alpha uP) + R'.$$

We conclude by induction on $\max(X')$ that Y is a weak left divisor of X' and thus of X. $\qquad\square$

Let $x_1, x_2, \ldots$ be a sequence of elements of a ring, of length at least n. We define the n-th *continuant polynomial* relative to this sequence by $p(x_1, \ldots, x_n)$, where $p(x_1, \ldots, x_n)$ is the 1, 1 coefficient of the matrix

$$\begin{pmatrix} x_1 & 1 \\ 1 & 0 \end{pmatrix} \begin{pmatrix} x_2 & 1 \\ 1 & 0 \end{pmatrix} \cdots \begin{pmatrix} x_n & 1 \\ 1 & 0 \end{pmatrix}.$$

It is a simple exercise to show that this matrix is actually equal to

$$\begin{pmatrix} p(x_1, \ldots, x_n) & p(x_1, \ldots, x_{n-1}) \\ p(x_2, \ldots, x_n) & p(x_2, \ldots, x_{n-1}) \end{pmatrix}. \tag{14.11}$$

Indeed, for the entry in position 2,1 for example, one sees that it is $p(x_2, \ldots, x_n)$ by computing the product of the first matrix by the product of the remaining ones and using induction.

For sake of coherence, the 0-th continuant polynomial is equal to 1, and the (-1)-th is equal to 0. From Equation (14.11), one deduces that

$$p(x_1, \ldots, x_n) = p(x_1, \ldots, x_{n-1})x_n + p(x_1, \ldots, x_{n-2}), \tag{14.12}$$

and

$$p(x_1, \ldots, x_n) = x_1 p(x_2, \ldots, x_n) + p(x_3, \ldots, x_n).$$

We often use the latter equation in the form

$$p(x_n, \ldots, x_1) = x_n p(x_{n-1}, \ldots, x_1) + p(x_{n-2}, \ldots, x_1). \tag{14.13}$$

By induction, one deduces the *Wedderburn relation*:

$$p(x_1, \ldots, x_n)p(x_{n-1}, \ldots, x_1) = p(x_1, \ldots, x_{n-1})p(x_n, \ldots, x_1). \tag{14.14}$$

To prove it, use Equation (14.12) for the left-hand side, Equation (14.13) for the right-hand side and induction.

The next result shows that, in essence, each relation $XY' = YX'$ in $K\langle A \rangle$ comes from a Wedderburn relation (14.14).

Theorem 14.3.3 *Let X, Y, X', Y' be nonzero polynomials in $K\langle A\rangle$ such that $XY' = YX'$. Then there exist $n \geq 1$ and polynomials $U, V, x_1, \ldots, x_n$ such that*

$$X = Up(x_1, \ldots, x_n), \qquad Y' = p(x_{n-1}, \ldots, x_1)V$$

$$Y = Up(x_1, \ldots, x_{n-1}), \quad X' = p(x_n, \ldots, x_1)V.$$

Furthermore, $x_1, \ldots, x_{n-1}$ have positive degree, and if $\deg(X) > \deg(Y)$, then x_n also has positive degree.

The proof is a simple noncommutative version of the Euclidean algorithm, obtained by iteration of the Euclidean division of Corollary 14.3.2.

Proof. The hypothesis and Corollary 14.3.2 imply that Y is a weak left divisor of X. Thus $X = YQ + Z$, for some polynomials Q and Z with $\deg(Z) < \deg(Y)$; note that if $\deg(X) > \deg(Y)$, then $\deg(Q) > 0$. From $XY' = YX'$, we have $(YQ + Z)Y' = YX'$. We set $Z' = X' - QY'$. This implies $ZY' = YZ'$; since $\deg(Z) < \deg(Y)$, we deduce that $\deg(Z') < \deg(Y')$. Note that $Z = 0 \Leftrightarrow Z' = 0$. In this case, the result follows with $n = 1$, $U = Y$, $x_1 = Q$ and $V = Y'$. We now assume that $Z \neq 0$.

Then we have $YZ' = ZY'$, and by induction, there exist polynomials $U, V, x_1, \ldots, x_n$ such that

$$Y = Up(x_1, \ldots, x_n), \qquad Z' = p(x_{n-1}, \ldots, x_1)V$$

$$Z = Up(x_1, \ldots, x_{n-1}), \quad Y' = p(x_n, \ldots, x_1)V.$$

Moreover, $x_1, \ldots, x_{n-1}$ have positive degrees and, since $\deg(Z) < \deg(Y)$, x_n also has positive degree. This, together with $X = YQ + Z$ and $X' = QY' + Z'$ gives

$$X = U(p(x_1, \ldots, x_n)Q + p(x_1, \ldots, x_{n-1})), \quad Y' = p(x_n, \ldots, x_1)V,$$

$$Y = Up(x_1, \ldots, x_n), \quad X' = (Qp(x_n, \ldots, x_1) + p(x_{n-1}, \ldots, x_1))V.$$

The result follows by (14.12) and (14.13) with $x_{n+1} = Q$ (recall that Q has positive degree if $\deg(X) > \deg(Y)$). $\qquad\square$

We shall also need the next result in the proof of Theorem 14.2.1 (with $\underline{A} - 1$ playing the role of the polynomial of degree 1). For polynomials X, X', Y we write $X \equiv X'$ modulo Y if Y is a weak left divisor of X, X' with the same remainder, that is if $X = YQ + R$ and $X' = YQ' + R$.

Theorem 14.3.4 *Let B be a polynomial of degree 1, and let $x_1, \ldots, x_n$ be polynomials such that $x_1, \ldots, x_{n-1}$ have positive degree. If B is a weak left divisor of $p(x_{n-1}, \ldots, x_1)$ and $p(x_n, \ldots, x_1)$ then $p(x_1, \ldots, x_i) \equiv p(x_i, \ldots, x_1)$ modulo B for each $i = 1, \ldots, n$.*

To prove this, we need a lemma.

Lemma 14.3.5 *Let* $x_1, \ldots, x_n$ *be polynomials.*

(i) $p(x_1, \ldots, x_n) = 0$ *if and only if* $p(x_n, \ldots, x_1) = 0$.

(ii) *If the degrees of* $x_1, \ldots, x_{n-1}$ *are strictly positive, then the polynomials* 1, $p(x_1)$, $\ldots, p(x_{n-1}, \ldots, x_1)$ *have strictly increasing degrees.*

Proof. Claim (i) is proved using the Wedderburn relation (14.14) if $p(x_{n-1}, \ldots, x_1)$ and $p(x_1, \ldots, x_{n-1})$ are both nonzero, and Equations (14.12) and (14.13) if they are both zero (by induction, if one is zero, so is the other).

Similarly (ii) is proved by induction, using Equation (14.13).   □

Proof of Theorem 14.3.4. The proof is by induction. The case $n = 1$ is obvious, so assume $n > 1$. If $p(x_{n-1}, \ldots, x_1)$ vanishes, then $p(x_1, \ldots, x_{n-1})$ also vanishes by Lemma 14.3.5 (i). Then by Equations (14.12) and (14.13), we have $p(x_1, \ldots, x_n) = p(x_1, \ldots, x_{n-2})$ and $p(x_n, \ldots, x_1) = p(x_{n-2}, \ldots, x_1)$. Thus we conclude the proof by induction in this case.

Suppose that $p(x_{n-1}, \ldots, x_1) \neq 0$. Then by (14.13),

$$x_n p(x_{n-1}, \ldots, x_1) + p(x_{n-2}, \ldots, x_1) = BQ + \alpha$$

for some polynomial Q and some scalar $\alpha \in K$.

By Lemma 14.3.5(ii), we have $\deg(p(x_{n-2}, \ldots, x_1)) < \deg(p(x_{n-1}, \ldots, x_1))$. Accordingly, by Theorem 14.3.1, the above equality implies that B is a weak left divisor of x_n. Hence, $x_n \equiv \gamma$ modulo B. By hypothesis, the left division of $p(x_1, \ldots, x_i)$ and $p(x_i, \ldots, x_1)$ by B have the same remainder denoted δ_i for $i \leq n - 1$. Since B has degree 1, γ and all the δ_i are scalars. Thus (14.12) implies that

$$p(x_1, \ldots, x_n) \equiv \delta_{n-1}\gamma + \delta_{n-2}$$

and (14.13) implies

$$p(x_n, \ldots, x_1) \equiv \gamma\delta_{n-1} + \delta_{n-2}.$$

This proves the claim.   □

We consider now polynomials over $\mathbb{Z}$ and $\mathbb{Q}$. A nonzero polynomial $P \in \mathbb{Z}\langle A \rangle$ is called *primitive* if the greatest common divisor of its coefficients is 1. The *content* of a nonzero $P \in \mathbb{Q}\langle A \rangle$ is the unique positive rational number $c(P)$ such that $P/c(P)$ is primitive; the latter polynomial is then denoted by $\bar{P}$. Hence $P = c(P)\bar{P}$. Actually, $\bar{P}$ is the unique primitive polynomial such that $P = q\bar{P}$ for some nonzero $q \in \mathbb{Q}_+$.

The next result is the analogue for noncommutative polynomials of *Gauss' lemma*.

Lemma 14.3.6 (Gauss' lemma)

(i) *If* P, Q *are primitive polynomials in* $\mathbb{Z}\langle A \rangle$, *then so is* PQ.

(ii) *If* P, Q *are polynomials in* $\mathbb{Q}\langle A \rangle$, *then* $c(PQ) = c(P)c(Q)$ *and* $\overline{PQ} = \bar{P}\bar{Q}$.

Proof. For (i), if PQ is not primitive, some prime number p divides all its coefficients. One obtains a contradiction by reducing coefficients in $\mathbb{Z}/p\mathbb{Z}$, since polynomials over a field do not have zero divisors. Now (ii) follows easily from (i). □

In the proof of the following statements, the exponent in the expressions like $PQ^{-1}R$ refers to the inverse in the ring of series, and not to the residual.

Theorem 14.3.7 *Let P, Q, R be nonzero polynomials in $\mathbb{Z}\langle A\rangle$ with $(Q, 1) \neq 0$. Then $PQ^{-1}R$ is a polynomial if and only if there exist polynomials P', S, T, Q' in $\mathbb{Z}\langle A\rangle$ such that $P = P'S$, $Q = TS$, $R = TR'$.*

Proof. The condition is of course sufficient. Conversely, we begin by proving the corresponding statement with $\mathbb{Z}$ replaced by $\mathbb{Q}$. Then we use Gauss' lemma to lift our conclusion to $\mathbb{Z}\langle A\rangle$.

1. Consider the set E of pairs of polynomials $V = (V_1, V_2)$ such that $V_1 = PQ^{-1}V_2$. Clearly E is a right $\mathbb{Q}\langle A\rangle$-module, that is if (V_1, V_2) is in E, then for any polynomial $U \in \mathbb{Q}\langle A\rangle$, the pair (V_1U, V_2U) is in E. Note that E contains the pairs (P, Q) and $(PQ^{-1}R, R)$. Note also that if the constant term of the second component of $V = (V_1, V_2) \in E$ is zero, then $Va^{-1} = (V_1a^{-1}, V_2a^{-1})$ is in E. Indeed, since $(V_2, 1) = 0$, we have $(PQ^{-1}V_2)a^{-1} = (PQ^{-1})(V_2a^{-1})$ and thus $PQ^{-1}(V_2a^{-1}) = V_1a^{-1}$. Choose $V = (V_1, V_2)$ to be nonzero in E and of minimal degree, where $\deg(V)$ is the maximum degree of its two components. Note that $V_1, V_2 \neq 0$ since otherwise $V = 0$. Suppose that the constant term of V_2 is zero. Let a be a letter such that $V_1a^{-1} \neq 0$. This exists because $V_1 \neq 0$. Then the pair (V_1a^{-1}, V_2a^{-1}) is in E and has degree less than V. This shows that the constant term of V_2 is nonzero.

We show that $E = V\mathbb{Q}\langle A\rangle$. For this, we prove by induction on $\deg(W)$ that every $W = (W_1, W_2)$ in E is of the form $W = VT$ for some polynomial T. We may assume that $\deg(W) \geq \deg(V)$. If W has constant term zero, then $W_i = \sum_{a\in A}(W_ia^{-1})a$ for $i = 1, 2$. Each pair $Wa^{-1} = (W_1a^{-1}, W_2a^{-1})$ is in E by the remark above, and by induction Wa^{-1} is in $V\mathbb{Q}\langle A\rangle$. Thus W is in $V\mathbb{Q}\langle A\rangle$. This shows the property when W has constant term zero.

Otherwise since every $W = (W_1, W_2)$ in E satisfies $(W_1, 1) = \gamma(W_2, 1)$ with $\gamma = (PQ^{-1}, 1)$, one has $(V_2, 1) \neq 0$ and $(W_2, 1) = \alpha(V_2, 1)$ with $\alpha = (W_2, 1)/(V_2, 1)$. It follows that $(W_1, 1) = \gamma(W_2, 1) = \gamma\alpha(V_2, 1) = \alpha(V_1, 1)$. This shows that the pair $W - \alpha V = (W_1 - \alpha V_1, W_2 - \alpha V_2)$ has zero constant term. Using the above argument, we have $W - \alpha V \in V\mathbb{Q}\langle A\rangle$ and thus $W \in V\mathbb{Q}\langle A\rangle$.

Since (P, Q) and $(PQ^{-1}R, R)$ are in $V\mathbb{Q}\langle A\rangle$, there exist polynomials S and R' such that $P = V_1S$, $Q = V_2S$ and $PQ^{-1}R = V_1R'$, $R = V_2R'$. This concludes this part with $P' = V_1$ and $T = V_2$.

2. By the first part, we have $P = P'S$, $Q = TS$, $R = TR'$ with $P', S, T, R' \in \mathbb{Q}\langle A\rangle$. By Lemma 14.3.6, we have $c(P) = c(P')c(S)$, $c(Q) = c(T)c(S)$, $c(R) = c(T)c(R')$. Since P, Q, R are in $\mathbb{Z}\langle A\rangle$, their contents are in $\mathbb{N}$. Now $PQ^{-1}R = P'R'$ is a polynomial and $c(PQ^{-1}R) = c(P')c(R')$. From the above, one has $c(P)c(R) = c(P')c(S)c(T)c(R') = c(PQ^{-1}R)c(Q)$. Since the four factors are integers, there exist

factorizations

$$c(P) = p's, \ c(R) = r't, \ c(PQ^{-1}R) = p'r', \ c(Q) = st$$

for integers p', s, r', t. This implies that

$$P = p'\bar{P}'s\bar{S}, \ Q = t\bar{T}s\bar{S}, \ R = t\bar{T}r'\bar{R}'$$

whence the result, since the polynomials $p'\bar{P}', \ s\bar{S}, \ t\bar{T}, \ r'\bar{R}'$ have integral coefficients.

□

We shall also need the following result.

Lemma 14.3.8 *Let B be a primitive polynomial of degree 1 which vanishes for some integer value of the variables. Let $P, Q \in \mathbb{Z}\langle A \rangle$ be such that B is a weak left divisor of PQ in $\mathbb{Z}\langle A \rangle$ with nonnull remainder α. Then B is a weak left divisor, in $\mathbb{Z}\langle A \rangle$, of P with remainder β and of Q with remainder γ, where $\beta\gamma = \alpha$.*

Proof. Set $PQ = BQ' + \alpha$ for some $Q' \in \mathbb{Z}\langle A \rangle$ and $\alpha \in \mathbb{Z}, \ \alpha \neq 0$. Since $Q \neq 0$ (because $\alpha \neq 0$), we may apply Theorem 14.3.1. Consequently, $P = BT + \beta, \ T \in \mathbb{Q}\langle A \rangle, \ \beta \in \mathbb{Q}$. Thus $BQ' + \alpha = \beta Q + BTQ$. We have $\beta \neq 0$ (since $\alpha \neq 0$, and $\deg(B) = 1$). Hence $Q = \gamma + BS$ for some $S \in \mathbb{Q}\langle A \rangle$, and $\gamma \in \mathbb{Q}$, with $\alpha = \beta\gamma$. Now, the assumption on B and the fact that $P, Q \in \mathbb{Z}\langle A \rangle$ imply that $\beta, \gamma \in \mathbb{Z}$. Since $BT = P - \beta$, we obtain by Gauss' lemma $c(B)c(T) = c(P - \beta) \in \mathbb{N}$, hence $c(T) \in \mathbb{N}$, because B is primitive. This shows that $T \in \mathbb{Z}\langle A \rangle$. Similarly we obtain $S \in \mathbb{Z}\langle A \rangle$.

□

Finally we prove the following lemma which will be used later.

Lemma 14.3.9 *If $a_1, \ldots, a_n \in \mathbb{Q}\langle A \rangle$, then $p(a_1, \ldots, a_n)$ and $p(a_n, \ldots, a_1)$ are both zero or have the same content.*

Proof. By induction on n. Recall the Wedderburn relation

$$p(a_1, \ldots, a_n)p(a_{n-1}, \ldots, a_1) = p(a_1, \ldots, a_{n-1})p(a_n, \ldots, a_1).$$

Assume $p(a_1, \ldots, a_n) = 0$. By the Wedderburn relation, either $p(a_1, \ldots, a_{n-1}) = 0$ or $p(a_n, \ldots, a_1) = 0$. If $p(a_1, \ldots, a_{n-1}) = 0$, then by (14.12), one has $p(a_1, \ldots, a_{n-2}) = 0$. By induction, this implies $p(a_{n-1}, \ldots, a_1) = 0$ and $p(a_{n-2}, \ldots, a_1) = 0$ which implies by (14.13) $p(a_n, \ldots, a_1) = 0$.

Assume now $p(a_1, \ldots, a_n) \neq 0$ and $p(a_n, \ldots, a_1) \neq 0$. If $p(a_1, \ldots, a_{n-1}) = 0$, we have also $p(a_{n-1}, \ldots, a_1) = 0$ by induction. By (14.12), $c(p(a_1, \ldots, a_n)) = c(p(a_1, \ldots, a_{n-2}))$ and by (14.13), $c(p(a_n, \ldots, a_1)) = c(p(a_{n-2}, \ldots, a_1))$. The conclusion follows by induction. Otherwise Gauss' Lemma and the Wedderburn relation give

$$c(p(a_1, \ldots, a_n))c(p(a_{n-1}, \ldots, a_1)) = c(p(a_1, \ldots, a_{n-1}))c(p(a_n, \ldots, a_1)).$$

By induction, $c(p(a_1, \ldots, a_{n-1})) = c(p(a_{n-1}, \ldots, a_1))$ and thus we obtain the conclusion.

□

14.4 Proof of the factorization theorem

Given a word u and a series $T \in \mathbb{Z}\langle\langle A \rangle\rangle$, the residual of T by u is defined by

$$u^{-1}T = \sum_{w \in A^*} (T, uw)w.$$

This is consistent with the definition given in Chapter 1. Observe that $(uv)^{-1}T = v^{-1}(u^{-1}T)$. The notation Tv^{-1} is defined symmetrically. Note that $u^{-1}(Tv^{-1}) = (u^{-1}T)v^{-1}$. Here, the exponent refers to the residual and not to the inverse.

Given a code X and words u, v, we define $S(u) = \{s \in A^* \mid us = x_1 \cdots x_n, x_i \in X, |s| < |x_n|\}$ and $P(v) = \{p \in A^* \mid pv = x_1 \cdots x_n, x_i \in X, |p| < |x_1|\}$. These are the sets $C_r(u)$ and $C_\ell(v)$ of strict right and left contexts of u and v already defined earlier.

Lemma 14.4.1 *Let X be a finite code. For each pair of words u, v, there exists a finite set $F(u, v)$ such that*

$$u^{-1}\underline{X^*}v^{-1} = \underline{S(u)}\,\underline{X^*}\,\underline{P(v)} + \underline{F(u, v)}. \tag{14.15}$$

Proof. The series $u^{-1}\underline{X^*}v^{-1}$ is the characteristic series of the set W of words w such that $uwv \in X^*$. Let $F(u, v)$ be the set of words w such that $uwv = xyz$ for some words $x, z \in X^*$ and $y \in X$ with x a prefix of u, z a suffix of v and w a proper factor of y. Since X is finite, this set is finite.

Let us verify that W is the disjoint union of $S(u)X^*P(v)$ and $F(u, v)$. Indeed, the sets $S(u)X^*P(v)$ and $F(u, v)$ are contained in W. They are disjoint since if w is a word in $S(u)X^*P(v) \cap F(u, v)$, then uwv has two distinct factorizations $x_1 x_2 \cdots x_n$ with $x_i \in X$, one in which w is a proper factor of some x_i and the other in which it is not.

Conversely, given a word w such that $uwv = x_1 \cdots x_n$, with $x_i \in X$, either there is an index i such that $x_i = swp$ with $x_1 \cdots x_{i-1}u' = u$, and $v = v'x_{i+1}\cdots x_n$, and both u', v' nonempty. In this case, $w \in F(u, v)$. Otherwise, $w \in S(u)X^*P(v)$. This proves Equation (14.15). □

Lemma 14.4.2 *Let X be a finite maximal code of degree d. Then there exist words $u_1, \ldots, u_d, v_1, \ldots, v_d$ with $u_1, v_1 \in X^*$, such that, for any $1 \le i, j \le d$,*

$$\underline{A^*} = \sum_{1 \le \ell \le d} u_i^{-1}\underline{X^*}v_\ell^{-1} = \sum_{1 \le k \le d} u_k^{-1}\underline{X^*}v_j^{-1}.$$

Proof. Let $\mathcal{A} = (Q, 1, 1)$ be an unambiguous automaton recognizing X^*, set $\varphi = \varphi_{\mathcal{A}}$ and let $M = \varphi_{\mathcal{A}}(A^*)$ be the transition monoid of $\mathcal{A}$. Let G be an $\mathcal{H}$-class of the minimal ideal of M that meets $\varphi(X^*)$, and let e be its neutral element. The set $H = G \cap \varphi(X^*)$ is a subgroup of index d of G. In particular, $e \in \varphi(X^*)$ and $\varphi^{-1}(e) \subset X^*$.

Let $u_1, \ldots, u_d, v_1, \ldots, v_d$ be words in $\varphi^{-1}(G)$ such that

$$G = \bigcup_{1 \le i \le d} \varphi(v_i)H = \bigcup_{1 \le j \le d} H\varphi(u_j).$$

We may assume that $\varphi(u_1) = \varphi(v_1) = e$, and that $\varphi(u_i)$ is the inverse of $\varphi(v_i)$ in G. It follows that $u_1, v_1 \in \varphi^{-1}(e) \subset X^*$. Fix j, $1 \le j \le d$. Let $w \in A^*$. Observe that $\varphi(v_j) \in G$, hence that $e\varphi(wv_j) = e\varphi(w)\varphi(v_j) = e\varphi(w)\varphi(v_j)e$ is in $eMe = G$. Thus $e\varphi(wv_j)$ is in some $\varphi(v_i)H$, for some uniquely determined i, depending on w. We show that

$$e\varphi(wv_j) \in \varphi(v_i)H \Leftrightarrow u_i w v_j \in X^*.$$

Indeed, $e\varphi(wv_j) \in \varphi(v_i)H \Leftrightarrow \varphi(u_i)e\varphi(wv_j) \in \varphi(u_i)\varphi(v_i)H \Leftrightarrow \varphi(u_i w v_j) \in H \Leftrightarrow u_i w v_j \in X^*$ (since $\varphi(u_i w v_j) = e\varphi(u_i w v_j)e \in G$).

Thus we obtain that for any w in A^*, there is a unique i such that $w \in u_i^{-1} X^* v_j^{-1}$, which implies the second equality in the lemma and the first one by symmetry. $\square$

The following lemma is easily derived.

Lemma 14.4.3 *Let X be a finite maximal code of degree d. There exist finite subsets P, S, P_1, S_1 of A^* with $1 \in P_1, S_1$, finite subsets L_1, R_1 of A^+ and a polynomial Q with coefficients in $\mathbb{N}$ such that*

(i) $d\underline{A}^* = Q + \underline{S}\,\underline{X}^*\underline{P}$.
(ii) $\underline{A}^* = \underline{L_1} + \underline{S}\,\underline{X}^*\underline{P_1} = \underline{R_1} + \underline{S_1}\,\underline{X}^*\underline{P}$.
(iii) *If $S_1 = \{1\}$ (resp. $P_1 = \{1\}$), then X is prefix (resp. suffix). Conversely, if X is prefix (resp. suffix), then one can chose $S_1 = \{1\}$ (resp. $P_1 = \{1\}$).*

Proof. According to Lemma 14.4.2, there exist words $u_1, \dots, u_d, v_1, \dots, v_d$ with u_1, v_1 in X^* such that

$$\underline{A}^* = \sum_{1 \le \ell \le d} u_i^{-1}\underline{X}^* v_\ell^{-1} = \sum_{1 \le k \le d} u_k^{-1}\underline{X}^* v_j^{-1}.$$

By Lemma 14.4.1

$$u_i^{-1}\underline{X}^* v_j^{-1} = \underline{S(u_i)}\,\underline{X}^*\underline{P(v_j)} + \underline{F(u_i, v_j)}$$

where $S(u_i), P(v_j), F(u_i, v_j)$ are finite sets. Thus, for any $i, j = 1, \dots, d$,

$$\underline{A}^* = \sum_{1 \le \ell \le d} \underline{S(u_i)}\,\underline{X}^*\underline{P(v_\ell)} + \sum_{1 \le \ell \le d} \underline{F(u_i, v_\ell)} \qquad (14.16)$$

$$= \sum_{1 \le k \le d} \underline{S(u_k)}\,\underline{X}^*\underline{P(v_j)} + \sum_{1 \le k \le d} \underline{F(u_k, v_j)}.$$

Let $P = \bigcup_{1 \le \ell \le d} P(v_\ell)$ and $S = \bigcup_{1 \le k \le d} S(u_k)$. Observe that, by (14.16), the unions are disjoint and therefore

$$\underline{P} = \sum_{1 \le \ell \le d} \underline{P(v_\ell)}, \quad \underline{S} = \sum_{1 \le k \le d} \underline{S(u_k)}.$$

Let $P_1 = P(v_1)$, $S_1 = S(u_1)$. Let $L_1 = \bigcup_{1 \le i \le d} F(u_i, v_1)$, $R_1 = \bigcup_{1 \le j \le d} F(u_1, v_j)$ which are again disjoint unions and finally $Q = \sum_{1 \le i, j \le d} \underline{F(u_i, v_j)}$.

Summing up both sides of Equation (14.16) for $i = 1, \ldots, d$, one gets assertion (i). Assertion (ii) is a reformulation of the equations for $i = 1$ (resp. $j = 1$).

Since $u_1, v_1 \in X^*$, one has $1 \in S(u_1)$ and $1 \in P(v_1)$. By (ii), we have $(\underline{L_1}, 1) + (\underline{S}\,\underline{X}^*\underline{P}_1, 1) = 1$. Since $1 \in S$ and $1 \in P_1$, this implies $1 \notin L_1$. This finishes the verification of the properties of the finite sets.

It remains to prove (iii).

If X is prefix, then X^* is right unitary. Thus the set of right contexts $S_1 = S(u_1)$ is reduced to the empty word.

Conversely, if $S_1 = \{1\}$, we have $\underline{A}^* = \underline{R}_1 + \underline{X}^*\underline{P}$. We show that X is right complete and hence, by Theorem 3.3.8, that X is a prefix code. Indeed, let w be a word, and let u be a word longer than any word in R_1 and in P. The word wu is not in R_1, therefore it is in X^*P. Consequently, w is a prefix of a word in X^*. This completes the proof. $\qquad\square$

Proof of Theorem 14.2.1. For convenience, we set $B = 1 - \underline{A}$. With the notation of Lemma 14.4.3, one has $\underline{A}^* = \underline{L}_1 + \underline{S}\,\underline{X}^*\underline{P}_1$. Thus $\underline{S}\,\underline{X}^*\underline{P}_1 = B^{-1}(1 - B\underline{L}_1)$. Hence

$$B\underline{S}\,\underline{X}^* = (1 - B\underline{L}_1)\underline{P}_1{}^{-1}.$$

By Lemma 14.4.3(i), we have $d - BQ = B\underline{S}\,\underline{X}^*\underline{P}$. Replacing $B\underline{S}\,\underline{X}^*$ gives $d - BQ = (1 - B\underline{L}_1)\underline{P}_1{}^{-1}\underline{P}$. This implies

$$\underline{P} = \underline{P}_1(1 - B\underline{L}_1)^{-1}(d - BQ).$$

We apply Theorem 14.3.7 to the last equality and we obtain the existence of E, F, G, H in $\mathbb{Z}\langle A\rangle$ such that $\underline{P}_1 = EF$, $1 - B\underline{L}_1 = GF$, $d - BQ = GH$, $\underline{P} = EH$. Lemma 14.3.8 implies that $G \equiv \pm 1$ (we write $P \equiv \alpha$ as a shorthand for saying that α is the remainder of the weak left division of P by B). Replacing if necessary E, F, G, H by their negatives, we may suppose that $G \equiv 1$. Then Lemma 14.3.8 again implies that $H \equiv d$. Thus

$$\underline{P} = E(d + BI) \tag{14.17}$$

for some $I \in \mathbb{Z}\langle A\rangle$.

By Lemma 14.4.3(ii), we have $B^{-1}(1 - BR_1) = \underline{A}^* - R_1 = \underline{S}_1\underline{X}^*\underline{P}$. Hence

$$1 - \underline{X} = \underline{P}(1 - BR_1)^{-1}B\underline{S}_1.$$

This is very close to Equation (14.7), but with a central inverted polynomial, which we must eliminate. For this, we use Theorem 14.3.7 again. There exist J, K, L, M in $\mathbb{Z}\langle A\rangle$ such that $\underline{P} = JK$, $1 - BR_1 = LK$, $B\underline{S}_1 = LM$, $1 - \underline{X} = JM$. Let π be a positive Bernoulli morphism. It extends linearly to an algebra homomorphism $\mathbb{Q}\langle A\rangle \to \mathbb{R}$.

We may assume that $\pi(K) \geq 0$. Then we deduce from Lemma 14.3.8 that $K = 1 + BK'$ and $L = 1 + BL'$ for some K', L' in $\mathbb{Z}\langle A\rangle$. Thus $B\underline{S}_1 = (1 + BL')M = M + BL'M$, which implies that $M = BM'$ for some M' in $\mathbb{Z}\langle A\rangle$. Therefore

$$1 - \underline{X} = JBM'. \tag{14.18}$$

Equation (14.18) will imply Equation (14.7), if we show that J is of the form $J_1(d + B J_2)$. This is the most technical part of the proof. It will follow from

$$E(d + BI) = J(1 + BK') \qquad (14.19)$$

(which holds in view of (14.17) and the fact that $\underline{P} = JK$ and $K = 1 + BK'$) and from the divisibility property of Theorem 14.3.3. The difficulty is that in this theorem, the polynomials involved have coefficients in $\mathbb{Q}$. Therefore a lot of additional work is required to draw the conclusion in $\mathbb{Z}$.

Theorem 14.3.3 applied to Equation (14.19) guarantees the existence of polynomials $x_1, \ldots, x_n, U, V$ in $\mathbb{Q}\langle A \rangle$ such that

$$E = Up(x_1, \ldots, x_n), \qquad d + BI = p(x_{n-1}, \ldots, x_1)V,$$

$$J = Up(x_1, \ldots, x_{n-1}), \qquad 1 + BK' = p(x_n, \ldots, x_1)V.$$

We write p_i, q_i for $p(x_1, \ldots, x_i)$ and $p(x_i, \ldots, x_1)$. We apply Theorem 14.3.1 to the two equalities at the right, and obtain that q_{n-1} and q_n are both congruent to a scalar modulo B. Thus Theorem 14.3.4 implies that p_{n-1} and q_{n-1} (resp. p_n and q_n) are congruent to the same scalar modulo B. Furthermore, by Lemma 14.3.9, $c(p_{n-1}) = c(q_{n-1})$ and $c(p_n) = c(q_n)$.

Observe that $1 - BR_1$ is primitive, since R_1 has coefficients $0, 1$. The equation $1 - BR_1 = LK$ implies that L, K are primitive, since they are in $\mathbb{Z}\langle A \rangle$. We have $K = 1 + BK' = q_n V$, hence by Gauss' lemma $c(q_n)C(V) = c(K) = 1$, and $\bar{q}_n \bar{V} = \bar{K} = K$. This equality together with Lemma 14.3.8 implies that $\bar{V} = \epsilon + BV'$, with $V' \in \mathbb{Z}\langle A \rangle$ and $\epsilon = \pm 1$. Now $1 - \underline{X} = JM$ and $1 - \underline{X}$ is primitive, hence J is primitive. Since $JK = E(d + BI)$, Gauss' lemma again implies that $d + BI$ is primitive. Since $d + BI = q_{n-1}V$, the same lemma implies that $d + BI = \bar{q}_{n-1}\bar{V}$. Lemma 14.3.8 now implies that $\bar{q}_{n-1} = \epsilon d + BN$ for some $N \in \mathbb{Z}\langle A \rangle$.

We have seen that p_{n-1} and q_{n-1} are congruent to the same scalar modulo B and that $c(p_{n-1}) = c(q_{n-1})$. Hence $\bar{p}_{n-1}$ and $\bar{q}_{n-1}$ are congruent to the same scalar modulo B, and we have $\bar{p}_{n-1} = \epsilon d + BH$ with $H \in \mathbb{Q}\langle A \rangle$. But $\bar{p}_{n-1} - \epsilon d = BH$ and B is primitive. By Gauss' lemma, $c(H) = c(\bar{p}_{n-1} - \epsilon d)$ is in $\mathbb{Z}$ and H is in $\mathbb{Z}\langle A \rangle$.

Now, J is primitive and $J = Up_{n-1}$, hence $J = \bar{J} = \bar{U}\bar{p}_{n-1}$, which implies $J = \bar{U}(\epsilon d + BH)$. Thus Equation (14.18) implies

$$1 - \underline{X} = \bar{U}(\epsilon d + BH)BM'.$$

This implies that for some polynomials W, Y, Z in $\mathbb{Z}\langle A \rangle$ (defined by $W = \pm\bar{U}$, $Y = \pm H, Z = \pm M'$) and $\epsilon_1 = \pm 1$, one has

$$1 - \underline{X} = W(\epsilon_1 dB + BYB)Z, \qquad (14.20)$$

with $\pi(W), \pi(Z) \geq 0$.

Now define the linear mapping $\lambda : \mathbb{Q}\langle A \rangle \to \mathbb{R}$ by $\lambda(w) = |w|\pi(w)$ for each word w in A^*. It is easily shown that $\lambda(\underline{P_1}\,\underline{P_2}) = \lambda(\underline{P_1})\pi(\underline{P_2}) + \pi(\underline{P_1})\lambda(\underline{P_2})$, for $\underline{P_1}, \underline{P_2}$ in $\mathbb{Q}\langle A \rangle$. Applying λ to (14.20) and observing that $\lambda(B) = -1$, we obtain $\lambda(\underline{X}) = \pi(W)\epsilon_1 d\pi(Z)$. Since $\lambda(\underline{X}) > 0$, this shows that $\epsilon_1 = 1$.

To conclude the proof of Theorem 14.2.1, observe that if X is prefix, then one can choose $\underline{S_1} = 1$ by Lemma 14.4.3(iii); since $B\underline{S_1} = LM$ and $M = BM'$, we obtain that $B = LBM'$. Thus $M' = \pm 1$. Since $\pi(Z) \geq 0$ and $Z = \pm M$, we deduce $Z = 1$.

On the other hand, if X is suffix, one can choose $\underline{P_1} = 1$ by Lemma 14.4.3(iii) again. Since $\underline{P_1} = EF$, we obtain $E = \pm 1$. Since $E = Up_n$, we obtain by Gauss' lemma, $\pm 1 = \overline{U}\,\overline{p}_n$, hence $W = \pm \overline{U} = \pm 1$. Since $\pi(X) \geq 0$, one has $W = 1$. $\qquad\square$

Remark 14.4.4 A closer look at the previous proof proves the following claim: under the hypothesis of Theorem 14.2.1, one has

$$\underline{X} - 1 = W(d(\underline{A} - 1) + (\underline{A} - 1)Y(\underline{A} - 1))Z,$$

and moreover

$$\underline{P_1} = W(1 + (\underline{A} - 1)W'), \quad \underline{S_1} = (1 + Z'(\underline{A} - 1))Z,$$

for some polynomials W, Y, Z, W', Z' in $\mathbb{Z}\langle A \rangle$, and in particular

$$\pi(W) = \pi(\underline{P_1}), \quad \pi(Z) = \pi(\underline{S_1}).$$

Recall that $\underline{P_1}, \underline{S_1}$ are as defined in Lemma 14.4.3 and its proof, and therefore satisfy:

$$u_1^{-1}\underline{X}^* = \underline{S_1}\,\underline{X}^*, \quad \underline{X}^* v_1^{-1} = \underline{X}^*\underline{P_1}$$

for some words u_1, v_1 in X^*. Note that the average length $\sum_{w \in X} \pi(w)|w|$ of X is equal to $\pi(W)d\pi(Z)$.

We prove the claim, by going through the proof of Theorem 14.2.1: first, we have $\underline{P_1} = EF, F \equiv 1$ (by Lemma 14.3.8 since $G \equiv 1$ and $GF \equiv 1$). Next, $E = \overline{U}\,\overline{p}_n$ (by Gauss' lemma, since $E = Up_n$, and E being primitive since $\underline{P_1}$ is and $\underline{P_1} = EF$). Furthermore $\overline{q}_n \equiv \pm 1$ (by Lemma 14.3.8, since $\overline{q}_n \overline{V} = K \equiv 1$), which implies, by an argument similar to that for p_{n-1} and q_{n-1} in the proof of Theorem 14.2.1, that $\overline{p}_n \equiv \pm 1$.

We obtain that $\overline{p}_n F \equiv \pm 1$, and $\underline{P_1} = \underline{P_1} = \overline{U}\,\overline{p}_n F$, which is the product of $\pm W$ with a polynomial which is $\equiv \pm 1$. Since $\pi(\underline{P_1}) > 0$ and $\pi(W) \geq 0$, we obtain finally that $\underline{P_1}$ is of the desired form $W(1 + (\underline{A} - 1)W')$.

On the other hand, $Z = \pm M'$, $M = BM'$, $B\underline{S_1} = (1 + BL')M$. Thus $B\underline{S_1} = (1 + BL')BM'$, which implies that $\underline{S_1} = (1 + L'B)M'$, and $\pi(\underline{S_1}) = \pi(M')$. Since $\pi(\underline{S_1}) > 0$ and $\pi(Z) \geq 0$, we have in fact $\underline{S_1} = (1 + L'B)Z$, which proves the claim.

14.5 Applications

Let π be a Bernoulli distribution. Recall that the *average length* (with respect to π) of a finite code X is the number $\sum_{w \in X} \pi(w)|w|$. The distribution is *positive* if $\pi(w) > 0$ for any word w.

The following statement is easily obtained from Remark 14.4.4. However, the same result holds for arbitrary thin complete codes, as proved in Corollary 13.5.2.

Corollary 14.5.1 *Let X be a finite maximal code and let π be a positive Bernoulli distribution. The average length of X is greater or equal to the degree of X, and equality holds if and only if X is bifix.*

Proof. With the notation of Remark 14.4.4, we have $\pi(W) = \pi(\underline{P_1})$ and $\pi(Z) = \pi(\underline{S_1})$. By Lemma 14.4.3, $\pi(\underline{S_1}) \geq 1$ (resp. $\pi(\underline{P_1}) \geq 1$), with equality if and only if $\underline{P_1} = 1$ (resp. $\underline{S_1} = 1$). Thus, since the average length of X is equal to $\lambda(\underline{X}) = \pi(\underline{W})d\pi(Z)$, we obtain that it is $\geq d$.

If equality holds, then we must have $\underline{P_1} = \underline{S_1} = 1$. Then the code X is bifix by Lemma 14.4.3(iii).   □

Let x be any word and X a finite code. Recall from Section 13.5 that a *strict context* of a word w with respect to X is a pair (p, s) such that either $pws = x_1 \cdots x_n, x_i \in X$, $n \geq 1$, with p a proper prefix of x_1 and s a proper suffix of x_n, or $pws = 1$. Thus, for $w \in X^*$, the pair $(1, 1)$ is a strict context. Observe that the set $C(w)$ of strict contexts of a word w is finite. The *measure* of $C(w)$ is by definition $\sum \pi(p)\pi(s)$, where the sum is over all strict contexts (p, s) of w.

The next result is easily obtained with the help of Theorem 14.2.1. The same result holds for an arbitrary thin complete code (Theorem 13.5.5).

Corollary 14.5.2 *Let X be a finite code over A, and let π be a positive Bernoulli distribution on A^*. For any word $w \in A^*$, the measure of the set $C(w)$ of strict contexts of w is equal to the average length of the code X.*

We prove in fact a noncommutative version of this result.

Proof. Fix a finite maximal code X and a word w. We define a mapping e from $\mathbb{Z}\langle\langle A \rangle\rangle$ into the complete tensor product $\mathbb{Z}\langle\langle A \rangle\rangle \otimes_{\mathbb{Z}} \mathbb{Z}\langle\langle A \rangle\rangle$, which is the set of series of the form $\sum_{u,v \in A^*} \alpha_{u,v} u \otimes v$ for integers $\alpha_{u,v}$. The mapping is defined by $e(z) = \sum_{uwv=z} u \otimes v$ for a word $z \in A^*$. It is easily seen that $e(\underline{A^*}) = \underline{A^*} \otimes \underline{A^*}$. Furthermore, the very definition of a strict context implies that $e(\underline{X^*}) = \sum_{p,s} \underline{X^*}p \otimes s\underline{X^*}$, where the sum is extended to all strict contexts (p, s) of w with respect to X. Thus $e(\underline{X^*}) = (\underline{X^*} \otimes 1)T(1 \otimes \underline{X^*})$, where $T = \sum p \otimes s$, summed over all strict contexts of w.

Suppose that w is nonempty; then we have for any words s, m, p:

$$e(smp) = (s \otimes 1)e(m)(1 \otimes p) + e(s)(1 \otimes mp) + (sm \otimes 1)e(p)$$
$$+ \sum_{u,v \neq 1, w=uv} (su^{-1} \otimes (v^{-1}m)p + s(mu^{-1}) \otimes v^{-1}p)$$
$$+ \sum_{u,v \neq 1} (umv, w)su^{-1} \otimes v^{-1}p,$$

where we use u^{-1} in the same way as the notation recalled at the beginning of Section 14.4, and where $(,)$ is the scalar product on $\mathbb{Z}\langle A \rangle$ that has A^* as an orthonormal basis.

The proof of this formula follows by inspection, once the six possibilities for the word w to be a factor of the word smp have been observed: either w appears as a factor of m, or of s or p, or w is an overlapping factor of the product sm or mp, or finally w is factor of smp which starts properly in s and ends properly in p.

Note that the previous formula is linear in each of s, m, p, so it extends to series S, M, P. Now we have by Corollary 14.2.3, $\underline{A}^* = S\underline{X}^*P$, where P, S are polynomials. Hence we obtain

$$\underline{A}^* \otimes \underline{A}^* = e(\underline{A}^*) = e(S\underline{X}^*P)$$

$$= (S \otimes 1)e(\underline{X}^*)(1 \otimes P) + e(S)(1 \otimes \underline{X}^*P) + (S\underline{X}^* \otimes 1)e(P)$$

$$+ \sum_{u,v\neq 1, w=uv} (Su^{-1} \otimes (v^{-1}\underline{X}^*)P + S(\underline{X}^*u^{-1}) \otimes v^{-1}P)$$

$$+ \sum_{u,v\neq 1} (u\underline{X}^*v, w)Su^{-1} \otimes v^{-1}P.$$

Note that the last sum is finite. Denote it by R. Observe that $e(\underline{X}^*) = (\underline{X}^* \otimes 1)T(1 \otimes \underline{X}^*)$. By the proof of Lemma 14.4.1, where $S(v)$ and $P(u)$ are defined, we thus have

$$\underline{A}^* \otimes \underline{A}^* = (S\underline{X}^* \otimes 1)T(1 \otimes \underline{X}^*P) + e(S)(1 \otimes \underline{X}^*P) + (S\underline{X}^* \otimes 1)e(P)$$

$$+ \sum_{u,v\neq 1, w=uv} (Su^{-1} \otimes S(v)\underline{X}^*P + S\underline{X}^*P(u) \otimes v^{-1}P) + R.$$

Let us multiply by $PB \otimes 1$ on the left and by $1 \otimes BS$ on the right. Since PBS is the inverse of $\underline{X}^*$, we obtain

$$P \otimes S = T + (PB \otimes 1)e(S) + e(P)(1 \otimes BS)$$

$$+ \sum_{u,v\neq 1, w=uv} (PB(Su^{-1}) \otimes S(v) + P(u) \otimes (v^{-1}P)BS)$$

$$+ (PB \otimes 1)R(1 \otimes BS).$$

Note that when w is the empty word, then the formula for $e(smp)$ has to be slightly modified: the Σ's are replaced by $-s \otimes mp - sm \otimes p$, and from here on the argument is similar and hence we omit it.

This shows that the sum of the strict contexts of the word w is equal to $P \otimes S$ modulo the two-sided ideal of $\mathbb{Z}\langle A \rangle \otimes \mathbb{Z}\langle A \rangle$ generated by $\underline{A} - 1 \otimes 1$ and $1 \otimes (\underline{A} - 1)$.

The homomorphism $\pi \otimes \pi : \mathbb{Z}\langle A \rangle \otimes \mathbb{Z}\langle A \rangle \to \mathbb{R}$ vanishes on this ideal. Thus the measure of the set of strict contexts is equal to $\pi(P)\pi(S)$. Now, using $\underline{X} - 1 = P(\underline{A} - 1)S$, we find that the average length of X is equal to $\lambda(\underline{X}) = \pi(P)\pi(S)$. $\quad\square$

A code of degree 1 is called synchronized, see Section 9.3. Recall that for a finite set X of words in A^*, we denote by $\alpha(\underline{X})$ the sum in $\mathbb{Z}[A]$ of the commutative images of the words in X.

Corollary 14.5.3 *Let X be a finite maximal code on the alphabet A. Then $\alpha(\underline{X}) - 1$ is a multiple of $\alpha(\underline{A}) - 1$. If the quotient of these two polynomials is irreducible in $\mathbb{Z}[A]$, then X has at least two of the following properties: prefix, suffix, synchronized.*

Proof. Let ρ the canonical homomorphism $\mathbb{Z}\langle A \rangle \to \mathbb{Z}[A]$. Then by Remark 14.4.4, we have $\alpha(\underline{X}) - 1 = \rho(W)\rho(Z)(d + \rho(Y)(\alpha(\underline{A}) - 1))(\alpha(\underline{A}) - 1)$, which proves the first assertion. If the quotient is irreducible, then we must have two of the three following equalities: $\rho(W) = \pm 1$, $\rho(Z) = \pm 1$, $d + \rho(Y)(\alpha(\underline{A}) - 1) = \pm 1$.

The equality $\rho(W) = \pm 1$ implies, by Remark 14.4.4, that $\pi(S_1) = 1$, hence $S_1 = 1$, and then that X is prefix (Lemma 14.4.3(vi)). We deal with the second equality similarly.

If the third equality holds, then we must have $\rho(Y) = 0$, and $d = \pm 1$, which implies $d = 1$, hence X is synchronized.   □

Observe that the first assertion is Theorem 2.5.30.

14.6 Commutative equivalence

Recall that the canonical morphism that associates to a formal power series its commutative image is denoted by $\alpha : \mathbb{Q}\langle\!\langle A \rangle\!\rangle \to \mathbb{Q}[[A]]$ and that $\alpha(A^*) = A^\oplus$ is the free commutative monoid on A. By definition, for each $\sigma \in \mathbb{Q}\langle\!\langle A \rangle\!\rangle$ and $w \in A^\oplus$,

$$(\alpha(\sigma), w) = (\sigma, \alpha^{-1}(w)) = \sum_{\alpha(v)=w} (\sigma, v).$$

Two series $\sigma, \tau \in \mathbb{Q}\langle\!\langle A \rangle\!\rangle$ are called *commutatively equivalent* if $\alpha(\sigma) = \alpha(\tau)$.

Two subsets X and Y of A^* are commutatively equivalent if their characteristic series $\underline{X}$ and $\underline{Y}$ are so, which means that $\alpha(\underline{X}) = \alpha(\underline{Y})$. In an equivalent manner, X and Y are commutatively equivalent if and only if there exists a bijection $\gamma : X \to Y$ such that $\gamma(x) \in \alpha^{-1}\alpha(x)$ for all $x \in X$.

A subset X of A^* is called *commutatively prefix* if there exists a prefix subset Y of A^* which is commutatively equivalent to X. It is conjectured that every finite maximal code is commutatively prefix. This is the *commutative equivalence conjecture*.

Example 14.6.1 Any suffix code X is commutatively prefix (since $\tilde{X}$ is prefix). More generally, any code obtained by a sequence of compositions of prefix and suffix codes is commutatively prefix. In particular, our friend $X = \{aa, ba, baa, bb, bba\}$ is commutatively prefix.

Example 14.6.2 Let $A = \{a, b\}$ and let

$$X = \{aa, ba, bb, abab, baab, bbab, a^3b^2, a^3ba^2, a^3b^2ab, a^3ba^3b, a^3babab\}.$$

This set is easily verified to be a code, by computing, for instance, the sets U_i of Section 2.3,

$$U_1 = \{abb, aba^2, ab^2ab, aba^3b, (ab)^3, ab\}, \qquad U_2 = \{ab\}, \qquad U_3 = \{ab\}.$$

Further, X is maximal since for $\pi(a) = \pi(b) = \frac{1}{2}$, we obtain $\pi(X) = 1$. Finally X is commutatively prefix since

$$Y = \{aa, ba, bb, abab, abba, abbb, abaab, aba^4, aba^3b^2, aba^3ba^2, aba^3bab\}$$

is a prefix code commutatively equivalent to X. Observe that

$$\underline{X} - 1 = (1 + a + b + a^3b + a^3ba)(a + b - 1)(1 + ab)$$

is a positive factorization for X. Actually, X belongs to the family of indecomposable finite maximal codes described in Exercise 14.1.7.

Proposition 14.6.3 *Let $A = \{a, b\}$ and let $X \subset a^*ba^*$. Then X is commutatively prefix if and only if, for all $n \geq 1$,*

$$\mathrm{Card}(X \cap A^{(n+1)}) \leq n. \tag{14.21}$$

Recall that $A^{(n+1)} = 1 \cup A \cup \ldots \cup A^n$.

Proof. The condition is necessary. Indeed, let Y be a prefix code commutatively equivalent to X. Since Y is prefix, the map π from $X \cap A^{(n+1)}$ to $\{0, 1, \ldots, n-1\}$ defined by $\pi(a^iba^j) = i$ is injective. This implies that we cannot have more than n words of length at most n in X. Conversely, suppose that the condition is satisfied. We show by induction on $n \geq 1$ that there is a prefix code Y commutatively equivalent to $X_1 \cup \ldots \cup X_n$ with $X_n = X \cap A^n$. This is true for $n = 1$. Assume that it is true for $n \geq 1$. Set $I = \{i \geq 0 \mid a^iba^* \cap Y \neq \emptyset\}$. Then $\mathrm{Card}(I) = \mathrm{Card}(X \cap A^{(n+1)})$ and thus $\mathrm{Card}(I) + \mathrm{Card}(X_{n+1}) \leq n + 1$. This shows that we can choose Z commutatively equivalent to X_{n+1} formed of words a^iba^j with distinct indices $i \in \{0, 1, \ldots, n\} \setminus I$. The code $Y \cup Z$ is prefix and commutatively equivalent to $X_1 \cup \ldots \cup X_{n+1}$. $\qquad\square$

Theorem 14.6.4 *For each subset X of A^* the following conditions are equivalent:*

(i) *X is commutatively prefix.*
(ii) *The series $(1 - \alpha(\underline{X}))/(1 - \alpha(\underline{A}))$ has nonnegative coefficients.*

The proof uses the following lemma.

Lemma 14.6.5 *Let $U \subset A^*$ and $V \in \mathbb{Z}\langle\langle A \rangle\rangle$ be such that $(\alpha(\underline{U}), w) \geq (\alpha(V), w) \geq 0$ for all $w \in A^\oplus$. Then there exists $U' \subset U$ such that $\alpha(\underline{U'}) = \alpha(V)$.*

Proof. Let $w \in A^\oplus$. Since $(\underline{U}, \alpha^{-1}(w)) \geq (\alpha(V), w) \geq 0$, there exists a subset U_w of $U \cap \alpha^{-1}(w)$ such that $(\alpha(\underline{U_w}), w) = (\alpha(V), w)$. Then $U' = \bigcup_{w \in A^\oplus} U_w$ is a subset of U and $(\alpha(\underline{U'}), w) = (\alpha(V), w)$. $\qquad\square$

Proof of Theorem 14.6.4. (i) $\Rightarrow$ (ii). First assume that X is commutatively equivalent to some prefix set Y. Let $P = A^* - YA^*$. Then $A^* = Y^*P$, hence $1 - \underline{Y} = \underline{P}(1 - \underline{A})$. Thus $1 - \alpha(\underline{X}) = \alpha(\underline{P})(1 - \alpha(\underline{A}))$. Clearly $\alpha(\underline{P}) = (1 - \alpha(\underline{X}))/(1 - \alpha(\underline{A}))$ has nonnegative integral coefficients.

(ii)$\Rightarrow$ (i). Let $X_n = X \cap A^n$ for $n \geq 0$. Set $Q = (1 - \underline{X})A^*$. Then $\alpha(Q) = (1 - \alpha(\underline{X}))/(1 - \alpha(\underline{A}))$ has nonnegative coefficients. Note that, since $Q(1 - \underline{A}) = 1 - \underline{X}$, we have for $1 \leq i \leq n$

$$Q_i = Q_{i-1}\underline{A} - \underline{X}_i, \tag{14.22}$$

where Q_i is the homogeneous component of degree i of Q.

We show by induction on $n \geq 1$ that there exists a prefix code Y commutatively equivalent to $X_1 \cup \ldots \cup X_n$. The property is true for $n = 1$ since $Y = X_1$ satisfies the condition.

Suppose that the property is true for $n \geq 1$. Let $P = A^* \setminus YA^*$. Thus $1 - \underline{Y} = \underline{P}(1 - \underline{A})$. Set $Y_i = Y \cap A^i$ and $P_i = P \cap A^i$ for $0 \leq i \leq n$. Since $1 - \alpha(\underline{X}) = \alpha(Q)(1 - \alpha(\underline{A}))$ and $1 - \alpha(\underline{Y}) = \alpha(\underline{P})(1 - \alpha(\underline{A}))$ coincide up to degree n, we have $\alpha(\underline{Q_i}) = \alpha(\underline{P_i})$ for $0 \leq i \leq n$. Since $Q_{n+1} = Q_n\underline{A} - \underline{X}_{n+1}$, the polynomial $Q_n\underline{A} - Q_{n+1}$ has nonnegative coefficients. This implies that $\alpha(P_n\underline{A}) - \alpha(Q_{n+1})$ also has nonnegative coefficients. In view of Lemma 14.6.5, we can choose a subset P_{n+1} of $P_n A$ in such a way that $\alpha(\underline{P_{n+1}}) = \alpha(\underline{Q_{n+1}})$.

We define $Y_{n+1} = P_n A \setminus P_{n+1}$. Then $Y \cup Y_{n+1}$ is prefix and commutatively equivalent to $X_1 \cup \ldots \cup X_{n+1}$. □

It is interesting to note the connection of this statement with Kraft's inequality given in (2.16) (see Exercise 14.6.2).

Corollary 14.6.6 *A positively factorizing code is commutatively prefix.*

Proof. Let $X \subset A^+$ be a factorizing code and let (P, Q) be a factorization of X. Then by definition $1 - \underline{X} = \underline{P}(1 - \underline{A})Q$. Passing to commutative variables gives $1 - \alpha(\underline{X}) = \alpha(\underline{P})(1 - \alpha(\underline{A}))\alpha(\underline{Q})$ or also $(1 - \alpha(\underline{X}))/(1 - \alpha(\underline{A})) = \alpha(\underline{P})\alpha(\underline{Q})$. Since $\alpha(\underline{P})\alpha(\underline{Q})$ has nonnegative coefficients, the conclusion follows from Theorem 14.6.4. □

Now we give an example of a code which is not commutatively prefix.

Example 14.6.7 Let $X \subset a^*ba^*$ be the set given in Table 14.2, with the convention that $a^i ba^j \in X$ if and only if the entry (i, j) contains a 1. Clearly $X \subset A^{(16)}$ and $\mathrm{Card}(X) = 16$. According to Proposition 14.6.3, X is not commutatively prefix.

Let us show that X is a code with deciphering delay 1. Let $x, y, z, t \in X$ be such that $xy \leq zt$. We may suppose $x \leq z$. Then (see Figure 14.4) we have

$$x = a^i ba^j, \quad y = a^k a^\ell ba^n, \quad z = xa^k, \quad t = a^\ell ba^n.$$

The 1's representing x and z are in the same row in Table 14.2. Necessarily $k \in \{0, 1, 2, 4, 6, 7, 12, 13, 14\}$ since these are the distances separating two 1's in the same row. Next, the 1's representing y and t are in rows whose difference of indices is k. Thus $k \in \{0, 3, 5, 8, 11\}$. This gives $k = 0$ and consequently $x = z$.

Corollary 14.6.6 shows that the factorization conjecture implies the commutative equivalence conjecture.

Table 14.2 *A code X which is not commutatively prefix.*

	0	1	2	3	4	5	6	7	8	9	10	11	12	13	14
0	1	1					1							1	1
1															
2															
3	1		1		1		1								
4															
6															
7															
8	1		1		1		1								
9															
10															
11	1	1	1												
12															
13															
14															
15															

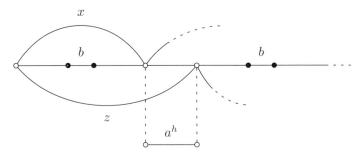

Figure 14.4 If X where not a code.

It is not known whether the code of Example 14.6.7 is included in a finite maximal code. It this were true, this would be a counterexample to the commutative equivalence conjecture and thus also to the factorization conjecture.

We use Theorem 14.6.4 to prove the following statement.

Theorem 14.6.8 *A circular code is commutatively prefix.*

We first prove the following lemma.

Lemma 14.6.9 *Let* $X \subset A^+$ *be a circular code. Then the series* $\log \alpha(\underline{A}^*) - \log \alpha(\underline{X}^*)$ *has nonnegative coefficients.*

Proof. We have

$$\log \underline{A}^* - \log \underline{X}^* = \log(1 - \underline{A})^{-1} - \log(1 - \underline{X})^{-1} = \sum_{n \geq 1} \frac{A^n}{n} - \sum_{n \geq 1} \frac{X^n}{n}.$$

Now, denoting by L the set of Lyndon words, and by L' the set of Lyndon words whose conjugacy class meets X^*, we have

$$\alpha(\underline{A})^n = \sum_{x \in L} \sum_{p \mid n} \frac{n}{p} \alpha(x)^p,$$

since the conjugacy class of x^p has $|x| = n/p$ elements. And, since X is circular, $\alpha(\underline{X})^n = \sum_{x \in L'} \frac{n}{p} \alpha(x)^p$. Thus

$$\log \alpha(\underline{A}^*) - \log \alpha(\underline{X}^*) = \sum_{x \in L} \sum_{p \geq 1} \frac{\alpha(x)^p}{p} - \sum_{x \in L'} \sum_{p \geq 1} \frac{\alpha(x)^p}{p}$$

$$= \sum_{x \in L \setminus L'} \sum_{n \geq 1} \frac{\alpha(x)^n}{n}.$$

This shows that the series $s = \log \alpha(\underline{A}^*) - \log \alpha(\underline{X}^*)$ has nonnegative coefficients.

□

Proof of Theorem 14.6.8 Let X be a circular code. Set $s = \log \alpha(\underline{A}^*) - \log \alpha(\underline{X}^*)$. By Lemma 14.6.9, the series s has nonnegative coefficients. We have

$$\exp(s) = \alpha(\underline{A}^*)/\alpha(\underline{X}^*) = (1 - \alpha(\underline{X})/(1 - \alpha(\underline{A})).$$

Since s has nonnegative coefficients, so does $\exp(s)$. Thus X is commutatively prefix by Theorem 14.6.4.

□

Note that a circular code is not always cyclically equivalent to a prefix code (see Exercise 14.6.1).

We now consider the problem of the commutative equivalence to synchronized codes. The *period* of a set of words is the greatest common divisor of the lengths of its elements. Two commutatively equivalent sets have the same period. If a finite maximal prefix code X has period p, then $X = Y \circ A^p$ and thus $d(X) = d(Y)p$ by Proposition 11.1.2. In particular, a finite maximal prefix code X of period $p \geq 2$ is not synchronized. The following result shows that this is the only obstruction.

Theorem 14.6.10 *Any finite maximal prefix code of period* 1 *is commutatively equivalent to a synchronized prefix code.*

The proof relies on three lemmas. Since the only maximal prefix code on one letter a of period 1 is the alphabet $\{a\}$ itself, we may assume that the alphabet has at least two letters.

For any nonempty finite set X of words, we denote by $\deg(X)$ the maximal length of the words of X and by $\widehat{X}$ the set of words of X of length $\deg(X)$. For a polynomial P, we write $\widehat{P}$ for the set of words of maximal length in $\mathrm{supp}(P)$.

Lemma 14.6.11 *If X is a finite maximal prefix code of period p such that*

$$\underline{X} - 1 = L(\underline{A} - 1)R$$

where $\widehat{R} = A^n$ for some $n \geq 1$, then R is a polynomial in $\underline{A}$ dividing $1 + \underline{A} + \cdots + \underline{A}^{p-1}$.

Proof. 1. Let $E = (\underline{A} - 1)R$. We first show that E is a polynomial in $\underline{A}$. Let us prove by descending induction on $m \leq n$ that

$$E = E' + \sum_{i=m+1}^{n+1} s_i \underline{A}^i \tag{14.23}$$

with $\deg(E') \leq m$. The property is true for $m = n$ since $\widehat{E} = A\widehat{R} = A^{n+1}$. Suppose that it holds for $m \leq n$. Let g be a word in $\widehat{L}$ and let h be a word of length m. For all words k of length $n - m + 1$ we have $ghk \in \widehat{L}\widehat{E} \subset X$ and thus $ghk \in X$. Since X is prefix and $k \neq 1$, we have $(LE, gh) = 0$.

But, by Formula (14.23) we have

$$(LE, gh) = (L, g)(E', h) + \sum_{i=0}^{t-1}(L, g_i)s_{t+m-i} \tag{14.24}$$

where g_i is the prefix of length i of g and $t = |g|$. Since $(LE, gh) = 0$, we deduce from (14.24) the formula

$$(E', h) = -\frac{1}{(L, g)}\sum_{i=0}^{t-1}(L, g_i)s_{t+m-i}.$$

It shows that (E', h) does not depend on the word h and proves that (14.23) is true for $m - 1$. Thus we have proved by induction that E is a polynomial in $\underline{A}$, that is

$$E = \sum_{i=0}^{n+1} s_i \underline{A}^i.$$

Consequently, R is also a polynomial in $\underline{A}$.

2. Let x be a word of X and let $q = |x|$. Let ℓ, s be the polynomials in the variable z defined by

$$\ell(z) = \sum_{i=0}^{q} \ell_i z^i, \quad s(z) = \sum_{i=0}^{n+1} s_i z^i,$$

where ℓ_i is the coefficient in L of the prefix x_i of length i of x. We have for each integer m such that $0 \leq m \leq q$

$$(LE, x_m) = \sum_{i+j=m} \ell_i s_j$$

(we set $s_i = 0$ for $j > n + 1$). Suppose that $0 < m < q$. Since X is prefix and $X - 1 = LE$, we have $(LE, x_m) = 0$ and thus

$$\sum_{i+j=m} \ell_i s_j = 0.$$

Since $(LE, x) = 1$ and $(LE, 1) = -1$, we therefore have $z^q - 1 = \ell(z)s(z)$. This shows that E divides $\underline{A}^q - 1$ and that R divides $1 + \underline{A} + \cdots + \underline{A}^{q-1}$ for each q such that X contains a word of length q. This proves the lemma. $\qquad\square$

The second lemma is a simple property of commutative equivalence.

Lemma 14.6.12 *Let Y be a maximal prefix code on the alphabet A with $\widehat{Y} = AR$ and $\deg(R) = n$. If $R \neq A^n$, then Y is commutatively equivalent to a prefix code Y' such that $\widehat{Y'}$ is not of the form AR' and, in particular $\widehat{Y'} \neq \widehat{Y}$.*

Proof. We use an induction on n to prove in a first step that for a nonempty set R strictly included in A^n, there exist a word h and letters a, b such that $(ha)^{-1}R \neq (hb)^{-1}R$ (note that one of the sides can be the empty set). The property holds trivially for $n = 0$ since then R is equal to $\{1\} = A^0$. Assume, for some $n \geq 1$, that it holds for $n - 1$. If for some $a \in A$, the set $S = a^{-1}R$ is nonempty and not equal to A^{n-1}, there exist, by induction hypothesis, a word g and letters b, c such that $(gb)^{-1}S \neq (gc)^{-1}S$. Then the assertion is proved with $h = ag$. Otherwise, we have $a^{-1}R = A^{n-1}$ or $a^{-1}R = \emptyset$ for each letter a. Since $R \neq \emptyset$ and $R \neq A^n$, the sets $a^{-1}R$ cannot be all equal. Thus, there exist letters a, b such that only one of the sets $a^{-1}R$, $b^{-1}R$ is empty. Then the conclusion holds with $h = 1$.

For h, a, b as above, let $U = (ahb)^{-1}Y$, $V = (bha)^{-1}Y$. Then $\widehat{U} = (hb)^{-1}R$ and $\widehat{V} = (ha)^{-1}R$. This implies that $\widehat{U} \neq \widehat{V}$. Let $Y = W \cup ahbU \cup bhaV$ with the three terms of the union disjoint. Then $Y' = W \cup ahbV \cup bhaU$ is commutatively equivalent to Y. Suppose that $\widehat{Y'} = AR'$. Since $V = (bha)^{-1}Y$, we have

$$\widehat{V} = (bha)^{-1}\widehat{Y} = (ha)^{-1}R = (aha)^{-1}\widehat{Y} = (aha)^{-1}\widehat{W} = (aha)^{-1}\widehat{Y'} = (ha)^{-1}R'.$$

On the other hand, we have

$$\widehat{U} = (bha)^{-1}\widehat{Y'} = (ha)^{-1}R'$$

and thus we obtain $\widehat{U} = \widehat{V}$, a contradiction. $\qquad\square$

For a finite maximal prefix code X, we denote by $e(X)$ the integer defined by

$$e(X) = \max\{e \geq 0 \mid \underline{X} - 1 = L(\underline{A} - 1)R, \; e = \deg(R)\}. \qquad (14.25)$$

Lemma 14.6.13 *Let X be a finite maximal prefix code such that*

$$\underline{X} - 1 = L(\underline{A} - 1)R \qquad (14.26)$$

with $\deg(R) = n \geq 1$ and $\widehat{R} \neq A^n$. Then there exists a prefix code X' commutatively equivalent to X such that

$$e(X') < e(X).$$

Proof. We first note that (14.26) implies that $\widehat{X} = \widehat{L}A\widehat{R}$. Observe that this also holds for the characteristic series of these sets. Let $g \in \widehat{L}$ and let $Y = g^{-1}X$. Then $\widehat{Y} = A\widehat{R}$.

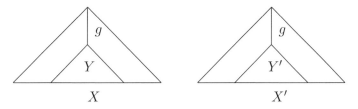

Figure 14.5 The codes X and X'.

Since $\widehat{R} \neq A^n$, there exists by Lemma 14.6.12, a prefix code Y' commutatively equivalent to Y such that $\widehat{Y}'$ is not of the form AR'.

Let X' be the prefix code commutatively equivalent to X defined by (see Figure 14.5)

$$X' = (X \setminus gY) \cup gY'.$$

In order to prove that $e(X') < e(X)$, consider a factorization

$$\underline{X}' - 1 = L'(\underline{A} - 1)R' \tag{14.27}$$

and suppose by contradiction that $\deg(R) \leq \deg(R')$.

Since Y' is commutatively equivalent to Y, we have $\deg(Y') = \deg(Y)$ and therefore $\deg(X) = \deg(X')$. This implies that $g\widehat{Y}' \subset \widehat{X}' = \widehat{L}'A\widehat{R}'$. Consider a word $y \in \widehat{Y}'$. Then $gy \in \widehat{L}'A\widehat{R}'$ implies that $gy = g'r$ with $g' \in \widehat{L}'$ and $r \in A\widehat{R}$. Since $\deg(L) \geq \deg(L')$, the word g' is a prefix of g. Let $g = g'h$. Then $\widehat{Y}' = g^{-1}\widehat{X}' = h^{-1}A\widehat{R}'$.

Suppose first that $h = 1$, that is that $g = g'$. Then $\widehat{Y}' = A\widehat{R}'$, a contradiction.

Thus $h \neq 1$. Let a be the first letter of h and set $h = ah'$. Let b be a letter distinct from a (recall that the alphabet is supposed to have at least two elements). We have

$$\widehat{Y}' = h^{-1}A\widehat{R}' = h'^{-1}\widehat{R}' = (bh')^{-1}A\widehat{R}' = (g'bh')^{-1}\widehat{L}'A\widehat{R}' = (g'bh')^{-1}\widehat{X}'.$$

Since the words of X and X' which do not begin with g are the same, this implies

$$\widehat{Y}' = (g'bh')^{-1}\widehat{X} = (g'bh')^{-1}\widehat{L}A\widehat{R}.$$

Since $\deg(Y') = \deg(Y)$, we have $\deg(Y') = \deg(R) + 1$. Thus the equality $\widehat{Y}' = (g'bh')^{-1}\widehat{L}A\widehat{R}$ with $|g'bh'| = |g| = \deg(L)$ implies $\widehat{Y}' = A\widehat{R}$, which is a contradiction. $\square$

Proof of Theorem 14.6.10. We use an induction on the integer $e(X)$. The property is true when $e(X) = 0$ since then X itself is synchronized. Indeed, we consider the factorization

$$\underline{X} - 1 = L(\underline{A} - 1)(d + D(\underline{A} - 1))$$

given by Theorem 14.2.1, knowing that X is prefix. Then $e(X) = 0$ implies $D = 0$ and thus $d = 1$.

Figure 14.6 The codes X and X'.

When $e(X) \geq 1$, we have $\underline{X} - 1 = L(\underline{A} - 1)R$ with $\deg(R) = n \geq 1$. If $\widehat{R} = A^n$, then by Lemma 14.6.11, R divides $1 + \underline{A} + \cdots + \underline{A}^{p-1}$ with p the period of X. Hence, $p \geq n + 1 \geq 2$ in contradiction with the hypothesis $p = 1$. Therefore, $\widehat{R} \neq A^n$ and by Lemma 14.6.13, there exists a prefix code X' commutatively equivalent to X such that $e(X') < e(X)$, whence the property by induction.   □

Example 14.6.14 Consider the maximal bifix code of degree 3 on the alphabet $A = \{a, b\}$

$$\underline{X} = aaa + aab\underline{A} + ab + baa + bab\underline{A} + bba + bbb.$$

We have $\underline{X} - 1 = (\underline{A} - 1)R$ with $R = 1 + a + b + b\underline{A} + ab\underline{A}$. We choose, with the notation of the proof of Lemma 14.6.13, $g = 1$ and therefore $Y = X$. We have $\widehat{R} = abA$. Then, with the notation of Lemma 14.6.12, we choose $h = a$, since $(aa)^{-1}\widehat{R} = \emptyset$ and $(ab)^{-1}\widehat{R} = A$. Thus we obtain

$$\underline{X'} = aaa + aab + ab + baa\underline{A} + bab\underline{A} + bba + bbb.$$

The code X' is commutatively equivalent to X and is synchronized since $baab$ is a synchronizing word (see Figure 14.6).

14.7 Complete reducibility

Let A be an alphabet and let $\sigma \in \mathbb{Q}\langle\langle A \rangle\rangle$ be a series. For each word $u \in A^*$, we define a series $\sigma \cdot u$ by $(\sigma \cdot u, w) = (\sigma, uw)$ for all $w \in A^*$. The following formulas hold :

$$\sigma \cdot 1 = \sigma, \quad (\sigma \cdot u) \cdot v = \sigma \cdot uv.$$

Let V_σ be the subspace of the vector space $\mathbb{Q}\langle\langle A \rangle\rangle$ generated by the series $\sigma \cdot u$ for $u \in A^*$. For each word $w \in A^*$, we denote by $\psi_\sigma(w)$ the linear function from V_σ into itself (acting on the right) defined by

$$\psi_\sigma(w) : \rho \mapsto \rho \cdot w.$$

The formula $(\rho \cdot u)\psi_\sigma(w) = \rho \cdot uw = \rho\psi_\sigma(uw)$ is straightforward. It follows that ψ_σ is a morphism

$$\psi_\sigma : A^* \rightarrow \mathrm{End}(V_\sigma)$$

from A^* into the monoid $\mathrm{End}(V_\sigma)$ of linear functions from V_σ into itself. The morphism ψ_σ is called the *syntactic representation* of σ.

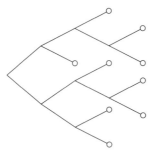

Figure 14.7 A bifix code.

Proposition 14.7.1 *Let Y be a subset of A^* and let $\sigma = \underline{Y}$. Let φ be the canonical morphism from A^* onto the syntactic monoid of Y. Then for all $u, v \in A^*$,*

$$\varphi(u) = \varphi(v) \Leftrightarrow \psi_\sigma(u) = \psi_\sigma(v).$$

In particular the monoid $\psi_\sigma(A^*)$ is isomorphic to the syntactic monoid of Y.

Proof. Assume first that $\psi_\sigma(u) = \psi_\sigma(v)$. Then for all $r \in A^*$,

$$\sigma \cdot ru = (\sigma \cdot r)\psi_\sigma(u) = (\sigma \cdot r)\psi_\sigma(v) = \sigma \cdot rv.$$

Thus also for all $s \in A^*$,

$$(\sigma, rus) = (\sigma \cdot ru, s) = (\sigma \cdot rv, s) = (\sigma, rvs).$$

This means that $rus \in Y$ if and only if $rvs \in Y$, which shows that $\varphi(u) = \varphi(v)$.

Conversely, assume $\varphi(u) = \varphi(v)$. Since the vector space V_σ is generated by the series $\sigma \cdot r$ $(r \in A^*)$, it suffices to show that for $r \in A^*$,

$$(\sigma \cdot r)\psi_\sigma(u) = (\sigma \cdot r)\psi_\sigma(v).$$

Now for all $s \in A^*$,

$$((\sigma \cdot r)\psi_\sigma(u), s) = (\sigma \cdot ru, s) = (\sigma, rus) = (\sigma, rvs) = ((\sigma \cdot r)\psi_\sigma(v), s). \qquad \square$$

The preceding result gives a relationship between the syntactic representation of the characteristic series σ of a set $Y \subset A^*$ and the syntactic monoid of Y. It should be noted that the dimension of the vector space V_σ can be strictly less than the number of states of the minimal automaton of Y (see Example 14.7.3). However, it can be shown that the vector space V_σ has finite dimension if and only if Y is recognizable (Exercise 14.7.2).

Example 14.7.2 Let $\sigma = \underline{A}^*$. Then $\sigma \cdot u = \sigma$ for all $u \in A^*$. Consequently $V_\sigma = \mathbb{Q}\sigma$ is a vector space of dimension 1.

Example 14.7.3 Let $A = \{a, b\}$ and let $X \subset A^+$ be the bifix code of Figure 14.7. Let $\sigma = \underline{X}^*$. We shall see that the vectors $\sigma, \sigma \cdot a, \sigma \cdot a^2$, and $\sigma \cdot b$ form a basis of the vector space V_σ. Indeed, the formulas

$$\sigma \cdot a^3 = \sigma \cdot ab = \sigma, \quad \sigma \cdot ba = \sigma \cdot a^2,$$

$$\sigma \cdot b^2 = \sigma \cdot a^2 b = \sigma \cdot a + \sigma \cdot a^2 - \sigma \cdot b,$$

show that the four vectors σ, $\sigma \cdot a$, $\sigma \cdot a^2$ and $\sigma \cdot b$ generate V_σ. A direct computation shows that they are linearly independent. The matrices of the linear mappings $\psi_\sigma(b)$ in this basis are

$$\psi_\sigma(a) = \begin{bmatrix} 0 & 1 & 0 & 0 \\ 0 & 0 & 1 & 0 \\ 1 & 0 & 0 & 0 \\ 0 & 0 & 1 & 0 \end{bmatrix}, \quad \psi_\sigma(b) = \begin{bmatrix} 0 & 0 & 0 & 1 \\ 1 & 0 & 0 & 0 \\ 0 & 1 & 1 & -1 \\ 0 & 1 & 1 & -1 \end{bmatrix}.$$

The relation between ψ_σ and the minimal automaton of X^* is now to be shown. The minimal automaton has five states which may be written as $1, 1 \cdot a, 1 \cdot a^2, 1 \cdot b, 1 \cdot b^2$. Let V be the $\mathbb{Q}$-vector space formed of formal linear combinations of these five states. The linear function $\alpha : V \to V_\sigma$ defined by $\alpha(1 \cdot u) = \sigma \cdot u$ satisfies the equality $\alpha(q \cdot u) = \alpha(q) \cdot u$ and moreover we have $\alpha(1 \cdot a + 1 \cdot a^2 - 1 \cdot b - 1 \cdot b^2) = 0$. Thus V has dimension 5 and V_σ has dimension 4.

Let V be a vector space over $\mathbb{Q}$ and let N be a submonoid of the monoid $\text{End}(V)$ of linear functions from V into itself. The action of elements in $\text{End}(V)$ will be written on the right.

A subspace W of V is *invariant* under N if for $\rho \in W, n \in N$, we have $\rho n \in W$. The submonoid N is called *reducible* if there exists a subspace W of V which is invariant under N and such that $W \neq \{0\}$, $W \neq V$. Otherwise, N is called *irreducible*.

The submonoid N is *completely reducible* if for any subspace W of V which is invariant under N, there exists a subspace W' of V which is a supplementary space of W and invariant under N.

If V has finite dimension, a completely reducible submonoid N of $\text{End}(V)$ has the following form. There exists a decomposition of V into a direct sum of invariant subspaces $W_1, W_2, \ldots, W_k$,

$$V = W_1 \oplus W_2 \oplus \cdots \oplus W_k$$

such that the restrictions of the elements of N to each of the W_i's form an irreducible submonoid of $\text{End}(W_i)$. In a basis of V composed of bases of the subspaces W_i, the matrix of an element n in N has a diagonal form by blocks,

$$n = \begin{bmatrix} n_1 & & & 0 \\ & n_2 & & \\ & & \ddots & \\ 0 & & & n_k \end{bmatrix}.$$

Let M be a monoid and let V be a vector space. A *linear representation* ψ of M over V is a morphism from M into the monoid $\text{End}(V)$. A subspace W of V is called *invariant* under ψ if it is invariant under $\psi(M)$. Similarly ψ is called reducible, irreducible, or completely reducible if this holds for $\psi(M)$.

The syntactic representation of a series σ is an example of a linear representation of a free monoid. The aim of this section is to study cases where this representation is completely reducible. We recall that all the vector spaces considered here are over the field $\mathbb{Q}$ of rational numbers. The following result is a classical one.

Theorem 14.7.4 (Maschke) *A linear representation of a finite group is completely reducible.*

Proof. Let V be a vector space over $\mathbb{Q}$. It suffices to show that each finite subgroup of the monoid $\text{End}(V)$ is completely reducible. Let G be a finite subgroup of $\text{End}(V)$ and let W be a subspace of V which is invariant under G. Let W_1 be any supplementary space of W in V. Let $\pi : V \to V$ be the linear function which associates to $\rho \in V$ the unique ρ_1 in W_1 such that $\rho = \rho_1 + \rho'$ with $\rho' \in W$. Then $\pi(\rho) = 0$ for all $\rho \in W$ and $\pi(\rho) = \rho$ for $\rho \in W_1$. Moreover, $\rho - \pi(\rho) \in W$ for all $\rho \in V$.

Let $n = \text{Card}(G)$. Define a linear function $\theta : V \to V$ by setting for $\rho \in V$,

$$\theta(\rho) = \frac{1}{n} \sum_{g \in G} \pi(\rho g) g^{-1}.$$

Let $W' = \theta(V)$. We shall see that W' is an invariant subspace of V under G which is a supplementary space of W. First, for $\rho \in W$,

$$\theta(\rho) = 0. \tag{14.28}$$

Indeed, if $\rho \in W$, then $\rho g \in W$ for all $g \in G$ since W is invariant under G. Thus $\pi(\rho g) = 0$ and consequently $\theta(\rho) = 0$. Next, for $\rho \in V$,

$$\rho - \theta(\rho) \in W. \tag{14.29}$$

Indeed

$$\rho - \theta(\rho) = \rho - \frac{1}{n} \sum_{g \in G} \pi(\rho g) g^{-1} = \frac{1}{n} \sum_{g \in G} (\rho g - \pi(\rho g)) g^{-1}.$$

By definition of π, each $\rho g - \pi(\rho g)$ is in W for $g \in G$. Since W is invariant under G, also $(\rho g - \pi(\rho g)) g^{-1} \in W$. This shows Formula (14.29).

By (14.28) we have $W \subset \text{Ker}(\theta)$ and by (14.29), $\text{Ker}(\theta) \subset W$ since $\rho \in \text{Ker}(\theta)$ implies $\rho - \theta(\rho) = \rho$. Thus

$$W = \text{Ker}(\theta).$$

Formula (14.28) further shows that $\theta^2 = \theta$. Indeed, $\theta(\rho) - \theta^2(\rho) = \theta(\rho - \theta(\rho))$. By (14.29), $\rho - \theta(\rho) \in W$. Hence $\theta(\rho) - \theta^2(\rho) = 0$ by (14.28). Since $\theta^2 = \theta$, the subspaces $W = \text{Ker}(\theta)$ and $W' = \text{im}(\theta)$ are supplementary. Finally, W' is invariant under G. Indeed, let $\rho \in V$ and $h \in G$. Then

$$\theta(\rho)h = \frac{1}{n} \sum_{g \in G} \pi(\rho g)g^{-1}h.$$

The function $g \mapsto k = h^{-1}g$ is a bijection from G onto G and thus

$$\theta(\rho)h = \frac{1}{n} \sum_{g \in G} \pi(\rho hk)k^{-1} = \theta(\rho h).$$

This completes the proof. $\qquad\square$

Theorem 14.7.5 *Let $X \subset A^+$ be a very thin bifix code. The syntactic representation of $\underline{X}^*$ is completely reducible.*

In the case of group codes, this theorem is a direct consequence of Theorem 14.7.4. For the general case, we need the following proposition in order to be able to apply Theorem 14.7.4.

Proposition 14.7.6 *Let $X \subset A^+$ be a very thin prefix code and let $\psi = \psi_{X^*}$ be the syntactic representation of $\underline{X}^*$. The monoid $M = \psi(A^*)$ contains an idempotent e such that*

(i) $e \in \psi(X^*)$.

(ii) *The set eMe is the union of the finite group $G(e)$ and of the element 0, provided $0 \in M$.*

Proof. Let S be the syntactic monoid of X^* and let $\varphi : A^* \to S$ be the canonical morphism. Consider also the minimal automaton $\mathcal{A}(X^*)$ of X^*. Since X is prefix, the automaton $\mathcal{A}(X^*)$ has a single final state which is the initial state (Proposition 3.2.5). Let $\mu = \varphi_{\mathcal{A}(X^*)}$ be the morphism associated with $\mathcal{A}(X^*)$. We claim that for all $u, v \in A^*$,

$$\mu(u) = \mu(v) \Leftrightarrow \psi(u) = \psi(v). \tag{14.30}$$

Indeed, in view of Proposition 1.4.5, we have

$$\mu(u) = \mu(v) \Leftrightarrow \varphi(u) = \varphi(v),$$

and by Proposition 14.7.1,

$$\varphi(u) = \varphi(v) \Leftrightarrow \psi(u) = \psi(v).$$

Formula (14.30) shows that there exists an isomorphism $\beta : \mu(A^*) \to \psi(A^*) = M$ defined by $\beta \circ \mu = \psi$. In particular, $\psi(X^*) = \beta(\mu(X^*))$. By Theorem 9.4.7, the monoid M has a unique 0-minimal or minimal ideal, say J, according to whether M does or does not have a zero. There exists an idempotent e in J which is also in $\psi(X^*)$. The $\mathcal{H}$-class of this idempotent is isomorphic to the group of X. $\qquad\square$

Proof of Theorem 14.7.5. For convenience, set $V = V_{X^*}$ and denote by ψ the syntactic representation ψ_{X^*}. Let $M = \psi(A^*)$. By Proposition 14.7.6, there exists an idempotent $e \in \psi(X^*)$ such that eMe is the union of 0 (if $0 \in M$) and of the group $G(e)$. The element 0 of the monoid M corresponds to the zero of $\psi(A^*)$. Let $L = Me$ and define $S = \{\rho e \mid \rho \in V\}$. Since $e^2 = e$, we have $\tau e = \tau$ for all $\tau \in S$. Next, for all $\ell \in L$, we have $\ell e = \ell$ since $\ell = me$ for some $m \in M$ and consequently $\ell e = me^2 = me = \ell$. Thus for all $\ell \in L$,

$$V\ell \subset S. \tag{14.31}$$

Let W be a subspace of V which is invariant under M. We shall see that there exists a supplementary space of W which is invariant under M. For this, set $T = W \cap S$ and $G = G(e)$.

The group G acts on S. The subspace T of S is invariant under G. Indeed, let $\tau \in T$ and let $g \in G$. Then $\tau g \in W$ since W is invariant under M and $\tau g \in S$ by (14.31) since $g = ge$. By Theorem 14.7.4, there exists a subspace T' of S which is supplementary of T in S and which is invariant under G. Set

$$W' = \{\rho \in V \mid \forall \ell \in L, \rho\ell \in T'\}.$$

We shall verify that W' is a supplementary space of W invariant under M. First observe that W' clearly is a subspace of V. Next it is invariant under M since for $\rho \in W'$ and $m \in M$, we have, for all $\ell \in L$, $(\rho m)\ell = \rho(m\ell) \in T'$ and consequently $\rho m \in W'$.

Next we show that

$$T' \subset W'. \tag{14.32}$$

Indeed, let $\tau' \in T'$. Then $\tau' \in S$ and thus $\tau' e = \tau'$. Hence $\tau'\ell = \tau'e\ell$ for all $\ell \in L$. Since $e\ell \in eMe$ and since T' is invariant under G, it follows that $\tau'\ell \in T'$. This shows that $\tau' \in W'$.

Now we verify that $V = W + W'$. For this, set $\sigma = \underline{X}^*$ and first observe that

$$\sigma e = \sigma. \tag{14.33}$$

(Note that $\sigma \in V$ and e acts on V.) Indeed, let $x \in X^*$ be such that $\psi(x) = e$. Since X^* is right unitary, we have for all $u \in A^*$ the equivalence $xu \in X^* \Leftrightarrow u \in X^*$. This shows that $(\sigma e, u) = (\sigma \cdot x, u) = (\sigma, xu) = (\sigma, u)$ and proves (14.33).

In view of (14.33), we have $\sigma \in S$. Since $S = T + T'$, there exist $\tau \in T$ and $\tau' \in T'$ such that $\sigma = \tau + \tau'$. Then for all $m \in M$, $\sigma m = \tau m + \tau'm$. For each $m \in M$, $\tau m \in Tm \subset Wm \subset W$, whence $\tau m \in W$. Using (14.32), also $\tau'm \in T'm \subset W'm$. Since W' is invariant under M, we obtain $\tau'm \in W'$. Thus $\sigma m \in W + W'$. Since V is generated by the vectors σm for $m \in M$, this proves that $V = W + W'$.

Finally, we claim that $W \cap W' = \{0\}$. Indeed, let $\rho \in W \cap W'$. Then for all $\ell \in L$,

$$\rho\ell = 0. \tag{14.34}$$

Indeed, let $\ell \in L$. Then $\rho\ell \in W$, since W is invariant under M and $\rho\ell \in S$ by Equation (14.31). This implies $\rho\ell \in W \cap S = T$. Further $\rho\ell \in T'$ by the definition of W' and by the fact that $\rho \in W'$. Thus $\rho\ell \in T \cap T' = \{0\}$.

Since V is generated by the series $\sigma \cdot u$ ($u \in A^*$), there exist numbers $\alpha_u \in \mathbb{Q}$ ($u \in A^*$), with only a finite number among them nonzero, such that

$$\rho = \sum_{u \in A^*} \alpha_u(\sigma \cdot u).$$

Again, let $x \in X^*$ be such that $\psi(x) = e$. Since X^* is left unitary, we have, as above, $(\sigma, w) = (\sigma, wx)$ for all $w \in A^*$. Consequently, for all $v \in A^*$,

$$(\rho, v) = \sum_{u \in A^*} \alpha_u(\sigma \cdot u, v) = \sum_{u \in A^*} \alpha_u(\sigma, uv) = \sum_{u \in A^*} \alpha_u(\sigma, uvx)$$

$$= \sum_{u \in A^*} \alpha_u(\sigma \cdot u, vx) = (\rho, vx) = (\rho \cdot vx, 1).$$

Setting $m = \psi(v)$, we have $(\rho, v) = (\rho m e, 1)$, and since $m e \in L$, we have $\rho m e = 0$ by (14.34). Consequently $(\rho, v) = 0$ for all $v \in A^*$. Thus $\rho = 0$. This shows that $W \cap W' = \{0\}$ and completes the proof. $\qquad\square$

Example 14.7.3 (*continued*) The subspace W of $V = V_\sigma$ generated by the vector $\rho = \sigma + \sigma \cdot a + \sigma \cdot a^2$ is invariant under ψ_σ. Indeed we have

$$\rho \cdot a = \rho, \quad \rho \cdot b = \rho.$$

We shall exhibit a supplementary space of W invariant under ψ_σ. It is the subspace generated by

$$\sigma - \sigma \cdot a, \ \sigma - \sigma \cdot a^2, \ \sigma - \sigma \cdot b.$$

Indeed, in the basis

$$\rho, \ \sigma - \sigma \cdot a, \ \sigma - \sigma \cdot a^2, \ \sigma - \sigma \cdot b,$$

the linear mappings $\psi_\sigma(a)$ and $\psi_\sigma(b)$ have the form

$$\alpha = \left[\begin{array}{c|ccc} 1 & 0 & 0 & 0 \\ \hline 0 & -1 & 1 & 0 \\ 0 & -1 & 1 & 0 \\ 0 & -1 & 1 & 0 \end{array}\right], \quad \beta = \left[\begin{array}{c|ccc} 1 & 0 & 0 & 0 \\ \hline 0 & 0 & 0 & 1 \\ 0 & 1 & 1 & -1 \\ 0 & 1 & 1 & -2 \end{array}\right].$$

We can observe that there are no other nontrivial invariant subspaces.

We now give a converse of Theorem 14.7.5 for the case of complete codes. The result does not hold in general if the code is not complete (see Example 14.7.5)

Theorem 14.7.7 *Let $X \subset A^+$ be a thin complete code. If the syntactic representation of X^* is completely reducible, then X is bifix.*

Proof. Let $\mathcal{A} = (Q, 1, 1)$ be a trim unambiguous automaton recognizing X^*. Let φ be the associated representation and let $M = \varphi(A^*)$.

Set $\sigma = \underline{X}^*$ and also $V = V_\sigma$, $\psi = \psi_\sigma$. Let μ be the canonical morphism from A^* onto the syntactic monoid of X^*. By Proposition 1.4.4, we have for $u, v \in A^*$, $\varphi(u) = \varphi(v) \Leftrightarrow \mu(u) = \mu(v)$. Thus we can define a linear representation $\theta : M \to \text{End}(V)$ by setting for $m \in M$, $\theta(m) = \mu(u)$ where $u \in A^*$ is any word such that $\varphi(u) = m$. If ψ is completely irreducible, then this holds also for θ.

For notational ease, we shall write, for $\rho \in V$ and $m \in M$, $\rho \cdot m$ instead of $\rho \cdot u$, where $u \in A^*$ is such that $\varphi(u) = m$. With this notation, we have for $m = \varphi(u)$,

$$\rho \cdot u = \rho\psi(u) = \rho\theta(m) = \rho \cdot m.$$

Observe further that with $m = \varphi(u)$,

$$(\sigma \cdot m, 1) = (\sigma \cdot u, 1) = (\sigma, u).$$

Hence

$$(\sigma \cdot m, 1) = \begin{cases} 1 & \text{if } u \in X^*, \\ 0 & \text{otherwise.} \end{cases} \tag{14.35}$$

Finally, we have for $\rho \in V$, $m, n \in M$, $(\rho \cdot m) \cdot n = \rho \cdot mn$. For $\rho \in V$ and for a finite subset K of M, we define

$$\rho \cdot K = \sum_{k \in K} \rho \cdot k.$$

In particular, (14.35) gives

$$(\sigma \cdot K, 1) = \text{Card}(K \cap \varphi(X^*)). \tag{14.36}$$

The code X being thin and complete, the monoid M has a minimal ideal J that intersects $\varphi(X^*)$. Further, J is a $\mathcal{D}$-class. Its $\mathcal{R}$-classes (resp. $\mathcal{L}$-classes) are the minimal right ideals (resp. minimal left ideals) of M (see Chapter 9, Section 9.4).

Let R be an $\mathcal{R}$-class of J and let L be an $\mathcal{L}$-class of J. Set $H = R \cap L$. For each $m \in M$, the function $h \mapsto hm$ induces a bijection from H onto the $\mathcal{H}$-class $Hm = Lm \cap R$. Similarly, the function $h \mapsto mh$ induces a bijection from H onto the $\mathcal{H}$-class $mH = L \cap mR$.

To show that X is suffix, consider the subspace W of V spanned by the series

$$\sigma \cdot H - \sigma \cdot K \tag{14.37}$$

for all pairs H, K of $\mathcal{H}$-classes of J contained is the same $\mathcal{R}$-class. We shall first prove that $W = \{0\}$.

The space W is invariant under M. Indeed, let H and K be two $\mathcal{H}$-classes contained in some $\mathcal{R}$-class R of J. Then for $m \in M$, $(\sigma \cdot H) \cdot m = \sigma \cdot (Hm)$ since, by Proposition 1.12.2, the right multiplication by m is a bijection from H onto Hm.

Thus $(\sigma \cdot H - \sigma \cdot K) \cdot m = \sigma \cdot (Hm) - \sigma \cdot (Km)$ and the right-hand side is in W since $Hm, Km \subset R$. Next for all $\rho \in W$ and $m \in J$,

$$\rho \cdot m = 0. \tag{14.38}$$

Indeed, let H and K be two $\mathcal{H}$-classes contained in an $\mathcal{R}$-class R of J. Then for $m \in J$, $Hm, Km \subset R \cap Rm$. Since $R \cap Mm$ is an $\mathcal{H}$-class, we have $Hm = Km = R \cap Mm$. This implies

$$(\sigma \cdot H - \sigma \cdot K) \cdot m = 0.$$

Since $p \in W$ is a linear combination of series of the form given in Equation (14.37). This proves Equation (14.38).

Since the representation of M over V is completely reducible, there exists a subspace W' of V which is complementary of W and invariant under M. Set $\sigma = \rho + \rho'$ with $\rho \in W, \rho' \in W'$. Let H, K be two $\mathcal{H}$-classes of J contained in an $\mathcal{R}$-class R. We shall prove that

$$\sigma \cdot H = \sigma \cdot K. \tag{14.39}$$

We have

$$\sigma \cdot H - \sigma \cdot K = (\rho \cdot H - \rho \cdot K) + (\rho' \cdot H - \rho' \cdot K).$$

Since $\rho \in W$ and $H, K \subset J$, it follows from (14.38) that

$$\rho \cdot H = \rho \cdot K = 0. \tag{14.40}$$

Next, there exist numbers $\alpha_m \in \mathbb{Q}$ $(m \in M)$ which almost all vanish such that $\rho' = \sum_{\in M} \alpha_m (\sigma \cdot m)$. Since the left multiplication is a bijection on $\mathcal{H}$-classes, we have

$$(\sigma \cdot m) \cdot H - (\sigma \cdot m) \cdot K = \sigma \cdot (mH) - \sigma \cdot (mK).$$

Thus, since $mH, mK \subset mR$, the right-hand side is in W and consequently also $\rho' \cdot H - \rho' \cdot K \in W$. Since W' is invariant under M, this element is also in W'. Consequently it vanishes and

$$\rho' \cdot H = \rho' \cdot K. \tag{14.41}$$

Consequently (14.39) follows from (14.40) and (14.41).

In view of (14.36), Formula (14.39) shows that if $\varphi(X^*)$ intersects some $\mathcal{H}$-class H in J, then it intersects all $\mathcal{H}$-classes which are in the $\mathcal{R}$-class containing H. In view of Proposition 9.4.9, this is equivalent to X being suffix.

We conclude by showing that X is prefix. Let T be the subspace of V composed of the elements $\rho \in V$ such that $(\rho \cdot H, 1) = (\rho \cdot K, 1)$ for all pairs H, K of $\mathcal{H}$-classes of J contained in a same $\mathcal{L}$-class.

The subspace T is invariant under M. Indeed if $\rho \in T$ and $H, K \subset L$, then for all $m \in M$,

$$(\rho \cdot m) \cdot H = \rho \cdot mH, \qquad (\rho \cdot m) \cdot K = \rho \cdot mK. \qquad (14.42)$$

Since mH, mK are in the $\mathcal{L}$-class L, we have by definition $((\rho \cdot m) \cdot K, 1) = ((\rho \cdot m) \cdot K, 1)$. It follows that $\rho \cdot m \in T$.

Next for all $m \in J$, and $\rho \in V$,

$$\rho \cdot m \in T. \qquad (14.43)$$

Indeed, let $m \in J$ and let H, K be two $\mathcal{H}$-classes contained in the L-class $L \subset J$. Then $mH = mK$. By (14.42), $((\rho \cdot m) \cdot H, 1) = ((\rho \cdot m) \cdot K, 1)$. Thus $\rho \cdot m \in T$.

Let T' be a supplementary space of T which is invariant under M. Again, set

$$\sigma = \rho + \rho'$$

this time with $\rho \in T$, $\rho' \in T'$. Let H, K be two $\mathcal{H}$-classes in J both contained in some $\mathcal{L}$-class L. Then

$$(\sigma \cdot H, 1) - (\sigma \cdot K, 1) = ((\rho \cdot H, 1) - (\rho \cdot K, 1)) + (\rho' \cdot H, 1) - (\rho' \cdot K, 1).$$

By definition of T, we have $(\rho \cdot H, 1) - (\rho \cdot K, 1) = 0$. In view of (14.43), we have $\rho' \cdot H, \rho' \cdot K \in T$ whence $\rho' \cdot H - \rho' \cdot K \in T \cap T' = \{0\}$. Thus $(\sigma \cdot H, 1) = (\sigma \cdot K, 1)$. Interpreting this equality using (14.36), it is shown that if $\varphi(X^*)$ meets some $\mathcal{H}$-class of J, it intersects all $\mathcal{H}$-classes contained in the same $\mathcal{L}$-class. By Proposition 9.4.9, this shows that X is prefix. □

Example 14.7.8 Let $A = \{a, b\}$ and let $X = \{a, ba\}$. The code X is prefix but not suffix. It is not complete.

Let $\sigma = \underline{X^*}$. The vectors σ and $\sigma \cdot b$ form a basis of the vector space V_σ since

$$\sigma \cdot a = \sigma, \qquad \sigma \cdot ba = \sigma, \qquad \sigma \cdot bb = 0.$$

In this basis, the matrices of $\psi_\sigma(a)$ and $\psi_\sigma(b)$ are

$$\psi_\sigma(a) = \begin{bmatrix} 1 & 0 \\ 1 & 0 \end{bmatrix}, \qquad \psi_\sigma(b) = \begin{bmatrix} 0 & 1 \\ 0 & 0 \end{bmatrix}.$$

The representation ψ_σ is irreducible. Indeed,

$$\psi_\sigma(ba) = \begin{bmatrix} 1 & 0 \\ 0 & 0 \end{bmatrix}, \qquad \psi_\sigma(a) - \psi_\sigma(ba) = \begin{bmatrix} 0 & 0 \\ 1 & 0 \end{bmatrix},$$

$$\psi_\sigma(b) = \begin{bmatrix} 0 & 1 \\ 0 & 0 \end{bmatrix}, \qquad \psi_\sigma(ab) - \psi_\sigma(b) = \begin{bmatrix} 0 & 0 \\ 0 & 1 \end{bmatrix}.$$

This shows that the matrices $\psi_\sigma(u)$, $u \in A^*$ generate the whole algebra $\mathbb{Q}^{2 \times 2}$. Thus no nontrivial subspace of V is invariant under A^*.

This example shows that Theorem 14.7.7 does not hold in general for codes which are not complete.

14.8 Exercises

Section 14.1

14.1.1 A code $X \subset A^+$ is called *separating* if there is a word $x \in X^*$ such that each $w \in A^*$ admits a factorization $w = uv$ with $xu, vx \in X^*$

(a) Show that a separating code is complete and synchronized.

(b) Show that a separating code is positively factorizing and that its positive factorization is unique.

14.1.2 Let $X \subset A^+$ be a synchronized code and let $\mathcal{A} = (Q, 1, 1)$ be a trim unambiguous automaton recognizing X^*. For $x \in X^*$ let

$$U(x) = \{p \in Q \mid 1 \xrightarrow{x} p\}, \quad V(x) = \{q \in Q \mid q \xrightarrow{x} 1\}.$$

Show that X is separating if and only if there is a word x such that $x A^* x \subset X^*$ and any path from a state in $U(x)$ to a state in $V(x)$ passes through state 1.

14.1.3 Let $X \subset A^+$ be a code. A pair (L, R) of subsets of A^* is called a *separating box* for X if for any word $w \in A^*$ there is a unique pair $(\ell, r) \in L \times R$ such that w admits a factorization $w = uv$ with $\ell u, vr \in X^*$.

Show that a code which has a separating box is positively factorizing.

14.1.4 Let $X \subset A^+$ be a synchronized code and let $\mathcal{A} = (Q, 1, 1)$ be a trim unambiguous automaton recognizing X^*. For sets $S, T \subset A^*$, let $\ell = \sum_{s \in S} \varphi_A(s)_{1*}$ and $c = \sum_{t \in T} \varphi_A(t)_{*1}$. Show that (S, T) is a separating box if and only if

(i) for each $w \in A^*$, one has $\ell \varphi_A(w) c = 1$, and

(ii) any path from a state of ℓ to a state of c passes through state 1.

14.1.5 Let $b \in A$ be a letter and let $X \subset A^+$ be a finite maximal code such that for all $x \in X, |x|_b \leq 1$. Let $A' = A \setminus b$. Let $X' = X \cap A'^*$. Show that there is a factorization (P, Q) of X' considered as a code over A' such that

$$X = X' \cup PbQ.$$

14.1.6 Let $A = \{a, b\}$. Use Exercise 14.1.5 to show that a finite code $X \subset a^* \cup a^*ba^*$ is maximal if and only if $X = a^n \cup PbQ$ with $n \geq 1$ and $P, Q \subset a^*$ satisfying $PQ = 1 + a + \cdots + a^{n-1}$.

14.1.7 Let $X, Y \subset A^+$ be two distinct finite maximal prefix codes such that $X \cap Y \neq \emptyset$. Let $P = A^* \setminus X A^*$, $Q = A^* \setminus Y A^*$ and let

$$R \subset (X \cap Y)^*$$

be a finite set satisfying $uv \in R, u \in (X \cap Y)^* \implies v \in R$. (This means that R is suffix-closed considered as a set over the alphabet $X \cap Y$.)

(a) Show that there is a unique finite code $Z \subset A^+$ such that

$$\underline{Z} - 1 = (\underline{X \cap Y} - 1)\underline{R}.$$

(b) Show that there exists a unique finite maximal code $T \subset A^+$ such that

$$\underline{T} - 1 = (\underline{P} + w\underline{Q})(\underline{A} - 1)\underline{R},$$

where w is a word of maximal length in Z.

(c) Show that the code T is indecomposable under the following three assumptions:

   (i) Z is separating.
   (ii) $\mathrm{Card}(P \cup wQ)$ and $\mathrm{Card}(R)$ are prime numbers.
   (iii) R is not suffix-closed (over the alphabet A).

(*Hint*: First prove that T is uniquely factorizing. For this, suppose that $\underline{T} - 1 = F(\underline{A} - 1)G$. Let $n = |w|$ and let m be the maximal length of words in G. Show that, for all $f \in F$, $|f| + m + \geq n$ implies $f \in wA^*$.)

(d) Compare with Example 14.1.3, by taking $P = \{1, a\}$, $Q = \{1, a, b\}$, $R = \{1, aa\}$, $w = abaa$.

14.1.8 Let $A = \{a, b\}$ and let

$$X = (A^2 \setminus b^2) \cup b^2 A, \qquad Y = A^2 a \cup b.$$

(a) Verify that X is a maximal prefix code and that Y is a maximal suffix code.
(b) Show that the code Z defined by $Z^* = X^* \cap Y^*$ satisfies

$$\underline{Z} - 1 = (1 + \underline{A} + b^2)((\underline{A} - 1)a(\underline{A} - 1) + \underline{A} - 1)(1 + a + \underline{A}a).$$

(*Hint*: Show that $\underline{Z} - 1 = (\underline{X} - 1)\underline{P} = \underline{Q}(\underline{Y} - 1)$ for some $P \subset X^*$, $Q \subset Y^*$.)
(c) Show that Z is synchronized but not separating.
(d) Show that Z has a separating box. (*Hint*: Show that $(\{b^3\}, \{1, a^5\})$ is a separating box.)

14.1.9 Let $X \subset A^+$ be a set. A word $x \in X$ is said to be a *pure square* for X if

(i) $x = w^2$ for some $w \in A^+$,
(ii) $X \cap wA^* \cap A^*w = \{x\}$.

(a) Let $X \subset A^+$ be a finite maximal prefix code and let $x = w^2$ be a pure square for X. Set $G = Xw^{-1}$, $D = w^{-1}X$. Show that the polynomial

$$\sigma = (1 + w)(\underline{X} - 1 + (\underline{G} - 1)w(\underline{D} - 1)) + 1$$

is the characteristic polynomial of a finite maximal prefix code denoted by $\delta_w(X)$. (*Hint*: Set $G_1 = G \setminus w$ and $D_1 = D \setminus w$. Show that $\sigma = (1 + w)R + w^4$ where

$$R = (\underline{X} - \underline{G_1}w - w\underline{D}) + \underline{G_1}w\underline{D} + w^2\underline{D_1}$$

is a prefix code.

Show that the polynomial

$$(\underline{X} - 1 + (\underline{G} - 1)w(\underline{D} - 1))(1 + w) + 1$$

is the characteristic polynomial of a finite maximal code denoted by $\gamma_w(X)$.)

(b) Let $X \subset A^+$ be a finite maximal prefix code. Show that if $x = w^2$ is a pure square for X, then x^2 is a pure square for $\delta_w(X)$ and $\gamma_w(X)$.

(c) Let $X \subset A^+$ be a finite maximal prefix code. Let $x = w^2$ be a pure square for X. Show that the codes $Y = \gamma_w(X)$ and $Z = \delta_w(X)$ have the same degree. (*Hint:* Show that there is a bijection between Y-interpretations and Z-interpretations of a word.)

(d) Let X be a finite maximal bifix code. Let $x = w^2$ be a pure square for X and $Y = \delta_w(X)$. Show that $d(X) = d(Y)$. (*Hint:* Show that $\underline{Y} - 1 = (1 + w)(\underline{A} - 1)\underline{L}$, where L is a disjoint union of $d(X)$ maximal prefix codes.)

(e) Let X be a finite maximal bifix code. Let $x = w^2$ be a pure square for X and let $Y = \delta_w(X)$. By (b) the word x^2 is a pure square for Y. Let $Z = \gamma_x(Y)$. Show that $d(Z) = d(X)$. (*Hint:* Set $T = \delta_x(Y)$. Show that $\underline{T} - 1 = (1 + w)(1 + w^2)(\underline{A} - 1)\underline{M}$, where $\underline{M}$ is a disjoint union of $d(X)$ prefix codes.)

(f) Show that if $d(X)$ is a prime number and $d(X) > 2$, the code Z of (e) does not admit any decomposition over a suffix or a prefix code.

(g) Use the above construction to show that for each prime number $d > 3$, there exist finite maximal codes of degree d which are indecomposable and are neither prefix nor suffix.

Section 14.3

14.3.1 Show that if Y is a weak left divisor of X, then one may find polynomials P, Q, satisfying the hypothesis of Theorem 14.3.1.

14.3.2 Show that if the $x_1, \ldots, x_n$ are elements of a field and if the fraction

$$x_1 + \cfrac{1}{x_2 + \cfrac{1}{\ddots + \cfrac{1}{x_n}}}$$

is defined, then it is equal to

$$\frac{p(x_1, \ldots, x_n)}{p(x_2, \ldots, x_n)}.$$

(*Hint:* Use an induction on n.)

14.3.3 Show that if $k \leq n$, then

$$p(a_1, \ldots, a_n)\, p(a_{n-1}, \ldots, a_k) - p(a_1, \ldots, a_{n-1})\, p(a_n, \ldots, a_k)$$

$$= (-1)^{n+k}\, p(a_1, \ldots, a_{k-2})$$

(*Hint:* Use descending induction on k.)

14.3.4 Show that $p(1, \ldots, 1)$ (n times) is the $n + 1$-th Fibonacci number.

Section 14.4

14.4.1 Show that $S(u)$ (resp. $P(u)$, $F(u, v)$) defined in the proof of Lemma 14.4.1 is a sum of proper suffixes (resp. prefixes, factors) of words of C.

14.4.2 If $S \in \mathbb{Z}\langle\langle A \rangle\rangle$ has constant term 0 and $a \in A$, show that $a^{-1}(S^*) = (a^{-1}S)S^*$.

Section 14.5

14.5.1 Show that if ℓ is the number of leaves of a finite complete a-ary tree, and i the number of its internal nodes then $\ell - 1 = i(a - 1)$. Deduce from the literal representation of a complete prefix code, the corresponding equality relating its cardinality to the number of its prefixes.

Section 14.6

14.6.1 Let X be the circular code $X = \{a, ab, c, acb\}$. Show that there is no bijection $\alpha : X \to Y$ of X onto a prefix code Y such that $\alpha(x)$ is a conjugate of x for all $x \in X$.

14.6.2 Let $u(z) = \sum_{n \geq 1} u_n z^n$ with $u_n \geq 0$. Let $k \geq 1$ be an integer. Show that $(1 - u(z))/(1 - kz)$ has nonnegative coefficients if and only if $u(1/k) \leq 1$.

Section 14.7

14.7.1 Let $\mathcal{A} = (Q, i, T)$ be a finite automaton. The aim of this exercise is to construct the syntactic representation of the series $\sigma = |\mathcal{A}|$.

Let φ be the representation associated with $\mathcal{A}$ and let $M = \varphi(A^*)$. We may assume $Q = \{1, 2, \ldots, n\}$ and $i = 1$.

Let E_0 be the subspace of $\mathbb{Q}^n$ generated by the vectors m_{1*}, for $m \in M$. Let E_1 be the subspace of E_0 composed of all vectors ℓ in E_0 such that for all $m \in M$, $\sum_{t \in T}(\ell m)_t = 0$.

Show that the linear function $\alpha : E_0 \to V_\sigma$ defined by $\alpha : \varphi(u)_{1*} \mapsto \sigma \cdot u$ has kernel E_1. Deduce from this fact a method for computing a basis of V_σ and the matrices of $\psi_\sigma(a)$ in this basis for $a \in A$.

14.7.2 Let $S \subset A^+$ and $\sigma = \underline{S}$. Show that V_σ has finite dimension if and only if S is recognizable (use Exercise 14.7.1).

14.7.3 Let K be a commutative field and let $\sigma \in K\langle\langle A \rangle\rangle$. The syntactic representation of σ over K is defined as in the case $K = \mathbb{Q}$. Recall that the characteristic of a field is the greatest common divisor of all integers n such that $n \cdot 1 = 0$ in K.

Let X be a very thin bifix code. Let K be a field of characteristic 0 or which is prime to the order of $G(X)$. Show that the syntactic representation of $\underline{X}^*$ over K is completely reducible.

14.7.4 Let X be a very thin bifix code. Show that if X is synchronizing, then $\psi_{X^*}(A^*)$ is irreducible.

14.9 Notes

The results in Section 14.2 and the proof in Section14.4 are from Reutenauer (1985). Theorem 14.2.1 extends a commutative factorization result by Schützenberger (1965b), see also Hansel *et al.* (1984). Theorem 14.3.1 and Corollary 14.3.2 are a particular case of Paul Cohn's weak algorithm, see Cohn (1985). For their proofs, we have followed a lexicographic argument from Melançon (1993). Theorem 14.3.3 and Theorem 14.3.7 are from Cohn (1985). Theorem 14.3.4, Lemmas 14.3.8 and 14.3.9 are from Reutenauer (1985). Corollary 14.5.1 is due to Schützenberger (1961b). Corollary 14.5.2 is due to Hansel and Perrin (1983). Corollary 14.5.3 is from Schützenberger (1965b).

Note that the relations (ii) and (iii) in Lemma 14.4.3 are each a weak form of the factorization conjecture, since L_1 is a finite sum of words (for the conjecture, one would need to have $L_1 = 0$). This form was also found by Zhang and Gu (1992). For partial results on the factorization conjecture, see Restivo (1977), Boë (1981), De Felice and Reutenauer (1986), De Felice (1992), De Felice (1993). For results involving constructions of factorizing codes and multiple factorizations, see Perrin (1977a), Vincent (1985), Bruyère and De Felice (1992).

New results on the polynomial of a finite code, evaluated in an algebric structure called the stochastic free field, appear in Lavallée *et al.* (2009). Theorem 14.6.10 is from Perrin and Schützenberger (1992). It solves the analogue, for commutative equivalence, of the road coloring problem (see Section 10.4).

The problem of characterizing commutatively prefix codes has an equivalent formulation in terms of optimality of prefix codes with respect to some cost functions, namely, the average length of the code for a given weight distribution on the letters. In this context, it has been treated in several papers, and in particular in Carter and Gill (1974), Karp (1961). The codes of Proposition 14.6.3 have been studied under the name of *bayonet codes* (Hansel (1982); Pin and Simon (1982); De Felice (1983)). Example 14.6.7 is due to Shor (1983). It is a counterexample to a conjecture of Perrin and Schützenberger (1981). A particular case of commutatively prefix codes is studied in Mauceri and Restivo (1981).

Results of Section 14.7 are due to Reutenauer (1981). The syntactic representation appears for the first time in Schützenberger (1961a). It has been developed more systematically in Fliess (1974) and in Reutenauer (1980).

Theorem 14.7.4 is Maschke's theorem. The property for an algebra of matrices to be completely reducible is equivalent to that of being semisimple (see, e.g., Herstein (1969)). Thus Theorem 14.7.5 expresses the fact that the syntactic algebra $\psi_\sigma(A^*)$ for $\sigma = \underline{X}^*$, X a thin bifix code, is semisimple. This theorem is a generalization of Maschke's theorem.

Solutions of exercises

Chapter 2

Section 2.1

2.1.1 Any word $w = a^{k_0}ba^{k_1}b\cdots ba^{k_r}$ with $k_1,\ldots k_r \geq 0$ has at most one factorization $w = a^{t_0 n}y_0 a^{t_1 n}y_1\cdots y_{r-1}a^{t_r n}$ where $y_u = a^{i_u}ba^{j_u}$ with $k_0 \equiv i_0 \bmod n$, $k_r \equiv j_{r-1} \bmod n$ and for $1 \leq u \leq r-1$, $k_u \equiv j_{u-1} + i_u \bmod n$.

Section 2.2

2.2.1 Suppose that $|x| \leq |y|$. If X is not a code, then x is a prefix of y. Let $y = xy'$. Then $X' = \{x, y'\}$ is not a code and we have, by induction hypothesis, $x, y' \in z^*$. Thus $x, y \in z^*$.

2.2.2 The map β is clearly surjective. To see that it is injective, consider a polynomial $P = \sum_{i=1}^{n} \alpha_i w_i$ for some $w_i \in B^*$, such that $\beta(P) = 0$, and set $\beta(w_i) = x_i$. For each x_j, one gets $0 = (\beta(P), x_j) = \sum \alpha_i(x_i, x_j)$. Since X is a code, $(x_i, x_j) = 1$ if $i = j$, and 0 otherwise. Thus $\alpha_j = 0$ for all j.

2.2.3 A stable submonoid satisfies this condition. Conversely, let $u, v, w \in M$ be such that $u, v, uw, wv \in N$. Then $n = vu$, $m = w$ satisfy $nm, n, mn \in N$ and thus $w \in N$. Thus N is stable.

2.2.4 A stable submonoid of a commutative monoid is right unitary: If $u, uv \in N$, then also $vu \in N$ and thus $v \in N$.

2.2.5 We proceed as in the proof of Proposition 2.2.16. Suppose that $y \in Y$ is not in $(Y^*)^{-1}X$. Then $Z = y^*(Y \setminus y)$ is such that $X \subset Z^* \subset Y^*$, $Z^* \neq Y^*$ and Z^* is right unitary, a contradiction. This proves (a). Statement (b) follows directly. For $X = \{a, ab\}$, we have $Y = \{a, b\}$ and thus $\mathrm{Card}(X) = \mathrm{Card}(Y)$ although X is not a prefix code.

2.2.6 We show by induction on $n \geq 0$ that if Y is a code such that $X \subset Y^*$, then $S_n \subset Y^*$. It is true for $n = 0$. Assuming the property true for n, let $w \in S_n^{-1}S_n \cap S_n S_n^{-1}$.

Let $u, v \in S_n$ be such that $uw, wv \in S_n$. Then $uw, wv \in Y^*$ by induction hypothesis and thus $w \in Y^*$ since Y^* is stable. Hence $S_n^{-1}S_n \cap S_n S_n^{-1} \subset Y^*$ and consequently $S_{n+1} \subset Y^*$. This shows that $S(X)$ is the free hull of X.

To prove the second statement, we introduce an intermediary statement. For any $Z \subset A^*$, define U_i and V_i by $U_0 = V_0 = \{1\}$ and for $i \geq 0$ by $U_{i+1} = U_i^{-1}Z \cup Z^{-1}U_i$, $V_{i+1} = ZV_i^{-1} \cup V_i Z^{-1}$. Let $U = \bigcup_{i \geq 0} U_i$ and $V = \bigcup_{i \geq 0} V_i$. Setting $Q = Z^*$, we prove that

$$(Q^{-1}Q \cap QQ^{-1})^* = (U \cap V)^*. \tag{15.1}$$

To prove (15.1), consider first $w \in U \cap V$. It is easy to see that $U \subset Q^{-1}Q$ and $V \subset QQ^{-1}$. Thus $w \in Q^{-1}Q \cap QQ^{-1}$. This proves one inclusion. Next, consider $w \in Q^{-1}Q \cap QQ^{-1}$. One may verify that $Q^{-1}Q \subset UQ$, and $QQ^{-1} \subset QV$. We have $w = uq$ and $wq' \in Q$ for some $u \in U$ and $q, q' \in Q$. Since $uqq' \in Q$, we have $u \in QQ^{-1}$. Since $u \in QQ^{-1}$ and $QQ^{-1} \subset QV$, we have $u = q''v$ for some $q'' \in Q$ and $v \in V$. Since $Q^{-1}U \subset U$, we have $v \in U$ and thus $w = q''vq \in Q(U \cap V)Q$. Since $Q \subset U \cap V$, this completes the proof of (15.1).

If X is recognizable, let $\varphi : A^* \to M$ be a morphism on a finite monoid M recognizing X. Then each submonoid S_n is generated by a set Z_n recognized by φ. Indeed, it is true for $n = 0$ since $S_0 = X^*$. Arguing by induction, let us suppose that $S_n = Z_n^*$ where Z_n is recognized by φ. Then, by (15.1), we have $S_{n+1} = (U \cap V)^*$ where U, V are recognized by φ. Then the free hull of X is generated by the union of all Z_n, which is also recognized by φ. Therefore it is recognizable.

2.2.7 This is a direct consequence of the closure of the family of recognizable sets by Boolean operations, product and star.

2.2.8 The conditions are obviously necessary. Conversely, let A be the set of elements which cannot be written bc with $b, c \neq 1$. Condition (i) shows that this set generates M. Indeed, if $m = bc$, with $b, c \neq 1$, then $\lambda(b), \lambda(c) < \lambda(m)$, so any m has a decomposition as a finite product of elements in A. Condition (ii) implies that the decomposition is unique. Thus M is isomorphic with A^*.

Section 2.3

2.3.1 We have $(u, v) \in \rho^*$ if and only if there exist $x_1, \ldots, x_n, y_1, \ldots, y_m \in X$ such that $ux_1 \cdots x_n = y_1 \cdots y_m v$ with u prefix of y_1, v suffix of x_n, $x_1 \neq y_1$, $x_n \neq y_m$.

Section 2.4

2.4.1 The fact that X is a code is checked like in Exercise 2.1.1. Let π be a Bernoulli distribution and set $p = \pi(a), q = \pi(b)$. Set $U = \{i + j \mid i \in, j \in j, i + j < n\}$. We have in characteristic series $a^U + a^V = (a^n - 1)/(a - 1)$ and $a^I a^J = a^U + a^n a^V$.

Thus

$$\pi(X) - 1 = \frac{p^I q p^J}{1 - q p^V} + p^n - 1$$

$$= \frac{q p^U + q p^n p^V}{1 - q p^V} + p^n - 1$$

$$= \frac{q p^U + q p^n p^V + p^n - 1 - p^n p^V q + p^V q}{1 - q p^V}$$

$$= \frac{q(p^n - 1)/(p - 1) + p^n - 1}{1 - q p^V} = 0,$$

which shows that X is maximal. Another approach consists in showing directly that X is complete.

2.4.2 We have $f_P(t) = t^2/(1 - t - f_P(t))$. Thus $f_P(t) = (1 - t - \sqrt{1 - 2t - 3t^2})/2$ whence the result.

2.4.3 A word $x \in D_a$ has a factorization $x = au_1 \cdots u_m \bar{a}$ with $u_i \in D$. If u_i is in $D_{\bar{a}}$, then $au_1 \cdots u_{i-1} \bar{a}$ is in D, a contradiction with the fact that D is a prefix code. Thus $D_a \subset a(D \setminus D_{\bar{a}})^* \bar{a}$. The converse inclusion is clear. Finally the products are all unambiguous since D is a code. Since all series $f_{D_a}(t)$ for $a \in A$ are equal, we have

$$f_{D_a}(t) = \frac{t^2}{1 - (2n - 1) f_{D_a}(t)}$$

or equivalently $(2n - 1) f_{D_a}^2 - f_{D_a} + t^2 = 0$ and thus

$$f_{D_a}(t) = \frac{1}{2(2n - 1)} \left(1 - \sqrt{1 - 4(2n - 1)t^2}\right).$$

From $f_D(t) = 2n f_{D_a}(t)$, it follows that

$$f_D(t) = \frac{n}{2n - 1} \left(1 - \sqrt{1 - 4(2n - 1)t^2}\right).$$

The probability generating series of D for the uniform Bernouilli distribution on A is $F_D(t) = f_D(t/(2n))$. Since $1 - \frac{4(2n-1)}{(2n)^2} = \left(\frac{n-1}{n}\right)^2$, we obtain $\pi(D) = F_D(1) = \frac{n}{2n-1}(1 - \frac{n-1}{n}) = \frac{1}{2n-1}$.

2.4.4 It is easy to check that the set Y is a bifix code generating U. Since the generating series of X^* is $f_X^*(t) = \sum_{n \geq 0} f_{n+1} t^n$, the generating series of U is $f_U(t) = \sum_{n \geq 0} f_{n+1}^2 t^n$. On the other hand, $f_Y(t) = t + t^2 + 2t^2/(1 - t)$ whence the identity.

Section 2.5

2.5.1 To check that X is complete, we compute the minimal automaton of X^* shown on Figure 15.1 and deduce that $bA^*b \subset X^*$. If one withdraws an element of X, it is

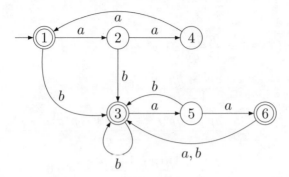

Figure 15.1 The minimal automaton of X^*.

not complete anymore. For example, if a^3 is withdrawn, the word a^4 is not a factor of $\{b, ab, ba^2, aba^2\}^*$, and similarly for the other words of X. Finally, X is not a code since $(b)(aaa)(b) = (baa)(ab)$.

2.5.2 The family $\mathcal{F}$ is closed under arbitrary union and intersection and $\emptyset \in \mathcal{F}$. We may thus consider the topology for which $\mathcal{F}$ is the family of open sets. Let P be dense in the sense that for any $m \in M$, there exist $u, v \in M$ such that $umv \in P$. Then any two-sided ideal has a nonempty intersection with P. Thus P is dense in the sense of the topology and conversely.

2.5.3 The first equality is clear since y is unbordered. The second one results from $V = U \cup X^*$, and thus $Vy = Uy \cup X^*y$. For the last identity, set $Z = y(Uy)^*$. Then $Y = X \cup Z$, and $(X^*y(Uy)^*)^* = (X^*Z)^* = 1 \cup (X^*Z)^*X^*Z = 1 \cup (X \cup Z)^*Z = 1 \cup Y^*Z$. Consequently, $A^* = (Uy)^*(X^*Z)^*V = (Uy)^*V \cup (Uy)^*Y^*ZV$. The fact that Y is a code follows from the equality $\underline{A}^* = \underline{R} + \underline{PY^*Q}$ with $R = (Uy)^*V$, $P = (Uy)^*$ and $Q = y(Uy)^*V$. The fact that Y is complete also follows easily.

2.5.4 Let X be a thin code. If X is complete, then it is maximal and there is nothing to prove. Otherwise we apply the construction of Proposition 2.5.25 to build $Y = X \cup y(Uy)^*$ starting with an unbordered word $y \notin F(X^*)$. Then $y^2 \notin F(Y)$ and thus Y is a thin maximal code containing X.

Section 2.6

2.6.1 Let us first suppose that X is decomposable, that is that $X \subset Y^*$ where Y is a code with $Y \neq A, X$. By Proposition 2.6.4, Y is bifix. We first prove that Y^* is also recognized by ψ. Let us consider $u \in Y^*$ and $v \in A^*$ such that $\psi(u) = \psi(v)$. Let $w \in A^*$ be such that $uw \in X^*$. Since Y is prefix, we also have $w \in Y^*$. Since $\psi(uw) = \psi(vw)$, we have $uw \in X^*$. Thus $u \in Y^*$. This shows that $\psi(Y^*)$ is a subgroup of G containing H and H is not maximal.

Conversely, if H is not maximal, then $H \subset K$, where K is a subgroup with $K \neq H, G$. Let Y be the bifix code such that $Y^* = \psi^{-1}(K)$. Since $X \subset Y^*$ and $Y \subset F(X^*)$, the code X is decomposable over Y.

2.6.2 If X is prefix, there is nothing to prove. Otherwise, one of the two words, say x is prefix of the other. Let $y = xy'$. Reasoning by induction, we may assume that $Z = \{x, y'\}$ is composed of prefix and suffix codes, whence the conclusion for X since $X = Y \circ Z$ with Y suffix.

2.6.3 Suppose that $X \subset Z^*$ with Z a prefix code. Then $a, aba \in X$ imply $ba \in Z^*$. Since $babaab \in X$, this forces $ab \in Z^*$ and finally $b \in Z^*$. Thus $Z = A$. Similarly, one proves that if $X \subset Z^*$ with Z a suffix, then $Z = A$.

The code Y is formed of 11 words:

$$Y = \{a, aba, babaaa, babaaaba, babaab, babaabba,$$

$$(ba)^4, bababb, bababbba, bb, bbba\}.$$

An easy computation shows that if $X \subset Z^*$ with Z prefix, then $Z = \{a, b\}$ and the same conclusion for Z suffix.

To obtain Y as in Exercise 14.1.7, choose $P = \{1, b\}$, $Q = \{1, a, b\}$, $R = \{1, ba\}$ and $w = baba$. The code Z defined by $Z - 1 = P(A - 1)R$ is separating because b is a separating word.

Chapter 3

Section 3.1

3.1.1 If P is infinite, there is at least one letter p_1 which is a prefix of an infinite number of elements of P. Then among this set, there is an infinite number of elements with the same prefix of length 2, and so on.

3.1.2 Indeed $XA^* \cap A^n$ is the disjoint union of the sets $(X \cap A^i)A^{n-i}$ for $1 \le i \le n - 1$. Thus $\mathrm{Card}(XA^* \cap A^n) \le \sum_{i=1}^{n} \alpha_i k^{n-i} \le k^n$. The desired inequality is obtained dividing both sides by k^n, and taking the limit for $n \to \infty$.

Section 3.2

3.2.1 Let $\rho(p) = i \cdot p$. Then ρ is surjective since $\mathcal{A}$ is trim. The identity $\rho(p \cdot a) = \rho(p) \cdot a$ is easy to verify in both cases $pa \in X$ and $pa \in P$. In the first case both sides are equal to i and in the second case, they are both equal to $i \cdot pa$.

3.2.2 (i) $\Longrightarrow$ (ii). By Proposition 3.2.6, $\mathrm{Stab}(i)$ is a right unitary submonoid. Its base, say Y, is a prefix code which is nonempty because $\mathrm{Stab}(i) \ne 1$. Let Z be the set of words defined as follows: $z \in Z$ if and only if $i \cdot z = t$ and $i \cdot z' \ne i$ for all proper nonempty prefixes z' of z. From $t \cdot A = \emptyset$, it follows that Z is a prefix code. Further $Y \cap Z \ne \emptyset$, by $i \ne t$. Finally $X = Y^*Z$. It remains to verify that $V = Y \cup Z$ is prefix. A proper prefix of a word in Z is neither in Z nor in Y, the latter by definition. A proper prefix w of a word y in Y cannot be in Z, since otherwise $i \cdot w = t$ whence $i \cdot y = \emptyset$. Thus V is prefix and X is a chain.

(ii) $\implies$ (iii). Assume that $X = Y^*Z$ with $V = Y \cup Z$ prefix and $Y \cap Z = \emptyset$. Consider a word $u \in Y$. The code V being prefix, we have $u^{-1}Z = \emptyset$. Thus $u^{-1}X = u^{-1}(Y^*Z) = u^{-1}Y^*Z = Y^*Z = X$.

(iii) $\implies$ (i). The automaton $\mathcal{A}(X)$ being minimal, the states of $\mathcal{A}(X)$ are in bijective correspondence with the nonempty sets $v^{-1}X$, where v runs over A^*. The bijection is given by associating the state $i \cdot v$ to $v^{-1}X$. Thus, the equality $u^{-1}X = X$ expresses precisely that $i \cdot u = i$. Consequently $u \in \mathrm{Stab}(i)$.

Section 3.3

3.3.1 Let $\lambda(X) = \min_{x \in X} |x|$. Then λ is clearly a morphism from the monoid of prefix subsets into the additive monoid $\mathbb{N}$. To be able to apply the result of Exercise 2.2.8, we have to prove first that $\lambda^{-1}(0) = 1$. Indeed, $\{1\}$ is the only prefix set containing 1. Next, let $X, Y, Z, T \subset A^*$ be prefix sets such that $XY = ZT$. Suppose that $\lambda(X) \leq \lambda(Z)$. Let $x \in X$ be of minimal length and let $U = x^{-1}Z$. For each $y \in Y$ there are $z \in Z, t \in T$ such that $xy = zt$. Then $z = xu$ and $y = ut$ for some $u \in U$. Thus $Y \subset UT$. Conversely, let $u \in U$ and $t \in T$. Then $xut \in ZT = XY$ hence $ut \in Y$. Thus $Y = UT$ and $XU = Z$. If X and XY are maximal prefix sets and if Y is prefix, then Y is also maximal. Thus the submonoid of maximal prefix sets is right unitary. The submonoid of recognizable prefix sets is also right unitary.

Section 3.4

3.4.1 To prove that L is the set of words w such that $\|w\| = -1$ and $\|u\| \geq 0$ for any proper prefix u of w, we note that it is easy to prove that the condition is necessary, by induction on the length of words in L. Conversely, let w satisfy the condition. If $|w| = 1$, then $w = b$. Otherwise, the first letter of w has to be a. Set $w = aw_1 \cdots w_k$ where $aw_1 \cdots w_i$ is, for $1 \leq i \leq k$, the shortest prefix of w such that $\|aw_1 \cdots w_i\| = k - i - 1$. Then w_i is in L by induction and thus w is in L.

Let w be such that $\|w\| = -1$. Let y be the minimal value of φ on the prefixes of w. Then the conjugate vu of $w = uv$ is in L if and only if u is the shortest prefix of w such that $\|u\| = y$.

A word of L with n letters a has length $n + (k-1)n + 1 = kn + 1$. The number of them is thus $\frac{1}{kn+1}\binom{kn+1}{n}$.

Finally, the map λ from prefix-closed sets on the alphabet $A_k = \{a_1, \ldots, a_k\}$ to $\{a, b\}^*$ which maps $\emptyset$ to b and $P = 1 \cup a_1 P_1 \cup \ldots \cup a_k P_k$ to $a\lambda(P_1) \cdots \lambda(P_k)$ is a bijection from the family of prefix-closed subsets of A_k^* to L such that $|\lambda(P)| = k \operatorname{Card}(P) + 1$.

3.4.2 Since XY is a maximal prefix code, X is right complete and Y is prefix. Let π be a positive Bernoulli distribution. Then $\pi(XY) = 1$ since XY is a maximal prefix code. Since the product XY is unambiguous, we have $\pi(XY) = \pi(X)\pi(Y)$. Thus $\pi(X)\pi(Y) = 1$ for any positive Bernoulli distribution. Let $p = \alpha(X)$ and $q = \alpha(Y)$. Then $\pi(pq) = 1$. Let $a \in A$ be a letter and let $\zeta_a(p)$ be the polynomial in the variables from $A \setminus a$ obtained by the substitution $a \mapsto 1 - \sum_{b \in A \setminus A} b$ in the polynomial p. By Proposition 2.5.29, $\pi(pq) = 1$ implies that $\zeta_a(pq) = 1$. Thus $\zeta_a(p) = \zeta_a(q) = 1$ and thus $\pi(p) = \pi(q) = 1$. Since X is right complete, the set $X' = X \setminus XA^+$ is a maximal prefix code. Since $\pi(X') = \pi(X) = 1$, we have $X = X'$. Thus X is a maximal prefix code. Since Y is prefix with $\pi(Y) = 1$, Y is also a maximal prefix code.

3.4.3 The set $Z = RA \setminus R$ is a prefix code because R is prefix-closed. To prove the formula $Z = (X \cap Q) \cup (X \cap Y) \cup (P \cap Y)$, we use that $X = PA \setminus P$ and $Y = QA \setminus Q$. Thus a word in $RA \setminus R$ is either in $X \cap Y$ or in X but not in Y and thus in $X \cap Q$ or in Y but not in X and thus in $P \cap Y$. If X, Y are maximal, P and Q are the sets of their prefixes. Then R is the set of prefixes of Z which is thus maximal.

3.4.4 The operations obviously preserve the family $\mathcal{F}$ of recognizable maximal prefix codes. To see that it contains all of them, consider an element Z of $\mathcal{F}$. Let $\mathcal{A}$ be the minimal deterministic automaton recognizing $Z \neq A$. We argue by induction on the number of edges in $\mathcal{A}$. We consider two cases. (i) There exists a nonempty word w such that $i \cdot w = i$. In this case, let X be the set of first returns to state i, and let Y be the set of words which are labels of paths from i to a terminal state that do not pass through i inbetween. Then $Z = X^*Y$. Next, $X \cup Y$ is in $\mathcal{F}$ in view of case (ii) below. (ii) Otherwise, let $Z = aX \cup Y$ for $a \in A$ such that $a \notin Z$. Then X and $a \cup Y$ are recognized by automata with strictly less edges than Z and the conclusion follows.

Section 3.5

3.5.1 Let us first assume (i). The code X is semaphore since $A^*X \subset XA^*$. If the property of the minimal set of semaphores $S = X \setminus A^+X$ stated in condition (ii) does not hold, there exist two overlapping words $s, t \in S$, that is such that $s = uv, t = vw$ with nonempty u, v, w. Then $sw = ut$ is in A^*X but not in X^+, a contradiction. Conversely, if X satisfies (ii), consider a word $w \in A^*$ and $x \in X$. Since two occurrences of words in S do not overlap, wx is a product of words in X.

3.5.2 The first inequality is clear since $x \in J^n$, $y \in J^m$ imply $xy \in J^{n+m}$. To see the second one, we observe that if $xy \in J^p$, there exist $u, v \in A^*$ and $n, m \geq 0$ such that $x \in J^n u, uv \in J, y \in vJ^m$, and $p = n + m + 1$. Since $x \in J^n u$ and J is an ideal, one has $x \in J^n$. Similarly for y. Then $n \leq \|x\|, m \leq \|y\|$ and thus $p \leq \|x\| + \|y\| + 1$.

Section 3.6

3.6.1 For any finite maximal prefix code X, there is an integer n such that $A^*a^n \subset X^*a^*$. Since $a \in X$, we have $A^*a^n \subset X^*$, showing that a^n is synchronizing.

3.6.2 Since X is synchronized, there are at least two states $p, q \in Q$ such that $p \cdot w = q \cdot w$ for some word w. If $|w| \geq n^2$, all the pairs $(p \cdot r, q \cdot r)$ for r running through the $|w| + 1 > n^2$ prefixes of w cannot be distinct. Thus there is a factorization of w in $w = rst$ such that $p \cdot r = p \cdot rs$ and $q \cdot r = q \cdot rs$. Then $p \cdot rt = q \cdot rt$ and thus we can choose a shorter w. We can therefore choose a word w_1 of length $\leq n^2$

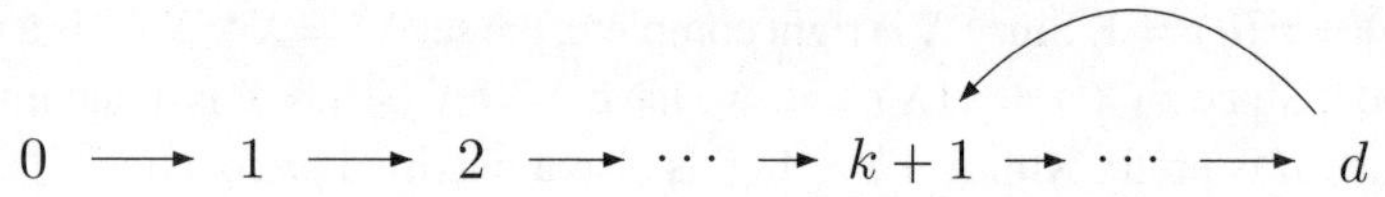

Figure 15.2 The action of $m''a$.

such that $\mathrm{Card}(Q \cdot w_1) \le n - 1$. Next, there is at least one word w_2 of length at most n^2 such that there exist two states $p, q \in Q \cdot w_1$ with $p \cdot w_2 = q \cdot w_2$. Continuing in this way, we obtain a word $w_1 w_2 \cdots$ of length at most n^3 which is synchronizing.

3.6.3 (a) for $m \in M_{d,e}$ and $i \in I_{d+j,e+j}$, we have $i - j \in I_{d,e}$ and $ia^{-j}ma^j = (i - j)ma^j = (i - j)a^j = i$. Thus $a^{-j}ma^j \in M_{d+j,e+j}$.

(b) We have
$$iba^{-1} = \begin{cases} j > i & \text{for } 0 \le i < n - t, \\ i & \text{for } n - t \le i < n. \end{cases}$$

Thus some power w of ba^{-1} is in $M_{n-t,n}$. Then $a^{-t}wa^t \in M_{0,t}$ by (a).

(c) Let $m \in M_{0,d}$ and let j be the least integer such that $jm \not\equiv j \bmod d$. Let $m' = a^{j-d}m$. We have for each $i \in I_{0,d}$, $im' = (i + j - d)m \equiv i + j \bmod d$. Thus $Qm' = I_{0,d}$ and m' is a permutation on $I_{0,d}$. This implies that m' has a power, say m'' which is in $M_{0,d}$. Moreover, since $dm' = km'$ for some $k \ne 0$ in $I_{0,d}$ we have $dm'' = km'' = k$ (that is we have shown that we might have chosen $j = d$). The map $m''a$ defines a cycle $(k + 1 \cdots d)$ and sends every element of $I_{0,d}$ ultimately into this cycle (see Figure 15.2). Thus $m''a =$ has a power in $M_{k+1,d+1}$. This implies by (a) that $M_{0,d-k} \ne \emptyset$ and contradicts the minimality of d.

(d) Arguing by contradiction, let $n = dq + r$ with $q \ge 1$ and $0 < r < d$. The unique element m in $M_{0,d}$ satisfies
$$ia^{n-r}m = \begin{cases} i & \text{for } 0 \le i < r, \\ i - r & \text{for } r \le i < d. \end{cases}$$

Thus some power of $a^{n-r}m$ is in $M_{0,r}$, a contradiction.

(e) Since ba^{-1} fixes each $i \in I_{n-t,n}$, we have $ba^{-1}m \in M_{n-d,n}$ and thus $ba^{-1}m = m$. For each $i \in Q$, we have $iba^{-1}m \equiv iba^{-1} \bmod d$ and $iba^{-1}m \equiv i \bmod d$. Thus $iba^{-1} \equiv i \bmod d$.

3.6.4 Let $\mathcal{A} = (Q, 1, 1)$ be the minimal automaton of X^*. Let $u \ge 1$ be such that $un \ge m$. Then for any $i \ge 0$, we have $1 \cdot a^i ba^{un} \in 1 \cdot a^*$ since a^{un} is not a factor of a word in X by condition (i). Let $j \le n - 1$ be such that $1 \cdot a^i ba^{un} = 1 \cdot a^j$. Then $|y_i| = i + 1 + n - j$. By condition (ii), we have $j \ge i + 1$ with equality if and only if $n - t \le i \le n - 1$. Identifying the state $1 \cdot a^i$ with the element $i \in \mathbb{Z}/n\mathbb{Z}$, we conclude that the maps $\alpha : i \to i + 1$ and $\beta : i \to j$ with $1 \cdot a^j = 1 \cdot a^i ba^{un}$ satisfy the hypotheses of Exercise 3.6.3. Thus, by (d), d divides n and by (e), $i\beta \equiv i + 1$

mod d for all $i \in \mathbb{Z}/n\mathbb{Z}$. This implies that $|y_i| \equiv 0 \mod d$ for $0 \le i \le n - 1$. By (iii), this forces $d = 1$.

3.6.5 We have to show that $Z'^* \subset U$. Let $w \in A^*$ be such that that $uw \in D$. There is a $v \in A^*$ such that $uwv \in X^*$. Since Z' is prefix, we have $wv \in Z'^*$. Since Y^* is right dense, there is some $s \in Z'^*$ such that $wvs \in X^*$. This shows that $w \in D$ and thus that $u^{-1}D \subset D$. Let then $w \in D$. There is some $v \in A^*$ such that $wv \in X^*$. Since $uwv \in Z'^*$ and since Y^* is right dense, there is an $s \in A^*$ such that $uwvs \in X^*$. This shows that $uw \in D$ and it follows that $D \subset u^{-1}D$. We have shown that $w \in U$ and thus that $Z'^* \subset U$.

Section 3.7

3.7.1 We have

$$H(X) - \lambda(X) = \sum_{x \in X} \pi(x) \log_k \frac{k^{-|x|}}{\pi(x)}.$$

Since $\log_k(t) \le (\log_k e)(t - 1)$ for all $t > 0$, we obtain

$$H(X) - \lambda(X) \le (\log_k e)\left(\left(\sum_{x \in X} k^{-|x|}\right) - 1\right) = 0$$

because $\sum_{x \in X} k^{-|x|} = 1$. Since $\log_k(t) < (\log_k e)(t - 1)$ unless $t = 1$ the equality $H(X) = \lambda(X)$ holds if and only if $\pi(x) = k^{-|x|}$ for all $x \in X$. Finally, if X has n elements,

$$H(X) - \log_k n = \sum_{x \in X} \pi(x) \log_k \frac{1}{n\pi(x)} \le (\log_k e)\left(\left(\sum_{x \in X} \frac{1}{n}\right) - 1\right) = 0.$$

Section 3.8

3.8.1 Let $u(z)$ be the generating series of a thin maximal prefix code on k letters. Then condition (i) holds since, by Theorem 2.5.16, we have $\pi(X) = 1$ for any positive Bernoulli distribution. Let w be a word which is not a factor of the words of X and let $p = |w|$. Let P be the set of proper prefixes of X. Then $v(z)$ is the generating series of P. Since no word of P can have w as a suffix, we have $v_{n+p} \le v_n(k^p - 1)$ for all $n \ge 1$. This proves (ii).

Conversely, let us build a maximal prefix code X as in the proof of Theorem 2.4.12 using the following strategy: Fix a letter a in A, and for each $n \ge 1$, choose the words of $X \cap A^n$ among those which have a suffix in a^* of maximal length. To prove that a^{2p} is not a factor of a word of X, it is enough to prove that for each $n \ge 1$, one has

$$v_n \le \sum_{i=1}^{2p} u_{n+i}.$$

Indeed, for each proper prefix q of length n there is a unique exponent $m(q)$ such that $qa^{m(q)}$ is in X. This gives v_n words in X, each of which has length $> n$. In view of the inequality, one may chose an exponent $m(q)$ between $n + 1$ and $n + 2p$ for each prefix q.

To prove the above inequality, we start from $v_{n+p} = v_n k^p - \sum_{i=1}^{p} u_{n+i} k^{p-i}$, which results from the definition of v. Using condition (ii), we obtain

$$v_n \le \sum_{i=1}^{p} u_{n+i} k^{p-i}. \tag{15.2}$$

Hence, using Equation (15.2) with n replaced by $n + p$,

$$v_n k^p - \sum_{i=1}^{p} u_{n+i} k^{p-i} = v_{n+p} \le \sum_{i=1}^{p} u_{n+p+i} k^{p-i}$$

and finally

$$v_n \le \sum_{i=1}^{p} u_{n+i} k^{-i} + \sum_{i=1}^{p} u_{n+p+i} k^{-i} \le \sum_{i=1}^{2p} u_{n+i}.$$

3.8.2 Except for the case where the sequence u_m is ultimately equal to one, we may choose the words of X in such a way that for some integer $n \ge 1$ and letters $a, b \in A$,

 (i) a^n does not appear as a proper factor in the words of X,
 (ii) the prefix code $Y = X \cap (a^* \cup a^* b a^*)$ has the form

$$Y = \{a^n, y_0, y_1, \ldots, y_{n-1}\}$$

where each $y_i = a^i b a^{\lambda_i - i - 1}$ is a word of length λ_i satisfying $i + 1 \le \lambda_i \le n$ and there is an integer t with $0 \le t \le n - 1$ such that $\lambda_i = n$ if and only if $i \ge t$ and finally the numbers λ_i are relatively prime.

Then the code X is synchronized by Exercise 3.6.4.

Finally, if the sequence u_n is ultimately equal to 1, we may choose X of the form $Y \cup a^n a^* b$ where Y is formed of words of length at most n. Then the word $a^n b$ is synchronizing.

Section 3.9

3.9.1 Indeed, (3.34) is equivalent with

$$p^m (1 + p) \le 1 < p^{m-1}(1 + p)$$

or equivalently

$$m \ge -\frac{\log(1 + p)}{\log p} > m - 1.$$

Set $Q = 1 - p^m$. By the choice of m, one has $p^{-1-m} \geq 1/Q > p^{1-m}$. We consider, for $k \geq -1$, the bounded alphabet

$$B_k = \{0, \ldots, k, \ldots, k + m\}.$$

In particular, $B_{-1} = \{0, \ldots, m - 1\}$. We consider on B_k the distribution

$$\pi(i) = \begin{cases} p^i q & \text{for } 0 \leq i \leq k, \\ p^i q/Q & \text{for } k < i \leq k + m. \end{cases}$$

Clearly $\pi(i) > \pi(k)$ for $i < k$ and $\pi(k + i) > \pi(k + m)$ for $1 < i < m$. Observe that also $\pi(i) > \pi(k + m)$ for $i < k$ since $\pi(k + m) = p^{k+m}q/Q \leq p^{k+m}q/p^{m+1} = \pi(k - 1)$. Also $\pi(k + i) > \pi(k)$ for $1 < i < m$ since indeed $\pi(k + i) > \pi(k + m - 1) = p^{k+m-1}q/Q > p^k q = \pi(k)$. As a consequence, the symbols k and $k + m$ are those of minimal weight. Huffman's algorithm replaces them with a new symbol, say k' which is the root of a tree with say left child k and right child $k + m$. The weight of k' is

$$\pi(k') = \pi(k) + \pi(k + m) = p^k q(1 + p^m/Q) = p^k q/Q.$$

Thus we may identify $B_k \setminus \{k, k + m\} \cup \{k'\}$ with B_{k-1} by assigning to k the new value $\pi(k) = p^k q/Q$. We get for B_{k-1} the same properties as for B_k and we may iterate.

After m iterations, we have replaced B_k by B_{k-m}, and each of the symbols $k - m + 1, \ldots, k$ now is the root of a tree with two children. Assume now that $k = (h + 1)m - 1$ for some h. Then after hm steps, one gets the alphabet $B_{-1} = \{0, \ldots, m - 1\}$, and each of the symbols i in B_{-1} is the root of a binary tree of height h composed of a unique right path of length h, and at each level one left child $i + m, i + 2m, \ldots, i + (h - 1)m$. This corresponds to the code $P_h = \{0, 10, \ldots, 1^{h-1}0, 1^h\}$. The weights of the symbols in B_{-1} are decreasing, and moreover $\pi(m - 2) + \pi(m - 1) > \pi(0)$ because $p^{m-2} + p^{m-1} > 1$. The optimal binary tree corresponding to such a sequence of weights has the heights of its leaves differing at most by one, as can be checked by induction on m. This shows that the code R_m is optimal for this probability distribution.

Thus we have shown that the application of Huffman's algorithm to the truncated source produces the code $R_m P_k$. When h tends to infinity, the sequence of codes converges to $R_m 1^* 0$. Since each of the codes in the sequence is optimal, the code $R_m 1^* 0$ is an optimal prefix code for the exponential distribution. The Golomb code $G_m = 1^* 0 R_m$ has the same length distribution and so is also optimal.

3.9.2 Consider a complete prefix code X_1 built by the algorithm. Assume it is not optimal, and consider a complete prefix tree X_2 which is optimal and which is closest to X_2 in the sense that the number of common elements of $X_1 \cup X_1 A^-$ and of $X_2 \cup X_2 A^-$ is maximal. There is a word x_1 in X_1 which is a proper prefix of a word in X_2. Otherwise every word in X_1 which is not in X_2 has a prefix which is in X_2, but then $\text{Card}(X_2) > \text{Card}(X_1)$. Symmetrically, there is a word x_2 in X_2 which is a proper prefix of a word in X_1.

Let p be a word that has x_1 as a prefix and such that $pa \in X_2$ for all $a \in A$. Since x_2 is a proper prefix of a of a word in X_1 and x_1 is a word of X_1, one has $c(x_2) \leq c(x_1)$. Next, $c(x_1) \leq c(p)$. Thus $c(x_2) \leq c(p)$. Let $X_3 = X_2 \setminus (pA \cup x_2) \cup p \cup x_2 A$. The difference of costs is

$$C_{X_3} - C_{X_2} = \sum_{a \in A} c(x_2 A)_c(x_2) + c(p) - \sum_{a \in A} c(pA) = (k-1)(c(x_2) - c(p)) \leq 0.$$

Thus X_3 is optimal and clearly, X_3 is closer to X_1 than X_2.

Chapter 4

Section 4.1

4.1.1 If M is recognizable and free, let X be the code such that $M = X^*$. Since $X = (M \setminus 1) \setminus (M \setminus 1)^2$, X is recognizable. Let $\mathcal{A}$ be a deterministic finite automaton recognizing X. Then the automaton $\mathcal{A}^* = (Q, 1, 1)$ is finite, trim and, by Proposition 1.10.5, it is an unambiguous automaton recognizing X^*. Conversely, let $\mathcal{A} = (Q, 1, 1)$ be an unambiguous trim finite automaton. The set M recognized by $\mathcal{A}$ is recognizable submonoid. By Proposition 4.1.5, M is free.

Section 4.2

4.2.1 The proof is the same as that of Proposition 4.2.3.

4.2.2 Any path $j \overset{w}{\to} q$ in $\mathcal{B}$ can be lifted to a path $j \overset{w}{\to} p$ in $\mathcal{A}$ such that $\rho(p) = q$. Thus such a path is unique.

Chapter 5

Section 5.1

5.1.1 The deciphering delay of a code X is infinite if and only if there is an infinite word that has two disjoint factorizations. This is equivalent to the existence of an infinite path in G_X. In the case X is finite, this is equivalent to the existence of a cycle accessible from some vertex in X.

5.1.2 (a) is straightforward.

(b) If the path e is empty ($n = 0$), then $s = t$, form (ii) holds and there is no crossing edge, so $c = 0$. Assume that for some n the form (i) holds and that c is odd. Let $e_{n+1} = (t, u)$ be a crossing edge. Setting $z = tu$, one has $z \in X$ and and one gets $sy_1 \cdots y_\ell z = x_1 \cdots x_k u$, so form (ii) is obtained and the number of crossing edges is now even. The same argument is valid when one starts with form (ii). This proves the hint.

The previous argument shows that all occurrences of crossing edges which are even contribute to $y_1 \cdots y_\ell$, and the other crossing edges to $x_1 \cdots x_k$. So the claim holds for crossing edges. It suffices to observe that the extending edges have the same parity as the closest preceding crossing edge.

(c) The graph having no cycle, the computation can be carried out bottom up from vertices without successors to vertices in X. For each vertex s, we maintain the pairs (ℓ, r) corresponding to paths of form (i) and (ii), and with maximal values: so there are four pairs for each vertex.

For a vertex without successor there is only the pair $(0, 0)$, and for other vertices u a computation of maxima is carried out for all edges (u, s). This gives the corresponding values in time proportional to the number of outgoing edges. For each $x \in X$, the deciphering delay is derived from these pairs according to (a).

5.1.3 Let $x \in X^*$, $y \in X^{d(Y)}$, $z \in X^{d(Z)}$ and $v \in A^*$ be such that $xyzv \in X^*$. Since $z \in Z^*$ and $|z|_Z \geq |z|_X$, we have $z \in S(Z)$, where $S(Z)$ is the set of simplifying words for Z, and so $zv \in Z^*$. Since y, viewed as a word on the alphabet of Y is in $S(Y)$, and since $zv \in Z^*$, we have $yzv \in X^*$. This proves that $yz \in S(X)$.

5.1.4 We prove the property by induction on $|x| + |y|$. If X is not prefix, we have, supposing that $|y| > |x|$, $y = xy'$. Then $X = Y \circ Z$ with $Z = \{x, y'\}$. Since Y and Z are two-element codes, they have finite deciphering delay by induction hypothesis. Thus, X also by the previous exercise.

5.1.5 (a) The code X being finite, there is only a finite number of codes T such that X decomposes over T. The smallest submonoid M generated by a code with finite deciphering delay such that $X^* \subset M$ is the intersection of the (finitely many) submonoids T^* containing X generated by a code T with finite deciphering delay.

It suffices to show that if Y, Z have finite deciphering delay, then $Y^* \cap Z^*$ is also generated by a code with finite deciphering delay. Indeed, let T be the code such that $T^* = Y^* \cap Z^*$. Then $S(Y) \cap S(Z) \subset S(T)$. If d is greater than the delays of Y and of Z, then $T^d \subset S(Y) \cap S(Z)$, and so T has delay d.

(b) Assume for instance that Y is not a subset of $X(Y^*)^{-1}$. There is $y \in Y$ which does not appear as the first factor of a factorization of a word in X as a product of words in Y. Set $Z = (Y \setminus y)y^*$. Then Z has finite deciphering delay, and moreover $X \subset Z^*$ and Z^* is strictly contained in Y^*.

Finally, assume that X does not have finite deciphering delay. Consider words $x \neq x'$, $y \in X^d$ and u such that $xyu \in x'X^*$. If d is greater than the deciphering delay of Y, then the Y-factorizations of x and x' start with the same word in Y. Thus the conclusion follows.

5.1.6 Let $Y = X^d$. Consider $x_1, \ldots, x_d, x_1', \ldots, x_d' \in X$, $y \in X^d$ and $u \in A^*$ such that $x_1 \cdots x_d yu \in x_1' \cdots x_d' Y^*$. If X has delay d, we have successively $x_1 = x_1'$, $x_2 = x_2'$, and finally $x_d = x_d'$. Thus $x_1 \cdots x_d = x_1' \cdots x_d'$, which shows that Y has delay 1. Conversely, suppose that Y has delay 1. Let $x, x' \in X$, $y \in X^d$ and $u \in A^*$ be such

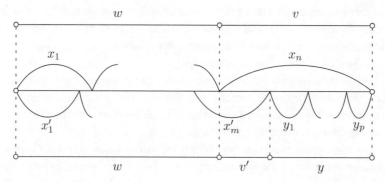

Figure 15.3 Factorization of $wv = wv'y$.

that $xyu \in x'X^*$. Then $x^d y$ is a prefix of a word of $x^{d-1}x'Y^*$ and thus $x^d = x^{d-1}x'$, whence $x = x'$.

5.1.7 Let us show first the inclusion $S(X) \subset E(X)$. Let $s \in S(X)$, $p \in E(X)$. Note that $pt \in X^*$ for some word t and that pt still is strongly right completable. Thus, we may assume that $p \in E(X) \cap X^*$. Consider any word $u \in A^*$. Since $p \in E(X)$, the word psu can be completed: there is a word $v \in A^+$ such that $psuv \in X^*$. But p is in X^* and s is simplifying. Thus, $suv \in X^*$, showing that s is strongly right completable.

Conversely, let $s \in S(X)$, $p \in E(X)$. To show that p is simplifying, let $x \in X^*$, $v \in A^*$ such that $xpv \in X^*$. Since the word pvs is right completable, we have $pvsw \in X^*$ for some $w \in A^*$. But then $xpvsw \in X^*$ also and since s is simplifying, we have $sw \in X^*$. Thus, finally, the four words x, $x(pv)$, $(pv)(sw)$, and sw are in X^*. The set X^* is stable, thus $pv \in X^*$. This shows that p is simplifying.

5.1.8 We first verify the following property (∗): if $vuz = v'u'$ for $v, v' \in C_r(w)$, $u, u' \in U$, and $z \in A^*$, then $v = v', u = u', z = 1$.

Indeed, first note that $u \in E(X)$. Thus, there exists $t \in A^*$ such that $uzt \in X^*$. Then

$$(wv)(uzt) = (wv')(u't). \qquad (15.3)$$

Each one of the first three parenthesized words is in X^*. Now the fourth word, namely $u't$, is also in X^*, because u' is simplifying. The set X being a code, we have $v = v'y$ or $v' = vy$ for some $y \in X^*$. This implies that $v = v'$ as follows: assume, for instance, that $v = v'y$, and set $wv = x_1 x_2 \cdots x_n$, $wv' = x_1' \cdots x_m'$, $y = y_1 \cdots y_p$, with $x_1, \ldots, x_n, x_1', \ldots, x_m', y_1 \ldots, y_p \in X$. Then $|x_n| > |v|$ and assuming $p > 0$, we have on the one hand (see Figure 15.3)

$$|y_p| \le |y| \le |v| < |x_n|,$$

and on the other hand, since $x_1 x_2 \cdots x_n = x'_1 \cdots x'_m y_1 \cdots y_p$, we have $x_n = y_p$. Thus $p = 0$, $y = 1$, and $v = v'$. Going back to (15.3), this gives $uz = u'$. Now U is prefix. Consequently $z = 1$ and $u = u'$. This proves property $(*)$.

It follows immediately from $(*)$ that $C_r(w)U$ is prefix, and also, taking $z = 1$, that the product $C_r(w)U$ is unambiguous. This proves 1 and 2. To prove 3, consider a word $t \in A^*$. The word wt is right completable, since $w \in E(X)$. Thus, $wtt' \in X^*$ for some $t' \in A^*$. Thus, tt' is in $w^{-1}X^*$. Consequently $tt' = vy$ for some $v \in C_r(w)$, $y \in X^*$. Now observe that $w \in E(X)$, and consequently also $yw \in E(X)$. Thus, $tt'w = vyw \in C_r(w)S(X)$. This shows that $C_r(w)S(X)$ is right dense. From $C_r(w)S(X) = C_r(w)UA^*$ it follows then by Proposition 3.3.3 that the prefix set $C_r(w)U$ is maximal prefix.

5.1.9 Let X be a maximal finite code with deciphering delay d. According to Propositions 5.1.5 and 5.2.3, both $S(X)$ and $E(X)$ are nonempty. Thus by Exercise 5.1.7, they are equal. Set $S = S(X) = E(X)$. Then $X^d \subset S$, further S is a right ideal, and the prefix set $U = S \setminus SA^+$ satisfies $S = UA^*$. We claim that U is a finite set. Indeed, set $\delta = d \max_{x \in X} |x|$ and let us verify that a word in U has length $\le \delta$. For this, let $s \in S$ with $|s| > \delta$. The word s being strongly right completable, there is a word $w \in A^*$ such that $sw \in X^*$. By the choice of δ, the word sw is a product of at least $d + 1$ words in X, and s has a proper left factor, say s', in X^d. From $X^d \subset S$, we have $s \in SA^+$. Thus, $s \notin U$. This proves the claim.

Now, fix a word $x \in X^d$, and consider the set $C_r(x)$ of right contexts of x. The set $C_r(x)$ is finite since each element of $C_r(x)$ is a right factor of some word in the finite set X.

By Exercise 5.1.8, the set $Z = C_r(x)U$ is a maximal prefix set, since $x \in X^d \subset S$. Further, Z is the unambiguous product of the finite sets $C_r(x)$ and U. By Exercise 3.4.2, both $C_r(x)$ and U are maximal prefix sets. Since $1 \in C_r(x)$, we have $C_r(x) = \{1\}$.

Thus, we have shown that $C_r(x) = \{1\}$ for $x \in X^d$. This implies as follows that X is prefix. Assume that $y, y' \in X$ and $yt = y'$ for some $t \in A^*$. Let $x = y^d$. Then $xt = y^d t = y^{d-1}y'$ and $|t| < |y'|$ show that $t \in C_r(x)$. Since $x \in X^d$, we have $t = 1$. Thus, X is a prefix code.

5.1.10 We first show that P is thin proving that for each $p \in P$ and $a \in A$, the word pa cannot be a factor of P. Indeed, if $upav \in P$, then up is also in P, a contradiction. Next, by Lemma 5.2.12, we have $S \subset \bigcup_{i=1}^{d-1} X^i P$, and thus S is thin. Since $R \subset XS$, we also have that R is thin. Finally, let us show that S^* is thin. Otherwise, since S is prefix by Lemma 5.2.15, S would be a maximal prefix code. Any element of R would then be comparable for the prefix order with an element of S, a contradiction with Lemma 5.2.16(i).

Section 5.3

5.3.1 It is clear that if $\mathcal{A}$ is a (d, d')-complete automaton with bidelay (d, d'), then with the pairs (U_p, V_p) chosen as indicated and the sets (U_e, V_e) defined by the compatibility conditions 2 and 4, the result satisfies conditions 1 and 3 and thus is an extended automaton without boundary edges. Conversely, we show that in an extended automaton with delay (d, d') without boundary edges, for $0 \le k \le d' + 1$, the set of labels of paths of length $\le k$ starting at p (resp. ending at q) is the set of prefixes of $V_p A$ (resp. $A U_q$) of length $\le k$. We prove the first alternative. The other one is symmetrical. The statement is true for $k = 0$. Assume that it holds for $k \le d'$. Let $p \xrightarrow{a} q \xrightarrow{u}$ be a path of length $\le k + 1$ with $a \in A$. Then, by induction hypothesis, u is a prefix of $V_q A$ and thus of V_q. By condition 1, au is a prefix of $V_p A$. This proves the property for $k + 1$ in one direction (observe that we did not use the hypothesis that there are no boundary edges). Conversely, if au is a prefix of $V_p A$, by the compatibility condition 1, there is an edge $e \in F(p)$ such that $a = \lambda(e)$ and $u \in V_e$. Since is not a boundary edge, we have $e = (p, a, q)$ for some state q. By condition 4, $u \in V_q$. By the induction hypothesis, there is a path $q \xrightarrow{u}$, hence a path $p \xrightarrow{au}$. Thus the property holds for $k + 1$ and the statement is proved by induction on k.

5.3.2 According to conditions 1 and 2, we have

$$\sum_{p \in Q} \underline{U_p} \underline{V_p} A = \sum_{e \in E_+} \underline{U_e} \lambda(e) \underline{V_e},$$

where E_+ is the set of edges which have an origin (that is which are not backward boundary edges). Similarly, $\sum_{p \in Q} A \underline{U_p} \underline{V_p} = \sum_{e \in E_-} \underline{U_e} \lambda(e) \underline{V_e}$ where E_- is the set of edges which have an end. This proves the formula.

5.3.3 The automaton $\mathcal{A}_0$ is clearly a (d, d')-complete automaton with bidelay (d, d') and thus an extended automaton (without boundary edges). For all $u \in A^d$ and $v \in A^{d'}$, there is a path $p \xrightarrow{u} q \xrightarrow{v} r$ in $\mathcal{A}_0$ if and only if $q = uv$.

It is not difficult to verify that $\mathcal{A}_{-x}$ and $\mathcal{A}_x$ still satisfy the four conditions defining extended automata. In $\mathcal{A}_{-x}$, the set of forward boundary edges is Ax and the set of backward boundary edges is xA. Thus $\sum_{e \in E} \partial(e) = \underline{A}x - x\underline{A} = -f_x$. The forward boundary edges of $\mathcal{A}_x$ are the backward boundary edges of $\mathcal{A}_{-x}$ and vice versa. This proves the last formula.

5.3.4 Suppose that e is a forward boundary edge from state p with label a such that U_e or V_e is not a singleton. We add a terminal state q to e with $U_q = A^- U_e a$ and $V_q = V_e$. For every word $w = a_1 \cdots a_{d'} a_{d'+1} \in V_e A$, we add a forward boundary edge e_w starting at q with label a_1, and with $U_{e_w} = A^- U_e a$, $V_{e_w} = \{a_2 \cdots a_{d'+1}\}$. In addition, for every word $w = a_1 \cdots a_{d+1}$ in $A(A^- U_e a)$ which is not in $U_e a$, we add a backward boundary edge e'_w ending at q with label a_{d+1} and with $U_{e'_w} = \{a_1 \cdots a_d\}$, $V_{e'_w} = V_e$. Iterating this transformation a finite number of times, we obtain an extended automaton in which all boundary edges are simple.

5.3.5 By Exercise 5.3.4 we may suppose that the extended automaton $\mathcal{A}$ is such that all boundary edges are simple. By Exercise 5.3.2, we have $\sum_{e \in E} \partial(e) \in \mathcal{L}$. Let us write

$$\sum_{e \in E} \partial(e) = \sum b_x f_x,$$

where the coefficients b_x are integers.

For each $x \in A^{d+d'}$ such that $b_x > 0$ (resp. $b_x < 0$), we add to the automaton $\mathcal{A}$ the disjoint union of b_x copies of $\mathcal{A}_{-x}$ (resp. $\mathcal{A}_x$). The resulting extended automaton $\bar{\mathcal{A}}$ is now such that $\sum \partial(e) = 0$. Each boundary edge e of $\bar{\mathcal{A}}$ is simple and thus $\partial(e) \in A^{d+d'+1}$. Thus, for each word $w \in A^{d+d'}$ we may define a bijection $\tau_w : \{e \in \bar{E} \mid \partial(e) = w\} \to \{e \in \bar{E} \mid \partial(e) = -w\}$. We now identify each forward boundary edge of $\bar{\mathcal{A}}$ with the backward boundary edge $\tau_w(e)$ where $w = \partial(e)$. The resulting extended automaton has no boundary edges.

5.3.6 For each state q, define U_q as the set of labels of paths of length d ending at q and V_q as the set of labels of paths of length d' starting at q. For each edge e from p to q, set $U_e = U_p$ and $V_e = V_q$. Since $\mathcal{A}$ has (right) delay d', for each state $q \in Q$, the sets aV_e for each edge e starting at q, with a the label of e, are disjoint. Thus we may attach forward boundary edges to state q to complete a partition of $V_q A$ as follows. For each $w = a_1 \cdots a_{d'+1} \in V_q A$ which is not in any of the sets aV_e, we define a boundary edge e with origin q and label a_1 with $U_e = U_q$ and $V_e = \{a_2 \ldots a_{d'+1}\}$. In a completely symmetric fashion, we attach backward boundary edges to each state q in order that the family of sets $U_e a$ is a partition of the set AU_q.

Thus we obtain, by adding boundary edges, an extended automaton $\mathcal{B}$ containing $\mathcal{A}$. By Exercise 5.3.5, there is an extended automaton $\mathcal{C}$ without boundary edges such that every edge of $\mathcal{A}$ is an edge of $\mathcal{C}$. Since $\mathcal{C}$ is (d, d')-complete, the stabilizer of 1 is generated by a code Y with bidelay (d, d') containing X.

5.3.7 We first add boundary edges as indicated on Figure 15.4 on the left (for each boundary edge e, we indicate the pair (U_e, V_e)). We have then $\sum_{e \in E} \partial(e) = abb - bba = -f_{bb}$. We thus add the automaton $\mathcal{A}_{bb}$ represented on the right in Figure 15.4. Merging the boundary edges by pairs which are compatible, we obtain the automaton on the right in Figure 5.18.

Chapter 6

Section 6.1

6.1.1 Let U be the set of parses of u. If $(L, u) = (L, uvu)$, then for each $(p, x, s) \in U$, there exists $(p', x', s') \in U$ such that $svp' \in X^*$ and conversely. Otherwise, there would be more parses for uvu than for u. This implies that $(L, (uv)^m u) = (L, u)$ for all $m \geq 0$.

6.1.2 Let $\mathcal{A} = (Q, 1, 1)$ be the minimal deterministic automaton of X^*. Suppose first that $\mathcal{A}$ is bideterministic. Let $t, u, v, w \in A^*$ be such that $tu, vu, vw \in X$. Then

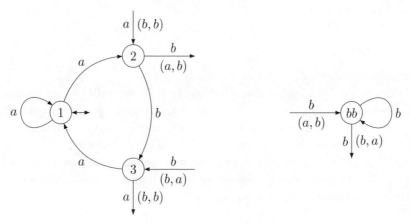

Figure 15.4 The construction of an extended automaton with delay $(1, 1)$.

$1 \cdot tu = 1$ and $1 \cdot vu = 1$ imply that $1 \cdot t = 1 \cdot v$. Since $1 \cdot vw = 1$, we obtain $1 \cdot tw = 1$. Thus $tw \in X^*$. This implies that tw has a prefix in X. Since t is a prefix of X, we have $w = w'w''$ with $tw' \in X$. For the same reason, we obtain $vw' \in X^*$ and thus $w = w'$. This proves that (ii) holds.

Next, if (ii) holds, consider $x \in H \cap A^*$. Then $x = h_1^{\epsilon_1} h_2^{\epsilon_2} \cdots h_n^{\epsilon_n}$ with $h_i \in X$ and $\epsilon_i = \pm 1$. Since $x \in A^*$, the words $h_i^{\epsilon_i}$ such that $\epsilon_i = -1$ cancel with their neighbors. Since X is bifix, h_i^{-1} cannot cancel completely with h_{i-1} or with h_{i+1}. This, if $\epsilon_i = -1$, we have $\epsilon_{i-1} = 1$, $\epsilon_{i+1} = 1$ and $h_{i-1} = tu$, $h_i = vu$, $h_{i+1} = vw$ for $t, u, v, w \in A^*$. But then $h_{i-1} h_i^{-1} h_{i+1} = tw$ is in X by (ii). This shows that $x \in X^*$. Thus (iii) holds.

Suppose finally that $H \cap A^* = X^*$. Let $p, q \in Q$ and $a \in A$ be such that $p \cdot a = q \cdot a$. Let $u, v \in A^*$ be such that $1 \cdot u = p$ and $1 \cdot v = q$. Let $w \in A^*$ be such that $p \cdot aw = q \cdot aw = 1$ in such a way that $uaw, vaw \in X^*$. Suppose that $p \cdot ax = 1$. Then $uax \in X^*$ and thus $vaw(uaw)^{-1}uax \in H$. Since $vaw(uaw)^{-1}uax = vax \in A^*$, the hypothesis implies that $vax \in X^*$ and thus $q \cdot ax = 1$. This shows that $p = q$. Thus $\mathcal{A}$ is bideterministic.

6.1.3   The definition of w being symmetrical, it is enough to show that w can be decoded from left to right. By construction, x_1 is a prefix of w and the first codeword can therefore be decoded with delay at most ℓ. But this also identifies the prefix of length $\ell + |x_1|$ of the second term of the right side of (6.57). Adding this prefix to the corresponding prefix of w gives a word beginning with $x_1 x_2$ and thus identifies x_2, and so on.

Section 6.2

6.2.1   (a) The existence of k follows from the fact that $w \in \bar{F}(X)$ since then $a_i \cdots a_n w \in X A^*$ for each $i \in \{1, \ldots, n\}$.

(b) If X is suffix, then clearly, ρ_w is injective. Conversely, if $v, uv \in X$, then the map ρ_w is not injective for any $w \in \bar{F}(X)$ with uv as a suffix. This proves assertion (b). The proof of (c) is similar.

(d) The proof results from the fact that a map of a finite set into itself is injective if and only if it is surjective.

6.2.2 (a) Set $X = P \setminus PA^+$. We prove that $X^* = P^*$. Let $x, y \in A^*$ be such that $x \in X$, $xy \in P^*$. We have $x = u\tilde{u}$, $xy = v\tilde{v}$. If $|x| \leq |v|$, then $v = xw$ and $xy = u\tilde{u}w\tilde{w}u\tilde{u}$. Thus $y \in P^*$. Otherwise, $x = vw$ and $\tilde{v} = wy$. Then, $x = \tilde{y}\tilde{w}w$ and thus $\tilde{x} = \tilde{w}wy$. Since $x = \tilde{x}$, this forces $y = 1$. This proves that P^* is right unitary. The proof that it is left unitary is symmetric.

(b) For each $u \in A^*$, $u\tilde{u}$ and $\tilde{u}u$ are in P.

6.2.3 If X is recognizable, then the sets G, D, G_0, D_0 are recognizable and thus also Y given by $Y = (X \cup w \cup G_1(wD_0)^*D_1) \setminus (Gw \cup wD)$. Conversely, $X = Y \setminus (w \cup G_1(xD_0)^*D_1) \cup Gw \cup wD$, and if Y is recognizable, then X is also recognizable.

6.2.4 By Exercise 6.1.2, the condition is satisfied if and only if the minimal deterministic automaton of X^* is bideterministic. Since X is maximal, the automaton is complete and the result follows.

Section 6.3

6.3.1 It is clear that each set Y_i is maximal prefix. They are disjoint because if $y \in Y_i \cap Y_j$ one of $p_i y$, $p_j y \in X$ is a suffix of the other. Any suffix s of X is in some Y_i since $ws \in A^*X$. This shows that S is the disjoint union of the sets Y_i.

6.3.2 The existence follows from Theorem 6.3.15 since the decomposition built in the proof satisfies this property. The uniqueness follows from the fact that a suffix s is in Y_i if and only if it has $i - 1$ proper prefixes which are in S.

Section 6.4

6.4.1 We may suppose that X is not maximal. Since, X is finite, $\mu(X) = \max\{(L_X, x) \mid x \in X\}$ is finite. By Theorem 6.4.3, for each $d \geq \mu(X) + 1$, X is the kernel of a maximal bifix code Z of degree d (which is unique by Theorem 6.4.2). Let us show that Z is recognizable. For a word w, we denote by $c(w)$ the pair (i, s) formed by the integer $i = (L_X, w)$ and the word s which is the longest suffix of w which is a prefix of X. It can be verified that $c(w) = c(w')$ implies $w^{-1}Z = w'^{-1}Z$. The number of possible pairs $c(w)$ is finite, and thus Z is recognizable.

6.4.2 By Proposition 6.3.14, the set P' of proper prefixes of the derived code is $P \cap H$. When X is recognizable, so are $P = XA^-$ and $H = A^-XA^-$. Thus P' is recognizable and so is $X' = P'A \setminus P'$.

6.4.3 If $|x| < |s|$, then x is in the kernel of X and so is in X'. Otherwise, let $s = ua$ with $a \in A$. Then $s \notin H = A^-XA^-$ since otherwise s would not be the longest prefix of w which is a proper suffix of X. Thus $s \in (HA \setminus H) \cap (AH \setminus H)$ which is contained in X' by Proposition 6.4.4.

6.4.4 The code Z is clearly (by Exercise 3.4.14) a thin maximal prefix code. To see that it is also suffix, suppose that a word of $X_1 \cap X_2A^-$ is a suffix of a word

Table 15.1 *The values of* $\lambda(k, d)$.

	1	2	3	4	5
2	1	2	4	8	22
3	1	2	5		
4	1	2	6		
5	1	2	7		

of $X_2 \cap X_1 A^-$. Then it belongs to the kernel of X_1, which the same as that of X_2, a contradiction. If X_1, X_2 are finite and have also the same degree d, then, by Proposition 6.5.1, a^d is in $X_1 \cap X_2$ for any letter $a \in A$. Thus a^d is also in Z. This implies that the degree of Z is also equal to d. But the degree of a finite maximal bifix code is also equal to its average length with respect to any positive Bernoulli distribution (Proposition 6.3.16). Since Z is formed of prefixes of the words of X_1 and X_2, this forces $Z = X_1 = X_2$.

6.4.5 Consider $X = a \cup ba^*b$ which is a maximal bifix code of degree 2 with kernel $\{a\}$. Let Y be the set of words formed of a and the words of the form $ba^i b$ for all integers $i \geq 0$ which are powers of 2. By Theorem 6.4.6, since $\{a\} \subset K \subsetneq X$, there exists a unique maximal bifix code Z of degree 3 such that $K(Z) = Y$. Moreover, X is the derived code of Z. Finally, Z is not rational since otherwise $Y = X \cap Z$ would be rational.

Section 6.5

6.5.1 Suppose $|p| < |r|$. Since $pwq = rws$ is chosen of maximal length, there is a prefix q' of q such that $rwq' \in X$. Thus $wq' \in H(X) \cap S$ and $wq' \in S'$ by Proposition 6.3.14 (3). This implies $w \in H(X')$.

6.5.2 Let $x = aub \in X$ with $a, b \in A$. If a word w of length $\ell(X') - 1$ has two occurrences in u, then $w \in H(X')$ by the previous exercise, which is impossible because the words in $H(X')$ have length at most $\ell(X') - 2$. Thus each word of length $\ell(X') - 1$ has at most one occurrence in u, whence $|u| \leq \ell(X') - 1 + k^{\ell(X')-1} - 1$ and finally $|x| \leq \ell(X') + k^{\ell(X')-1}$. The second formula follows directly. Some values of $\lambda(k, d)$ are given in Table 15.1.

For $d = 3$, the formula gives the exact value. Actually $\lambda(k, 2) = 2$ and one may verify that $\lambda(k, 3) = k + 2$. For $k = 4$, one has $\lambda(2, 4) = 8$ but the bound given by the formula is $\lambda(2, 4) \leq 12$.

6.5.3 The function φ is injective because X is suffix and therefore also surjective (the latter is also a consequence of the fact that X is maximal suffix).

6.5.4 For each finite maximal bifix code X of degree d, AX and XA are finite maximal bifix codes of degree $d + 1$. Since $AX \neq XA$ unless $X = A^d$, we obtain $\beta_k(d + 1) \geq 2\beta_k(d) - 1$. Since $\beta(k, 3) \geq 2$ for $k \geq 2$, the conclusion follows. Some values of $\beta_k(d)$ are represented in Table 15.2.

Table 15.2 *The values of $\beta_k(d)$.*

	1	2	3	4	5
2	1	1	3	73	50 567 83
3	1	1	25		
4	1	1	543		
5	1	1	29 281		

Table 15.3 *The 3 finite maximal binary bifix codes of degree 3.*

	kernel	length distribution	symmetry class
1	Ø	0 0 8	1
2	ab	0 1 4 4	2

6.5.5 A word of length α_n has two non overlapping factors of length α_n which are equal. Thus it has a factor of the form uvu where u is of length α_n. The claim follows by induction.

6.5.6 Let us suppose that X contains a word x of length $\alpha_{d-1} + 2$. By the previous exercise, x contains an internal factor which is a quasipower of order $d - 1$. Since, by Exercise 1.1, $(L, uvu) > (L, u)$ for any internal factor uvu with $u \neq 1$, we obtain $(L, x) > d$ which is impossible. The bound is less accurate than the one given by Exercise 6.5.2.

6.5.7 We will describe the 73 finite maximal binary bifix codes of degree 4 according to their derived code. The 3 finite maximal binary bifix codes of degree 3 are given by Table 15.3. The table is made of 3 columns describing the code. The first one gives the kernel of the code, the second one its length distribution. The third column gives the number of codes obtained by the symmetries consisting either in the exchange of the letters a, b or the reversal of words. There can be either 1, 2 or 4 such symmetrical codes. In this way we reduce the number of codes to be listed and and we list only one representative of each symmetry class, the third column giving the number of elements of the class. For example, there is just one code with empty internal part, namely A^3. There is one code with kernel $\{ab\}$ and one with kernel $\{ba\}$. The symmetry class has two elements, in correspondence with the fact that ab and ba are both obtained one from the other by reversal or exchange of a, b.

There are 39 bifix codes with derived code A^3 listed on Table 15.4. We may observe that the length distribution can be read from the internal part as follows. The fact that the code X on line 5 has 4 words of length 6 corresponds to the fact that the internal words aab and aba overlap on ab. Thus, $aaba$ is an internal factor of X and $\{a, b\}aaba\{a, b\} \subset X$.

Table 15.4 *The 39 finite maximal binary bifix codes of degree 4 with derived code* A^3.

	kernel	length distribution	symmetry class
0	∅	0 0 0 16	1
1	*aab*	0 0 1 12 4	4
2	*bab*	0 0 1 12 4	2
3	*aab, bab*	0 0 2 8 8	4
4	*aab, bba*	0 0 2 8 8	2
5	*aab, aba*	0 0 2 9 4 4	4
6	*aab, abb*	0 0 2 9 4 4	2
7	*aab, baa*	0 0 2 9 4 4	2
8	*aab, bab, baa*	0 0 3 5 8 4	2
9	*aab, aba, bba*	0 0 3 5 8 4	4
10	*aab, aba, abb*	0 0 3 6 4 8	4
11	*aab, abb, bba*	0 0 3 6 5 4 4	4
12	*aab, aba, abb, bba*	0 0 4 3 5 8 4	4

The remaining 34 bifix codes have a derivative with kernel $\{ab\}$ or $\{ba\}$ (there are 17 of each kind). They are listed in Table 15.5. The fact that the code X on line 23 has 4 words of length 8 can be read as follows on its internal part. The word *abbaab* has two interpretations, namely $(ab)(baa)b$ and $a(bba)(ab)$. Thus it is an internal factor and $\{a, b\}abbaab\{a, b\} \subset X$.

We have represented on Figure 15.5, the generation of the finite maximal bifix codes of degree 4 by internal transformation. The labels of the nodes are the indices of the first column in Tables 15.4 and 15.5. Each edge corresponds to an internal transformation. The label of the edge is the prefix used. We have only represented a part of the acyclic graph of internal transformations which is actually a covering tree of this graph. There are only three nodes without successor in the complete graph, which are 18, 20, and 23.

6.5.8 The formula is a direct consequence of $\underline{X} - 1 = (\underline{A} - 1)(d + T(\underline{A} - 1))$, where $T = \sum_{i=1}^{d} \underline{R_i}$.

6.5.9 The variance is $v_X = \sum_{n \geq 1} n^2 u_n k^{-n} - d^2$. Since $u(z) = \sum_{n \geq 1} u_n z^n$, we have $zu'(z) = \sum_{n \geq 1} n u_n z^n$, whence $u'(z) + zu''(z) = \sum_{n \geq 1} n^2 u_n z^{n-1}$. Finally, by Problem 6.5.8, $u(z) - 1 = (kz - 1)d + (kz - 1)^2 t(z)$. Differentiating twice, we obtain $u''(1/k) = 2k^2 t(1/k)$.

Table 15.5 *The remaining 34 finite maximal binary bifix codes of degree* 4.

	kernel	length distribution							symmetry class
13	ab	0	1	0	5	12	4		2
14	$ab, aabb$	0	1	0	6	8	8		2
15	$ab, aaba$	0	1	0	6	9	4	4	4
16	$ab, aaba, aabb$	0	1	0	7	5	8	4	4
17	$ab, aaba, babb$	0	1	0	7	6	5	4 4	2
18	$ab, aaba, aabb, babb$	0	1	0	8	2	9	4 4	2
19	ab, baa	0	1	1	3	9	8	4	4
20	$ab, baa, babb$	0	1	1	4	6	8	8	4
21	$ab, baa, aabb$	0	1	1	4	6	9	4 4	4
22	$ab, bba, aaba, aabb$	0	1	1	5	3	9	8 4	4
23	ab, baa, bba	0	1	2	2	4	9	12 4	2

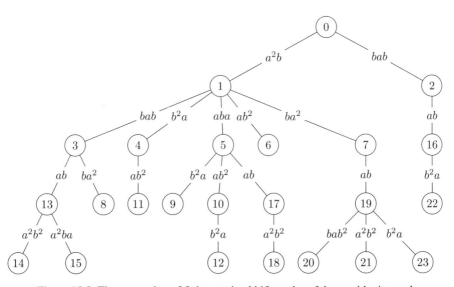

Figure 15.5 The generation of finite maximal bifix codes of degree 4 by internal transformations.

Section 6.6

6.6.1 Since $\overline{X} = I(X) \setminus I(X)A^+$ is the set of words of $I(X)$ which are minimal for the prefix order, it is prefix. Since it is contained in $I(X)$, the union $Y = X \cup \overline{X}$ is prefix. If X is rational, the set $A^-X \cup XA^*$ of words comparable to X is rational. The set $I(X)$ is the complement of this set, and so is rational too. Finally, the set $\overline{X} = I(X) \setminus I(X)A^+$ is rational. The code Y is right complete. Indeed, if a word is not comparable to a word in X, then it belongs to $I(X)$, and so it has a prefix in (X). This shows that the code Y is maximal.

Chapter 7

Section 7.1

7.1.1 Let $X = \{ab, ba\}$. Let $x \in A^*$ and $n \geq 2$ be such that $x^n \in X^*$. If $x \notin X^*$, then x has more than one X-interpretation. This forces $x \in F(ab)^*$, and thus $x^n \in (ab)^*$ or $x^n \in (ba)^*$, but then $x \in X^*$, a contradiction.

7.1.2 Set $p = |x|$ and $q = |y|$ with $p \geq q$ and $d = \gcd(p, q)$. Let $w = a_1 a_2 \cdots a_n$ with $n \geq p + q - 1$ be a prefix of a power of x and of a power of y. This means that p and q are periods of w, in the sense that $a_i = a_{i+p}$ for $1 \leq i \leq n - p$ and $a_i = a_{i+q}$ for $1 \leq i \leq n - q$. We want to prove that d is a period of w. First suppose that $d = 1$. Consider $i \leq n - p + q$. If $i \leq n - p$, we have $a_i = a_{i+p} = a_{i+p-q}$. Otherwise, we have $i > n - p$ and thus $i > q - 1$. Thus $a_i = a_{i-q} = a_{i+p-q}$. Thus $p - q$ is a period of w. This shows that $\gcd(p, q) = 1$ is a period of w. The general case follows by considering w as a word on the alphabet A^d.

Section 7.2

7.2.1 Suppose first that $M = \varphi(A^*)$ is aperiodic. Let $x \in A^*$ and $n \geq 1$ be such that $x^n \in X^*$. Let e be the idempotent in $\varphi(x^+)$. Then $\varphi(x)e = e\varphi(x) = e$ and thus $x \in X^*$. Thus X^* is pure. Conversely, let $e \in M$ be an idempotent and let G be its $\mathcal{H}$-class. Let $w \in \varphi^{-1}(G)$. We may suppose that $w \notin F(X)$. There is an $n \geq 1$ such that $\varphi(w^n) = e$. Let p be a fixed point of e. Since X is finite, there is a factorization $w^n = uv$ such that

$$p \overset{u}{\to} 1 \overset{v}{\to} p$$

and thus such that $vu \in X^*$. We have $vu = (rs)^n$ with r, s such that $w = sr$. Since X^* is pure we have $rs \in X^*$. Thus p is also a fixed point of w. This shows that the group containing e is trivial.

7.2.2 (a) Let us suppose for instance that X is $(1, 1)$-constrained. Suppose that $u_0 u_1, u_1 u_2 \in X^*$. We may assume $u_0, u_1, u_2 \neq 1$. If they belong to X, then $u_1 \in X^*$. Otherwise, if for example $u_0 u_1 \notin X$, then $u_0 = xu, u_1 = vy$ with $x, y \in X^*$ and $uv \in X$. Then $v \in X^*$ and thus $u_1 \in X^*$.

(b) X is $(3, 0)$-constrained since $u_0 u_1, u_1 u_2, u_2 u_3 \in X$ imply $u_0 = u_2 = 1$ or $u_1 = u_3 = 1$. It is not $(3, 0)$-limited since it is not prefix.

7.2.3 Let X be a recognizable circular code. Let $\varphi : A^* \to M$ be the morphism on the syntactic monoid of X^*. We show that X is (p, p)-limited with $p = \mathrm{Card}(M) + 1$. Let indeed $u_0, u_1, \ldots, u_{2p} \in A^*$ with $u_{i-1} u_i \in X^*$ for $1 \leq i \leq p + q$. We first observe that for any i, j such that $0 \leq i < j \leq 2p$, if $u_i, u_j \in X^*$, then $u_k \in X^*$ for $i \leq$

$k \leq j$ since X is a code. Now, since $\varphi(u_0), \ldots, \varphi(u_p)$ cannot be all distinct, there are indices j, k with $0 \leq j < k \leq p$ such that $\varphi(u_j) = \varphi(u_k)$. Then, since X is circular $u_j, u_{j+1}, \ldots, u_k \in X^*$. In the same way, there exist two indices ℓ, m with $p + 1 \leq \ell < m \leq 2p$ such that $\varphi(u_\ell) = \varphi(u_m)$ and thus $u_\ell, u_{\ell+1}, \ldots, u_m \in X^*$. This implies $u_p \in X^*$, proving the claim.

Section 7.3

7.3.1 We have by Proposition 3.7.17, with $P = XA^-$, $Ps = XR$ whence $t^p f_P(t) = f_X(t) f_R(t)$. Since $\underline{P}(\underline{A} - 1) = \underline{X} - 1$, we have $(kt - 1) f_P(t) = f_X(t) - 1$. The formula for $f_X(t)$ follows. The second formula also follows easily from $t^p + kt f_X(t) = f_X(t) + f_U(t) t^p$.

7.3.2 This is a direct consequence of Formula (7.15).

7.3.3 Let X be a circular code on a suitable alphabet B such that $u_n = \text{Card}(X \cap B^n)$ (the alphabet may be infinite). One may define a one-to-one correspondence $\alpha : A \to X$ between A and X such that the weight $w(a)$ of a is the length of $\alpha(a)$. Then the result follows from the fact that for any $z \in A^*$

  (i) z is primitive if and only if $\alpha(z)$ is primitive,
  (ii) $w(z) = |\alpha(z)|$,
 (iii) $y \in A^*$ is conjugate to z if and only if $\alpha(y)$ is conjugate to $\alpha(z)$.

7.3.4 Let A and B be two weighted alphabets such that A (resp. B) has u_n (resp. v_n) letters of weight n for each $n \geq 1$. Since $u_n \leq v_n$, we may suppose that $A \subset B$. Then the set of primitive necklaces of weight n on A is a subset of those on B.

7.3.5 One has

$$\sum_{n \geq 1} \frac{p_n}{n} z^n = \sum_{n \geq 1} \sum_{d \mid n} \frac{d v_d^{\frac{n}{d}}}{n} z^n = \sum_{d, e \geq 1} \frac{(v_d z^d)^e}{e} = \sum_{d \geq 1} \log(1 - v_d z^d)^{-1}$$

whence the formula by taking the exponential of both sides.

Chapter 8

Section 8.1

8.1.1 The unique factorization of a word $w \in \{1, 2, \ldots, n\}$ is obtained as follows. Let i be the least letter of w and let $w = uiv$ where all letters of u are at least equal to $i + 1$. Then $iv \in X_i^*$. We factorize in the same way u and obtain the factorization of w.

8.1.2 The factorization of a word $w = a_1 a_2 \cdots a_n$ corresponds to the convex hull of the graph of points $(i, \varphi(a_1 \cdots a_i))$.

8.1.3 Let $m = \ell_1 \ell_2 \cdots \ell_n$ be the factorization of m in a nonincreasing product of Lyndon words. Arguing by contradiction, suppose that $n > 1$. If $\ell \prec \ell_1$, then $\ell \ell_1 \in L$, a contradiction with the definition of ℓ. Thus $\ell_1 \preceq \ell$, showing that w has a nonincreasing factorization in Lyndon words of length $n + 1$, a contradiction with the fact that $w \in L$. Thus $n = 1$ and $m \in L$. Since $\ell \prec w$ and $w \prec m$, we have also $\ell \prec m$.

If $\ell \prec p$, then $\ell p \in L$ and ℓ is not the longest proper prefix of w which is in L. Thus $p \preceq \ell$.

8.1.4 We show by induction on $i \geq 1$ that Z_i contains all z_r such that $\pi(z_r) = (z_s, z_t)$ and $s < i \leq r$. It is true for $i = 1$. Suppose that it is true for $j \leq i - 1$ and consider z_r such that $\pi(z_r) = (z_s, z_t)$ with $s < i \leq r$. If $s < i - 1$, then $z_r \in Z_{i-1}$ by the induction hypothesis, and thus $z_r \in Z_i$. Otherwise, $\pi(z_r) = (z_{i-1}, z_t)$. Suppose first $z_t \in A$. Since $r < t$, we have $z_t \in Z_i$ and thus $z_r \in Z_i$. Otherwise, let $\pi(z_t) = (z_u, z_v)$. By the previous exercise, we have $u \leq s$ and thus $u < i$. We can thus repeat the same discussion with z_t replacing z_r. Iterating this argument, we can suppose that $z_r = z_{i-1}^k z_t$ with $k \geq 0$, $i - 1 < t$ and $z_t \in A$ or $\pi(z_t) = (z_u, z_v)$ with $u < i - 1$. We have, as above, $z_t \in Z_i$ and thus $z_r \in Z_i$.

8.1.5 Suppose that for $x_1, \ldots, x_k \in L_n$ and $y_1, \ldots, y_k \in L_n$ we have $x_1 \cdots x_k = s y_2 \cdots y_k p$ and $y_1 = ps$ with $ps \neq 1$. Then $x_1 < y_2 < x_2 < \cdots < x_k < y_1 < x_1$, a contradiction. Thus L_n is circular.

The set L_2 is comma-free only if $k \leq 3$ since for $k = 4$, $(ab)(cd) = a(bc)d$ with $ab, bc, cd \in L_2$. The sets L_3, L_4 are not comma-free for $k \geq 3$ since $(aab)(bbc) = a(abb)bc$ and $(aaab)(bbbc) = a(aabb)bc$.

8.1.6 We argue by contradiction and suppose that x, y, z are primitive and distinct.

First observe that $|x| \leq |z|$. Indeed, otherwise x would have two distinct z-interpretations, which is impossible for a primitive word. In the same way, $|y| \leq |z|$.

Let us first prove that the conclusion holds if $p \geq 3$. We consider the conjugate z' of z which is a Lyndon word. Then z' is either a factor of x^m or of y^n. In both cases, since z' is longer than x and y, this implies that z' is bordered. This is a contradiction since a Lyndon word is unbordered (Proposition 8.1.11).

Let us finally consider the case $p = 2$. We may suppose that $|x^m| > |y^n|$. Then we have $x^m = zu$, $z = uy^n$ for some word u. Thus $x^m = uy^n u$. But this implies that, changing x by some conjugate x', the equality $x'^m = u^2 y^n$. By induction, we have $x', u, y \in t^*$ whence the contradiction.

8.1.7 We suppose $|x| \geq |y|$. We may also suppose that x and y are primitive (since otherwise $y^* x \cup x^* y$ contains an imprimitive word). If X^* is not pure, there exists $u \notin X^*$ such that $u^n \in X^*$. Let $w = u^n$. We may suppose that $w \notin x^* \cup y^*$ since otherwise x or y is not primitive. Set $w = u^n = x_1 \cdots x_m$ with $x_i \in X$, and let j be the index such that $u^{n-1} = x_1 \cdots x_{j-1} k$, $x_j = kh$, $h x_{j+1} \cdots x_m = u$. Then $wh = hw'$ for $w' = x_{j+1} \cdots x_m x_1 \cdots x_j$. Note that $h \notin X^*$ since $u \notin X^*$.

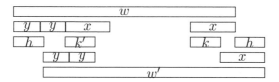

Figure 15.6 Case 1: $w' \in yX^*x$.

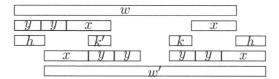

Figure 15.7 Case 2: $w' \in xX^*x$ and $|hx| > y^i$.

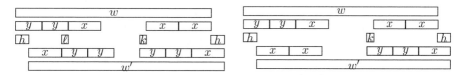

Figure 15.8 Case 2: $w' \in xX^*x$ and $|hx| < |y^i|$.

We consider the least integer $i \geq 1$ such that $w^2 \in X^*xy^ixX^*$. Replacing w by an X-conjugate, we may suppose that y^ix is a prefix of w and x a suffix of w. We distinguish several cases.

Case 1. $w' \in yX^*x$ (see Figure 15.6). By definition of the integer i, one has $w' \in y^i X^*x$. Let k, k' be such that $xh = kx$ and $y^ik' = hy^i$. Since k and k' are prefixes of x of the same length, $k = k'$. Thus $y^ixh = y^ikx = y^ik'x = hy^ix$ which shows that y^ix is not primitive.

Case 2. $w' \in xX^*x$. Suppose first that $|hx| > y^i$ (see Figure 15.7). We have in fact $w' \in xy^iX^* \cap X^*y^ix$, since otherwise x would be a nontrivial factor of x^2, a contradiction with the hypothesis that x is primitive. Since y^ix is a suffix of w', there exists k such that $y^ix = kxh$. Since y^ix is a prefix of w, there exists k' such that $y^ix = hxk'$. Since $|k| = |k'|$, and both are prefixes of y^i, we have $k = k'$. Thus $y^ix^2 = hxkx = kxhx$ is imprimitive. Since x, y are not powers of a common root, we have $i = 1$ by the Lyndon–Schützenberger theorem and yx^2 is imprimitive.

If $|hx| < |y^i|$, then $i > 1$ (see Figure 15.8). We have $w' \in X^*y^2x$ since otherwise x is a nontrivial factor of x^2. And $w \in X^*x^2$ since otherwise y is a nontrivial factor of y^2. Thus, there is a prefix k of y^i such that $y^ix = kx^2h$.

If $w' \in xy^2X^*$, then there is a prefix ℓ of y^i such that $y^ix = hx\ell x$. Since $|k| = |\ell|$, we have $k = \ell$. Thus $y^ix^2 = hxkx^2 = kx^2hx$ is not primitive, which is impossible since $i > 1$.

Thus $w' \in x^2X^*$. If $|hx^2| < y^i$, then x is a factor of y^* with two y-interpretations, a contradiction with the fact that x is primitive. Thus $|hx^2| > y^i$. Since x has only

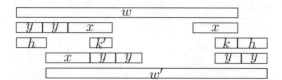

Figure 15.9 Case 3: $w' \in X^*y$ and $|hx| > |y^i|$.

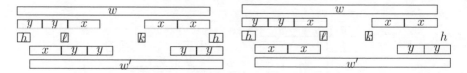

Figure 15.10 Case 3: $w' \in X^*y$ and $|hx| < |y^i|$.

one y-interpretation, we have $h = k$. Thus $y^i x^3 = (hx^2)^2$, which is impossible since $i > 1$.

Case 3. $w' \in X^*y$. Suppose first that $|hx| > |y^i|$ (see Figure 15.9). Then there is a suffix k of x such that $y^i = kh$ and a suffix k' of x such that $y^i x = hxk'$. Since $|k| = |k'|$, we have $k = k'$. Thus $y^i x = hxk = khx$ is not primitive.

Suppose now that $|hx| < |y^i|$ (see Figure 15.10). Then $i > 1$ and there is a prefix k of y^i such that $y^i = kxh$. If $w \in xy^i X^*$, then there is a prefix ℓ of y^i such that $y^i = hx\ell$. Since $|\ell| = |k|$, we have $k = \ell$. Thus $y^i x = kxhx = hxkx$ is imprimitive.

Finally, suppose that $w' \in x^2 X^*$. If $|hx^2| < |y^i|$, then x has two y-interpretations, which is impossible since x and y are primitive. Thus $|hx^2| > |y^i|$. We cannot have $w' \in x^3 X^*$ since otherwise x has two x-interpretations. Thus $w' \in x^2 y^i$. Let ℓ be the prefix of y^i such that $y^i x = hx^2 \ell$. Since $|k| = |\ell|$, we have $k = \ell$. Thus $y^i x^2 = hx^2 kx = kxhx^2$ is imprimitive, which is impossible since $i > 1$.

8.1.8 Suppose that X^* is not pure. Then $x^*y \cup y^*x$ contains a word which is not primitive. Suppose that $x^n y = z^m$ for some $n \geq 1$ and $m \geq 2$. If $(n-1)|x| \geq |z|$ then z^m and x^n have a common prefix of length $n|x| \geq |x| + |z|$. Thus x and z are powers of a common word by Fine–Wilf's theorem, a contradiction. Otherwise, we have $(n-1)|x| < |z|$. Since $|x| = |y|$ we have $(n+1)|x| = m|z|$. Thus $(n-1)m < n+1$ or equivalently $(n-1)(m-1) < 2$. The only case remaining to check is $n = m = 2$. Suppose that $|x| + |u| > |z|$. Then $u = rs$ with $z = uvr = svuu$. It follows that $|r| = |v| + |s|$. Thus $rsvr = svvrs$ implies $svr = vrs$ and we obtain that s and vr are powers of the same word, a contradiction with the fact that $y = vrs$ is primitive. The case $|x| + |u| < |z|$ is similar.

8.1.9 The right-hand side of Equation (7.17) may be rewritten as $\prod (1 - z^{|v|})^{-1}$ where the product is over all primitive necklaces v meeting X^*, in some fixed decreasing ordering of these necklaces. This in turn is equal to $\prod \sum_{n \geq 0} z^{n|v|}$, which is the sum of all monomials $z^{n_1 |v_1|} \cdots z^{n_k |v_k|}$, for all integers $k, n_1, \ldots, n_k$ and necklaces as above with $v_1 > \cdots > v_k$. For the second solution, one uses the fact that a free monoid has the complete factorization of Lyndon words, that these are in bijection with primitive necklaces, and that primitive necklaces within X^* coincide with primitive necklaces of A^* meeting X^*, since X^* is a very pure submonoid.

8.1.10 The last factorization is proved by induction on n, together with the fact that each C_i is contained in A^i and that X_{n+1} has only words of length at least $n + 1$. The case $n = 0$ is clear. If it is true for n, then define C_{n+1}, X_{n+2} as indicated and verify the previous properties, using the bisection $H^* = K^*((H \setminus K)K^*)^*$ where $K \subset H$. The finite factorization above leads to the infinite factorization $X^* = C_1^* C_2^* \cdots C_n^* \cdots$. To deduce the nonnegativity of the integers v_n, apply the homomorphism sending each letter in A onto z.

8.1.11 If X is rational, then X^* too, and it is easy to show that the closure under conjugacy of a rational language is rational, by using the syntactic monoid of the language. Since X^* is very pure, its closure under conjugacy is a cyclic language. Now, the generating function of X^* is by Equation (7.13) equal to the zeta function of its closure under conjugacy.

To show that the zeta function of a cyclic language L has the indicated expansion, proceed as in the proof of Proposition 7.3.4: first, one has Equation (7.16); then one shows by taking the logarithmic derivative that the equality of the zeta function with the right-hand side of Equation (7.17) is equivalent to Equation (7.16).

Section 8.2

8.2.1 We prove the statement by induction on n. Let $A^* = X_{n-1}^* \cdots X_1^*$ be a factorization obtained by composition of bisections and $X_i^* = Y^* Z^*$ be a bisection of X_i^*. Then, by induction hypothesis, X_i is an $(i - 1, n - i - 1)$-limited code. We consider the factorization $A^* = Y_n^* \cdots Y_1^*$ with $Y_n = X_{n-1}, \ldots, Y_{i+2} = X_{i+1}$, $Y_{i+1} = Y, Y_i = Z$ and $Y_{i-1} = X_{i-1}, \ldots, Y_1 = X_1$. Then Y_j is a $(j - 1, n - j)$-limited code for $1 \le j \le i - 1$ and for $i + 2 \le j \le n$. Let us show that $Y_{i+1} = Y$ is $(i, n - i - 1)$-limited. Let $u_0, \ldots, u_{n-1}$ be such that $u_{j-1} u_j \in Y^*$ for $1 \le j \le n - 1$. Since $Y \subset X_i^*$ and since X_i is $(i - 1, n - i - 1)$-limited, we have $u_{i-1}, u_i \in X_i^*$. Since Y is $(1, 0)$-limited, we have $u_i \in Y^*$. Thus Y is $(i, n - i - 1)$-limited. The proof that $Y_i = Z$ is $(i - 1, n - i)$-limited is similar.

8.2.2 The submonoid M satisfies $C(1, 0)$ and thus U is $(1, 0)$-limited. Consequently, there exists a bisection of the form (U, Z). Let $u, v \in U^*$ be such that $uv \in X^*$. Let $u = u_1 \cdots u_n$ with $u_1, u_2, \ldots, u_n$ suffixes of X. Since X is $(2, 0)$-limited, we have successively $u_2 \cdots u_n v \in X^*, \ldots, u_n v \in X^*$, and finally $v \in X^*$. Thus, considered as a code on U, X is $(1, 0)$-limited, which implies the existence of a bisection (X, Y) of U^*.

8.2.3 An easy inspection shows that Y is $(1, 1)$-limited. Suppose that (X, Y, Z) is a trisection of A^*. Since $ged \in Y$ and since $X^* Y^*$ is suffix-closed (by Proposition 8.2.9), $ed \in X^* Y^*$, which implies $ed \in X$. Similarly, since $dac \in Y$ and since $Y^* Z^*$ is prefix-closed, $da \in Z$. But then $eda \in X^2 \cap Z^2$, which is impossible.

8.2.4 The submonoid M generated by the suffixes of y clearly satisfies the condition $C(1, 0)$. Let X' be the code generating M. Since X' is $(1, 0)$-limited, there exists a bisection of A^* of the form (X', Z). Since y is unbordered, we have $y \in X'$. Thus $X'^* = X^* y^*$ with $X = y^*(X' \setminus y)$.

Chapter 9

Section 9.1

9.1.1 Since e is an idempotent, one has also $p \xrightarrow{e} r$, and by Proposition 9.1.6(ii), there is a fixed point s of e such that $p \xrightarrow{e} s \xrightarrow{e} r$. By unambiguity, we get $q = s$.

9.1.2 For any $(u, v), (u'v') \in D$ such that $(u, v)\rho(u', v')$, one has also $(u, v'), (u', v) \in D$ and $(u, v)\rho(u, v')\rho(u', v)$. Indeed, since $(u, v)\rho(u', v')$ there are $n, n', m, m' \in N$ such that

$$nu = n'u', \quad vm = v'm'.$$

Multiplying the first equality by v' on the right and the second one on the left by u', we obtain $nuv' = n'u'v' \in N$ and $uv'm' = uvm \in N$. Since N is stable, this implies $uv' \in N$. Thus $(u, v') \in D$ and $(u, v)\rho(u, v')$. A similar proof holds for (u', v).

Since $(1, n)\rho(1, n)\rho(1, 1)$ for any $n \in N$, $N \times N$ is the class of $(1, 1)$.

All we have to verify is that φ is well defined, in the sense that $(U, V)\varphi(m)(U', V')$ if and only if there are $u \in U$, $v' \in V'$ such that $um \in U'$ and $mv' \in V$. Let us consider $r \in U$ and $s \in V'$. Then $(u, mv')\rho^*(r, mv')$ and thus $rmv' \in N$. Moreover, since $(u, mv') = (u_0, v_0)\rho(u_1, v_1)\rho \cdots \rho(u_k, v_k) = (r, mv')$, we obtain $(um, v') = (u_0 m, v')\rho(u_1 m, v')\rho \cdots \rho(u_k m, v') = (rm, v')$. Thus $rm \in U'$. The proof that $ms \in V$ is similar. Thus $\varphi(m)$ is well defined.

If $M = A^*$ and $N = X^*$, the classes of ρ^* are the sets $X^* u \times v X^*$ for $u, v \neq 1$ such that $uv \in X$. Thus the classes are in bijection with the states of the flower automaton. The action also coincides (by Proposition 4.2.3).

9.1.3 The condition is obviously sufficient. Conversely, let c be a $n \times p$ matrix such that its columns form a basis of the columns of m. Then $m = \ell r$ in a unique way. The matrix $n = r\ell$ is invertible and satisfies $n^3 = n^2$. Thus n is the identity.

9.1.4 (a) Choose

$$R = \begin{bmatrix} 0 & -1 & 1 & 0 & 0 & 0 & 0 & 0 \\ 0 & -1 & 0 & 1 & 0 & 0 & 0 & 0 \\ 1 & 0 & 0 & 0 & 0 & 0 & 0 & 0 \\ 0 & 1 & 0 & 0 & 0 & 0 & 0 & 0 \\ 0 & 0 & 0 & 0 & 1 & 1 & 1 & 1 \\ 0 & 0 & 0 & 0 & 0 & 1 & 0 & 0 \\ 0 & 0 & 0 & 0 & 0 & 0 & 1 & 0 \\ 0 & 0 & 0 & 0 & 0 & 0 & 0 & 1 \end{bmatrix}, \quad R^{-1} = \begin{bmatrix} 0 & 0 & 1 & 0 & 0 & 0 & 0 & 0 \\ 0 & 0 & 0 & 1 & 0 & 0 & 0 & 0 \\ 1 & 0 & 0 & 1 & 0 & 0 & 0 & 0 \\ 0 & 1 & 0 & 1 & 0 & 0 & 0 & 0 \\ 0 & 0 & 0 & 0 & 1 & -1 & -1 & -1 \\ 0 & 0 & 0 & 0 & 0 & 1 & 0 & 0 \\ 0 & 0 & 0 & 0 & 0 & 0 & 1 & 0 \\ 0 & 0 & 0 & 0 & 0 & 0 & 0 & 1 \end{bmatrix}.$$

(b) Set $M = \varphi_A(A^*)$. For $q \in Q$, let $u_q \in A^*$ be such that $q \overset{u_q}{\to} q$ and that $e_q = \varphi_A(u_q)$ is an idempotent of minimal rank of M. Since ρ is a reduction, there exist $p, p' \in \rho^{-1}(q)$ such that $p \overset{u_q}{\to} p'$. Since e_q is idempotent, there is a fixed point s_q of e_q such that $p \overset{u_q}{\to} s_q \overset{u_q}{\to} p'$. By unambiguity, we have $\rho(s_q) = q$. Let $e_q = \boldsymbol{\ell}_q \boldsymbol{r}_q$ be the column-row decomposition of e_q. Define $\lambda(p) = q$ if $\rho(p) = q$ and $(s_q, p) \in \boldsymbol{r}_q$. Next, define $\mu(p) = q$ if $\rho(p) = q$ and there is a fixed point s of e_q such that $(p, s) \in \boldsymbol{\ell}_q$. Let $q \overset{w}{\to} q'$ be a path in $\mathcal{B}$ and let $m = \varphi_A(w)$. Then $(q, q') \in e_q m e_{q'}$ and thus $e_q m e_{q'} \neq 0$. By Proposition 9.1.9, the relation $\boldsymbol{r}_q m \boldsymbol{\ell}_{q'}$ is a bijection from the set of fixed points of e_q on the set of fixed points of $e_{q'}$. This shows that the pair (λ, μ) is an unambiguous realization of ρ.

Section 9.2

9.2.1 We have

$$(H * m)S_{H'K} = r_H m \ell_{H'} r_{H'} \ell_K$$
$$= r a_H m a'_{H'} \ell r a_{H'} \ell_K$$
$$= r e a_H m a'_{H'} a_{H'} \ell_K$$
$$= r_H m \ell_K.$$

The last equality comes from the fact that the right multiplication by $a'_{H'} a_H$ is the identity on H' and $e a_H m \in H'$. The proof that $S_{HK'}(m * H)$ reduces to the same expression is similar.

Consider the map ρ from D to $\Lambda \times G_e \times \Gamma$ associating to $m \in D$ the triple $\rho(m) = (K, g, H)$ defined by $m \in KM \cap MH$ and $g = r b'_K m a'_H \ell$. It is one-to-one because $m = \ell_K g r_H$. It is a morphism since for $m \in KM \cap MH \cap D$ and $m' \in K'M \cap MH' \cap D$, we have

$$\rho(m)\rho(m') = (K, r b'_K m a'_H \ell, H)(K', r b'_{K'} m' a'_{H'} \ell, H')$$
$$= (K, r b'_K m a'_H r a_H b_K \ell r b'_{K'} m' a'_{H'} \ell, H')$$
$$= (K, r b'_K m m' a'_{H'} \ell, H') = \rho(mm').$$

Section 9.3

9.3.1 For each s in the set S of fixed points of e, there exists a unique $t \in T$ such that sut and tvs. Define $s\varphi$ to be this element t. Suppose that for $s, s' \in S$, we have $s\varphi = s'\varphi = t$. Then $s \overset{u}{\to} t \overset{v}{\to} s \overset{u}{\to} t \overset{v}{\to} s$ and $s \overset{u}{\to} t \overset{v}{\to} s' \overset{u}{\to} t \overset{v}{\to} s$. Since the product ee is unambiguous, we have $s = s'$. Since $\mathrm{Card}(T) = \mathrm{Card}(S)$, this implies that φ is a bijection.

Thus we may suppose that φ is the identity on S, and we are reduced to the case $e = uv$ with $u : Q \to S, v : S \to Q$ and sus, svs for any $s \in S$. We prove that $u = \ell$ and $v = r$. Let us show that qus if and only if qes for $q \in Q$ and $s \in S$.

Assume qus. Then qes because sds, and similarly svq implies seq. Thus $vu = I_S$ as in the last part of the implication (ii) $\Longrightarrow$ (iii) of Proposition 9.1.6. Finally qes

implies qus since $u = uvu = eu$. Similarly, seq implies svq. This proves that $u = \ell$ and $v = r$.

9.3.2 If m has rank r in the sense of linear algebra, then we can write $m = cl$ with c a $n \times r$ matrix whose columns form a basis of the columns of m. Conversely, if $m = cl$ with $c \in K^{n \times r}$ and $l \in K^{r \times n}$ then the columns of c generate the columns of m.

9.3.3 One has

$$
m = \begin{bmatrix} 1 & 0 & 0 \\ 1 & 1 & 0 \\ 0 & 1 & 1 \\ 0 & 0 & 1 \end{bmatrix} \begin{bmatrix} 1 & 0 & 0 & 1 \\ 0 & 1 & 0 & -1 \\ 0 & 0 & 1 & 1 \end{bmatrix}
$$

and thus the rank over $\mathbb{Z}$ is 3. It can be verified that there is no such decomposition with nonnegative coefficients.

9.3.4 We treat the case where M does not have a zero. Since $R \cap L \cap N$ is a subgroup, it contains the idempotent e of $R \cap L$. In the same way the idempotent e' of $R' \cap L'$ is in N. Thus ee' is in $N \cap R \cap L'$.

9.3.5 (i) implies (ii). Indeed, let m be of minimal rank and such that $v = m_{p*}$ is a row of m. For any $n \in M$, since the right ideal mM is minimal, there is an $m' \in M$ such that $mnm' = m$. Since $vnm' = v$, we have $vn \neq 0$.

(ii) implies (iii). Suppose that v is not maximal and let $v' > v$ be a row of an element of M. Let $q \in Q$ be such that $(v' - v)_q = 1$. Let $m \in M$ be such that $w = m_{p*}$ is a maximal row. Let $n \in M$ be such that $n_{qp} = 1$. Then $v'nm$ is a row of an element of M which is $\geq w$ and thus equal to w. This forces $vnm = 0$ and thus $0 \in vM$.

(iii) implies (iv). Let $v = m_{p*}$ be a maximal row. Let $m' \in M$ have a minimal number of distinct nonzero rows. Let $q, s \in Q$ be such that $m'_{qs} = 1$. Let $n \in M$ be such that $n_{sp} = 1$. Then $m'nm$ has a minimal number of distinct nonzero rows and $(m'n)_{qp} = 1$. Thus v is the row of index q of $m'nm$.

(iv) implies (ii). Let $v = m_{p*}$ where m has a minimal number of distinct nonzero rows. If $vn = 0$, then mn has less distinct nonzero rows than n.

(iii) implies (i). Let $v = m_{p*}$ be a maximal row. Let n be of minimal rank with $n_{qp} = 1$. Then $(nm)_{q*} \geq v$ and thus $(nm)_{q*} = v$. This shows that v is a row of an element of minimal rank.

Observe that a matrix of minimal rank r has r distinct nonzero rows and thus a matrix has a minimal number of distinct nonzero rows if and only if it has minimal rank. Indeed, let e be an idempotent of minimal rank d. Let $e = \ell r$ be the column row decomposition of e. Then the rows of e are sums of rows of r. But since the columns of ℓ are in particular columns of e, they are maximal. Thus all rows of e are rows of r.

9.3.6 (a) The statement is a simple consequence of the fact that a word u is right completable if and only if $\varphi(u)_{1*} \neq 0$.

(b) By Exercise 9.3.5, the vector $\varphi(w)_{1*}$ is maximal and $0 \notin \varphi(w)_{1*}M$. Thus w is strongly right completable by (a). Let $x \in X^*$ and $u \in A^*$ be such that $xwu \in X^*$. Then $\varphi(xw)_{1*} \geq \varphi(w)_{1*}$ implies $\varphi(xw)_{1*} = \varphi(w)_{1*}$. Thus $\varphi(xwu)_{1*} = \varphi(wu)_{1*}$, showing that $wu \in X^*$.

9.3.7 The first statement is clear. To see the converse, first observe that R and L contain singletons and thus, for any $q \in Q$ there is $r \in R$ (resp. $\ell \in L$) such that $r_q = 1$ (resp. $\ell_q = 1$). Next, for any $r \in R$ and $m \in M$, we have $rm \in R$. Similarly, for any $m \in M$ and $\ell \in L$, we have $m\ell \in L$. Let now $m, n \in M$. For any $r \in R$ and $\ell \in L$, we have $rm \in R$ by the previous remark and thus $rmn\ell = (rm)n\ell \leq 1$. Hence $mn \in M$, which shows that M is a monoid. For any $p, q \in Q$, let $r \in R$ and $\ell \in L$ be such that $r_p = \ell_q = 1$. Then $1 \geq rmn\ell \geq (mn)_{pq}$. This shows that M is unambiguous. Any product ℓr for $\ell \in L$ and $r \in R$ is in M since for any $\ell', r' \in L \times R, r'\ell r\ell' = (r'\ell)(r\ell') \leq 1$. Thus M is additionally transitive. This proves (a).

To prove (b), consider a transitive unambiguous monoid of relations on the set Q. Let R (resp. L) be the set of rows (resp. columns) of the elements of M. Let R' be the set of all row vectors r in $\{0, 1\}^Q$ such that $r\ell \leq 1$ for all $\ell \in L$. Then $R'M = R'$. Indeed, for any $r \in R', m \in M$ and $\ell \in L$, we have $rm\ell = r(m\ell) \leq 1$ because $ML = L$. Thus $rm \in R'$. Next, let L' be the set of column vectors ℓ in $\{0, 1\}^Q$ such that $r\ell \leq 1$ for all $r \in R'$. Then R' and L' satisfy the condition (9.25). Let N be the transitive unambiguous monoid of relations formed of all n such that $rn\ell \leq 1$ for all $r \in R'$ and $\ell \in L'$. For any $r \in R', m \in M$ and $\ell \in L'$, we have $rm\ell = (rm)\ell \leq 1$ since $rm \in R'$. Thus M is a submonoid of N.

9.3.8 Let e be an idempotent of M of minimal rank with column-row decomposition $e = \ell r$ such that u is the sum of the rows of r and v is the first column of ℓ. Then $rm\ell$ is a permutation and thus $umv = 1$.

The rest of the proof is the same as that of Exercise 9.3.7.

9.3.9 (a) is clear since G acts transitively on the set Q.

(b) The first equality comes from the two ways to express the set of pairs (q, w) for $q \in Q$ and $w \in U$. The second one is analogous. The first equality of the second group corresponds to the one-to-one correspondence between an element $w \in U$ and the set of pairs $(q, \ell) \in Q \times V$ such that $w \cap \ell = q$.

(c) For each pair $(w, \ell) \in U \times V$ there is a unique pair (p, q) in $Q \times Q$ such that $w_p = m_{pq} = \ell_q = 1$. We conclude that

$$t = \frac{pq}{hk} = rs = \frac{n^2}{rs} = n.$$

9.3.10 Let M be a transitive unambiguous monoid of relations on Q. By Exercise 9.3.7 there is a pair R, L of row and column vectors in $\{0, 1\}^Q$ satisfying Equations (9.25) such that $rm\ell \leq 1$ for all $r \in R$ and $\ell \in L$. Let U (resp. V) be the set of maximal elements of R (resp. L). We consider the set P obtained by adding to Q a set p_u of elements in one-to-one correspondence with U. We form the set U' of subsets of P obtained by adding to each $u \in U$ the element p_u. We also denote by U' the set of characteristic vectors of the sets $u \in U'$. Let V' be the subset of $\{v \in \{0, 1\}^P \mid uv \leq 1 \text{ for all } u \in U'\}$ which are maximal. One has actually $uv = 1$ for all $v \in V'$ and $u \in U'$ since v contains either an element of u or the element p_u.

Let us show that for any $m \in \{0, 1\}^{P \times P}$ such that $umv \leq 1$ for all $u \in U'$ and $v \in V'$ and which is maximal for this property, one has actually $umv = 1$ for all $u \in U'$ and $v \in V'$. Suppose indeed that $umv = 0$. For any $q \in v$, there is a pair $(r, s) \in m$ and a pair $(u', v') \in U \times V'$ such that $r \in U$ and $q, s \in V'$. When q runs through v, the set of states s forms a set u' which is such that $u'v \leq 1$ for all $v \in V'$. Suppose that u' and v have a common element k. Then, choosing $q = k$, we obtain that $u'v' \geq 2$, a contradiction. Thus $u'v = 0$, which is also a contradiction. This proves the claim.

9.3.11 If ℓ is a clique and r is stable, then $\text{Card}(\ell \cap r) \leq 1$. Conversely, let ℓ be a set of vertices such that $\text{Card}(\ell \cap r) \leq 1$ for any stable set r. Let s, t be in ℓ. If (s, t) is not an edge of G, then $r = \{s, t\}$ is stable and $\text{Card}(\ell \cap r) = 2$, a contradiction. Thus ℓ is a clique. This shows that the pair (L, R) satisfies the the first equality. The proof for the second one is analogous.

The second assertion can be verified easily.

9.3.12 Suppose that $m'_{pq} = 1$ for some $p, q \in Q$. Since M is transitive and does not contain zero, there exists a maximal row r such that $r_p = 1$. Let us assume that $r = n_{s*}$ for some $n \in M$. Then $nm \leq nm'$ and $(nm)_{s*} = n_{s*}m$ is a maximal row by Exercise 9.3.5. Thus $(nm)_{s*} = (nm')_{s*}$. This forces $m_{pq} = 1$ since $m \leq m'$.

9.3.13 Let $p \in Q$ and $u \in A^*$, be such that $\varphi(u)_{p*}$ is not a maximal row. Since $\mathcal{A}$ is strongly connected, there exists a maximal row r such that $r_p = 1$. There is at least a state p' distinct of p such that $r_{p'} = 1$ and $\varphi(u)_{p'*} \neq 0$ since otherwise $r\varphi(u)$ is not maximal. Hence there is a state $q \in Q$ and a word v of length at most $n(n-1)/2$ such that $q \xrightarrow{v} p$ and $q \xrightarrow{v} p'$. Then $\varphi(u)_{p*} < \varphi(vu)_{q*}$. This proves the claim.

By the claim and its symmetric form, there exist pairs (p_1, u_1), (p_2, u_2), $\ldots, (p_s, u_s)$ in $Q \times A^*$ and $(v_1, q_1), (v_2, q_2), \ldots, (v_t, q_t)$ in $A^* \times Q$ such that, with $x_i = \varphi(u_i \cdots u_1)_{p_i*}$ and $y_j = \varphi(v_1 \cdots v_j)_{*q_i}$,

  (i) $u_1 = v_1 = 1$ and $p_1 = q_1$.
  (ii) for $2 \leq i \leq s$, the word u_i has length at most $n(n-1)/2$ and $x_i > x_{i-1}$.
  (iii) for $2 \leq j \leq t$, the word v_j has length at most $n(n-1)/2$ and $y_j > y_{j-1}$.
  (iv) x_s is a maximal row and y_t is a maximal column.

Let $u = u_s \cdots u_1$ and $v = v_1 \cdots v_t$. We have $|u| \leq (s-1)n(n-1)/2$ and $|v| \leq (t-1)n(n-1)/2$. Thus $|uv| \leq (s+t-2)n(n-1)/2$. Since $\mathcal{A}$ is unambiguous, we have $x_s y_t = 1$.

Thus $s + t \leq \sum_{q \in Q}(x_s)_q + \sum_{q \in Q}(y_t)_q \leq n + 1$. Let finally $z \in A^*$ be such that $q_t \xrightarrow{z} p_s$ with $|z| \leq n - 1$. Then $w = vzu$ is such that $y_t x_s \leq \varphi(w)$. By Exercise 9.3.12, this implies $\varphi(w) = y_t x_s$, whence the conclusion.

Section 9.4

9.4.1 We treat the case where the code is complete. Let $\mathcal{A} = (Q, 1, 1)$ be an unambiguous trim automaton recognizing X^*. Let K be the set of minimal rank of $M' = \varphi_{\mathcal{A}}(A^*)$. There exists a morphism ψ from M' onto M such that $\varphi = \psi \varphi_{\mathcal{A}}$. Then $J = \psi(K)$ is the minimal ideal of M and the other properties follow from the fact that they hold for K.

9.4.2 If $\mu(m) = \mu(n)$, then for any $H \in \Lambda$, we have $H \cdot m = H \cdot n$ and $H * m = H * n$. Let $H' = H \cdot m$. Since $H * m = r_H m \ell_{H'}$ and $H * n = r_H n \ell_{H'}$, we obtain $r_H m a'_{H'} \ell = r_H n a'_{H'} \ell$. Multiplying on the right by $r a_H$ we have $r_H m a'_{H'} \ell r a_H = r_H n a'_{H'} \ell r a_H$ whence $r_H m = r_H n$ since $x a'_{H'} a_H = x$ for all $x \in H$. This proves the equivalence concerning μ. The other one is proved in the same way. To prove that the function $m \mapsto (\mu(m), \nu(m))$ is injective, let $m, n \in M$ be such that $\mu(m) = \mu(n)$ and $\nu(m) = \nu(n)$. Let $p, q \in Q$ be such that $m_{p,q} = 1$. Let $H \in \Lambda$ be such that p is a fixed point of the idempotent of H and let $K \in \Gamma$ be such that q is a fixed point of the idempotent of K. Since $e a_H \in H$, there is an $s \in Q$ such that $s \xrightarrow{r_H} p$ and since $b_K e \in K$, there is a $t \in Q$ such that $q \xrightarrow{\ell_K} t$. Since $r_H m = r_H n$, there is an $u \in Q$ such that $s \xrightarrow{r_H} u \xrightarrow{n} p$. Since $m \ell_K = n \ell_K$, there is a $v \in Q$ such that $q \xrightarrow{n} v \xrightarrow{\ell_K} t$. Then $p = u$ and $q = v$ since otherwise the product $r_H n \ell_K$ is ambiguous. Thus $n_{p,q} = 1$. This shows that $m = n$.

9.4.3 Let X be a prefix code and let e be an idempotent of J. Suppose that $Me \cap \varphi(X^*) \neq \emptyset$. Let $f \in Me$ be an idempotent in $\varphi(X^*)$. Then $fe = f$ implies $e \in \varphi(X^*)$ since $\varphi(X^*)$ is right unitary.

Conversely, let $u, v \in M$ be such that $u, uv \in \varphi(X)$. We may assume, multiplying u on the left by an element of $J \cap \varphi(X^*)$ that $u \in J$. For any $n \geq 0$, we have $(uv)^{n+1} \in X^*$ and thus $(vu)^n \neq 0$. Let e be the idempotent in $(vu)^+$. Since the left ideal Mu is minimal and since $e \in Mu$, we have $u \in Me$. Thus $Me \cap \varphi(X^*) \neq \emptyset$, which implies $e \in \varphi(X^*)$. Since X is a code $u, uv, e \in \varphi(X^*)$ imply $v \in X^*$ by stability. Thus X is prefix.

9.4.4 Let C be a maximal class. For any $a \in A$, the set $a^{-1}C = \{q \in Q \mid q \cdot a \in C\}$ is again a maximal class. We have, for any maximal class C, the equality $MC = \sum_{a \in A} a^{-1}C$ where we identify a class with its characteristic column vector. Multiplying on the left by w, we obtain $\mathrm{Card}(A)wC = \sum_{a \in A} w(a^{-1})C$. Since $wC = w(C)$, we obtain

$$\sum_{a \in A} w(a^{-1}C) = \mathrm{Card}(A)w(C). \tag{15.4}$$

This implies that $w(C)$ is a constant. Indeed, the action of A on the maximal classes is transitive. Thus, if w is not constant on the set of maximal classes, there is a maximal class C such that the value $w(C)$ is maximal and a letter $a \in A$ such that $w(a^{-1}C) < w(C)$. Thus, by (15.4), there is a letter $b \in A$ such that $w(b^{-1}C) > w(C)$, a contradiction.

Let $u \in A^*$ be a word of minimal rank r. Then $w(Q) = \sum w(C)$ where the sum is on the classes of the nuclear equivalence of u. Thus $w(Q) = rw(C)$ since $w(C)$ is the same for each class. This shows that r divides $w(Q)$.

Section 9.5

9.5.1 We treat the case where the code is complete. Let $\mathcal{A} = (Q, 1, 1)$ be an unambiguous trim automaton recognizing X^*. Let K be the set of minimal rank of $M' = \varphi_{\mathcal{A}}(A^*)$. There exists a morphism ψ from M' onto M such that $\varphi = \psi\varphi_{\mathcal{A}}$. Then $J = \psi(K)$ is the minimal ideal of M. Let G' be an $\mathcal{H}$-class in K such that $\psi(G') = G$. Let $H' = G' \cap \varphi_{\mathcal{A}}(X^*)$. The restriction of ψ to G' is one-to-one and $\psi(H') = H$. This proves the claim since $G(X)$ equivalent to G' viewed as a permutation group acting on the right cosets of H'.

9.5.2 Let e be an idempotent of D. By assumption $D \cap \varphi((\bar{F}(X)) \neq \emptyset$. Let $m \in D \cap \varphi(\bar{F}(X))$. Since m is in D, we have $e \in MmM$ and thus $e \in \varphi(\bar{F}(X))$. Since $D \neq \{0\}$ the relation e has at least one fixed point s. Let $w \in \bar{F}(X) \cap \varphi^{-1}(e)$. Since s is a fixed point of e, there is a path $s \xrightarrow{w} s$ in $\mathcal{A}$. Since $w \in \bar{F}(X)$, there exist $u, v \in A^*$ such that $w = uv$ with $s \xrightarrow{u} 1 \xrightarrow{v} s$. Then vu is in X^* since $1 \xrightarrow{v} s \xrightarrow{u} 1$. Moreover, $\varphi(vu)^4 = \varphi(v)\varphi(w)^2\varphi(u) = \varphi(v)\varphi(w)\varphi(u) = \varphi(vu)^2$ and thus $\varphi(vu)^2$ is an idempotent. It belongs to D because $uv \mathcal{R} uvu \mathcal{L} vuvu$.

Suppose that X is finite. Let D be a regular $\mathcal{D}$-class. If $1 \in \varphi^{-1}(D)$, the conclusion holds. Otherwise $\varphi^{-1}(D)$ meets $\bar{F}(X)$ since it contains arbitrary long words. The conclusion thus follows from the previous case.

9.5.3 Let $u \in A^*$ be a word which is not a factor of X. Then, for each integer $i \geq 1$, there is a prefix p_i of u and a suffix s_i of u such that $s_i z^i, z^i p_i \in X^*$. Since there is a finite number of pairs (s_i, p_i), there exist integers $i < j$ such that $p_i = p_j$ and $s_i = s_j$. Then $s_i z^{i+j} p_i = (s_i z^i)(z^j p_j) = (s_j z^j)(z^i p_i)$ imply $z^{j-i} \in X^*$.

9.5.4 If $Z = X \wedge Y$ is thin maximal, there exists, by Exercise 9.3.6, a word $x \in Z^*$ which is strongly right completable in Z^* (and thus in X^*) and symmetrically a word $y \in Z^*$ which is strongly left completable in Z^* (and thus in Y^*), which proves that the condition is satisfied.

Conversely, the existence of $y \in Y^*$ strongly right completable in X^* shows that X is complete. Thus, there exists $x' \in A^*$ strongly left completable in X^*. Similarly, there exists $y' \in A^*$ strongly right completable in Y^*. Let $u = x'x$ and $v = yy'$. Then u is strongly left completable in both X^* and Y^* and v is strongly right completable in both X^* and Y^*. Thus, for any $w \in A^*$, the word vwu is both strongly right and left completable in X^* and Y^*. It follows from Exercise 9.5.3 that some power of

uwv is in Z^*. Thus Z is complete. It is moreover thin since Z^* is recognized by the direct product of automata $\mathcal{A}$ and $\mathcal{B}$ recognizing X^* and Y^* (which has finite minimal rank as $\mathcal{A}$ and $\mathcal{B}$). It is thus a maximal code.

9.5.5 We may suppose that Z is not maximal. Let T be a rational (resp. thin) code containing Z built using Theorem 2.5.24 (resp. Exercise 2.5.4). Let u, v be two distinct words in T which are not in Z (the method used to build T adds an infinite number of words). Let

$$X = Z \cup u \cup (T \setminus (Z \cup u))(T \setminus u)^* u,$$

$$Y = Z \cup v \cup (T \setminus (Z \cup v))(T \setminus v)^* v.$$

Then X and Y are obtained by composition as maximal rational (resp. thin) codes. Clearly $Z^* \subset X^* \cap Y^*$. To show the converse, let $w = t_1 \cdots t_n \in X^* \cap Y^*$ with $t_i \in T$. Suppose that $w \notin Z^*$. Then u and v appear among the t_i and the uniqueness of the factorization forces $u = v$, a contradiction.

Chapter 10

Section 10.2

10.2.1 The inclusion from left to right is clear. Conversely, let $x \in (X^s A^* \cap A^* X^s) \setminus W$. Since $x \notin W$, there exist $u, v \in A^*$ such that $uxv \in X^*$. Let $x = ry = zt$ with $r, t \in A^*$ and $y, z \in X^s$. Then $uztv \in X^*$ implies $ztv \in X^*$. And $ztv = ryv \in X^*$ implies $x = ry \in X^*$. This proves (10.9).

To prove (10.10), consider a word $v \in V$. Suppose that $v \notin W$. Let n be the least integer such that v is a factor of X^n. Then $uvw = x_1 x_2 \cdots x_n$ for some $u, w \in A^*$ and $x_i \in X$. By the definition of V we have $n \geq s + 2$ and by the minimality of n, u is a prefix of x_1 and v is a suffix of x_n. Thus $x_2 \cdots x_{n-1}$ is a factor of v, a contradiction with the fact that v does not have a factor in X^{s+1}.

To prove the opposite inclusion, let w be a word in W without any proper factor in W. We have to prove that w does not have a factor in X^{s+1}. If $w \in A$, the conclusion holds. Otherwise, let $w = ahb$ with $a, b \in A$ and $h \in A^*$. Let us first suppose that h has a factor in X^s. Since $ah, hb \notin W$, there exist $u_1, u_2, u_3, u_4 \in A^*$ such that $u_1 ahu_2, u_3 hbu_4 \in X^*$. Since h has a factor in X^s, we obtain by synchronization $u_1 ahbu_4 \in X^*$ a contradiction. Suppose now that w has a factor in X^{s+1}. Since h does not have a factor in X^s, the only possibility is $w \in X^{s+1}$, a contradiction.

10.2.2 Assume first that $X^s A^* \cap A^* X^s \subset X^*$. Then by Proposition 10.1.13, every pair of words in X^s is synchronizing. Completion follows from the inclusion $X^s w X^s \subset X^*$ for all w, and from the fact that X is nonempty.

Conversely, let X be a complete code with synchronization delay s. Again by Proposition 10.1.13, every pair (x, y) of words in X^s is such that $yA^* \cap A^* x \subset X^*$.

10.2.3 Set $Y' = X \cup (T \setminus W)$. We show first that $Y' \subset Y$. Let $y \in Y'$ and suppose that $y \notin Y$. Then, since $Y' \subset M$, one has $y = y_1 \cdots y_n$, with $y_i \in Y$ and $n \geq 2$.

At least one of the y_i is not in X. Take $y_i \notin X$ with i minimum. Then $y_1, \ldots, y_{i-1} \in X$, and $y_i \in X^s A^*$ by definition of M. Hence $y \in X^{i-1+s} A^*$, which is possible only if $i = 1$ in view of the definition of T. Thus $y_1 \notin X$ and similarly $y_n \notin X$.

Now $y_1 \in A^* X^s$. Choose $i \in \{2, \ldots, n\}$ minimum with $y_i \notin X$. Then y_i is in $X^s A^*$, hence $y_1 \cdots y_i \in A^* X^{2s} A^*$ and so is also y, a contradiction. Thus $y \in Y$. This proves the inclusion.

Conversely, let $y \in Y$. If $y \in X^*$, then $y \in X$ and hence $y \in Y'$. Suppose now that $y \notin X^*$. Then $y \in X^s A^* \cap A^* X^s$, since $y \in M$.

If we assume that $y \in X^{s+1} A^*$, then $y = xzr$ with $x \in X$, $z \in X^s$. We cannot have $zr \in A^* X^s$, otherwise $zr \in M$ and y is decomposable in M, a contradiction. But $y = r'z'$ with $z' \in X^s$. It follows that zr is a proper suffix of z'. Since z' is a synchronizing word, we obtain $zr \in X^*$ and thus $y \in X^*$, a contradiction.

Symmetrically, $y \notin A^* X^{s+1}$. Thus $y \in T$. Suppose $y \in A^* X^{2s} A^*$. Then $y = rzz'r'$, with $z, z' \in X^s$. Since y is indecomposable in M, either rz or $z'r'$ is not in M. We may suppose that $rz \notin M$. Then $y = z''s$ with $z'' \in X^s$ and rz is a proper prefix of z''. Since z is synchronizing, we obtain $rz \in X^*$, a contradiction. Thus $y \notin W$, showing the inclusion $Y \subset Y'$.

10.2.4 Let φ be the representation associated with the flower automaton $\mathcal{A}_D^*(X)$ of X. Let $e \in \varphi(A^+)$ be an idempotent with positive minimal rank. According to Proposition 7.1.5, the rank of e is 1. Thus $d(X) = 1$.

10.2.5 (i) and (ii) are clearly equivalent. To prove that (ii) implies (iii), we first have that X is a semaphore code since (ii) implies $A^* X \subset X A^*$. Let $S = X \setminus A^* X$. If $uv, vw \in S$, then $uv, uvw \in A^* X$ imply $w \in X^*$. This forces $v = 1$, thus proving (iii). Conversely (iii) implies clearly (ii).

Section 10.3

10.3.1 It is clear that a strictly locally testable set is locally testable and that the family of locally testable sets is a Boolean algebra. Thus a finite Boolean combination of strictly locally testable sets is locally testable. Conversely, a locally testable language is a finite union of classes of $\sim_s$ and such a class is a Boolean combination of sets of the form yA^*, A^*y, and A^*yA^*, which are either strictly locally testable or complements of strictly locally testable sets.

10.3.2 Let Y be a strictly locally testable set. Let $\varphi : A^+ \to S$ be the morphism from A^+ onto the syntactic semigroup of Y. Let e be an idempotent of S and let $w \in \varphi^{-1}(e)$. We may assume that w is longer than any word of the sets T, U, V, W defining Y by (10.8). Then it is easy to verify that w is a constant.

10.3.3 Let s be the order of Y. Let $\varphi : A^+ \to S$ be the morphism from A^+ onto the syntactic semigroup of Y. Let e be an idempotent of S and let w be a word of $\varphi^{-1}(e)$ of length larger than s. Then for any words p, u, v, q, we have $pwuwuwq \sim_s pwuwq$ and $pwuwvwq \sim_s pwvwuwq$. Thus eSe is idempotent and commutative.

10.3.4 By Proposition 10.3.5, we need to prove only one direction. We use the characterization of strictly locally testable sets given by Exercise 10.3.2. Let $\varphi : A^+ \to S$ be the morphism onto the syntactic semigroup of the locally testable set X^*. Let e be an idempotent of S. Suppose that $p, q, r, s \in S$ are such that $peq, res \in \varphi(X^*)$. Since X^* is locally testable, the semigroup eSe is idempotent and commutative. Thus setting $m = peqres$, one gets that

$$m = mm = pespeqreqres = pesm = mpes$$

is an element of $\varphi(X^*)$. Since $\varphi(X^*)$ is stable, this implies $pes \in \varphi(X^*)$. Thus e is a constant.

Chapter 11

Section 11.1

11.1.1 Let $u, uv \in R$ and let $x \in X^*$. Since X^* is right dense, there exists $w \in A^*$ such that $vxw \in X^*$. Since $u \in R$, there exists $y \in X^*$ such that $uvxwy \in X^*$. Since X^* is right dense, there is $s \in A^*$ such that $wys \in X^*$. Since $uv \in R$ there is $z \in X^*$ such that $uvxwysz \in X^*$. Finally, since X^* is right unitary, we have $sz \in X^*$. Thus $vxwysz \in X^*$ with $wysz \in X^*$ and this shows that $v \in R$, completing the proof of (a).

(b) The fact that Y is synchronized results from Proposition 3.6.6.

(c) Let $z \in Z'^*$ and let $x \in X^*$. Since Y' is synchronized, there exists $y \in X^*$ such that $zxy \in X^*$. Thus $z \in Z^*$.

(d) Let first $r \in R$. We may restrict to $m = \varphi(x)$ with $x \in X^*$. Then, there is $y \in X^*$ such that $rxy \in X^*$. Thus $1 \cdot rxy = 1$ and thus, by maximality of $\mathrm{Ker}(\varphi(x))$, $1 \cdot rx = 1$. Conversely, if r satisfies the condition, let $x \in X^*$. Let $y \in X^*$ be such that $\varphi(xy)$ is of minimal rank. Then $1 \cdot rxy = 1$ and thus $r \in R$.

Section 11.3

11.3.1 This follows from Theorem 11.3.1 applied to the subset of the alphabet formed of letters $a \in A$ such that $\varphi_A(a)$ is invertible.

11.3.2 It is clear that X is finite since $X \subset F(Y^2)$ and it is bifix since $X \subset Z$. Let φ be the representation associated with the minimal automaton $\mathcal{A}(X^*)$. Let e be an idempotent in $\varphi(Y^*)$. Let us show that G_e is equivalent to G. Indeed, let $w \in \varphi^{-1}(e) \cap Y^*$ and let U be the set of words in wA^*w of rank in $\mathcal{A}(X^*)$ equal to the degree d of G. Then, $\psi(U) = G$ since U contains wY^*w. Further, for $u, u' \in U$, $\psi(u) = \psi(u')$ implies $\varphi(u) = \varphi(u')$. Indeed, set $u = wyw$ and $u' = wy'w$. Let $r, t \in A^*$ be such that $rut \in X^*$. Then $w = ps = p's'$ with $rp, syp', s't \in X^*$. Since $X \subset Z$, we have $rut \in Z^*$ and thus $ru't \in Z^*$. This implies $sy'p' \in X^*$ since otherwise the rank of $\varphi(u')$ would be less than d. Thus $ru't \in X^*$. Thus $\varphi(u) = \varphi(u')$. This shows that $\psi^{-1}\varphi$ defines a morphism from G onto the $\mathcal{H}$-class of e. It is clearly bijective. Moreover, $\psi(u) \in H$ if and only if $u \in X^*$. This shows that G and G_e are equivalent.

Section 11.4

11.4.1 Let $w \in \varphi^{-1}(e) \cap \bar{F}(X)$. Let p, p' be fixed points of e such that $p \cdot wtw = p' \cdot wtw \neq \emptyset$ for some $t \in A^*$. Since $w \in \bar{F}(X)$, we have $w = uv = u'v'$ with v, v' prefixes of X and $p \cdot u = p' \cdot u' = 1$ and $1 \cdot vtw = 1 \cdot v'tw$. Since one of v, v' is suffix of the other, $1 \cdot vtw = 1 \cdot v'tw$ forces $v = v'$ by Proposition 6.1.14. Since $\varphi(w) = e$, we have $p \cdot w = p$ and $p' \cdot w = p'$. Thus $p = 1 \cdot v = p'$.

Section 11.5

11.5.1 Let d be the degree of X. If $w \in \bar{H}_X$, then w has d interpretations $w = s_i x_i p_i$ with $s_i \in A^- X$, $x_i \in X^*$, and $p_i \in XA^-$. Thus $\varphi_D(w) = \bigcup_{i=1}^{d}(Xs_i^{-1}, s_i) \times (p_i, p_i^{-1}X)$ which shows that $\varphi_D(w)$ has rank d and thus $\varphi_D(w) \in J_D$.

Conversely, if $w \in H(X)$, let $u, v \in A^+$ be such that $uwv \in X$. Then the row of index (u, wv) of $\varphi_D(w)$ is reduced to $\{(uw, v)\}$ and is not maximal because the row of index $(1, 1)$ of $\varphi_D(uw)$ contains all the (uw, v') for $v' \in (uw)^{-1}X$. Thus $\varphi_D(w) \notin J_D$ by Exercise 9.3.5.

11.5.2 The states $1, 1 \cdot a, \ldots, 1 \cdot a^{n-1}$ are the fixed points of the idempotent in $\varphi(a^+)$, which has thus rank n. If $n \geq d(X) + 1$, the idempotent in $\varphi(a^+)$ is not in the minimal ideal of $\varphi(A^+)$, which is therefore not nil-simple.

11.5.3 Let $B = \{a \in A \mid aA^* \cap X^* \neq \emptyset\}$ and $C = \{a \in A \mid A^*a \cap X^* \neq \emptyset\}$. Then the submonoid $BA^* \cap A^*C$ is generated by a code Z such that $X \subset Z^*$. Each word in Z^* has a power in X^*. Indeed, let $n \geq 1$ be such that $\varphi_A(z^n)$ is idempotent. Then z^n is left and right completable and thus in X^*. Thus $X = Y \circ Z$ with Y elementary bifix.

11.5.4 The sequence of equivalences θ_i, with θ_0 being the equality, is increasing. If $\theta_i = \theta_{i+1}$, then $\theta_i = \theta_{i+k}$ for all $k \geq 1$. There is an i such that θ_i has one class. The smallest such integer i is the depth d of $\varphi(A^+)$. This forces the sequence $\theta_0, \ldots, \theta_d$ to be strictly increasing from θ_0 to θ_d, whence $d \leq \mathrm{Card}(Q) - 1$.

11.5.5 Let $w \in \psi^{-1}(J)$. Then for any $L, L' \in \Lambda$, $L \cdot w = L' \cdot w$. Thus w has rank one. Conversely, if w has rank 1, then it is in $\psi^{-1}(J)$. Thus $\psi^{-1}(I) = \psi^{-1}(I)$. As a direct consequence of Exercise 11.5.4, the depth of $\varphi(A^+)$ is at most $\mathrm{Card}(\Lambda)$.

Section 11.6

11.6.1 Let $j = i(u *_a a^k)$. There is a path labeled ua^k from i to $j \cdot a^k$. If this path does not pass by 1, the finiteness of X imposes $j + k \geq i + 1$.

11.6.2 If X is a group code, we have $k = 0$ and by the previous exercise, for each letter $b \in A$ we have $i \cdot b \geq i + 1$ for all i except one. This forces $X = A^d$.

11.6.3 With $k = 1$ and $u = b$, we obtain $i(b *_a a) \geq i$ for all i provided $a^{i-1}ba$ does not have a prefix in X, that is except when $a^{i-1}b \in X$.

11.6.4 Let $\pi \in G$ not a power of α. Let $[d] = \{1, 2, \ldots, d\}$, $E = \{i \in [d] \mid i\pi \leq i\}$, and $F = [d] - E$. Let $d - 1 = ku + v$ with $u \geq 0$ and $0 \leq v < k$. Let N be the set formed of the $(u - 2)k$ first elements of F ordered by increasing value of $i\pi - i$. Let $I_1 + \cdots + I_{u-2}$ be a partition of N in consecutive intervals with respect to the value of $i\pi - i$. Let us show by induction on r, $1 \leq r \leq u - 2$ that for each $i \in I_r$, $i\pi - i \geq r$. It is true for $r = 1$. Suppose now that the element j of I_r with minimal value of $i\pi - i$ is such that $j\pi - j \leq r - 1$. Then by induction, we have $i\pi - i = r - 1$ for each $i \in I_{r-1}$. But then π coincides with α^{r-1} on the $r + 1$ elements of $I_{r-1} \cup j$, which implies by definition of k that $\pi = \alpha^{r-1}$, a contradiction. Thus

$$S = \sum_{i \in F} (i\pi - i) \geq \sum_{r=1}^{u-2} \sum_{i \in I_r} (i\pi - i) \geq \sum_{r=1}^{u-2} kr = k(u - 1)(u - 2)/2.$$

On the other hand

$$S = \sum_{i \in E} (i - i\pi) \leq (2k + 1)(d - 1).$$

Comparing the two inequalities, we obtain $k(u - 1)(u - 2)/2 \leq (2k + 1)(d - 1)$. Since $d - 1 \leq (u + 1)k$, this implies $k(u - 1)(u - 2)/2 \leq (2k + 1)(u + 1)k$ or $(u - 1)(u - 2)/2 \leq (2k + 1)(u + 1)$. Since $(u - 1)(u - 2) \geq (u + 1)(u - 5)$ for $u \geq 0$, we obtain $2(2k + 1) \geq u - 5$ and finally $d \leq 4k^2 + 8k + 1$.

11.6.5 (a) We write as usual i for $1 \cdot a^{i-1}$. Thus $\alpha = (12 \cdots d)$. According to Theorem 11.4.3, one has $a^k \in J(X)$. Thus the permutation $\pi = w *_a a^k$ is defined by $i \cdot wa^k = i\pi \cdot a^k$ for $1 \leq i \leq d$. Let $\sigma = \pi\alpha^k$. Then $i\sigma = 1 \cdot a^{i+1}wa^k$. There are exactly $2k$ values of i such that $a^{i-1}wa^k$ has a prefix in X. Otherwise, $a^{i-1}wa^k$ is a prefix of X and i is an excedance of σ. Thus σ has at least $d - 2k$ excedances. This implies, by Exercise 11.6.4, that σ belongs to the subgroup generated by α.

(b) We show that if X^* contains a word t of length at most k, then it contains all the conjugates of t. This is a contradiction since all the powers of t would have $k < d$ interpretations. Let $t = a_1 \cdots a_\ell$ with $a_i \in A$ and $\ell \leq k$. We show by descending induction on i that $t_i = a_i \cdots a_\ell a_1 \cdots a_{i-1} \in X^*$. Assume that $t_{i-1} \in X^*$. We apply statement 1 with $a = a_{i-2}$ and $w = t_{i-1}a^{k-\ell}$. Thus $\pi = t_{i-1}a^{k-\ell} *_a a^d$ is in the subgroup generated by $\alpha = (12 \ldots d)$. Since $1\pi = 1 \cdot a^{k-\ell}$, we have $\pi = \alpha^{k-\ell}$. Thus $1 \cdot t_{i-2}a^d = 1 \cdot at_{i-2}a^{d-1} = 2 \cdot t_{i-1}a^{d-1} = 1$. This shows that $t_{i-2} \in X^*$ and concludes the proof.

11.6.6 Assume by contradiction that $k \leq \sqrt{d}/2 - 2$. Then $d \geq 4k^2 + 16k + 16$. By Exercise 11.6.5, X does not contain words of length less than or equal to k. Thus, by Theorem 11.5.2, the depth of the syntactic semigroup of X^* is at most equal to k. Let Y be the base of the right ideal $J(X)$. For any $a \in A$ and $y \in Y$, the permutation $\sigma = (ay *_a a^k)$ has at least $d - 2k - 1$ excedances. By Exercise 11.6.1, this implies that σ is the subgroup generated by α. Since $ay *_a a^k = (a *_a y)(y *_a a^k)$ and since $G(X)$ is generated by the permutations $a *_a y$ for $a \in A$ and $y \in Y$, we obtain that $G(X)$ is cyclic and thus that $X = A^d$.

Section 11.7

11.7.1 It can be verified that the conditions stated on β and γ are equivalent to:

1. $1\beta^{-1} = 1\gamma^{-1}$,
2. for each $i \neq 1\beta^{-1}$, one has $i\beta \geq i$.
3. γ is an n-cycle such that $1\gamma^i \leq i + 2$ for all i.
4. for all $i \neq 1\beta^{-1}, 1\beta^{-1}\gamma^{-1}$, one has $i\gamma\beta \geq i$.

and that in turn, these conditions are necessary and sufficient for the code to be finite. The first condition is necessary and sufficient for the code to be bifix.

11.7.2 We use the following facts concerning the group $PGL_2(5)$. It is sharply 3-transitive on six points, of order $120 = 6 \times 5 \times 4$. As an abstract group it is isomorphic with the symmetric group S_5. Let $\alpha = (123456)$, $\beta = b *_a a$, $\gamma = b *_a b$. Since all the elements of order 6 of $PGL_2(5)$ are internally conjugate, we may suppose that the identification of $\{1, 2, 3, 4, 5, 6\}$ with the projective line $\mathbb{Z}/5\mathbb{Z} \cup \infty$ is the same as the bijection ρ used in Example 11.7.7, with $\alpha^\rho = (\infty01423)$ realized by the homography $\zeta \mapsto 2/(\zeta + 2)$. By Exercise 11.7.1, β and γ are such that $\beta = (i_1 \cdots i_k)$ with $i_1 < \cdots < i_k$ and $\gamma = \alpha^\tau$ where τ is a product of cycles of the form $(k, k + 1, \ldots, k + m)$ with $k\beta \geq k + m$ or $k\beta = 1$.

If β has no fixed points, then $\beta = \alpha$. The permutation γ is conjugate of α by an involution which is a product of two cycles. The only solution is $\gamma = (132546)$. This gives the finite maximal bifix code X_1 of Example 11.7.7 (Table 11.5).

If β has one fixed point, then it coincides with α on four points, which is impossible.

If β has two fixed points, these cannot be consecutive since otherwise β would coincide with α on three points. These two points cannot either form an orbit of α^3, since otherwise β would commute with α^3, in contradiction with the fact that the stabilizer of two points is, in $PGL_2(5)$ its own centralizer. Thus, the possible sets of fixed points are $(2, 4)$, $(2, 6)$, $(4, 6)$, and $(3, 5)$, corresponding to

$$\beta_1 = (1356), \quad \beta_2 = (1345), \quad \beta_3 = (1235), \quad \beta_4 = (1246).$$

Each of them generates, together with α, the group $PGL_2(5)$. As for γ, we have either $\gamma = \alpha$ or $\gamma = \alpha^\tau$ where τ is a product of two transpositions. This gives the two solutions

1. $\gamma_1 = (132546)$ with $\tau = (23)(45)$ compatible with $\beta = \beta_4$.
2. $\gamma_2 = (124365)$ with $\tau = (34)(56)$ compatible with $\beta = \beta_2$ or $\beta = \beta_3$.

Thus, in the case where β has two fixed points, the code X is one of the five possible codes.

1. The code X_2 corresponding to $\beta = \beta_1$ and $\gamma = \alpha$ whose minimal automaton is described in Table 15.6.
2. The code $X_3 = \bar{X}_1$ symmetric of X_1 by the exchange of a, b with $\beta = \beta_2, \gamma = \gamma_2$.
3. The code $X_4 = \widetilde{X}_2$ which is the reversal of X_2 with $\beta = \beta_3$ and $\gamma = \gamma_2$.

Table 15.6 *The transitions of the minimal automaton of X_2^*.*

	1	2	3	4	5	6	7	8	9	10
a	2	3	4	5	6	1	4	3	6	5
b	7	8	9	10	6	1	8	9	10	6

Table 15.7 *The vector space $(\mathbb{Z}/2\mathbb{Z})^3$.*

1	1	0	0
2	0	1	0
3	0	0	1
4	1	1	0
5	0	1	1
6	1	1	1
7	1	0	1

4. The code $\bar{X}_4$ with $\beta = \beta_4$ and $\gamma = \alpha$.
5. The code $\bar{X}_2$ with $\beta = \beta_4$ and $\gamma = \gamma_1$.

Note that $X_1 = \tilde{X}_1$, so X_1 is equal to its reversal.

11.7.3 The identification of $(\mathbb{Z}/3\mathbb{Z})^3$ with $\{1, 2, \ldots, 7\}$ is shown in Table 15.7. In this way, the permutation $\alpha = (1234567)$ corresponds, via the identification, to the matrix

$$\begin{bmatrix} 0 & 1 & 0 \\ 0 & 0 & 1 \\ 1 & 1 & 0 \end{bmatrix}$$

which represents the multiplication by x in the basis $1, x, x^2$. The group $G(X)$ is generated by $\alpha = (1234567)$ and the permutations:

$$b *_a a^2 = (1236)(45)(7), \quad b *_a ab = (146)(235)(7), \quad b *_a b = (1254376)$$

correspond, via the identification, to the matrices:

$$b *_a a^2 = \begin{bmatrix} 0 & 1 & 0 \\ 0 & 0 & 1 \\ 1 & 1 & 1 \end{bmatrix}, \quad b *_a ab = \begin{bmatrix} 1 & 1 & 0 \\ 0 & 0 & 1 \\ 0 & 1 & 1 \end{bmatrix}, \quad b *_a b = \begin{bmatrix} 0 & 1 & 0 \\ 0 & 1 & 1 \\ 1 & 0 & 1 \end{bmatrix},$$

which generate the group $GL_3(2)$.

11.7.4 The images of a^2, ab, b are of minimal rank and thus the group $G(X)$ is generated by $\alpha = (1\,2\cdots 11), \beta = b *_a a^2, \gamma = b *_a ba$, and $\delta = b *_a b$. We compute from the transitions of the automaton

$$\beta = (1\ 2\ 3\ 6\ 5\ 4\ 7\ 10)(8\ 9)(11),$$

$$\gamma = (1\ 4\ 7\ 9\ 6\ 3\ 8\ 10)(2\ 5)(11) = \beta\alpha^2\beta^{-1}\alpha^{-1},$$

$$\delta = (1\ 2\ 5\ 6\ 3\ 4\ 9\ 8\ 7\ 11\ 10) = \beta\alpha\beta^{-1}.$$

Let us show that α and β generate the Mathieu group M_{11} (see the Notes for a reference). Let $h(x)$ be the polynomial with coefficients in the field $\mathbb{F}_3 = \mathbb{Z}/3\mathbb{Z}$

$$h(x) = -1 + x^2 - x^3 + x^4 + x^5.$$

The columns of the matrix K below are the remainders of the polynomials $1, x, \ldots, x^{10}$ modulo $h(x)$.

$$K = \begin{bmatrix} 1 & 0 & 0 & 0 & 0 & 1 & -1 & -1 & -1 & 1 & 0 \\ 0 & 1 & 0 & 0 & 0 & 0 & 1 & -1 & -1 & -1 & 1 \\ 0 & 0 & 1 & 0 & 0 & -1 & 1 & -1 & 0 & 1 & -1 \\ 0 & 0 & 0 & 1 & 0 & 1 & 1 & 0 & 1 & 1 & 1 \\ 0 & 0 & 0 & 0 & 1 & -1 & -1 & -1 & 1 & 0 & 1 \end{bmatrix}$$

Multiplying the last one by x, one obtains that $x^{11} - 1 \equiv 0$ modulo $h(x)$. We consider the vector space $V = \mathbb{F}_3[x]/(x^{11} - 1)$. Let H be the subspace of V formed by the multiples of $h(x)$. Since $h(x)$ has degree 5, H has dimension $11 - 5 = 6$. The Mathieu group M_{11} is the group of permutations of $\{1, 2, \ldots, 11\}$ which leave invariant the support of the vectors in H (that is the set of coordinates with nonzero coefficient). A basis of the orthogonal of H is made of the rows of the matrix K above.

The columns of K are the components in the basis $\{1, \xi, \xi^2, \xi^3, \xi^4\}$ of the powers of a root ξ of the polynomial $h(x)$. Thus the group M_{11} contains α, which corresponds to the multiplication by ξ, and also β whose action on the columns of K corresponds to the matrix

$$\begin{bmatrix} 0 & 0 & -1 & -1 & 0 \\ -1 & 0 & 0 & 1 & 0 \\ 0 & -1 & 1 & 1 & 0 \\ 0 & 0 & -1 & 1 & -1 \\ 0 & 0 & 1 & -1 & 0 \end{bmatrix}$$

One may verify that α, β generate G by showing that they they generate a 4-transitive group.

Chapter 12

Section 12.1

12.1.1 Let $m, n \geq 1$ be integers. We show that if n is not a Hajós number, then neither is mn. Let $G = \mathbb{Z}/mn\mathbb{Z}$ and $H = \{0, m, \ldots, (n - 1)m\}$. Thus H is a subgroup of G and $H \simeq \mathbb{Z}/n\mathbb{Z}$. Let $H = K + L$ be a factorization of H where neither K

nor L is periodic. Let $M = \{0, 1, \ldots, m - 1\}$ and $N = L + M$. Since M is a set of representatives of the cosets of H, $G = H + M$ is a factorization of G. Thus $G = K + N$ is a factorization of G. We show that N is not periodic. Assume by contradiction that p is a period of N and consider $i \in M$. We have $p + i + L \subset p + N = N$ and $p + i + L \subset p + i + H = j + H$ for some appropriate $j \in M$. Thus $p + i + L \subset N \cap (j + H) = j + L$. Since L is not periodic, we have $p + i = j$. Thus we have proved that $p + M \subset M$, a contradiction since M is not periodic.

12.1.2 The proof is by induction on n. Let $G = \mathbb{Z}/n\mathbb{Z}$ and let $G = L + R$ be a factorization. Since n is a Hajós number, L or R is periodic. We may suppose that R is periodic. Then, we can write $R = H + S$ where H is a nontrivial subgroup of G and the sum is direct. We have a factorization $G/H = (L + H)/H + (S + H)/H$. Since G/H has the Hajós property by Exercise 12.1.1, it has the Rédei property by induction hypothesis. Thus either $\langle (L + H)/H \rangle \neq G/H$ and thus $\langle L \rangle \neq G$, or $\langle (S + H)/H \rangle \neq G/H$ and thus $\langle R \rangle \neq G$.

12.1.3 For $x, y \in \mathbb{Z}$ with $x \leq y$, denote $[x, y] = \{z \in \mathbb{Z} \mid x \leq z \leq y\}$. We may suppose that $L \subset [0, d]$ for some $d \geq 0$. Let $x, y \in \mathbb{Z}$ with $x \leq y$ be such that $R \cap [x, x + d] = R \cap [y, y + d]$. Then $R \cap [x + kd, x + (k + 1)d] = R \cap [y + kd, y + (k + 1)d]$ for all $k \geq 0$, as one may verify by induction on k. Thus R is periodic of period at most 2^d.

Section 12.2

12.2.1 Suppose that $\mathbb{Z}/n\mathbb{Z} = L + R$ is a factorization. For each $i \in \{0, 1, \ldots, n - 1\}$ there is exactly one pair $(\ell, r) \in L \times R$ such that $i \equiv \ell + r \bmod n$. Since $0 \leq \ell + r \leq 2n - 2$, we have actually $\ell + r = i$ or $\ell + r = i + n$. Thus $a^\ell a^r = a^i$ or $a^\ell a^r = a^i a^n$. This shows that

$$a^L a^R \equiv 1 + a + \cdots + a^{n-1} \quad \bmod a^n - 1, \tag{15.5}$$

and thus $a^L a^R (a - 1) \equiv 0 \bmod a^n - 1$ as claimed.

12.2.2 We first prove the preliminary remark. If $p' \leq p$ and $q' \leq q$, then $p' + q' \leq p + q$, a contradiction. Suppose $p' > p$. Then $q' \leq q$ since otherwise $p' + q' \geq p + q + 2$. If p' is not the successor of p in P, then there exists p'' such that $p < p'' < p'$, then $p + q < p'' + q < p' + q'$, a contradiction. The other case is handled in an analogous way.

We have $0 \in P \cap Q$ and we may assume that $1 \in P$. If $Q = \{0\}$, there is nothing to prove. Otherwise let m be the least nonzero element of Q. Then $\{0, 1, \ldots, m - 1\} \subset P$.

Let r in $\{0, 1, \ldots, n - 1\}$. We claim that

(i) if $r \in Q$, then $m \mid r$,
(ii) if r is in P, then $s, s + 1, \ldots, s + m - 1$ are in P, where s is the unique integer such that $m \mid s$ and $s \leq r < s + m$.

The proof is by induction on r. The property holds for $r = 0$ since $0 \in Q$ and $\{0, 1, \ldots, m - 1\} \subset P$. Assume that it holds for $s < r$. Set $r = um + v$ with $u \geq 0$ and $0 \leq v < m$. Let $um = p + q$ with $p \in P$ and $q \in Q$. We distinguish three cases.

Case 1. $p < r$ and $q < r$. Then by (i) $m|q$ and thus $m|p$. By (ii), we have $p + v \in P$ and thus $r = (p + v) + q$ is the decomposition of r in $P + Q$. We cannot have $r \in Q$ since otherwise $p = v = 0$ and thus $q = r$. If r is in P, then $q = 0$ and $p = um$. By the induction hypothesis, $p, p + 1, \ldots, p + m - 1$ are in P. Thus (ii) is satisfied with $s = um$.

Case 2. $p = r$ and thus $v = q = 0$. Set $r + 1 = p' + q'$ with $p' \in P$ and $q' \in Q$. By the preliminary remark, we have either $p' = r + 1$ or $q' = m$. If $q' = m$, then $m|(p' - 1)$ and thus $p' - 1 \in P$ by (ii). Therefore $r = (p' - 1) + m$ is another decomposition of r, a contradiction. Thus $p' = r + 1$ and $r + 1$ is in P. One proves in the same way that $r + 2, \ldots, r + m - 1$ are in P. Thus r satisfies also (ii).

Case 3. $q = r$ and thus $p = v = 0$. In this case, $m|r$ and thus (i) holds.

We have shown that there exist sets P' and Q' such that $P = \{0, 1, \ldots, m - 1\} + P'$ and that $Q = mQ'$. Thus $\{0, 1, \ldots, n/m - 1\} = P' + Q'$. This proves the statement taking $n_1 = m$.

Section 12.3

12.3.1 Let $m \geq 1$ be such that $x = b^m$ is not a proper factor of a word in X. Then, since $b \in X$, the pair (x, x) is synchronizing. Suppose that $\ell \in L$, that is $a^\ell b^+ \cap X \neq \emptyset$. Then $a^\ell x \in X^*$ and thus ℓ is in the set $L(x)$ defined in Proposition 12.2.9. Conversely, if $\ell \in L(x)$, then $\ell = kn + \ell'$ with $a^{\ell'} b^+ \cap X \neq \emptyset$. Thus the set of residues modulo n of L and $L(x)$ are the same. The same holds for R and $R(x)$. Thus Theorem 12.3.1 follows from Proposition 12.2.9.

12.3.2 The property is a simple consequence of the fact that the sums $\bar{H} + \bar{K}$ and $\bar{S} + \bar{T}$ are direct.

12.3.3 Let $Y \subset \{a, b\}^*$ be a finite maximal code containing X. Let L, R be as in Proposition 12.3.7. We cannot have $X \cap (a^*b^* \cup b^*a^*) = Y \cap (a^*b^* \cup b^*a^*)$ since otherwise $\text{Card}(L) = \text{Card}(K)\,\text{Card}(T) = t\,\text{Card}(T)$ and $\text{Card}(R) = \text{Card}(H)\,\text{Card}(S) = d\,\text{Card}(S)$. The pair (L, R) would thus be a dt-factorization of $\mathbb{Z}/n\mathbb{Z}$ and thus not an m-factorization.

Assume first that there is an $h \notin R$ such that $b^+ a^h \cap Y \neq \emptyset$. Let us show the multiplicity of h in $L + (R \cup h)$ is larger than m. Indeed, since (S, T) is a factorization of $\mathbb{Z}/n\mathbb{Z}$, there is a pair $(r, \ell) \in S \times T$ such that $h \equiv \ell + r \bmod n$. Thus the value h is represented modulo n in t ways as the sum $h + n$ and in dt ways as the sum $\ell + r$. Thus the multiplicity of h is $dt + t > m$. The proof that the same property holds for $(L \cup h) + R$ is symmetrical.

12.3.4 Use Solution 12.3.3 with $m = 10, n = 2$ and $H = \{1, 2, 10\}$, $K = \{3, 6, 10\}$ and $S = \{2\}$, $T = \{1, 2\}$.

12.3.5 The proof is by induction on $n \geq 1$. The property is true for $n = 1$ since $a \cup a^\ell b a^r$ is composed of a prefix and a suffix code. Consider next an integer $n \geq 2$.

Since n has the Hajós property, either L or R is periodic. We may assume that L is periodic of period p. Then $n = pq$ and $L = L' + \{0, p, \ldots, p(q-1)\}$. The pair (L', R) is a factorization of $\mathbb{Z}/q\mathbb{Z}$. By the induction hypothesis, the code $Z = a^q \cup a^{L'} b a^R$ is composed of prefix and suffix codes. Then $X \subset a^n \cup \{1, a^p, \ldots, a^{p(q-1)}\} a^{L'} b a^R$ has the same property.

Chapter 13

Section 13.1

13.1.1 Let μ be the matrix representation of $\mathcal{A}$. We have for any $w \in A^*$

$$\sum_{a \in A} \pi(aw) = \sum_{a \in A} I\mu(aw)T = IP\mu(w)T = I\mu(w)T = \pi(w).$$

Section 13.2

13.2.1 Let φ be the representation associated with $\mathcal{A}$. The hypotheses imply that each X_p is rational and a maximal prefix code. Thus, by Theorem 13.2.9, we have $\delta(X_p^*) = 1/\lambda(X_p)$.

For $p, q \in Q$, let $L_{p,q}$ be the set defined by $L_{p,q} = \{w \in A^* \mid p \cdot w = q\}$. Set $Y_{p,q} = L_{p,q} \setminus L_{p,q} A^+$. Since $L_{p,q} = Y_{p,q} X_q^*$, and since each $Y_{p,q}$ is a rational maximal prefix code, we have for each $p, q \in Q$, by Proposition 13.4.3, $\delta(L_{p,q}) = \delta(X_q^*) = 1/\lambda(X_q)$.

First assume that I is given by $I_q = 1/\lambda(X_q)$ for each $q \in Q$. Since $\mathcal{A}$ is deterministic and complete, the family of sets $L_{i,q}$ for $q \in Q$ forms a partition of A^*. Thus $\sum_{q \in Q} I_q = \sum_{q \in Q} \delta(L_{i,q}) = \delta(A^*) = 1$.

For each $q \in Q$, the sets $L_{i,q}$ and $\bigcup_{p \cdot a = q} L_{i,p} a$ differ at most by the empty word and thus $\delta(L_{i,q}) = \sum_{p \cdot a = q} \delta(L_{i,p}) \pi(a)$. Since $\delta(L_{i,q}) = \delta(X_q^*)$, this shows that $(IP)_q = \sum_{p \in Q} I_p P_{p,q} = \sum_{p \in Q} (I_p (\sum_{p \cdot a = q} \pi(a)) = \sum_{p \in Q} \sum_{p \cdot a = q} \delta(L_{i,p}) \pi(a) = \delta(L_{i,q}) = I_q$. Thus I is stationary.

Conversely, suppose that $\sum_{q \in Q} I_q = 1$ and that $IP = I$. We have also $IP^n = I$ for all $n \geq 0$. But $P_{p,q}^n = \pi(L_{p,q} \cap A^n)$ and thus the sequence of matrices $(S^{(n)})$ defined by $S^{(n)} = 1/n \sum_{i < n} P^i$ converges to the matrix S with coefficients $S_{p,q} = \delta(L_{p,q}) = 1/\lambda(X_q)$. Since $IS^{(n)} = I$, we obtain $IS = I$, and for each $q \in Q, I_q = \sum_{p \in Q} I_p S_{p,q} = \sum_{p \in Q} I_p/\lambda(X_q) = (\sum_{p \in Q} I_p)/\lambda(X_q)$. This shows that $I_q = 1/\lambda(X_q)$ for each $q \in Q$.

Section 13.3

13.3.1 Since $X^* \subset Y^* \subset F(X^*)$, one has $h(X^*) \leq h(Y^*) \leq h(F(X^*))$, where h denotes the entropy. By Proposition 13.3.1, one has $h(X^*) = h(F(X^*))$. Thus $h(X^*) = h(Y^*)$. Set $h(X^*) = -\log r$. By Theorem 13.3.3, we have $f_X(r) = f_Y(r) = 1$, which implies $X = Y$.

Section 13.4

13.4.1 (a) is clear by bounded convergence.

(b) We have

$$\pi^{(n)} * \pi^{(1)}(L) = \sum_{u \in L} \pi^{(n)} * \pi^{(1)}(u) = \sum_{u \in L \cap A^{n+1}} \sum_{va=u} \pi(v)\pi(a)$$

$$= \sum_{u \in L \cap A^{n+1}} \pi(u) = \pi^{(n+1)}(L).$$

(c) Let $\mu_n = \frac{1}{n} \sum_{i=0}^{n-1} \pi^{(n)} \varphi^{-1}$. Then $\nu = \lim \mu_n$ and thus

$$\nu * \nu = (\lim \mu_n) * (\lim \mu_m) = \lim(\mu_{n+m}) = \nu.$$

13.4.2 We verify that the vector K defined by $K_q = \frac{1}{d} \sum_{E \in \mathcal{E}_q} J_E$ is stationary and satisfies $\sum K_q = 1$. Since every minimal image has d elements, we have $\sum_{q \in Q} K_q = \frac{1}{d} \sum_{E \in \mathcal{E}} d J_E = \sum_{E \in \mathcal{E}} J_E = 1$. Next,

$$\sum_{p \cdot a = q} K_p \pi(a) = \sum_{p \cdot a = q} \frac{1}{d} \sum_{E \in \mathcal{E}_p} J_E \pi(a) = \sum_{F \in \mathcal{E}_q} \frac{1}{d} \sum_{E \in \mathcal{E}_p, E \cdot a = F} J_E \pi(a) = \sum_{F \in \mathcal{E}_q} \frac{1}{d} J_F = K_q.$$

Section 13.5

13.5.1 We rely on the fact that for a thin maximal code, the sets $E(X)$ and $S(X)$ are nonempty and equal (see Exercises 5.1.7 and 9.3.6). Thus (i) and (ii) are equivalent.

If $C_r(w)$ is maximal, then $w \in S(X)$. Indeed, suppose that $xwv \in X^*$ for some $x \in X^*$. Since $C_r(w) \subset C_r(xw)$, we have $C_r(w) = C_r(xw)$. Thus $wv \in X^*$.

If $C_r(w)$ is not maximal, then $w \notin E(X)$. Suppose indeed that $C_r(w) \subset C_r(u)$ with $v \in C_r(u) \setminus C_r(w)$. Let $s \in S(X)$ and suppose that for some $t \in A^*$, we have $wvst \in X^*$. Since $C_r(w) \subset C_r(u)$, we have $uvst \in X^*$. Since $v \in C_r(u)$ we have $uv \in X^*$ and consequently $st \in X^*$. Let $v' \in C_r(w)$ be such that $vst = v'x$ with $x \in X^*$. Then $uv'x = uvst$ forces $v = v'$ by unambiguity, a contradiction. Thus there is no t as above and $w \notin E(X)$.

13.5.2 Let $U = S(X) \setminus S(X)A^+$. Since $S(X)$ is a right ideal, we have $S(X) = UA^*$. Thus $\delta(S(X)) = \pi(U)$. Moreover, we have $E(X) \cap \varphi^{-1}(K) = D_X \cap \varphi^{-1}(K)$. Indeed, let $u \in D_X \cap \varphi^{-1}(K)$.

Since the right ideal $\varphi(uA^*)$ is minimal, for any $v \in A^*$ there is a $w \in A^*$ such that $\varphi(uvw) = \varphi(u)$. Since $u \in D_X$ there is a $w' \in A^*$ such that $uvww' \in X^*$. Thus $u \in E(X)$. The other inclusion is clear. Thus $\delta(D_X) = \delta(S(X))$. For any $w \in \varphi^{-1}(K) \cap D_X$, by Exercise 5.1.8, the set $C_r(w)U$ is a maximal prefix code and the product is unambiguous. Thus

$$\pi(C_r(w))\delta(D_X) = \pi(C_r(w))\delta(S(X)) = \pi(C_r(w))\pi(U) = \pi(C_r(w)U) = 1.$$

13.5.3 First, we have $\pi^\alpha(1) = \frac{1}{\lambda(\alpha)} \sum_{uv \in X} \pi\alpha^{-1}(uv) = \frac{1}{\lambda(\alpha)} \sum_{x \in X} |x| \pi\alpha^{-1}(x) = 1$.
Next,

$$\sum_{a \in A} \pi^\alpha(wa) = \frac{1}{\lambda(\alpha)} \sum_{(u,v) \in C(wa)} \pi\alpha^{(-1)}(uwav)$$

$$= \frac{1}{\lambda(\alpha)} \left(\sum_{\substack{(u,v) \in C(wa) \\ v \neq 1}} \pi\alpha^{(-1)}(uwv) + \sum_{\substack{(u,1) \in C(w) \\ x \in X}} \pi\alpha^{(-1)}(uwx) \right)$$

$$= \frac{1}{\lambda(\alpha)} \left(\sum_{\substack{(u,v) \in C(wa) \\ v \neq 1}} \pi\alpha^{(-1)}(uwv) + \sum_{(u,1) \in C(w)} \pi\alpha^{(-1)}(uw) \right) = \pi^\alpha(w).$$

A symmetric argument shows that $\sum_{a \in a} \pi^\alpha(aw) = \pi^\alpha(w)$. The contextual probability corresponds to the case where π is a Bernoulli distribution on B^*.

Chapter 14

Section 14.1

14.1.1 A word $x \in X^*$ as in the statement is called *separating*.

(a) A separating code is complete and synchronized since for any $w \in A^*$, one has $xwx \in X^*$.

(b) Let P be the set of strict left contexts of x and let S be the set of strict right contexts of x. Then $A^* = SX^*P$ unambiguously. Suppose that $A^* = S'X^*P'$ unambiguously. Let us first verify that the product $S'X^*P$ is unambiguous. Suppose indeed that $syp = s'y'p'$ for some $s, s' \in S'$, $y, y' \in X^*$ and $p, p' \in P$. Then $sypx = s'y'p'x$ are two factorizations in $S'X^*$ which is unambiguous and thus $s = s'$, $yp = y'p'$. Since X^*P is unambiguous, $y = y'$ and $p = p'$.

Let now R be the set such that $\underline{A}^* = \underline{S'X^*P} + \underline{R}$. Then $\underline{SX^*P} = \underline{S'X^*P} + \underline{R}$ and multiplying on the right both sides by $(1 - \underline{A})\underline{S}$, we obtain

$$\underline{S'} = \underline{S} - \underline{R}(1 - \underline{A})\underline{S}. \tag{15.6}$$

One can show symmetrically that the product SX^*P' is unambiguous and that the set T such that $\underline{A}^* = \underline{SX^*P'} + \underline{T}$ satisfies

$$\underline{P'} = \underline{P} - \underline{P}(1 - \underline{A})\underline{T}. \tag{15.7}$$

Substituting the expressions for $\underline{P'}$ and $\underline{S'}$ given by Equations (15.6) and (15.7) in the equality $\underline{S'X^*P'} = \underline{SX^*P}$, we obtain

$$\underline{S'X^*P'} = (\underline{S} - \underline{R}(1 - \underline{A})\underline{S})\underline{X^*}(\underline{P} - \underline{P}(1 - \underline{A})\underline{T})$$

$$= \underline{SX^*P} - \underline{R} - \underline{T} + \underline{R}(1 - \underline{A})\underline{T}.$$

Thus $\underline{R} + \underline{T} + \underline{R}\underline{A}\underline{T} = \underline{R}\underline{T}$ which forces $\underline{R} = \underline{T} = 0$, by considering the terms of lowest degree of both sides. Thus $\underline{S'}\underline{X}^*\underline{P} = \underline{S}\underline{X}^*\underline{P}$ and $\underline{S}\underline{X}^*\underline{P'} = \underline{S}\underline{X}^*\underline{P}$, which implies $P = P'$ and $\underline{S} = \underline{S'}$.

14.1.2 If x satisfies the conditions, for any word $w \in A^*$ there is a path $1 \xrightarrow{x} p \xrightarrow{w} q \xrightarrow{x} 1$. Then p is in $U(x)$ and q is in $V(x)$. The hypothesis on $U(x)$, $V(x)$ implies that $w = uv$ with $p \xrightarrow{u} 1 \xrightarrow{v} q$, showing that xu, $vx \in X^*$. Thus X is separating. The converse is clear.

14.1.3 Let (L, R) be a separating box. Let P be the set of right contexts of words in L and let Q be the set of left contexts of the words in R. Then $A^* = PX^*Q$ unambiguously.

14.1.4 Suppose that S, T satisfy the hypotheses. For $w \in A^*$, there is a unique pair $(s, t) \in S \times T$ such that $\varphi_A(swt)_{11} = 1$, and thus such that there is a path $1 \xrightarrow{s} p \xrightarrow{w} q \xrightarrow{t} 1$. Since $\varphi(s)_{1p} = 1$, p is in the set ℓ. Since $\varphi(t)_{q1} = 1$, q is in c. By condition (ii), we have $w = uv$ with $p \xrightarrow{u} 1 \xrightarrow{v} q$. We obtain su, $vt \in X^*$. Thus S, T is a separating box.

The converse implication is similar.

14.1.5 Let P (resp. Q) be the set of left (resp. right) contexts of b. Then $\underline{A}^* = \underline{Q}\underline{X}_u^*(\underline{P})$ and thus $\underline{X} - 1 = \underline{P}(\underline{A} - 1)\underline{Q} = \underline{X'} - 1 + \underline{P}b\underline{Q}$.

14.1.6 One has $a^n - 1 = \underline{P}(a - 1)\underline{Q}$ if and only if $\underline{P}\,\underline{Q} = 1 + a + \cdots + a^{n-1}$.

14.1.7 (a) is clear since Z is a suffix code on the alphabet X.

(b) Let V be the code defined by $\underline{V} - 1 = \underline{Q}(\underline{A} - 1)\underline{R}$. We have

$$\underline{P}(\underline{A} - 1)\underline{R} + w\underline{Q}(\underline{A} - 1)\underline{R} = \underline{Z} - 1 + w\underline{V} - w.$$

Since w is of maximal length in Z, the right-hand side has the form $\underline{T} - 1$ for a subset T of A^* which is a code by Proposition 14.1.1.

(c) We first show that T is uniquely factorizing. Suppose that $\underline{T} - 1 = \underline{F}(\underline{A} - 1)\underline{G}$. Let $n = |w|$ and m be the maximal length of words in G. It is possible to show that, for all $f \in F$, $|f| + m + 1 > n$ implies $f \in wA^*$.

This is shown by descending induction on the length of f. If f is of maximal length, then $fAg \subset wV$ for $|g| = m$ and thus $f \in wA^*$. Consider next $f \in F$, $a \in A$, and $g \in G$ such that $|fag| > n$ with $|g| = m$. We first rule out out the case $|f| < n$. If this were the case, we first suppose that $fag \in wV$. Then, for $b \neq a$, we have $fbg \notin wV$ and thus $fbg = f_1 g_1$ for some $f_1 \in F$ and $g_1 \in G$. Since $|g|$ is maximal, we have $|f_1| > |f|$, whence $f_1 \in wA^*$ by the induction hypothesis, a contradiction. Suppose next that $fAg \cap wV = \emptyset$. Using the same argument as above, we conclude that fa and fb are prefixes of w for $a \neq b$, a contradiction. Thus $|f| \geq n$. If $fag \notin wV$, then $fag = f_1g_1$ for some $f_1 \in F$ and $g_1 \in G$. Then $|f_1| > |f|$ implies $f_1 \in wA^*$ by induction hypothesis and finally $f \in wA^*$.

Let F_1 be the set of $f \in F$ such that $|fag| \le |w|$ for all $a \in A$ and $g \in G$ and let $F_2' = F \setminus F_1$. Then, as we have seen, $F_2' = wF_2$ and $F_1AG \cap wF_2G = \{w\}$. We thus obtain $\underline{P}(\underline{A} - 1)\underline{R} = \underline{F_1}(\underline{A} - 1)\underline{G}$ and $\underline{Q}(\underline{A} - 1)\underline{R} = \underline{F_2}(\underline{A} - 1)\underline{G}$. Since Z is separating, it is uniquely factorizing, and thus $R = G$. Thus T is uniquely factorizing.

The three-factor expression of $\underline{T} - 1$ does not correspond to a decomposition of T since $P \cup wQ$ is not prefix-closed and R is not suffix-closed. Since $\underline{P} + w\underline{Q}$ and $\underline{R}$ cannot be factorized into products of nontrivial characteristic polynomials, these are the only possible decompositions of T. Thus T is indecomposable.

(d) Z is separating. Let indeed $z = b$. We have for any word $w \in A^*$, $wb \in X^*$. Since $X^* = RZ^*$, we have either $wb \in Z^*$ or $wb = aav$ with $v \in Z^*$. In the first case we have $b, wb \in Z^*$ and in the second one $baa, vb \in Z^*$. Thus condition (i) is satisfied. Next, we have Card$(P \cup wQ) = 5$ and Card$(R) = 2$. Thus condition (ii) is satisfied. Finally, R is not suffix-closed since $a \notin R$ and thus condition (iii) is also satisfied.

14.1.8 (a) is a direct verification.

(b) We show that the code Z defined by the expression satisfies $Z^* = X^* \cap Y^*$. We have

$$\underline{Z} - 1 = (1 + \underline{A} + b^2)(\underline{A} - 1)(a(\underline{A} - 1) + 1)(1 + a + \underline{A}a)$$

$$= (\underline{X} - 1)(a(\underline{A} - 1) + 1)(1 + a + \underline{A}a)$$

$$= (\underline{X} - 1)(1 + a\underline{A} + ba + a\underline{A}^2a)$$

and thus $Z \subset X^*$, since $\underline{Z} - 1 = (\underline{X} - 1)\underline{P}$ with $P \subset X^*$. In the same way

$$\underline{Z} - 1 = (1 + \underline{A} + b^2)((\underline{A} - 1)a + 1)(\underline{A} - 1)(1 + a + \underline{A}a)$$

$$= (1 + a + b + +b^2)(1 - a + a^2 + ba)(\underline{Y} - 1)$$

$$= (1 + a\underline{A}a + b + b^2 + ba^2 + b^2\underline{A}a)(\underline{Y} - 1)$$

and $\underline{Z} - 1 = \underline{Q}(\underline{Y} - 1)$ with $Q \subset Y^*$. Thus Z decomposes on X and Y and consequently $Z \subset \overline{X}^* \cap Y^*$. The other inclusion follows from the fact that these are the only possible decompositions of Z.

(c) Z is synchronized since X and Y are. Let $x, y \in Z^*$ be such that $yA^*x \subset Z^*$. Then $yA^* \subset Y^*$ since Y is suffix and $A^*x \subset X^*$ since X is prefix. Consider the word xay. We cannot have $ya \in Z^*$ (since $a \notin X^*$) and neither $ax \in Z^*$ (since $a \notin Y^*$). Thus Z is not separating.

(d) Consider the automaton recognizing Z^* represented on Figure 15.11 (it can be computed either from the list of words forming Z or using the direct product of automata recognizing X^* and Y^*). Let us verify that $(\{b^3\}, \{1, a^5\})$ is a separating box. Indeed, the set of states q such that $0 \xrightarrow{b^3} q$ is $\ell = \{1, 3, 6\}$. It is a maximal row of the transition monoid of the automaton appearing as the first column of Table 15.8. The other maximal rows are $\{2, 4, 5\}$ and $\{4, 7\}$. Each of these sets

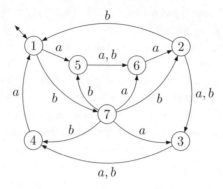

Figure 15.11 An automaton recognizing Z^*.

Table 15.8 *The maximal rows and columns.*

1	2	7
3	5	7
6	4	4

intersects in exactly one point the set $\{1, 5, 7\} = \{1\} \cup \{q \in Q \mid q \xrightarrow{a^5} 1\}$. This shows that condition (i) of Exercise 14.1.4 is satisfied for the pair $(\{1, 3, 6\}, \{1, 5, 7\})$. It can be checked that condition (ii) is also satisfied and thus the pair is a separating box. The corresponding factorization is $\underline{Z} - 1 = (\underline{X} - 1)\underline{P}$. Another separating box is $(\{1, a^4, a^4b\}, \{1, aba\}$. Indeed, the set of states q such that $q \xrightarrow{aba} 1$ is $\{2, 7\}$. But the set $\{1, 2, 7\}$ is a maximal column of the transition monoid of the automaton, appearing as the first row of Table 15.8. The other maximal columns are $\{3, 5, 7\}$ and $\{4, 6\}$. Each of them intersects in exactly one point the set $\{1, 3, 4\}$ which is the set of states q such that $1 \xrightarrow{u} q$ for $u = 1$, a^4 or a^4b. Thus the pair $(\{1, 3, 4\}, \{1, 2, 7\})$ satisfies condition (i). Since condition (ii) is also satisfied, the pair is a separating box. It corresponds to the other factorization $\underline{Z} - 1 = \underline{Q}(\underline{Y} - 1)$.

14.1.9 (a) We have

$$\sigma = (1 + w)(\underline{X} - 1 + \underline{G}_1 w \underline{D}_1 + \underline{G}_1 w^2 - \underline{G}_1 w + w^2 \underline{D}_1 + w^3 - w^2 - w\underline{D} + w) + 1$$
$$= (1 + w)\underline{R} + (1 + w)(w^3 - w^2 + w - 1) + 1 = (1 + w)\underline{R} + w^4.$$

It is easy to verify that R is a prefix code, that w is not a prefix of R, and that σ is the characteristic polynomial of a maximal prefix code.

The polynomial $\tau = (\underline{X} - 1 + (\underline{G} - 1)w\underline{D}(\underline{D} - 1))(1 + w) + 1$ satisfies $\tau = \underline{R}(1 + w) + w^4$ and thus τ has nonnegative coefficients. We have also $\tau - 1 = (\underline{P} + (\underline{G} - 1)w\underline{Q})(\underline{A} - 1)(1 + w)$ where P is the set of prefixes of X and Q the set of prefixes of D. Thus τ is the characteristic polynomial of a finite maximal code.

(b) We have $\gamma_w(X) \cap w^2 A^* = w^2 D_1 \cup w^2 D w$ and $\gamma_w(X) \cap A^* w^2 = G_1 w \cup G w^3$. Thus $\gamma_w(X) \cap w^2 A^* \cap A^* w^2 = \{w^4\}$, which shows that $x^2 = w^4$ is a pure square for $\gamma_w(X)$.

(c) It follows from the fact that $\underline{Y} = (1 + w)\underline{R} + w^4$ and $\underline{Z} = \underline{R}(1 + w) + w^4$ that for each $y \in Y^*$, we have either $y \in Z^*$ or $y = wz$ with $z, zw \in Z^*$. Indeed, if $y = y_1 y_2 \cdots y_n$, we have for each $i = 1, \ldots, n$, $y_i \in R$ or $y_i \in wR$ or $y_i = w^4$. We then glue each prefix w with the previous element of the factorization, except perhaps for the first one. Thus a word with d disjoint interpretations in Y^* has also d disjoint interpretations in Z^*.

(d) Let S be the set of suffixes of X and T be the set of suffixes of G. We have $\underline{X} - 1 = (\underline{A} - 1)\underline{S}$ and $\underline{G} - 1 = (\underline{A} - 1)\underline{T}$. Thus

$$\underline{Y} - 1 = (1 + w)(\underline{X} - 1 + (\underline{G} - 1)w(\underline{D} - 1)) = (1 + w)(\underline{A} - 1)(\underline{S} + \underline{T}w(\underline{D} - 1))$$

$$= (1 + w)(\underline{A} - 1)\underline{L}$$

with $L = (S \setminus Tw) \cup TwD$. Thus, equivalently $A^* = \underline{L}\underline{Y}^*(1 + w)$ is a factorization. Since S is a disjoint union of $d(X)$ maximal prefix codes and $Tw \subset S$, the set L is a disjoint union of $d(X)$ maximal prefix codes. Thus any word has $d(X)$ disjoint interpretations in Y^*.

(e) Let $G' = Yw^{-2}$ and $D' = w^{-2}Y$. We have $\underline{G'} = (1 + w)\underline{G_1} + w^2$ and $\underline{D'} = (1 + w)\underline{D_1} + w^2$. Thus $\underline{G'} - 1 = (1 + w)(\underline{G} - 1)$ and $\underline{D'} - 1 = (1 + w)(\underline{D} - 1)$. We have then the factorization

$$\underline{T} - 1 = (1 + w^2)(\underline{Y} - 1 + (\underline{G'} - 1)w(\underline{D'} - 1))$$

$$= (1 + w^2)(1 + w)(\underline{X} - 1 + (\underline{G} - 1)w(\underline{D} - 1) + (\underline{G} - 1)w(\underline{D'} - 1))$$

$$= (1 + w^2)(1 + w)(\underline{A} - 1)(\underline{S} + \underline{T}w(1 + w + w^2)(\underline{D} - 1))$$

$$= ((1 + w^2)(1 + w)(\underline{A} - 1)\underline{M},$$

where M is a disjoint union of $d(X)$ maximal prefix codes (observe that $(1 + w + w^2)(\underline{D} - 1) = \underline{E} - 1$ where E is a maximal prefix code). This shows that $d(T) = d(X)$. By (c) we obtain the conclusion $d(Z) = d(X)$.

(f) Suppose that $Z \subset V^*$ where V is a prefix code. Fix a letter $a \in A$. Set $d = d(X)$ and let $e < d$ be such that $a^e \in D$. Since $d > 2$, we have $w \neq a$. Since Z contains a^d and $a^d w$, we have $w \in V^*$. Since $Gw^3 D \setminus w^5$ is a subset of Z, we have $D_1 \subset V^*$ and thus $a^e \in V^*$. We conclude, since d is prime that $a \in V$. The case where V is a suffix code is symmetric.

(g) The set X defined by $\underline{X} = \underline{A}^d + (\underline{A} - 1)a^{n-1}ba^n(\underline{A} - 1)$ with $d = 2n + 1$ is a maximal bifix code. The word $(a^n b)^2$ is a pure square for X. Thus we may apply the above construction for any prime number $d > 2$.

Section 14.3

14.3.1 Take $P = 1$ and $Q = 0$.

14.3.2 By induction on n. It is clear for $n = 1$. Assume it holds for n. Then

$$x_1 + \cfrac{1}{x_2 + \cfrac{1}{\ddots + \cfrac{1}{x_n + \cfrac{1}{x_{n+1}}}}}$$

is equal to

$$\frac{p(x_1, \ldots, x_n + 1/x_{n+1})}{p(x_2, \ldots, x_n + 1/x_{n+1})}.$$

Next,

$$p(x_1, \ldots, x_n + 1/x_{n+1}) = p(x_1, \ldots, x_{n-1})(x_n + 1/x_{n+1}) + p(x_1, \ldots, x_{n-2})$$

$$= p(x_1, \ldots, x_{n-1})x_n + p(x_1, \ldots, x_{n-2}) + p(x_1, \ldots, x_{n-1})\frac{1}{x_{n+1}}$$

$$= p(x_1, \ldots, x_n) + p(x_1, \ldots, x_{n-1})\frac{1}{x_{n+1}}$$

$$= \frac{1}{x_{n+1}} p(x_1, \ldots, x_n, x_{n+1}),$$

Thus, the fraction is equal to

$$\frac{p(x_1, \ldots, x_n + 1/x_{n+1})}{p(x_2, \ldots, x_n + 1/x_{n+1})} = \frac{\frac{1}{x_{n+1}} p(x_1, \ldots, x_n, x_{n+1})}{\frac{1}{x_{n+1}} p(x_2, \ldots, x_n, x_{n+1})}.$$

14.3.3 The formula holds for $k = n$ since it reduces to $p(a_1, \ldots, a_n) - p(a_1, \ldots, a_{n-1})a_n = p(a_1, \ldots, a_{n-2})$. It also holds for $k = n - 1$. Indeed, the left-hand side is equal to $p(a_1, \ldots, a_n)a_{n-1} - p(a_1, \ldots, a_{n-1})(a_n a_{n-1} + 1)$, and since

$$p(a_1, \ldots, a_{n-1})(a_n a_{n-1} + 1) = p(a_1, \ldots, a_{n-1})a_n a_{n-1} + p(a_1, \ldots, a_{n-1})$$

$$= p(a_1, \ldots, a_{n-1})a_n a_{n-1} + p(a_1, \ldots, a_{n-2})a_{n-1} + p(a_1, \ldots, a_{n-3})$$

$$= p(a_1, \ldots, a_n)a_{n-1} + p(a_1, \ldots, a_{n-3})$$

we get

$$p(a_1, \ldots, a_n)a_{n-1} - p(a_1, \ldots, a_{n-1})(a_n a_{n-1} + 1)$$

$$= p(a_1, \ldots, a_n)a_{n-1} - p(a_1, \ldots, a_n)a_{n-1} - p(a_1, \ldots, a_{n-3})$$

$$= -p(a_1, \ldots, a_{n-3})$$

as required. Arguing by induction on decreasing values of k, we have, using the formula $p(a_n, \ldots, a_k) = p(a_n, \ldots, a_{k+1})a_k + p(a_n, \ldots, a_{k+2})$

$p(a_1, \ldots, a_n)\, p(a_{n-1}, \ldots, a_k) - p(a_1, \ldots, a_{n-1})\, p(a_n, \ldots, a_k)$

$$= p(a_1, \ldots, a_n)\, p(a_{n-1}, \ldots, a_{k+1})a_k + p(a_1, \ldots, a_n)\, p(a_{n-1}, \ldots, a_{k+2})$$

$$- p(a_1, \ldots, a_{n-1})\, p(a_n, \ldots, a_{k+1})a_k - p(a_1, \ldots, a_{n-1})\, p(a_n, \ldots, a_{k+2})$$

$$= (-1)^{n+k+1}\, p(a_1, \ldots, a_{k-1})a_k + (-1)^{n+k+2}\, p(a_1, \ldots, a_k)$$

$$= (-1)^{n+k}\, p(a_1, \ldots, a_{k-2}).$$

14.3.4 Set $f_{n+1} = p(1, \ldots, 1)$ (n times). Then $f_0 = 0$, $f_1 = 1$, and by the definition, one gets $f_{n+1} = f_n + f_{n-1}$.

Section 14.4

14.4.1 This is clear for $S(u)$, $P(u)$, and $F(u, v)$ by definition.

14.4.2 This results from the formula $a^{-1}(ST) = a^{-1}(S)T + (S, 1)a^{-1}(T)$ and from the fact that $S^* = 1 + SS^*$.

Section 14.5

14.5.1 The proof is easy by induction on the number of nodes of the tree and the number of states of the literal automaton.

Section 14.6

14.6.1 Since $a, c \in Y$, we have $ba \in Y$. But then all conjugates of acb have a prefix in Y.

14.6.2 Set $p(z) = (1 - u(z)/(1 - kz))$ with $p(z) = \sum_{i \geq 0} p_i z^i$. Then for each $n \geq 1$

$$p_n/k^n = 1 - u_1/k - \cdots - u_n/k^n,$$

whence the result.

Section 14.7

14.7.1 Any $\ell \in E_0$ is a linear combination $\sum \lambda_u i\varphi(u)$, where **i** denotes the characteristic row vector of I and **T** denotes the characteristic column vector of T. For $v \in A^*$, we have $(\gamma(\ell), v) = (\sum \lambda_u(\sigma \cdot u), v) = \sum \lambda_u(\sigma, uv) = \sum \lambda_u i\varphi(uv)\mathbf{T} = \ell\varphi(v)\mathbf{T}$. Thus $\gamma(\ell) = 0$ if and only if $\ell \in E_1$.

14.7.2 If S is recognizable, there is a finite automaton $\mathcal{A} = (Q, i, T)$ recognizing S. Then, by Exercise 14.7.1, the dimension of V_σ is at most equal to $\text{Card}(Q)$.

14.7.3 Theorem 14.7.4 can be stated more generally as: A linear representation of a finite group G over a field of characteristic 0 or prime to the order of G is completely reducible. The same proof applies with the observation that the map θ is well defined under the hypothesis. The rest of the proof of Theorem 14.7.5 remains unchanged.

14.7.4 Suppose, as in the proof of Theorem 14.7.5, that W is an invariant subspace of V. Let W' be the supplementary subspace of W defined in the proof. Since X is synchronized, the idempotent e has rank 1 and therefore S has dimension 1. Thus either $T = \{0\}$ or $T = S$. In the first case, $T' = S$, which implies $W' = V$ and thus $W = \{0\}$. In the second case, $W' = \{0\}$ and thus $W = V$. Thus, the representation is irreducible.

Appendix: Research problems

In this appendix, we gather, for the convenience of the reader, the conjectures mentioned in the book and present some additional open problems. We take this opportunity to discuss some of them in more detail.

The inclusion problem Recall from Chapter 2 that the *inclusion problem* for a finite code X is the existence of a finite maximal code containing X. The *inclusion conjecture* is that this problem is decidable.

The smallest integer k for which a k element code is known which is not included in a finite maximal code is $k = 4$. Such an example is the code $X = \{a^5, ba^2, ab, b\}$ of Example 2.5.7. Proposition 12.3.3 describes an infinite family of codes to which X belongs. It is not known whether every code with three elements is included in a finite maximal code.

For a finite bifix code X, the existence of a finite maximal bifix code containing X is decidable. Indeed, if X is insufficient, then any maximal bifix code with kernel X is finite by Proposition 6.5.6. On the contrary, if X is sufficient, then the degree of a finite maximal code containing X must be equal to the common value (L_X, w) of the indicator L_X of X for any full word w whose length exceeds the maximal length of the words of X. Since there is a finite number of finite maximal bifix codes with given degree, this gives a decision procedure (although it is not a very practical one).

Complexity of unique decipherability The precise complexity of the test for unique decipherability is still unknown. The same holds for the property of completeness. The length of the shortest word w such that w is not a factor of X^* for a finite set X has been studied by Restivo (1981). The bound proposed in Restivo (1981) is $2k^2$ where $k = \max_{x \in X} |x|$. A counterexample has been obtained by a computer-aided search using the software Vaucanson. It is believed that the conjecture is true with a larger value of the constant.

Černý's conjecture Recall from Chapter 3 that *Černý's conjecture* asserts that any synchronized strongly connected deterministic automaton with n states has a synchronizing word of length at most $(n-1)^2$. The conjecture is known to be true in several particular cases. For example, the conjecture holds if there is a letter which acts as an n-cycle on the set of states, see Dubuc (1998). This result has been generalized to so-called strongly transitive automata by Carpi and D'Alessandro (2008).

The best upper bound known is $(n^3 - n)/6$, far from the lower bound. For an n-state so-called monotonic automaton over a k-letter input alphabet there exists an algorithm that finds a synchronizing word in $O(n^3 + n^2 k)$ time and $O(n^2)$ space; for this subclass of automata, an upper bound of $(n - 1)^2$ on the length of a synchronizing word can be proven. It has also been proved that finding the minimum length synchronizing word is an NP-complete problem. For a recent survey, see Volkov (2008).

The same conjecture can be formulated for unambiguous automata instead of deterministic ones. The cubic bound which is easy to obtain for deterministic automata can still be proved by a result of Carpi (1988) (Exercise 9.3.13).

Bifix codes   Recall from Chapter 6 that it is conjectured that for any sequence of nonnegative integers u_n such that $\sum_{n \geq 0} u_n k^{-n} \leq 3/4$, there exists a bifix code X on k letters with length distribution $(u_n)_{n \geq 0}$. Among the partial results obtained so far, we mention that for $k = 2$, the conjecture holds with $3/4$ replaced by $5/8$, as shown by Yekhanin (2004).

Groups of codes   The first problem is simply to study whether Proposition 11.1.6 holds for arbitrary thin maximal codes.

Next, let $X \subset A^+$ be a finite code with n elements and let $\mathcal{A} = (Q, 1, 1)$ be a trim unambiguous automaton recognizing X^*. Let $\varphi = \varphi_{\mathcal{A}}$ and let $M = \varphi(A^*)$. Let e be an idempotent in the transition monoid of the automaton $\mathcal{A}$ and let H be the $\mathcal{H}$-class of e. Schützenberger (1979a) has proved that either $\varphi^{-1}(H)$ is cyclic or the group G_e has degree at most $2n$. This bound can be reduced to n by using the *critical factorization theorem* (see Lothaire (1997)). It is conjectured that actually, the degree of G_e is at most $n - 1$ if $\varphi^{-1}(H)$ is not cyclic. This is known to be true if X is prefix (Perrin and Rindone (2003)).

Finally, it is not known whether Theorem 11.6.5 holds more generally for finite maximal prefix codes. For example, it is not known if there exists a finite maximal prefix code X such that $G(X)$ is the dihedral group D_5.

Finite factorizations   Given a factorization $A^* = X_n^* X_{n-1}^* \cdots X_1^*$ with n factors, are the codes X_i always limited? This is true if the factorization is obtained by iterating bisections (Exercise 8.2.1). It is true for factorizations with up to four factors by a result of Krob (1987). A conjecture in relation to factorizations is the following. If $A^* = M_1 \cdots M_n$ where $M_1, \ldots, M_n$ are submonoids, then the M_i are free submonoids. This is known to hold up to $n = 4$, see Krob (1987).

Probability distributions   Let $X \subset A^*$ be a finite maximal code and let π be a probability distribution on A^*. It is conjectured that if π is invariant and multiplicative on X^*, then it is a Bernoulli distribution. This has been proved to hold for a finite maximal prefix code by Langlois, as reported in Hansel and Perrin (1989).

Factorization conjecture   Recall from Chapter 14 that the *factorization conjecture* states that any finite maximal code is positively factorizing and that the *commutative equivalence conjecture* states that any finite maximal code is commutatively prefix. By Corollary 14.6.6, the factorization conjecture implies the commutative equivalence conjecture. There are relations between the factorization conjecture and factorizations of cyclic groups. These have been described in a series of papers, see De Felice (2007). It is not known whether every finite maximal code has a separating box (see Exercise 14.1.3). A positive answer would solve the factorization conjecture. It is conjectured that the polynomial Q in Equation (14.7) has the form $Q = \sum_{i=1}^{d-1} \underline{U_i}$, where each U_i is a nonempty prefix-closed set.

Noncommutative polynomials   Let K be a field and let A be an alphabet. A subring R of $K\langle A \rangle$ is *free* if it is isomorphic to $K\langle B \rangle$ for some alphabet B. A subring R of $K\langle A \rangle$ is called an *anti-ideal* if for any $u \in K\langle A \rangle$ and nonzero $v, w \in R$, $uv, wu \in R$ implies $u \in R$. By a theorem of Kolotov (1978), a free subring of $K\langle A \rangle$ is an anti-ideal, see also Lothaire (2002). Thus the subring generated by a submonoid M of A^* is an anti-ideal if and only if it is free. Indeed, if $K\langle M \rangle$ is an anti-ideal, then M is stable and therefore is free. This is not true for arbitrary subrings of $K\langle A \rangle$, Cohn (1985), Exercise 6.6.11 gives a counterexample which he credits to Dicks. It is not known whether the property that $K\langle Y \rangle$ is free, for a finite set Y of $K\langle A \rangle$, is decidable.

Some of the problems presented in this appendix were already mentioned in Berstel and Perrin (1986). They are also discussed in Bruyère and Latteux (1996) and Béal *et al.* (2009).

References

Roy L. Adler and Benjamin Weiss (1970). *Similarity of Automorphisms of the Torus.* Memoirs of the American Mathematical Society, No. 98. American Mathematical Society. 396

Roy L. Adler, L. Wayne Goodwyn, and Benjamin Weiss (1977). Equivalence of topological Markov shifts. *Israel J. Math.*, **27** (1): 48–63. 396

Roy L. Adler, Donald Coppersmith, and Martin Hassner (1983). Algorithms for sliding block codes. *IEEE Trans. Inform. Theory*, **IT-29**: 5–22. 175

Rudolf Ahlswede, Bernhard Balkenhol, and Levon H. Khachatrian (1996). Some properties of fix-free codes. In *Proc. 1st Int. Sem. on Coding Theory and Combinatorics, Thahkadzor, Armenia*, pp. 20–33. 273

Alfred V. Aho and Margaret J. Corasick (1975). Efficient string matching: An aid to bibliographic search. *Communications of the ACM*, **18**: 335–40. 105

Alfred V. Aho, John E. Hopcroft, and Jeffrey D. Ullman (1974). *The Design and Analysis of Computer Algorithms.* Addison-Wesley. 104

Alberto Apostolico and Raffaele Giancarlo (1984). Pattern-matching implementation of a fast test for unique decipherability. *Inform. Process. Lett.*, **18**: 155–8. 105

Robert B. Ash (1990). *Information Theory.* Dover Publications Inc. Corrected reprint of the 1965 original. 104

Jonathan Ashley, Brian Marcus, Dominique Perrin, and Selim Tuncel (1993). Surjective extensions of sliding block codes. *SIAM J. Discrete Math.*, **6**: 582–611. 224

G. Bandyopadhyay (1963). A simple proof of the decipherability criterion of Sardinas and Patterson. *Inform. and Control*, **6**: 331–6. 104

Evelyne Barbin-Le Rest and Michel Le Rest (1985). Sur la combinatoire des codes à deux mots. *Theoret. Comput. Sci.*, **41** (1): 61–80. 325

Frédérique Bassino, Marie-Pierre Béal, and Dominique Perrin (2000). A finite state version of the Kraft-McMillan theorem. *SIAM J. Comput.*, **30** (4): 1211–30 (electronic). 175

Marie-Pierre Béal (1993). *Codage symbolique.* Masson. 104

Marie-Pierre Béal and Dominique Perrin (2005). Codes and sofic constraints. *Theoret. Comput. Sci.*, **340**(2): 381–93. 105

Marie-Pierre Béal and Dominique Perrin (2006). Complete codes in a sofic shift. In *STACS 2006*, volume 3884 of *Lecture Notes in Computer Science*, pp. 127–36. Springer-Verlag. 105

Marie-Pierre Béal, Olivier Carton, and Christophe Reutenauer (1996). Cyclic languages and strongly cyclic languages. In C. Puech and R. Reischuk, editors, *STACS'96*, volume 1046 of *Lecture Notes in Computer Science*, pp. 49–59. Springer-Verlag. 326

Marie-Pierre Béal, Sylvain Lombardy, and Jacques Sakarovitch (2005). On the equivalence of ℤ-automata. In *ICALP'05*, volume 3580 of *Lecture Notes in Computer Science*, pp. 397–409. Springer-Verlag. 175

Marie-Pierre Béal, Eugen Czeizler, Jarkko Kari, and Dominique Perrin (2008). Unambiguous automata. *Math. Comput. Sci.*, **1** (4): 625–38. 370

Marie-Pierre Béal, Jean Berstel, Brian H. Marcus, Dominique Perrin, Christophe Reutenauer, and Paul H. Siegel (2009). Variable length-codes and finite automata. In Isaac Woungang, editor, *Selected Topics in Information and Coding Theory*. World Scientific. 104, 593

Jean Berstel (1979). *Transductions and Context-Free Languages*. Teubner. 198

Jean Berstel and Dominique Perrin (1986). Trends in the theory of codes. *Bull. Eur. Assoc. Theor. Comput. Sci. EATCS*, **29**: 84–95. 593

Jean Berstel and Dominique Perrin (2007). The origins of combinatorics on words. *European J. Combin.*, **28** (3): 996–1022. 105

Jean Berstel and Christophe Reutenauer (1988). *Rational Series and their Languages*. Springer-Verlag. 53, 370

Jean Berstel and Christophe Reutenauer (1990). Zeta functions of formal languages. *Trans. Amer. Math. Soc.*, **321**: 533–46. 326

Jean Berstel, Dominique Perrin, Jean François Perrot, and Antonio Restivo (1979). Sur le théorème du défaut. *J. Algebra*, **60**: 169–80. 104, 224

Marek Tomasz Biskup (2008). Shortest synchronizing strings for Huffman codes. In E. Ochmanski and J. Tyszkiewicz, editors, *Mathematical Foundations of Computer Science 2008, Torun, Poland, August 25-29, 2008*, volume 5162 of *Lecture Notes in Computer Science*, pp. 120–31. Springer-Verlag. 175

François Blanchard and Georges Hansel (1986). Systèmes codés. *Theoret. Comput. Sci.*, **44** (1): 17–49. 491

François Blanchard and Dominique Perrin (1980). Relèvement d'une mesure ergodique par un codage. *Z. Wahrsch. Verw. Gebiete*, **54**: 303–11. 491

Edward K. Blum (1965). Free subsemigroups of a free semigroup. *Michigan Math. J.*, **12**: 179–82. 104

Jean-Marie Boë (1976). Représentations des monoïdes: Applications à la théorie des codes. PhD thesis, Montpellier. 370

Jean-Marie Boë (1981). Sur les codes synchronisants coupants. In *Noncommutative Structures in Algebra and Geometric Combinatorics (Naples, 1978)*, volume 109 of *Quad. "Ricerca Sci."*, pp. 7–10. 534

Jean Marie Boë (1991). Les boîtes. *Theoret. Comput. Sci.*, **81** (1, (Part A)): 17–34. 370

Jean-Marie Boë, Jeanine Boyat, Jean-Pierre Bordat, and Yves Cesari (1979). Une caractérisation des sous-monoïdes libérables. In D. Perrin, editor, *Théorie des Codes (actes de la septième École de Printemps d'Informatique Théorique)*, LITP, pp. 9–20. 370

Jean-Marie Boë, Aldo de Luca, and Antonio Restivo (1980). Minimal completable sets of words. *Theoret. Comput. Sci.*, **12**: 325–32. 104

Phillip G. Bradford, Mordecai J. Golin, Lawrence L. Larmore, and Wojciech Rytter (2002). Optimal prefix-free codes for unequal letter costs: dynamic programming with the Monge property. *J. Algorithms*, **42** (2): 277–303. 175

Véronique Bruyère (1987). Maximal prefix products. *Semigroup Forum*, **36**: 147–57. 174

Véronique Bruyère (1992). Automata and codes with bounded deciphering delay. In *LATIN '92 (São Paulo, 1992)*, volume 583 of *Lecture Notes in Computer Science*, pp. 99–107. Springer-Verlag. 224

Véronique Bruyère (1998). On maximal codes with bounded synchronization delay. *Theoret. Comput. Sci.*, **204**: 11–28. 395

Véronique Bruyère and Clelia De Felice (1996). Any lifting of a trace coding is a word coding. *Inform. and Comput.*, **130**(2): 183–193, 1996. 105

Véronique Bruyère and Clelia De Felice (1992). Synchronization and decomposability for a family of codes. *Internat. J. Algebra Computation*, **2**: 367–93. 534

Véronique Bruyère and Michel Latteux (1996). Variable-length maximal codes. In *ICALP'96*, volume 1099 of *Lecture Notes in Computer Science*, pp. 24–47. Springer-Verlag. 224, 593

Véronique Bruyère and Dominique Perrin (1999). Maximal bifix codes. *Theoret. Comput. Sci.*, **218** (1): 107–21. 274

Véronique Bruyère, Li Min Wang, and Liang Zhang (1990). On completion of codes with finite deciphering delay. *European J. Combin.*, **11** (6): 513–21. 224

Véronique Bruyère, Denis Derencourt, and Michel Latteux (1998). The meet operation in the lattice of codes. *Theoret. Comput. Sci.*, **191** (1-2): 117–29. 176, 371

John A. Brzozowski (1967). Roots of star events. *J. Assoc. Comput. Mach.*, **14**: 466–77. 104

Renato M. Capocelli and Christoph M. Hoffmann (1985). Algorithms for factorizing semigroups. In A. Apostolico and Z. Galil, editors, *Combinatorial Algorithms on Words (Maratea, 1984)*, volume 12 of *NATO Adv. Sci. Inst. Ser. F*, pp. 59–81. Springer-Verlag. 105

Renato M. Capoceli, Alfredo A. De Santis, Luisa Gargano, and Ugo Vaccaro (1992). On the construction of statistically synchronizable codes. *IEEE Trans. Inform. Theory*, **38**(2, part 1): 407–14. 175

Arturo Carpi (1987). On unambiguous reductions of monoids of unambiguous relations. *Theoret. Comput. Sci.*, **51** (1-2): 215–20. 370

Arturo Carpi (1988). On synchronizing unambiguous automata. *Theoret. Comput. Sci.*, **60** (3): 285–96. 370, 592

Arturo Carpi and Flavio D'Alessandro (2008). The synchronization problem for strongly transitive automata. In Masami Ito and Masafumi Toyama, editors, *Developments in Language Theory, 12th International Conference, DLT 2008,*

Kyoto, Japan, September 16-19, 2008, volume 5257 of *Lecture Notes in Computer Science*, pp. 240–51. Springer-Verlag. 592

Larry Carter and John Gill (1974). Conjectures on uniquely decipherable codes. *IRE Trans. Inform. Theory*, **IT-20**: 394–96. 534

Ján Černý (1964). Poznamka k homogenym s konecnymi automati. *Mat.-fyz. cas. SAV.*, **14**: 208–15. 175

Yves Césari (1972). Sur un algorithme donnant les codes bipréfixes finis. *Math. Systems Theory*, **6**: 221–25. 273

Yves Césari (1974). Sur l'application du théorème de Suschkevitch à l'étude des codes rationnels complets. In *Automata, Languages and Programming*, volume 14 of *Lecture Notes in Computer Science*, pp. 342–50. Springer-Verlag. 370

Yves Césari (1979). Propriétes combinatoires des codes bipréfixes. In D. Perrin, editor, *Théorie des Codes (actes de la septième École de Printemps d'Informatique Théorique)*, pp. 20–46. LITP. 273, 274

Christian Choffrut (1979). Une caractérisation des codes à délai borné par leur fonction de décodage. In D. Perrin, editor, *Théorie des Codes (actes de la septième École de Printemps d'Informatique Théorique)*, pp. 47–56. LITP. 224

Alfred H. Clifford and Gordon B. Preston (1961). *The Algebraic Theory of Semigroups*, volume 1. American Mathematical Society. 53, 370

Paul M. Cohn (1962). On subsemigroups of free semigroups. *Proc. Amer. Math. Soc.*, **63**: 347–51. 104

Paul M. Cohn (1985). *Free Rings and their Relations*, volume 19 of *London Mathematical Society Monographs*. Academic Press, second edition. (First edition 1971). 104, 534, 593

John H. Conway (1971). Three lectures on exceptional groups. In *Finite Simple Groups (Proc. Instructional Conf., Oxford, 1969)*, pp. 215–47. Academic Press. 434

Karel Culik, Juhani Karhumäki, and Jarkko Kari (2002). A note on synchronized automata and the road coloring problem. In W. Kuich, editor, *Developments in Language Theory (Vienna, 2001)*, volume 2295 of *Lecture Notes in Computer Science*, pp. 175–85. Springer-Verlag. 396

Nicholaas Govert De Bruijn (1953). On the factorization of cyclic groups. *Indag. Math.*, **15**: 258–64. 449

Clelia De Felice (1983). A note on the triangle conjecture. *Inform. Process. Lett.*, **14**: 197–200. 534

Clelia De Felice (1992). On the factorization conjecture. In *STACS'92*, volume 577 of *Lecture Notes in Computer Science*, pp. 545–56. Springer-Verlag. 534

Clelia De Felice (1993). A partial result about the factorization conjecture for finite variable-length codes. *Discrete Math.*, **122**: 137–52. 534

Clelia De Felice (1996). An application of Hajós factorizations to variable-length codes. *Theoret. Comput. Sci.*, **164**: 223–52. 449

Clelia De Felice (2007). Finite completions via factorizing codes. *Internat. J. Algebra Computation*, **17** (4): 715–60. 593

Clelia De Felice and Antonio Restivo (1985). Some results on finite maximal codes. *RAIRO Informat. Theor.*, **19**: 383–403. 449

Clelia De Felice and Christophe Reutenauer (1986). Solution partielle de la conjecture de factorisation des codes. *C. R. Acad. Sci. Paris*, **302**: 169–70. 534

Aldo de Luca (1976). A note on variable length codes. *Inform. and Control*, **32**: 263–71. 104

Aldo de Luca and Antonio Restivo (1980). On some properties of very pure codes. *Theoret. Comput. Sci.*, **10**: 157–70. 298, 396

Xiaotie Deng, Guojun Li, and Wenan Zang (2004). Proof of Chvátal's conjecture on maximal stable sets and maximal cliques in graphs. *J. Combin. Theory Ser. B*, **91** (2): 301–25. 370

Xiaotie Deng, Guojun Li, and Wenan Zang (2005). Corrigendum to: "Proof of Chvátal's conjecture on maximal stable sets and maximal cliques in graphs" [J. Combin. Theory Ser. B **91** (2004), no. 2, 301–25; mr2064873]. *J. Combin. Theory Ser. B*, **94** (2): 352–53. 370

Christian Deppe and Holger Schnettler (2006). On q-ary fix-free codes and directed deBrujin graphs. In *IEEE International Symposium on Information Theory*, pp. 1482–5. 273

Denis Derencourt (1996). A three-word code which is not prefix-suffix composed. *Theoret. Comput. Sci.*, **163**: 145–60. 106

John S. Devitt and David M. Jackson (1981). Comma-free codes: An extension of certain enumerative techniques to recursively defined sequences. *J. Combin. Theory Ser. A*, **30**: 1–18. 299

Volker Diekert and Anca Muscholl (1996). Code problems on traces. In *Mathematical Foundations of Computer Science 1996 (Cracow)*, volume 1113 of *Lecture Notes in Computer Science*, pp. 2–17. Springer-Verlag. 105

Louis Dubuc (1998). Sur les automates circulaires et la conjecture de Černý. *RAIRO Inform. Théor. Appl.*, **32** (1-3): 21–34. 592

Williard L. Eastman (1965). On the construction of comma-free codes. *IEEE Trans. Inform. Theory*, **IT-11**: 263–7. 299

Andrei Ehrenfeucht and Gregorz Rozenberg (1978). Elementary homomorphisms and a solution to the D0L sequence equivalence problem. *Theoret. Comput. Sci.*, **7**: 169–84. 104

Andrei Ehrenfeucht and Gregorz Rozenberg (1983). Each regular code is included in a regular maximal code. *RAIRO Informat. Theor.*, **20**: 89–96. 104

Samuel Eilenberg (1974). *Automata, Languages and Machines*, volume A. Academic Press. 52, 53, 104, 198

Samuel Eilenberg (1976). *Automata, Languages and Machines*, volume B. Academic Press. 299, 396

Peter Elias (1975). Universal codeword sets and representations of the integers. *IEEE Trans. Inform. Theory*, **21**(2): 194–203. 174

William Feller (1968). *An Introduction to Probability Theory and Its Applications*. Wiley, third edition. 175, 491

Michel Fliess (1974). Matrices de Hankel. *J. Math. Pures Appl.*, **53**: 197–222. 534

Dominique Foata and Guo Niu Han (1994). Nombres de Fibonacci et polynômes orthogonaux. In M. Morelli and M. Tangheroni, editors, *Leonardo Fibonacci:*

il tempo, le opere, l'eredità scientifica, pp. 179–200, Pisa, 23–25 March. Pacini Editore (Fondazione IBM Italia). 106

Christopher F. Freiling, Douglas S. Jungreis, Francois Théberge, and Kenneth Zeger (2003). Almost all complete binary prefix codes have a self-synchronizing string. *IEEE Trans. Inform. Theory*, **49**(9):2219–25. 175

Joel Friedman (1990). On the road coloring problem. *Proc. Amer. Math. Soc.*, **110** (4): 1133–5. 371, 396

Zvi Galil (1985). Open problems in stringology. In A. Apostolico and Z. Galil, editors, *Combinatorial Algorithms on Words (Maratea, 1984)*, volume 12 of *NATO Adv. Sci. Inst. Ser. F*, pp. 1–8. Springer-Verlag. 105

Robert G. Gallager and David C. van Voorhis (1975). Optimal source codes for geometrically distributed integer alphabets. *IEEE Trans. Inform. Theory*, **21**: 228–30. 176

Felix R. Gantmacher (1959). *The Theory of Matrices* Vols 1, 2. Chelsea. Translated from the Russian original. 52

Adriano M. Garsia and Michelle L. Wachs (1977). A new algorithm for minimum cost binary trees. *SIAM J. Comput.*, **6** (4): 622–42. 175

Israel M. Gel′fand and Vladimir S. Retakh (1991). Determinants of matrices over noncommutative rings. *Funktsional. Anal. i Prilozhen.*, **25** (2): 13–25, 96. 198

Edgar N. Gilbert (1960). Synchronization of binary messages. *IRE Trans. Inform. Theory*, **IT-6**: 470–77. 299

Edgar N. Gilbert and Edward F. Moore (1959). Variable length binary encodings. *Bell System Tech. J.*, **38**: 933–67. 104, 175, 223, 224, 273

David Gillman and Ronald Rivest (1995). Complete variable length fix-free codes. *Designs, Codes and Cryptography*, **5**: 109–14. 273

Bernd Girod (1999). Bidirectionally decodable streams of prefix code words. *IEEE Communications Letters*, **3** (8): 245–7. 274

Mordecai J. Golin and Günter Rote (1998). A dynamic programming algorithm for constructing optimal prefix-free codes with unequal letter costs. *IEEE Trans. Inform. Theory*, **44** (5): 1770–81. 175

Mordecai J. Golin, Claire Kenyon, and Neal E. Young (2002). Huffman coding with unequal letter costs. In *Proceedings of the Thirty-Fourth Annual ACM Symposium on Theory of Computing*, pp. 785–91 (electronic). ACM. 176

Solomon W. Golomb (1966). Run-length encodings. *IEEE Trans. Inform. Theory*, **IT-12**: 399–401. 174

Solomon W. Golomb and Basil Gordon (1965). Codes with bounded synchronization delay. *Inform. and Control*, **8**: 355–72. 395

Solomon W. Golomb, Basil Gordon, and Lloyd R. Welch (1958). Comma free codes. *Canad. J. Math.*, **10**: 202–9. 299

Ian P. Goulden and David M. Jackson (2004). *Combinatorial Enumeration*. Dover Publications Inc. Reprint of the 1983 original. 105

Ulf Greenander (1963). *Probabilities on Algebraic Structures*. Wiley. 491

Yannick Guesnet (2003). On maximal synchronous codes. *Theoret. Comput. Sci.*, **307**(1):129–38. 395

Leonidas J. Guibas and Andrew M. Odlyzko (1978). Maximal prefix synchronized codes. *SIAM J. Appl. Math.*, **35**: 401–18. 299

Paul R. Halmos (1950). *Measure Theory*. Van Nostrand. 490

Georges Hansel (1982). Baïonettes et cardinaux. *Discrete Math.*, **39**: 331–5. 534

Georges Hansel and Dominique Perrin (1983). Codes and Bernoulli partitions. *Math. Systems Theory*, **16**: 133–57. 491, 534

Georges Hansel and Dominique Perrin (1989). Rational probability measures. *Theoret. Comput. Sci.*, **65** (2): 171–88. 592

Georges Hansel, Dominique Perrin, and Christophe Reutenauer (1984). Factorizing the polynomial of a code. *Trans. Amer. Math. Soc.*, **285**: 91–105. 534

Tero Harju and Dirk Nowotka (2004). The equation $x^i = y^j z^k$ in a free semigroup. *Semigroup Forum*, **68** (3): 488–90. 325

Kosaburo Hashiguchi and Namio Honda (1976a). Homomorphisms that preserve star-height. *Inform. and Control*, **30**: 247–66. 104

Kosaburo Hashiguchi and Namio Honda (1976b). Properties of code events and homomorphisms over regular events. *J. Comput. System Sci.*, **12**: 352–67. 299

Tom Head and Andreas Weber (1993). Deciding code related properties by means of finite transducers. In R. Capocelli, A. De Santis, and U. Vaccaro, editors, *Sequences, II (Positano, 1991)*, pp. 260–72. Springer-Verlag. 105

Tom Head and Andreas Weber (1995). Deciding multiset decipherability. *IEEE Trans. Inform. Theory*, **41** (1): 291–7. 105

Israel N. Herstein (1969). *Non-commutative Rings*. Carus Mathematical Monographs. Wiley. 534

Christoph M. Hoffmann (1984). A note on unique decipherability. In *Math. Foundations Comput. Sci. (MFCS)*, volume 176 of *Lecture Notes in Computer Science*, pp. 50–63. Springer-Verlag. 105

Te Chiang Hu and Alan C. Tucker (1971). Optimal computer search trees and variable-length alphabetical codes. *SIAM J. Appl. Math.*, **21**: 514–32. 176

Te Chiang Hu and Paul A. Tucker (1998). Optimal alphabetic trees for binary search. *Inform. Process. Lett.*, **67** (3): 137–40. 176

Te Chiang Hu and Man-Tak Shing (2002). *Combinatorial Algorithms*. Dover Publications Inc., second edition. 176

David A. Huffman (1952). A method for the construction of minimum redundancy codes. *Proceedings of the Institute of Electronics and Radio Engineers*, **40** (10): 1098–101. 175

David A. Huffman (1959). Notes on information-lossless finite-state automata. *Nuovo Cimento (10)*, **13** (supplemento): 397–405. 198

Bertram Huppert (1967). *Endliche Gruppen*. Springer-Verlag. 434

Bertram Huppert and Norman Blackburn (1982). *Finite Groups II and III*, volume 242 and 243 of *Grundlehren der Mathematischen Wissenschaften [Fundamental Principles of Mathematical Sciences]*. Springer-Verlag. 434

Alon Itai (1976). Optimal alphabetic trees. *SIAM J. Comput.*, **5** (1): 9–18. 176

Masami Ito and Gabriel Thierrin (1994). Congruences, infix and cohesive prefix codes. *Theoret. Comput. Sci.*, **136**(2):471–85. 274

Masami Ito, Helmut Jürgensen, Huei-Jan Shyr, and Gabriel Thierrin (1991). Outfix and infix codes and related classes of languages. *J Comput. Syst. Sci.*, **43**(3):484–508. 274

B. H. Jiggs (1963). Recent results in comma-free codes. *Canad. J. Math.*, **15**: 178–87. Jiggs is a pseudonym for Baumert, Hales, Jewett, Golomb, Gordon, Selfridge; the i is imaginary. 299

Juhani Karhumäki (1984). A property of three element codes. In *STACS'84*, volume 166 of *Lecture Notes in Computer Science*, pp. 305–13. Springer-Verlag. 224

Richard M. Karp (1961). Minimum redundancy codes for the discrete noiseless channel. *IRE Trans. Inform. Theory*, **IT-7**: 27–38. 175, 534

Gerhard Keller (1991). Circular codes, loop counting, and zeta-functions. *J. Combin. Theory Ser. A*, **56** (1): 75–83. 299

Jeffrey H. Kingston (1988). A new proof of the Garsia-Wachs algorithm. *J. Algorithms*, **9** (1): 129–36. 175

Bruce Kitchens (1981). Continuity properties of factor maps in ergodic theory. PhD thesis, University of North Carolina. 224

Donald E. Knuth (1971). Optimum binary search trees. *Acta Informatica*, **1**: 14–25. 175

Donald E. Knuth (1985). Dynamic Huffman coding. *J. Algorithms*, **6** (2): 163–80. 175

Donald E. Knuth (1998). *The Art of Computer Programming, Volume III: Sorting and Searching*. Addison-Wesley, second edition. 175

Zvi Kohavi (1978). *Switching and Automata Theory*. McGraw-Hill, second edition. 198, 224

Alexander T. Kolotov (1978). Free subalgebras of free associative algebras. *Sibirsk. Mat. Ž.*, **19** (2): 328–35. 593

Marc Krasner and Britt Ranulac (1937). Sur une propriété des polynômes de la division du cercle. *C. R. Acad. Sci. Paris*, **240**: 397–9. 449

Daniel Krob (1987). Codes limites et factorisations finies du monoïde libre. *RAIRO Inform. Théor. Appl.*, **21** (4): 437–67. 325, 592

Michal Kunc (2004). Undecidability of the trace coding problem and some decidable cases. *Theoret. Comput. Sci.*, **310**(1-3):393–459. 105

Gérard Lallement (1979). *Semigroups and Combinatorial Applications*. Wiley. 53, 370

Gérard Lallement and Dominique Perrin (1981). A graph covering construction of all the finite complete biprefix codes. *Discrete Math.*, **36**: 261–71. 433

Nguyen Huong Lam (1996). A property of finite maximal codes. *Acta Mathematica Vietnamica*, **21**: 279–88. 449

Nguyen Huong Lam (1997). Hajós factorizations and completion of codes. *Theoret. Comput. Sci.*, **182**: 245–56. 449

Nguyen Huong Lam (2000). Finite maximal infix codes. *Semigroup Forum*, **61** (3):346–56. 274

Nguyen Huong Lam (2001). Finite maximal solid codes. *Theorect. Comput. Sci.*, **262**(1-2):333–47. 299

Nguyen Huong Lam (2003). Completing comma-free codes. *Theor. Comput. Sci.*, **1-3**(301):399–415. 299

Serge Lang (1965). *Algebra*. Addison-Wesley. 299

Jean-Louis Lassez (1973). Prefix codes and isomorphic automata. *Internat. J. Comput. Math.*, **3**: 309–14. 176

Jean-Louis Lassez (1976). Circular codes and synchronization. *Internat. J. Computer System Sciences*, **5**: 201–8. 298

Sylvain Lavallée, Dominique Perrin, Vladimir Retakh, and Christophe Reutenauer (2009). Codes and noncommutative stochastic matrices. *J. Noncommutative Geometry*, to appear. 534

André Lentin (1972). *Equations dans les monoïdes libres*. Gauthier-Villars. 104

André Lentin and Marcel-Paul Schützenberger (1969). A combinatorial problem in the theory of free monoids. In *Combinatorial Mathematics and its Applications (Proc. Conf., Univ. North Carolina, Chapel Hill, N.C., 1967)*, pp. 128–44. Univ. North Carolina Press, Chapel Hill, N.C. 325

Evelyne Lerest and Michel Lerest (1980). Une representation fidèle des groupes d'un monoïde de relations sur un ensemble fini. *Semigroup Forum*, **21**: 167–72. 370

Martine Léonard (1988). A property of biprefix codes. *RAIRO Inform. Théor. Appl.*, **22**(3):311–18. 273

Vladimir I. Levenshtein (1964). Some properties of coding and self-adjusting automata for decoding messages. *Problemy Kirbernet.*, **11**: 63–121. 104, 223

Frank W. Levi (1944). On semigroups. *Bull. Calcutta Math. Soc.*, **36**: 141–6. 104

Benjamin Lewin (1994). *Genes V*. Oxford University Press. 299

Douglas A. Lind and Brian H. Marcus (1995). *An Introduction to Symbolic Dynamics and Coding*. Cambridge University Press. 175, 198, 224, 298, 370, 396, 491

Yun Liu (2009). Composition of maximal codes. *Theoret. Comput. Sci.*, 2009. 104

Dongyang Long (1996). On group codes. *Theoret. Comput. Sci.*, **163** (1-2): 259–67. 274

M. Lothaire (1997). *Combinatorics on Words*. Cambridge University Press, second edition. (First edition 1983). 52, 176, 299, 325, 434, 592

M. Lothaire (2002). *Algebraic Combinatorics on Words*, volume 90 of *Encyclopedia of Mathematics and its Applications*. Cambridge University Press. 104, 224, 593

M. Lothaire (2005). *Applied Combinatorics on Words*, volume 105 of *Encyclopedia of Mathematics and its Applications*. Cambridge University Press. 198

Jean-Gabriel Luque and Jean-Yves Thibon (2007). Noncommutative symmetric functions associated with a code, Lazard elimination, and Witt vectors. *Discrete Math. Theor. Comput. Sci.*, **9** (2): 59–72 (electronic). 299

Roger C. Lyndon and Marcel-Paul Schützenberger (1962). The equation $a^m = b^n c^p$ in a free group. *Michigan Math. J.*, **9**: 289–98. 325

Ian G. Macdonald (1995). *Symmetric Functions and Hall Polynomials*. Oxford University Press. 298

F. Jessie MacWilliams and Neil J. Sloane (1977). *The Theory of Error Correcting Codes*. North-Holland. 103

Wilhelm Magnus, Abraham Karrass, and Donald Solitar (2004). *Combinatorial Group Theory*. Dover, second edition. 104

Gennady S. Makanin (1976). On the rank of equations in four unknowns in a free semigroup. *Mat. Sb. (N.S.)*, **100**: 285–311. 104

Anthony Manning (1971). Axiom A diffeomorphisms have rational zeta functions. *Bull. London Math. Soc.*, **3**: 215–20. ISSN 0024-6093. 298

Roberto Mantaci (1991). Anti-exceedences in permutation groups. *Europ. J. Combinatorics*, **12**: 237–44. 434

Sabrina Mantaci and Antonio Restivo (2001). Codes and equations on trees. *Theoret. Comput. Sci.*, **255**(1-2):483–509. 105

Brian H. Marcus (1979). Factors and extensions of full shifts. *Monatsh. Math*, **88**: 239–47. 175

Stuart W. Margolis (1982). On the syntactic transformation semigroup of a language generated by a finite biprefix code. *Theoret. Comput. Sci.*, **21**: 225–30. 434

Aleksandr A. Markov (1962). On alphabet coding. *Soviet. Phys. Dokl.*, **6**: 553–4. 224

Per Martin-Löf (1965). Probability theory on discrete semigroups. *Z. Wahrsch. Verw. Gebiete*, **4**: 78–102. 491

Silvana Mauceri and Antonio Restivo (1981). A family of codes commutatively equivalent to prefix codes. *Inform. Process. Lett.*, **12**: 1–4. 534

Robert J. McEliece (2004). *The Theory of Information and Coding*, volume 86 of *Encyclopedia of Mathematics and its Applications*. Cambridge University Press, student edition. With a foreword by Mark Kac. 104

Brockway McMillan (1956). Two inequalities implied by unique decipherability. *IRE Trans. Inform. Theory*, **IT-2**: 115–16. 104

Robert McNaughton and Seymour Papert (1971). *Counter Free Automata*. MIT Press. 396

Guy Melançon (1993). Constructions des bases standards des $K\langle A\rangle$-modules à droite. *Theoret. Comput. Sci.*, **117**: 255–72. 534

Nicolas C. Metropolis and Gian-Carlo Rota (1983). Witt vectors and the algebra of necklaces. *Advances in Math.*, **50**: 95–125. 299

Edward F. Moore (1956). Gedanken-experiments on sequential machines. In C. E. Shannon and J. McCarthy, editors, *Automata Studies*, volume 34 of *Ann. of Math. Stud.*, pp. 129–153. 175

Donald J. Newman (1977). Tesselations of integers. *J. Number Theory*, **9**: 107–11. 449

Maurice Nivat (1992). Binary tree codes. In M. Nivat and A. Podelski, editors, *Tree Automata and Languages*, pp. 1–20. North-Holland. 105

Maurice Nivat (1966). Éléments de la théorie générale des codes. In E. Caianiello, editor, *Automata Theory*, pp. 278–94. Academic Press. 104, 224

George L. O'Brien (1981). The road coloring problem. *Israel J. Math*, **39**: 145–54. 396

Yehoshua Perl, Michael R. Garey, and Shimon Even (1975). Efficient generation of optimal prefix codes: equiprobable words using unequal cost letters. *J. Assoc. Comput. Mach.*, **22** (2): 202–14. 175

Micha Perles, Michael Rabin, and Eliahu Shamir (1963). The theory of definite automata. *IEEE Trans. Electronic Computers*, **12**: 233–43. 434

Dominique Perrin (1975). Codes bipréfixes et groupes de permutations. PhD thesis, Universite Paris 7. 434

Dominique Perrin (1977a). Codes asynchrones. *Bull. Soc. Math. France*, **105**: 385–404. 175, 273, 534

Dominique Perrin (1977b). La transitivité du groupe d'un code bipréfixe fini. *Math. Z.*, **153**: 283–7. 434

Dominique Perrin (1978). Le degré minimal du groupe d'un code bipréfixe fini. *J. Combin. Theory Ser. A*, **25**: 163–73. 434

Dominique Perrin (1979). La représentation ergodique d'un automate fini. *Theoret. Comput. Sci.*, **9**: 221–41. 433

Dominique Perrin (1981). Sur les groupes dans les monoïdes finis. In *Noncommutative Structures in Algebra and Geometric Combinatorics (Naples, 1978)*, volume 109 of *Quad. "Ricerca Sci."*, pp. 27–36. CNR. 434

Dominique Perrin and Jean-Éric Pin (2004). *Infinite Words, Automata, Semigroups, Logic and Games*. Elsevier. 395, 396

Dominique Perrin and Giuseppina Rindone (2003). On syntactic groups. *Bull. Belg. Math. Soc. Simon Stevin*, **10** (suppl.): 749–59. 434, 592

Dominique Perrin and Marcel-Paul Schützenberger (1977). Codes et sous-monoïdes possédant des mots neutres. In H. Tzschach, H. Waldschmidt, and Hermann K.-G. Walter, editors, *Theoretical Computer Science, 3rd GI Conference, Darmstadt*, volume 48 of *Lecture Notes in Computer Science*, pp. 270–81. Springer-Verlag. 370, 449

Dominique Perrin and Marcel-Paul Schützenberger (1981). A conjecture on sets of differences of integer pairs. *J. Combin. Theory Ser. B*, **30**: 91–93. 534

Dominique Perrin and Marcel-Paul Schützenberger (1992). Synchronizing words and automata and the road coloring problem. In P. Walters, editor, *Symbolic Dynamics and its Applications*, pp. 295–318. American Mathematical Society. Contemporary Mathematics, vol. 135. 396, 534

Jean-François Perrot (1972). Contribution à l'étude des monoïdes syntaxiques et de certains groupes associés aux automates finis. Thèse d'État, Université de Paris. 175, 433

Jean-Éric Pin (1978). Le problème de la synchronisation et la conjecture de Černy. PhD thesis, Université Paris 6. 175

Jean-Éric Pin (1986). *Varieties of Formal Languages*. Foundations of Computer Science. Plenum Publishing Corp. With a preface by M.-P. Schützenberger, Translated from the French by A. Howie. 299

Jean-Éric Pin and Imre Simon (1982). A note on the triangle conjecture. *J. Combin. Theory Ser. A*, **32**: 106–9. 534

Vera S. Pless, W. Cary Huffman, and Richard A. Brualdi, editors, (1998). *Handbook of Coding Theory. Vol. I, II*. North-Holland. 103

Lázló Rédei (1965). Ein Überdeckungssatz für endliche abelsche Gruppen im Zusammenhang mit dem Hauptsatz von Hajós. *Acta Sci. Math. Szeged*, **26**: 55–61. 449

Clive Reis and Gabriel Thierrin (1979). Reflective star languages and codes. *Inform. and Control*, **42**: 1–9. 433

Antonio Restivo (1974). On a question of McNaughton and Pappert. *Inform. and Control*, **25**: 1. 299

Antonio Restivo (1975). A combinatorial property of codes having finite synchronization delay. *Theoret. Comput. Sci.*, **1**: 95–101. 395

Antonio Restivo (1977). On codes having no finite completions. *Discrete Math.*, **17**: 309–16. 104, 449, 534

Antonio Restivo (1981). Some remarks on complete subsets of a free monoid. In *Noncommutative Structures in Algebra and Geometric Combinatorics (Naples, 1978)*, volume 109 of *Quad. "Ricerca Sci."*, pp. 19–25. CNR, Rome. 591

Antonio Restivo (1990). Codes and local constraints. *Theoret. Comput. Sci.*, **72** (1): 55–64. 105, 491

Antonio Restivo, Sergio Salemi, and Tecla Sportelli (1989). Completing codes. *RAIRO Inform. Théor. Appl.*, **23**: 135–47. 106, 449

Christophe Reutenauer (1980). Séries formelles et algèbres syntaxiques. *J. Algebra*, **66**: 448–83. 534

Christophe Reutenauer (1981). Semisimplicity of the algebra associated to a biprefix code. *Semigroup Forum*, **23**: 327–42. 371, 534

Christophe Reutenauer (1985). Noncommutative factorization of variable-length codes. *J. Pure and Applied Algebra*, **36**: 157–86. 534

Christophe Reutenauer (1986). Ensembles libres de chemins dans un graphe. *Bull. Soc. Math. France*, **114**(2):135–52. 105

Christophe Reutenauer (1997). $\mathbb{N}$-rationality of zeta functions. *Adv. Appl. Math.*, **18**: 1–17. 326

Robert F. Rice (1979). Some pratical universal noiseless coding techniques. Technical report, Jet Propulsion Laboratory. 174

Iain Richardson (2003). *H.264 and MPEG-4 Video Compression: Video Coding for Next-generation Multimedia*. Wiley. 174

John A. Riley (1967). The Sardinas-Patterson and Levenshtein theorems. *Inform. and Control*, **10**: 120–36. 104

Giuseppina Rindone (1983). Groupes finis et monoïdes syntaxiques. PhD thesis, Université Paris 7. 434

Michael Rodeh (1982). A fast test for unique decipherability based on suffix trees. *IEEE Trans. Inform. Theory*, **IT-28**: 648–51. 105

Jacques Sakarovitch (2009). *Elements of Theory of Automata*. Cambridge University Press. 198

Arto Salomaa (1981). *Jewels of Formal Language Theory*. Computer Science Press. 224

David Salomon (2007). *Variable-length Codes for Data Compression*. Springer-Verlag. 174, 274

Arthur D. Sands (2000). Replacement of factors by subgroups in the factorization of Abelian groups. *Bull. London Math. Soc.*, **32** (3): 297–304. 449

Arthur D. Sands (2007). A question concerning the factorization of cyclic groups. *Internat. J. Algebra Comput.*, **17** (8): 1573–5. 449

August Albert Sardinas and George W. Patterson (1953). A necessary and sufficient condition for the unique decomposition of coded messages. *IRE Internat. Conv. Rec.*, **8**: 104–8. 104

Robert A. Scholtz (1969). Maximal and variable length comma-free codes. *IEEE Trans. Inform. Theory*, **IT-15**: 300–6. 299

Marcel-Paul Schützenberger (1955). Une théorie algébrique du codage. In *Séminaire Dubreil-Pisot 1955-56*, p. Exposé N°. 15. 104

Marcel-Paul Schützenberger (1956). On an application of semigroup methods to some problems in coding. *IRE Trans. Inform. Theory*, **IT-2**: 47–60. 273, 433

Marcel-Paul Schützenberger (1961a). On the definition of a family of automata. *Inform. and Control*, **4**: 245–70. 534

Marcel-Paul Schützenberger (1961b). On a special class of recurrent events. *Ann. Math. Statist.*, **32**: 1201–13. 273, 534

Marcel-Paul Schützenberger (1961c). On a family of submonoids. *Publ. Math. Inst. Hungar. Acad. Sci. Ser. A*, **VI**: 381–91. 273, 274

Marcel-Paul Schützenberger (1961d). A remark on finite transducers. *Inform. and Control*, **4**: 185–96. 198

Marcel-Paul Schützenberger (1964). On the synchronizing properties of certain prefix codes. *Inform. and Control*, **7**: 23–36. 174, 433

Marcel-Paul Schützenberger (1965a). On a factorization of free monoids. *Proc. Amer. Math. Soc.*, **16**: 21–4. 325

Marcel-Paul Schützenberger (1965b). Sur certains sous-monoïdes libres. *Bull. Soc. Math. France*, **93**: 209–23. 198, 491, 534

Marcel-Paul Schützenberger (1965c). Sur une question concernant certains sous-monoïdes libres. *C. R. Acad. Sci. Paris*, **261**: 2419–20. 298, 299

Marcel-Paul Schützenberger (1966). On a question concerning certain free submonoids. *J. Combin. Theory*, **1**: 437–22. 224

Marcel-Paul Schützenberger (1967). On synchronizing prefix codes. *Inform. and Control*, **11**: 396–401. 175, 176

Marcel-Paul Schützenberger (1975). Sur certaines opérations de fermeture dans les langages rationnels. In *Symposia Mathematica, Vol. XV (Convegno di Informatica Teorica, INDAM, Roma, 1973)*, pp. 245–53. Academic Press. 395, 396

Marcel-Paul Schützenberger (1979a). A property of finitely generated submonoids of free monoids. In G. Pollak, editor, *Algebraic Theory of Semigroups*, pp. 545–76. North-Holland. 370, 592

Marcel-Paul Schützenberger (1979b). Codes à longueur variable. In D. Perrin, editor, *Théorie des codes (actes de la septième École de Printemps d'Informatique Théorique)*, pp. 247–71. LITP. Reproduction of the notes for a NATO school, Royan. 449

Louis W. Shapiro (1981). A combinatorial proof of a Chebyshev polynomial identity. *Discrete Math.*, **34** (2): 203–6. 106

Lev N. Shevrin (1960). On subsemigroups of free semigroups. *Soviet. Math. Dokl.*, **1**: 892–4. 104

Paul C. Shields (1996). *Ergodic Theory of Discrete Sample Paths*. Springer-Verlag. 490

Peter W. Shor (1983). A counterexample to the triangle conjecture. *J. Combin. Theory Ser. A.*, **38**: 110–12. 534

Huei-Jan Shyr and Shyr-Shen Yu (1990). Solid codes and disjunctive domains. *Semigroup Forum*, **41**(1):23–37. 299

Jean-Claude Spehner (1975). Quelques constructions et algorithmes relatifs aux sous-monoïdes d'un monoïde libre. *Semigroup Forum*, **9**: 334–53. 104

Jean-Claude Spehner (1976). Quelques problèmes d'extension, de conjugaison et de présentation des sous-monoïdes d'un monoïde libre. PhD thesis, Université Paris 7. 104

Richard P. Stanley (1997). *Enumerative Combinatorics. Vol. 1*, volume 49 of *Cambridge Studies in Advanced Mathematics*. Cambridge University Press. With a foreword by Gian-Carlo Rota, Corrected reprint of the 1986 original. 106, 298, 299

Lubert Stryer (1975). *Biochemistry*. Freeman. 299

Sandor Szabó (1985). A type of factorization of finite Abelian groups. *Discrete Mathematics*, **54**: 121–4. 449

Sandor Szabó (2004). *Topics in the Factorisation of Abelian Groups*. Birkhaüser. 449

Sandor Szabó (2006). Completing codes and the Rédei property of groups. *Theoret. Comput. Sci.*, **359**: 449–54. 449

Sandor Szabó and Arthur D. Sands (2009). *Factoring Groups into Subsets*, volume 257 of *Lecture Notes in Pure and Applied Mathematics*. CRC Press. 449

Yasuhiro Takishima, Masahiro Wada, and Hitomi Murakami (1995). Reversible variable length codes. *IEEE Trans. Comm.*, **43**:158–62. 273

Jukka Teuhola (1978). A compression method for clustered bit-vectors. *Inform. Process. Lett.*, **7** (6): 308–11. 174

Bret Tilson (1972). The intersection of free submonoids is free. *Semigroup Forum*, **4**: 345–50. 104

Avraham N. Trahtman (2008). The road coloring problem. *Israel J. Math.*, to appear. 396

Jacobus H. van Lint (1982). *Introduction to Coding Theory*. Springer-Verlag. 104

Ben Varn (1971). Optimal variable length codes (arbitrary symbol cost and equal code word probabilities). *Inform. and Control*, **19**: 289–301. 175

Gérard Viennot (1974). Algèbres de Lie libres et monoïdes libres. PhD thesis, Université Paris 7. 325, 326

Gérard Viennot (1978). *Algèbres de Lie et monoïdes libres*, volume 691 of *Lecture Notes in Mathematics*. Springer-Verlag. 325

Max Vincent (1985). Construction de codes indécomposables. *RAIRO Informatique Théorique*, **19**: 165–78. 534

Mikhail V. Volkov (2008). Synchronizing automata and the Cerny conjecture. In Carlos Martín-Vide, Friedrich Otto, and Henning Fernau, editors, *Language and Automata Theory and Applications, Second International Conference, LATA 2008, Tarragona, Spain, March 13-19. Revised Papers*, volume 5196 of *Lecture Notes in Computer Science*, pp. 11–27. Springer-Verlag. 592

Helmut Wielandt (1964). *Finite Permutation Groups*. Academic Press. 53, 434

Chunxuan Ye and Raymond W. Yeung (2001). Some basic properties of fix-free codes. *IEEE Trans. Inform. Theory*, **47** (1): 72–87. 273

Sergey Yekhanin (2004). Improved upper bound for the redundancy of fix-free codes. *IEEE Trans. Inform. Theory*, **50** (11): 2815–18. 273, 592

Liang Zhang and Changkang Gu (1992). On factorization of finite maximal codes. In M. Ito, editor, *Words, Languages and Combinatorics*, pp. 534–41. World Scientific. 534

Liang Zhang and Zhong Hui Shen (1995). Completion of recognizable bifix codes. *Theoret. Comput. Sci.*, **145** (1-2): 345–55. 274

Index of notation

Index